# Radiological Risk Assessment and Environmental Analysis

# Radiological Risk Assessment and Environmental Analysis

*Edited by*
**JOHN E. TILL**
**HELEN A. GROGAN**

UNIVERSITY PRESS
2008

# OXFORD
UNIVERSITY PRESS

Oxford University Press, Inc., publishes works that further
Oxford University's objective of excellence
in research, scholarship, and education.

Oxford New York
Auckland Cape Town Dar es Salaam Hong Kong Karachi
Kuala Lumpur Madrid Melbourne Mexico City Nairobi
New Delhi Shanghai Taipei Toronto

With offices in
Argentina Austria Brazil Chile Czech Republic France Greece
Guatemala Hungary Italy Japan Poland Portugal Singapore
South Korea Switzerland Thailand Turkey Ukraine Vietnam

Copyright © 2008 by Oxford University Press, Inc.

Published by Oxford University Press, Inc.
198 Madison Avenue, New York, New York 10016
www.oup.com

Oxford is a registered trademark of Oxford University Press

All rights reserved. No part of this publication may be reproduced,
stored in a retrieval system, or transmitted, in any form or by any means,
electronic, mechanical, photocopying, recording, or otherwise,
without the prior permission of Oxford University Press.

Library of Congress Cataloging-in-Publication Data
Radiological risk assessment and environmental analysis / edited by John E. Till
and Helen A. Grogan.
p. ; cm.
Includes bibliographical references and index.
ISBN 978-0-19-512727-0
1. Radiation dosimetry. 2. Radiation—Safety measures. 3. Health risk
assessment. I. Till, John E. II. Grogan, Helen A.
[DNLM: 1. Radioactive Pollutants—adverse effects. 2. Accidents, Radiation—
prevention & control. 3. Environmental Exposure—prevention & control.
4. Environmental Monitoring—methods. 5. Radiation Injuries—prevention & control.
6. Risk Assessment. WN 615 R1292 2008]
RA569.R328 2008
363.17'99—dc22        2007036918

9 8 7 6 5 4 3 2

Printed in the United States of America
on acid-free paper

For

Susan S. Till, PhD,

and

R. Scott Yount,

Our spouses, who have been so supportive and patient as we worked together on this book.

In Memoriam

Todd V. Crawford

A dear friend and professional colleague who contributed to chapter 3 and who passed away before the book's publication.

# Preface

This textbook is an update and major revision of *Radiological Assessment: A Textbook on Environmental Dose Analysis*, published by the U.S. Nuclear Regulatory Commission in 1983. The earlier book was widely used as a graduate-level text and as a reference book at universities, in special courses, and by individuals who perform radiological assessment. Although the previous book made a unique contribution in bringing together different elements of radiological assessment as a science, a number of deficiencies were difficult to resolve at the time it was written. For example, there was considerable disparity in the level of detail among the chapters and in the information that each provided. In this new book, we have tried to address some of these deficiencies. It is written more specifically as a textbook, and it includes examples and sample problems throughout.

We have worked hard to improve the editing so there is better consistency among the chapters and greater cohesiveness in the different subjects presented. Nevertheless, we recognize some differences still exist in the way material is presented. These are due in large part to the multiple authors who contributed to this work. We do not believe, however, that an individual author could have adequately captured the state-of-the-art science and effectively conveyed the in-depth concepts required in such a textbook. That is why, as with the 1983 edition, we asked other scientists to contribute to the text. It is an honor and privilege to have the participation of these respected scientists, and we are indebted to their efforts and expertise. This book would not have been possible without them.

There have been many significant changes in radiological assessment over the past 20 years, and we have tried to capture them. Some changes were caused by the natural evolution and improvements of the underlying sciences that make up radiological assessment, and other changes resulted from events such as the

Chernobyl accident, phenomenal advances in computer technology, and a vastly different interest in understanding the health implications of radioactive materials that are or may be released to the environment by nuclear facilities.

It is apparent that interest in radiological assessment will continue to grow. There is a renewed emphasis on nuclear power as an energy source. Many nuclear power plants around the world have matured and will, at some point, require significant upgrades to their design or need to be decommissioned altogether. Government authorities in many countries are working to find the right balance between ecological destruction and remediation of environmental sites that were contaminated with radioactive materials during decades of deliberate disposal of residues from the production of nuclear weapons. We hope that this textbook will provide a reliable reference document for teaching and will set a standard for how radiological assessment should be performed.

Unfortunately, there are some elements of radiological assessment that we could not include. One example, briefly discussed in chapter 1, is communication of radiological assessment results and the participation of stakeholders. Another example is how to screen sources, materials, and pathways of exposure in order to focus the assessment on those elements that are the most important. These subjects had to be omitted to keep to a reasonable length.

We did not want the book to be simply a compilation of papers by authors; rather, we worked hard with the contributors to have the chapters fit together as cohesive elements to cover the entire science of radiological assessment. Undertaking this effort to employ the talents of a diverse group of individuals who are recognized as experts in their own right is a considerable challenge. We hope we have achieved success at merging the materials so that this textbook is useful, readable, and applicable to a wide variety of users.

John E. Till
Helen A. Grogan

# Acknowledgments

We are very grateful to the contributors to this book who provided their chapters while diligently and thoroughly responding to our editing suggestions and ideas. We especially thank them for their cooperation and persistence because this effort has taken such a long time to complete. Some chapters were revised significantly over the time we have been working together, and the patience that our contributors showed during this time is especially appreciated. We have been truly fortunate to have some of the best scientists in our profession to assist us.

Editing and proofreading this book has been the primary responsibility of Ms. Cindy Galvin and Ms. Julie Wose. Cindy works as the editor for our research team. In addition to her routine duties trying to keep our technical reports and publications in top quality, the book has been a responsibility she willingly took on for almost two years. She has had to work with the different contributors and accommodate their individual styles and writing mannerisms. She has accomplished this task in a pleasant and professional way. Julie Wose assisted with proofreading and checking text, references, tables, and figures. Julie has a superb ability to pour through hundreds of numbers in tables and figures looking for errors or items that seem incongruous with other information being presented. She has been of immense help throughout the course of this effort.

Ms. Shawn Mohler assisted us with graphics in some chapters. Shawn has a superb talent for creating new graphic art or revising old or outdated artwork.

We especially acknowledge our entire research team, Risk Assessment Corporation (RAC), who agreed to contribute to the book or who helped us in other ways. RAC team members Art Rood, Jim Rocco, Lisa Stetar, Lesley Hay Wilson, and Paul Voillequé contributed chapters to the book. We fully understand and appreciate the extra effort required to contribute to this book while they were meeting deadlines for other projects. A considerable amount of the work presented in our chapters was taken from work performed by the RAC team as a whole. Therefore, everyone on our team deserves credit and special recognition for their contribution to this high-quality and unique volume.

A special thanks and recognition to Dr. Bob Meyer, who co-edited the first book, published in 1983. Bob was instrumental in keeping the idea of an updated version alive over the years. Bob's new work responsibilities, intense commitments, and busy schedule prevented him from continuing this collaboration. Nevertheless, his contributions and involvement in radiological assessment over the past 30 years have helped shape the science.

Finally, we acknowledge the patience of Peter Prescott of Oxford University Press, who has exemplified an extraordinary patience with us in delivering the manuscript. Peter has worked with us from the beginning. Since we began assembling and editing this book, our lives have undertaken a number of turns, both positive and negative. Our research always had to come first because of commitments to customers, which meant that we often lost our focus on the book. Nevertheless, Peter always maintained his confidence in the book and its importance. We appreciate this dedication to our effort by both Peter Prescott and his entire staff at Oxford University Press.

# Contents

| | | |
|---|---|---|
| Contributors | | xxv |
| **1** | **The Radiological Assessment Process** | **1** |
| | John E. Till | |
| | Radiological Assessment Process | 2 |
| |    Source Term | 3 |
| |    Environmental Transport | 5 |
| |       Environmental Transport of Plutonium in Air During the 1957 Fire at Rocky Flats | 6 |
| |    Exposure Factors | 8 |
| |       Rocky Flats Representative Exposure Scenarios | 9 |
| |       Hanford Site Scenarios for Native Americans | 10 |
| |    Conversion to Dose | 12 |
| |       Uncertainty in Dose Coefficients | 12 |
| |       Appropriate Use of Dose Coefficients as a Function of Age | 13 |
| |    Conversion of Dose to Risk | 13 |
| |       Why Risk? | 14 |
| |       Risk Coefficients | 14 |
| |    Uncertainty Analysis | 15 |
| |       Use of Uncertainty for Determining Compliance with Standards | 17 |
| |    Validation | 18 |
| |    Communication of Dose and Risk and Stakeholder Participation | 21 |
| |       Communication of Results from Radiological Assessment | 21 |
| |       Stakeholder Participation | 24 |
| | Conclusion | 28 |
| | References | 28 |

## 2  Radionuclide Source Terms                                     31
Paul G. Voillequé

    Radionuclides of Interest and Their Properties                    32
    Situations That Do Not Require Source Terms                       37
    Human Activities Producing Releases of Radionuclides               38
        Uranium Mining                                                 39
        Uranium Milling                                                39
        Uranium Conversion                                             40
        Uranium Enrichment                                             40
        Weapon Component and Fuel Fabrication                          41
        Reactors                                                       42
            Source Terms for Normal Operations                         44
            Source Terms for Accidents                                 56
        Fuel Processing Plants                                         58
        Solid Waste Disposal                                           61
    Source Term Development for Facilities                             62
        Source Terms for Prospective Analyses                          62
        Source Terms for Retrospective Analyses                        64
    Problems                                                           66
    References                                                         69

## 3  Atmospheric Transport of Radionuclides                        79
Todd V. Crawford, Charles W. Miller, and Allen H. Weber

    The Atmosphere                                                     80
        Composition                                                    80
        Vertical Extent Important for Atmospheric Releases             80
        Scales of Motion                                               81
            Macroscale                                                 84
            Mesoscale                                                  84
            Microscale                                                 87
    Input Data for Atmospheric Transport
            and Diffusion Calculations                                 89
        Source                                                         89
        Winds                                                          90
        Turbulence and Stability                                       93
        Atmospheric Stability Categories                               94
            Pasquill-Gifford Stability Categories                      95
            Richardson Number                                          99
            Mixing Height                                             106
            Meteorological Data Quality                               108
    Modeling of Transport and Diffusion                               108
        Gaussian Diffusion Models                                     109
            Instantaneous Point Source                                109
            Continuous Point Source                                   110
            Continuous Line Source                                    110
            Continuous Point Source Release from a Stack             111

|  |  |
|---|---|
| Sector Averaging | 112 |
| Modifications Based on Source Characteristics | 113 |
| Special Considerations | 116 |
| Summary of Gaussian Plume Model Limitations | 119 |
| Puff-Transport and Diffusion Models | 120 |
| Puff Transport | 120 |
| Puff Diffusion | 121 |
| Sequential Puff-Trajectory Model | 122 |
| Multibox Models | 125 |
| Calculation Grid | 125 |
| Calculation Methods and Limitations | 126 |
| Particle-in-Cell Models | 126 |
| Screening Models | 127 |
| Atmospheric Removal Processes | 128 |
| Fallout | 129 |
| Dry Deposition | 129 |
| Wet Deposition | 130 |
| Model Validation | 132 |
| Model Uncertainty | 134 |
| Guidelines for Selecting Models | 136 |
| Regulatory Models | 137 |
| AERMOD Model | 137 |
| CALPUFF Model | 138 |
| CAP88 Model | 138 |
| Conclusions | 139 |
| Problems | 139 |
| References | 140 |

**4** **Surface Water Transport of Radionuclides** — 147
Yasuo Onishi

|  |  |
|---|---|
| Basic Transport and Fate Mechanisms | 149 |
| Transport | 149 |
| Water Movement | 149 |
| Sediment Movement | 150 |
| Bioturbation | 151 |
| Intermedia Transfer | 151 |
| Adsorption and Desorption | 151 |
| Precipitation and Dissolution | 152 |
| Volatilization | 153 |
| Physical Breakup | 153 |
| Degradation/Decay | 153 |
| Radionuclide Decay | 153 |
| Transformation | 153 |
| Yield of Daughter Products | 153 |
| Radionuclide Contributions from Other Environmental Media | 154 |

xiv Contents

|  |  |
|---|---|
| Radionuclide Transport Models | 154 |
| *Accidental Radionuclide Releases* | *155* |
| *Routine Long-Term Radionuclide Releases* | *156* |
| *Rivers* | *159* |
| Basic River Characteristics | 159 |
| Screening River Model | 163 |
| *Estuaries* | *167* |
| Basic Estuarine Characteristics | 167 |
| Screening Estuary Methodology | 171 |
| *Coastal Waters and Oceans* | *176* |
| Basic Coastal Water and Ocean Characteristics | 176 |
| Coastal Water Screening Model | 179 |
| *Lakes* | *181* |
| Basic Lake Water Characteristics | 181 |
| Small Lake Screening Model | 183 |
| Large Lake Screening Model | 185 |
| *Sediment Effects* | *187* |
| *Numerical Modeling* | *190* |
| Governing Equations | 190 |
| Some Representative Models | 191 |
| Chernobyl Nuclear Accident Aquatic Assessment | 192 |
| *Radionuclide Transport in Rivers* | *194* |
| *Aquatic Pathways and Their Radiation Dose Contributions* | *197* |
| *New Chernobyl Development* | *200* |
| Problems | 200 |
| References | 203 |

## 5 Transport of Radionuclides in Groundwater — 208

Richard B. Codell and James O. Duguid

|  |  |
|---|---|
| Applications of Groundwater Models for Radionuclide Migration | 209 |
| *Geologic Isolation of High-Level Waste* | *209* |
| *Shallow Land Burial* | *210* |
| *Uranium Mining and Milling* | *210* |
| *Nuclear Power Plant Accidents* | *211* |
| Types of Groundwater Models | 211 |
| *Groundwater Models for High-Level Waste Repositories* | *211* |
| Near-Field Performance | 212 |
| Far-Field Performance | 213 |
| *Groundwater Models for Shallow Land Burial of Low-Level Waste* | *215* |
| *Groundwater Models for Mill Tailings Waste Migration* | *215* |
| Equations for Groundwater Flow and Radioactivity Transport | 216 |
| *Groundwater Flow* | *216* |
| Saturated Flow | 218 |

| | |
|---|---|
| Unsaturated Flow | 219 |
| Mass Transport | 219 |
| Chain Decay of Radionuclides | 220 |
| Percolation of Water into the Ground | 221 |
| Parameters for Transport and Flow Equations | 222 |
|   Diffusion and Dispersion in Porous Media | 222 |
|     Molecular Diffusion | 222 |
|     Dispersion | 222 |
|     Macrodispersion | 223 |
|     Determination of Dispersion | 224 |
|   Porosity and Effective Porosity | 224 |
|   Hydraulic Conductivity for Saturated Flow | 226 |
|   Sorption, Retardation, and Colloids | 228 |
|     Transport Based on Assumption of Equilibrium (Retardation Factor) | 228 |
|     Transport Based on Geochemical Models | 231 |
|     Colloid Migration | 232 |
| Methods of Solution for Groundwater Flow and Transport | 233 |
|   Numerical Methods | 233 |
|     Finite Difference | 233 |
|     Finite Element | 234 |
|     Method of Characteristics | 234 |
|     Random Walk Method | 234 |
|     Flow Network Models | 235 |
|     Advection Models | 235 |
|     Analytic Elements | 235 |
|   Analytical Solutions of the Convective-Dispersive Equations | 236 |
|     Point Concentration Model | 237 |
|     Flux Model | 240 |
|     Generalization of Instantaneous Models | 243 |
|     Superposition of Solutions | 243 |
|   Simplified Analytical Methods for Minimum Dilution | 243 |
|   Models for Population Doses | 246 |
| Source Term Models for Low-Level Waste | 250 |
| Model Validation and Calibration | 251 |
| Misuse of Models | 253 |
| Problems | 254 |
| References | 254 |
| **6 Terrestrial Food Chain Pathways: Concepts and Models** | **260** |
| F. Ward Whicker and Arthur S. Rood | |
| Conceptual Model of the Terrestrial Environment | 262 |
| Strategies for Evaluating Food Chain Transport | 268 |
|   Predictive Approaches | 268 |
|     Direct Measurements | 268 |
|     Statistical Models | 269 |

| | |
|---|---:|
| Mechanistic Models | *270* |
| Choosing a Predictive Approach | *271* |
| Model Attributes | *271* |
| Mechanistic Models: The Mathematical Foundations for Single Compartments | *273* |
| Concepts and Terminology of Tracer Kinetics | *273* |
| Single-Compartment, First-Order Loss Systems | *275* |
| Source and Sink Compartments | *275* |
| Single Compartments with Constant Input Rates | *277* |
| Single Compartments with Time-Dependent Input Rates | *280* |
| Single-Compartment, Non–First-Order Loss Systems | *284* |
| The Convolution Integral | *284* |
| Borel's Theorem | *285* |
| Derivation of Rate Constants Involving Fluid Flow Compartments | *286* |
| Numeric Solutions | *287* |
| Individual Transport Processes: Concepts and Mathematical Formulations | *287* |
| Types of Processes | *288* |
| Continuous Processes | *288* |
| Discrete Processes | *288* |
| Stochastic Processes | *288* |
| Deposition from Air to Soil and Vegetation | *289* |
| Gravitational Settling | *289* |
| Dry Deposition | *290* |
| Wet Deposition | *292* |
| Soil–Vegetation Partitioning of Deposition | *294* |
| Transport from Soil to Vegetation | *295* |
| Suspension and Resuspension | *296* |
| Root Uptake | *303* |
| Transport from Vegetation to Soil | *307* |
| Weathering | *307* |
| Senescence | *308* |
| Transport within the Soil Column | *309* |
| Percolation | *310* |
| Leaching | *310* |
| Other Natural Processes Producing Vertical Migration in Soil | *317* |
| Tillage | *318* |
| Transport from Vegetation to Animals | *318* |
| Transport from Soil to Animals | *320* |
| Ingestion | *320* |
| Inhalation | *321* |
| Transfers to Animal-Derived Human Food Products | *321* |
| Ingestion Pathways to Humans | *323* |
| Dynamic Multicompartment Models: Putting It All Together | *324* |
| Conclusions | *331* |

|  |  |  |
|---|---|---|
|  | Problems | 333 |
|  | References | 334 |

## 7 Aquatic Food Chain Pathways    340
Steven M. Bartell and Ying Feng

|  |  |
|---|---|
| Aquatic Ecosystem Classification | 341 |
| Conceptual Model for an Aquatic Environment | 342 |
| *Physicochemical Processes* | *343* |
| Radionuclide Uptake and Concentration Factors | 344 |
| *Examples of Bioconcentration Factors* | *348* |
| *Bioconcentration Factors in Screening-Level Risk Estimations* | *351* |
| *Bioaccumulation Factors in Estimating Exposure* | *352* |
| *Bioaccumulation under Nonequilibrium Conditions:* | |
|     *The Chernobyl Cooling Pond Example* | *354* |
| *Initial $^{137}$Cs Contamination in the Chernobyl Cooling* | |
|     *Pond Water* | *355* |
| *The Chernobyl Cooling Pond Ecosystem* | *356* |
| *Chernobyl Cooling Pond Model Structure* | *357* |
| *Food Web Structure* | *360* |
| *Population Dynamics and Biomass Distributions* | *360* |
| *Spatial and Temporal Radionuclide Ingestion Rates* | *362* |
| *Radionuclide Transport and Distribution* | *363* |
| *Case Studies in Exposure and Bioaccumulation* | *364* |
|   *Case 1: Homogeneous and Steady-State Exposures* | *364* |
|   *Case 2: Homogeneous and Dynamic Radioactive Environment* | *364* |
|   *Case 3: Homogeneous and Dynamic Radioactive Environment* | |
|     *with Dynamic Population Biomass* | *366* |
|   *Case 4: Heterogeneous and Dynamic Radioactive Environment* | *366* |
|   *Case 5: Heterogeneous and Dynamic Radioactive Environment* | |
|     *with Varying Biomass* | *367* |
|   *Case 6: Dynamic Exposures and Variations in Feeding Rates* | *368* |
|   *Case 7: Dynamic Exposures and Multiple Prey* | *368* |
| *Discussion of the Chernobyl Modeling Results* | *368* |
| *Temporally and Spatially Dependent Ecological Factors* | *370* |
| Problems | 371 |
| References | 372 |

## 8 Site Conceptual Exposure Models    376
James R. Rocco, Elisabeth A. Stetar, and Lesley Hay Wilson

|  |  |
|---|---|
| Evaluation Area | 377 |
| Interested Party Input | 377 |
| Exposure Pathways | 378 |
| *Sources and Source Areas* | *379* |
| *Radionuclides* | *380* |
| *Exposure Areas* | *381* |
| *Potentially Exposed Persons* | *382* |

|  |  |  |
|---|---|---|
| | *Behaviors and Activities* | *383* |
| | *Exposure Media* | *383* |
| | *Exposure Routes* | *383* |
| | *Transport Mechanisms* | *384* |
| | *Transfer Mechanisms* | *385* |
| | Exposure Scenarios | 385 |
| | *Exposure Factors* | *386* |
| | Problems | 387 |
| | References | 388 |

## 9  Internal Dosimetry — 389
John W. Poston, Sr., and John R. Ford

|  |  |
|---|---|
| External versus Internal Exposure | 389 |
| *Internal Dose Control* | *392* |
| Regulatory Requirements | 394 |
| *ICRP Publication 26 Techniques* | *394* |
| Tissues at Risk | 397 |
| *ICRP Publication 30 Techniques* | *399* |
| Determination of the Tissue Weighting Factors | 399 |
| Secondary and Derived Limits | 400 |
| Other Definitions | 402 |
| *Calculation of the Committed Dose Equivalent* | *402* |
| *Dosimetric Models Used in the ICRP 30 Calculations* | *409* |
| Model of the Respiratory System | 409 |
| Model of the Gastrointestinal Tract | 417 |
| Dosimetric Model for Bone | 419 |
| Submersion in a Radioactive Cloud | 421 |
| Recent Recommendations | 422 |
| *ICRP Publication 60* | *422* |
| Dosimetric Quantities | 423 |
| Dose Limits | 429 |
| *Age-Dependent Doses to the Public (ICRP Publications 56, 67, 69, 71, and 72)* | *430* |
| ICRP Publication 56 | 432 |
| ICRP Publication 67 | 434 |
| ICRP Publication 69 | 437 |
| ICRP Publication 71 | 437 |
| ICRP Publication 72 | 442 |
| *ICRP Publication 89* | *443* |
| *ICRP Publications 88 and 95* | *444* |
| Summary | 444 |
| Problems | 445 |
| References | 445 |

| 10 | **External Dosimetry** | 447 |
|---|---|---|
| | David. C. Kocher | |
| | Dose Coefficients for External Exposure | 448 |
| | *Definition of External Dose Coefficient* | 449 |
| | *Compilation of External Dose Coefficients* | 449 |
| | *Description of Dose Coefficients in Current Federal Guidance* | 450 |
| | *Applicability of Dose Coefficients* | 452 |
| | *Effective Dose Coefficients* | 453 |
| | *Dose Coefficients for Other Age Groups* | 454 |
| | Corrections to Dose Coefficients for Photons | 454 |
| | *Shielding during Indoor Residence* | 455 |
| | *Effects of Ground Roughness* | 455 |
| | *Exposure during Boating Activities* | 456 |
| | *Exposure to Contaminated Shorelines* | 456 |
| | Point-Kernel Method | 457 |
| | *Description of the Point-Kernel Method* | 457 |
| | *Point-Kernel Method for Photons* | 459 |
| | *Applications of the Point-Kernel Method for Photons* | 459 |
| | *Point-Kernel Method for Electrons* | 461 |
| | Problems | 461 |
| | References | 462 |
| 11 | **Estimating and Applying Uncertainty in Assessment Models** | 465 |
| | Thomas B. Kirchner | |
| | Why Perform an Uncertainty Analysis? | 468 |
| | Describing Uncertainty | 469 |
| | *Probability Distributions* | 471 |
| | *Descriptive Statistics* | 471 |
| | *Statistical Intervals* | 476 |
| | Confidence Intervals | 476 |
| | Tolerance Intervals | 479 |
| | *Typical Distributions* | 483 |
| | *Correlations and Multivariate Distributions* | 485 |
| | Assigning Distributions | 487 |
| | *Deriving Distributions from Data* | 490 |
| | Estimating Parameters of a Distribution | 491 |
| | Using Limited Data | 491 |
| | Using Expert Elicitation | 493 |
| | Methods of Propagation | 497 |
| | *Analytical Methods* | 497 |
| | Sum and Difference of Random Variables | 498 |
| | Product of Random Variables | 499 |
| | Quotient of Random Variables | 500 |
| | *Formulas for Normal and Lognormal Distributions* | 501 |
| | Linear Operations | 501 |
| | Geometric Means and Standard Deviations | 501 |

xx Contents

| | |
|---|---|
| Mathematical Approximation Techniques | 502 |
| Mean | 502 |
| Variance | 502 |
| Propagation Using Interval Estimates | 504 |
| Sum and Difference | 504 |
| Products and Quotients | 504 |
| Other Functions | 505 |
| Covariance and the Order of Operations | 505 |
| Monte Carlo Methods | 506 |
| Generating Random Numbers | 508 |
| Potential Problems with Monte Carlo Methods | 508 |
| Sampling Designs | 509 |
| Simple Random Sampling | 509 |
| Latin Hypercube Sampling | 509 |
| Importance Sampling | 510 |
| Sampling Designs to Partition Variability and True Uncertainty | 511 |
| Number of Simulations | 511 |
| Interpretation of the Output Distributions | 513 |
| Sensitivity Analysis | 517 |
| Local Sensitivity Analysis | 518 |
| Global Sensitivity Analysis | 520 |
| Statistics for Ranking Parameters | 521 |
| Uncertainty and Model Validation | 522 |
| Summary | 524 |
| Problems | 525 |
| References | 526 |

**12  The Risks from Exposure to Ionizing Radiation** — 531
Roger H. Clarke

| | |
|---|---|
| Radiobiological Effects after Low Doses of Radiation | 533 |
| Biophysical Aspects of Radiation Action on Cells | 533 |
| Chromosomal DNA as the Principal Target for Radiation | 535 |
| Epigenetic Responses to Radiation | 535 |
| Effects at Low Doses of Radiation | 537 |
| Dose and Dose-Rate Effectiveness Factor | 538 |
| Genetic Susceptibility to Cancer | 538 |
| Heritable Diseases | 539 |
| Cancer Epidemiology | 540 |
| Japanese A-Bomb Survivors | 541 |
| Other Cohorts | 542 |
| In Utero Exposures | 543 |
| Uncertainties in Risk Estimates Based on Mortality Data | 544 |

| | | |
|---|---|---|
| | Risk Coefficients for Cancer and Hereditary Effects | 545 |
| | *Cancer Risk Coefficients* | *545* |
| | *Hereditary Risk* | *546* |
| | Overall Conclusions on Biological Effects at Low Doses | 547 |
| | Problems | 549 |
| | References | 549 |
| **13** | **The Role of Epidemiology in Estimating Radiation Risk: Basic Methods and Applications** | **551** |
| | Owen J. Devine and Paul L. Garbe | |
| | Measures of Disease Burden in Populations | 552 |
| | *Estimating Disease Risk* | *552* |
| | *Estimating Disease Rate* | *554* |
| | *Estimating Disease Prevalence* | *555* |
| | Measures of Association between Disease Risk and Suspected Causative Factors | 557 |
| | *Risk Ratio* | *557* |
| | *Risk Odds Ratio* | *559* |
| | *Exposure Odds Ratio* | *559* |
| | Study Designs Commonly Used in Epidemiologic Investigations | 563 |
| | *Cohort Designs* | *563* |
| | *Case–Control Designs* | *565* |
| | *Nested Designs* | *566* |
| | Assessing the Observed Level of Association between Disease and Exposure | 566 |
| | *Interpreting Estimates of Disease Exposure Association* | *567* |
| | *Confidence Intervals* | *570* |
| | Issues in Radiation Epidemiology | 580 |
| | Conclusion | 584 |
| | Problems | 584 |
| | References | 587 |
| **14** | **Model Validation** | **589** |
| | Helen A. Grogan | |
| | Validation Process | 590 |
| | *Model Composition* | *591* |
| | *Model Performance* | *593* |
| | *Calibration* | *594* |
| | Tests of Model Performance | 596 |
| | *Testing for Bias* | *596* |
| | *Measures of Scatter* | *598* |
| | *Correlation and Regression* | *599* |
| | *Visual Display of Information* | *599* |

|    | Reasons for Poor Model Performance | 603 |
|---|---|---|
|    | *User Error* | *604* |
|    | *The Model* | *605* |
|    | *The Assessment Question* | *607* |
|    | Conclusions | 608 |
|    | Problems | 608 |
|    | References | 609 |
| **15** | **Regulations for Radionuclides in the Environment** | **613** |
|    | David C. Kocher | |
|    | Principal Laws for Regulating Exposures to Radionuclides and Hazardous Chemicals in the Environment | 614 |
|    | Institutional Responsibilities for Radiation Protection of the Public | 617 |
|    | *Responsibilities of U.S. Governmental Institutions* | *617* |
|    | U.S. Environmental Protection Agency | *617* |
|    | U.S. Nuclear Regulatory Commission | *617* |
|    | U.S. Department of Energy | *618* |
|    | State Governments | *618* |
|    | *Role of Advisory Organizations* | *619* |
|    | Standards for Controlling Routine Radiation Exposures of the Public | 619 |
|    | *Basic Approaches to Regulating Exposure to Radionuclides in the Environment* | *619* |
|    | *Radiation Paradigm for Risk Management* | *620* |
|    | *Chemical Paradigm for Risk Management* | *622* |
|    | *Linear, Nonthreshold Dose–Response Hypothesis* | *622* |
|    | *Radiation Protection Standards for the Public* | *624* |
|    | Guidance of the U.S. Environmental Protection Agency | *624* |
|    | Radiation Protection Standards of the U.S. Nuclear Regulatory Commission | *625* |
|    | Radiation Protection Standards of the U.S. Department of Energy | *626* |
|    | State Radiation Protection Standards | *627* |
|    | Current Recommendations of the ICRP, NCRP, and IAEA | *627* |
|    | Summary of Radiation Protection Standards for the Public | *628* |
|    | *Standards for Specific Practices or Sources* | *629* |
|    | Operations of Uranium Fuel-Cycle Facilities | *630* |
|    | Radioactivity in Drinking Water | *631* |
|    | Radioactivity in Liquid Discharges | *635* |
|    | Uranium and Thorium Mill Tailings | *636* |
|    | Other Residual Radioactive Material | *639* |
|    | Radioactive Waste Management and Disposal | *648* |
|    | Airborne Emissions of Radionuclides | *660* |
|    | Indoor Radon | *662* |
|    | Risks Associated with Radiation Standards for the Public | *664* |

| | |
|---|---:|
| *Consistency of Radiation Standards for the Public* | 666 |
| *Importance of ALARA Objective to Consistent Regulation* | 669 |
| Exemption Levels for Radionuclides in the Environment | 671 |
| *Concepts of Exemption* | 671 |
| De Minimis Level | 671 |
| Exempt or Below Regulatory Concern Level | 671 |
| *Recommendations of Advisory Organizations* | 672 |
| Recommendations of the NCRP | 672 |
| Recommendations of the IAEA | 672 |
| *Exemptions Established by the U.S. Nuclear Regulatory Commission* | 673 |
| Exemptions in U.S. Nuclear Regulatory Commission Regulations | 673 |
| U.S. Nuclear Regulatory Commission Guidance on Disposal of Thorium or Uranium | 674 |
| Protective Action Guides for Accidents | 674 |
| *Purpose and Scope of Protective Action Guides* | 675 |
| *Time Phases for Defining Protective Actions* | 675 |
| *Protective Action Guides Established by Federal Agencies* | 676 |
| Recommendations of the U.S. Environmental Protection Agency | 676 |
| Recommendations of the U.S. Food and Drug Administration | 676 |
| Proposed Recommendations of the U.S. Department of Homeland Security | 677 |
| *U.S. Nuclear Regulatory Commission's Reactor Siting Criteria* | 679 |
| *ICRP Recommendations on Responses to Accidents* | 680 |
| *IAEA Guidelines for Intervention Levels in Emergency Exposure Situations* | 681 |
| Conclusions | 682 |
| References | 683 |
| Index | 689 |

# Contributors

Steven M. Bartell, PhD
Principal Scientist and Manager
E2 Consulting Engineers, Inc.
339 Whitecrest Drive
Maryville, Tennessee 37801

Roger H. Clarke
Emeritus Member, International
   Commission on Radiological
   Protection
Corner Cottage, Woolton Hill
Newbury, RG209XJ
United Kingdom

Richard B. Codell, PhD
Consultant to the U.S. Nuclear
   Regulatory Commission
4 Quietwood Lane
Sandy, Utah 84092


Todd V. Crawford, PhD[a]
Consultant

Owen J. Devine, PhD
National Center on Birth Defects
and Developmental Disabilities
MS E-87
Centers for Disease Control and
   Prevention
1600 Clifton Road
Atlanta, Georgia 30333

James O. Duguid, PhD
JK Research Associates
29 Touchstone Lane
Amissville, Virginia 20106

Ying Feng, PhD
7047 Dean Farm Road
New Albany, Ohio 43504

John R. Ford, PhD
Department of Nuclear Engineering
Texas A&M University
3133 TAMU

---

[a] Deceased.

College Station,
  Texas 77843-3133

Paul L. Garbe, DVM
National Center for Environmental
  Health MS E-17
Centers for Disease
  Control and Prevention
1600 Clifton Road
Atlanta, Georgia 30333

Helen A. Grogan, PhD
Cascade Scientific, Inc.
1678 NW Albany Avenue
Bend, Oregon 97701

Thomas B. Kirchner, PhD
Carlsbad Environmental Monitoring
  and Research Center
New Mexico State University
1400 University Drive
Carlsbad, New Mexico 88220

David C. Kocher, PhD
SENES Oak Ridge, Inc.
102 Donner Drive
Oak Ridge, Tennessee 37830

Charles W. Miller, PhD
Chief, Radiation Studies Branch
Division of Environmental
  Hazards and Health Effects
National Center for
  Environmental Health
Centers for Disease Control
  and Prevention
2400 Century Parkway
Atlanta, Georgia 30345

Yasuo Onishi, PhD
Yasuo Onishi Consulting, LLC
Adjunct Full Professor,
  Washington State University
144 Spengler Street
Richland, Washington 99354

John W. Poston, Sr., PhD
Department of Nuclear Engineering

Texas A&M University
3133 TAMU
College Station, Texas 77843-3133

James R. Rocco
Sage Risk Solutions, LLC
360 Heritage Road
Aurora, Ohio 44202

Arthur S. Rood, MS
K-Spar, Inc.
4835 W. Foxtrail Lane
Idaho Falls, Idaho 83402

Elisabeth A. Stetar, CHP
Performance Technology Group, Inc.
1210 Seventh Avenue North
Nashville, Tennessee 37208-2606

John E. Till, PhD
Risk Assessment Corporation
417 Till Road
Neeses, South Carolina 29107

Paul G. Voillequé, MS
MJP Risk Assessment, Inc.
P.O. Box 200937
Denver, Colorado 80220-0937

Allen H. Weber, PhD
Consultant
820 Jackson Avenue
North Augusta, Georgia 29841

F. Ward Whicker, PhD
Department of Radiological
  Health Sciences
Colorado State University
Fort Collins, Colorado 80523

Lesley Hay Wilson, PhD
Sage Risk Solutions, LLC
3267 Bee Caves Road, Suite 107
PMB 96
Austin, Texas 78746

# 1

# The Radiological Assessment Process

John E. Till

R adiological assessment is defined as the process of estimating dose and risk to humans from radioactive materials in the environment. These radioactive materials are generally released from a source that may be either man-made or natural. The materials may be transported through the environment and appear as concentrations in environmental media. These concentrations can be converted to dose and risk by making assumptions about exposure to people.

The chapters in this book explain the basic steps of radiological assessment that are typically followed. There is some logic to the order of the chapters and to the sequence of steps generally undertaken in radiological assessment. Some of this logic comes from my own experience over the years, and some of it is defined by the calculation process in radiological assessment because certain information must be known before proceeding to the next step. This logic is explained in the sections that follow.

Over the years, some scientists have suggested that the term "radiological assessment" does not quantitatively express the intense level of computational science that is necessary to estimate dose or risk. As a result, scientists have instead used the terms "environmental risk assessment" or "environmental risk analysis" to describe the process. Regardless of what it is called, radiological assessment has matured significantly over the past three decades. It has become the foundation of many regulations and legal cases and has provided a means for decision makers to take action on important issues such as cleanup of contaminated sites and control of emissions to the environment from nuclear facilities. Additionally, radiological assessment has increasingly become a fundamental element in communicating information to stakeholders about exposure to radioactive materials in the environment. Stakeholders are people who have an interest in the assessment process and the policies or recommendations that result from it. This term is defined in greater detail later in this chapter.

Radiological assessment requires the merging of a number of scientific disciplines to provide quantitative estimates of risk to humans. The book focuses on humans as an end point because decision makers typically use that end point to allocate resources and resolve issues. More specifically, the targeted person is a member of the public. The public is usually the objective in the assessment because when radioactive materials are released to the environment, it is members of the public who are or will be exposed. Although the primary target of exposure in the book is a member of the public, many of the technical methods described here also address occupational exposures.

Although the book focuses on how we estimate dose and risk to a member of the public, it is becoming more evident that impacts to the environment must also be taken into account. Whicker et al. (2004) stress that care must be taken to avoid destroying ecological systems in the interest of reducing inconsequential human health risks. This is an essential point for everyone to understand. Although we do not address ecological impacts in this book, many of the same principles described could be used to consider these impacts. In the end, both impacts on humans and impacts on the environment must be taken into account before good decisions can be made.

A number of examples are used in this chapter to illustrate key points being made about specific areas of radiological assessment. These examples are taken from work I and my research team, Risk Assessment Corporation, have performed over the years. Although specific reports are cited, it must be emphasized that radiological assessment can rarely be performed by a single individual. It generally requires the skills of people across several scientific disciplines.

## Radiological Assessment Process

Contemporary radiological assessment began with the testing of nuclear weapons as scientists tried to predict the path of radioactive fallout and the dose to people who lived downwind. Early research in this area, more than any other, laid the foundation for the methods we still use today to estimate risk to people from radioactive materials in the environment. More recently, research to reconstruct historical releases of radionuclides to the environment from atmospheric nuclear weapons testing and from nuclear weapons facilities resulted in significant improvements in methods to estimate risk (Till 1990; Till et al. 2000, 2002). This research included many new areas of investigation, such as estimation of source terms, transport of radioactive materials in the environment, uptake of radionuclides by humans and biota, and the development of dose and risk coefficients.

Radiological assessment is not confined to a specific time frame; it can address the past, present, or future. Dose and risk can be estimated for possible future releases of materials (prospective), for present-day releases, or for releases that occurred in the past (retrospective). Risk can be estimated for present-day or potential future releases of materials at existing or planned facilities. Such assessments are typically designed to demonstrate compliance with standards. Risk can also be estimated for releases that occurred in the past to help understand the impact of those releases. The dose reconstruction studies conducted on the weapons complex facilities in the United States provide excellent examples of retrospective risk assessments, as do the studies of populations exposed following the Chernobyl reactor accident. Although the risk assessments may be undertaken somewhat differently in their methods, there are many similarities in the techniques applied to each.

The components that comprise radiological assessment today evolved from individual sciences that have been merged gradually (and lately, more frequently) to form the computational methods we now use to estimate dose and risk to humans. In explaining the process of radiological assessment to colleagues and to the public, I often use the following illustrative equation to express the interdisciplinary nature of this research:

$$\text{Risk} = (S \cdot T \cdot E \cdot D \cdot R)_{uvcp} \tag{1.1}$$

where

$S$ = source term
$T$ = environmental transport
$E$ = exposure factors
$D$ = conversion to dose
$R$ = conversion of dose to risk
$u$ = uncertainty analysis
$v$ = validation
$c$ = communication of results
$p$ = participation of stakeholders

In the sections that follow, each of these components of radiological assessment is discussed, with emphasis on several key concepts that are important to keep in mind.

## *Source Term*

The source term is the characterization and quantification of the material released to the environment. It is the heart of a risk assessment. We frequently give too little attention to the derivation of the source term, and yet this step is where the greatest potential lies for losing scientific and stakeholder credibility. This is also the component of radiological assessment that typically requires the most resources relative to the other steps. Therefore, it is important that development of the source term be given highest priority and that the source term be carefully estimated before moving to the next step of radiological assessment.

Chapter 2 of this book, contributed by Paul Voillequé, addresses source terms. The chapter covers an expansive scope of which nuclear materials are typically released to the environment and how to quantify them. Issues such as chemical form, particle size, temporal trends, and estimating releases when measurement data are not available are discussed.

**4** Radiological Risk Assessment and Environmental Analysis

Two key points need to be mentioned about how source terms should be derived in radiological assessment. The first point is that uncertainties should be included with the estimates of releases if a realistic estimate of the source is to be used. Methods for estimating uncertainties in the source term and other aspects of radiological assessment are discussed in chapter 11. This aspect has been overlooked in the past, with release estimates being reported as point values when in reality we know there is a *range* of possible values that exist even when good monitoring data are available on which to base the source term.

An alternative approach to addressing uncertainties in the source term that may be useful for screening or providing preliminary estimates to determine the significance of a particular source is the use of an upper bound, deterministic value. The upper bound approach is designed to provide doses and risks that are significantly greater than what is expected to occur. This approach may be useful for screening or making preliminary comparisons of the impact of different sources.

The second key point is that the source term should be derived using as many different independent approaches as possible to increase the confidence that a credible estimate has been made. This is especially important in historical dose reconstruction, where sources that may have occurred many years ago are being estimated. This point is illustrated in the work of Meyer et al. (1996) that reconstructed source terms for the Fernald Feed Materials Production Center (FMPC), near Cincinnati, Ohio, which was formerly a part of the U.S. nuclear weapons complex that processed uranium ore. The facility has now been decommissioned and cleaned up. This study estimated the release of uranium from the FMPC using two methods. The first method, which could be called the "inside-out" approach, considered the amounts of material being processed at the site and estimated the fractional release of uranium to the atmosphere through various effluent treatment systems (primarily scrubbers and dust collectors). Using this approach, it was determined that the median quantity of uranium released to the atmosphere was 310,000 kg, with the 5th and 95th percentiles ranging between 270,000 kg and 360,000 kg, respectively. These results are shown in table 1.1.

An alternative calculation, called the "outside-in" approach, was performed as a check to verify the calculation, looking at the amount of uranium deposited on soil within 7.5 km of the site based on soil samples that had been collected over time. Taking into account environmental removal of some of the uranium and the amount of uranium that would have been deposited from the atmosphere, it was estimated that the source term for uranium released from the site to the atmosphere

Table 1.1 Uranium and radon source terms for the Fernald Feed Materials Production Center for 1951–1988[a] (Voillequé et al. 1995)

| Source: uranium to atmosphere | Median release estimate | 5th percentile | 95th percentile |
| --- | --- | --- | --- |
| Primary estimate | 310,000 | 270,000 | 360,000 |
| Alternative calculation | 212,000 | 78,000 | 390,000 |

[a] Values are in kilograms of uranium.

would lie between 78,000 kg and 390,000 kg, with a median value of 212,000 kg. This alternative calculation, although the uncertainties are large, provides additional confidence that the source term estimate for uranium is reasonable.

Without a defensible estimate of the source term, it is not possible to provide a defensible estimate of dose or risk, and the credibility of the assessment is lost. Therefore, it is critical to carefully and thoroughly address this first step in radiological assessment.

## Environmental Transport

Once the source term has been estimated, the next step is to determine where in the environment the radioactive materials go and what are the resulting concentrations in environmental media. This step is called environmental transport.

One of the first tasks in evaluating environmental transport is to identify the relevant exposure pathways. Figure 1.1 illustrates possible pathways typically considered in radiological assessment. However, not all pathways shown in the diagram typically apply to every site or radionuclide. Special pathways of concern may also exist that are not shown here. Determining important pathways and eliminating those that are not important is a critical step that can help focus resources. This process can be accomplished using screening models or other techniques that are easy

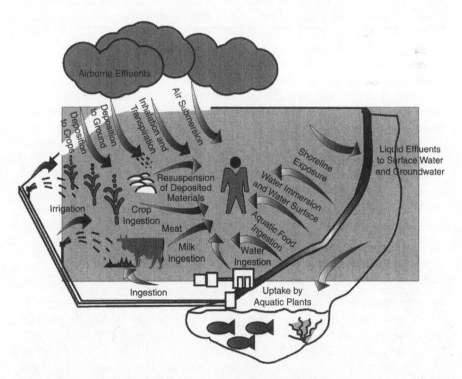

Figure 1.1  Diagram illustrating pathways typically considered in radiological assessment.

to use and help set priorities for the focus of the assessment (NCRP 1996; Mohler et al. 2004).

Environmental transport necessarily involves individuals from a number of scientific disciplines because different pathways of exposure have to be considered. Consequently, environmental transport of materials in the environment makes up a large part of this book. Chapter 3, contributed by Todd Crawford, Charles Miller, and Allen Weber, focuses on the transport of radioactive materials through the atmosphere. Chapter 4, contributed by Yasuo Onishi, discusses the transport of radioactive materials in surface water. Chapter 5, contributed by Richard Codell and James Duguid, looks at transport in groundwater. Chapter 6, contributed by Ward Whicker and Arthur Rood, provides methods for evaluating transport of radioactive materials in terrestrial food chain pathways. Chapter 7, contributed by Steven Bartell and Ying Feng, considers the transport of materials in aquatic food chain pathways. These chapters present a comprehensive look at the state-of-the-art science today in estimating environmental transport techniques used in radiological assessment.

Transport of radioactive materials in the environment can be determined in several ways. If there are measurement data in the environment that are sufficiently thorough, these measurements may be used directly to determine concentrations in media. The more data that are available to characterize the environment around a site, the more defensible will be the estimates of dose and risk. In fact, measurements of environmental concentrations are always preferable to modeling. It is rare, however, that measured data can be used in place of models. This is especially true when radiological assessment is being undertaken for a new facility where releases of materials will occur at some point in the future. In most cases, environmental transport is determined using a combination of both modeling and measurement data.

To illustrate environmental transport, I use the work performed by our research team during the historical dose reconstruction for the Rocky Flats Environmental Technology Site near Denver, Colorado. This site has been decommissioned and is now a wildlife refuge. The goal of the project was to reconstruct risks to members of the public from releases of plutonium and other materials at the site. Most of the plutonium was released to the atmosphere, and the most significant release was during a fire that occurred in September 1957. Understanding the risks associated with this source of plutonium and where it went in the environment was crucial to the success of the study (Rood et al. 2002; Till et al. 2002).

## Environmental Transport of Plutonium in Air During the 1957 Fire at Rocky Flats

In order to determine environmental transport of plutonium during and after the 1957 fire at Rocky Flats, several critical pieces of information had to be obtained. First, a source term was needed that estimated the amount of plutonium released, the distribution of the release over time, the heat generated by the fire (to account for the rise of the plume), and the size of the particles released. This important part of the puzzle controlled the concentrations of plutonium in the plume as it moved downwind from the site. Since inhalation was determined to be the only major pathway of exposure, air concentration coupled with where people were located during the fire and their breathing rate determined the resulting risk. The source term was reconstructed (Voillequé 1999) by reviewing historical records detailing the fire and interviewing fire experts. Quantities of plutonium released to the atmosphere were estimated for 15-min intervals during the time of the fire, along with the physical and chemical form of plutonium that was dispersed.

The next critical piece of information we needed was data describing the meteorological conditions during the fire. Information such as wind direction, wind velocity, and atmospheric conditions was required if the transport of the plutonium through air was to be understood. Fortunately, these data were collected and could be found in historical records. This information was used as input to RATCHET, an atmospheric dispersion model (Ramsdell 1994) that could take advantage of the resolution of meteorological data and the temporal distribution of the source.

Once these steps were taken, time-integrated concentration values were combined with scenario exposure information and risk coefficients to yield the incremental lifetime cancer incidence risk to hypothetical individuals in the model domain. Plutonium released during the 1957 fire was modeled as puffs that entered the atmosphere every 15 min from 10:00 P.M. September 11 until 2:00 A.M. September 12, 1957 (Rood and Grogan 1999). The transport calculations were continued until 7:00 A.M. September 12, 1957, to allow all the released plutonium to disperse throughout the model domain. The computer code simulations performed using RATCHET covered a 9-h period. Because the effluent release temperature was estimated to be near 400° C, there was significant plume rise, and maximum plutonium concentrations in ground-level air were estimated some distance southeast of the Rocky Flats Plant, not adjacent to it. The concentration in air at ground level, typically at a height of 1 m, represents the air concentration to which people would have been exposed.

At the time the fire started, the plume was transported in a westerly direction for a few kilometers. Around 10:45 P.M., the wind direction at the Rocky Flats plant shifted so that it blew out of the northwest and continued to blow from that direction until about 4:00 A.M., September 12. Those winds transported the bulk of the airborne plutonium to the suburb of Arvada and toward the Denver metropolitan area. Near southern Arvada, the air mass converged with air flowing from the southwest in the Platte River Valley, which resulted in a northeasterly plume trajectory. Figure 1.2 shows the median (50th percentile) estimated time-integrated plutonium concentrations in air near ground level.

This example of environmental transport of plutonium during the 1957 fire at Rocky Flats illustrates a very important point. First, without meteorological data collected at the time of the event, it would have been difficult to understand where the plume carried the plutonium and who may have been exposed. Obtaining these data that characterized atmospheric conditions at the precise time and location of the accident was essential to the assessment. In radiological assessment, a significant amount of time will be spent obtaining site-specific data that characterize the situation being investigated.

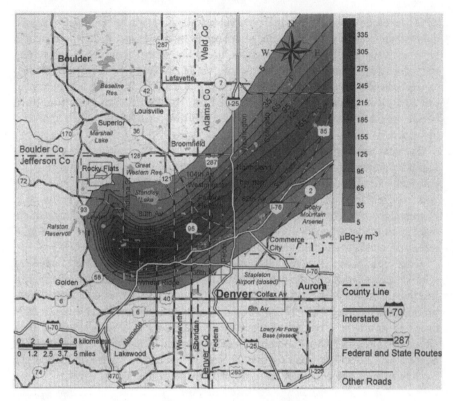

Figure 1.2 Estimated nine-hour average plutonium concentration in air one meter above ground at the 50th percentile level during the 1957 fire (Till et al. 2002).

## Exposure Factors

The dose or risk to a person depends upon a number of characteristics, called exposure factors, such as time, location, transport of radionuclides through the environment, and the traits of the individual. These traits include physiological parameters (e.g., breathing rate), dietary information (e.g., consumption rate of various foods), residence data (e.g., type of dwelling), use of local resources (e.g., agricultural resources), recreational activities (e.g., swimming), and any other individual-specific information that is necessary to estimate dose or risk. In radiological assessment, a specific set of these characteristics is referred to as an exposure scenario.

The target of radiological assessment may be real individuals or representative individuals. Real individuals are those who are or were actually exposed. Their characteristics should be defined as closely as possible to those that actually exist. Representative, or hypothetical, individuals are not characterized by specific persons but have characteristics similar to people in the area who are or were exposed in the past or who may be exposed in the future.

Exposure scenarios are described in a site conceptual exposure model (SCEM) that contains information about exposure factors specific for a given source and location. Chapter 8, contributed by James Rocco, Elisabeth Stetar, and Lesley Hay Wilson, addresses exposure factors and the SCEM.

There is no prescribed approach for defining and presenting scenarios of exposure in radiological assessment. This decision must fit the particular assessment being undertaken, the type of individual (real or representative) being evaluated, and the goals of the assessment. Two examples follow that come from studies we performed at Rocky Flats and at the Hanford Site, a nuclear weapons production facility in Washington State.

## Rocky Flats Representative Exposure Scenarios

A key component of the Rocky Flats dose reconstruction work was estimating the health impacts to representative individuals in the model domain. In this case, the cancer risk to people depended upon a number of factors, such as where the person lived and worked, when and how long that person lived near the site, the age and gender of the person, and lifestyle. It was not possible to create an exposure scenario that fit every person in the exposed population. To consider the many factors that influence exposure, exposure scenarios were developed for residents for whom representative risk estimates could be made, incorporating typical lifestyles, ages, genders, and times in the area. The scenarios provided a range of potential profiles and included a laborer, an office worker, a homemaker, an infant-child, and a student. The infant-child scenario represented a single individual who matured during the exposure period. Table 1.2 lists key features of the exposure scenarios used in the analysis.

The five exposure scenarios were organized according to occupational and nonoccupational activities. Occupational activities included work, school, and extracurricular activities away from the home. Nonoccupational activities included time spent at home doing chores, sleeping, and leisure activities (e.g., watching television). In these calculations, the receptor was assumed to perform occupational and nonoccupational activities at the same location. The age of the individual during which exposure occurred was also considered when calculating risk.

Risks were reported for these scenarios at various locations in the domain as illustrated in figure 1.3, which shows risks estimated for the laborer scenario.

Table 1.2  Exposure scenario descriptions for Rocky Flats

| Exposure scenario | Gender | Year of birth | Year beginning exposure | Year ending exposure | Days per year exposed |
|---|---|---|---|---|---|
| Laborer | Male | 1934 | 1953 | 1989 | 365 |
| Homemaker | Female | 1934 | 1953 | 1989 | 350 |
| Office worker | Female | 1940 | 1965 | 1989 | 350 |
| Infant-child | Female | 1953 | 1953 | 1960 | 350 |
| Student | Male | 1957 | 1964 | 1974 | 350 |

**10**  Radiological Risk Assessment and Environmental Analysis

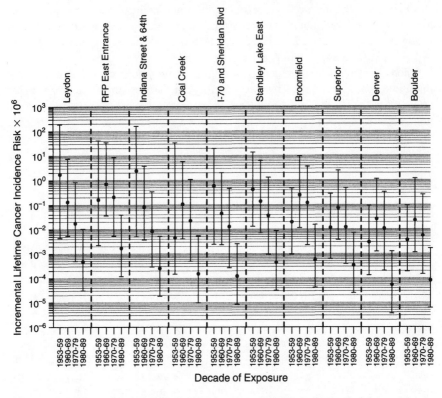

Figure 1.3  Lifetime cancer incidence risk from plutonium inhalation for the laborer scenario at selected locations in the model domain. Dots represent the 50th percentile value; horizontal bars represent the 5th and 95th percentile range. Cancer risks have been sorted by decade of exposure.

### Hanford Site Scenarios for Native Americans

In almost every risk assessment, there are special population groups who do not fit the usage factors for the general public. One example of this occurred in the dose reconstruction project for the Hanford Site. The Hanford facility released large amounts of radionuclides, $^{131}$I in particular, to the atmosphere. Significant quantities of materials were also released directly into the Columbia River, which was used for cooling the production reactors at the site (Farris et al. 1994a).

Pathways of exposure from the river were investigated thoroughly. Members of the general public who lived near the river received relatively small doses (estimated to be 15 mSv over about 40 years) from consumption of river water, consumption of fish from the river (150 kg of fish per year), and activities in and around the river. Special attention, however, had to be given to Native Americans who relied on the river for fish, a major component of their food (Grogan et al. 2002).

There was concern by Native Americans that because their unique lifestyles relied more heavily on natural sources of local foods and materials and because they had unique pathways of exposure, their risk may have been significantly greater than that of non-Native American people. Working with nine tribes in the region, scientists collected and summarized data, which allowed specific exposure factor information on diet, lifestyle, and special cultural ceremonies to be included to assess risk.

There were several pathways for which few data were available to estimate exposure and for which Native Americans were concerned about risk. Examples of these included exposure from shoreline sediment used for paints and medicinal purposes, sweat lodges using Columbia River water, and inhalation of river water spray during fishing. The pathway of most concern was that of fish consumption, not only because of the large quantities of fish consumed but also because they consumed the whole fish, which could significantly increase the dose and risk since some radionuclides concentrate in the bones of fish. Table 1.3 shows fish consumption data gathered with the involvement of Native Americans in the area and used in our risk estimates.

The results of the study indicated that except for the consumption of fish from the river, the risks from all other pathways would be small. In the case of consumption of fish, risks to Native Americans could have been substantially greater than those of non-Native American people. Since this study was a screening analysis, it is evident that the only pathway that deserved more detailed analysis, if quantitative estimates of risks were warranted, was consumption of fish from the Columbia River.

This discussion related to exposure factors illustrates several important points. The individual who is the target of exposure must be clearly defined in the beginning of the assessment. This step will help determine the scenarios of exposure and help identify specific exposure factors needed for the assessment. It must also be decided how the scenarios of exposure will be presented in the end so that people can understand what dose or risk they may have received.

The design of exposure scenarios and the data used to describe them are important to the credibility of the study. In some cases, generic information will be sufficient to characterize individuals for whom dose or risk is being calculated. It may be necessary, however, to undertake surveys or other methods for collecting exposure factor data when generic information is not available for specific groups of individuals with uncommon habits.

Table 1.3 Fish consumption of Native Americans for the Columbia River near the Hanford Site as reported by Walker and Pritchard (1999)

| Fish category[a] | Jan | Feb | Mar | Apr | May | Jun | Jul | Aug | Sept | Oct | Nov | Dec | Total | Holdup[b] (days) |
|---|---|---|---|---|---|---|---|---|---|---|---|---|---|---|
| Omnivore | 4 | 4 | 4 | 2 | 2 | 2 | 2 | 2 | 2 | 2 | 4 | 4 | 34 | 3 |
| First-order predator | — | — | — | — | — | — | — | — | — | — | — | — | 0 | 0 |
| Second-order predator | 4 | 4 | 4 | 2 | 2 | 2 | 2 | 2 | 2 | 2 | 4 | 4 | 34 | 3 |
| Salmon | 3 | 3 | 3 | 22 | 22 | 22 | 22 | 22 | 22 | 22 | 3 | 3 | 169 | 14 |

[a] Omnivorous fish include bullhead, catfish, suckers, whitefish, chiselmouth, chub, sturgeon, minnows, and shiners. First-order predators include perch, crappie, punkinseed, and bluegill. Second-order predators include bass, trout, and squawfish.
[b] The time between obtaining fish from the river and consuming it.

## Conversion to Dose

The conversion of radioactive materials taken into the body or the conversion of external radiation to dose has become a routine process because of the large effort put into deriving and publishing dose coefficients over the past several decades. Two chapters in the book address conversion to dose. Chapter 9, contributed by John Poston and John Ford, describes concepts of internal dosimetry. Chapter 10, contributed by David Kocher, focuses on external dosimetry.

There are two brief issues about conversion to dose I wish to make in this introductory chapter. The first is the importance of uncertainties related to dose coefficients and when, or when not, to take this uncertainty into account. The second issue is relatively new (ICRP 2007) and is related to the appropriate use of dose coefficients for compliance as a function of age.

## Uncertainty in Dose Coefficients

Until recently, little was understood about uncertainties associated with dose coefficients, and these values were typically used as single point values even when the radiological assessment was performed probabilistically. The use of single values probably came about because dose coefficients were first introduced as a means for determining compliance with a regulatory standard rather than for determining dose to individuals in a population. However, as more emphasis was placed on studies of populations where dose to specific individuals for use in epidemiology was the objective, it became clear that more information was needed on the uncertainty of these coefficients to properly address uncertainties in the calculation. As a result, considerable attention has been given to this important area of dosimetry over the past 10 years.

In the Hanford Environmental Dose Reconstruction Project (Farris et al. 1994b), it was determined that one of the two most important contributors to overall uncertainty was the dose coefficient for $^{131}$I; the other key component of uncertainty was the feed-to-milk transfer coefficient. In this analysis, it was pointed out that the uncertainty in the iodine dose coefficient was due primarily to variability in the mass of the thyroid, uptake of iodine in the gastrointestinal tract, transfer of iodine to the thyroid, and the biological half-time of iodine.

It is generally assumed that uncertainties associated with external dose coefficients are much less than those for internal dose coefficients and that there is little variability in dose per unit of exposure with age (Golikov et al. 1999, 2000). One reason for this low variability is because external radiation fields can be measured, and if measurements are carried out properly, the variability is small for a given location. Determining uncertainty of internal dose coefficients is a much more complicated process because radionuclides disperse after being taken into the body, and it is not possible to quantify precisely where they go and to measure the resulting dose. Therefore, it becomes an intensive computational process involving many assumptions. Nevertheless, much progress is being made in this area of dosimetry, and it will continue to be a viable area for research in the future.

## Appropriate Use of Dose Coefficients as a Function of Age

The International Commission on Radiological Protection (ICRP) has issued age-specific dose coefficients (dose per unit intake, $Sv\ Bq^{-1}$) for members of the public in six age ranges covering the time period from the newborn infant to 70 years of age (ICRP 1995, 1996a, 1996b). Additional refinements of these coefficients are also available for the embryo/fetus (ICRP 2001, 2005). The ICRP (2007) points out that "application of dose coefficients for the six age groups should be weighed in relation to the ability to predict concentrations in the environment from a source and the ability to account for uncertainties in habit data for individuals exposed." This is an important statement to consider in radiological assessment, especially when the assessment is being made for prospective calculations. It implies that a careful balance is needed between the resolution of dose coefficients being applied and the overall uncertainty in assessment.

Most likely, scientists will continue to refine dose coefficients into more discrete categories of age; however, this increased resolution will not likely give a better estimate of dose. As a result, ICRP (2007) recommends that some consolidation of dose coefficients is justified when the coefficients are being used for the purpose of determining compliance. There are a number of reasons the ICRP changed its policy, including the idea that compliance is generally determined by a dose standard that is typically set at a level to protect individuals from exposure to a continuing source over the lifetime of an individual. Table 1.4 lists the three age groups now recommended by the ICRP for compliance calculations.

The ICRP does recommend the use of specific age-group categories for retrospective calculations of dose and in addressing accidents. The reason for this recommendation is that specific information about age, diet, lifestyle, and other habit data is generally known.

Internal dosimetry and external dosimetry continue to be important areas of research. Too frequently, we assume that work in this area of radiological assessment is essentially complete; this assumption is not correct.

## *Conversion of Dose to Risk*

If the objective of radiological assessment is to estimate risk, then converting dose to risk is the next step. This step is generally accomplished by applying risk coefficients to doses that have been calculated for individuals. Increasingly, the intermediate

Table 1.4  Dose coefficients recommended by ICRP (2007) for compliance calculations

| Age category (years) | Name of age category | Dose coefficient and habit data to be used |
|---|---|---|
| 0–5 | Infant | 1-year-old |
| 6–15 | Child | 10-year-old |
| 16–70 | Adult | Adult |

step of calculating dose is subsumed into the calculation that converts exposure to risk. For example, Federal Guidance Report 13 (Eckerman et al. 1999) presents risk coefficients in terms of risk per unit intake via inhalation or ingestion. Risk coefficients and their foundation are covered in chapters 12 and 13. Chapter 12, contributed by Roger Clarke, addresses exposure standards, risk coefficients, and how these coefficients were developed over the years. Chapter 13, contributed by Owen Devine and Paul Garbe, explains how epidemiological studies to investigate the effects of exposure on populations can be designed to help determine if there are effects in populations following radiation exposure and if those effects can be attributed to the exposure.

Until the past decade, the end point of radiological assessment was typically dose, and conversion to risk was not routinely undertaken. Converting dose to risk, however, is becoming more important and useful for several reasons that are discussed below.

## Why Risk?

In the context of this chapter and this book, "risk" refers to risk of adverse health effects, primarily cancer, to humans from exposure to radioactive materials in the environment. Unfortunately, in radiological assessment, people are exposed not only to radioactive materials but also to chemicals. By using risk as an end point for the calculation in radiological assessment, one can compare the effects of radioactive materials with chemicals that may be present. Risk is the most fundamental common denominator in an assessment that can be estimated to help people understand current and prospective effects on humans and the environment from both radioactive materials and chemicals. If people have a better understanding of the risk imposed from exposure to these materials, it gives them a starting point for making decisions about potential cleanup or remediation.

There are other reasons to estimate risk in radiological assessment. The term "risk" is becoming more common in our language today. Medications are often described as having a risk of side effects. We discuss the risk posed by potential bad weather. Farmers refer to the risk of investing in an expensive crop. Of course, the type of risk referred to in this book could be described as a chance of harm from being exposed to radioactive materials in the environment. More specifically, risk is quantified in radiological assessment as a risk of the incidence of, or dying from, cancer following exposure.

## Risk Coefficients

As with conversion of intake or external exposure to dose, conversion of dose to risk is a straightforward process involving risk factors published by a number of different groups (UNSCEAR 2000). The current risk estimates of cancer following exposure to ionizing radiation are based primarily upon analyses of Japanese survivors of the atomic bombings at Hiroshima and Nagasaki. These risk estimates essentially relate to uniform whole-body exposures to predominantly low linear energy transfer radiation doses ranging from 0.01 Gy to 4 Gy delivered at high dose rate.

Risk coefficients and dose coefficients are similar with regard to our lack of understanding about the uncertainties associated with them. Typically, risk coefficients are applied with a single, deterministic value. We know that such a value is not valid and that the range of uncertainty associated with risk coefficients often may be quite large. Little work has been done to try to quantify this uncertainty, although scientists are working to quantify uncertainties in risk coefficients and to apply these uncertainties in their results.

Uncertainty in the risk factors for radiation was described by Sinclair (1993) and investigated more thoroughly by Grogan et al. (2001) as having five primary components:

- Epidemiological uncertainties
- Dosimetric uncertainties
- Projection to lifetime
- Transfer between populations
- Extrapolation to low dose and dose rate

Epidemiological uncertainties include statistical uncertainties associated with quantifying the relatively small number of excess cancers attributable to ionizing radiation from the background cancers resulting from all causes. Also included in epidemiological uncertainties are uncertainties from underreporting of cancers per unit population and nonrepresentativeness of populations used to determine risk. Dosimetric uncertainties include those from random errors in individual dose estimates arising from errors in the input parameters used to compute doses, and systematic errors due to the presence of more thermal neutrons at Hiroshima than originally estimated. Risk projection includes uncertainties associated with extrapolating beyond the time period covered by the observed population. Transfer of estimates of risk from one population (Japanese) to another introduces an additional source of uncertainty that must be considered. Finally, since the exposures for the A-bomb population were at relatively high dose rate, uncertainty is introduced when we extrapolate estimates of risk to low-dose, low-dose-rate situations common in most risk assessments. This area of risk assessment research is very important for the future, and the ideas introduced by Sinclair (1993) must be pursued. Indeed, we may find that the risk factors themselves introduce more uncertainty into the overall estimate of risk than does any other single component.

## *Uncertainty Analysis*

Uncertainty has been mentioned frequently up to this point, but it has not been explained or tied to the other components of radiological assessment. Uncertainty is covered in chapter 11, contributed by Thomas Kirchner. Uncertainty analysis is an essential element of risk assessment. Of all the steps in radiological assessment, this is the area where the greatest progress has been made over the past three decades. This success has been partly due to advances in techniques that are used to propagate uncertainties in calculations, but it is mainly due to the rapid improvements in computer technology. Today, uncertainties can be readily estimated with off-the-shelf software and laptop computers. This success was not imaginable even a decade ago.

Methods for quantification of uncertainty have been well established. Today, it is expected that when one carries out a risk assessment, the best estimate of risk is reported along with associated uncertainties.

The most common method for uncertainty analysis uses Monte Carlo statistical techniques incorporating a random sampling of distributions of the various models and parameters involved (see figure 1.4). In this simplified illustration, $A$ is an input

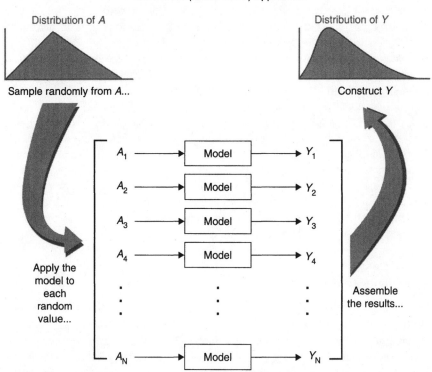

Figure 1.4 Schematic presentation of Monte Carlo methods for propagating a parametric uncertainty distribution through a model to its results.

parameter to the model, and $Y$ is the result, or output, corresponding to $A$. For each specific value of $A$, the model produces a unique output $Y$. Such an application of the model is deterministic because $A$ determines $Y$. But $A$ may not be known with certainty. If uncertainty about $A$ is represented by a distribution, such as the triangular one in the figure, repeatedly sampling the distribution at random and applying the model to each of the sample input values $A_1, A_2 \ldots$ gives a set of outputs $Y_1, Y_2, \ldots$, which can be arranged into a distribution for $Y$. The distribution of $Y$ is then the estimate of the uncertainty in $Y$ that is attributable to the uncertainty in $A$. This is a stochastic, or probabilistic, application of the model.

Proposed distributions may be based on measurements or on scientific judgment when data are not available. Site-specific data are used when such measurements exist for relevant times, locations, and processes, but often surrogate data based on other times or locations must be used. The most difficult aspect of uncertainty analysis is the selection of parameters and distributions to be used in the analysis.

## Use of Uncertainty for Determining Compliance with Standards

Little attention has been given to how uncertainties might be considered when radiological assessment was being used to determine compliance with environmental regulations. Until recently, deterministic calculations were used as the comparison value without regard to the uncertainties associated with them. ICRP (2007) clarifies this matter for exposures to the public in prospective situations.

The difficulty in applying uncertainties in determining compliance with a standard arises from the fact that, in almost all cases, some members of the population exposed will exceed the dose benchmark (e.g., 50th percentile, 95th percentile) that is used as the basis for comparison. The number of people who exceed the criterion for comparison and the level of dose they may receive are important to consider. As a result, ICRP (2007) recommends the following:

> In a prospective probabilistic assessment of dose to individuals, whether from a planned facility or an existing situation, the ICRP recommends that the representative individual be defined such that the probability is less than about 5% that a person drawn at random from the population will receive a greater dose. In a large population, many individuals will have doses greater than that of the representative individual, because of the nature of distributions in probabilistic assessments. This need not be an issue if the doses are less than the relevant dose constraint. However, if such an assessment indicates that a few tens of persons or more could receive doses above the relevant constraint, then the characteristics of these people need to be explored. If, following further analysis, it is shown that doses to a few tens of persons are indeed likely to exceed the relevant dose constraint, actions to modify the exposure should be considered.

This recommendation by the ICRP illustrates some of the problems that will be encountered as uncertainties are accounted for in future radiological assessments. Other difficult issues will be encountered, as well. These include the acceptance and understanding of uncertainty by the public and the misuse of uncertainty to argue the presence of an upper bound (e.g., 99th percentile) dose or risk to an individual as being the basis for a legal decision. Regardless of the difficulties introduced when uncertainties are accounted for in radiological assessment, the benefits far outweigh the problems, and the result is a more realistic understanding of dose and risk.

## Validation

The term "validation" is used here to mean efforts taken to verify the estimates made in radiological assessment. Since direct measurements of dose to people exposed cannot be readily taken, validation typically involves comparing predicted concentrations in the environment with measurement data. Validation in radiological assessment is discussed in chapter 14, contributed by my co-editor, Helen Grogan. Validation can be expensive and is very difficult to accomplish, especially in prospective assessments. However, retrospective assessments in historical dose reconstructions conducted over the past two decades have provided a unique opportunity to validate the predictions of environmental transport models. The historical records contain a vast resource of measurement data, much of it collected many years ago, that have been used as the basis for comparisons between estimated environmental concentrations and values that were measured. These comparisons have given scientists good indications of the reliability of many of the mathematical models being developed and applied by scientists in radiological risk assessment.

Before beginning the assessment process, methods for validating the results should be considered. In the case of prospective assessments, this could be identifying areas where deposition of materials is expected to occur and locating measurements to be collected in the future to determine the amount of material actually deposited. These future data could confirm the accuracy of model predictions or lead to changes in the model. This type of planned validation is becoming more common at facilities today.

More frequently, however, validation has been performed in radiological assessments after the release of a source (an accident or routine release), often many years after the release occurred. Comparing environmental measurements with predicted environmental concentrations is one way of gaining confidence in the methods used to estimate concentrations in the environment. When the source term is well known (i.e., has little uncertainty), the validation is primarily checking how well the environmental model works. However, when the source term is not well known, as in the case of historical releases to the environment, the validation cannot discern between the source term and the environmental transport to know which introduces the most uncertainty.

It is important to look for data that could be used for validation and to keep these data independent from other data that may be used as part of the calculation. As an example of how validation can help improve confidence in modeling, two examples are discussed below that were taken from the Fernald Historical Dose Reconstruction Project (Till et al. 2000).

Unfortunately, in this dose reconstruction, very little information existed to validate radon releases from the on-site radium storage facilities during the period of highest releases (before 1979). Most of the environmental monitoring during the site's early history focused on uranium. Two examples of these comparisons are shown below.

Table 1.5 compares predicted and observed concentrations of uranium at air monitoring stations around the site. There was no routine monitoring of uranium in air in the mid-1950s when uranium releases from the Fernald FMPC were highest. Because predicted concentrations were made on an annual basis, the comparisons with observations were also annual. There were 14 pairs of comparisons at each of the perimeter stations, and 17 pairs at each of the boundary stations, with the exception of boundary station 7, for which 11 years of data were available. The observed annual concentrations were computed from compilations of weekly measurements. The measurements were adjusted for inefficiency of the samplers in collecting larger particles that were predicted to be present at each sampler. The geometric bias (the geometric mean of the annual predicted-to-observed ratios) at both the perimeter and boundary stations was 1.0, which is excellent overall agreement. The average correlation between the predicted and observed concentrations at perimeter stations was more than 0.65, but it was less at four of the boundary stations. The lower correlation at the boundary stations could be partially due to the smaller range of concentrations at the stations. In general, concentrations were overpredicted to the east of the site and underpredicted to the north and west. An exception was boundary station BS-6, to the west, which had a geometric bias of 1.3.

Figure 1.5 illustrates time trends for the perimeter and boundary stations northeast and southeast of the FMPC from 1958–1988. Data from routine air monitoring were

Table 1.5 Comparison of predicted (P) and observed (O) uranium concentrations in air at monitoring stations at the Fernald Feed Materials Production Center (Till et al. 2000)

| Monitoring station | Distance from FMPC center(km) | Correlation between $\ln(P)$ and $\ln(O)$ | Geometric bias | Long-term average concentration (mBq m$^{-3}$) | |
|---|---|---|---|---|---|
| *Perimeter (1958–1971)* | | | | Predicted | Observed |
| SW | 0.5 | 0.65 | 0.57 | 5.6 | 8.1 |
| NW | 0.5 | 0.86 | 0.64 | 3.0 | 4.1 |
| NE | 0.5 | 0.85 | 1.86 | 16 | 7.4 |
| SE | 0.5 | 0.79 | 1.42 | 9.6 | 5.6 |
| *Boundary (1972–1988)* | | | | | |
| BS–1 | 0.9 | 0.62 | 0.64 | 0.48 | 0.63 |
| BS–2 | 1.3 | 0.72 | 1.57 | 0.96 | 0.48 |
| BS–3 | 0.7 | 0.50 | 1.46 | 1.4 | 0.74 |
| BS–4 | 1.4 | 0.30 | 1.28 | 0.41 | 0.24 |
| BS–5 | 1.3 | 0.18 | 0.81 | 0.32 | 0.30 |
| BS–6 | 1.1 | 0.67 | 1.30 | 0.85 | 0.44 |
| BS–7 | 1.6 | 0.90 | 0.52 | 0.16 | 0.33 |
| **Perimeter group** | | | 1.0 | | |
| **Boundary group** | | | 1.0 | | |

available for comparison beginning in 1958 at the four perimeter stations. In 1972, air monitoring ceased at the perimeter and began at the seven boundary stations. Each plot in figure 1.5 contains the monitoring data record from one perimeter station and one boundary station. There is a clear decrease in uranium concentration over time, which is consistent with decreasing releases from FMPC facilities. The lower detection limit in the 1970s and 1980s also permitted the lower concentrations to be measured accurately.

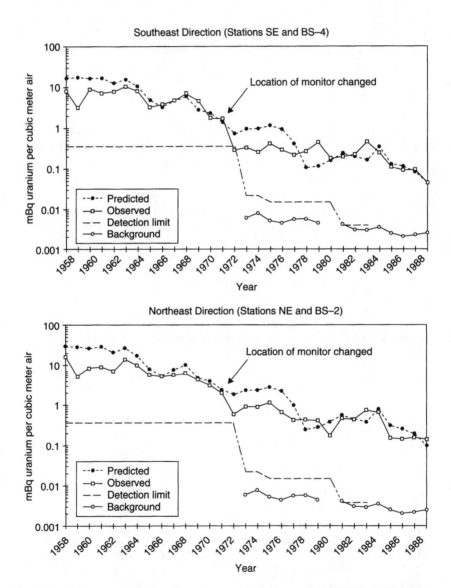

Figure 1.5 Time trend in observed and predicted concentrations of uranium in air at the Fernald Feed Materials Production Center from 1958 to 1988 (Till et al. 2000).

Natural background concentrations of uranium, mainly because of particles of soil suspended in the air, were much less than the uranium concentrations in air historically measured near the FMPC. For this reason, there was no need to subtract background concentrations from the observations before making comparisons to the predicted concentrations.

Validation, like uncertainty analysis, must be an integral part of radiological assessment. It must be planned into the assessment process from the beginning. In most cases, it is only possible to perform validation on one or two aspects of the results. Often the results of validation do not support the calculations being made and will require us to rethink the source term, the environmental transport mechanism, or both. On the other hand, being able to show that measurement data of concentrations in the environment confirm predicted concentrations adds enormous credibility (both scientific and public) to the results.

## *Communication of Dose and Risk and Stakeholder Participation*

The final two steps in radiological assessment are

- Communication of dose and risk
- Public participation

Unfortunately, these two aspects of the equation are not addressed in this book, but they may be the two most important elements of radiological assessment. Scientists involved in the field of radiological assessment are generally not very competent at communicating to the public what they do or recognizing the importance of stakeholder participation in the process.

During the years of my work in radiological assessment, the term "stakeholder" has become increasingly used. Because there are many definitions for this term depending on who is using it, it is helpful to provide a definition of stakeholder that seems to consistently fit situations in which I have been involved. For the purposes of radiological assessment, stakeholders are defined as "[i]ndividuals who have a personal, financial, health, or legal interest in policy or recommendations that affect their well-being or that of their environment" (Till 2002).

If the results of the assessment cannot be effectively communicated to the individuals who are exposed (typically the decision makers and the public, i.e., the stakeholders), the work may not be accepted as credible. Participation of the people who are exposed is also critical to the success of the assessment process. This section examines some fundamental principles that are important in achieving effective communication of the results and public participation in the radiological assessment.

## Communication of Results from Radiological Assessment

In almost all situations, we are required to estimate risk to members of the public who have been or who may be exposed to releases of radionuclides to the environment and to consider how these estimates compare with regulations for exposure. Regulations are a very important aspect of radiological assessment. A history of regulations and a summary of current regulations related to environmental exposures to radioactive materials are provided in chapter 15, contributed by David Kocher.

In the end, we must achieve both public and scientific credibility in radiological assessment (Till 1995). As scientists, we have come a long way in establishing the technical methods described in the sections above. On the whole, however, we are quite inept in communicating our results with the public and working with the public in conducting risk assessments. This is an area where the most progress could be made with a small investment of resources.

Risk communication has become a branch of radiological and chemical assessment, with individuals who specialize in effective methods to communicate dose and risk to people who are exposed (National Research Council 1989, 1996; Lundgren and McMakin 1998). Many techniques have been devised to explain to the public the meaning of doses and risks. Comparisons have been proposed between risks from exposure to radiation and risks from other sources common in life. We may be able to help people understand the potential importance of risks by referring to the overall risks posed by natural background, fallout, and other involuntary exposures and comparing these to our own estimates of risk for a particular situation. Although these comparisons help, there is no replacement for scientists personally conveying the results of their calculations to the public who is at risk. Communication of the results of radiological assessment is primarily the responsibility of those conducting the assessment. We would all like to think that society is ready to agree on a common set of risk levels for decision making. Indeed, it would make life simpler, but we are not at that point with either the technology to compute risks or the public's understanding and acceptance of risk. Trying to impose a level of risk as being "significant" or "insignificant" leads to a serious loss of credibility. During my past decade of work on dose reconstruction projects, I have learned that the best way to respond in communicating the significance of a level of risk is to respond personally. We as scientists can tell people how *we* feel about the significance of a level of risk, but we cannot tell members of the public what *they* should feel about that level of risk.

What we can do, however, is to take steps to help people understand the assessment process so that they will have a better understanding of how the results were achieved in an assessment and how to interpret them. The next section discusses several principles that, if followed during radiological assessment, can greatly help communicate the results and earn credibility in the process.

*Principles to Improve Communication of Radiological Assessment Results*
Table 1.6 lists four principles that would, if implemented, greatly enhance the communication and understanding of results in radiological assessment. These principles should be followed regardless of whether the materials being released (or potentially released) to the environment are radioactive materials or chemicals.

The first principle is that environmental data related to the facility being evaluated should be readily available to all stakeholders. Many facilities currently follow this practice today by issuing annual environmental reports that explain what radioactive materials were released and how these releases compare to environmental regulations and by providing information about background concentrations of these

Table 1.6  Principles for improving communication and understanding of results in radiological assessment

- Environmental data should be readily accessible to decision makers and the public
- Risk should be a fundamental end point of the assessment if possible
- User-friendly tools should be available to aid in decision making
    - Transparent
    - Flexible
    - Peer reviewed
- There should be a process to work with those who are exposed (or may be exposed if assessment is prospective) during the assessment and during decision making

materials and other benchmarks to help put the releases into perspective. The best way to make environmental data available to stakeholders is to put it on the Internet, where people have access to the data as they become available. This practice is relatively new but will become the standard approach for disseminating information in the future.

The second principle is to use human health risk as an end point in radiological assessment whenever possible. Risk gives a better perspective to help people understand the meaning of exposure from radionuclides in the environment. It also allows comparisons with risks from exposure to chemicals and with hazards of other exposures and activities in life.

A third principle is to provide readily accessible tools that help people clearly follow the assessment process and provide interpretation of the results. These tools can be made available in different formats, from software to reports that are written specifically for the public or decision makers, and they can be made available on the Internet. Software tools should have several important features: (1) transparency, so users know how calculations are being made and what values are being used; (2) flexibility, so individuals can readily make changes to input data to see the effect on results; and (3) independent peer review, to ensure scientific and public credibility.

There is always concern that making radiological assessment tools available to individuals who are not trained in the field will lead to misuse of information. This misuse is certainly possible and is inevitable to some degree. From my experience, however, the positive benefits far outweigh the negative aspects of this principle, and taking this approach will ultimately widen people's knowledge about radiological assessment and the meaning of exposures to radiation.

The fourth principle is that there must be a mechanism for decision makers to consider input from the people who are exposed. There are many different mechanisms to achieve this feedback, such as from public meetings, through the Internet, and by actively engaging a group of stakeholders to participate in the radiological assessment.

Regardless of the mechanism used, it is critical to success that decision makers and the people exposed (or potentially exposed) have an ongoing dialogue throughout the process. This principle may increase the cost of the work in the short term, but it will save significant resources in the long term if the completed assessment is accepted as thorough and credible. How to achieve this participation by stakeholders is discussed next.

## Stakeholder Participation

There have been many successful radiological assessment studies over the past two decades where stakeholders have been involved in the process. There have also been many unsuccessful attempts. However, there is no question that radiological assessment is more successful if stakeholders are actively engaged in the process from the beginning.

Probably the most groundbreaking radiological assessment project where stakeholder involvement was critical to the outcome and acceptance of the work was the Hanford Environmental Dose Reconstruction Project (Till 1995). This study established a number of precedents for stakeholder participation that are in use today. These include elements such as openness of the technical project work, documentation of responses to comments received from stakeholders, and the participation of Native American tribes in gathering information on diet and lifestyle needed to estimate doses and risks.

This lesson about public involvement at Hanford was expanded further in another dose reconstruction project that focused on the Savannah River Site, another facility in the U.S. nuclear weapons complex. The goal was to begin the project with as much stakeholder involvement as possible. The first phase for this project was to review and catalogue all of the historical records on site, amounting to almost 50,000 boxes of historical records in total (Till 1997). Each box had to be reviewed for content and entered into a database for future retrieval in subsequent phases of the study. With the cooperation of management at the site, we invited stakeholders representing several groups of citizens to watch as we reviewed historical records during the early stages of the project. This step clearly showed the stakeholders what was involved in reviewing historical records and how tedious the process really was. Although the interest by stakeholders in watching us review historical records soon dwindled, providing them an opportunity to watch first-hand helped achieve credibility and led to a positive and fruitful interaction with the public during the course of the project.

There are three basic questions that need to be answered to have effective stakeholder participation (Till 2002). The first question is, "Do you believe stakeholders can play a role in making policy recommendations and can help make better decisions about protecting the environment?"

A good example of an answer to this question is the Rocky Flats Radionuclide Soil Action Level Study (Till and Meyer 2001). A panel of stakeholders directed the study. The panel was composed of a cross-section of the community, which consisted of a mix of technical specialists and people with no technical experience drawn from public interest groups, local governments, and the general pubic. The three responsible agencies (U.S. Department of Energy, U.S. Environmental Protection Agency, and Colorado Department of Public Health and Environment) were each represented by one ex-officio member, making the number of people on the panel total 16. The panel met monthly for work sessions with my research team from October 1998 to March 2000, with all meetings open to the public. During this time, the stakeholders became familiar with terminology such as uncertainty, concentration factors, breathing rates, resuspension, and many other technical terms that were crucial to determining a level for cleanup.

What is important to recognize in this example is that although the stakeholders may not have had the detailed knowledge of the science that was being conducted, they firmly grasped the significance of crucial components of the analysis and how changes in these components affected the final result. For example, figure 1.6 shows how we presented uncertainty, which was as a probability of exceeding the dose limit as a function of the cleanup level. Distributions were developed for different scenarios of exposure. The stakeholders decided that they wanted a 10% probability level, that is, a 90% probability of not exceeding the dose limit. In setting this level, the stakeholders recognized the dose limit could be exceeded 10% of the time, therefore acknowledging that dose limits can be exceeded due to the uncertainties of the real world. The concept of uncertainty is often difficult to explain, but this approach was readily accepted by the stakeholders.

What evolved was a strong endorsement by the panel for a plutonium cleanup level of about 1,300 Bq kg$^{-1}$. We made it clear that there could not be a single value for soil cleanup, but that any value within a general range of between 1,000 and 2,300 Bq kg$^{-1}$ would likely be acceptable.

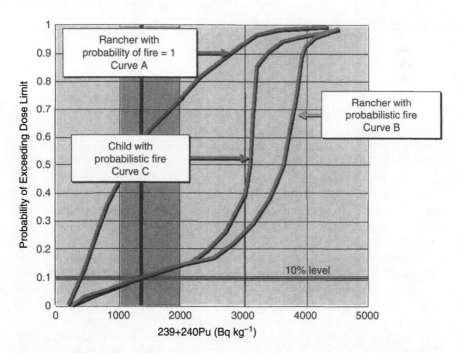

Figure 1.6 Illustration of the bounding scenarios that shows a range of possible soil action levels centered around about 1,300 Bq kg$^{-1}$, which is the 10% probability level for the rancher and child scenarios.

The example from Rocky Flats clearly demonstrates effective stakeholder involvement in radiological assessment. It is certainly a perfect example showing that stakeholder involvement, if conducted with a deliberate set of ground rules to follow, can be effective in making recommendations and decisions about risk.

The second question is "Do you believe stakeholders can help us conduct better science to be used in making recommendations about protection of the environment?"

For this example, I refer the reader back to table 1.3 and fish consumption by Native Americans from the Columbia River near the Hanford Site in Washington State. Early in the historical dose reconstruction at Hanford, it was realized that the scientists conducting the study had little knowledge about the habits and diets of Native Americans who lived along and extensively used the Columbia River. As discussed above, there was concern by Native Americans that because their unique lifestyles relied more heavily on natural sources of local foods and materials and because they had unique pathways of exposure, their risks may have been significantly greater than that of non-Native American people. The data in table 1.3 (along with other important diet and lifestyle information) were collected by involving the affected tribes in the study.

The point of this example is to emphasize that stakeholders can provide critical information in risk analysis and can help us conduct more defensible science. Although this is just one example of the stakeholders' role in radiological assessment, it has been our experience that a better product has evolved in every case where stakeholders are engaged in the scientific process.

The third question is "Do you believe that by involving stakeholders in the decisions and recommendations we make today, those decisions and recommendations will be more enduring and better accepted in the future?"

Of course, there is no way to answer this question conclusively, and only time will tell us the truth. The following example, however, helps provide some insight.

In 1995, a lawsuit was filed by a local citizens' group alleging that the Los Alamos National Laboratory was not in compliance with the requirements prescribed by the Clean Air Act for radionuclides. The suit was settled through an agreement that called for a series of independent audits. Our research team was asked by both parties to be the independent auditor. The question posed to us was, "Did Los Alamos National Laboratory meet requirements for compliance with 40 CFR 61, subpart H (U.S. EPA 1989), for the year 1996?" The settlement agreement also provided that representatives of the citizen's group were to observe my team during the audit process to verify the audit's integrity.

Because of the nature of research that is undertaken at Los Alamos National Laboratory and its mission, the laboratory must maintain a high degree of security and therefore neither is open to the public nor has traditionally encouraged public participation in many site-related activities. The audit process included site visits and tours of facilities, document review and retrieval, and interviews with staff.

It is easy to imagine the complex situation that existed, not only having to carry out the tasks required to conduct the audit, but also being observed during the process, as well as communicating what we were doing to representatives of the citizens group that had filed the lawsuit.

We quickly became overwhelmed with questions being asked by the stakeholders' representatives. Therefore, we decided to require that all questions be written down, and we promised to try to answer them as part of the audit. Ultimately, 75 questions were submitted by the stakeholders that were responded to in our report. The documentation of stakeholder questions was critical to our success and added significantly to the stakeholders' understanding of the audit process and its credibility.

Regular work sessions were held where we allowed the stakeholders to join us and look at how we were evaluating compliance. Although we found the process tedious at first, once we laid out a plan for the audit, announced our schedule of visits, and clearly established the rules to be followed, it was an orderly and rather quick process. We conducted a total of three audits over a period of five years.

What does the Los Alamos audit tell us about how involving stakeholders today can help us make better and more enduring decisions for the future? What is so remarkable about this example is that when we began our first audit of Los Alamos National Laboratory, the laboratory's attitude toward any involvement of the public in compliance issues was defensive and closed. The laboratory did not see any role that the public should play in its complying with the regulation, and viewed it as strictly a regulatory, technical matter.

One outcome of the audits is that local stakeholder groups, Los Alamos scientists, and the regulator (the New Mexico Environment Department) now meet regularly to discuss issues openly and to improve communication about the audit process. This example at Los Alamos illustrates how stakeholders can help us make better and more enduring decisions.

Stakeholder involvement is not public relations. Nor is it an excuse for investing fewer resources into science, thinking you can get by with a lower quality product. Indeed, the stakeholders' expectations of the product, whether it is recommendations on protection of the environment or demonstration of compliance, are even tougher to meet.

So, how do we involve stakeholders in developing environmental radiation protection policy and recommendations? First, we must believe the answer to each of the three questions above is, "Yes." Second, we must have some guidelines within which to work. Although these are still evolving as we learn more with each study, six key guidelines to follow are presented in table 1.7.

Table 1.7 Guidelines to follow in working with stakeholders

- Recognize the difficulty of this commitment
- Understand that short-term costs are greater
- Clearly define the role and authority of stakeholders
- Develop a plan for receiving and responding to stakeholder input
- Have a well-defined schedule and end product
- Recognize that stakeholder involvement cannot be retracted once the commitment is made

## Conclusion

It should now be evident that radiological assessment requires the skills of many different sciences because it is a discipline that brings these different skills together to provide us with an understanding of the meaning of radioactive materials when they are released to the environment. Each step of the assessment process builds upon the steps that precede it. To be conducted properly, it takes a team of individuals who have these skills working together to perform radiological assessments in a defensible and proper manner. From source terms to transport, to exposure, dose, and risk, each step is unique and important.

The purpose of this chapter is to set the stage for the remainder of the book, where you will find greater detail and more explanation about each step of the science of radiological assessment.

### References

Eckerman, K.F., R.W. Leggett, C.B. Nelson, J.S. Puskin, and A.C.B. Richardson. 1999. *Cancer Risk Coefficients for Environmental Exposure to Radionuclides*. Federal Guidance Report 13. EPA 402-R-99–001. Office of Radiation and Indoor Air, U.S. Environmental Protection Agency, Washington, DC.

Farris, W.T., B.A. Napier, J.C. Simpson, S.F. Snyder, and D.B. Shipler. 1994a. *Columbia River Pathway Dosimetry Report, 1944–1992*. PNWD-2227 HEDR. Battelle Pacific Northwest Laboratories, Richland, WA.

Farris, W.T., B.A. Napier, P.W. Eslinger, T.A. Ikenberry, D.B. Shipler, and J.C. Simpson. 1994b. *Atmospheric Pathway Dosimetry, 1944–1992*. PNWD-2228 HEDR. Battelle Pacific Northwest Laboratories, Richland, WA.

Golikov, V., M. Balonov, V. Erkin, and P. Jacob. 1999. "Model Validation for External Doses Due to Environmental Contaminations by the Chernobyl Accident." *Health Physics* 77(6): 654–661.

Golikov, V., M. Balonov, and P. Jacob. 2000. "Model of External Exposure of Population Living in the Areas Contaminated after the Chernobyl Accident and Its Validation." In *Harmonization of Radiation, Human Life, and the Ecosystem*. Proceedings of the 10th International Congress of the IRPA, International Conference Centre, Hiroshima, Japan, 746-T-19(1)-2.

Grogan, H.A., W.K. Sinclair, and P.G. Voillequé. 2001. "Risks of Fatal Cancer from Inhalation of Plutonium-239,240 by Humans: A Combined Four Method Approach with Uncertainty Evaluation." *Health Physics* 80(5): 447–461.

Grogan, H.A., A.S. Rood, J.W. Aanenson, E.B. Liebow, and J.E. Till. 2002. "A Risk-Based Screening Analysis for Radionuclides Released to the Columbia River from Past Activities at the U.S. Department of Energy Nuclear Weapons Site in Hanford, Washington." RAC Report 3-CDC-Task Order 7–2000-FINAL. Risk Assessment Corporation, Neeses, SC.

ICRP (International Commission on Radiological Protection). 1995. "Age-Dependent Doses to Members of the Public from Intake of Radionuclides: Part 3, Ingestion Dose Coefficients." ICRP Publication 69. *Annals of the ICRP* 25(1).

ICRP. 1996a. "Age-Dependent Doses to Members of the Public from Intake of Radionuclides: Part 4, Inhalation Dose Coefficients." ICRP Publication 71. *Annals of the ICRP* 25(3).

ICRP. 1996b. "Age-Dependent Doses to Members of the Public from Intake of Radionuclides: Part 5, Compilation of Ingestion and Inhalation Dose Coefficients." ICRP Publication 72. *Annals of the ICRP* 26(1).

ICRP. 2001. "Doses to the Embryo and Fetus from Intakes of Radionuclides by the Mother." ICRP Publication 88. *Annals of the ICRP* 31(1–3).

ICRP. 2005. "Doses to Infants from Ingestion of Radionuclides in Mothers' Milk." ICRP Publication 95. *Annals of the ICRP* 34(3–4).

ICRP. 2007. "Assessing Dose of the Representative Person for the Purpose of Radiation Protection of the Public and the Optimization of Radiological Protection: Broadening the Process." ICRP Publication 101. *Annals of the ICRP* 36(3).

Lundgren, R., and A. McMakin. 1998. *Risk Communication: A Handbook for Communicating Environmental, Safety, and Health Risks.* Battelle Press, Columbus, OH.

Meyer, K.R., P.G. Voillequé, D.W. Schmidt, S.K. Rope, G.G. Killough, B. Shleien, R.E. Moore, M.J. Case, and J.E. Till. 1996. "Overview of the Fernald Dosimetry Reconstruction Project and Source Term Estimates for 1951–1988." *Health Physics* 71(4): 425–437.

Mohler, H.J., K.R. Meyer, H.A. Grogan, J.W. Aanenson, and J.E. Till. 2004. "Application of NCRP Air Screening Factors for Evaluating Both Routine and Episodic Radionuclide Releases to the Atmosphere." *Health Physics* 86(2).

National Research Council. 1989. *Improving Risk Communication.* National Academy Press, Washington, DC.

National Research Council. 1996. *Understanding Risk: Informing Decisions in a Democratic Society.* National Academy Press, Washington, DC.

NCRP (National Council on Radiation Protection and Measurements). 1996. *Screening Models for Releases of Radionuclides to Atmosphere, Surface Water, and Ground.* NCRP report 123. National Council on Radiation Protection and Measurements, Bethesda, MD.

Ramsdell, J.V., Jr., C.A. Simonen, and K.W. Burk. 1994. *Regional Atmospheric Transport Code for Hanford Emission Tracking (RATCHET).* PNWD-2224 HEDR. Battelle Pacific Northwest Laboratories, Richland, WA.

Rood, A.S., and H.A. Grogan. 1999. *Estimated Exposure and Lifetime Cancer Incidence Risk from Plutonium Released from the 1957 Fire at the Rocky Flats Plant.* RAC Report 2-CDPHE-RFP-1999-FINAL. Prepared for the Colorado Department of Public Health and Environment by Radiological Assessments Corporation, Neeses, SC.

Rood, A.S., H.A. Grogan, and J.E. Till. 2002. "A Model for a Comprehensive Assessment of Exposure and Lifetime Cancer Incidence Risk from Plutonium Released from the Rocky Flats Plant, 1953–1989." *Health Physics* 82(2): 182–212.

Sinclair, W.K. 1993. "Science, Radiation Protection, and the NCRP," Lauriston S. Taylor Lectures in Radiation Protection and Measurements, Lecture 17. National Council on Radiation Protection and Measurements, Bethesda, MD.

Till, J.E. 1990. "Reconstructing Historical Exposures to the Public from Environmental Sources." In *Radiation Protection Today—the NCRP at 60 Years.* Proceedings of the 25th annual meeting of the NCRP. National Council on Radiation Protection and Measurements, Bethesda, MD.

Till, J.E. 1995. "Building Credibility in Public Studies." *American Scientist* 83(5).

Till, J.E. 1997. "Environmental Dose Reconstruction." In *Proceedings of the Thirty First Annual Meeting of the National Council on Radiation Protection and Measurements (NCRP), Washington, D.C., April 12–13, 1995.* National Council on Radiation Protection, Bethesda, MD.

Till, J.E. 2002. "Stakeholder Involvement in Developing Environmental Radiation Protection Policy and Recommendations." In *Radiological Protection of the Environment, the Path Forward to a New Policy*. Proceedings from the NEA Forum in Collaboration with the International Commission on Radiological Protection, Taormina, Sicily, Italy.

Till, J.E., and K.R. Meyer. 2001. "Public Involvement in Science and Decision-Making." *Health Physics* 80(4): 370–379.

Till, J.E., G.G. Killough, K.R. Meyer, W.S. Sinclair, P.G. Voillequé, S.K. Rope, and M.J. Case. 2000. "The Fernald Dosimetry Reconstruction Project." *Technology* 7: 270–295.

Till, J.E., A.S. Rood, P.G. Voillequé, P.D. Mcgavran, K.R. Meyer, H.G. Grogan, W.K. Sinclair, J.W. Aanenson, H.R. Meyer, S.K. Rope, and M.J. Case. 2002. "Risk to the Public from Historical Releases of Radionuclides and Chemicals at the Rocky Flats Nuclear Weapons Plant." *Journal of Exposure, Analysis, and Epidemiology* 12(5): 355–372.

UNSCEAR (U.N. Scientific Committee on the Effects of Atomic Radiation). 2000. *Sources and Effects of Ionizing Radiation*. Report to the General Assembly, with annexes. United Nations, New York.

U.S. EPA (U.S. Environmental Protection Agency). 1989. "40 CFR 61.90–61.97, Subpart H, National Emission Standards for Emissions of Radionuclides Other Than Radon from Department of Energy Facilities." U.S. Environmental Protection Agency, Washington, DC.

Voillequé, P.G. 1999. *Estimated Airborne Releases of Plutonium During the 1957 Fire in Building 71*. RAC Report 10-CDPHE-RFP-1999-FINAL. Prepared for the Colorado Department of Public Health and Environment by Radiological Assessments Corporation, Neeses, SC.

Voillequé, P.G., K.R. Meyer, D.W. Schmidt, G.G. Killough, R.E. Moore, V.I. Ichimura, S.K. Rope, B.S. Shleien, and J.E. Till. 1995. *The Fernald Dosimetry Reconstruction Project, Tasks 2 and 3, Radionuclide Source Terms and Uncertainties*. RAC Report CDC-5. Radiological Assessments Corporation, Neeses, SC.

Walker, D.E., Jr., and L.A. Pritchard. 1999. *Estimated Radiation Doses to Yakima Tribal Fishermen*. Walker Research Group Ltd., Boulder, CO.

Whicker, F.W., T.G. Hinton, M.M. MacDonnell, J.E. Pinder, and L.J. Habegger. 2004. "Avoiding Destructive Remediation at DOE Sites." *Science* 303(5664): 1615–1616.

# 2

# Radionuclide Source Terms

Paul G. Voillequé

The first step in an assessment of risks due to releases of radionuclides into the environment is often a description of the "source term," a shorthand expression that refers to the quantities and compositions of radioactive materials released, locations of the release points, and the rates of release during the times considered in the assessment. The released radionuclides may be gaseous, associated with airborne particles, or dissolved or suspended in aqueous or other liquids. Operating facilities typically have routine releases of radionuclides to air via stacks and chimneys and to water bodies via liquid discharge outfalls. Waste materials that are stored onsite or at disposal facilities may be released to the air or water, or to the soil and then to groundwater. In all cases, the particle size and chemical form or solubility of the released activity can be important for the proper estimation of radionuclide transport in the environment (discussed in chapters 3–7).

The next section contains a brief discussion about the radionuclides of interest for source term development for nuclear facilities. This is followed by a section that identifies situations when source term estimation was found to be unnecessary or was not the best approach for particular dose and risk assessments. These can generally be identified as circumstances when reliable information exists on levels of environmental or personal contamination. The fourth section describes the various activities and facilities for which source terms may be needed for dose and risk assessment purposes. These cover the range of nuclear fuel cycle facilities and other human activities. The last section provides guidance for source term development for prospective and retrospective analyses.

## Radionuclides of Interest and Their Properties

Estimates of source terms for nuclear facilities typically will include several radionuclides. These may be roughly categorized as naturally occurring and "man-made," although the distinction is not clear-cut. Some of the so-called man-made radionuclides are also produced naturally by activation and fission. Radionuclides such as $^3$H (also called tritium) and $^{14}$C, which result from the operation of nuclear reactors, are also produced by cosmic ray interactions in the atmosphere (NCRP 1987). Spontaneous fission of uranium isotopes also occurs (rarely) in nature, but measurements of fission and activation products have identified the location of natural nuclear chain reactions that continued for an extended period long ago in African uranium deposits (Cowan 1976).

The principal naturally occurring radionuclides used in nuclear facilities are the isotopes of uranium and, to a much lesser extent, those of thorium. The half-lives (ICRP 1983) of the three principal uranium isotopes ($4.468 \times 10^9$ yr for $^{238}$U, $2.446 \times 10^5$ yr for $^{234}$U, and $7.038 \times 10^8$ yr for $^{235}$U) are so long that their current relative abundances (0.99275, 0.000054, and 0.0072, respectively) have not changed in the past century during which they have been a focus of human activity. Also present with the uranium isotopes are radioactive decay products that have accumulated due to the decay of uranium isotopes. The decay chain for $^{238}$U includes isotopes of thorium (Th), protactinium (Pa), radium (Ra), and radon (Rn). The first part of the decay sequence after $^{238}$U, with half-lives (ICRP 1983) in parentheses, is $^{234}$Th (24.10 days), $^{234m}$Pa (1.17 min), $^{234}$Pa (6.7 h), $^{234}$U, $^{230}$Th (77,000 yr), $^{226}$Ra (1,600 yr), and $^{222}$Rn (3.8235 days). Additionally, the short-lived isotopes of polonium (Po), lead (Pb), and bismuth (Bi) produced as the result of decay of radon gas that remains with the ore are also present. These include $^{218}$Po (3.05 min), $^{214}$Pb (26.8 min), $^{214}$Bi (19.9 min), $^{214}$Po (164.3 µs), $^{210}$Pb (22.3 yr), $^{210}$Bi (5.012 days), and $^{210}$Po (138.38 days). The last isotope in the chain is $^{206}$Pb, which is stable. There is a similar decay chain that begins with $^{235}$U, but it is less important for most risk assessments and is not listed here (for details, see Shleien 1992).

The discovery of neutron-induced fission of uranium led to research and large-scale production operations that produced atomic bombs prior to the end of World War II (Smyth 1945). Fission of uranium or plutonium produces pairs of "fission fragments" or "fission products," most of which are radioactive and decay by emitting beta particles, often followed by gamma rays. As shown in figure 2.1, the masses of fission products are distributed unevenly.

Only combinations that preserve both mass and charge occur, so the distributions due to fission of $^{235}$U and $^{239}$Pu differ. For both, there are two major mass yield peaks with fission yields in the range of 6–8% separated by a deep trough. For $^{235}$U, fission yields of 1% or more occur for atomic mass numbers in the ranges of 84 to 104 and 130 to 149. The fission products formed during detonation of an atomic weapon constitute the vast majority of the radioactivity released by the explosion. Both fission and activation products are produced during controlled fission in a nuclear power plant.

Activation products are the other main category of man-made radionuclides. They are produced when the nucleus of an atom absorbs a neutron, a proton,

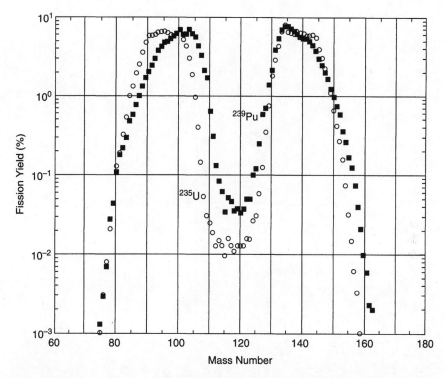

**Figure 2.1** Fission yields for $^{235}$U and $^{239}$Pu. Plotted values are based on General Electric Company (1989).

a deuteron, a helium nucleus (alpha particle), or other combination of nucleons. Accelerators can produce a wide variety of radionuclides by various reactions, depending upon the nature of the particle beam and other conditions (Patterson and Thomas 1973; Sullivan 1992; NCRP 2003). Production of radionuclides due to neutron absorption is common in reactors (e.g., $^{59}$Co absorbing a neutron to produce $^{60}$Co, a high-energy gamma emitter that has been used for radiotherapy). In a reactor, the activation process produces $^{134}$Cs as the result of neutron absorption by the stable fission product $^{133}$Cs. The $^{133}$Cs produced in the fuel can accumulate and absorb neutrons, producing $^{134}$Cs. Although the precursors of $^{133}$Cs are produced by weapons detonations, there is little production of $^{134}$Cs. Perkins et al. (1990) illustrate the difference between $^{134}$Cs levels from fallout and from the Chernobyl accident. Detailed methods for calculating the activities of individual radionuclides in fission product decay chains and of radionuclides produced by nuclear bombardment are given in Evans (1955) and Glasstone (1955). The equations include a production term appropriate for the process and consider losses due to radioactive decay and other processes as appropriate for the isotope.

There are several sources of information about the properties of radionuclides. One of these is ICRP Publication 38 (ICRP 1983), which shows the decay schemes of many radionuclides of interest for dose and risk assessment. Some radionuclides

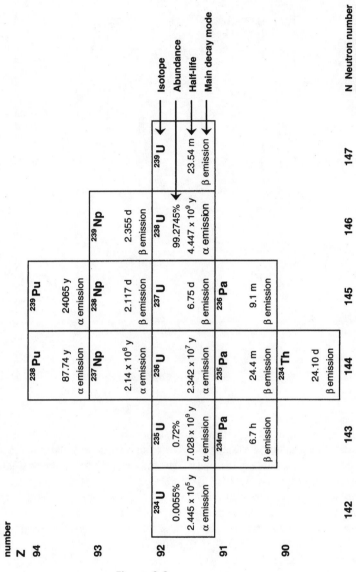

Figure 2.2

**(b)**

**Atomic Number Z**

| Z | N=78 | N=79 | N=80 | N=81 | N=82 | N=83 | N=84 |
|---|---|---|---|---|---|---|---|
| 55 | 133 Cs — 100% | 134m Cs, 2.9 h, IT \| 134 Cs, 2.062 y, β | 135m Cs, 53 m, IT \| 135 Cs, 2×10⁶ y, β | 136m Cs, 19 s, IT \| 136 Cs, 13.1 d, β | 137 Cs, 30.0 y, β emission | 138m Cs, 2.9 m, IT \| 138 Cs, 32.2 m, β | 139 Cs, 9.3 m, β emission |
| 54 | 132 Xe — 26.9% | 133m Xe, 2.188 d, IT \| 133 Xe, 5.245 d, β | 134 Xe — 10.4% | 135m Xe, 15.79 m, IT \| 135 Xe, 9.09 h, β | 136 Xe — 8.9% | 137 Xe, 3.82 m, β emission | 138 Xe, 14.17 m, β emission |
| 53 | 131 I, 8.04 d, β emission | 132 I, 2.3 h, β | 133 I, 20.8 h, β | 134m I, 3.7 m, IT \| 134 I, 52.6 m, β | 135 I, 6.61 h, β emission | 136m I, 47 s, β \| 136 I, 1.39 m, β | 137 I, 24.5 s, β emission |
| 52 | 130 Te — 33.87% | 131m Te, 80 h, IT \| 131 Te, 25.0 m, β | 132 Te, 78.2 h, β emission | 133m Te, 55.4 m, β \| 133 Te, 12.45 m, β | 134 Te, 41.8 m, β emission | 135 Te, 19.0 s, β emission | 136 Te, 17.5 s, β emission |
| 51 | 129m Sb, 17.7 m, β \| 129 Sb, 4.32 h, β | 130m Sb, 6.5 m, β \| 130 Sb, 40 m, β | 131 Sb, 2.3 m, β emission | 132m Sb, 2.8 m, β \| 132 Sb, 4.2 m, β | 133 Sb, 2.53 m, β emission | 134m Sb, 10.4 m, β \| 134 Sb, 0.8 s, β | 135 Sb, 1.71 s, β emission |

**Neutron number**

| Mass number | 131 | 132 | 133 | 134 | 135 | 136 | 137 | 138 |
|---|---|---|---|---|---|---|---|---|
| $^{235}$U fission yield | 2.89% | 4.31% | 6.69% | 7.87% | 6.54% | 6.32% | 6.19% | 6.71% |

(a) Relationships between some isotopes of thorium (Th), protactinium (Pa), uranium (U), neptunium (Np), and plutonium (Pu). Natural abundances for the long-lived uranium isotopes are shown. Half-lives and the main modes of decay are shown for all isotopes. (b) Relationships between some isotopes of antimony (Sb), tellurium (Te), iodine (I), xenon (Xe), and cesium (Cs) showing the beta-particle emission decay sequences. The scheme is the same as in (a), with the natural abundances, half-lives, and principal decay modes shown. Some complex decay schemes are only partially indicated. Numerical values of $^{235}$U fission yields are listed below for mass numbers 131–138.

with half-lives less than 10 min are included in the compilation, but that half-life is often used as a cutoff for tabulations of radionuclides of interest for dosimetry. The encyclopedic *Table of Isotopes* (Firestone and Shirley 1998) provides a comprehensive listing of all stable and unstable isotopes whose properties were known at the time of publication. Over time, high-energy physics research has expanded the periodic table by creating isotopes of new elements; a complete and current list would be longer, but not more interesting for most radiation protection purposes. Focusing on radiation protection, Shleien (1992) has selected alpha, beta, and gamma emitters of particular interest and presented basic information for those radioisotopes. Data on the energies of the emitted radiations and on radionuclide half-lives are provided as an aid to identifying unknown radionuclides based on those properties. The sequences of radionuclides produced by decay of naturally occurring uranium and thorium isotopes and their progeny are clearly presented in Shleien (1992). The booklet form of the *Chart of the Nuclides* (General Electric Company 1989) is a compact source of basic information about stable and radioactive isotopes of interest for most radiation protection situations. Also included in the charts are the $^{235}$U and $^{239}$Pu fission yields for the mass number chains produced.

Figure 2.2 presents a simplified picture of the relationships between a few of the many radioactive isotopes that have been studied. The atomic number (Z) varies along the vertical axis, and the neutron number (N) increases along the horizontal axis. Figure 2.2a includes some of the isotopes of plutonium, neptunium, uranium, and thorium, including the isotopes involved in plutonium production. Absorption of a neutron by $^{238}$U yields $^{239}$U followed by gamma-ray emission [$^{238}$U (n,γ) $^{239}$U]. Uranium-239 decays to $^{239}$Np, and beta decay of that radionuclide produces $^{239}$Pu, which is fissionable. A second important set of transitions begins with absorption of a neutron by $^{238}$U followed by emission of two neutrons to produce $^{237}$U [$^{238}$U (n,2n) $^{237}$U], which decays by beta-particle emission to $^{237}$Np. This long-lived isotope can undergo an (n,γ) reaction to produce $^{238}$Np and subsequently $^{238}$Pu, used in radionuclide thermogenerators to produce electricity. The first steps in the previously mentioned long sequence of radioactive transformations following the decay of $^{238}$U can also be seen in figure 2.2a. Alpha-particle emission by $^{238}$U removes two protons and two neutrons, resulting in an atom of $^{234}$Th. Beta decay of $^{234}$Th and then $^{234m}$Pa produces $^{234}$U, and the sequence continues from there.

Figure 2.2b illustrates the transformations of some typical beta-emitting fission products as they decay along an isobaric (constant mass) chain to a stable nuclide. The beta decay sequence that produces the stable cesium isotope includes $^{133}$In (half-life = 0.18 s), $^{133}$Sn (half-life = 1.44 s), which are not shown in figure 2.2, followed by the mass 133 isotopes of antimony (Sb), tellurium (Te), iodine (I), and xenon (Xe). Generally, the more distant an isotope is from the stable destination nuclide, the shorter its half-life, and that pattern is seen in this sequence. Also provided in figure 2.2b are numerical values of the $^{235}$U fission yields for the mass chains 131–138. Included in those chains are a number of fission products important for health assessments.

## Situations That Do Not Require Source Terms

There are situations in which definition of a source term is not the logical first step in a risk assessment. As a general rule, one begins the assessment as close to those receiving the radiation doses as possible. In some cases, personal measurements may be available; these could include chromosome aberrations for high-dose situations (Maletskos 1991; Guskova et al. 1988), in vivo measurements of radionuclides (e.g., Maletskos 1991), or in vitro bioassay measurements (Maletskos 1991). Estimating the release of $^{137}$Cs from an originally encapsulated source and subsequent dispersion of the material in Goiania, Brazil (Maletskos 1991), could not have provided dose estimates as reliable as those obtained from measures of environmental and personal contamination or external exposure. When bioassay data are not available, reliable measurements of radionuclide concentrations in air, drinking water, and foodstuffs are generally preferred over modeled concentrations that were based upon estimated release amounts and mathematical models, which may be complex and have greater uncertainties.

Even though the radionuclide releases from detonation of nuclear weapons could be calculated, researchers chose environmental measurements as starting points in dose assessments for those exposed to fallout from explosions at the Nevada Test Site (Gesell and Voillequé 1990; National Cancer Institute 1997) because the calculation of atmospheric dispersion and fallout deposition was a less reliable option. Similarly, evaluations of the consequences of nuclear weapons testing in the Marshall Islands relied on measurements of ground contamination, external exposure rates, radionuclide concentrations found in many local foods, in vivo measurements, and in vitro bioassays (Simon and Vetter 1997).

Although the amounts of certain radionuclides released during the Chernobyl accident were estimated in several different ways (e.g., International Nuclear Safety Advisory Group 1986; Gudiksen et al. 1989; Devell et al. 1995; Talerko 2005), that information was not an important input to the dose assessment process. Evaluations of thyroid doses relied upon in vivo measurements of thyroid activity (e.g., Likhtarev et al. 1994; Gavrilin et al. 1999) or the measured $^{137}$Cs depositions on the ground (summarized by DeCort et al. 1998) and ratios of $^{131}$I to $^{137}$Cs (e.g., Stepanenko et al. 2004). Zvonova and Balonov (1993) employed measurements of $^{131}$I contamination in milk, the main source of intakes, as well as measured thyroid activities in dose assessment. External radiation fields were estimated using $^{137}$Cs depositions on the ground, ratios of other gamma-emitting isotopes to $^{137}$Cs, and downward migration and surface roughness (e.g., Golikov et al. 1993; Likhtarev et al. 2002) and were compared with measured doses (e.g., Erkin and Lebedev 1993). Internal doses to all body tissues, largely due to intakes of $^{137}$Cs and $^{134}$Cs, were assessed using models based on ground contamination and modeled transfer to food, measurements of foodstuffs, and whole-body counting of members of the population (e.g., Balonov and Travnikova 1993; Likhtarev et al. 1996, 2000). The models generally predicted higher body burdens than were measured in the populations, indicating that persons exercised control over their intakes.

Assessing the consequences of the malicious dispersal of radionuclides in an urban setting will involve approaches that are similar to those used under the circumstances described above. Initially, a variety of source terms may be assumed in order to make some dispersion calculations that can be compared with the first environmental measurements, but as the range and scope of measurements increases, the utility of source-term–driven calculations will decrease. Management of terrorist events ranging from detonation of a nuclear weapon to explosion of a radiological dispersal device is discussed in detail in National Council on Radiation Protection and Measurements Report No. 138 (NCRP 2001). Particular methods for retrospective assessment of radiation exposures have been examined and compared in International Commission on Radiation Units and Measurements Report 68 (ICRU 2002).

## Human Activities Producing Releases of Radionuclides

A range of human activities results in releases of radionuclides to the environment. Some of them have already been mentioned, but in this section they are discussed sequentially beginning with the extraction of naturally occurring material from the earth, proceeding through its processing and uses, and ending with disposal of wastes containing natural activity and man-made radionuclides. The latter category includes radionuclides produced with a specific purpose in mind (e.g., $^{239}$Pu for weapons) and those produced incidentally as the result of other activities (e.g., fission and activation products produced in a nuclear power reactor). The primary focus of the subsections that follow is on activities in the United States. Ergorov et al. (2000) provide information regarding similar enterprises in the former Soviet Union that were frequently quite similar to those in the United States. Bradley (1997) examined the waste disposal aspects of Soviet industrial sites, the releases that occurred, and, in some cases, the estimated doses.

Production of radioactive chemicals that are used for research and in medical diagnosis and treatment is an activity that involves only man-made radionuclides. Ongoing research in high-energy physics is another source of releases of radionuclides, primarily gaseous activation products, to the environment.

The U.S. Environmental Protection Agency (EPA) estimated source terms for generic facilities that constituted the uranium fuel cycle in the early 1970s (EPA 1973a, 1973b, 1973c). Subsequently, using information from the U.S. Nuclear Regulatory Commission (NRC) and the U.S. Department of Energy (DOE), the EPA also performed a more detailed evaluation of the impact of radionuclide emissions into the atmosphere (EPA 1979). These source term estimates are unlikely to be relevant to prospective source term assessments for new facilities, but are indicative of potential releases during operations of such facilities at that time.

New environmental standards established for normal operations of fuel cycle facilities became effective for most facilities in December 1979 and for uranium mills in December 1980. The standards limited annual dose equivalents to the whole body and individual organs, except the thyroid, to 25 mrem (Title 40 Code of Federal Regulations, part 190 [40 CFR 190]). Meeting the new dose standard typically required reductions in environmental discharges, making previous source term estimates or measurements out of date for future operations. Under the authority of the Clean Air Act, the EPA established National Emission Standards for Hazardous Air Pollutants (NESHAPs) for radionuclides (EPA 1989) that apply to particular source categories, including nuclear fuel cycle activities.

## Uranium Mining

Uranium mining was mentioned only in passing in an analysis of the uranium fuel cycle conducted by the EPA (1973a), suggesting that it was not considered to be a significant source of radionuclide releases. Some radioactivity is no doubt present in water released from a uranium mine, but it is unlikely that those releases would be an important source of public exposure. Air is purposefully discharged from operating mines to reduce in-mine concentrations of radon ($^{222}$Rn) and radon progeny (principally $^{218}$Po, $^{214}$Pb, $^{214}$Bi, $^{214}$Po, and $^{210}$Pb), thereby reducing radiation doses to the lungs of miners in the workplace. This practice leads to exposure of the public to the discharged radionuclides. On the basis of its authority under the Clean Air Act, the EPA established a NESHAP for underground uranium mines limiting the effective dose equivalent to the maximally exposed individual to 10 mrem (EPA 1989). To comply with the regulation, it is necessary to define a source term, based on measurements of the releases, and to calculate the effective dose equivalent to the most exposed person. The EPA (1989) decided it was not necessary to establish a NESHAP for surface uranium mines. Important factors in the decision were that surface mining activities were already governed by many other laws and regulations, estimated risks of exposure to radon in that setting were low, and it was considered unlikely that new surface mining projects would be initiated.

## Uranium Milling

The ore from current uranium mining operations typically contains only a fraction of 1% uranium. Before the uranium can be utilized, it is necessary to separate it from other minerals in the ore and to produce a concentrated form, which typically is $U_3O_8$, called yellowcake. To accomplish this goal, the ore is ground and leached with an acid or base, and the uranium fraction is separated from the other material, called tailings. Merritt (1971) described the range of processes that were in use at a time when many milling facilities were operating in the United States. Releases from uranium milling facilities included particles from crushing, grinding, and conveyance of ground ore, as well as releases during yellowcake drying and packaging (EPA 1973a; NRC 1980a). The mill tailings contain most of the activities of the radionuclides listed above that were originally present in the ore and small amounts of uranium isotopes. Waterborne effluents are associated with tailings impoundments and processes that produced waste liquids. Releases from tailings due to radon emanation and suspension of radioactive particles by winds depend on the management of the tailings ponds.

The Manhattan Project employed numerous facilities to process uranium obtained from a variety of sources and in several forms. The first plants that could be considered uranium mills did not begin operation until after the war, when the arms race began. Treatment of gaseous effluents from uranium mills has changed over time, so source terms for planned new facilities will differ substantially from those measured during operations that occurred years ago. Prospective development of source terms for a new uranium mill and associated tailings disposal must take into account current guidance and regulations as well as effluent treatment systems incorporated in the design of the mill. The NRC (1987) established methods for estimating releases from uranium mills and considered some effluent treatment systems. Most historic mill operations were not affected by the EPA NESHAP for existing and new tailings piles (EPA 1989).

## Uranium Conversion

Several processes were developed during the 1940s to increase the fraction of $^{235}$U above the 0.72% present naturally. All the enrichment processes (i.e., separation by diffusion, centrifugation, and electromagnetic means) utilized gaseous forms of uranium. Uranium hexafluoride ($UF_6$) was used most frequently, although uranium chloride ($UCl_4$) was also employed. Yellowcake is converted to $UF_6$ through a series of chemical processes with intermediate products of $UO_3$, $UO_2$, and $UF_4$. In the war years, Harshaw Chemical produced $UF_6$. Production facilities for $UF_6$ were added at the K-25 plant in Oak Ridge, Tennessee, in 1947 and later at other gaseous diffusion plants. Conversion of $U_3O_8$ to $UF_6$ was performed by the private sector after 1962 at facilities in Illinois and Oklahoma.

Historic information about uranium releases from Oak Ridge facilities has been investigated as part of the Oak Ridge Dose Reconstruction Project (Buddenbaum et al. 1999). The information relates primarily to historic activities; it would not be representative of activities expected from a new facility. Information about the source term for an accidental release from the Sequoyah Fuels Corporation plant in Oklahoma is provided in reports of investigations of the accident (NRC 1986a, 1986b).

## Uranium Enrichment

The electromagnetic separation method, employed at the Y-12 facility in Oak Ridge, Tennessee, was most productive during the war years, but gaseous diffusion (with plants at Oak Ridge; Paducah, Kentucky; and Piketon, Ohio) was the predominant method employed by the government agencies responsible for producing enriched uranium. The name (and mission) of the responsible agency changed with time, from 1947 when the Atomic Energy Commission (AEC) was established, to the Energy Research and Development Agency (1974–1976), and then to the DOE. Highly enriched uranium (HEU), containing more than 20% $^{235}$U, and uranium products with lower amounts of $^{235}$U (LEU) were both produced. The HEU with the highest levels (>90%) of $^{235}$U was used for weapons production, and various grades of LEU were used for weapons complex and, later, commercial reactor fuels.

Following passage of the Energy Policy Act in 1992, the DOE has leased enrichment facilities in Ohio and Kentucky to the U.S. Enrichment Corporation (USEC) for production of commercial-reactor fuel-grade LEU (DOE 1997). More recently, a new uranium enrichment facility using centrifuge technology has been proposed for Piketon, Ohio, by USEC. A safety evaluation report and an environmental impact statement (EIS) have been prepared for that facility (NRC 2006a, 2006b). The EIS notes an anticipated increase in demand for enriched uranium for new nuclear generating capacity expected prior to 2025. The airborne activity release is estimated to be about 5 GBq yr$^{-1}$, and air concentrations to which the public could be exposed are anticipated to be well below established standards (NRC 2006b).

Louisiana Energy Services has plans to build a plant in New Mexico that would also employ the gas centrifuge enrichment process (NRC 2005). The highest proposed enrichment levels are 5–10% $^{235}$U. Releases of uranium to the atmosphere from the facility are estimated to be less than 10 g yr$^{-1}$; releases of hydrogen fluoride are estimated to be less than 1 kg yr$^{-1}$. Both releases are well below the NRC and EPA limits. Liquid waste releases are also estimated to be quite small.

## Weapon Component and Fuel Fabrication

Enriched uranium hexafluoride was converted to UF$_4$, which could then be reduced to uranium metal suitable for casting into weapons components or the fuel elements used in early reactors. Initially, these operations were performed at Los Alamos, New Mexico, and at the Hanford Site in Washington State. After the war, conversion of HEU and weapon component production were carried out at the Y-12 Plant at Oak Ridge. This facility became the repository of HEU used in nuclear weapons; HEU recovered from naval propulsion reactor fuel elements was also shipped to the Y-12 plant. Other uranium processing and fuel element fabrication activities involving LEU were dispersed to several locations, including the Feed Materials Production Center at Fernald, Ohio. Source terms for uranium and other radionuclides released from this facility have been studied in detail (Meyer et al. 1996), but they would not be representative of new facilities built for the same purposes. Deworm and Tedder (1991) describe effluent treatment systems and generic release estimates for different types of fuel fabrication facilities.

Plutonium produced during the Manhattan Project was also reduced to a metallic form for casting and machining nuclear weapons components. Initially, the work was performed at Los Alamos, but later it was performed primarily at the Rocky Flats facility in Colorado. Source terms for plutonium, uranium, and other releases during routine operations and accidents at the Rocky Flats facility were a focus of the first phase of studies following shutdown of the facility in 1989 (Ripple et al. 1996; Mongan et al. 1996a, 1996b). Revised estimates of the plutonium releases from routine operations, two fires, and wind suspension of contaminated soils, developed during the second phase of the Rocky Flats Health Studies, are given in the comprehensive risk assessment by Rood et al. (2002).

## Reactors

After controlled nuclear fission was demonstrated at the University of Chicago in December 1942, the potential for plutonium production from neutron capture by $^{238}$U was established (Smyth 1945). A prototype reactor, the Clinton Pile (later called the Graphite Reactor), was built at the Oak Ridge Reservation in Tennessee. A pilot plant plutonium separation facility was also constructed there. The Oak Ridge facilities were originally conceived as temporary to develop processes that would be used at industrial-scale plants built elsewhere.

The first large plutonium production reactors were built at the Hanford Reservation in southeast Washington State in 1944, as were the first large-scale chemical processing facilities needed to separate plutonium from uranium fuel elements removed from the reactors. The reactors were situated along the Columbia River, from which water was withdrawn for cooling. The cooling water flowed through the core and, after a brief holdup period, back into the river, carrying radionuclides produced by neutron activation of elements in the water and fission products released from the reactor fuel elements because of cladding failures. Initially, the fuel elements removed from the reactors were stored only one to two months, so the fuel inventories of even relatively short-lived radionuclides (e.g., $^{131}$I, half-life of ~8 days) were substantial at the time of processing, and large releases of radioactive gases occurred during fuel dissolution. Radionuclide releases to the environment and doses to the public were estimated for both the reactor and processing plant sources in a major dose reconstruction effort, of which source term development was an important component (Shipler et al. 1996). Voillequé et al. (2002) estimated airborne releases of other short-lived gases and particulate radionuclides from these facilities. Those releases were found to be generally less important for public health than were the releases of $^{131}$I.

A total of nine production reactors were constructed at Hanford between 1944 and 1963. Their operational periods varied; most were shut down by 1971, but the N Reactor operated until 1987. Five more plutonium production reactors were built at the Savannah River Site (SRS) in South Carolina, with reactor operations beginning in 1954 and continuing until 1988. Estimates of releases from the SRS reactors were performed as part of a different dose reconstruction project (Till et al. 2001).

A demonstration of electricity production using steam generated by nuclear reactor operation was accomplished in 1951 when the Experimental Breeder Reactor No. 1 at the National Reactor Testing Station in Idaho provided electric power to the nearby town of Arco, Idaho (Glasstone 1955). In 2004, the International Atomic Energy Agency (IAEA) noted the 50th anniversary of nuclear power production. On June 26, 1954, the first nuclear power plant in Obninsk, a town with many scientific institutes in Kaluga Oblast southwest of Moscow, began providing electricity to local residences and businesses (Wedekind 2004).

The nuclear-powered submarine *Nautilus* was launched in January 1954, commissioned in September of that year, and undertook her first voyage in January 1955. The ship was powered by a pressurized water reactor built by Westinghouse Electric Company, which later constructed an electrical power plant of similar design at Shippingport, Pennsylvania (Murray 1961). After President Eisenhower's Atoms for

Peace speech at the United Nations in 1953 and the first Geneva Conference in 1955, information that was previously kept secret was made available to train and prepare engineers to build facilities for the expected growth of a nuclear power industry. The text on nuclear reactor engineering prepared by Glasstone (1955) with the assistance of staff of the Oak Ridge National Laboratory, which contains drawings of many early reactor designs and experimental facilities, is an example.

Over time, various types and designs of reactors were built, demonstrated, and operated in several countries. Eichholz (1976) summarized the characteristics of the principal reactor types in operation during a period when the nuclear power industry was in full flower. The facilities in many countries are still dominated by the same three plant types that were developed more than 30 years ago: the pressurized water reactor (PWR), the boiling water reactor (BWR), and the pressurized heavy-water–moderated reactor (HWR). (Heavy water contains deuterium, called heavy hydrogen. It is symbolized by D or $^2$H, which replaces the normal hydrogen, H or $^1$H, in the water molecule.) This reactor type is exemplified by the CANDU reactor, which was developed in Canada but has also been constructed and operated in India and other countries. Other globally important reactor types are the gas-cooled reactor (GCR), operating mainly in the United Kingdom, and the light-water–cooled graphite-moderated reactor (LWGR), which is similar to the original Hanford reactors, located at sites in the Russian Federation and republics of the former Soviet Union.

In 1997, electrical energy generated at nuclear power plants totaled about 260 gigawatts (GW yr) (UNSCEAR 2000). Nearly two-thirds (64.7%) was produced by PWRs, and another 23.8% was generated by BWRs. The HWRs produced about 4.8% of the electricity that year, while the GCRs contributed 3.5% of the total and the LWGRs produced 3.0% of the total. Fast breeder reactor (FBR) plants have operated in various locations since the startup of Experimental Breeder Reactor No. 1 in 1951. In 1997, they contributed only a small fraction (0.2%) of the electrical power generated. The contribution of the FBR units was greater in 1985 (0.4%), but an overwhelming majority of production at that time was by PWRs (59.3%) and BWRs (23.5%). The HWRs, GCRs, and LWGRs accounted for, respectively, 5.1%, 4.5%, and 7.1% of the total electrical generation (160.4 GW yr) in 1985 (UNSCEAR 1993). In 1973, the total electrical generation from nuclear power was just 21.6 GW yr, and the contributions of the various reactor types were somewhat different. At that time, the PWR and BWR generation fractions were similar (36.6% and 30.7%, respectively) and the GCRs contributed 20.6% of the total electricity generated. The HWRs, LWGRs, and FBRs accounted for the remainder with 8.4%, 8.8%, and 0.02%, respectively (UNSCEAR 1993).

The Nuclear Energy Agency (NEA) reported that 346 reactors were connected to power grids in countries belonging to the Organisation for Economic Co-operation and Development (OECD) in 2006. The electrical production capacity of those reactors is about 308 GW (electric) or 83.6% of the world's nuclear generating capacity. The OECD includes 17 countries, but not Russia, China, or India, where additional generating capacity is located. The contribution of nuclear energy to electricity production in the member countries ranged as high as 78% (in France). The average contribution was 23%, somewhat higher than in Great Britain and the

United States, where the fraction was about 19%. Although several countries have plans to phase out nuclear power, concerns about $CO_2$ emissions from fossil fuel plants and other factors may lead to construction of new nuclear power reactors (OECD 2007).

## Source Terms for Normal Operations

Table 2.1 lists some gaseous and particulate radionuclides that are produced during reactor operation. Some of these radionuclides are released during reactor operation or when spent fuel from the reactor is processed. Not all of the radionuclides listed are of equal importance, and the radionuclide source terms vary according to reactor type. This is illustrated in table 2.2, which shows global average normalized releases in airborne effluents for the six types of reactors during three recent periods (UNSCEAR 1993, 2000). In addition to the tabled values, the U.N. Scientific Committee on the Effects of Atomic Radiation reports (UNSCEAR 1993, 2000) include normalized releases of $^{14}C$ to the atmosphere. The normalized rates have varied little over time; they are generally between 0.1 and 1 TBq per GW yr of electrical energy, except for HWRs, whose normalized rates averaged 4.2 TBq per GW yr during the period 1985–1997. Table 2.3 shows comparable information for $^3H$ and radionuclides other than $^3H$ in liquid wastes for the same periods. It should be noted that tables 2.2 and 2.3 present global average values and, as indicated in the table 2.2 notes, normalized releases from individual plants or groups of plants may differ substantially from these values. Plots of the distribution of releases from individual reactors (UNSCEAR 2000) show a very broad range of values. Also, not all of the monitoring results are equally reliable, and there are periods for which some facilities did not report releases or reported only normalized mean values. In general, monitoring techniques have improved with time, and good practices are now more widespread. Therefore, the normalized releases based on more recent effluent measurements are likely more reliable than estimates for earlier periods. These caveats notwithstanding, general trends of normalized radionuclide releases can be seen, as shown in tables 2.1–2.3.

The PWR and the BWR are the reactor types that generate the largest amounts of electrical energy. The PWR is characterized by a system where the heat generated by nuclear fission in the reactor core is transferred to water (called the "reactor coolant" or "primary coolant") circulating under pressure through the core. This superheated water transfers its heat content via a heat exchanger to secondary cooling water, which is converted to steam. This steam drives a turbine generator, producing electricity, and is then cooled, condensed, and recirculated. A major benefit of this system is the isolation of the reactor coolant loop from the environment. In BWRs, steam is generated within the reactor vessel and is carried directly to the turbine generator. Afterward it is condensed, demineralized, and returned to the reactor. This arrangement results in simpler piping and somewhat more efficient heat utilization, but radioactive gases in the coolant can also follow the same path as the steam.

Figure 2.3 shows the history of normalized noble gas releases from PWRs and BWRs around the world during a 27-year period (UNSCEAR 2000). The large difference in releases from the two reactor types was due to the path noted above and the

Table 2.1  Fission and activation products generated by nuclear power plant operation

| Product | Half-life | Product | Half-life |
|---|---|---|---|
| *Noble gases* | | *Major corrosion and activation products* | |
| $^{83m}$Kr | 1.83 h | | |
| $^{85m}$Kr | 4.48 h | $^{3}$H[a] | 12.35 y |
| $^{85}$Kr | 10.72 y | $^{14}$C | 5,730 y |
| $^{87}$Kr | 76.3 m | $^{24}$Na | 15.00 h |
| $^{88}$Kr | 2.84 h | $^{110m}$Ag | 249.9 d |
| $^{89}$Kr | 3.15 m | $^{134}$Cs | 2.062 y |
| $^{131m}$Xe | 11.9 d | | |
| $^{133m}$Xe | 2.188 d | *Some important fission products* | |
| $^{133}$Xe | 5.245 d | | |
| $^{135m}$Xe | 15.29 m | $^{88}$Rb | 17.8 m |
| $^{135}$Xe | 9.09 h | $^{89}$Rb | 15.2 m |
| $^{137}$Xe | 3.82 m | $^{89}$Sr | 50.5 d |
| $^{138}$Xe | 14.17 m | $^{90}$Sr | 29.12 y |
| | | $^{90}$Y | 64.0 h |
| *Radioiodines and precursors* | | $^{91}$Sr | 9.5 h |
| | | $^{91}$Y | 58.51 d |
| $^{129}$I | $15.7 \times 10^7$ y | $^{92}$Sr | 2.71 h |
| $^{131}$I | 8.04 d | $^{92}$Y | 3.54 h |
| $^{132}$I | 2.3 h | $^{95}$Zr | 63.98 d |
| $^{133}$I | 20.8 h | $^{95}$Nb | 35.15 d |
| $^{134}$I | 52.6 m | $^{97}$Zr | 16.9 h |
| $^{135}$I | 6.61 h | $^{97}$Nb | 72.1 m |
| $^{131m}$Te | 30 h | $^{99}$Mo | 66.0 h |
| $^{132}$Te | 78.2 h | $^{103}$Ru | 39.28 d |
| | | $^{106}$Ru | 368.2 d |
| *Major corrosion and activation products* | | $^{124}$Sb | 60.2 d |
| | | $^{125}$Sb | 2.77 y |
| $^{51}$Cr | 27.7 d | $^{136}$Cs | 13.1 d |
| $^{54}$Mn | 312.5 d | $^{137}$Cs | 30.0 y |
| $^{56}$Mn | 2.58 h | $^{138}$Cs | 32.2 m |
| $^{55}$Fe | 2.7 y | $^{140}$Ba | 12.74 d |
| $^{59}$Fe | 44.53 d | $^{140}$La | 40.272 h |
| $^{57}$Co | 270.9 d | $^{141}$Ce | 30.501 d |
| $^{58}$Co | 70.80 d | $^{144}$Ce | 284.3 d |
| $^{60}$Co | 5.27 y | $^{144}$Pr | 17.28 m |
| $^{65}$Zn | 243.9 d | | |

[a] Also a product of ternary fission.

simple treatment system employed initially. The first BWRs utilized only a 30-min holdup of offgases prior to discharge. Over time, this simple gaseous effluent treatment system has been modified to reduce releases of noble gases from the BWRs. Noble gas releases from PWRs, which were never as high as those from BWRs, were not reduced as dramatically. As shown in figure 2.4, the releases of $^{131}$I and particulate radionuclides from BWRs also were substantially reduced by better effluent

Table 2.2  Normalized releases of noble gases, tritium, $^{131}$I, and particulate radionuclides to the atmosphere from nuclear power production of electricity (UNSCEAR 1993, 2000)

| Reactor type[a] | Normalized release rates (TBq per GW yr) | | | Normalized release rates (GBq per GW yr) | | |
|---|---|---|---|---|---|---|
| | 1985–89 | 1990–94 | 1995–97 | 1985–89 | 1990–94 | 1995–97 |
| *Noble gases* | | | | $^{131}I$ | | |
| PWR[b] | 81 | 27 | 13 | 0.93 | 0.33 | 0.17 |
| BWR | 290 | 354 | 171 | 1.8 | 0.81 | 0.33 |
| HWR | 191 | 2,050[c] | 252[c] | 0.19 | 0.35 | 0.11 |
| GCR | 2,150 | 1,560 | 1,240 | 1.4 | 1.4 | 0.42 |
| LWGR | 2,000 | 1,720 | 465 | 14 | 6.8 | 6.9 |
| FBR | 150 | 380 | 209 | _[d] | _[d] | _[d] |
| *Tritium ($^3H$)* | | | | *Particulate radionuclides* | | |
| PWR[b] | 2.8 | 2.3 | 2.4 | 2.0 | 0.18 | 0.13 |
| BWR | 2.5 | 0.94 | 0.86 | 9.1 | 178[e] | 351[e] |
| HWR | 480 | 650 | 329 | 0.23 | 0.051 | 0.048 |
| GCR | 9.0 | 4.7 | 3.9 | 0.68 | 0.30 | 0.17 |
| LWGR | 26[f] | 26[f] | 26[f] | 12 | 14 | 8.4 |
| FBR | 96 | 49 | _[g] | 0.19 | 12 | 1.0 |

[a] PWR, pressurized water reactor; BWR, boiling water reactor; HWR, heavy water reactor; GCR, gas cooled reactor; LWGR, lightwater–cooled, graphite–moderated reactor; FBR, fast breeder reactor.
[b] In reports of releases from PWRs operating in France, releases of tritium are included with those for noble gases, and releases of $^{131}$I are included with those for particulate radionuclides.
[c] Elevated average normalized release rate for 1990–1994 due to large releases reported for reactors in India; no releases from those reactors were reported for the period 1995–1997.
[d] No information reported or zero releases reported.
[e] These unusual normalized releases were due to elevated releases at a single reactor (Ringhals 1) during 1993–1997. Excluding releases from that reactor, normalized releases for the periods 1990–1994 and 1995–1997 were estimated to be approximately 5.3 and 1.4 GBq per GW yr, respectively.
[f] Only average normalized releases were reported.
[g] No power production at the FBR reporting releases of $^3$H during the period.

treatment systems during the same period. Discharges of these radionuclides in PWR gaseous effluents were not as high initially but were also reduced by more than a factor of 10 between 1970–1974 and 1995–1997, as shown in figures 2.3 and 2.4.

The UNSCEAR (1993) report provides a snapshot of the mix of isotopes in the "other radionuclides" section of table 2.3. The information is for PWRs and BWRs operating in the United States during 1988. Table 2.4 includes the radionuclides whose normalized release rate exceeded 0.5 GBq per GW yr of electrical energy produced. As is the case for the gaseous radionuclides, the amounts released depend upon the effluent treatment system, reactor power level, and other factors. These data for 1988 may not be indicative of long-term average radionuclide mixtures and normalized releases.

Critical elements of the design and operation of nuclear power plants are fuel rod integrity, the systems for radionuclide removal from the primary coolant (and secondary coolant in the PWRs), and monitoring and control of the water chemistry in

Table 2.3 Normalized releases of tritium and other radionuclides in liquid wastes from nuclear power production of electricity (UNSCEAR 1993, 2000)

| Reactor type[a] | Normalized release rates (TBq per GW yr) | | | Normalized release rates (GBq per GW yr) | | |
|---|---|---|---|---|---|---|
| | 1985–89 | 1990–94 | 1995–97 | 1985–89 | 1990–94 | 1995–97 |
| *Tritium* ($^3H$) | | | | *Other radionuclides* | | |
| PWR | 25 | 22 | 19 | 45 | 19 | 8.1 |
| BWR | 0.79 | 0.94 | 0.87 | 36 | 43 | 11 |
| HWR | 374 | 490 | 330 | 30 | 130 | 44 |
| GCR | 120 | 220 | 280 | 960 | 510 | 700 |
| LWGR | 11 | _[b] | _[b] | _[b] | 4.8 | 5.8 |
| FBR | 3 | 1.8 | 1.7 | 30 | 49 | 23 |

[a]PWR, pressurized water reactor; BWR, boiling water reactor; HWRb, heavy water reactor; GCR, gas-cooled reactor; LWGR, light-water-cooled, graphite-moderated reactor; FBR, fast breeder reactor.
[b]No information.

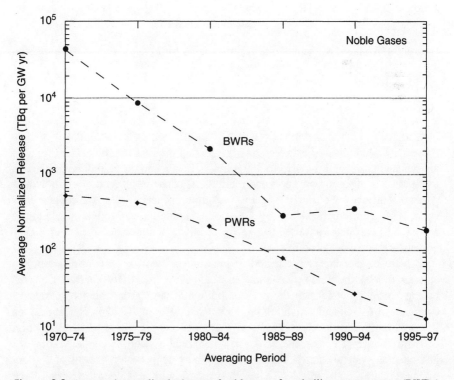

Figure 2.3 Averaged normalized releases of noble gases from boiling water reactors (BWRs) and pressurized water reactors (PWRs) between 1970 and 1997 (UNSCEAR 2000).

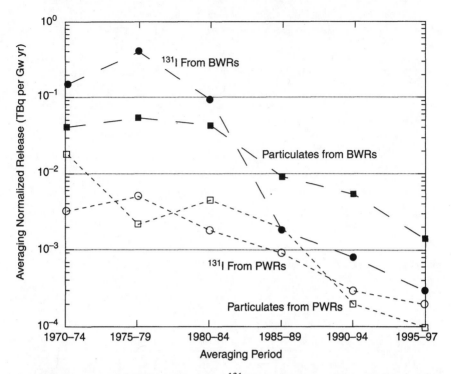

Figure 2.4 Averaged normalized releases of $^{131}$I and particulate radionuclides from boiling water reactors (BWRs) and pressurized water reactors (PWRs) between 1970 and 1997 (UNSCEAR 2000).

the coolant loops. The fuel rod consists of cylindrical sintered uranium oxide pellets placed in sealed tubes of a zirconium alloy, called the fuel cladding. The fuel rods are placed in assemblies called fuel bundles. The fuel and the cladding are barriers to the release of fission products and retain most of the fission products, including gases that diffuse out of the fuel pellets. Nevertheless, leakage into the coolant does occur through microcracks and by diffusion of noble gases through the cladding. Ion exchange systems are used to remove reactive contaminants from the primary coolant and secondary cooling water.

Careful monitoring is conducted to measure pH, corrosion control additives, and, in PWRs, the amount of boric acid added for reactivity control. Concentrations of radionuclides in the coolant fluids are the starting points for release estimates of fission and activation products from the plant. Coolant concentrations of corrosion products, which are metal elements that are activated by absorption of neutrons, depend upon the choices of materials as well as coolant chemistry and cleanup systems. The activated particles may deposit on the walls of coolant piping, creating high-radiation fields. Resuspension of such deposits can produce short-term spikes in the concentrations of corrosion products (e.g., $^{60}$Co) in the coolant.

Table 2.4  Normalized releases of radionuclides[a] (excluding $^3$H) in liquid effluents from U.S. power reactors in 1988 (UNSCEAR 1993)

| Radionuclide | Normalized releases (GBq per GW yr) | |
| --- | --- | --- |
| | PWRs | BWRs |
| $^{24}$Na | 0.12 | 32.0 |
| $^{51}$Cr | 2.8 | 21 |
| $^{54}$Mn | 1.9 | 5.0 |
| $^{55}$Fe | 5.6 | 1.9 |
| $^{59}$Fe | 0.22 | 1.7 |
| $^{58}$Co | 26.7 | 2.8 |
| $^{60}$Co | 7.6 | 11 |
| $^{65}$Zn | 0.016 | 1.3 |
| $^{95}$Nb | 0.56 | 0.016 |
| $^{110m}$Ag | 2.4 | 0.060 |
| $^{124}$Sb | 0.85 | 0.018 |
| $^{125}$Sb | 3.9 | 0.018 |
| $^{131}$I | 2.3 | 0.32 |
| $^{133}$I | 0.75 | 0.047 |
| $^{134}$Cs | 2.3 | 1.3 |
| $^{137}$Cs | 4.6 | 2.9 |

[a] Included are radionuclides whose normalized releases from either reactor type exceeded 0.5 GBq per GW yr.

Radionuclide concentration measurements in coolants of operating reactors and measurements of the performance of cleanup systems form the basis for a consensus standard developed in the United States (ANS 1999). Operational parameters for the reference BWR and for two reference PWRs are shown in table 2.5 (ANS 1999). The two PWRs differ in the type of steam generator employed: once-through or U-tube. Plants with specific parameters that are within about 10% of the tabled values are considered to be in the range covered by the standard (NRC 1976a, 1976b). For example, reactors with power levels between 3,000 and 3,800 MW (thermal) are within the identified range.

Table 2.6 shows the radionuclide concentration in reactor water and steam for the reference BWR (ANS 1999). The tabulated values are intended to reflect plant lifetime average concentrations and are based upon data for BWRs available at the time the standard was prepared. The radionuclides are divided into six classes: noble gases, halogens, rubidium and cesium isotopes, volatile activation products, tritium, and other particulate activation and fission products. All of the noble gases are carried out of the reactor vessel with the steam. The water-to-gas partition factor for the radioiodine isotopes is indicated to be 50, but the standard notes that a higher value (200) may be appropriate for plants with higher concentrations of copper in the coolant. Radionuclide concentrations based on historic operating data are much

Table 2.5  Operational parameters for reference reactors (ANS 1999)[a,b]

| Parameter | Reference BWR | Reference PWRs according to steam generator type | |
| --- | --- | --- | --- |
| | | U-tube | Once-through |
| Thermal power (MWt) | $3.4 \times 10^3$ | $3.4 \times 10^3$ | $3.4 \times 10^3$ |
| Steam flow rate (kg s$^{-1}$) | $1.9 \times 10^3$ | $1.9 \times 10^3$ | $1.9 \times 10^3$ |
| Weight of water in reactor vessel (kg) | $1.7 \times 10^5$ | NA | NA |
| Cleanup demineralizer flow rate (kg s$^{-1}$) | $1.6 \times 10^1$ | NA | NA |
| Weight of water in reactor coolant system (kg) | NA | $2.5 \times 10^5$ | $2.5 \times 10^5$ |
| Weight of secondary coolant in all steam generators (kg) | NA | $2.0 \times 10^5$ | $2.0 \times 10^5$ |
| Reactor coolant letdown flow rate (purification) (kg s$^{-1}$) | NA | 4.7 | 4.7 |
| Reactor coolant letdown flow rate (yearly average for boron control) (kg s$^{-1}$) | NA | $6.3 \times 10^{-2}$ | $6.3 \times 10^{-2}$ |
| Steam generator blowdown rate (kg s$^{-1}$) | NA | 9.5 | NA |
| Fraction of radioactivity in blowdown stream that is not returned to the secondary coolant system | NA | 1.0 | NA |
| Flow through the purification system cation demineralizer (kg s$^{-1}$) | NA | $4.7 \times 10^{-1}$ | $4.7 \times 10^{-1}$ |
| Ratio of condensate demineralizer flow rate to steam flow rate | 1.0 | 0.0 | 0.65 |
| Fraction of noble gas activity in the letdown stream that is not returned to the reactor coolant system (not including the boron recovery system) | NA | 0.0 | 0.0 |

[a] Extracted from American National Standard ANSI/ANS-18.1-1999, with permission of the publisher, the American Nuclear Society.
[b] Weights in pounds and flow rates in pounds per hour are also given in the standard.

lower than were those estimated years ago using the assumption of 1% failed fuel (e.g., EPA 1973b).

Figure 2.5 (Eichholz 1983, figure 1.12) is a schematic diagram of the main elements of a BWR gaseous waste system with possible components and discharge points. The gases in the steam are removed via the air ejector, pass through the hydrogen recombiner, and are then subject to treatments of various types depending upon plant design. Appendix A of Regulatory Guide 1.112 (NRC 2007) lists information requirements for BWR gaseous effluent source term calculations. Besides the basic operating parameters (shown in table 2.5), information such as the holdup time in the delay line for gases leaving the condenser and the parameters of the off-gas treatment systems for the air ejector and mechanical vacuum pump air streams is needed. The size, operating parameters, and effectiveness of charcoal retention systems and cryogenic distillation systems, as appropriate for the system design, are also needed. Other necessary information includes the source and flow rate of steam for the turbine gland seal and, if it is primary steam, appropriate partition coefficients, decontamination factors, and other data related to expected releases. Similar data are needed for other parts of the plant that have separate ventilation systems (typically the reactor building, turbine building, and

Table 2.6 Nuclide concentrations ($\mu$Ci g$^{-1}$) in principal fluid streams of the reference BWR [a,b]

| Nuclide | Reactor water[c] | Reactor steam[c] | Nuclide | Reactor water[c] | Reactor steam[c] |
|---|---|---|---|---|---|
| *Class 1* | | | *Class 6* | | |
| $^{83m}$Kr | | $5.9 \times 10^{-4}$ | $^{24}$Na | $2.0 \times 10^{-3}$ | $2.0 \times 10^{-6}$ |
| $^{85m}$Kr | | $1.0 \times 10^{-3}$ | $^{32}$P | $4.0 \times 10^{-5}$ | $4.0 \times 10^{-8}$ |
| $^{85}$Kr | | $4.0 \times 10^{-6}$ | $^{51}$Cr | $3.0 \times 10^{-3}$ | $3.0 \times 10^{-6}$ |
| $^{87}$Kr | | $3.3 \times 10^{-3}$ | $^{54}$Mn | $3.5 \times 10^{-5}$ | $3.5 \times 10^{-8}$ |
| $^{88}$Kr | | $3.3 \times 10^{-3}$ | $^{56}$Mn | $2.5 \times 10^{-2}$ | $2.5 \times 10^{-5}$ |
| $^{89}$Kr | | $2.1 \times 10^{-2}$ | $^{55}$Fe | $1.0 \times 10^{-3}$ | $1.0 \times 10^{-6}$ |
| $^{131m}$Xe | | $3.3 \times 10^{-6}$ | $^{59}$Fe | $3.0 \times 10^{-5}$ | $3.0 \times 10^{-8}$ |
| $^{133m}$Xe | | $4.9 \times 10^{-5}$ | $^{58}$Co | $1.0 \times 10^{-4}$ | $1.0 \times 10^{-7}$ |
| $^{133}$Xe | | $1.4 \times 10^{-3}$ | $^{60}$Co | $2.0 \times 10^{-4}$ | $2.0 \times 10^{-7}$ |
| $^{135m}$Xe | | $4.4 \times 10^{-3}$ | $^{63}$Ni | $1.0 \times 10^{-6}$ | $1.0 \times 10^{-9}$ |
| $^{135}$Xe | | $3.8 \times 10^{-3}$ | $^{64}$Cu | $3.0 \times 10^{-3}$ | $3.0 \times 10^{-6}$ |
| $^{137}$Xe | | $2.6 \times 10^{-2}$ | $^{65}$Zn | $1.0 \times 10^{-4}$ | $1.0 \times 10^{-7}$ |
| $^{138}$Xe | | $1.5 \times 10^{-2}$ | $^{89}$Sr | $1.0 \times 10^{-4}$ | $1.0 \times 10^{-7}$ |
| | | | $^{90}$Sr-Y[d] | $7.0 \times 10^{-6}$ | $7.0 \times 10^{-9}$ |
| *Class 2* | | | $^{91}$Sr | $4.0 \times 10^{-3}$ | $4.0 \times 10^{-6}$ |
| $^{131}$I | $2.2 \times 10^{-3}$ | $4.4 \times 10^{-5}$ | $^{92}$Sr | $1.0 \times 10^{-2}$ | $1.0 \times 10^{-5}$ |
| $^{132}$I | $2.2 \times 10^{-2}$ | $4.4 \times 10^{-4}$ | $^{91}$Y | $4.0 \times 10^{-5}$ | $4.0 \times 10^{-8}$ |
| $^{133}$I | $1.5 \times 10^{-2}$ | $3.0 \times 10^{-4}$ | $^{92}$Y | $6.0 \times 10^{-3}$ | $6.0 \times 10^{-6}$ |
| $^{134}$I | $4.3 \times 10^{-2}$ | $8.6 \times 10^{-4}$ | $^{93}$Y | $4.0 \times 10^{-3}$ | $4.0 \times 10^{-6}$ |
| $^{135}$I | $2.2 \times 10^{-2}$ | $4.4 \times 10^{-4}$ | $^{95}$Zr-Nb[d] | $8.0 \times 10^{-6}$ | $8.0 \times 10^{-9}$ |
| | | | $^{99}$Mo-$^{99m}$Tc[d] | $2.0 \times 10^{-3}$ | $2.0 \times 10^{-6}$ |
| *Class 3* | | | $^{103}$Ru-$^{103m}$Rh[d] | $2.0 \times 10^{-5}$ | $2.0 \times 10^{-8}$ |
| $^{89}$Rb | $5.0 \times 10^{-3}$ | $5.0 \times 10^{-6}$ | $^{106}$Ru-Rh[d] | $3.0 \times 10^{-6}$ | $3.0 \times 10^{-9}$ |
| $^{134}$Cs | $3.0 \times 10^{-5}$ | $3.0 \times 10^{-8}$ | $^{110m}$Ag | $1.0 \times 10^{-6}$ | $1.0 \times 10^{-9}$ |
| $^{136}$Cs | $2.0 \times 10^{-5}$ | $2.0 \times 10^{-8}$ | $^{129m}$Te | $4.0 \times 10^{-5}$ | $4.0 \times 10^{-8}$ |
| $^{137}$Cs-$^{137m}$Ba[d] | $8.0 \times 10^{-5}$ | $8.0 \times 10^{-8}$ | $^{131m}$Te | $1.0 \times 10^{-4}$ | $1.0 \times 10^{-7}$ |
| $^{138}$Cs | $1.0 \times 10^{-2}$ | $1.0 \times 10^{-5}$ | $^{132}$Te | $1.0 \times 10^{-5}$ | $1.0 \times 10^{-8}$ |
| | | | | | |
| *Class 4* | | | $^{140}$Ba-La[d] | $4.0 \times 10^{-4}$ | $4.0 \times 10^{-7}$ |
| $^{16}$N | $6.0 \times 10^{1}$ | $5.0 \times 10^{1}$ | $^{141}$Ce | $3.0 \times 10^{-5}$ | $3.0 \times 10^{-8}$ |
| | | | $^{144}$Ce-Pr[d] | $3.0 \times 10^{-6}$ | $3.0 \times 10^{-9}$ |
| *Class 5* | | | $^{187}$W | $3.0 \times 10^{-4}$ | $3.0 \times 10^{-7}$ |
| $^{3}$H | $1.0 \times 10^{-2}$ | $1.0 \times 10^{-2}$ | $^{239}$Np | $8.0 \times 10^{-3}$ | $8.0 \times 10^{-6}$ |

[a] Extracted from American National Standard ANSI/ANS-18.1-1999, with permission of the publisher, the American Nuclear Society.
[b] The numerical concentration values in ANS (1999) are given to two significant figures in the mixed units shown here. To convert, use 1 $\mu$Ci = $3.7 \times 10^4$ Bq (e.g., $5.9 \times 10^{-4} \mu$Ci g$^{-1}$ = 22 Bq g$^{-1}$).
[c] Reactor water concentrations are specified at the nozzle where it leaves the reactor vessel; reactor steam concentrations are specified at $t = 0$.
[d] Nuclides in secular equilibrium with concentration of the progeny equal to that shown for the precursor; mass numbers are the same unless indicated.

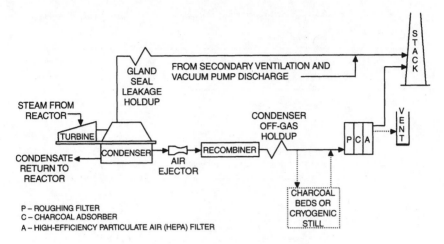

Figure 2.5 Elements of the gaseous waste system for an example pressurized water reactor (PWR) with release points and possible effluent treatment components.

radwaste building) that may be transfer paths for releases of radioactivity to the environment.

Information from a previous version (ANSI 1976) of the consensus standard was incorporated into computer codes used by the NRC to estimate gaseous and liquid effluent (GALE) releases (Cardile et al. 1979; Chandrasekaran et al. 1985). Regulatory Guide 1.112 (NRC 2007) allows use of the fluid concentrations in the new standard (ANS 1999) as an alternative starting point for release calculations. Tables 2.7 and 2.8 provide the expected lifetime average concentrations in primary and secondary coolants for the two reference PWRs. The radionuclide categories are the same as discussed above for BWRs. The numerical values are based on measurements at operating PWR stations that were available at the time the standard was prepared. The standard also provides reference values for demineralizers that treat the coolant and condensate and methods for calculating removal rates from coolants for facilities whose basic parameters do not fall within the ranges associated with the reference values (table 2.5).

Figure 2.6 (Eichholz 1983, figure 1.13) is a schematic diagram of the main elements of a PWR gaseous waste treatment system. Appendix B of Regulatory Guide 1.112 (NRC 2007) describes procedures for estimating radionuclide source terms for gaseous effluents from normal operations of PWRs. The PWR GALE code document (Chandrasekaran et al. 1985) examines the basis for estimating airborne releases from a variety of sources. These include leakage into the containment building, the internal containment cleanup system, and treatment of the exhaust during periodic containment purges. These features are shown schematically in figure 2.6. Also shown are the release paths for uncondensed gases in steam, and storage of gases released from the primary system and held in waste gas decay tanks. Details about all the relevant systems are needed to estimate the releases from individual pathways.

**Table 2.7** Nuclide concentrations ($\mu\text{Ci g}^{-1}$) in principal fluid streams of the reference PWR with U-tube steam generators[a,b]

| Nuclide | Reactor coolant[c] | Secondary coolant Water[d] | Secondary coolant Steam[d] | Nuclide | Reactor coolant[c] | Secondary coolant Water[d] | Secondary coolant Steam[d] |
|---|---|---|---|---|---|---|---|
| *Class 1* | | | | *Class 6* | | | |
| $^{85m}$Kr | $1.6 \times 10^{-2}$ | nil | $3.4 \times 10^{-9}$ | $^{24}$Na | $4.7 \times 10^{-2}$ | $1.5 \times 10^{-6}$ | $7.5 \times 10^{-9}$ |
| $^{85}$Kr | $4.3 \times 10^{-1}$ | nil | $8.9 \times 10^{-8}$ | $^{51}$Cr | $3.1 \times 10^{-3}$ | $1.3 \times 10^{-7}$ | $6.3 \times 10^{-10}$ |
| $^{87}$Kr | $1.7 \times 10^{-2}$ | nil | $1.0 \times 10^{-8}$ | $^{54}$Mn | $1.6 \times 10^{-3}$ | $6.5 \times 10^{-8}$ | $3.3 \times 10^{-10}$ |
| $^{88}$Kr | $1.8 \times 10^{-2}$ | nil | $3.8 \times 10^{-9}$ | $^{55}$Fe | $1.2 \times 10^{-3}$ | $4.9 \times 10^{-8}$ | $2.5 \times 10^{-10}$ |
| $^{131m}$Xe | $7.3 \times 10^{-1}$ | nil | $1.5 \times 10^{-7}$ | $^{59}$Fe | $3.0 \times 10^{-4}$ | $1.2 \times 10^{-8}$ | $6.1 \times 10^{-11}$ |
| $^{133m}$Xe | $7.0 \times 10^{-2}$ | nil | $1.5 \times 10^{-8}$ | $^{58}$Co | $4.6 \times 10^{-3}$ | $1.9 \times 10^{-7}$ | $9.4 \times 10^{-10}$ |
| $^{133}$Xe | $2.9 \times 10^{-2}$ | nil | $6.0 \times 10^{-9}$ | $^{60}$Co | $5.3 \times 10^{-4}$ | $2.2 \times 10^{-8}$ | $1.1 \times 10^{-10}$ |
| $^{135m}$Xe | $1.3 \times 10^{-1}$ | nil | $2.7 \times 10^{-8}$ | $^{65}$Zn | $5.1 \times 10^{-4}$ | $2.1 \times 10^{-8}$ | $1.0 \times 10^{-10}$ |
| $^{135}$Xe | $6.7 \times 10^{-2}$ | nil | $1.4 \times 10^{-8}$ | $^{89}$Sr | $1.4 \times 10^{-4}$ | $5.7 \times 10^{-9}$ | $2.9 \times 10^{-11}$ |
| $^{137}$Xe | $3.4 \times 10^{-2}$ | nil | $7.1 \times 10^{-9}$ | $^{90}$Sr | $1.2 \times 10^{-5}$ | $4.9 \times 10^{-10}$ | $2.4 \times 10^{-12}$ |
| $^{138}$Xe | $6.1 \times 10^{-2}$ | nil | $1.3 \times 10^{-8}$ | $^{91}$Sr | $9.6 \times 10^{-4}$ | $2.8 \times 10^{-8}$ | $1.4 \times 10^{-10}$ |
| | | | | $^{91m}$Y | $4.6 \times 10^{-4}$ | $3.2 \times 10^{-9}$ | $1.6 \times 10^{-11}$ |
| *Class 2* | | | | $^{91}$Y | $5.2 \times 10^{-6}$ | $2.1 \times 10^{-10}$ | $1.1 \times 10^{-12}$ |
| $^{84}$Br | $1.6 \times 10^{-2}$ | $7.5 \times 10^{-8}$ | $7.5 \times 10^{-10}$ | $^{93}$Y | $4.2 \times 10^{-3}$ | $1.2 \times 10^{-7}$ | $6.1 \times 10^{-10}$ |
| $^{131}$I | $2.0 \times 10^{-3}$ | $8.1 \times 10^{-8}$ | $8.1 \times 10^{-10}$ | $^{95}$Zr | $3.9 \times 10^{-4}$ | $1.6 \times 10^{-8}$ | $7.9 \times 10^{-11}$ |
| $^{132}$I | $6.0 \times 10^{-2}$ | $8.9 \times 10^{-7}$ | $8.9 \times 10^{-9}$ | $^{95}$Nb | $2.8 \times 10^{-4}$ | $1.1 \times 10^{-8}$ | $5.7 \times 10^{-11}$ |
| $^{133}$I | $2.6 \times 10^{-2}$ | $9.0 \times 10^{-7}$ | $9.0 \times 10^{-9}$ | $^{99}$Mo | $6.4 \times 10^{-3}$ | $2.5 \times 10^{-7}$ | $1.2 \times 10^{-9}$ |
| $^{134}$I | $1.0 \times 10^{-1}$ | $7.2 \times 10^{-7}$ | $7.2 \times 10^{-9}$ | $^{99m}$Tc | $4.7 \times 10^{-3}$ | $1.1 \times 10^{-7}$ | $5.7 \times 10^{-10}$ |
| $^{135}$I | $5.5 \times 10^{-2}$ | $1.4 \times 10^{-6}$ | $1.4 \times 10^{-8}$ | $^{103}$Ru | $7.5 \times 10^{-3}$ | $3.1 \times 10^{-7}$ | $1.6 \times 10^{-9}$ |
| | | | | $^{106}$Ru | $9.0 \times 10^{-2}$ | $3.7 \times 10^{-6}$ | $1.8 \times 10^{-8}$ |
| *Class 3* | | | | $^{110m}$Ag | $1.3 \times 10^{-3}$ | $5.3 \times 10^{-8}$ | $2.7 \times 10^{-10}$ |
| $^{89}$Rb | $1.9 \times 10^{-1}$ | $5.3 \times 10^{-7}$ | $2.6 \times 10^{-9}$ | $^{129m}$Te | $1.9 \times 10^{-4}$ | $7.8 \times 10^{-9}$ | $3.9 \times 10^{-11}$ |
| $^{134}$Cs | $3.7 \times 10^{-5}$ | $1.7 \times 10^{-9}$ | $9.0 \times 10^{-12}$ | $^{129}$Te | $2.4 \times 10^{-2}$ | $2.2 \times 10^{-7}$ | $1.1 \times 10^{-9}$ |
| $^{136}$Cs | $8.7 \times 10^{-4}$ | $4.0 \times 10^{-8}$ | $2.0 \times 10^{-10}$ | $^{131m}$Te | $1.5 \times 10^{-3}$ | $5.4 \times 10^{-8}$ | $2.7 \times 10^{-10}$ |
| $^{137}$Cs-$^{137m}$Ba[e] | $5.3 \times 10^{-5}$ | $2.5 \times 10^{-9}$ | $1.2 \times 10^{-11}$ | $^{131}$Te | $7.7 \times 10^{-3}$ | $2.9 \times 10^{-8}$ | $1.5 \times 10^{-10}$ |
| | | | | $^{132}$Te | $1.7 \times 10^{-3}$ | $6.6 \times 10^{-8}$ | $3.3 \times 10^{-10}$ |
| *Class 4* | | | | $^{140}$Ba | $1.3 \times 10^{-2}$ | $5.2 \times 10^{-7}$ | $2.6 \times 10^{-9}$ |
| $^{16}$N | $4.0 \times 10^{1}$ | $1.0 \times 10^{-6}$ | $1.0 \times 10^{-7}$ | $^{140}$La | $2.5 \times 10^{-2}$ | $9.3 \times 10^{-7}$ | $4.6 \times 10^{-9}$ |
| | | | | $^{141}$Ce | $1.5 \times 10^{-4}$ | $6.1 \times 10^{-9}$ | $3.1 \times 10^{-11}$ |
| *Class 5* | | | | $^{143}$Ce | $2.8 \times 10^{-3}$ | $1.0 \times 10^{-7}$ | $5.1 \times 10^{-10}$ |
| $^{3}$H | 1.0 | $1.0 \times 10^{-3}$ | $1.0 \times 10^{-3}$ | $^{144}$Ce | $4.0 \times 10^{-3}$ | $1.6 \times 10^{-7}$ | $8.2 \times 10^{-10}$ |
| | | | | $^{187}$W | $2.5 \times 10^{-3}$ | $8.7 \times 10^{-8}$ | $4.4 \times 10^{-10}$ |
| | | | | $^{239}$Np | $2.2 \times 10^{-3}$ | $8.4 \times 10^{-8}$ | $4.2 \times 10^{-10}$ |

[a] Extracted from American National Standard ANSI/ANS-18.1-1999, with permission of the publisher, the American Nuclear Society.
[b] The numerical concentration values in (ANS 1999) are given to two significant figures in the mixed units shown here. To convert, use $1~\mu\text{Ci} = 3.7 \times 10^{4}~\text{Bq}$ (e.g., $1.6 \times 10^{-2}~\mu\text{Ci g}^{-1} = 0.59~\text{kBq g}^{-1}$).
[c] Reactor coolant concentrations are given for coolant entering the letdown line.
[d] Secondary coolant concentrations are based on a primary-to-secondary leakage rate of $3.9 \times 10^{-4}~\text{kg s}^{-1}$; values are for water in a steam generator and steam leaving a steam generator.
[e] Nuclides in secular equilibrium with concentration of the progeny are equal to that shown for the precursor.

**Table 2.8** Nuclide concentrations ($\mu$Ci g$^{-1}$) in principal fluid streams of the reference PWR with once-through steam generators[a,b]

| Nuclide | Reactor coolant[c] | Secondary coolant[d] | Nuclide | Reactor coolant[c] | Secondary coolant[d] |
|---|---|---|---|---|---|
| *Class 1* | | | *Class 6* | | |
| $^{85m}$Kr | $1.6 \times 10^{-1}$ | $3.4 \times 10^{-8}$ | $^{24}$Na | $4.7 \times 10^{-2}$ | $1.1 \times 10^{-7}$ |
| $^{85}$Kr | $4.3 \times 10^{-1}$ | $8.9 \times 10^{-8}$ | $^{51}$Cr | $3.1 \times 10^{-3}$ | $6.9 \times 10^{-9}$ |
| $^{87}$Kr | $1.7 \times 10^{-2}$ | $3.4 \times 10^{-9}$ | $^{54}$Mn | $1.6 \times 10^{-3}$ | $3.5 \times 10^{-9}$ |
| $^{88}$Kr | $1.8 \times 10^{-2}$ | $3.8 \times 10^{-9}$ | $^{55}$Fe | $1.2 \times 10^{-3}$ | $2.7 \times 10^{-9}$ |
| $^{131m}$Xe | $7.3 \times 10^{-1}$ | $1.5 \times 10^{-7}$ | $^{59}$Fe | $3.0 \times 10^{-4}$ | $6.7 \times 10^{-10}$ |
| $^{133m}$Xe | $7.0 \times 10^{-2}$ | $1.5 \times 10^{-8}$ | $^{58}$Co | $4.6 \times 10^{-2}$ | $1.0 \times 10^{-8}$ |
| $^{133}$Xe | $2.9 \times 10^{-2}$ | $6.0 \times 10^{-9}$ | $^{60}$Co | $5.3 \times 10^{-4}$ | $1.2 \times 10^{-9}$ |
| $^{135m}$Xe | $1.3 \times 10^{-1}$ | $2.7 \times 10^{-8}$ | $^{65}$Zn | $5.1 \times 10^{-4}$ | $1.1 \times 10^{-9}$ |
| $^{135}$Xe | $8.5 \times 10^{-1}$ | $1.8 \times 10^{-7}$ | $^{89}$Sr | $1.4 \times 10^{-4}$ | $3.1 \times 10^{-10}$ |
| $^{137}$Xe | $3.4 \times 10^{-2}$ | $7.1 \times 10^{-9}$ | $^{90}$Sr | $1.2 \times 10^{-5}$ | $2.7 \times 10^{-11}$ |
| $^{138}$Xe | $6.1 \times 10^{-2}$ | $1.3 \times 10^{-8}$ | $^{91}$Sr | $9.6 \times 10^{-4}$ | $2.1 \times 10^{-9}$ |
| | | | $^{91m}$Y | $4.6 \times 10^{-4}$ | $9.7 \times 10^{-10}$ |
| *Class 2* | | | $^{91}$Y | $5.2 \times 10^{-6}$ | $1.2 \times 10^{-11}$ |
| $^{84}$Br | $1.6 \times 10^{-2}$ | $1.8 \times 10^{-8}$ | $^{93}$Y | $4.2 \times 10^{-3}$ | $9.3 \times 10^{-9}$ |
| $^{131}$I | $2.0 \times 10^{-3}$ | $2.3 \times 10^{-9}$ | $^{95}$Zr | $3.9 \times 10^{-4}$ | $8.7 \times 10^{-10}$ |
| $^{132}$I | $6.0 \times 10^{-2}$ | $6.9 \times 10^{-8}$ | $^{95}$Nb | $2.8 \times 10^{-4}$ | $6.2 \times 10^{-10}$ |
| $^{133}$I | $2.6 \times 10^{-2}$ | $3.0 \times 10^{-8}$ | $^{99}$Mo | $6.4 \times 10^{-3}$ | $1.4 \times 10^{-8}$ |
| $^{134}$I | $1.0 \times 10^{-1}$ | $1.1 \times 10^{-7}$ | $^{99m}$Tc | $4.7 \times 10^{-3}$ | $1.0 \times 10^{-8}$ |
| $^{135}$I | $5.5 \times 10^{-2}$ | $6.4 \times 10^{-8}$ | $^{103}$Ru | $7.5 \times 10^{-3}$ | $1.7 \times 10^{-8}$ |
| | | | $^{106}$Ru | $9.0 \times 10^{-2}$ | $2.0 \times 10^{-7}$ |
| *Class 3* | | | $^{110m}$Ag | $1.3 \times 10^{-3}$ | $2.9 \times 10^{-9}$ |
| $^{89}$Rb | $1.9 \times 10^{-1}$ | $6.0 \times 10^{-7}$ | $^{129m}$Te | $1.9 \times 10^{-4}$ | $4.2 \times 10^{-10}$ |
| $^{134}$Cs | $3.7 \times 10^{-5}$ | $1.6 \times 10^{-10}$ | $^{129}$Te | $2.4 \times 10^{-2}$ | $5.1 \times 10^{-8}$ |
| $^{136}$Cs | $8.7 \times 10^{-4}$ | $3.6 \times 10^{-9}$ | $^{131m}$Te | $1.5 \times 10^{-3}$ | $3.3 \times 10^{-9}$ |
| $^{137}$Cs-$^{137m}$Ba[e] | $5.3 \times 10^{-5}$ | $2.2 \times 10^{-10}$ | $^{131}$Te | $7.7 \times 10^{-3}$ | $1.5 \times 10^{-8}$ |
| | | | $^{132}$Te | $1.7 \times 10^{-3}$ | $3.8 \times 10^{-9}$ |
| *Class 4* | | | $^{140}$Ba | $1.3 \times 10^{-2}$ | $2.9 \times 10^{-8}$ |
| $^{16}$N | $4.0 \times 10^{1}$ | $1.0 \times 10^{-6}$ | $^{140}$La | $2.5 \times 10^{-2}$ | $5.6 \times 10^{-8}$ |
| | | | $^{141}$Ce | $1.5 \times 10^{-4}$ | $3.3 \times 10^{-10}$ |
| *Class 5* | | | $^{143}$Ce | $2.8 \times 10^{-3}$ | $6.2 \times 10^{-9}$ |
| $^{3}$H | 1.0 | $1.0 \times 10^{-3}$ | $^{144}$Ce | $3.9 \times 10^{-3}$ | $8.7 \times 10^{-9}$ |
| | | | $^{187}$W | $2.5 \times 10^{-3}$ | $5.6 \times 10^{-9}$ |
| | | | $^{239}$Np | $2.2 \times 10^{-3}$ | $4.9 \times 10^{-9}$ |

[a] Extracted from American National Standard ANSI/ANS-18.1-1999, with permission of the publisher, the American Nuclear Society.
[b] The numerical concentration values in (ANS 1999) are given to two significant figures in the mixed units shown here. To convert, use 1 $\mu$Ci = $3.7 \times 10^4$ Bq (e.g., $1.6 \times 10^{-1}$ $\mu$Ci g$^{-1}$ = 5.9 kBq g$^{-1}$).
[c] Reactor coolant concentrations are given for coolant entering the letdown line.
[d] Secondary coolant concentrations are given for steam leaving a steam generator and are based on a primary-to-secondary leakage rate of $3.9 \times 10^{-4}$ kg s$^{-1}$.
[e] Nuclides in secular equilibrium with concentration of the progeny are equal to that shown for the precursor.

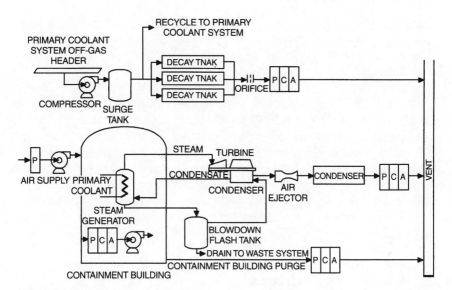

Figure 2.6 Elements of the gaseous waste system for an example pressurized water reactor (PWR) with release points and possible effluent treatment components.

Gilbert (1991) describes systems used to reduce emissions of radioactive noble gases from both BWRs and PWRs. Delay lines and gas holdup tanks, which are used to take advantage of radioactive decay, as well as longer term charcoal delay systems are examined. The design and effectiveness of such systems affect the final source term estimates for most of the noble gases released from reactors.

The details of the liquid waste processing system, including the decontamination factors for each step, must be considered when calculating liquid waste releases. Appendixes A and B to Regulatory Guide 1.112 identify information needs for the liquid waste calculations for BWRs and PWRs, respectively. The two GALE code documents (Cardile et al. 1979; Chandrasekaran et al. 1985) include estimates of basic parameters (e.g., liquid waste volumes and flow rates) that were established at that time. Much of the information needed for the liquid waste system, such as regeneration frequency, decontamination factors, and number, size, and type of demineralizers also can be used to estimate volumes of solid wastes that will be generated during operations. The GALE code reports (Cardile et al. 1979; Chandrasekaran et al. 1985) provide example calculations of radioactivity releases, but source term estimates for particular new facilities will likely differ in at least some respects from the reference BWR or the appropriate reference PWR. Prospective estimates of source terms for any new reactor must also reflect the planned effluent treatment systems that will be used at the facility.

Figures 2.3 and 2.4 show that the source terms for airborne effluents have changed substantially with time and that the introduction of new effluent treatment systems can have a dramatic impact on the releases. Estimates of historic releases of radionuclides from power reactors must address the same parameters discussed, as well as procedures that were in place and materials that were used for valves and piping during the operational period of interest. Numerous special studies of radionuclide releases at U.S. nuclear power plants were performed in the 1970s (Kahn et al. 1970, 1971, 1974; Blanchard et al. 1976; Marrero 1976; Pelletier et al. 1978a, 1978b; Cline et al. 1979; Voillequé 1979; Mandler et al. 1981) during the period when appendix I to 10 CFR 50, defining "as low as practicable" release criteria, was developed and refined, and more detailed information about reactor radionuclide releases was needed. These efforts employed state-of-the-art sampling techniques and parallel collection of plant operating data during the measurement periods. Results of those efforts would be useful for atmospheric source term calculations for either reactor type in the early years of power reactor operation in the United States (prior to the requirements of appendix I and subsequent EPA regulations for offsite dose limits for power reactors [40 CFR 190]). Wilhelm and Deuber (1991) discuss the effectiveness of systems for removal of short-lived radioiodines (particularly in organic form) from effluents, which could have been employed after implementation of those regulations.

There is evidence of a renewed interest in nuclear power plant construction. At least two applications for early site permits have been filed with the NRC Division of New Reactor Licensing (NRC 2006c, 2006d). These applications represent first steps toward building additional power reactors in close proximity to existing plants. In both applications, two more nuclear power reactors with total power levels of up to 8,600 MW (thermal) are considered.

## Source Terms for Accidents

Calculation of radionuclide releases from reactor facilities following postulated accidents is a complex process that has evolved from a simple set of assumptions. These assumptions were considered to provide cautious overestimates of the consequences of the accident. In spite of the fact that reactor accident source term analyses are now highly complex and require resources likely to be found only in regulatory bodies, large utilities, large engineering firms, and national laboratories, it is worth reviewing the history of such assessments.

In the beginning, it was assumed that a dramatic failure of the reactor coolant system, called a loss of coolant accident (LOCA), would release large amounts of radioactivity into the containment (AEC 1957). All of the noble gases and half of the halogens in the core inventory were assumed to enter the containment area immediately, together with 1% of the inventories of the particulate radionuclides. (Core inventory activities can be estimated from the fission yields; results that are more accurate can be obtained from detailed computer calculations that address all aspects of buildup and removal of fission and activation products in the reactor core. The ORIGEN code [Croff 1982] is often used for such calculations.) The halogens of greatest concern were the radioiodines. Half of the radioiodines in the containment (25% of the core inventories of those isotopes), together with the other radionuclides present, were assumed to be available for release according to a constant containment leakage rate of 0.1% per day. This set of assumptions was called a "maximum credible accident" and was believed to be highly unlikely to occur. The term "design basis accident" was also applied to this and other limiting accident types.

Criteria for reactor sites are given in 10 CFR 100. The exclusion area for the site is sufficiently large that a person located at the boundary for 2 h immediately following the start of a postulated accidental release of fission products would not receive a whole-body dose greater than 25 rem or a thyroid dose from radioiodines greater than 300 rem. The low population zone was defined using the same dose criteria for a person at the outer boundary who remained there throughout the duration of the postulated release. A nominal population center distance was defined as at least 1.33 times greater than the distance to the outer boundary of the low population zone. If a large city were involved, an even greater distance to the population center could be required. With the establishment of the NRC in January 1975, 10 CFR 100, other regulations (e.g., 10 CFR 50), and regulatory guidance documents previously developed by the AEC were transferred to the NRC.

Using the assumptions presented in the WASH-740 report (AEC 1957) and the criteria in 10 CFR 100, AEC Report TID-14844 by DiNunno et al. (1962) evaluated the radii of the exclusion zone and low population zone as a function of the thermal power level of the reactor. Besides the generic calculations, analyses for particular facilities also were performed. The analyses showed that the assumed releases of particulate radionuclides did not make a substantial contribution to the external gamma dose. Reference to the DiNunno et al. report in the regulation (10 CFR 100) provides it with a special status not shared by many other AEC/NRC reports related to accident analyses.

Among the first safety guides prepared in 1970 (later called regulatory guides) were those related to calculations of doses from reactor accidents. Revised Regulatory Guides 1.3 and 1.4 (AEC 1974a, 1974b) deal with LOCAs at BWRs and PWRs, respectively, and include additional assumptions about iodine ·chemical forms, breathing rates, and meteorological dispersion calculations to be used in the assessments.

Since the time the relatively simple assumptions about bounding reactor accidents were made, much effort has been devoted to detailed studies of the behavior of reactor systems and fission products, particularly radioiodines. Notable among these is the Reactor Safety Study (report WASH-1400, NRC 1975a), an independent study of risks of commercial reactor operations, commissioned by the AEC in 1972 and published by the newly formed NRC in 1975. The study, directed by Professor Norman Rasmussen of MIT, involved scientists from many organizations around the country who contributed to assessments of possible failures leading to reactor accidents, analyses of the relevant physical processes, the associated releases of radioactivity, and the consequent health effects in exposed populations. These evaluations, which employed probabilistic risk assessment (PRA) methods, were performed for a broad range of possible accident sequences and were used in preparing overall estimates of health risks from the operation of 100 power reactors. Accidents that resulted in core melting with releases of large fractions of noble gases and radioiodines in the core were evaluated, and their probabilities were estimated.

In March 1979, there was a serious accident at the Three Mile Island Unit 2 reactor in Pennsylvania (NRC 1979; Kemeny et al. 1979; Rogovin et al. 1980). Although there was substantial fuel damage (Toth et al. 1986), there was no failure of the containment building, and little radioactivity was released to the environment. Nonetheless, the accident and subsequent analyses of it led to needed improvements in the design and operation of U.S. nuclear power reactors (NRC 1980b).

The Three Mile Island Unit 2 accident led to reassessments of the releases of fission products and their behavior during reactor accidents and of the complex computer codes used in such analyses (e.g., NRC 1981). A subsequent report by Silberberg et al. (1985) discussed the history of reactor accident source terms in more detail and identified substantial improvements that had been made since the Reactor Safety Study. The authors and outside reviewers examined in detail the approaches used in the new computer codes, assessed their validity, and pointed to areas where additional research was needed.

A more recent report on reactor accident risks (NRC 1990) contains the results of detailed analyses of risks from five reactors operating in the United States utilizing the new approach to source term calculations described by Silberberg et al. (1985). PRA techniques were used to estimate ranges of results that could occur considering uncertainties in accident frequencies, estimated source terms, and containment building behavior and leakage. The accident at the Chernobyl nuclear power plant in Ukraine occurred while the document was being prepared (April 26, 1986). The Chernobyl reactor differed in important ways from BWRs and PWRs. Perhaps the most important difference was that the Chernobyl reactor did not have a containment building. The graphite moderator, which burned at Chernobyl, is not present in U.S. power reactor cores. The release estimates for the Chernobyl accident, comparable to those predicted for severe fuel damage and early loss of containment in a light-water reactor, were in the range of those predicted in the study (NRC 1990).

As a matter of policy, the NRC (1998) has taken the position that "use of PRA technology should be increased in all regulatory matters to the extent supported by the state of the art in PRA methods and data and in a manner that complements the NRC's deterministic approach and supports the NRC's traditional defense-in-depth philosophy." The relevant regulations have not been changed to address use of PRA methods, but there is a policy emphasis on reducing unnecessary conservatism and making realistic assessments.

New regulatory guides (NRC 2003a, 2003b) have been published that deal with the evaluation of accident consequences and the important corollary concern of control room habitability during an accident. The guidance in those documents can be used instead of Regulatory Guides 1.3 and 1.4 (AEC 1974a, 1974b) for LOCA analyses. For power reactors whose initial operating license was issued before January 10, 1997, Regulatory Guide 1.183 (NRC 2000) can be used to voluntarily revise the design basis accident source term for the facility.

## Fuel Processing Plants

As noted above, a pilot plant for fuel processing was established in Oak Ridge during the Manhattan Project, and two full-scale facilities to recover plutonium from irradiated uranium fuel were built at the Hanford Site soon thereafter. Initial fuel processing at Hanford began after the fuel had been stored a relatively short time, and substantial amounts of short-lived radionuclides were released during processing. The most important of these was $^{131}$I, whose releases were not measured at the time. The releases of $^{131}$I at Hanford from 1944 to 1972 were estimated as part of the Hanford Environmental Dose Reconstruction Project (Heeb et al. 1996).

Releases of $^{131}$I also occurred during processing of short-cooled fuel elements at the X-10 site (near Oak Ridge, Tennessee) during a project (1944–1956) to recover and purify the fission product $^{140}$La for use in plutonium weapons development at Los Alamos, New Mexico. Apostoaei et al. (1999) developed source term estimates for the so-called RaLa (radioactive lanthanum) processing performed during those years.

Other fuel processing plants were built at the National Reactor Testing Station (now the Idaho National Laboratory) and at the Savannah River plant. Iodine-131 releases were also important for these facilities, particularly during the early years of operation. The RaLa processing effort was moved from Oak Ridge to the more remote location in Idaho, where a source of irradiated fuel elements was readily available. Releases from those facilities were also estimated as part of other dose reconstruction projects (Meyer et al. 2000).

Substantial effort has been devoted to the development of effluent treatments to reduce releases of $^{131}$I and other radionuclides from fuel processing plants. The simple technique of greater fuel storage time to allow $^{131}$I to decay was implemented at Hanford after the wartime pressure for plutonium production was relieved. Even so, further efforts to remove $^{131}$I from offgas streams at Hanford were needed, and a variety of cleanup systems (scrubbers, filters, and silver reactors) was installed after the war. The largest releases from the two fuel processing plants occurred between late 1944 and the end of 1949 (Heeb et al. 1996).

The AEC (and later the DOE) organized periodic meetings of scientists and engineers who were working on various aspects of filtration, other methods of removal, and measurements of radioactive contaminants in gaseous effluents. These meetings became known as the "air cleaning conferences." Staff members from various facilities reported the status of their efforts to collect airborne particles and gases (e.g., $^{131}$I in its various forms) and discussed common problems (e.g., AEC 1954; DOE 1985). First (1991), a leader of these efforts for many years, summarized the development of aerosol filtration techniques.

Two IAEA publications document progress in understanding radioiodine behavior and the development of radioiodine effluent control (IAEA 1973, 1980). Removal of iodine in effluents from fuel processing plants is a focus of the later publication (IAEA 1980). Jubin and Counce (1991) address the joint problem of removal of both radioiodine and nitrogen oxides from the exhausts of fuel processing facilities.

Once releases of short-lived and particulate radionuclides were better controlled, attention turned to the volatile longer lived isotopes $^3$H, $^{14}$C, $^{85}$Kr, and $^{129}$I (OECD 1980). The World Health Organization developed environmental health criteria for these and other isotopes, including a discussion of the sources (World Health Organization 1983). Kovach (1991) examines the control technologies for these four radionuclides in fuel processing facilities.

As part of its survey of the nuclear fuel cycle, the EPA considered fuel reprocessing in some detail (EPA 1973c). At that time, one plant had been built and had been in operation but was in an outage for modifications to increase the processing capacity from 1 metric ton of uranium (MTU) day$^{-1}$ to 3 MTU day$^{-1}$. Cochran et al. (1970) performed some measurements of the airborne releases, including $^{129}$I, from the facility. Planning had begun for construction of two other commercial facilities, a 3-MTU day$^{-1}$ plant in Illinois and a 5-MTU day$^{-1}$ plant in South Carolina.

The EPA analysis of future doses due to releases from fuel processing plants focused on releases of $^3$H, $^{85}$Kr, $^{129}$I, and actinides (e.g., $^{239}$Pu) from 40 years of operation of a model facility (EPA 1973c). In the end, such analyses were less important than concerns about proliferation of weapons capabilities resulting from potential access to plutonium recovered by processing spent fuel. Presidents Ford and Carter shared this concern and did not believe that fuel processing should go forward without assurance that proliferation risks would be small. In 1977, President Carter announced a policy to defer indefinitely commercial reprocessing of fuel and the recycling of plutonium for nuclear power production, and vetoed the bill that would have authorized funding for a breeder reactor and reprocessing plant (Andrews 2006).

Fuel processing had begun in other countries, however, and has continued since that time. UNSCEAR (2000) has summarized information about fuel processing activities in France at Cap de la Hague, at Sellafield in the United Kingdom, and at Tokai in Japan from the 1970s through 1997. For source term estimates, the most useful results are the normalized release rates, which are presented in table 2.9. The normalization is to the amount of electricity produced. According to estimates in the EPA report (1973c), the amount of fuel discharged per gigawatt can range from about 10 MTU for light-water reactors to 20 MTU or more for mixes of reactor types. Considering the reactor types in the three countries, the ratio seems likely to fall in the

Table 2.9 Normalized releases from fuel processing facilities in France, the United Kingdom, and Japan (UNSCEAR 2000)

| Period | Fuel processed (GW yr) | Normalized release (TBq per GW yr) in effluents | | | | | | | |
|---|---|---|---|---|---|---|---|---|---|
| | | $^3$H | $^{14}$C | $^{129}$I | $^{137}$Cs | $^{85}$Kr | $^{131}$I | $^{90}$Sr | $^{106}$Ru |
| *Airborne effluents* | | | | | | | | | |
| 1970–1979 | 29.2 | 93 | 7.3 | 0.006 | 0.09 | 13,920 | 0.12 | | |
| 1980–1984 | 36.6 | 48 | 3.5 | 0.007 | 0.04 | 11,690 | 0.03 | | |
| 1985–1989 | 62.5 | 24 | 2.1 | 0.003 | 0.002 | 7,263 | 0.0003 | | |
| 1990–1994 | 131 | 24 | 0.4 | 0.001 | 0.00008 | 6,300 | 0.00009 | | |
| 1995–1997 | 160 | 9.6 | 0.3 | 0.001 | 0.00001 | 6,900 | 0.00005 | | |
| *Liquid effluents* | | | | | | | | | |
| 1970–1979 | 29.2 | 399 | 0.4 | 0.04 | 1,020 | | | 131 | 264 |
| 1980–1984 | 36.6 | 376 | 0.3 | 0.04 | 252 | | | 45 | 112 |
| 1985–1989 | 62.5 | 378 | 0.8 | 0.03 | 7.4 | | | 7.5 | 33 |
| 1990–1994 | 131 | 270 | 0.8 | 0.03 | 1.0 | | | 2.0 | 2.1 |
| 1995–1997 | 160 | 255 | 0.4 | 0.04 | 0.2 | | | 0.8 | 0.5 |

10–20 MTU GW$^{-1}$ range. The columns containing the four radionuclides common to both airborne and liquid effluent lists are aligned to facilitate comparisons. Liquid effluents carried most of the $^3$H, $^{129}$I, and $^{137}$Cs that were discharged, while $^{14}$C releases were primarily to the atmosphere except in the last two periods. Substantial decreases can be seen in discharges of the particulate radionuclides $^{137}$Cs, $^{90}$Sr, and $^{106}$Ru in liquid wastes.

As is the case for releases from reactors, the normalized release rates have changed with time, reflecting changes in plant operation and effluent treatment systems. Source term assessments for a fuel processing facility (or a facility to do research and development on fuel processing technology) must reflect the conditions at the time being considered and the plant features that most strongly affect releases. The NRC did not develop regulatory guides for analyses of routine releases from fuel processing facilities before such releases were no longer an issue; however, some related regulatory guides were adopted (AEC 1973, 1974c, 1974d; NRC 1975b). A regulatory guide addressing releases due to an accidental nuclear criticality was prepared but was later withdrawn.

There is evidence of an effort to overcome concerns about proliferation and to establish fuel processing as a viable activity in the United States. The Advanced Fuel Cycle Initiative undertaken by the DOE is intended to develop a proliferation-resistant approach that will recover the valuable materials in spent fuel, lead to further development of nuclear power, and reduce the toxicity and amounts of high-level nuclear waste requiring geologic disposal (Andrews 2006; Sandia 2007).

## *Solid Waste Disposal*

Management of solid radioactive wastes is largely based on the current source-based system used for waste classification, which means that wastes posing similar risks are treated differently because different laws apply to them. The NCRP proposed a risk-based classification of wastes (radioactive, chemical, or mixed) in which the wastes are separated according to the risks to public health posed by disposal of the wastes (NCRP 2002). Croff, who chaired the NCRP committee, has suggested a modified approach in which public health risk is still the primary determinant for classification, but other aspects of the problem are also considered; this is described as risk-informed waste classification (Croff 2006). The National Academy of Sciences (NAS) has prepared reports dealing with the DOE's high-level and transuranic wastes (NAS/NRC 2005) and low-activity wastes (NAS/NRC 2006), both of which endorse risk-informed decision making for radioactive waste management. The future will reveal whether the NAS committees' recommendations, which recognize the complexities created by more than 50 years of patchwork legislation, could be implemented to improve the situation.

Estimation of releases from wastes that are disposed by burial near the land surface or placed in a geologic repository is a complex problem that involves many elements and processes. Initially, the inventory of radionuclides in the waste, their relative toxicities, and the form(s) of the waste material must be considered. Both the waste forms and the containers used for burial affect the movement of the radionuclides. The containers have finite lifetimes, which, for wastes buried years ago, may already have been exceeded. The presence of water may facilitate chemical changes in the wastes and provides a mechanism for movement of the radionuclides. The rate of release of radionuclides from the disposal location, the source term, depends on these and other factors (e.g., low-frequency but dramatic natural events).

The NCRP has prepared a comprehensive report that examines factors affecting the potential releases of radionuclides from near-surface disposal facilities (NCRP 2005). It treats, in detail, elements of the prospective release analysis as part of the overall performance assessment of such facilities. In some disposal facilities, there may be other barriers, not mentioned above, such as concrete cell structures and covers to reduce infiltration of water. The report also discusses a range of issues related to the level of detail needed in an assessment and is recommended to anyone whose task includes evaluation of source terms resulting from waste disposal. Many of the issues for geologic repositories are similar, with the added complexity of addressing events that may occur very long times after disposal.

## Source Term Development for Facilities

Source term estimates are needed when dose estimates cannot be based on measurements of exposed individuals or on environmental measurements that can be used to estimate doses with some confidence. Use of incomplete environmental data on long-lived radionuclides may be a better choice than trying to model environmental transport over long distances or widely dispersed exposure locations. On the other hand, for releases of short-lived radionuclides that produce transient contamination, the desired environmental data are not generally available and estimation of a source term for the facility may be required. Two types of analyses, prospective and retrospective, are considered in the next subsections.

### Source Terms for Prospective Analyses

Estimates of releases to the environment are always needed for prospective analyses. These often focus on a representative year of future operation of a planned facility. The expected releases to air and water are estimated, and estimated doses are compared with applicable annual limits or goals that are lower than the limits. If the comparison is with regulatory limits, an assessment that is somewhat conservative, tending to overestimate the doses may be appropriate, but the reasons for and degree of overestimation may differ from case to case.

The process of prospective dose estimation for routine operation of power reactors, based on NRC regulatory guidance, has already been described in some detail. As noted, the estimate begins with the concentrations of radionuclides in reactor coolant. Processes that lead to releases are known, and methods of reducing those releases have been identified. Data on the performance of various types of coolant cleanup systems have been collected from operating reactors, and the effects of effluent treatment options have also been measured. Checks on predicted performance can be obtained using good-quality effluent monitoring data that have been collected at operating plants. The approach used for power reactors can serve as a model for other facilities if sufficient information is available to clearly identify the release paths, to estimate the efficacy of methods used for effluent reduction, and to check predicted releases.

Prospective analyses of the source term for routine operation can be checked after operations begin and facility monitoring data have been collected for a reasonable period of steady operation. The effluent and environmental monitoring data collected during that period will provide information that can be used to determine whether the prospective source term analysis was realistic. Such checks can be conducted periodically as operations continue and the database of experience grows. In-plant tests of effluent treatment systems can also help to determine whether the decontamination factors used for source term estimation are being achieved during operation of the facility.

The complex process of estimating releases to the atmosphere from reactor accidents has also been discussed in the previous section. For other types of nuclear facilities, the focus is also on airborne releases, but a different approach has been adopted. The American National Standards Institute has approved a procedure for estimating airborne release fractions (ARFs) for accident source term assessment at nonreactor nuclear facilities (ANS 1998). The ARF is the fraction of material involved in a short-term event that becomes airborne and available for release. The standard itself defines criteria for selection of an ARF for various situations. Appendix A to the standard provides bounding ARFs for various types of events (e.g., explosions, spills) together with an indication of the reliability of the information and references upon which the selection is based. For processes that continue for an extended period (e.g., suspension of contaminants), an airborne release rate is defined as the fraction released per unit time (e.g., per hour). Appendix B provides a source term calculation process that employs a number of factors to estimate the release of respirable material to the atmosphere for a particular accident. These appendixes are not part of the standard, but provide useful guidance to the user.

One of the principal references for appendix A to the standard is the handbook of release fractions prepared for the DOE, with contributions from many scientists (DOE 1994). Large volumes of experimental data, related to many situations that could lead to airborne releases, were analyzed and summarized according to the form (i.e., gas, liquid, solid, contaminated surfaces) of the material of interest. Experimental conditions and original data are presented in an extensive appendix that is part of the handbook. Accidental nuclear criticalities are treated separately. The handbook contains a listing of known criticalities for various material configurations, their causes, and the physical damage to the fissile material involved.

Quoting the NRC conclusion that the inhalation exposure pathway will be the most important source of dose following accidental releases from fuel cycle facilities, the DOE handbook (DOE 1994) presents a five-part source term equation, later adopted in the standard (ANS 1998). The components of the simple multiplicative equation are the amount of material at risk, also referred to as the inventory, in the location where the event occurs, the fraction of the material actually affected (damage ratio), the airborne release fraction (ARF), the respirable fraction based on the portion of the particles produced that have aerodynamic diameters of 10 µm or less, and the leak path factor. The latter reflects deposition along the leak path as well as losses from the air stream due to the presence of intact filters in the ventilation system.

The NRC has also published an updated accident analysis handbook for fuel cycle facilities (SAIC 1998) that considers a broader variety of facilities and accidents than were considered in the original version. The first NRC accident analysis handbook (NRC 1988) focused more on fuel processing and plutonium fuel production facilities. The more recent version (SAIC 1998) considers a number of other facilities and processes in the nuclear fuel cycle and also discusses chemical release. It includes an updated basis for computer calculation of the leak path factor and also updates release estimates from nuclear criticalities, replacing information formerly in regulatory guides that were withdrawn.

## Source Terms for Retrospective Analyses

Retrospective estimates of historic releases usually consider many years of operation of a particular plant or multiplant complex. Several evaluations of this type have been performed for facilities that were part of the nuclear weapons complex, with periods of interest dating back to the 1940s in some cases. Some of the source term reconstructions have been mentioned in preceding sections with references to relevant publications and technical reports. The source term estimates have typically been used as part of dose and risk assessments for representative persons exposed during previous operations of the facilities. For the $^{131}$I releases from the Hanford fuel processing plants, dose assessments were made for specific individuals who were members of a cohort for an epidemiologic study of thyroid disease (Kopecky et al. 2004; Davis et al. 2004). Dose reconstruction for epidemiologic studies was the subject of a workshop and subsequent report developed by an NAS advisory committee to the Centers for Disease Control and Prevention (NAS/NRC 1995).

When long time periods are considered, as in the Hanford study, many changes occurred in the processing facilities where the releases occurred and some of these affected the method of source term estimation. In general, close attention should be paid to any basic plant process modifications, installations of new exhaust treatment systems, and the introduction or improvement of monitoring capabilities. Scoping calculations can be used to identify the most important radionuclide releases, which may have varied during the different phases of operation.

Monitoring improvements may mean that the releases of interest will have been measured more reliably during later years of operation and that relevant environmental data are available for checking the source term estimates during some time periods. Particularly desirable would be a program of environmental monitoring data collection by an organization that is independent from the facility operator. Changes in modes of operation and effluent processing may also affect the chemical and physical forms of the materials discharged, which should be indicated to the extent possible. Such changes are particularly important to source term estimates and will warrant a detailed investigation to define blocks of time when the operations and associated releases reflected relatively consistent operating conditions.

The overall source term estimates will reflect the sequence of such periods during the life of the facility.

For each time period, the objective is to identify the relevant release points and to make unbiased estimates of the releases, with their associated uncertainties, from those locations. It is desirable to know the flow rates of the airborne or waterborne releases and radionuclide concentrations in those streams as well as the physical and chemical forms of the radionuclides that were discharged. Other factors that affect the dispersion of the released material should also be specified. Examples are the stack height, stack diameter, the temperature of airborne discharges, and the pH and temperature of liquid effluents.

Several documents contribute information useful for making source term estimates. These include routine reports of effluent measurements, analytical data sheets for effluent samples, reports of data from pilot plant operations and from special measurement campaigns, reports about incidents and of accident investigations, production records (useful for estimating normalized release rates), and daily waste processing logs. While the historical record will probably be incomplete, all data sources related to important releases from the facility should be examined.

The reported results of effluent measurements are quite useful, but it is also necessary to examine the basis for such data. Key issues are whether the samples that were extracted were representative of the discharge, whether the collection media employed were appropriate for all chemical forms of the radionuclides likely to be found in the discharge, and whether the analytic procedures and measurement techniques were appropriate. It is likely that documentation of such information improved with time and that deficiencies were recognized and corrected as understanding of monitoring problems increased. Possible reasons for incorrect estimates of effluent quantities are nonrepresentative sampling and bias, uneven distribution of contaminants in the effluent stream, losses during sample extraction and deposition in long sampling lines, incomplete sample collection, losses during chemical analysis (tracers not used), and inadequate correction for detector efficiency and absorption during sample counting. Samples of batch releases of liquid will not be representative unless the liquid is well mixed prior to sampling and discharge. Grab samples of liquid discharges may not accurately reflect releases; proportional samples, in which the volume sampled varies with the release flow rate at the point of discharge, are preferred.

It is necessary to estimate the releases of important radionuclides during periods when they were not measured. Brief periods may be caused by a temporary malfunction of a sampling system or loss of a filter prior to analysis. An extended period may be due to lack of an effluent monitoring program during initial operations. In the first case, review of a variety of reports is useful to ensure that there were no unusual events that could lead to elevated release rates during the period. If that was the case, then simple interpolation between measured releases before and after could provide a reasonable estimate of releases for the brief period without measurements. To make estimates of releases for an extended period when there was no effluent monitoring, it is necessary to relate the releases to a characteristic of plant operation, usually to a production or processing rate.

## 66  Radiological Risk Assessment and Environmental Analysis

The concept of a normalized release rate is illustrated in the tables and figures above presenting UNSCEAR summaries of reactor and fuel processing plant releases. Those releases were related to electrical energy generated. For some facilities, normalization of the release rate $Q$ (kBq day$^{-1}$) to a production or processing rate $P$ (kg day$^{-1}$) may be appropriate. The normalized release rate is defined as $QP^{-1}$ and must be estimated for periods when reliable monitoring and production data are available. The choice of the unit of time employed obviously depends upon the level of detail of the available information, which may range from hourly to annual. Once the pattern, trend, or distribution of normalized release rates is established, a choice of the values most representative of operations during the period without effluent measurements can be made. This approach assumes that production data are available for the time period when releases were not measured. Otherwise, some index for production, which must be defined during periods with and without effluent monitoring results, would be used to develop an appropriate normalized release rate for application during the period without monitoring. Because releases to the environment may be highest when maintenance is performed at the facility, the duration of various plant operating modes is important and must be considered in the evaluation of normalized release rates (illustrated in Pelletier et al. 1978a, 1978b).

It is important, whenever possible, to compare estimated environmental concentrations with independent measurements of those concentrations (NAS/NRC 1995). The data for comparison could come from routine programs in which the relevant medium (e.g., air, water, vegetation, food) is well monitored. Integrated depositions of long-lived radionuclides of interest can be measured in soils (e.g., Beck and Krey 1983) and sediments (e.g., Krey et al. 1990; Whicker et al. 1994). Besides quantifying the local deposition density and time history, such measurements can provide a check on cumulative releases. Killough et al. (1999) studied the measured deposition and migration of uranium in soils around the Fernald facility. Estimation of the total deposition based on that work provided a measure of confirmation of the airborne uranium source term that had been developed from incomplete effluent monitoring data supplemented by estimates based on production records.

## Problems

1. Early reactor fuel consisted of cylindrical "slugs" of natural uranium metal (enclosed in aluminum cans) that were placed in tubes in the pile and could be pushed out the discharge end of the tube after irradiation. Estimate the $^{131}$I inventory in a fuel slug that was 10 cm long and 3 cm in diameter and had spent 220 days in a fuel channel location where the average power was 3 MW. The handbook value for the density of uranium is 19.05 g cm$^{-3}$. The nominal energy release per fission of $^{235}$U is ~200 MeV, but ~11 MeV is carried away by neutrinos. The fission yield for the atomic mass 131 chain and the half-life of $^{131}$I are given in figure 2.2b.

2. Estimate the total release rate of noble gases from a 3,400-MW (thermal) BWR whose turbine air ejector exhaust passed through a 30-min delay line. Assume the steam flow rate given in table 2.5 and the radionuclide concentrations given in table 2.6. What are the principal radionuclides in the exhaust from the delay line? What reduction in total release rate is provided by the 30-min delay?
3. One way to reduce noble gas releases is to use massive carbon beds to adsorb the gases and delay their transport to the stack. Chandrasekaran et al. (1985) give the following formula for estimating the holdup time ($T$, days) for such a system:

$$T = \frac{0.011MK}{F}$$

where $M$ is the mass of the carbon absorber in thousands of pounds, $K$ is the dynamic absorption coefficient (cm$^3$ g$^{-1}$), $F$ is the system flow rate (ft$^3$ min$^{-1}$), and the numerical factor results from conversion of the mixture of units. For a system operating at 25°C and a dew point of –40°C, the dynamic absorption coefficients are estimated to be 70 and 1,160 cm$^3$ g$^{-1}$ for krypton and xenon, respectively. What are the expected holdup times for the two gases in a system containing 90,000 pounds of carbon that treats an offgas flow rate of 30 ft$^3$ min$^{-1}$? What are the release rates for the noble gases considered in problem 2 after passing through the 30-min delay line and this additional offgas treatment system? Littlefield et al. (1975) reported on the design and operation of such an augmented offgas system for Vermont Yankee. Their initial results indicated holdup times of more than 3.5 days for krypton and about two months for xenon. The difference in calculated and observed holdup times may be due to a difference between the design flow rate and the actual flow rate.
4. Radioiodines leave BWR reactor coolant in steam, are removed from the coolant by the cleanup demineralizer, and are lost due to radioactive decay. For a reference BWR (tables 2.5 and 2.6), determine which of these processes removes the most $^{131}$I. Compute the total removal rate constant assuming that both the cleanup and condensate demineralizers have efficiencies of 90% for halogens and that all the condensate is returned to the reactor vessel. What is the rate of leakage of $^{131}$I from the fuel into the coolant?
5. Several types of leakage contribute to the radionuclide source term for PWR liquid wastes. These include primary coolant system equipment drains, auxiliary building floor drains, and leakage of primary coolant in the reactor containment building. If the weighted average concentration in these liquids is 0.076 times the primary coolant activity, estimate the releases of $^{134}$Cs, $^{137}$Cs, $^{54}$Mn, and $^{58}$Co in liquid waste. Calculations by Kirchner and Noack (1988) indicated that, for Chernobyl, the ratio of $^{134}$Cs to $^{137}$Cs in the core increased at the rate of ∼0.0411 per GW-day ton$^{-1}$. Assuming this rate of increase applies, what is the approximate burnup of the fuel of the standard 3,400-MW (thermal) PWR described in tables 2.5 and 2.7?

6. A uranium processing facility began operation in 1953. Production records, which gave the annual amounts of uranium processed (MTU), were available for all years of operation. Files of effluent monitoring and stack flow rate data were located, and the annual uranium releases (kg U) to the atmosphere were computed for most years of operation. The normalized release rate is defined as the amount of uranium released during a year divided by the production rate for that year (kg U MTU$^{-1}$). The computed normalized release rates for the facility are shown in the following table.

| Year | Normalized release rate (kg U [MTU]$^{-1}$) | Year | Normalized release rate (kg U [MTU]$^{-1}$) |
|---|---|---|---|
| 1955 | 2.5 | 1963 | 0.017 |
| 1956 | 0.69 | 1964 | $-^a$ |
| 1957 | 0.093 | 1965 | 0.052 |
| 1958 | 0.061 | 1966 | 0.035 |
| 1959 | 0.14 | 1967 | 0.051 |
| 1960 | 0.020 | 1968 | 0.073 |
| 1961 | 0.032 | 1969 | 0.011 |
| 1962 | 0.071 | 1970 | 0.025 |

$^a$Effluent monitoring results not located.

Calculate the geometric mean and geometric standard deviation of the normalized release rates. There was no effluent monitoring during the first years of operation (1953 and 1954), when the production rates were 760 and 2,400 MTU, respectively. Estimate the uranium release rates for those years. Estimate the uranium releases to the atmosphere during 1964 in two different ways; the production rate during that year was 4,400 MTU.

7. Spontaneous combustion of plutonium (Pu) metal led to fire damage to a line of plastic gloveboxes. A prefire inventory indicated that there were eight Pu metal items in the fire area with a total mass of 54 kg. Another accounting, made after the fire, identified six intact Pu metal items weighing a total of 40 kg. The airborne release fraction (ARF) depends on the temperature and the ratio of surface area to mass. For combustion at the ignition temperature or higher, the ARF for plutonium is estimated to be $5 \times 10^{-4}$, and the respirable fraction (RF) of the aerosol produced is 0.5. If the temperature reached is below the ignition temperature, the ARF and RF are estimated to be $3 \times 10^{-5}$ and 0.04, respectively. For Pu pieces with a surface area-to-mass ratio of 100 cm$^2$ g$^{-1}$, the observed ignition temperature is 160°C. For bulky items, with surface area-to-mass ratios < 10 cm$^2$ g$^{-1}$, the ignition temperature is ~500°C (SAIC 1998). Estimate the release of respirable particles of Pu into the room during the fire assuming the temperatures of the burned items exceeded 500°C. What physical forms of Pu metal could have surface area-to-mass ratios ~100 cm$^2$ g$^{-1}$?

8. Predicting deposition densities of $^{131}$I due to wet and dry deposition of releases from the Chernobyl accident in areas where no measurements were

made is not a simple task. The rates and elevations of the release are not well known, nor are the forms of $^{131}$I released. Also not known are the changes in airborne species that occurred as the plume moved to downwind locations. One possible way to overcome the lack of information about deposition of $^{131}$I is to measure soil concentrations of the long-lived isotope $^{129}$I. Estimate the $^{131}$I and $^{129}$I inventories in the Chernobyl core. Devell et al. (1995) give a set of core inventory estimates for $^{131}$I from various investigators and quote burnup estimates of 10,921 and 11,660 ± 650 MW-day ton$^{-1}$. Straume et al. (2006) report on the status of such work in Belarus.

9. A uranium mill processed ore containing 0.12% $U_3O_8$ at the rate of 1,500 MT day$^{-1}$. If the mill had an average uranium recovery efficiency of 92%, operated 325 days per year, and produced a yellowcake product that is 89% $U_3O_8$, how much yellowcake and how much $U_3O_8$ were produced during a year? An important source of releases from uranium mills was the stack serving the yellowcake drying and packaging area of the mill. Estimate releases of $U_3O_8$ for these operations using an average normalized release rate (NRR) of 0.04 kg h$^{-1}$ (MT day$^{-1}$)$^{-1}$ that was based on the average $U_3O_8$ production rate and measurements of the drying and packaging area exhaust during a one-week period. If the performance of the scrubber in the offgas system deteriorated, the NRR could increase to 0.2–0.3 kg h$^{-1}$ (MT day$^{-1}$)$^{-1}$. In the absence of other measurements of the releases from this stack, how would you estimate the release during a year of operation? (a) Suppose that the operational parameters for the scrubber were recorded daily, but may not have been reviewed until several days later. (b) Suppose that changes in scrubber performance were estimated by observing the exhaust from the yellowcake area stack.

10. Estimate releases of radon from a tailings pond that covers an area of 60 hectares (∼150 acres). The NRC (1987) employs a generic release factor of 1 pCi m$^{-2}$ s$^{-1}$ of $^{222}$Rn per pCi g$^{-1}$ of $^{226}$Ra in the tailings. This factor reflects a combination of differing $^{226}$Ra concentrations in slimes and sands (small and large tailings particle sizes), radon diffusion coefficients, moisture contents, and thicknesses of the tailings pile. Information for more complex calculations that reflect mote detailed descriptions of tailings impoundment conditions and covered tailings piles can be found in the environmental impact statement (NRC 1980a).

References

AEC (U.S. Atomic Energy Commission). 1954. *Third Atomic Energy Commission Air Cleaning Conference Held at Los Alamos Scientific Laboratory, September 21, 22, and 23 1953*. Report WASH-170 (Del.). U.S. Atomic Energy Commission, Washington, DC.

AEC. 1957. *Theoretical Possibilities and Consequences of Major Accidents in Large Nuclear Power Plants*. Report WASH-740. U.S. Atomic Energy Commission, Washington, DC.

AEC. 1973. *Content of Technical Specifications for Fuel Reprocessing Plants*. Regulatory Guide 3.6. U.S. Atomic Energy Commission, Washington, DC.

AEC. 1974a. *Assumptions Used for Evaluating the Potential Radiological Consequences of a Loss of Coolant Accident for Boiling Water Reactors.* Regulatory Guide 1.3, Revision 2. U.S. Atomic Energy Commission, Washington, DC.

AEC. 1974b. *Assumptions Used for Evaluating the Potential Radiological Consequences of a Loss of Coolant Accident for Pressurized Water Reactors.* Regulatory Guide 1.4, Revision 2. U.S. Atomic Energy Commission, Washington, DC.

AEC. 1974c. *Confinement Barriers and Systems for Fuel Reprocessing Plants.* Regulatory Guide 3.18. U.S. Atomic Energy Commission, Washington, DC.

AEC. 1974d. *Process Offgas Systems for Fuel Reprocessing Plants.* Regulatory Guide 3.20. U.S. Atomic Energy Commission, Washington, DC.

Andrews, A. 2006. *Nuclear Fuel Reprocessing: U.S. Policy Development.* Congressional Research Service Report. Order Code RS22542.

ANS (American Nuclear Society). 1998. *American National Standard for Airborne Release Fractions at Non-reactor Nuclear Facilities.* ANSI/ANS-5.10–1998. American Nuclear Society, La Grange Park, IL.

ANS. 1999. *Radioactive Source Term for Normal Operation of Light Water Reactors, an American National Standard.* ANSI/ANS-18.1–1999. American Nuclear Society, La Grange Park, IL.

ANSI (American National Standards Institute). 1976. *Source Term Specification N237, Radioactive Materials in Principal Fluid Streams of Light-Water-Cooled Nuclear Power Plants.* ANS-18.1. American National Standards Institute, New York.

Apostoaei, A.I., R.E. Burns, F.O. Hoffman, T. Ijaz, C.J. Lewis, S.K. Nair, and T.E. Widner. 1999. *Iodine-131 Releases from Radioactive Lanthanum Processing at the X-10 Site in Oak Ridge, Tennessee (1944–1956)—an Assessment of Quantities Released, Off-Site Radiation Doses, and Potential Excess Risks of Thyroid Cancer.* Reports of the Oak Ridge Dose Reconstruction, Vol.1. The Report of Project Task 1. ChemRisk, Alameda, CA.

Balonov, M.I., and I.G. Travnikova. 1993. "Importance of Diet and Protective Actions on Internal Dose from Cs Radionuclides in Inhabitants of the Chernobyl Region." In S.E. Merwin and M.I. Balonov, eds. *The Chernobyl Papers.* Vol. 1: *Doses to the Soviet Population and Early Health Effects Studies.* Research Enterprises Publishing Segment, Richland, WA. Pages 127–166.

Beck, H.L., and P.W. Krey. 1983. "Radiation Exposures in Utah from Nevada Nuclear Tests." *Science* 220: 18–24.

Blanchard, R.L., W.L. Brinck, H.E. Kolde, H.L. Krieger, D.M. Montgomery, S. Gold, A. Martin, and B. Kahn. 1976. *Radiological Surveillance Studies at the Oyster Creek BWR Nuclear Generating Station.* Report EPA-520/5-76-003. Environmental Protection Agency, Cincinnati, OH.

Bradley, D.J. 1997. *Behind the Nuclear Curtain: Radioactive Waste Management in the Former Soviet Union.* Battelle Press, Columbus, OH.

Buddenbaum, J.E., R.E. Burns, J.K. Cockroft, T. Ijaz, J.J. Shonka, and T.E. Widner. 1999. *Uranium Releases from the Oak Ridge Reservation—a Review of the Quality of Historical Effluent Monitoring Data and a Screening Evaluation of Potential Off-Site Exposures.* Task 6 Report. ChemRisk, Alameda, CA.

Cardile, F.P., R.R. Bellamy (eds.), R.L. Bangart, L.G. Bell, J.S. Boegli, W.C. Burke, J.T. Collins, J.Y. Lee, J.L. Minns, P.G. Stoddard, and R.A. Weller. 1979. *Calculation of Releases of Radioactive Materials in Gaseous and Liquid Effluents from Boiling Water Reactors (BWR-GALE Code).* Report NUREG-0016, Revision 1. U.S. Nuclear Regulatory Commission, Washington, DC.

Chandrasekaran, T., J.Y. Lee, and C.A. Willis. 1985. *Calculation of Releases of Radioactive Materials in Gaseous and Liquid Effluents from Pressurized Water Reactors (PWR-GALE Code)*. Report NUREG-0017, Revision 1. U.S. Nuclear Regulatory Commission, Washington, DC.

Cline, J.E., C.A. Pelletier, J.P. Frederickson, P.G. Voillequé, and E.D. Barefoot. 1979. *Airborne Particulate Releases from Light-Water Reactors*. Report EPRI NP-1013. Electric Power Research Institute, Palo Alto, CA.

Cochran, J.A., D.G. Smith, P.J. Magno, and B. Shleien. 1970. *An Investigation of Airborne Radioactive Effluents from an Operating Nuclear Fuel Reprocessing Plant*. Report BRH/NERHL 70–3. National Technical Information Service, Springfield, VA.

Cowan, G.A. 1976. "A Natural Fission Reactor." *Scientific American* 235: 36–47.

Croff, A.G. 1982. "ORIGEN2: A Versatile Computer Code for Calculating the Nuclide Compositions and Characteristics of Nuclear Materials." *Nuclear Technology* 62: 335–352.

Croff, A.G. 2006. "Risk-Informed Radioactive Waste Classification and Reclassification." *Health Physics* 91: 449–460.

Davis, S., K.J. Kopecky, T.E. Hamilton, and L.E. Onstad. 2004. "The Hanford Thyroid Disease Study: Thyroid Neoplasia and Autoimmune Thyroiditis in Persons Exposed to I-131 from the Hanford Nuclear Site." *Journal of the American Medical Association* 292: 2600–2613.

DeCort, M., G. Dubois, Sh.D. Fridman, M.G. Germenchuk, Yu.A. Izrael, A. Janssens, A.R. Jones, G.N. Kelly, E.V. Kvasnikova, I.I. Maveenko, I.M. Nazarov, Yu.M. Pokumeiko, V.A. Sitak, E.D. Stukin, L.Ya. Tabachny, Yu.S. Tsaturov, and S.I. Avdyushin. 1998. *Atlas of Caesium Deposition on Europe after the Chernobyl Accident*. European Communities, Luxembourg.

Devell, L., S. Guntay, and D.A. Powers. 1995. "The Chernobyl Reactor Accident Source Term—Development of a Consensus View." Report NEA/CSNI/R(95)24. OECD Nuclear Energy Agency, Issy-les-Moulineaux, France.

Deworm, J.P., and Tedder, D.W. 1991. "Nuclear Fuel Fabrication Plants." In W.R.A. Goossens, G.G. Eichholz, and D.W. Tedder, eds. *Treatment of Gaseous Effluents at Nuclear Facilities*. Harwood Academic Publishers, New York.

DiNunno, J.J., F.D. Anderson, R.E. Baker, and R.L. Waterfield. 1962. *Calculation of Distance Factors for Power and Test Reactor Sites*. Report TID-14844. U.S. Atomic Energy Commission, Washington, DC.

DOE (U.S. Department of Energy). 1985. *Proceedings of the 18th DOE Nuclear Airborne Waste Management and Air Cleaning Conference, Held in Baltimore, Maryland 12–16 August 1984*. Report CONF-840806 (2 vols.). Harvard Air Cleaning Laboratory, Boston, MA.

DOE. 1994. *Airborne Release Fractions/Rates and Respirable Fractions for Nonreactor Nuclear Facilities*. DOE-HDBK-3010–94. U.S. Department of Energy, Washington, DC.

DOE. 1997. *Linking Legacies, Connecting the Cold War Nuclear Weapons Production Processes to Their Environmental Consequences*. DOE/EM-0319. U.S. Department of Energy, Washington, DC.

Eichholz, G.G. 1976. *Environmental Aspects of Nuclear Power*. Ann Arbor Science Publishers, Ann Arbor, MI.

Eichholz, G.G. 1983. "Source Terms." In J.E. Till and H.R. Meyer, eds. *Radiological Assessment: A Textbook on Environmental Dose Assessment*. NUREG/CR-3332. U.S. Nuclear Regulatory Commission, Washington DC. Pages 1-1–1-54

EPA (U.S. Environmental Protection Agency). 1973a. *Environmental Analysis of the Uranium Fuel Cycle, Part I—Fuel Supply*. EPA-520/9-73-003-B. U.S. Environmental Protection Agency, Washington, DC.

EPA. 1973b. *Environmental Analysis of the Uranium Fuel Cycle, Part II—Nuclear Power Reactors*. EPA-520/9-73-003-C. U.S. Environmental Protection Agency, Washington, DC.

EPA. 1973c. *Environmental Analysis of the Uranium Fuel Cycle, Part III—Nuclear Fuel Reprocessing*. EPA-520/9-73-003-D. U.S. Environmental Protection Agency, Washington, DC.

EPA. 1979. *Radiological Impact Caused by Emissions of Radionuclides into Air in the United States, Preliminary Report*. EPA 520-/7-79-006. U.S. Environmental Protection Agency, Washington, DC.

EPA. 1989. "40 CFR 61 National Emission Standards for Hazardous Air Pollutants; Radionuclides, Final Rule and Notice of Reconsideration." *Federal Register* 54(240): 51654–51715.

Egorov, N.N., V.M. Novikov, F.L. Parker, and V.K. Popov, eds. 2000. *The Radiation Legacy of the Soviet Nuclear Complex*. Earthscan Publications, London.

Erkin, V.G., and O.V. Lebedev. 1993. "Thermoluminescent Dosimeter Measurements of External Doses to the Population of the Bryansk Region after the Chernobyl Accident." In S.E. Merwin and M.I. Balonov, eds. *The Chernobyl Papers*. Vol. 1: *Doses to the Soviet Population and Early Health Effects Studies*. Research Enterprises Publishing Segment, Richland, WA. Pages 289–311.

Evans, R.D. 1955. *The Atomic Nucleus*. McGraw-Hill, New York.

Firestone, R.B., and V.S. Shirley, eds. 1998. *Table of Isotopes*. Updated 8th ed. John Wiley & Sons, New York.

First, M.W. 1991. "Removal of Airborne Particles." In W.R.A. Goossens, G.G. Eichholz, and D.W. Tedder, eds. *Treatment of Gaseous Effluents at Nuclear Facilities*. Harwood Academic Publishers, New York.

Gavrilin, Yu.I., V.T. Khrouch, S.M. Shinkarev, N.A. Krysenko, A.M. Skryabin, A. Bouville, and L.R. Anspaugh. 1999. "Chernobyl Accident: Reconstruction of Thyroid Dose for Inhabitants of the Republic of Belarus." *Health Physics* 76(2): 105–119.

General Electric Company. 1989. *Nuclides and Isotopes, Chart of the Nuclides*, 14th ed. General Electric Company, San Jose, CA.

Gesell, T.F., and P.G. Voillequé, eds. 1990. "Evaluation of Environmental Radiation Exposures from Nuclear Testing in Nevada." *Health Physics* 59(5): 503–746.

Gilbert, R. 1991. "The Removal of Short-Lived Noble Gases." In W.R.A. Goossens, G.G. Eichholz, and D.W. Tedder, eds. *Treatment of Gaseous Effluents at Nuclear Facilities*. Harwood Academic Publishers, New York.

Glasstone, S. 1955. *Principles of Nuclear Reactor Engineering*. D. Van Nostrand Company, Princeton, NJ.

Golikov, V.Yu., M.I. Balonov, and A.V. Ponomarev. 1993. "Estimation of External Gamma Radiation Doses to the Population after the Chernobyl Accident." In S.E. Merwin and M.I. Balonov, eds. *The Chernobyl Papers*. Vol. 1: *Doses to the Soviet Population and Early Health Effects Studies*. Research Enterprises Publishing Segment, Richland, WA. Pages 247–288.

Gudiksen, P.H., T.F. Harvey, and R. Lange. 1989. "Chernobyl Source Term, Atmospheric Dispersion, and Dose Estimation." *Health Physics* 57(5): 697–706.

Guskova, A.K., A.V. Barabanova, A.Y. Baranov, G.P. Gruszdev, Y.K. Pyatkin, N.M. Nadezhina, N.A. Metlyaeva, G.D. Selidovkin, A.A. Moiseev, I.A. Gusev, E.M. Dorofeeva, and I.E. Zykova. 1988. "Acute Radiation Effects in Victims of the Chernobyl

Nuclear Power Plant Accident" (appendix to annex G). In *Sources, Effects and Risks of Ionizing Radiation.* 1988 Report to the General Assembly, with annexes. United Nations, New York.

Heeb, C.M., S.P. Gydesen, J.C. Simpson, and D.J. Bates. 1996. "Reconstruction of Radionuclide Releases from the Hanford Site, 1944–1972." *Health Physics* 71(4): 545–555.

IAEA (International Atomic Energy Agency). 1973. *Control of Iodine in the Nuclear Industry.* Technical Report Series No. 148. International Atomic Energy Agency, Vienna, Austria.

IAEA. 1980. *Radioiodine Removal in Nuclear Facilities, Methods and Techniques for Normal and Emergency Situations.* Technical Reports Series No. 201. International Atomic Energy Agency, Vienna, Austria.

ICRP (International Commission on Radiological Protection). 1983. "Radionuclide Transformations, Energy and Intensity of Emissions." ICRP Publication 38. *Annals of the ICRP* Vols. 11–13.

ICRU (International Commission on Radiation Units and Measurements). 2002. "Retrospective Assessment of Exposures to Ionizing Radiation." ICRU Report 68. *Journal of the ICRU* 2(2).

International Nuclear Safety Advisory Group. 1986. *Summary Report on the Post-accident Review Meeting on the Chernobyl Accident.* Safety Series No. 75-INSAG-1. International Atomic Energy Agency, Vienna, Austria.

Jubin, R.T., and R.M. Counce. 1991. "Retention of Iodine and Nitrogen Oxides in Reprocessing Plants." In W.R.A. Goossens, G.G. Eichholz, and D.W. Tedder, eds. *Treatment of Gaseous Effluents at Nuclear Facilities.* Harwood Academic Publishers New York.

Kahn, B., R.L. Blanchard, H.L. Krieger, H.E. Kolde, D.B. Smith, A. Martin, S. Gold, W.J. Averett, W.L. Brinck, and G.J. Karches. 1970. *Radiological Surveillance Studies at a Boiling Water Nuclear Power Reactor.* Report BRH/DER 70–1. U.S. Public Health Service, Cincinnati, OH.

Kahn, B., R.L. Blanchard, H.E. Kolde, H.L. Krieger, S. Gold, W.L. Brinck, W.J. Averett, D.B. Smith, and A. Martin. 1971. *Radiological Surveillance Studies at a Pressurized Water Nuclear Power Reactor.* Report EPA 71–1. U.S. Environmental Protection Agency, Cincinnati, OH.

Kahn, B., R.L. Blanchard, W.L. Brinck, H.L. Krieger, H.E. Kolde, W.J. Averett, S. Gold, A. Martin, and G.L. Gels. 1974. *Radiological Surveillance Study at the Haddam Neck PWR Nuclear Power Station.* Report EPA-520/3-74-007. U.S. Environmental Protection Agency, Cincinnati, OH.

Kemeny, J.G., B. Babbitt, P.E. Haggerty, C. Lewis, P.A. Marks, C.B. Marrett, L. McBride, H.C. McPherson, R.W. Peterson, T.H. Pigford, T.B. Taylor, and A.D. Trunk. 1979. *Report of the President's Commission on the Accident at Three Mile Island.* President's Commission on the Accident at Three Mile Island, Washington, DC.

Killough, G.G., S.K. Rope, B. Shleien, and P.G. Voillequé. 1999. "Nonlinear Estimation of Weathering Rate Parameters for Uranium in Surface Soil near a Nuclear Facility." *Journal of Environmental Radioactivity* 45: 95–118.

Kirchner, G., and C.C. Noack. 1988. "Core History and Nuclide Inventory of the Chernobyl Core at the Time of the Accident." *Nuclear Safety* 29: 1–5.

Kopecky, K.J., S. Davis, T.E. Hamilton, M.S. Saporito, and L.E. Onstad. 2004. "Estimation of Thyroid Radiation Doses for the Hanford Thyroid Disease Study: Results and Implications for Statistical Power of the Epidemiological Analyses." *Health Physics* 87(1): 15–32.

Kovach, J.L. 1991. "Off-Gas Treatment for Fuel Reprocessing Plants." In W.R.A. Goossens, G.G. Eichholz, and D.W. Tedder, eds. *Treatment of Gaseous Effluents at Nuclear Facilities*. Harwood Academic Publishers, New York.

Krey, P.W., M. Heit, and K.M. Miller. 1990. "Radioactive Fallout Reconstruction from Contemporary Measurements of Reservoir Sediments." *Health Physics* 59(5): 541–554.

Likhtarev, I.A., G.M. Gulko, I.A. Kairo, I.P. Los, K. Henrichs, and H.G. Paretzke. 1994. "Thyroid Doses Resulting from the Ukraine Chernobyl Accident—Part I: Dose Estimates for the Population of Kiev." *Health Physics* 66(2): 137–146.

Likhtarev, I.A., L.N. Kovgan, S.E. Vavilov, R.R. Gluvchinsky, O.N. Perevoznikov, L.N. Litvinets, L.R. Anspaugh, J.R. Kercher, and A. Bouville. 1996. "Internal Exposure from the Ingestion of Foods Contaminated by $^{137}$Cs after the Chernobyl Accident. Report 1. General Model: Ingestion Doses and Countermeasure Effectiveness for the Adults of Rovno Oblast of Ukraine." *Health Physics* 70(3): 297–317.

Likhtarev, I.A., L.N. Kovgan, S.E. Vavilov, O.N. Perevoznikov, L.N. Litvinets, L.R. Anspaugh, P. Jacob, and G. Pröhl. 2000. "Internal Exposure from the Ingestion of Foods Contaminated by $^{137}$Cs after the Chernobyl Accident—Report 2. Ingestion Doses of the Rural Population of Ukraine up to 12 y after the Accident (1986–1997)." *Health Physics* 79(4): 341–357.

Likhtarev, I.A., L.N. Kovgan, P. Jacob, and L.R. Anspaugh. 2002. "Chernobyl Accident: Retrospective and Prospective Estimates of External Dose of the Population of Ukraine." *Health Physics* 82(3): 290–303.

Littlefield, P.S., S.R. Miller, and H. DerHagopian. 1975. "Vermont Yankee Advanced Off-gas System (AOG)." In First, M.W., ed. *Proceedings of the AEC Air Cleaning Conference Held at San Francisco, California, on 12–15 August 1974*. Report CONF740807V1. National Technical Information Service, Springfield, VA. Pages 99–124.

Maletskos, C.J., ed. 1991. "The Goiania Radiation Accident" [special issue]. *Health Physics Journal* 60(1).

Mandler, J.W., A.C. Stalker, S.T. Croney, C.V. McIsaac, G.A. Soli, J.K. Hartwell, L.S. Loret, B.G. Motes, T.E. Cox, D.W. Akers, N.K. Bihl, S.W. Duce, J.W. Tkachyk, C.A. Pelletier, and P.G. Voillequé. 1981. *In-Plant Source Term Measurements at Four PWRs*. Report NUREG/CR-1992. Prepared by EG and G Idaho, Inc., Idaho Falls, ID, for Office of Nuclear Regulatory Research, U.S. Nuclear Regulatory Commission, Washington, DC.

Marrero, T.R. 1976. *Airborne Releases from BWRs for Environmental Impact Evaluations*. Licensing Topical Report NEDO-21159. General Electric Company, San Jose, CA.

Merritt, R.C. 1971. *The Extractive Metallurgy of Uranium*. Colorado School of Mines Research Institute, Golden, CO.

Meyer, K.R., P.G. Voillequé, D.W. Schmidt, S.K. Rope, G.G. Killough, B. Shleien, R.E. Moore, M.J. Case, and J.E. Till. 1996. "Overview of the Fernald Dosimetry Reconstruction Project and Source Term Estimates for 1951–1988." *Health Physics* 71(4): 425–437.

Meyer, K.R., H.J. Mohler, J.W. Aanenson, and J.E. Till. 2000. *Identification and Prioritization of Radionuclide Releases from the Idaho National Engineering and Environmental Laboratory*. RAC Report No. 3-CDC-Task Order 5–2000-FINAL. Risk Assessment Corporation, Neeses, SC.

Mongan, T.R., S.R. Ripple, G.P. Brorby, and D.G. diTommaso. 1996a. "Plutonium Releases from the 1957 Fire at Rocky Flats." *Health Physics* 71(4): 510–521.

Mongan, T.R., S.R Ripple, and K.D. Winges. 1996b. "Plutonium Releases from the 903 Pad at Rocky Flats." *Health Physics* 71(4): 522–531.

Murray, R.L. 1961. *Introduction to Nuclear Engineering*, 2nd ed. Prentice-Hall, Englewood Cliffs, NJ.

NAS/NRC (National Academy of Sciences/National Research Council). 1995. *Radiation Dose Reconstruction for Epidemiologic Uses*. National Academy Press, Washington, DC.

NAS/NRC. 2005. *Risk and Decisions about Disposition of Transuranic and High-Level Radioactive Waste*. National Academies Press, Washington, DC.

NAS/NRC. 2006. *Improving the Regulation and Management of Low-Activity Radioactive Wastes*. National Academies Press, Washington, DC.

National Cancer Institute. 1997. *Estimated Exposures and Thyroid Doses Received by the American People from Iodine-131 in Fallout Following Nevada Atmospheric Nuclear Bomb Tests*. National Cancer Institute, National Institutes of Health, U.S. Department of Health and Human Services, Bethesda, MD.

NCRP (National Council on Radiation Protection and Measurements). 1987. *Exposure of the Population in the United States and Canada from Natural Background Radiation*. NCRP Report No. 94. National Council on Radiation Protection and Measurements, Bethesda, MD.

NCRP. 2001. *Management of Terrorist Events Involving Radioactive Material*. NCRP Report No. 138. National Council on Radiation Protection and Measurements, Bethesda, MD.

NCRP. 2002. *Risk-Based Classification of Radioactive and Hazardous Chemical Wastes*. NCRP Report No. 139. National Council on Radiation Protection and Measurements, Bethesda, MD.

NCRP. 2003. *Radiation Protection for Particle Accelerator Facilities*. NCRP Report No. 144. National Council on Radiation Protection and Measurements, Bethesda, MD.

NCRP. 2005. *Performance Assessment of Near-Surface Facilities for Disposal of Low-Level Radioactive Waste*. NCRP Report No. 152. National Council on Radiation Protection and Measurements, Bethesda, MD.

NRC (U.S. Nuclear Regulatory Commission). 1975a. *Reactor Safety Study, an Assessment of Accident Risks in U.S. Commercial Nuclear Power Plants*. Report WASH-1400 (NUREG-75/014). U.S. Nuclear Regulatory Commission, Washington, DC.

NRC. 1975b. *General Design Guide for Ventilation Systems for Fuel Reprocessing Plants*. Regulatory Guide 3.32 (for Comment). U.S. Nuclear Regulatory Commission, Washington, DC.

NRC. 1976a. *Calculation of Releases of Radioactive Materials in Gaseous and Liquid Effluents from Boiling Water Reactors (BWR-GALE Code)*. Report NUREG-0016. U.S. Nuclear Regulatory Commission, Washington, DC.

NRC. 1976b. *Calculation of Releases of Radioactive Materials in Gaseous and Liquid Effluents from Pressurized Water Reactors (PWR-GALE Code)*. Report NUREG-0017. U.S. Nuclear Regulatory Commission, Washington, DC.

NRC. 1979. *Investigation into the March 28, 1979 Three Mile Island Accident by Office of Inspection and Enforcement*. Investigative Report No. 50–320/79–10. Report NUREG-0600. U.S. Nuclear Regulatory Commission, Washington, DC.

NRC. 1980a. *Final Generic Environmental Impact Statement on Uranium Milling*. Report NUREG-0706. U.S. Nuclear Regulatory Commission, Washington, DC.

NRC. 1980b. *Clarification of TMI Action Plan Requirements*. Report NUREG-0737. U.S. Nuclear Regulatory Commission, Washington, DC.

NRC. 1981. *Technical Bases for Estimating Fission Product Behavior During LWR Accidents*. Report NUREG-0772. U.S. Nuclear Regulatory Commission, Washington, DC.

NRC. 1986a. *Assessment of the Public Health Impact from the Accidental Release of $UF_6$ at the Sequoyah Fuels Corporation Facility at Gore, Oklahoma.* NUREG-1189 (2 vols.). National Technical Information Service, Springfield, VA.

NRC. 1986b. *Release of $UF_6$ from a Ruptured Model 48Y Cylinder at Sequoyah Fuels Corporation Facility: Lessons-Learned Report.* NUREG-1198 U.S. Nuclear Regulatory Commission, Washington, DC.

NRC. 1987. *Methods for Estimating Radioactive and Toxic Airborne Source Terms for Uranium Milling Operations.* Regulatory Guide 3.59. U.S. Nuclear Regulatory Commission, Washington, DC.

NRC. 1988. *Nuclear Fuel Cycle Facility Accident Analysis Handbook.* NUREG-1320. U.S. Nuclear Regulatory Commission, Washington, DC.

NRC. 1990. *Severe Accident Risks: An Assessment for Five U.S. Nuclear Power Plants.* Report NUREG-1150. U.S. Nuclear Regulatory Commission, Washington, DC.

NRC. 1998. *An Approach for Using Probabilistic Risk Assessment in Risk-Informed Decisions on Plant-Specific Changes to the Licensing Basis.* Regulatory Guide 1.174. U.S. Nuclear Regulatory Commission, Washington, DC.

NRC. 2000. *Alternative Radiological Source Terms for Evaluating Design Basis Accidents at Nuclear Power Reactors.* Regulatory Guide 1.183. U.S. Nuclear Regulatory Commission, Washington, DC.

NRC. 2003a. *Methods and Assumptions for Evaluating Radiological Consequences of Design Basis Accidents at Light-Water Nuclear Power Reactors.* Regulatory Guide 1.195. U.S. Nuclear Regulatory Commission, Washington, DC.

NRC. 2003b. *Control Room Habitability at Light-Water Nuclear Power Reactors.* Regulatory Guide 1.196. U.S. Nuclear Regulatory Commission, Washington, DC.

NRC. 2005. *Safety Evaluation Report for the National Enrichment Facility in Lea County, New Mexico, Louisiana Energy Services.* NUREG-1827. U.S. Nuclear Regulatory Commission, Washington, DC.

NRC. 2006a. *Safety Evaluation Report for the American Centrifuge Plant in Piketon, Ohio.* NUREG-1851. U.S. Nuclear Regulatory Commission, Washington, DC.

NRC. 2006b. *Environmental Impact Statement for the Proposed American Centrifuge Plant in Piketon, Ohio.* NUREG-1834. U.S. Nuclear Regulatory Commission, Washington, DC.

NRC. 2006c. *Safety Evaluation of Early Site Permit Application in the Matter of System Energy Resources, Inc., a Subsidiary of Entergy Corporation, for the Grand Gulf Early Site Permit Site.* NUREG-1840. U.S. Nuclear Regulatory Commission, Washington, DC.

NRC. 2006d. *Safety Evaluation for an Early Site Permit (ESP) at the Exelon Generation Company, LLC (EGC) ESP Site.* NUREG-1844. U.S. Nuclear Regulatory Commission, Washington, DC.

NRC. 2007. *Calculation of Releases of Radioactive Materials in Gaseous and Liquid Effluents from Light-Water-Cooled Nuclear Power Plants.* Regulatory Guide 1.112. U.S. Nuclear Regulatory Commission, Washington, DC.

OECD (Organisation for Economic Co-operation and Development). 1980. *Radiological Significance and Management of Tritium, Carbon-14, Krypton-85, and Iodine-129 Arising from the Nuclear Fuel Cycle.* Organisation for Economic Co-operation and Development, Paris, France.

OECD. 2007. *NEA 2006 Annual Report.* Organisation for Economic Co-operation and Development, Paris, France.

Patterson, H.W., and R.H. Thomas. 1973. *Accelerator Health Physics.* Academic Press, New York.

Pelletier, C.A., E.D. Barefoot, J.E. Cline, R.T. Hemphill, W.A. Emel, and P.G. Voillequé. 1978a. *Sources of Radioiodine at Boiling Water Reactors.* Report EPRI NP-495. Electric Power Research Institute, Palo Alto, CA.

Pelletier, C.A., J.E. Cline, E.D. Barefoot, R.T. Hemphill, P.G. Voillequé, and W.A. Emel. 1978b. *Sources of Radioiodine at Pressurized Water Reactors.* Report EPRI NP-939. Electric Power Research Institute, Palo Alto, CA.

Perkins, R.W., D.E. Robertson, C.W. Thomas, and J.A. Young. 1990. "Comparison of Nuclear Accident and Nuclear Test Debris." In *Environmental Contamination Following a Major Nuclear Accident.* International Atomic Energy Agency, Vienna, Austria.

Ripple, S.R., T.E. Widner, and T.R. Mongan. 1996. "Past Radionuclide Releases from Routine Operations at Rocky Flats." *Health Physics* 71(4): 502–509.

Rogovin, M., G.T. Frampton, Jr., E.K. Cornell, R.C. DeYoung, R. Budnitz, and P. Norry. 1980. *Three Mile Island: A Report to the Commissioners and to the Public.* Report NUREG/CR-1250. U.S. Nuclear Regulatory Commission Special Inquiry Group, Washington, DC.

Rood, A.S., H.A. Grogan, and J.E. Till. 2002. "A Model for a Comprehensive Assessment of Exposure and Lifetime Cancer Incidence Risk from Plutonium Released from the Rocky Flats Plant, 1953–1989." *Health Physics* 82(2): 182–212.

SAIC (Science Applications International Corporation). 1998. *Nuclear Fuel Cycle Facility Accident Analysis Handbook.* Report NUREG/CR-6410. National Technical Information Service, Springfield, VA.

Sandia. 2007. *Advanced Fuel Cycle Initiative.* Available: afci.sandia.gov, accessed June 2007.

Shipler, D.B., B.A. Napier, W.T. Farris, and M.D. Freshley. 1996. "Hanford Environmental Dose Reconstruction Project—an Overview." *Health Physics* 71: 532–544.

Shleien, B., ed. 1992. *The Health Physics and Radiological Health Handbook*, rev. ed. Scinta, Inc., Silver Spring, MD.

Silberberg, M., J.A. Mitchell, R.O. Meyer, W.F. Pasedag, C.P. Ryder, C.A. Peabody, and M.W. Jankowski. 1985. *Reassessment of the Technical Bases for Estimating Source Terms.* Report NUREG-0956. U.S. Nuclear Regulatory Commission, Washington, DC.

Simon, S.L., and R.J. Vetter, eds. 1997. "Consequences of Nuclear Testing in the Marshall Islands." *Health Physics Journal* 73(1): 1–269.

Smyth, H.D. 1945. *A General Account of the Development of Methods of Using Atomic Energy for Military Purposes under the Auspices of the United States Government 1940–1945.* U.S. Government Printing Office, Washington, DC.

Stepanenko, V.F., P.G. Voillequé, Yu.I. Gavrilin, V.T. Khrouch, S.M. Shinkarev, M.Yu. Orlov, A.E. Kondrashov, D.V. Petin, E.K. Iaskova, and A.F. Tsyb. 2004. "Estimating Individual Thyroid Doses for a Case-Control Study of Childhood Thyroid Cancer in Bryansk Oblast, Russia." *Radiation Protection Dosimetry.* 108: 143–160.

Straume, T., L.R. Anspaugh, A.A. Marchetti, G. Voigt, V. Minenko, F. Gu, P. Men, S. Trofimik, S. Tretyakevich, V. Drozdovitch, E. Shagalova, O. Zhukova, M. Germenchuk, and S. Berlovich. 2006. "Measurement of $^{129}$I and $^{137}$Cs in Soils from Belarus and Reconstruction of $^{131}$I Deposition from the Chernobyl Accident." *Health Physics* 91: 7–19.

Sullivan, A.H. 1992. *A Guide to Radiation and Radioactivity Levels near High Energy Particle Accelerators.* Nuclear Technology Publishing, Ashford, Kent, UK.

Talerko, N. 2005. "Mesoscale Modelling of Radioactive Contamination Formation in Ukraine Caused by the Chernobyl Accident." *Journal of Environmental Radioactivity* 78: 311–329.

Till, J.E., J.W. Aanenson, P.J. Boelter, M.J. Case, M. Dreicer, H.A. Grogan, M.O. Langan, P.D. McGavran, K.R. Meyer, H.R. Meyer, H.J. Mohler, A.S. Rood, R.C. Rope, S.K.

Rope, L.A. Stetar, P.G. Voilleque, T.F. Winsor, and W. Yang. 2001. *SRS Environmental Dose Reconstruction Project Phase II*. RAC Report No. 1-CDC-SRS-1999-FINAL, Risk Assessment Corporation, Neeses, SC.

Toth, L.M., A.P. Malinauskas, G.R. Eidam, and H.M. Burton, eds. 1986. *The Three Mile Island Accident, Diagnosis, and Prognosis*. ACS Symposium Series 293. American Chemical Society, Washington, DC.

UNSCEAR (U.N. Scientific Committee on the Effects of Atomic Radiation). 1993. *Sources and Effects of Ionizing Radiation. Report to the General Assembly, with Scientific Annexes*. United Nations, New York.

UNSCEAR. 2000. *Sources and Effects of Ionizing Radiation, Report to the General Assembly, with Scientific Annexes*. Vol. 1: *Sources*. United Nations, New York.

Voilleque, P.G. 1979. *Iodine Species in Reactor Effluents and in the Environment*. Report EPRI NP-1269. Electric Power Research Institute, Palo Alto, CA.

Voilleque, P.G., G.G. Killough, S.K. Rope, and J.E. Till. 2002. *Methods for Estimating Radiation Doses from Short-Lived Gaseous Radionuclides and Radioactive Particles Released to the Atmosphere during Early Hanford Operations*. RAC Report No. 2-CDC-Task Order 3-2002-FINAL. Risk Assessment Corporation, Neeses, SC.

Wedekind, L. 2004. "Too Cheap to Meter What?" *IAEA Bulletin* 46: 1.

Whicker, J.J., F.W. Whicker, and S. Jacobi. 1994. "$^{137}$Cs in Sediments of Utah Lakes and Reservoirs: Effects of Elevation, Sedimentation Rate, and Fallout History." *Journal of Environmental Radioactivity* 23: 265–283.

Wilhelm, J.G., and H. Deuber. 1991. "The Removal of Short-Lived Iodine Isotopes." In W.R.A. Goossens, G.G. Eichholz, and D.W. Tedder, eds. *Treatment of Gaseous Effluents at Nuclear Facilities*. Harwood Academic Publishers, New York.

World Health Organization. 1983. *Environmental Health Criteria 25, Selected Radionuclides: Tritium, Carbon-14, Krypton-85, Strontium-90, Iodine, Caesium-137, Radon, Plutonium*. World Health Organization, Geneva, Switzerland.

Zvonova, I.A., and M.I. Balonov. 1993. "Radioiodine Dosimetry and Prediction of Consequences of Thyroid Exposure of the Russian Population Following the Chernobyl Accident." In S.E. Merwin and M.I. Balonov, eds. *The Chernobyl Papers*. Vol. 1: *Doses to the Soviet Population and Early Health Effects Studies*. Research Enterprises Publishing Segment, Richland, WA. Pages 71–125.

# 3

# Atmospheric Transport of Radionuclides

Todd V. Crawford
Charles W. Miller
Allen H. Weber

Nuclear facilities can release radionuclides to the atmosphere under routine operating conditions and through unintentional events. As a result, people who live and work around nuclear facilities could be exposed to radiation from a number of pathways:

- External exposures because of direct radiation from radionuclides in the plume or deposited on the ground
- Internal exposures because of inhalation of radionuclides in the air or ingestion of foods or other materials (e.g., soil) that have been contaminated by radioactive materials

It is expected that releases to the atmosphere will contribute more to the (nonnatural) dose received by people than will releases to other environmental media (Little 1984).

The magnitude of the exposure to atmospheric releases of radionuclides depends on atmospheric transport, diffusion, and deposition processes. Thus, it is necessary to determine *where* the pollutants are going, *when* they will arrive at a particular location, and *how much* of the pollutant concentration is in the air and on the ground at that location. Ideally, dose assessments would be based on measurements of air concentration, ground concentration, or radiation fields resulting from these releases. However, this is not always possible. Instead, mathematical models must be used to estimate the impact of radionuclide releases to the atmosphere.

A number of models have been developed for this purpose. All models and procedures, including those published by regulatory authorities (e.g., NRC 1977; EPA 1989a), must be used carefully or they can produce incorrect and misleading results.

Great differences exist in model assumptions and the data that are used to develop and test models; therefore, models should be selected that best fit the situation being assessed.

This chapter assists professionals in selecting and properly using atmospheric transport, diffusion, and deposition models. After presenting meteorological fundamentals applicable to atmospheric processes, we discuss different types of models and modeling approaches. Although the models presented in regulatory guidance are included, this chapter is *not* designed to teach how to implement regulatory guidance. Instead, we present fundamental principles and guidelines to help professionals select the most appropriate model for a given assessment problem and sensitize model users to the need for using appropriate and accurate site data in model calculations.

## The Atmosphere

The earth is surrounded by an invisible mixture of gases and water vapor, held in place by the earth's gravity. The nature of this mixture greatly influences the fate of various pollutants, including radionuclides, released into it.

### Composition

Ninety-nine percent of the atmosphere consists of two gases: oxygen (21%) and nitrogen (78%). The rest includes a number of trace gases, each at much less than 1%. These trace gases include two very important constituents: water vapor and carbon dioxide. Although the oxygen and nitrogen percentages are relatively uniform over the globe, the other percentages can vary widely with location. The condensation of water vapor forms clouds and precipitation, and clouds have a very important role in the energy balance of the atmosphere. Carbon dioxide is the "greenhouse" gas of concern. Solar radiation from the sun reaches the earth after passing through the atmospheric carbon dioxide, but infrared radiation from the earth's surface is absorbed by the atmospheric carbon dioxide and radiated back to the earth. The higher the carbon dioxide concentration, the greater is this radiation. Because of its role in photosynthesis and subsequent growth of plants, however, carbon dioxide is also necessary for life on earth.

### Vertical Extent Important for Atmospheric Releases

Most of the solar radiation reaching the earth's surface is absorbed and then radiated back into the atmosphere. One result of the complex interactions between the incoming solar energy, the surface of the earth, and the atmosphere is that, on average, the atmosphere develops a distinct vertical temperature structure. A thermometer rising above the surface at mid-latitudes would indicate that the temperature of the atmosphere decreases up to a height of about 12 km (Battan 1974). At this height, called the tropopause, the temperature begins to increase with height. The tropopause separates the lowest layer of the atmosphere, the troposphere, from the next higher layer, the stratosphere. Exchanges do occur between the troposphere and the stratosphere, but they occur on time scales of a few years. Nearly all weather phenomena and consequences of surface-based activities, such as releases from nuclear facilities, occur in the troposphere. However, the consequences of some very energetic radionuclide releases, such as nuclear weapon detonations, can extend into the stratosphere.

The bottom of the troposphere includes an interface extending upward from the surface of the earth to the "free atmosphere." This interface is called the atmospheric boundary layer (ABL). The height of the ABL varies spatially and temporally, but it generally ranges between a few meters and 2,000 m (Brenk et al. 1983) depending on the time of day and many other factors. The ABL is that "layer of air directly above the earth's surface in which the effects of the surface (friction, heating, and cooling) are felt directly on time scales less than a day, and in which significant fluxes of momentum, heat or matter are carried by turbulent motions on a scale of the order of the depth of the boundary layer or less" (Garratt 1994). The exact character of the ABL greatly influences the dispersion of radionuclides and the subsequent exposure to people.

The diurnal cycle of the ABL, caused by the rise and set of the sun, is clearly very important on the structure of the ABL as a function of time. Other cycles, such as seasonal and annual cycles, are important when examining climatological data as input to long-term release calculations. Variations also occur from year to year at a particular location; therefore, several years of data must be used to obtain a meaningful average.

## Scales of Motion

The most important meteorological parameter for atmospheric transport and diffusion is the wind. The mean wind (including direction and speed) provides the basis for answering the *where* and *when* questions. The intensity of the atmospheric turbulence, which includes the fluctuations in wind speed and direction, provides the basis for answering the *how much* question (turbulence consists of irregular and apparently random fluctuations that can only be characterized by statistical properties). Although motion is three dimensional in the atmosphere, the major motions are horizontal; therefore, discussions of winds in the meteorological literature usually refer to the horizontal wind (Battan 1974).

Winds are the result of different processes in the atmosphere. At the large (macro) scale (1,000–2,000 km), it is the synoptic weather patterns with pressure gradients, low- and high-pressure regions, weather fronts, and the rotation of the earth that determine wind speed and directions. As major synoptic weather systems move around the world, these winds change at a particular location with a periodicity of three to seven days. These patterns are more evident in winter months with well-defined storms than in summer months with weakly defined storms.

At a distance scale of about 100 km (the mesoscale), the winds are influenced by a number of local phenomena caused by differential heating of the earth's surface:

- Sea breezes (onshore in the afternoon and offshore at night) caused by land and sea temperature differences

- Gravity winds (downslope at night and upslope in the daytime) caused by radiation heating and cooling in regions of significant topography change
- Local convective activity caused by daytime heating of the earth's surface within a thermodynamically unstable atmosphere

Typical periodicity for such phenomena is one day. These flows are superimposed upon and modified by the larger-scale synoptic winds, which are all changing with time and space.

At the distance scale of about 1 km or less (the microscale), winds are influenced by very local topographical changes or, perhaps more important, changes in the surface roughness. For example, building structures, the transition from forest to grass cover, and the placement of a wind instrument with respect to obstructions all become important. You have to be concerned with the representativeness of the data when performing assessment calculations. Clearly, you need to know the locations of the wind instruments with respect to the contaminant release locations and surrounding features.

Winds vary in direction and speed with increasing height above the ground under all conditions. Typically, the wind speed increases with height above the ground, and wind direction "veers" in the ABL (e.g., shifts clockwise, coming from the southeast toward south or coming from the southwest toward west). The rate of change of wind with height is a function of the thermal structure of the atmosphere, and this function is modified by all of the processes that affect the horizontal winds.

Average winds and turbulence values used for transport and diffusion calculations use different scales depending on the size of the pollutant cloud in the atmosphere and the resolution of the input data. For instance, an entire pollutant cloud is moved by the average winds, but the cloud of pollutants itself is most efficiently diffused (spread out and concentrations reduced) by turbulent eddies in the atmosphere of about the same size as the cloud. Much smaller eddies just "fuzz" the cloud edges, and much larger eddies just move the cloud. Figure 3.1 (from Pendergast 1984) is a pictorial example of these phenomena.

On the scale of a plume from a smokestack, a snapshot of the plume reveals the small eddies causing the diffusion; however, a pollutant cloud the size of a continent would be most efficiently diffused by the large-scale eddies that we call synoptic storms. Are the systems that we see in satellite views of the earth just turbulent eddies, or are they causing the average wind at a location? The answer depends on the time and space scales of interest to the viewer.

Another scale problem is associated with input data for models used in transport and diffusion. You cannot deterministically calculate transport on scales smaller than that for which input wind data are available. Thus, turbulence parameterization is necessary for scales smaller than the data network spacing. Wind data can be used directly from stations or interpolated from observing locations to grid points. Often, it is necessary to use wind speed and direction forecasts from large-scale numerical models in estimating a cloud's transport.

Sampling time is also important. Depending on the sophistication of the observing station, wind data are typically reported as a spot reading on the hour from synoptic weather stations. This spot reading on the hour is then assumed to represent the

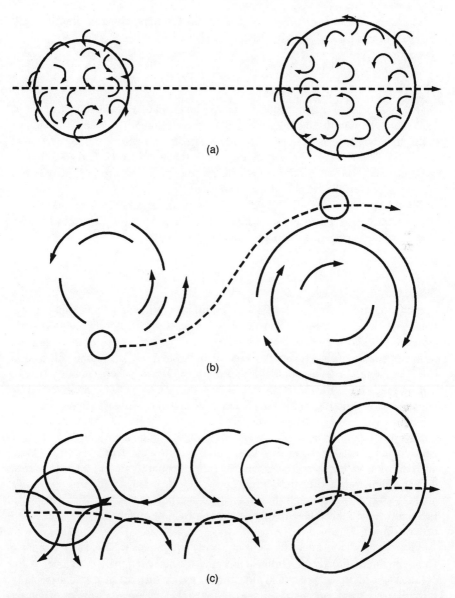

Figure 3.1 Idealized dispersion patterns. (a) A large cloud in a uniform field of small eddies. (b) A small cloud in a uniform field of large eddies. (c) A cloud in a field of eddies of the same size as the cloud. From Pendergast (1984).

whole hour. Atmospheric diffusion studies, which have provided empirical input for model parameterization, have had many different sampling times. Keep this in mind when selecting a model (and its embedded parameterization) for application to a particular problem.

The scales of motion outlined above are critical in determining the transport and diffusion of material released into the atmosphere.

## Macroscale

On the large scale, winds transport pollutant clouds that are small compared to the 1,000- to 2,000-km distance. Eddies of this size are important in the diffusion of pollutant clouds several days after release or after traveling part of the way around the world. The atmospheric concentration measurements in the United States of radioactive debris from the Chernobyl reactor event in the Ukraine in 1986 reflected the consequences of diffusion on this large scale.

Most frequently, you will be dealing with transport and diffusion on smaller scales, and it is useful to review some examples. Keep these examples in mind when applying a particular model to your local situation.

## Mesoscale

Complex terrain, such as hills and valleys, influences the path and diffusion of a pollutant plume. Furthermore, depending on the meteorological conditions, the same terrain obstacle can affect a plume in different ways. For example, if an elevated source of contaminant is located upwind of an isolated hill, the plume will tend to rise over the hill or go around the hill, depending on atmospheric stability (Drake and Barrager 1979; Snyder et al. 1985). Near-surface local winds are affected by the orientation of the terrain to the general flow at higher altitudes (figure 3.2).

Differential heating during the day can create local wind patterns that influence airflow in a mountain and valley system. At night, the generally heavier cool air tends to flow down the mountainside and into and down the valley. As the sun rises and heats the valley floor, the air in the valley becomes warmer and less dense than the higher air and begins to rise up the slope. At sunset, the upslope winds begin to reverse, and the nighttime drainage winds begin to reestablish themselves. Pollutants released into these regimes can be profoundly affected by such local wind patterns (figure 3.3).

Very high pollutant concentrations can occur in complex terrain, such as when a plume impinges directly on a hillside. Fortunately, however, there is a tendency for pollutants to disperse more quickly in complex terrain than in flat terrain, especially for pollutants from long-term releases (Egan 1975). The overall effect of terrain features on dispersion (transport and diffusion) for a given site can be complex and difficult to predict, especially for short-term pollutant releases. As a result, it is often necessary to use site-specific tracer studies or physical modeling techniques to study the airflow at a given location (e.g., see Widner et al. 1991).

Many nuclear and other industrial facilities are located along the coast of large bodies of water, such as the Atlantic or Pacific Ocean, the Gulf of Mexico, or one of the Great Lakes. Because bodies of water change temperature at a much slower rate than do land surfaces, large differences in land and water temperatures can develop and create local airflows along coasts that can significantly affect pollutant plumes.

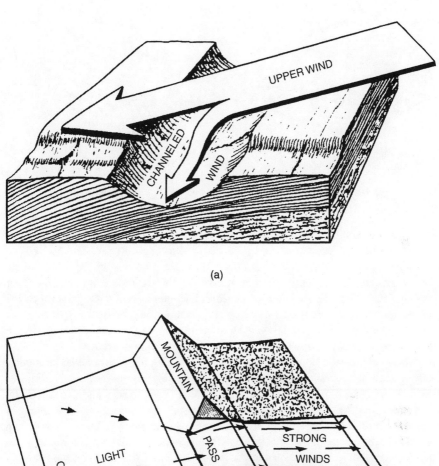

Figure 3.2 Distortion of the wind flow by topographic obstacles. (a) Channeling of the wind by a valley. (b) The effect of a mountain pass on the wind flow. From Pendergast (1984).

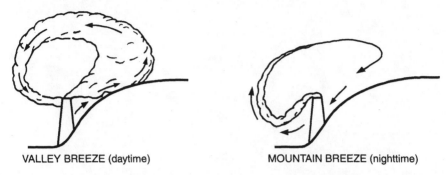

VALLEY BREEZE (daytime)        MOUNTAIN BREEZE (nighttime)

Figure 3.3   Effect of mountain valley breezes. From Pendergast (1984).

If the temperature of the land is warmer than the temperature of the water, air over the land will become heated (less dense) and begin to rise. This air will be replaced by cooler, heavier air from over the water, which is referred to as a lake or sea breeze. For example, this breeze may occur when the sun rises on a clear summer day. If the temperature of the land is cooler than the temperature of the water, the opposite circulation is set up (i.e., warmer, less dense air rises over the water and is replaced by cooler, heavier air originating over the land surface). Such a land breeze may occur when the sun sets on a clear summer day.

Abrupt changes in wind direction can occur within a few kilometers of a coast during sea breeze conditions (Lyons 1975). The leading edge of the cool air is known as the sea breeze front. At this front, the air can begin to rise, and the horizontal direction can change as the onshore flow interacts with the general movement of the air over the land. At night, or at other times when the flow of air is reversed and a land breeze occurs, pollutants move offshore and often into a much more stable atmospheric environment over the cooler water (Lyons 1975). Such plumes may travel relatively long distances, such as across Lake Michigan, without experiencing much diffusion. Furthermore, if the sea breeze returns when the sun rises, some of the material blown out over the water during the night may return to land with the new onshore flow (figure 3.4). Changes in atmospheric stability are also associated with the coastal atmospheric environment. When a sea breeze occurs, a thermal internal boundary layer (TIBL) is created between the cool, stable air moving onshore and the warm, unstable air rising from the land surface (Lyons 1975). Pollutants emitted into the TIBL will be trapped within that layer, and higher concentrations than otherwise expected may develop. Pollutants emitted above the TIBL will remain aloft until they intersect with the TIBL. Then rapid downward diffusion will occur, which results in increased pollutant concentrations at the surface (Barr and Clements 1984). This movement is shown in figure 3.5.

It is clear from this limited discussion that airflow in coastal environments can be difficult to measure or predict. Furthermore, large variations in airflow patterns exist between coastal sites. As with nuclear facilities located in other complex terrains, site-specific tracer studies or physical modeling may be required to accurately predict dispersal patterns of potential releases from nuclear sites in coastal environments (Ogram and Wright 1991).

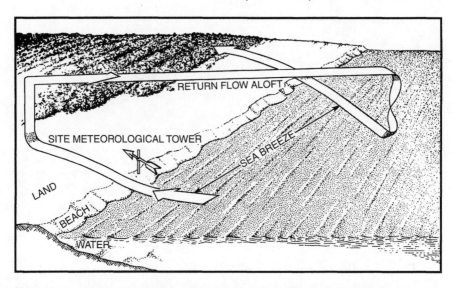

Figure 3.4 Example of wind flow onshore, return flow aloft, and the returning of the air with the sea breeze at a different location. From McKenna et al. (1987).

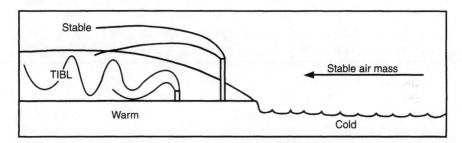

Figure 3.5 Example of pollutant behavior for emissions within and above the thermal internal boundary layer (TIBL). From Barr and Clements (1984).

## Microscale

The presence of buildings and other structures will disturb the flow of air over the earth's surface. Figure 3.6 shows the idealized flow around a simple building. The three main zones of flow around a building are as follows (Pasquill and Smith 1983):

- The upwind displacement zone, where the approaching air is deflected around the building
- The relatively isolated cavity zone immediately on the leeward side of the building
- The highly disturbed wake zone farther downwind from the building

The wake zone may extend downwind for some distance, with the distance depending upon the source configuration and meteorological conditions (Hosker 1984).

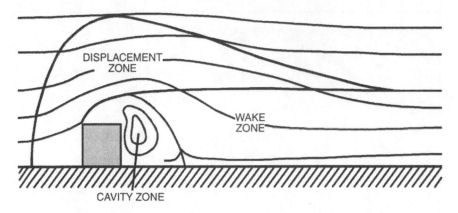

Figure 3.6 Conceptual characteristic flow zones around large structures (Thuilliere and Mancuso 1980).

If the effluent from a roof vent release penetrates into the displacement zone, the effects of the building are minimal and the release can be considered as coming from an elevated point source. If the plume fails to penetrate the displacement zone, it is drawn into the wake or cavity zone and should be treated as coming from a ground-level volume source for purposes of further air concentration calculations (Thuillier and Mancuso 1980). Plumes that fail to penetrate the displacement zone pose increased risk for exposures to people on the ground and in the building itself. The placement of intake vents for the building ventilation system is a major concern if plumes can be drawn into the cavity zone (Briggs 1974).

One obvious question is when can a release be expected to penetrate into the displacement zone. One often-used rule of thumb is that if the release point is at least 2.5 times the height of the building under consideration, the plume will penetrate into the displacement zone and may be considered as coming from an elevated point source (Hanna et al. 1982). Another means of determining when a plume may become entrained in the wake of a building is to consider the ratio between the vertical exit velocity of the plume and the horizontal wind speed at the height of the release. The U.S. Nuclear Regulatory Commission (NRC) has suggested a series of modeling guidelines for different values of this ratio (NRC 1977):

- If the exit stack velocity ($w_0$) divided by the horizontal wind speed at stack height ($u$) is greater than or equal to 5, the release may be assumed to penetrate into the displacement zone and can be considered an isolated point source
- If $w_0/u$ is less than 1 or is unknown, the plume should be assumed to be entrained into the cavity zone and subsequently treated as a ground-level volume source
- If $w_0/u$ is between 1 and 5, the plume is assumed to be elevated part of the time and entrained part of the time

The NRC (1977) provides modeling guidelines for each of these release conditions.

Table 3.1 Representative values of surface roughness length, $z_0$, for different types of surfaces (Randerson 1984)

| Type of surface | $z_0$ (cm) |
|---|---|
| Very smooth (ice, mud flats) | 0.001 |
| Smooth sand | 0.01 |
| Grass lawn up to 1 cm high | 0.1 |
| Grass lawn up to 5 cm high | 1.0 |
| Thin grass up to 50 cm high | 5.0 |
| Thick grass up to 50 cm high | 10.0 |
| Woodland forest | 20.0 |
| Urban areas | 100–300 |

In most considerations of building wake effects, the building from which the release occurs is assumed to be the one that most influences the resulting dispersion of the plume. This is not always the case, however. For example, if the release point is on one building and a much larger building is in the immediate vicinity, the larger building may exert more influence on the dispersion of the plume than the building from which the release originates (NCRP 1993).

Surface roughness affects the small-scale flows and the turbulence of the air within the ABL. In general, the greater the surface roughness, the greater the degree of mechanical turbulence. Surface roughness increases from a very smooth ice surface, to sand, to grass lawns, to field crops, to forests, and to urban buildings. For a given surface, turbulence increases with increases in wind speed. The greater the turbulence, the greater the rate of diffusion. Some representative values of surface roughness, $z_0$, are given in table 3.1.

## Input Data for Atmospheric Transport and Diffusion Calculations

You will need a variety of input information before you consider performing any type of atmospheric transport and diffusion calculation. The quantity and quality of that data will play a significant role in determining the type of calculation that you can perform and the accuracy of your results.

### Source

You need the following information concerning a release to the atmosphere before considering a calculation:

- Geographic location
- Characteristics of released material (identity, physical form, size and weight of particles, chemical stability in atmosphere, and radioactivity)

- Time and rate of release and duration
- Height of release above the ground
- Exit velocity from a stack
- Thermal content of release
- Location of nearby significant buildings

## Winds

You need to know the nearest location of a meteorological tower (or other source of meteorological data), the height of the measurements, and the representativeness of the data for the path of the pollutant.

The wind direction determines the path of the radionuclides (the *where* question) and, therefore, which downwind inhabitants may be exposed. Unfortunately, specifying the wind direction is not always an easy task. The wind direction can shift significantly in an hour or less, and the wind direction measured at the nuclear facility site may or may not be representative of the wind direction at other locations over which released radionuclides will travel. Figure 3.7 gives an example of time variations, and figure 3.8 gives an example of horizontal variations for different weather conditions.

To assess the potential impact of radionuclide releases on people living downwind of a site, you need to know the direction toward which the wind is blowing. However, wind vanes, the meteorological instruments used to measure wind direction, are read in terms of the direction from which the wind is blowing, and most meteorologists report wind direction as "direction from." This is a potential source of serious confusion when assessing the impact of radionuclide releases. The vectors in figures 3.7 and 3.8 are pointing toward the direction in which the wind is blowing. Another source of potential error is misalignment of the wind direction instrument itself. For example, at the Savannah River Site in the early construction period, the plant was aligned with an existing railroad track through the site, and this was called plant north. Unfortunately, this is 37 degrees off of true north, and it was a source of confusion in the early years of the plant's history. In addition, if compasses are used to align wind vanes, then corrections need to be made for the magnetic declination. You also need to remember that the accuracy of a wind vane is usually never better than about 5 degrees.

Wind speed refers to the rate at which the air is moving horizontally. Wind velocity refers to the vector quantity, which consists of both speed and direction. Like wind direction, wind speed can vary significantly. Rising in the ABL, the wind speed tends to increase in response to the lessening of the frictional effects of the earth's surface. Wind speed is important in assessing the impact of radionuclide releases because the speed of the wind determines the travel time and dilution between the point of release (the *when* question) and the location of any particular receptor. For a continuous release of contaminants, the concentration of any contaminant released into the atmosphere is inversely proportional to the wind speed (a part of the *how much* question). As a result, the wind speed is

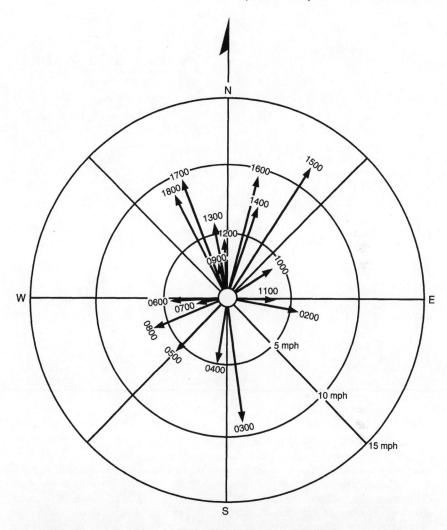

**Figure 3.7** The hourly wind vector as a function of time on March 28, 1979, at the Three-Mile Island site (McKenna et al. 1987). Arrows indicate direction toward which the onsite wind was blowing at the local time indicated. Circles represent varying wind speeds.

a key parameter in any assessment of the impact of radionuclide releases to the atmosphere. You must remember, however, that the instruments used are accurate only to about 5% of the observed speed, and that instrument stalling speeds vary from about 1 to 5 m s$^{-1}$ depending on the particular instrument and its maintenance.

Ideally, the wind instrumentation providing the data for your assessment should be located at the release height of the contaminant. If not, then you should correct the data to the height of the release using one of the formulas available to depict

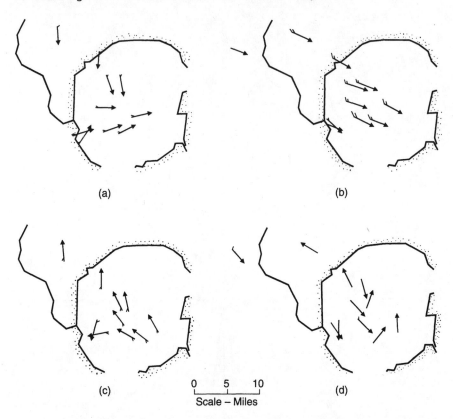

Figure 3.8 Different wind patterns at the Savannah River site, South Carolina, from the operational network of eight meteorological towers, each 61 m high; an offsite television tower; and the National Weather Service station at Augusta, Georgia. The wind field is at 61 m except for a second instrument at a lower level near the Savannah River and for Augusta. (a) A significant wind shift within the network. (b) A relatively uniform midafternoon flow. (c) An example of gravity flow affecting the winds near the Savannah River on an otherwise relatively light wind at nighttime with southeasterly flow over the site. (d) An illustration of a chaotic flow during a transition from northerly to southwesterly flow in midmorning. From Crawford (1990).

these changes with height. The simplest, reasonably accurate formula is the Power Law for speed change with height as a function of stability (Sutton 1953):

$$u = u_1 \left(\frac{z}{z_1}\right)^m, \tag{3.1}$$

where

$u$ = wind speed at height $z$
$u_1$ = wind speed at reference height $z_1$
$m$ = a variable that changes with atmospheric stability and surface roughness $\geq 0$

Table 3.2 Values of the Power Law variable $m$ as a function of surface roughness and atmospheric stability (Sutton 1953)

| Surface roughness | Atmospheric stability | | | |
|---|---|---|---|---|
| | Thermally Unstable | Near neutral | Stable | Very stable |
| Rural | 0.07 | 0.14 | 0.35 | 0.55 |
| Urban | 0.15 | 0.23 | 0.40 | 0.60 |

Table 3.2 shows the values of $m$ as a function of surface roughness and atmospheric stability.

## *Turbulence and Stability*

One of the most important characteristics of the ABL is its turbulent nature (Garratt 1994). Mechanical turbulence is generated by the frictional drag of the earth's surface on the lower layers of the atmosphere. This turbulence increases in proportion to the wind speed and the roughness of the underlying surface (NCRP 1984). Thermal turbulence occurs when the surface of the earth is heated by incoming solar radiation and buoyant thermals (large rising bubbles of warm air) are created. The depth of the ABL at any particular time is directly proportional to the strength of these processes. For example, on a calm, clear night, turbulence in the ABL is at a minimum, and the top of the ABL may be very close to the ground.

The intensity of thermal turbulence in the ABL depends on the stability of the atmosphere. The stability of the ABL can be related to the negative of the temperature change with height, called lapse rate, which is related to thermally induced turbulence. If a parcel of dry air rises vertically without exchanging heat with its surroundings (i.e., adiabatically), its temperature will decrease by 0.98 °C for every 100 m of rise (i.e., $-0.98$ °C $[100\ \mathrm{m}]^{-1}$). This is known as the dry adiabatic lapse rate. If the lapse rate of the atmosphere in which this parcel is rising is less than the dry adiabatic lapse rate, the displaced parcel will be warmer than its surroundings, and buoyant forces will tend to keep the parcel rising vertically. When this occurs, the atmosphere is unstable. If the lapse rate of the atmosphere is greater than the dry adiabatic lapse rate, the parcel will be cooler than its surroundings and tend to return to its original position. Under these conditions, the atmosphere is stable. One of the most stable atmospheric conditions is when the lapse rate of the atmosphere is positive. This occurs on clear, cold winter nights or when a center of high pressure settles over a large area of the earth's surface. When the lapse rate of the atmosphere is equal to the dry adiabatic lapse rate, the atmosphere is neutral. Figure 3.9 illustrates these concepts.

Figure 3.10 illustrates the effect of stability on plume behavior. The largest amount of diffusion (mixing) occurs under unstable conditions when a looping

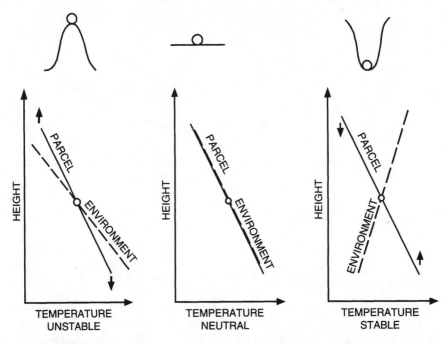

Figure 3.9 Illustration of the relationship between the adiabatic lapse rate (parcel), the environmental lapse rate (environment), and three classes of atmospheric stability (unstable, neutral, and stable). From Hanna et al. (1982).

plume is produced. Figure 3.10 shows that under these conditions, quite high instantaneous ground-level concentrations can occur when the looping plume touches the ground relatively close to the source. Under stable conditions, a plume will disperse very little, and it may travel great distances before touching the ground. When such a plume does touch the ground, however, the concentration of radionuclides in the plume may still be quite high because of lack of mixing. Note that if the plume contains energetic gamma-ray–emitting radionuclides, inhabitants under the plume might be exposed to radiation even though the plume has not touched the ground.

Radioactive materials released into the atmosphere are transported by the winds and diffused by the turbulence. These processes are treated separately in most models. Characterizing the effect of the turbulence on the diffusion process is very important.

## Atmospheric Stability Categories

To perform diffusion calculations, it is necessary to devise a scheme for characterizing the turbulence in a way that is consistent with observations of actual diffusion. The most common approach is to combine some atmospheric measurements of winds and temperatures in a way that theory indicates should be related to diffusion rates and then to correlate these measurements to the spread of pollutant plumes

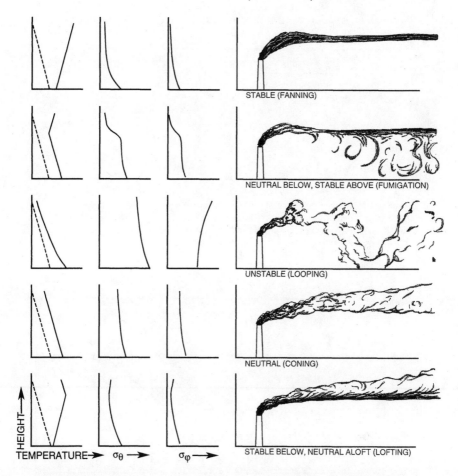

Figure 3.10   The effect of atmospheric stability on plumes. From Slade (1968).

or clouds in the atmosphere. Under the assumption that the plume or cloud has a Gaussian distribution of pollutants in crosswind directions, the distance from the peak concentration to the location of a concentration of 1/10 of the peak concentration is usually correlated with the stability category. This is 2.5 times the standard deviation of a Gaussian distribution fitted to the cloud's averaged crosswind concentration profile.

## Pasquill-Gifford Stability Categories

The Pasquill-Gifford stability categories are by far the most commonly used turbulence classification scheme in applied problems. Pasquill (1961) proposed the categories in table 3.3 and the relationships in figure 3.11 based on measurements of plume spread. Here $h$ is the vertical distance to the plume edge (considered to be 1/10 of the peak concentration); $\sigma_z$ is the standard deviation of a Gaussian

Table 3.3  Relation of atmospheric turbulence types to weather conditions (Pasquill 1961)

| Surface wind speed (m s$^{-1}$) | Daytime insolation[a] | | | Nighttime conditions[a] | |
|---|---|---|---|---|---|
| | Strong | Moderate | Slight | Thin overcast[b] or >3/8 cloudiness | ≤3/8 cloudiness |
| <2 | A | A | B | | |
| 2 | A–B | B | C | E | F |
| 4 | B | B–C | C | D | E |
| 6 | C | C–D | D | D | D |
| >6 | C | D | D | D | D |

[a]A, extremely unstable conditions; B, moderately unstable conditions; C, slightly unstable conditions; D, neutral conditions (applicable to heavy overcast, day or night); E, slightly stable conditions; F, moderately stable conditions.
[b]The degree of cloudiness is defined as that fraction of the sky above the local apparent horizon that is covered by clouds.

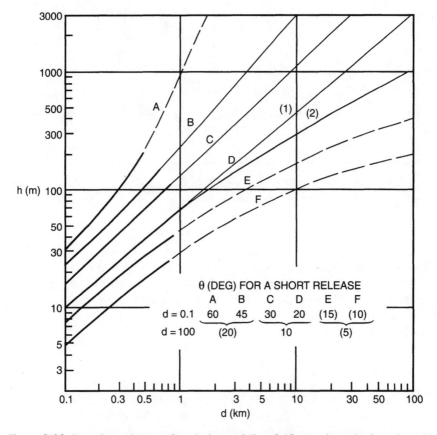

Figure 3.11  Tentative estimates of vertical spread ($h \approx 2.15\sigma_z$) and angular lateral spread ($\Theta \approx 4.3\sigma_y/x$) for a source in open country. From Pasquill (1974).

distribution in the vertical dimension; $d$ is downwind distance; $\Theta$ is the angular spread of the plume in the horizontal, crosswind direction; $\sigma_y$ is the standard deviation of a Gaussian distribution in the horizontal, crosswind direction; and $x$ is downwind distance. The light lines in figure 3.11 indicate the tentativeness of the extrapolations to distances greater than those for which plume measurements existed. These categories were slightly modified in Gifford (1976) and result in the widely used relationships shown in figures 3.12 and 3.13. Gifford (1976) also reviewed other similar schemes developed at Brookhaven National Laboratory by Cramer (1957) and at the Tennessee Valley Authority by Briggs (1974).

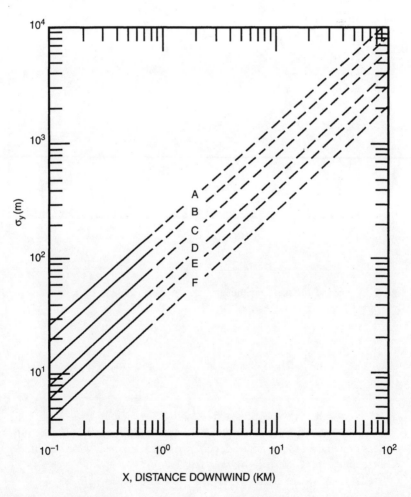

Figure 3.12 Lateral diffusion coefficient, $\sigma_y$, versus downwind distance for Pasquill-Gifford turbulence types. From Gifford (1976).

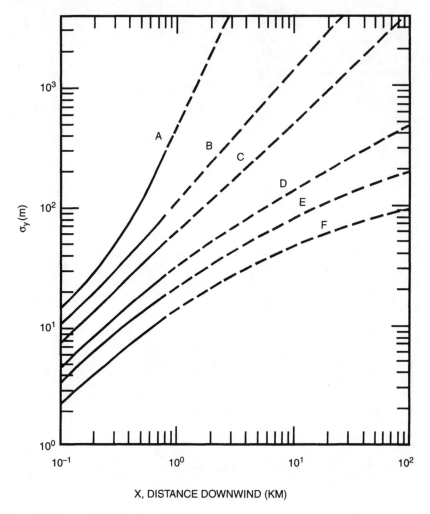

Figure 3.13 Vertical diffusion coefficient, $\sigma_z$, versus downwind distance for Pasquill-Gifford turbulence types. From Gifford (1976).

The actual Pasquill stability categories, A through F, may be determined by the following:

- General weather scheme indicated in table 3.3, which uses sun angle and measurements routinely made by first-order weather stations (Turner 1967)
- Measurements of the standard deviation of the horizontal wind direction fluctuations, $\sigma_\theta$ (Gifford 1976; see table 3.4)
- Vertical temperature gradient, $\Delta T/\Delta Z$ (NRC 1972; see table 3.5)

These methods are summarized in Bailey (2000).

Table 3.4 Relationships between Pasquill-Gifford stability category and the standard deviation of the horizontal wind direction ($\sigma_\theta$) (Gifford 1976)

| Pasquill-Gifford stability category | $\sigma_\theta$ (Degrees) |
| --- | --- |
| A | 25.0 |
| B | 20.0 |
| C | 15.0 |
| D | 10.0 |
| E | 5.0 |
| F | 2.5 |

Table 3.5 Classification of Pasquill-Gifford stability categories on the basis of vertical temperature gradient (NRC 1972)

| Pasquill-Gifford stability category | Vertical temperature difference $\Delta T/\Delta Z$ (°C/100 m) |
| --- | --- |
| A | $\Delta T/\Delta Z \leq -1.9$ |
| B | $-1.9 < \Delta T/\Delta Z \leq -1.7$ |
| C | $-1.7 < \Delta T/\Delta Z \leq -1.5$ |
| D | $-1.5 < \Delta T/\Delta Z \leq -0.5$ |
| E | $-0.5 < \Delta T/\Delta Z \leq +1.5$ |
| F | $+1.5 < \Delta T/\Delta Z \leq +4.0$ |
| G[a] | $+4.0 < \Delta T/\Delta Z$ |

[a] G, extremely stable conditions.

The most theoretically satisfying scheme for specifying atmospheric stability class is to use the standard deviation of the vertical wind direction fluctuations, $\sigma_\phi$, and the standard deviation of the horizontal wind direction fluctuations, $\sigma_\theta$, directly to classify the stability (Hanna et al. 1977). However, it is rare that vertical wind directions are available. Therefore, the temperature gradient approach (over the appropriate heights for the problem) could be used for vertical diffusion, and the standard deviation of the horizontal wind direction (if available) could be used for horizontal diffusion.

After the stability category has been determined, then the $\sigma_y$ and $\sigma_z$ of the plume with distance, $x$, can be determined from the curves in figures 3.12 and 3.13 or the equations developed by Briggs (1974) in tables 3.6 and 3.7.

## Richardson Number

The Richardson number (Ri) is another way of characterizing atmospheric stability. Richardson (1920) proposed it as a measure of whether turbulence should increase or decrease with time. He argued that turbulence would continue as long as the rate of energy removed from the mean flow by the turbulence is at least as great as the

Table 3.6 Formulas recommended for σ_y and σ_z under open country conditions ($10^2 < x < 10^4$ m) where $x$ is the downwind distance (Briggs 1974)

| Pasquill-Gifford stability category | σ_y (m) | σ_z (m) |
|---|---|---|
| A | $0.22x(1 + 0.0001x)^{-1/2}$ | $0.20x$ |
| B | $0.16x(1 + 0.0001x)^{-1/2}$ | $0.12x$ |
| C | $0.11x(1 + 0.0001x)^{-1/2}$ | $0.08x(1 + 0.0002x)^{-1/2}$ |
| D | $0.08x(1 + 0.0001x)^{-1/2}$ | $0.06x(1 + 0.0015x)^{-1/2}$ |
| E | $0.06x(1 + 0.0001x)^{-1/2}$ | $0.03x(1 + 0.0003x)^{-1}$ |
| F | $0.04x(1 + 0.0001x)^{-1/2}$ | $0.016x(1 + 0.0003x)^{-1}$ |

Table 3.7 Formulas recommended for σ_y and σ_z under urban conditions ($10^2 < x < 10^4$ m) where $x$ is the downwind distance (Briggs 1974)

| Pasquill-Gifford stability category | σ_y (m) | σ_z (m) |
|---|---|---|
| A–B | $0.32x(1 + 0.0004x)^{-1/2}$ | $0.24x(1 + 0.001x)^{-1/2}$ |
| C | $0.22x(1 + 0.0004x)^{-1/2}$ | $0.20x$ |
| D | $0.16x(1 + 0.0004x)^{-1/2}$ | $0.14x(1 + 0.0003x)^{-1/2}$ |
| E–F | $0.11x(1 + 0.0004x)^{-1/2}$ | $0.08x(1 + 0.00015x)^{-1/2}$ |

work that must be done to maintain turbulence against gravity. This nondimensional number is defined as

$$\text{Ri} = \frac{g \left( \frac{\partial T}{\partial z} + \Gamma \right)}{T \left( \frac{\partial u}{\partial z} \right)^2}, \tag{3.2}$$

where

- $g$ = acceleration of gravity ($\cong 9.8$ m s$^{-2}$)
- $T$ = temperature (K)
- $z$ = vertical axis
- $\Gamma$ = dry adiabatic lapse rate in the atmosphere (0.01 °C m$^{-1}$)
- $u$ = horizontal wind speed (m s$^{-1}$)

Slightly turbulent motion should remain turbulent as long as Ri < 1 and decay to laminar (smooth, no fluctuations) flow if Ri > 1. Wind tunnel tests and ABL studies suggest that the actual critical Ri for transition to laminar flow lies between Ri = 0.15 and 0.5.

Golder (1972) related the Pasquill-Gifford categories to ranges in Ri as shown in table 3.8. Pendergast and Crawford (1974) showed that stability category frequency distributions derived using Ri measurements corresponded more closely to stability

Table 3.8  A relationship between Pasquill-Gifford stability categories and Richardson number (Golder 1972)

| Pasquill-Gifford stability category | Richardson number (Ri) |
| --- | --- |
| A | $\text{Ri} < -3.5$ |
| B | $-3.5 \leq \text{Ri} < -0.75$ |
| C | $-0.75 \leq \text{Ri} < -0.1$ |
| D | $-0.1 \leq \text{Ri} < +0.15$ |
| E | $+0.15 \leq \text{Ri} < +0.75$ |
| F | $+0.75 \leq \text{Ri} < +3.5$ |
| G | $+3.5 < \text{Ri}$ |

category frequency distributions based on $\sigma_\theta$ (standard deviation of the horizontal wind direction) than those based on the vertical temperature gradient method (NRC 1972).

Implicit in the use of Ri is the assumption that the vertical exchange coefficients for momentum ($K_M$) and sensible heat ($K_H$) are equal. If they are not, a flux form of the Richardson number (Rf) can be written as

$$\text{Rf} = \frac{K_H \text{Ri}}{K_M}. \tag{3.3}$$

For unstable conditions, the (gradient) Richardson number can be shown equal to

$$\text{Rf} = \frac{z}{L}, \tag{3.4}$$

where $L$ is the Monin-Obukhov (1954) length scale.

Randerson (1984) discusses the derivation of $L$. It can be considered as the height below which turbulence is dominated by mechanically generated turbulence and above which turbulence is dominated by thermally generated turbulence. It is defined as

$$L = -\frac{\rho c_p u_*^3 T}{kgH}, \tag{3.5}$$

where

- $\rho$ = density of air (g m$^{-3}$)
- $c_p$ = specific heat of air at constant pressure (cal g$^{-1}$ K$^{-1}$)
- $u_*$ = friction velocity, which can be determined from the log law wind profile under constant momentum flux (m s$^{-1}$)
- $T$ = mean temperature of the ABL (K)
- $k$ = a constant from log law wind profile (the Von Karman constant, $\cong 0.4$)
- $g$ = acceleration of gravity ($\cong 9.8$ m s$^{-2}$)
- $H$ = vertical sensible heat flux (cal m$^{-2}$ s$^{-1}$)

If flux measurements of momentum and sensible heat are available (they rarely are in practical applications), then $L$ could be directly calculated. Usually, $L$ is approximated using wind and temperature profiles, which means we are no better off using $z/L$ as an indicator of stability categories than we are by just using the Ri.

*Eddy Diffusion Coefficients* The concept of eddy diffusion coefficients for turbulent flow arises from an analogy to the molecular transfer of momentum and heat for gases. The idea is that if an elemental box is formed, the time change of concentration of pollutants in the box is the result of divergence or convergence of the fluxes of the pollutant in each of three directions. This is represented in the following equation:

$$F_x = K_x \frac{\partial \chi}{\partial x}, \qquad (3.6)$$

where

$F_x$ = flux of pollutant in the $x$ direction (g m$^{-2}$ s$^{-1}$)
$K_x$ = eddy diffusivity in the $x$ direction (m$^2$ s$^{-1}$)
$\chi$ = concentration of the pollutant (g m$^{-3}$)

In the molecular kinematic case, exchanges occur over the mean free mixing length of the gas molecules. In turbulent flow, the mixing occurs over the mean distance between turbulent eddies. In equation 3.6, the molecular transfer of the pollutant has been ignored because it is always much smaller than that for turbulent flow. In molecular problems, the kinematic viscosity is a constant property of the media. The three-dimensional diffusion equation results from the changes in fluxes in each of three directions:

$$\frac{\partial \chi}{\partial t} = \frac{\partial}{\partial x}\left[K_x \left(\frac{\partial \chi}{\partial x}\right)\right] + \frac{\partial}{\partial y}\left[K_y \left(\frac{\partial \chi}{\partial y}\right)\right] + \frac{\partial}{\partial z}\left[K_z \left(\frac{\partial \chi}{\partial z}\right)\right], \qquad (3.7)$$

where

$x$, $y$, and $z$ = the downwind, crosswind, and vertical directions, respectively
$K_x$, $K_y$, and $K_z$ = the eddy diffusivities in the x-, y-, and z- directions, respectively

Assuming continuity of mass, solutions to the diffusion equation 3.7 vary with initial and boundary conditions and result in Gaussian distributions of pollutant, $\chi$, in the x-, y-, and z-directions. This results in

$$\sigma^2 = 2Kt \qquad (3.8)$$

for each of the three directions (x, y, and z), with $\sigma$ being the standard deviation of the Gaussian distribution.

However, to obtain the solutions, it has to be assumed that the eddy diffusion coefficients are constant in time and space. This is true for molecular processes but not in the turbulent atmosphere. Therefore, the empirical approaches discussed in the preceding section are used for obtaining values of $\sigma$ as a function of distance and stability conditions in the real atmosphere.

In the following discussion of numerical models and resulting numerical solutions to the diffusion equation, it is necessary to specify some reasonable values for $K$.

In the vertical direction, $K_z$ has to be related to height above the ground, which determines the maximum vertical eddy size. Sutton (1953) derived a formulation for $K_z$ for vertical transfer of momentum in a turbulent boundary layer regime for which the Power Law wind profile is a reasonable approximation (see equation 3.1):

$$K_z = K_l \left(\frac{z}{z_l}\right)^{1-m}, \tag{3.9}$$

where $K_l$ is the eddy viscosity at reference height $z_l$.

Sutton (1953) argued from an analysis of the Log Law wind profile that, for vertical transfer of momentum,

$$K_z = u_* l = k u_* z. \tag{3.10}$$

Because $k$ is about 0.4 and $u_*$ is 10–20 cm s$^{-1}$ over very smooth surfaces, $K_l$ is about $10^3$ cm$^2$ s$^{-1}$ at a $z_l$ of 100 cm.

For the horizontal direction, it can be argued that the $K$ values should be related to the size of the turbulent eddies that are the most efficient for diffusing the pollutant. Richardson (1926) studied the relative diffusion of particles on Loch Long, Scotland, and found that

$$K \cong 0.07 l^{1.4}. \tag{3.11}$$

For the atmosphere and using relatively crude data, Richardson (1926) found that

$$K \cong 0.2 l^{4/3}, \tag{3.12}$$

where $l$ is one of the following:

- Horizontal mixing length
- Mean distance between two particles
- Standard deviation of particles around a center point (i.e., the standard deviation of a Gaussian distribution)
- Characteristic dimension of the eddies most responsible for the diffusion

The data used by Richardson (1926) had a distance range of $10^3$ to $10^8$ cm. This approach is appealing because it implies that the "most efficient eddies for dispersion" are of a similar size as the particle separation. In fact, it was Richardson (1926) that introduced the concept of a continuous range of eddy sizes, where turbulent energy was being handed down from large to smaller eddies and ultimately dissipated in viscous action.

*Similarity Theory* Further development of this concept came with the work of Kolmogorov (1941) from his work on similarity theory and from Obukhov (1941) on the basis of evaluating the energy balance in the turbulent spectrum. Using the similarity theory assumption that turbulent energy flowed unchanged from the large eddies to smaller eddies and using dimensional analysis, Taylor (1959) showed that the turbulent dissipation rate $\varepsilon$ is the key parameter to consider in atmospheric diffusion. Taylor showed that $K \propto l^{4/3}$, which is the same as Richardson (1926) determined empirically.

Batchelor (1950) applied these concepts to the separation of two particles with time (or the spread of a cloud of particles about its center point) and found the following:

$$\frac{d\sigma^2(t)}{dt} = 2.66 C_l t (\varepsilon \sigma_0)^{2/3}, \qquad (3.13)$$

where $C_l$ is a constant of order unity that results from the dimensional analysis, and $\sigma_0$ is the initial separation of the particles (or the initial standard distribution for a Gaussian distribution). Integration over time gives

$$\sigma^2 = \sigma_0^2 + 1.33 C_l (\varepsilon \sigma_0)^{2/3} t^2. \qquad (3.14)$$

Equation 3.14 is for early times when the initial size of the dispersing cloud is important. For longer times, when $\sigma^2(t) \gg \sigma_0^2$, the diffusion processes are independent of the way the particles were released. The initial separation no longer has any influence on the way the particles are diffusing. Then dimensional analysis indicates that

$$\frac{d\sigma^2(t)}{dt} = C_2 \varepsilon t^2, \qquad (3.15)$$

where $C_2$ is another nondimensional constant of order unity. Integrating from $t_1$ (the time at which it is no longer necessary to consider the influence of $\sigma_0$) to $t$ gives

$$\sigma^2(t) = \sigma_1^2 + \left(\frac{C_2}{3}\right) \varepsilon t^3. \qquad (3.16)$$

Dimensional arguments give the time of transition, $t_1$, as

$$t_1 = C_3 \sigma_0^{2/3} \varepsilon^{-1/3}, \qquad (3.17)$$

where $C_3$ is another nondimensional constant of order unity.

The inertial subrange refers to a range of eddy sizes in the atmosphere through which turbulent energy is handed down to successively smaller eddies at a constant rate. The pertinent variables in this dimensional analysis of turbulent diffusion for the inertial subrange are $\sigma_0$, $t$, and $\varepsilon$. In the inertial subrange, eddy diffusivity is not dependent upon the time of release, so the dimensional arguments yield

$$K = C_4 \varepsilon^{1/3} \sigma^{4/3}, \qquad (3.18)$$

where $C_4$ is another nondimensional constant of order unity.

The above predictions from similarity theory should apply only in the inertial subrange where all turbulence is isotropic. However, Crawford (1966, 1967) extensively reviewed the literature and concluded that it applies reasonably well to atmospheric data out to the large eddy size of the synoptic storm (i.e., about 1,000 km or a travel time of a few days). At that point, the cloud of pollutants becomes larger than the largest horizontal eddies, and diffusion rates are characterized by a Fickian-like process with the eddy diffusivity in the horizontal direction having values of about $10^{10}$ cm$^2$ s$^{-1}$. Figure 3.14 summarizes relevant data for long-range diffusion. The

## Atmospheric Transport of Radionuclides 105

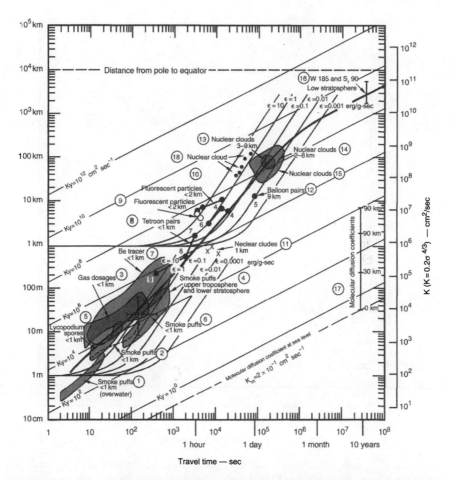

**Figure 3.14** The standard deviation of the horizontal particle distribution, normal to the direction of motion of the cloud, as a function of travel time from the point of injection of the cloud of particles into the atmosphere. From Hage et al. (1966) and modified by Crawford (1966).

figure was taken from Hage et al. (1966), and the lines for similarity predications were added by Crawford (1966, 1967).

Wilkins (1963) summarized most of the existing measurements of atmospheric dissipation at different heights for a normalized wind speed of 5 m s$^{-1}$. Wilkins's results are represented approximately by

$$\varepsilon = \frac{300}{z}\left(\frac{u}{5}\right)^3, \qquad (3.19)$$

where

$\varepsilon$ = dissipation (cm$^2$ s$^3$)
$u$ = wind speed (m s$^{-1}$)
$z$ = height above the ground (m)

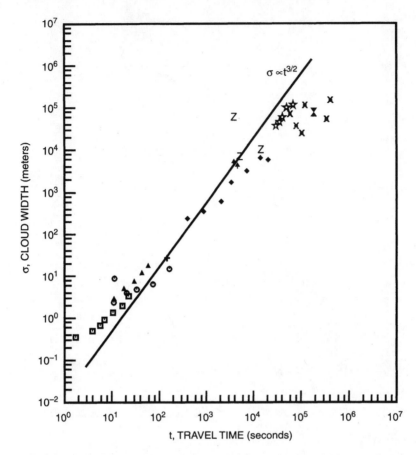

Figure 3.15 A composite summary of the results from 10 tropospheric experiments on relative diffusion. From Gifford (1977), as cited in Hanna et al. (1982).

Additional data are given in figures 3.15 and 3.16 (Hanna et al. 1982), showing that $\sigma^2 \propto t^3$ (equation 3.16) for puffs. Table 3.9 presents a summary of the comparisons between puff and plume diffusion.

### Mixing Height

Often, especially in the presence of strong thermal turbulence, the top of the ABL may be well defined by the presence of a stable layer (also known as a capping temperature inversion) (Garratt 1994). Turbulent motions, and the contaminants they may be transporting, have difficulty penetrating into this layer. This condition results in the contaminants being effectively trapped between the ground surface and the top of the ABL layer. (See figure 3.10 for an illustration of a stable lapse rate.) Because turbulent diffusion and the mixing of air with the contaminant plume are restricted, the distance from the surface to the top of the ABL may be called the mixing height (or depth).

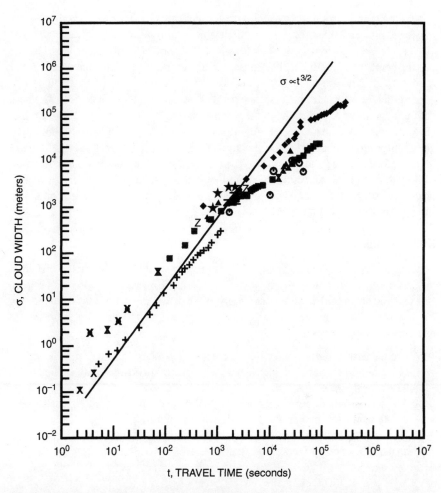

Figure 3.16 A composite summary of the results from 12 tropospheric experiments on relative diffusion. From Gifford (1977), as cited in Hanna et al. (1982).

Table 3.9 Comparison of puff and plume diffusion (Hanna et al. 1982)

|  | Short times | Long times |
|---|---|---|
| Puff | $\sigma \propto t$ (very short $t$) | $\sigma \propto t^{1/2}$ |
|  | $\sigma \propto t^{3/2}$ (intermediate $t$) |  |
| Plume | $\sigma \propto t$ | $\sigma \propto t^{1/2}$ |

Mixing heights are quite variable. Under unstable conditions the mixing height may be as high as 2,000 m or more. Under stable conditions the mixing height may be 100 m or less and extremely difficult to define (Garratt 1994). At a given location, the mixing height changes diurnally, generally being highest in midafternoon and lowest at night. There is also wide spatial variation in mixing heights (Holzworth 1972). Mixing-height information for some specific locations in the United States, where data are taken twice per day, can be obtained from the National Weather Service. However, site-specific mixing-height information is difficult to obtain without upper air monitoring equipment at the location of interest (Pendergast 1984). Mixing-height information can be determined from an analysis of temperatures as a function of height, from acoustic sounders, from Doppler radars, and from lidars (a special laser). To follow the diurnal evolution of the ABL, you need measurements as a function of time throughout the day.

## Meteorological Data Quality

The emphasis throughout this chapter is on the atmospheric transport and diffusion models to be used for radiological dose assessments. However, it is also very important that you are confident about the quality of the data being used in the calculations. As they say in the computer world, garbage in gives garbage out. Thus, you should ask the following questions:

- Are the data appropriate for the problem?
- Is the height of the wind and turbulence measurements the same as the height of the release?
- Are the sampling frequencies and the length of the data record appropriate?
- Are the data accurate and with sufficient resolution?
- Do adequate instrumentation, installation, and calibration records exist to ensure accuracy and representativeness?

## Modeling of Transport and Diffusion

Mathematical models play a key role in evaluating the actual or potential impact of radionuclides released into the atmosphere. The basic purpose of any diffusion model is to account for all of the material released to the atmosphere from a particular source. The model must address three fundamental questions (Barr and Clements 1984):

- *Where* does the released material go and *when* does it arrive at a particular location?
- *How* fast is the material diluted as it travels?
- *How* fast and by what mechanism is the material ultimately removed from the atmosphere?

The answers to these questions are often difficult to obtain, and they depend on interactions among the many fundamental meteorological properties described in preceding sections (see "The Atmosphere" and "Input Data for Atmospheric Transport and Diffusion Calculations," above).

Much additional research needs to be done before atmospheric turbulence and diffusion theory is complete. Nevertheless, several diffusion theories have been proposed that can serve as the basis for practical models of atmospheric dispersion. Some of these theories and their limitations have been extensively discussed elsewhere (Gifford 1968; NCRP 1984; Pasquill and Smith 1983). When using a model developed from any of these sources, however, you must always remember that any model, no matter how simple or complex, is an approximation of reality, and predictions made by the model contain some level of uncertainty. Models are very useful tools for radiological assessment purposes, but they must be used carefully and with some indication or explanation of their inherent uncertainties to be most useful to the assessor.

## Gaussian Diffusion Models

Gaussian diffusion models arise from the analytical solution to the three-dimensional conduction equation. The difference in fluxes into and out of an elemental cube is equal to the change in the content of the cube with time. This works well for molecular processes (e.g., conduction of heat in metal) because thermal conductivity is constant everywhere in the metal. The case of a constant conductivity or diffusivity is often called Fickian diffusion. There are limitations to this concept when applied to atmospheric diffusion because molecular processes are small compared to turbulent transfer processes, and the effective turbulent eddy diffusivity is not invariant with space or time. In any case, with suitable time averaging, the analytical solutions do give Gaussian-shaped solutions and mass is conserved. Despite their theoretical limitations, the simple Gaussian models for atmospheric diffusion have not been discarded, primarily because they produce results that often agree fairly well with measured experimental data and because their results are obtained quickly. Gaussian models yield solutions where the concentrations of material released into the atmosphere are described by the Gaussian distribution in all three directions (along-wind direction [$x$], crosswind direction [$y$], and vertical direction [$z$]). Solutions can be obtained for many different scenarios depending on the initial and boundary conditions. A few useful solutions for atmospheric problems are given in the next few sections.

### Instantaneous Point Source

The solution for an instantaneously generated point source in the free (away from any boundaries) atmosphere is

$$\chi = \frac{Q}{8(\pi t)^{3/2}(K_x K_y K_z)^{1/2}} \exp\left[-\frac{1}{4}t\left(\frac{x^2}{K_x} + \frac{y^2}{K_y} + \frac{z^2}{K_z}\right)\right], \qquad (3.20)$$

where

$\chi$ = air concentration of the contaminant as a function of $x$, $y$, $z$, and $t$
$Q$ = total amount of contaminant released
$K_x$, $K_y$, and $K_z$ = eddy diffusivities in the x-, y-, and z-directions, respectively, which are assumed to be equal and constant for the Gaussian solution

In the case of Fickian diffusion (necessary for the Gaussian solution), we have

$$\sigma^2 = 2Kt, \quad (3.21)$$

where $\sigma$ is the standard deviation of the Gaussian distribution. This form would be true for each orientation. For a perfect sphere, $x$, $y$, and $z$ can be replaced by the radius of the sphere, $r$, and the $\sigma$s in the x-, y-, and z-directions by $\sigma_r$ in the Gaussian distribution. Thus, the new equation form for the instantaneous point source is

$$\chi = \frac{Q}{(2\pi)^{3/2}\sigma_r^3} \exp\left[-\frac{1}{2}\left(\frac{r^2}{\sigma_r^2}\right)\right]. \quad (3.22)$$

## Continuous Point Source

The solution for a continuous point source located far from any boundary is obtained by integrating the point source solution along the x-axis. The concentrations are only a function of $y$ and $z$ at different locations along the x-axis. The solution then is given in equation 3.23 using the Fickian relationships between the eddy diffusivities ($K$ values) and the standard deviations ($\sigma$ values) of the Gaussian distribution:

$$\chi = \frac{Q'}{2\pi u \sigma_y \sigma_z} \exp\left[-\frac{1}{2}\left(\frac{y^2}{\sigma_y^2} + \frac{z^2}{\sigma_z^2}\right)\right], \quad (3.23)$$

where

$u$ = wind speed in the x- direction
$Q'$ = amount of contaminant released per unit time

When the continuous point source is on the ground, the ground surface is assumed to act as a perfect reflector. To maintain mass continuity, the right-hand side of equation 3.23 is then multiplied by 2.

## Continuous Line Source

The solution to an infinite continuous line source, which is far from any boundary and is perpendicular to the wind (i.e., the continuous line source lies along the y-axis), is obtained by integrating the continuous point source in the y-direction. The same assumptions on constant $K$ values and Fickian diffusion apply. The resulting equation is

$$\chi = \frac{Q''}{(2\pi)^{1/2} u \sigma_z} \exp\left[-\frac{1}{2}\left(\frac{z^2}{\sigma_z^2}\right)\right], \quad (3.24)$$

where $Q''$ is the amount of contaminant released per unit time per unit line length.

For a continuous line source on the ground (e.g., a highway perpendicular to the wind direction), the right-hand side of equation 3.24 has to be multiplied by 2 to maintain mass continuity.

## Continuous Point Source Release from a Stack

When material is released from an elevated source (i.e., a stack), the resulting plume will spread downward until it eventually encounters the ground. When this happens, it is assumed that the plume is reflected and dispersed back up into the air. This system is illustrated in figure 3.17. The equation for the Gaussian model under these conditions is given by Gifford (1968). It is the same as for the continuous point source far from any boundary presented above (equation 3.23); however, the stack height term is added to fix the origin point of the coordinate system at the stack top:

$$\chi = \frac{Q'}{2\pi u \sigma_y \sigma_z} \exp\left\{-\frac{1}{2}\left[\frac{y^2}{\sigma_y^2} + \frac{(z-h_p)^2}{\sigma_z^2}\right]\right\}, \quad (3.25)$$

where $h_p$ is the effective release height of the plume (see "Modifications Based on Source Characteristics," below).

Reflection by the ground is equivalent to a virtual source located at a distance $h_p$ below the surface of the ground. Under these circumstances, equation 3.25 becomes

$$\chi = \frac{Q'}{2\pi u \sigma_y \sigma_z} \exp\left[-\frac{1}{2}\left(\frac{y^2}{\sigma_y^2}\right)\right] \cdot \left\{\exp\left(-\frac{1}{2}\left[\frac{(z-h_p)^2}{\sigma_z^2}\right]\right) + \exp\left(-\frac{1}{2}\left[\frac{(z+h_p)^2}{\sigma_z^2}\right]\right)\right\}. \quad (3.26)$$

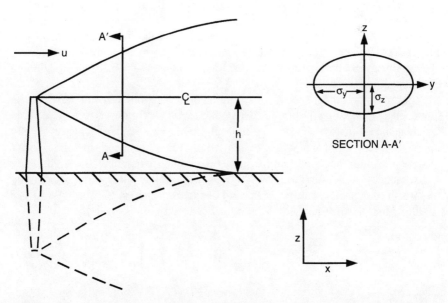

Figure 3.17 Coordinate system of the straight-line Gaussian plume model, including representations of the distribution of concentrations within the plume in the horizontal and vertical directions and the location of the image source. From Hanna et al. (1982).

As described above under "Mesoscale," the top of the ABL may be well defined in the presence of a stable layer of air. When a dispersing plume encounters this layer of air, it may be assumed to reflect off the layer rather than penetrate it. The result of assuming multiple reflections of the plume from both the ground and the top of the ABL is that the air concentration at any point downwind may be calculated by summing the contributions from many virtual sources (e.g., Miller et al. 1986). Equation 3.26 does not include this process, but a good approximation is to let $\sigma_z = h_i$ (the height of the base of a trapping inversion, i.e., the height of the top of the ABL) in the denominator of equation 3.26 and to eliminate the vertical distribution of $\chi$ in the equation (concentration is now uniform with height). This occurs at about the distance downwind where $\sigma_z$ becomes larger than twice the height of the ABL. When this occurs, the material in the plume is effectively distributed uniformly throughout the ABL (Miller et al. 1986).

### Sector Averaging

The various forms of the continuous point source releases above apply to a situation in which the duration of the release results in a distribution of the radionuclides that may be considered to be in the form of a plume. However, one of the most common uses of the Gaussian model is to calculate the average ground-level air concentration around a source for releases that occur over an extended period of time, such as a year (e.g., NRC 1977). As a result, the plume equation is modified to eliminate its dependence on $y$. The most common modification is based on the type of long-term meteorological records archived in the United States. These data are usually expressed in terms of a joint frequency distribution of wind direction, wind speed class, and atmospheric stability class. The average concentration of material across the sectors defined by the available wind directions is calculated. In the United States, 16 wind direction sectors are normally available; therefore, average air concentrations are calculated for 22.5-degree sectors, usually beginning with and centered on north.

The crosswind-integrated air concentration from a continuous point source can be obtained by integrating equation 3.26 over the variable $y$ from $-\infty$ to $+\infty$. You can calculate an estimate of the average concentration over a period that is very long (a month or a year) compared with the time period over which the mean wind velocity is determined (an hour) by multiplying the crosswind-integrated concentration by the frequency with which the wind blows toward a given sector and dividing by the width of that sector at the downwind distance of interest (Gifford 1968). The resulting long-term ground-level ($z = 0$) air concentration is given by

$$\chi_{is}(x) = \sum_{p=1}^{N_p} \sum_{r=1}^{N_r} \frac{N_f f_{prs} Q'_i}{\sqrt{2}\pi^{3/2} x u_r \sigma_{zp}(x)} \exp\left\{-\frac{1}{2}\left[\frac{h_p^2}{\sigma_{zp}(x)^2}\right]\right\}, \qquad (3.27)$$

where

$\chi_{is}(x)$ = ground-level air concentration of contaminant $i$ in wind direction $s$ at downwind distance $x$

$N_p$ = number of atmospheric stability classes (unitless)
$N_r$ = number of wind speed classes (unitless)
$N_s$ = number of wind directions (generally 16; unitless)
$f_{pr}$ = joint frequency of occurrence of atmospheric stability class $p$, wind speed class $r$, and wind direction $s$ (unitless)
$Q'_i$ = release rate of contaminant $i$ per unit time
$u_r$ = mean wind speed associated with wind speed category $r$
$\sigma_{zp}(x)$ = the specific value of $\sigma_z$ associated with stability class $p$ and downwind distance $x$

Reflection from the top of the ABL is ignored in equation 3.27. Another limitation associated with this equation is that wind velocity and stability class persistence are not considered. That is, for each stability class, the wind does not actually blow steadily in one direction with a given speed for a long period of time, as is implicit in the equation. Therefore, uniform vertical mixing will not be obtained under all conditions even at very large downwind distances, and concentrations at very large distances will be unduly influenced by stable classes with low wind speeds.

Various formulations of the Gaussian model are illustrated previously and can be applied under many different conditions, including both episodic releases (0–50 km distances) and long-term releases (50–500 km distances) and a variety of source configurations. All of these formulations require the same basic set of input data, however, and are subject to the same general limitations.

### Modifications Based on Source Characteristics

Basic input requirements are discussed under "Input Data for Atmospheric Transport and Diffusion Calculations," above. This section discusses ways to modify the calculations based upon the source characteristics.

*Gravitational Effects* Particles with diameters of 10 μm or more may have significant gravitational settling speeds, $v_g$, in the atmosphere. For particles with diameters greater than 100 μm, the Gaussian plume model may not apply because the particles are falling so fast that diffusion is no longer significant (Van der Hoven 1968). For particles with diameters of 100 μm or less, however, the Gaussian model may still be used by further modifying the effective release height, $hp$. This is accomplished by subtracting $xv_g u^{-1}$ from the effective stack height. Values of $v_g$ can be calculated by using the basic Stokes equation (Van der Hoven 1968).

*Effective Release Height* Another important input parameter for the Gaussian plume model is the value selected for the effective release height of the plume, $h_p$. This parameter is actually composed of three components:

$$h_p = h_{ph} + h_{pr} - h_{sr}, \quad (3.28)$$

where

$h_{ph}$ = physical height of the stack above the ground
$h_{pr}$ = height that the plume rises above the stack because of momentum and buoyancy effects
$h_{sr}$ = reduction in the plume height because of stack downwash or other source effects

The physical height of a release point is relatively easy to measure from plant drawings or by direct measurement. The other portions of $h_p$ are not always determined so easily. A plume may continue to rise above its physical release point if it is warmer than the surrounding air (buoyancy effects) or if it has an appreciable exit velocity, such as that caused by an exhaust fan (momentum effects). For most routine releases from nuclear facilities, momentum effects are likely to be more significant than buoyancy effects in causing plumes to rise above their release points. Numerous models have been developed to estimate plume rise for a number of source and meteorological conditions. Briggs (1984) provides an extensive review of such models. In some later work, Overcamp and Ku (1988) examined what happens when two or more large buoyant plumes from adjacent stacks merge. Their studies showed enhanced plume rise in such circumstances.

As noted above under "Microscale," the configuration of the source (e.g., the presence of building structures) can greatly influence the dispersion of a plume. The ratio between the exit velocity of the plume from the stack, $w_0$, and wind speed, $u$, can be used to judge whether a released plume is likely to be entrained in the wake of a building. The stack itself, even if it is considered an isolated point source, can have an effect on the plume. If the ratio of $w_0$ to $u$ is less than 1.5, the plume may be drawn downward behind the stack (NRC 1977). The amount of this reduction in plume height may be estimated by

$$h_{sr} = 2\left(\frac{w_0}{u} - 1.5\right)D, \qquad (3.29)$$

where $D$ is the internal diameter of the stack (Briggs 1974). Snyder and Lawson (1991) examined equation 3.29 in detail for neutrally buoyant plumes in the wind tunnel. They found that these predictions are overly conservative for most real stacks. By characterizing the flow as subcritical or supercritical (where criticality refers to turbulent flow in the boundary layer on the stack cylinder), they were able to use the ratio of effluent speed to wind speed to predict the extent and nature of the downwash. Most full-scale stacks are supercritical, and the ratio of the plume centroid to the stack diameter is given by

$$\frac{z_c}{D} = 2.2\left(\frac{W}{U} - 0.3\right)^{0.67}, \qquad (3.30)$$

when

$$0.3 \leq \frac{W}{U} \leq 2.0,$$

where

$z_c$ = plume centroid
$D$ = stack diameter
$W$ = vertical velocity of the plume
$U$ = wind speed at the stack tip

See Snyder and Lawson (1991) for complete details.

In the Gaussian plume model, the maximum ground-level air concentration is roughly proportional to the inverse square of the stack height (Hanna et al. 1982). As a result, $h_p$ must be specified as accurately as possible. A factor of 10 difference in $h_p$ can, under some circumstances, lead to as much as a factor of 100 difference in the predicted maximum ground-level air concentration (Hanna et al. 1982).

*Building Wake Considerations* Another complex airflow condition for which the standard Gaussian plume model may not directly apply is flow around buildings (see "Microscale," above). One common way of addressing this condition is to assume that the plume is released from ground level and then to modify the values of the standard diffusion parameters $\sigma_y$ and $\sigma_z$ that are used in the model. A number of different approaches to modifying $\sigma_y$ and $\sigma_z$ have been proposed, one of which is suggested in Regulatory Guide 1.111 (NRC 1977). Sometimes the receptor of interest may be on or near the building from which the release originates. For example, unintentional releases from building surfaces near building air intake vents could expose plant personnel inside the facility. Models have been proposed for such situations (e.g., Ramsdell and Fosmire 1995). A relatively simple model for pollutant drawn into the downwind cavity caused by airflow over the building for the Gaussian plume equation is as follows:

$$\chi = \frac{Q'}{(\pi\sigma_y\sigma_z + cWH)u}, \qquad (3.31)$$

where

$c$ = a constant empirically determined to be between 0.5 and 2.0
$W$ = width of the building
$H$ = height of the building

Site-specific field studies or physical modeling, or both, may be required to understand the airflow and resultant dispersion around individual building configurations (e.g., Widner et al. 1991).

*Mixing Height* Site-specific values of mixing height (or top of the ABL) are not routinely measured; however, they can be important constraints on vertical mixing, with very low mixing heights near the source of pollutant emission or at longer distances. Holzworth (1972) presents a climatological summary of annual average morning and afternoon mixing heights that can be used to calculate stability-dependent values of mixing height for annual average air concentration calculations in the absence of site-specific data (e.g., Miller et al. 1986).

*Wind Direction* Wind direction is an implicit, rather than an explicit, parameter in the Gaussian model because once the material enters the atmosphere, it is assumed to travel in the same direction throughout the assessment process. One of the major limitations of any Gaussian plume model is that spatial and temporal variations in wind direction are not considered.

The Gaussian plume model is mathematically unstable if $u = 0$. Fortunately, however, although anemometers may report a wind speed of zero, the winds in the ABL seldom stop entirely; they just drop to a speed below the measuring capability of the anemometer (Hanna et al. 1982). Calm winds are generally assigned a value of $u = 0.5$ m s$^{-1}$ in dispersion calculations. See also "Winds," above, for a discussion on wind input to transport and diffusion modeling.

*Atmospheric Stability Classes* Traditionally, atmospheric stability is specified in terms of a finite number of categories. The Pasquill-Gifford stability classes range from A, most unstable, to F or G, most stable. The stability class of the atmosphere is inferred from one or more readily measurable atmospheric quantities (see "Atmospheric Stability Categories," above). The atmospheric stability class is used in the Gaussian plume model as the starting point and is fundamental to the reliability of the results. The diffusion parameters $\sigma_y$ and $\sigma_z$ that are an integral part of the model are chosen on the basis of stability class. The Gaussian plume model has several theoretical prerequisites, assumptions, and boundary conditions that are rarely completely fulfilled in the atmosphere (Brenk et al. 1983). In practice, however, many of these limitations can be compensated for by judiciously selecting the diffusion parameters incorporated into the model. Not surprisingly, changes in the diffusion parameters strongly affect the calculated air concentrations. You should select a set of diffusion parameters that is based on measurements and theoretical assumptions that most closely coincide with the conditions expected in the model application (e.g., a predominately rural setting or a predominately urban setting). If appropriate, you should also use the stability classification methods used in deriving the standard set to derive the stability classes to be used in the modeling.

## Special Considerations

The Gaussian plume model equations presented above are based on certain assumptions. These assumptions are not always met in release conditions. As a result, there will be times when you will need to modify the form of the Gaussian plume model you use based on the specific problem you are considering.

*Averaging Time* The continuous plume equation above defines the concentration of an atmospheric contaminant at a given point in space over some period of time. As this time period increases, very high and low values of the pollutant concentration are not seen, and the standard deviations of the pollutant concentration within the plume tend to increase (Hanna et al. 1982). This phenomenon can be seen when just viewing the smoke plume (a snap shot vs. a time exposure) and is illustrated in figure 3.18. As a result, the averaging times of the field measurements from which the Gaussian plume model diffusion parameters $\sigma_y$ and $\sigma_z$ are derived

Atmospheric Transport of Radionuclides 117

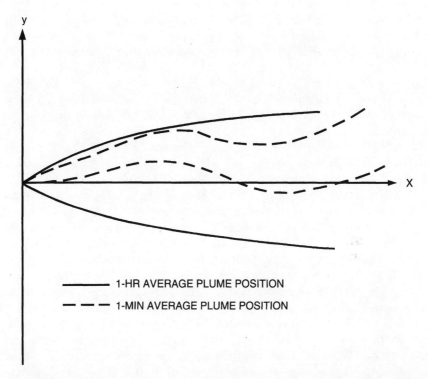

Figure 3.18 Comparison of short- and long-term average plume positions. From Hanna et al. (1982).

are an important consideration. For example, the standard Pasquill-Gifford values for $\sigma_y$ and $\sigma_z$ are associated with a sampling time of 10 min. Other sets of diffusion parameters are based on averaging times as long as 1 h (Brenk et al. 1983).

One factor to be considered in selecting a set of diffusion parameters for a given application is to select a set with an averaging time as close to the desired prediction time as possible. If this cannot be done, Gifford (1975) suggests the following empirical formula for adjusting the values of $\sigma_y$ and $\sigma_z$ for differences in averaging time:

$$\frac{\sigma_{y1}}{\sigma_{y2}} = \left(\frac{T_{s1}}{T_{s2}}\right)^q, \tag{3.32}$$

where

$\sigma_{y1}$ = standard deviation for sample time case 1
$\sigma_{y2}$ = standard deviation for sample time case 2
$T_{s1}$ = sample time for case 1
$T_{s2}$ = sample time for case 2
$q$ = an empirical constant in the range of 0.25 to 0.3 for $1\,h < T_{s1} < 100\,h$ and is approximately 0.2 for $3\,\min < T_{s1} < 1\,h$

The Pasquill-Gifford curves of $\sigma_y$ as a function of distance downwind and stability category were determined using sampling times of about 10 min. Thus, $\sigma_y$ for a sampling time of 1 h equals $6^{0.2}$ or 1.43 times the $\sigma_y$ for 10 min (Hanna et al. 1982).

*Complex Terrain Considerations* Flat terrain is an implicit assumption of the Gaussian plume model. Complex terrain, however, is not uncommon in association with nuclear facilities, and it can significantly affect the dispersion of material released from a plant into the atmosphere (see "Mesoscale," above). Whenever possible, site-specific information should be used in estimating dispersion in complex terrain (Pasquill and Smith 1983). In the absence of such information, approximations may be incorporated into the Gaussian plume model.

A plume released upwind of a terrain obstacle (e.g., a hill) will likely either directly affect the obstacle or rise and ride over the obstacle. A common way of accounting for such release conditions is by making further adjustments in the effective plume height $h_p$ used in the model. Under unstable and neutral conditions (i.e., Pasquill-Gifford categories A–D), a plume will tend to rise over the downwind obstacle but lose part of its effective height (Hanna et al. 1982). Under these conditions, Briggs (1974) suggests reducing $h_p$ by the height of the terrain obstacle, $h_t$, or by one-half of the effective plume height, whichever is smaller. Egan (1975) proposes a similar approach except that he suggests a reduction in $h_p$ of $h_t/2$ rather than $h_t$ when $h_t < h_p/2$. Under stable conditions (i.e., Pasquill-Gifford categories E–G), Briggs (1974) and Egan (1975) assume that the plume remains at a constant elevation. As a result, $h_p$ is effectively reduced by $h_t$. If $h_t > h_p$, the plume may impinge on the terrain obstacle.

Snyder et al. (1985) did a comprehensive study of these phenomena in the wind tunnel. They emphasized the dividing streamline concept, which enables the plume behavior to be rather readily predicted. The dividing streamline height separates the horizontal airflow approaching a terrain obstacle (hill or mountain) into two separate regions: a lower one where the flow is blocked and flows around the terrain obstacle, and an upper one where the flow moves vertically to flow over the terrain obstacle.

The Froude number is used to predict the height of the dividing streamline. The Froude number is the ratio of the inertial force to the force of gravity and is dimensionless:

$$\text{Fr} = \frac{V^2}{gL} = \frac{V}{N_{\text{BV}}L},$$

where

$V$ = mean speed
$g$ = acceleration of gravity
$L$ = width of the flow obstacle
$N_{\text{BV}}$ = Brunt-Vaisala frequency

The dividing streamline height is $H_c = H(1 - \text{Fr})$ where $H$ is the height of the obstacle and Fr is the Froude number. More stable flows where Fr is small imply that the dividing streamline height is near the top of the obstacle, and most of the horizontally moving air will move around rather than over the obstacle. On the other hand, when there is considerable kinetic energy in the horizontal air stream, the Froude number will be larger and the dividing streamline height, smaller; thus, more of the horizontal airflow moves over rather than around the obstacle. These concepts and wind tunnel studies also were tested in the field and found to be valid (Snyder et al. 1980).

Clearly, the above procedures are only approximations to a very complex problem. If such a simple procedure is deemed inappropriate for a given study, site-specific field studies, physical modeling (i.e., wind tunnel studies), or complex numerical modeling may be required to understand the airflow and resultant diffusion (Pasquill and Smith 1983).

*Finite Plumes*  The ground-level external dose to people at a given location resulting from their immersion in air contaminated by a radionuclide emitting gamma rays or X-rays is generally computed by multiplying the air concentration by a dose-rate conversion factor (Sv yr$^{-1}$ per Bq m$^{-3}$). External dose-rate conversion factors for various radionuclides have been compiled (e.g., Eckerman and Ryman 1993). Such compilations are based on the assumption that the contamination consists of a semi-infinite plume of photon emitters. At times, especially at locations near the release point for elevated emitters, the assumption of a semi-infinite plume at ground level is not met. Instead, you must calculate the photon dose from a finite plume passing over the ground. Failure to account for this overhead plume dose can lead to an underestimate of the near-in dose for an elevated release (Lahti et al. 1981). Healy and Baker (1968) present a basic methodology for calculating a finite external dose with the Gaussian plume model that is still widely used (e.g., Miller et al. 1986).

Overcamp and Fjeld (1983, 1987) provided a simple approximation for estimating centerline gamma absorbed dose rates from a continuous Gaussian plume. They found an exact solution for the Gaussian cloud approximation for long-term averaged estimates of gamma absorbed dose due to a ground-level release of radioactive gases and particles. Their solution is more realistic than the uniform cloud approximation of the earlier work by Healy and Baker (1968). More recently, Overcamp (2007) found an exact analytic solution for the dose rate for the Gaussian cloud approximation for an elevated source for one of the two complex integrals that result during the formulation of the problem. This exact solution will enable more rapid computation of the complex mathematical expressions that result during the computation of the gamma dose rate for a Gaussian cloud.

### Summary of Gaussian Plume Model Limitations

Throughout this section, a number of limitations associated with the straight-line Gaussian plume model have been mentioned. Theoretically, the model is only valid when certain basic assumptions are completely met (Brenk et al. 1983). In the real atmosphere, some of these basic assumptions are never met, such as the assumption of constant eddy diffusivity in time and space implicit in the derivation of the equations. Keep in mind that the Pasquill-Gifford curves were really only determined for distances out to about 1 km, yet they are commonly extrapolated to

100 km. In practical terms, the Gaussian plume model should not be applied under conditions of (a) low wind speeds, (b) complex terrain, (c) spatial and temporal changes in wind velocity, or (d) deposition and transformation (e.g., radiological decay) within the plume during travel. As described above, however, empirically based adjustments to the basic Gaussian plume formulation allow it to be applied in many situations in which it theoretically should not apply. Nevertheless, any system based extensively on empirical adjustments has a severely limited capacity to produce generalizations that may be applied to conditions not specifically measured.

## *Puff-Transport and Diffusion Models*

You have to answer the question of whether a particular release is a puff or a plume before deciding which model to use. For instance, the statistics for puff diffusion are different than are those for plume diffusion. A useful rule of thumb is that if the release duration of the effluent is short compared to the travel time of interest, then it is a puff (Pasquill and Smith 1983).

### Puff Transport

Determining the path taken by a puff (its trajectory) is based on the wind field over the domain that the model is designed to simulate. This wind field can be based on observations or the output of a wind-field model (e.g., Sullivan et al. 1993), and it is generally time dependent. The simplest model for interpolating wind observations to a grid is to assume a $1/r^2$ ($r$ is radial distance) weighting from points where wind data are available to grid points used in the trajectory model. Mass-consistent wind field analysis techniques have been useful tools for analyzing sparse wind data in situations of complex terrain and temperature inversions restricting vertical diffusion (Dickerson 1973). Puff trajectory models are most applicable to downwind distances beyond 10 km, and such models have been applied even on a synoptic (500–5,000 km) scale (e.g., Heffter et al. 1990; Crawford 1967, 1968, 1969, 1970). An example of trajectory calculations in the midtroposphere for an effluent cloud is given in figure 3.19. The actual path of this same cloud, as determined from instrumented aircraft tracking of the cloud of debris, is given in figure 3.20.

A combination of trajectory calculations and puff diffusion has been used for many long-range events of interest. The worldwide tracking of debris from the Chernobyl reactor accident in 1986 is a classic example.

If observed winds are used in the model, the resolution of the input data is determined by the spacing of the wind observation points. For example, the National Weather Service radiosonde stations have spacings on the order of 300 km, but a mesoscale grid (for research or special operational reasons) may have meteorological towers with spacings on the order of 20 km (Hanna et al. 1982; Crawford 1990). One of the chief disadvantages of the puff trajectory model is that it is very sensitive to errors in selecting values for the wind field, which drives the entire trajectory analysis (NCRP 1984).

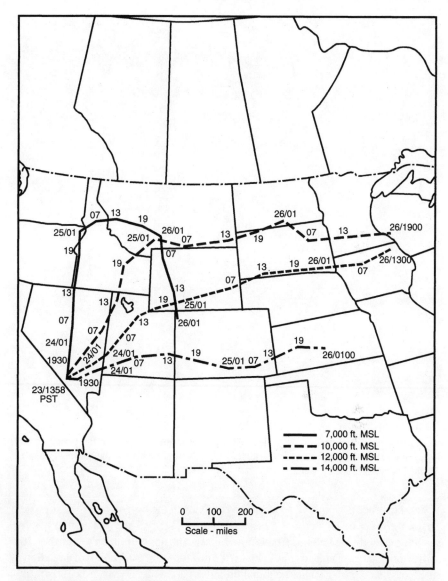

Figure 3.19 Long-range trajectories calculated from observed winds for the effluent cloud from the ground-level release of material associated with the test of a nuclear rocket engine. MSL, mean sea level. From Crawford (1968).

## Puff Diffusion

Although the Pasquill-Gifford curves are often used for microscale to mesoscale distances to estimate the puff $\sigma$ values, other schemes can be used, such as the similarity theories discussed under "Atmospheric Stability Categories," above. At very long distances, when the pollutant cloud reaches the size of the largest horizontal eddy size characteristic of the atmosphere (of the order of 1,000 km),

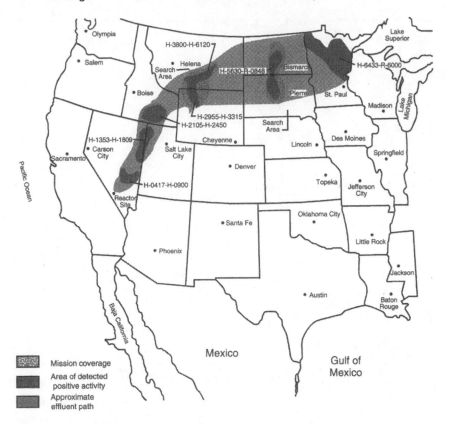

Figure 3.20 The effluent trajectory and diffusion pattern measured from tracking aircraft between the altitudes of 2,879 m and 3,182 m above mean sea level for the effluent cloud, from the ground-level release of material associated with the test of a nuclear rocket engine. From Crawford (1968).

a Fickian process would take over. The K values in the horizontal direction would be on the order of $10^{10}$ cm$^2$ s$^{-1}$ (see "Similarity Theory," above, and figure 3.14). Cylindrically shaped clouds in the horizontal plane approximate the transport and diffusion of large clouds to continental scale distances. The $\sigma_r$ (standard deviation of the Gaussian distribution in the radial direction) is calculated as a function of time. In the vertical direction, the diffusion process can be parameterized by a variety of methods depending on the particular problem. For mid-troposphere clouds of debris, Crawford (1966) input the vertical diffusivity, $K_z$, as a function of time and height along the trajectory. This input was based on analyses of wind and temperature structure along the trajectory.

### Sequential Puff-Trajectory Model

Wind velocity is seldom constant in time or space in the real atmosphere. One of the basic practical limitations of the straight-line Gaussian plume model is that it cannot simulate such varying conditions. The sequential Gaussian puff trajectory model is popular because it can reflect variations in wind velocity. In such a model, a continuous plume is approximated by a series of puffs released in succession. This is illustrated in figure 3.21.

The two fundamental assumptions associated with any sequential puff model (Ramsdell et al. 1994) are that (1) a plume may be represented by a sequence of puffs and (2) puff movement can be considered separately from puff diffusion. The air concentration at any given point in time and space over the assessment domain $(x, y, z, t)$ is assumed to be equal to the sum of the concentrations from all of the puffs at that location (Ramsdell et al. 1994). This can be represented by the following equation:

$$\chi(x, y, z, t) = \sum_{i=1}^{N} \chi_i(x, y, z, t), \tag{3.33}$$

where

$i$ = puff number
$N$ = total number of puffs

The concentration of material within each individual puff is assumed to be Gaussian in shape. Furthermore, each puff is assumed to be symmetrical in the horizontal plane, so $\sigma_r$ is used to specify the distribution in both the downwind ($x$) and crosswind ($y$) directions. The concentration distributions are expressed in terms of the

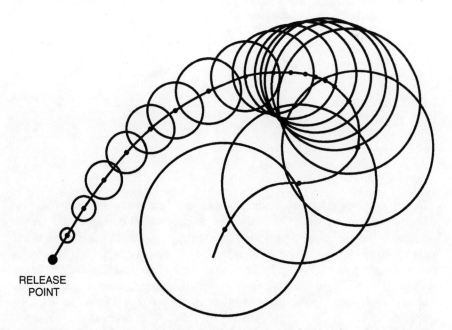

Figure 3.21 Simulation of a plume composed of a series of puffs released at equal time intervals and $\sigma \propto t$ (Hanna et al. 1982).

radial distance, $r$, between the center of the puff and the location for which the calculation is being made. This is true also for the puff discussion above under "Puff Diffusion." Assuming total reflection of the puffs from the ground and the top of the ABL ($h_L$), the concentration at location $x$, $y$, $z$ at time $t$ due to puff $i$ is given by Ramsdell et al. (1994) as

$$\chi_i(r,z,t) = \frac{Q(t)}{(2\pi)^{3/2}\sigma_r^2\sigma_z} \left\{\exp\left[-\frac{1}{2}\left(\frac{r^2}{\sigma_r^2}\right)\right]\right\} G(z), \tag{3.34}$$

where

$Q(t)$ = material in the puff at time $t$
$r^2 = (x - x_0)^2 + (y - y_0)^2$, where $x$ and $y$ are the horizontal location of the receptor of interest and $x_0$ and $y_0$ are the horizontal location of the puff center

In this formulation, the reflection term, $G(z)$, is given by

$$G(z) = 2 \sum_{m=-\infty}^{\infty} \exp\left\{-\frac{1}{2}\left[\frac{(2mh_L - h_p)^2}{\sigma_z^2}\right]\right\}, \tag{3.35}$$

where

$m$ = number of reflections from the ground and the top of the ABL
$h_L$ = height of the top of the ABL
$h_p$ = effective height of the plume

When the material may be assumed to be uniformly distributed within the ABL and $h_p \leq h_L$, then

$$G(z) = \frac{(2\pi)^{1/2}\sigma_z}{2h_L}. \tag{3.36}$$

When $h_p > h_L$, then

$$G(z) = \frac{(2\pi)^{1/2}\sigma_z}{2h_p}. \tag{3.37}$$

A discussion of the input requirements for the Gaussian plume model is presented above under "Special Considerations." Much of the discussion applies for the Gaussian puff model, as well. One set of critical selections, of course, is the values of $\sigma_r$ and $\sigma_y$ to be used in the model. An important conceptual difference between puff and plume models is the averaging time for the simulations. Plume models generally calculate time-averaged concentrations with an averaging time on the order of tens of minutes or longer. In contrast, puff models are designed to estimate instantaneous spreading about the center of the puff (NCRP 1984). This difference is one reason that puff trajectory models have found wide use in emergency response applications (e.g., NRC 1992).

## Multibox Models

Multibox models are used in regional pollution studies when there is a defined bottom of the box (the ground), a defined top of the box (the top of ABL), and most commonly when chemical processes have to be modeled at the same time (Knox and Walton 1984). The equation governing the time evolution of air pollutant concentration is the mass conservation equation and is written as

$$\frac{\delta \chi}{\delta t} + u\frac{\delta \chi}{\delta x} + v\frac{\delta \chi}{\delta y} + w\frac{\delta \chi}{\delta z} = \frac{\delta}{\delta x}\left(K_x \frac{\delta \chi}{\delta x}\right) + \frac{\delta}{\delta y}\left(K_y \frac{\delta \chi}{\delta y}\right)$$
$$+ \frac{\delta}{\delta z}\left(K_z \frac{\delta \chi}{\delta z}\right) + \frac{S_a}{V} + \frac{P_a}{V}, \quad (3.38)$$

where

$\chi$ = pollutant concentration as a function of $(x, y, z, t)$ for each pollutant, $a$, (subscript "a" left off for clarity) of interest

$u$, $v$, and $w$ = wind velocity components in the x-, y-, and z-direction, respectively

$K_x$, $K_y$, and $K_z$ = eddy diffusion coefficients in the x-, y-, and z-direction, respectively

$S_a$ = source field for pollutant, $a$, as a function of $(x, y, z, t)$

$P_a$ = expression allowing for chemical change of the pollutant, $a$, through possible interaction with other pollutants and the solar radiation field (in the case of photochemical smog)

$V$ = volume of the elemental "box"

This equation accounts for the advective fluxes and diffusion into and out of an elemental volume, but it is nonlinear and it cannot be solved analytically. For a multibox model of regional pollution, it can be simplified to eliminate the vertical flux and diffusion terms under the assumption of uniform mixing between the earth's surface and the top of the ABL. That still leaves the problem of specifying the input data for winds and, particularly, the turbulent eddy diffusivities in the horizontal plane.

## Calculation Grid

Because input has to be supplied on an x-y grid fixed to the earth's surface over the domain of interest, the multibox model is also sometimes referred to as a grid model. Source terms (topography, height of ABL, $u$, and $v$) have to be supplied for each grid point. This is also true for $K_x$ and $K_y$. Because these K values have to represent diffusion on the subgrid scale, it can be argued that they should be related to the grid size used in the model. For this reason, the Richardson (1926) relationship ($K \cong 0.2l^{4/3}$) discussed under "Eddy Diffusion Coefficients" above has often been used with $l$ set equal to the grid spacing.

## Calculation Methods and Limitations

With the advent of computers and numerical methods of solving differential equations, solutions of multibox models are possible. The advantage of this approach to pollution problems over a region is that the equations can be formulated so that you can also simultaneously solve the chemical rate equations for chemical reactions taking place within the "box." A disadvantage is that the solutions can be numerically unstable depending on the grid size chosen and the wind speeds. Decreasing grid size minimizes the instability problem but increases the computational demands. Another problem is that the solutions themselves tend to introduce numerical diffusion into the solution, an artifact of the numerical method chosen to solve the equations. Advances continue to be made in solution methods (Pepper et al. 1982; Pepper and Baker 1988; Pepper 1995; Pepper and Marino 1996). These types of models are not widely applicable to radiological assessments, so they are not discussed further here.

### Particle-in-Cell Models

Particle-in-cell (PIC) models assume that a plume can be represented by a large number of particles being continuously released. The concentration at any given location and time is simply the sum of the particles present in each box, defined by the three-dimensional grid system in which the calculations are being made, divided by the volume of the box. To overcome the numerical stability and numerical diffusion problems associated with multibox models, PIC models calculate a pseudovelocity field. This field is composed of the actual wind field, either observed or modeled, plus a diffusion velocity (Lange 1978). Using K-theory to define the diffusion velocity, $u_D$, and assuming that the wind field is nondivergent, the air concentration field in a PIC model can be defined as

$$\frac{\delta \chi}{\delta t} + u(\nabla \chi) = \nabla(K \nabla \chi), \tag{3.39}$$

where

$u$ = three-dimensional wind field
$K$ = eddy diffusivities in the x-, y-, and z-directions
$\nabla$ = derivative with respect to $x$, $y$, and $z$

This equation is in vector format, for simplicity, and is the same as the advection-diffusion equation 3.38 minus the source and chemical change terms. Under the assumption of incompressibility, the following equation is true:

$$u(\nabla \chi) = \nabla(\chi u). \tag{3.40}$$

Substituting equation 3.40 into equation 3.39, it can be shown that

$$\frac{\delta \chi}{\delta t} + \nabla(u_p \chi) = 0, \tag{3.41}$$

where

$u_p = u - u_D$ = pseudo-transport velocity
$u_D = K(\nabla \chi / \chi)$ = diffusion velocity

An approach to obtaining the necessary $K$ values (assume $K_x = K_y$) is outlined under "Similarity Theory," above.

The PIC model is initiated by releasing a large number of particles as a function of time from a point source or specifying a large number in a region to indicate an initial large source term (e.g., from an explosion). These particles are then transported to new locations over time. Advantages of this model are that (a) it is numerically stable, (b) there is no artificial numerical diffusion, (c) it can handle radioactive decay in transport, (d) it can handle deposition processes, and (e) it can incorporate terrain effects directly through adjustments in the three-dimensional grid system. The disadvantages are that (a) it suffers some of the resolution problems common to other models, (b) it has the same basic limitations of any model involving the use of eddy diffusivities ($K$ values), (c) it requires significant amounts of input data, and (d) it requires significant computer resources to keep track of all of the particles as a function of time and position. The NRC allows reactor licensees to use this model for analysis of routine radionuclide releases only after detailed justification (NRC 1977). In the event of an actual or potential release of radioactive material, a PIC model will likely be used to evaluate the mesoscale impact of the release (Sullivan et al. 1993). PIC models may also find use in dispersion calculations for special conditions, such as lake or sea breeze effects (Uliasz and Pielke 1991).

## Screening Models

Most of the discussion about models presented above assumes that the user is attempting to calculate as realistic an answer as possible. Another type of model that is finding increased application today, however, is the screening model. One application of screening is in the prioritization of contaminated sites for possible cleanup activities (e.g., Hoffman et al. 1993). Two different screening calculations can be performed in the prioritization process:

1. Low-priority contaminants can be identified by using models and assumptions deliberately designed to give conservative estimates of concentration or dose (i.e., models designed so it is unlikely they will underpredict actual values).
2. High-priority contaminants can be identified by using models and assumptions deliberately designed to be nonconservative so that if calculated values for a particular contaminant exceed a predetermined threshold value, the contaminant should be examined more carefully.

In principle, any of the mathematical formulations presented above can become an atmospheric screening model if the proper input parameters and modeling assumptions are selected. The National Council on Radiation Protection and Measurements (NCRP) has published a set of conservative screening models designed for assessing the impact of atmospheric emissions of radionuclides (NCRP 1996).

The NCRP screening approach is designed to produce predictions that are unlikely to underestimate actual values by more than a factor of 10. A subsequent uncertainty review indicated that this is likely to be achieved for most, but not necessarily all, atmospheric release conditions (NCRP 1993).

For annual average ground-level concentrations in the atmosphere from continuous releases, the NCRP has recommended several different screening levels (NCRP 1996). The simplest one is to assume that the atmospheric concentrations are the same as those in the stack. This is represented by

$$\chi = \frac{fQ'}{V}, \tag{3.42}$$

where

$f$ = fraction of time the wind blows toward the receptor of interest
$Q'$ = amount released per unit time
$V$ = volumetric flow rate at point of release

Additional screening models are based on the Gaussian plume equation presented above. These assume that the receptor is located at the plume centerline. The screening concentration is given by

$$\chi = \frac{fQ}{\pi u \sigma_y \sigma_z} \exp\left[-\frac{1}{2}\left(\frac{h_p}{\sigma_z}\right)^2\right], \tag{3.43}$$

if the release is at ground level, $h_p = 0$, and the exponential term above goes to 1. For purposes of screening annual average concentrations, neutral conditions are assumed and Pasquill-Gifford category D is used. If the release is from a stack high enough to be out of the influence of building wake effects (i.e., $h_p > 2.5H_b$ [building height]), then equation 3.43 can be used directly with $h_p$ specified. Various modifications of the above are given for the case of $h_p \leq 2.5H_b$ (NCRP 1996).

## Atmospheric Removal Processes

One of the fundamental questions that atmospheric dispersion models often address is how and when material emitted into the atmosphere is removed from it. For large particles (i.e., $>10$ μm), gravitational effects are important. These particles are often referred to as fallout. In the absence of precipitation processes, gaseous and small particles ($<10$ μm) may also be removed by coming in contact with the ground surface or with objects on the surface (e.g., leaves, grass, and structures). For these small particles, you can visualize a turbulent impaction process causing them to deposit on the surface. Gaseous materials diffuse into the surface. These processes are known as dry deposition. During precipitation, contaminant material may be incorporated into the precipitation-forming process within a cloud (rainout or snowout) and subsequently removed, or the contaminant may be washed out of the atmosphere by the falling precipitation. These two processes are known collectively as wet deposition. Radionuclides removed from the atmosphere by these processes become available to expose people via other exposure routes.

## Fallout

The influence of fallout on the behavior of a Gaussian plume model is discussed above under "Special Considerations." It should be noted, however, that a number of fallout models were developed to account for the distribution of debris following large explosions near the earth's surface. Usually, these models started with a cloud of debris stabilized at some height above the ground shortly after the explosion. The debris was assumed to have a range of particle sizes falling from each of several heights appropriate to the height and size of the debris cloud. These falling particle size classes were transported by the winds (as a function of height and time) and the falling speed of the particle size class until they reached the ground. Comparisons of the resulting patterns on the ground with measured patterns (e.g., from atmospheric nuclear weapons tests) yielded an appropriate activity distribution to associate with the debris cloud and its particle size distribution. Subsequent tests were then used to validate the fallout models (Knox et al. 1970).

## Dry Deposition

In the absence of precipitation, small particles and reactive gases may deposit on various surfaces via impingement, electrostatic interactions, chemical reactions, and other processes. These dry deposition processes are often parameterized in terms of a dry deposition velocity, $v_d$. This parameter is defined as the ratio between the rate at which material is deposited on the surface of the earth and the air concentration measured at some reference height, usually 1 m, above the depositing surface. It must be stressed that $v_d$ is merely a proportionality constant and not a true velocity, although it has the units of velocity.

Values of $v_d$ are expected to be a function of numerous meteorological variables and characteristics of both the depositing material and the deposition surface (Sehmel 1984). A review of field measurements of $v_d$ (Sehmel 1984), however, shows such a variation in $v_d$ values (often by more than two orders of magnitude) and such sparse data that a rigorous analysis of the relationship between $v_d$ and various controlling parameters is not possible (Hanna et al. 1982). Sehmel (1984) includes graphs of predicted values of $v_d$ for particles; these graphs are based primarily on theoretical considerations and wind tunnel studies rather than on actual field data. Murphy and Sigmon (1990) used extensive electrical resistance analogues to model the uptake $SO_2$, $NO$, $NO_2$, and $HNO_3$ by forest ecosystems. Field measurements indicated reasonable agreement with predictions. Ramsdell et al. (1994) uses a resistance-analogy model to calculate $v_d$ for both particles and gases in their puff trajectory computer code, but one of the three terms in their model is an empirical factor designed to place a lower bound on the calculation.

When material is deposited on the surface of the earth, it is, of course, simultaneously removed from the atmosphere. One traditional way of accounting for plume depletion in Gaussian-based models is to assume that as material is deposited on the ground, it is replenished from above at such a rate that the Gaussian profile in the vertical is maintained (i.e., that depletion occurs uniformly throughout the plume). This is the so-called source depletion model (Van der Hoven 1968). Other models have been proposed in which the vertical concentration profile is not necessarily maintained. Such surface depletion models appear more realistic to many modelers, but they are generally more difficult to use than the source depletion models. The greatest differences between concentrations calculated by these two models are expected to occur under stable atmospheric conditions, but such differences may not be a serious problem under realistic meteorological and deposition conditions (Jensen 1981). Regulatory Guide 1.111 (NRC 1977) includes plume depletion curves based on a surface depletion model, but these curves appear to contain calculation errors (Miller and Hoffman 1979).

The following values of $v_d$ have been suggested for radiological assessment purposes (Brenk et al. 1983):

- 1 cm s$^{-1}$ for reactive gases such as elemental molecular iodine
- 0.1 cm s$^{-1}$ for aerosols (1 μm in diameter)
- 0.01 cm s$^{-1}$ for relatively unreactive gases such as organic iodine

A $v_d$ of zero may be appropriate for nonreactive gases such as argon and krypton (Miller et al. 1986). Table 3.10 presents a summary of measured values of $v_d$ for both particles and reactive gases.

As much site-specific information as possible should be incorporated into selecting a measured or modeled value of $v_d$ for an assessment. In the absence of sufficient information, however, the values listed in table 3.10 may be adequate for most assessment purposes. Nevertheless, no matter how $v_d$ is determined, the uncertainties associated with it are large. Sehmel (1984) provides an extensive review of the dry deposition literature, including many measurements of $v_d$.

## Wet Deposition

Dry deposition processes are continuous, and wet deposition processes are more episodic. Wet deposition processes, however, can contribute significantly to the impact of radionuclide releases upon the environment, as measurements around the world following the Chernobyl event in 1986 clearly demonstrated (Hoffman 1991).

One way of modeling the wet deposition process is to use a scavenging rate, $\Lambda$ ($t^{-1}$) (Slinn 1984). This factor may be thought of as the fraction of the airborne

Table 3.10  Summary of measured values for dry and wet deposition parameters (Hoffman et al. 1984)

| Parameter | Number of observations | Geometric mean | Geometric standard deviation | Range |
|---|---|---|---|---|
| $v_d$ for particles | 98 | 0.33 | 7.7 | 0.024 to 20 |
| $v_d$ for reactive gases | 99 | 0.73 | 3.2 | 0.069 to 7.8 |
| $W_r$ | 52 | $3.3 \times 10^5$ | 3.8 | $4.3 \times 10^4$ to $2.5 \times 10^6$ |

material that is removed per unit time. The concentration of pollutant in the atmosphere decreases exponentially with time since the precipitation started. This is expressed as

$$\chi = \chi_0 e^{-\Lambda t}, \tag{3.44}$$

where

$\chi_0$ = air concentration at the time the precipitation started
$t$ = length of time that the precipitation has been occurring

Values of $\Lambda$ should theoretically be a function of the size of the precipitation droplets, the precipitation rate, and the physical and chemical characteristics of the particle or gas through which the precipitation is falling (Hanna et al. 1982). In this case, impaction is important for particulate scavenging by falling raindrops, and the use of Henry's Law is important for uptake of gases by falling raindrops. One form of Henry's Law is given by

$$\gamma_w = \alpha \chi_a, \tag{3.45}$$

where

$\gamma_w$ = equilibrium concentration of gas in water using volume measurements (i.e., g cm$^{-3}$ or mol L$^{-1}$)
$\alpha$ = a dimensionless Henry's constant, which is a function of gas on a per-volume basis (available in table form)
$\chi_a$ = equilibrium concentration of gas in the atmosphere using volume measurements (same units as $\gamma_w$)

Slinn (1984) presents details on how to use Henry's Law for calculating gaseous uptake by falling raindrops.

When small particles act as condensation nuclei within the cloud, this process is often referred to as rainout; when rain falls through a particulate debris cloud, the process is referred to as washout. Limited data exist on scavenging by snow. Because of significant variations found in measured values of $\Lambda$, Hanna et al. (1982) suggests that the values of $\Lambda$ for wet removal modeling may be, at best, an order-of-magnitude estimate of the amount of material removed by wet deposition processes. This method is more applicable to episodic releases than long-term averages because the geometry of the particular situation can be directly incorporated into the calculation. However, large uncertainty exists in the amounts deposited on the ground.

Another method of estimating wet deposition is through the washout ratio, $W_r$. For a given reference height, $W_r$ can be defined as (Hanna et al. 1982)

$$W_r = \frac{k_0}{\chi_0}, \tag{3.46}$$

where

$k_0$ = concentration of the pollutant in the rain
$\chi_0$ = concentration of the pollutant in the atmosphere

The flux of material to the earth's surface, $F_w$, is then defined as

$$F_w = \chi_0 W_r J_0, \tag{3.47}$$

where $J_0$ is the rainfall rate in length per unit time (e.g., mm h$^{-1}$). This expression can then be used to define a wet deposition velocity, $v_w$, which is analogous to the dry deposition velocity, $v_d$:

$$v_w = \frac{F_w}{\chi_0} = W_r J_0. \tag{3.48}$$

The parameter $v_w$ can now be used to model wet deposition and plume depletion in the same manner as dry deposition is modeled. Furthermore, it can be shown that $\Lambda$ is approximately equal to $v_w$ divided by the depth of the plume layer through which the precipitation falls.

Measured values of $W_r$ are also summarized in table 3.10. By necessity, these measurements include both rainout and washout processes. No matter whether these data are used to determine a value of $\Lambda$ or a value of $v_w$ for assessment modeling, the result can be expected to have an uncertainty of two orders of magnitude or more. The washout ratio method is most applicable to long-term averages.

Some models actually combine $v_d$ and $v_w$ into a total deposition velocity, $v_T$ (e.g., NCRP 1989). This is generally more convenient for assessment purposes than treating the two separately. For example, the use of $v_T = 1,000$ m day$^{-1}$ appears to be consistent with values reported for a number of locations using data for $^{131}$I and $^{137}$Cs deposition from the Chernobyl event (Kohler et al. 1991). It should be noted that these aerosols were well mixed in the atmosphere by the time measurements were made. This is true of washout ratios reported in the literature (Briggs et al. 1968). Materials deposited near the point of release will likely be less well mixed than those that are farther downwind. This further suggests that $v_w$ and $v_T$ are appropriate parameters for long-term mesoscale depletion and deposition calculations, whereas $\Lambda$ may be more appropriate for calculations of microscale or episodic releases. An extensive review of wet deposition processes and data is given in Slinn (1984).

## *Model Validation*

After a model is developed, two main steps need to be taken to assess its accuracy. The first step is verification against an analytical solution. In the case of an extremely complex model where an analytical solution is not possible, the term "verification" sometimes means that the model is tested against its published documentation to determine whether the model reproduces a test case or behaves as expected. This is particularly important for a complex numerical model. The second step is validation, which is a comparison against field data (Hoffman and Miller 1983). It is best if the validations are made over the range of conditions for which the model is intended to be used. Truly "valid" models, however, rarely exist (IAEA 1989). It is next to impossible to test a model under all conditions in which it might be used. Therefore, you must exercise scientific judgment when determining whether a model has been sufficiently validated for a given assessment.

The atmospheric dispersion model that has been subjected to the most extensive validation efforts is the straight-line Gaussian plume model. Typically, the Pasquill-Gifford stability categories, whose standard deviations were determined from field tests out to about 1 km, are used. Miller and Hively (1987) reviewed several reported validation studies for this Gaussian model. Their results are summarized in table 3.11. The accuracy of this model depends on the conditions under which it is applied. In general, the longer the averaging time, the flatter the terrain, the less complex the terrain and meteorology, and the more complete the meteorological instrumentation at the site, the better the model will perform. The usefulness of the model, however, will depend on what level of uncertainty is acceptable for the assessment problem, not on the absolute accuracy of the model.

Table 3.11 indicates a rather large range of uncertainty when calculating air concentration for a specific location with a short averaging time (hourly rather than annual averaging) and for episodic releases. Very useful decisions can be made, however, because a large part of the error when calculating for a specific point is associated with the error in the trajectory calculations. The typical wind direction instrument is only accurate to 5 degrees, and trajectory predictions themselves have some error. Thus, in taking protective actions during an episodic release, you must allow for some deviation in the path and then use the diffusion predictions without as large an error as indicated in table 3.11. When wind direction errors were removed, 80–88% of the predicted concentrations were within a factor of 10 of the measured concentrations during 37 experiments involving 15-min releases of $SF_6$ conducted at the Savannah River Site in South Carolina between 1981 and 1985 (Fast et al. 1991). Samples were collected along roads at distances of 25–40 km from the release

Table 3.11 Summary of the estimated uncertainty associated with predictions made by the straight-line Gaussian plume model (Miller and Hively 1987)

| Conditions | Range of the Ratio (predicted concentrations/observed concentrations) |
|---|---|
| Highly instrumented site, ground-level, centerline concentrations; within 10 km of a continuous point source | 0.65–1.35 |
| Specific hour and receptor point, flat terrain, steady meteorological conditions; within 10 km of the release point | 0.1–10 |
| Annual average for a specific point, flat terrain; within 10 km of the release point | 0.5–2 |
| Annual average for a specific point, flat terrain; 10–150 km downwind of the release point | 0.25–4 |
| Complex terrain or meteorological conditions (e.g., sea breeze regimes); episodic releases | 0.01–100 |
| Episodic surface-level releases; wind speeds less than 2 m s$^{-1}$ | 1–100 |

point, and the models were Gaussian sequential puff or puff-in-time and/or time- and space-varying wind fields.

A similar survey of validation studies for non-Gaussian models has not been developed; however, numerous validation studies have been reported for individual models (e.g., Sullivan et al. 1993; Weber et al. 1982). It is not yet clear that enough validation studies have been performed to specify when more complex models should replace the Gaussian plume model in radiological assessment calculations (Miller and Hively 1987). One major problem associated with nearly all non-straight-line models is that they require a spatially and temporally varying wind field as input. Lack of enough information to specify the wind field accurately appears to be a limitation of many models in field tests (e.g., Weber et al. 1982). On the other hand, there are clearly situations or sites where it is obvious that a straight-line model is not applicable. You must use scientific judgment to identify those sites and to select a more appropriate model. If possible, you should perform some validation tests with that model at that site before relying on it completely for the radiological assessment.

You may be required to use atmospheric dispersion models in real-time forecasting of the impact of unintentional radionuclide releases. In 1981, the NRC sponsored a study conducted at the Idaho National Engineering and Environmental Laboratory in which a nonradioactive tracer ($SF_6$) was released and the resulting downwind air concentrations were measured (Lewellen et al. 1985). It should be noted that the Idaho site is relatively flat, and several meteorological towers were used to monitor atmospheric conditions across the site. Figure 3.22 shows the observed and calculated 1-h ground-level air concentration for one of these tests. Clearly, none of the models very accurately reproduced the measured data. Again, treat all model results with caution.

## Model Uncertainty

As stated above, all models, no matter how complex, are only approximations of reality. As a result, some level of uncertainty is always associated with any model prediction. Hoffman and Miller (1983) placed the sources of model uncertainty into four major categories:

1. Incorrect model formulation
2. Incorrect parameter estimation
3. Errors in programming, computation, calculation, and report writing
4. Failure to account for parameter variability

Category 1 uncertainty can never be completely eliminated because no model will ever duplicate reality exactly. Category 2 uncertainty can be reduced by using as much site-specific information as possible in selecting the various parameter values to be used in the model. A vigorous quality control program can address category 3 uncertainty; however, even with such a program in place, any complex computer code is likely to contain some computational or coding errors. Finally, addressing category 4 by allowing for uncertainty is the state-of-the-art in radiological assessment modeling.

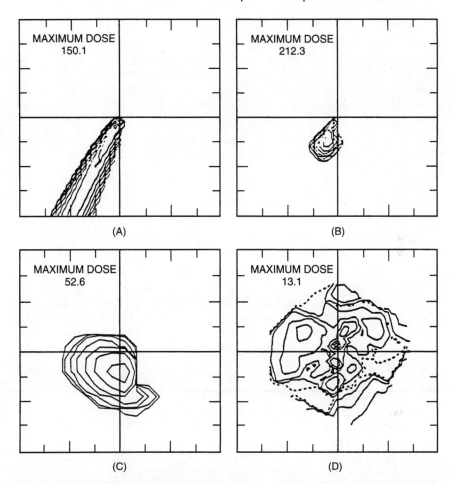

Figure 3.22 One-hour surface doses predicted by (A) a straight-line Gaussian plume model, (B) a Gaussian puff trajectory model, (C) a particle-in-cell model, and (D) 1-h surface doses actually observed (NRC 1987).

*Stochastic Modeling* Stochastic modeling explicitly addresses the issue of accounting for parameter variability in model output. In such a modeling procedure, all of the uncertain parameters are treated as random variables that have a probability distribution associated with them. The distribution associated with each parameter is sampled, and a distribution of values of model output is produced. Further details on this procedure may be found in IAEA (1989).

One of the most difficult tasks associated with stochastic procedures is selecting the distributions for each of the uncertain input parameters. For example, table 3.10 includes statistical parameters that might be used to specify a model input distribution for $v_d$. The statistical parameters presented in table 3.10, however, were developed from a noncritical review of measured values of $v_d$ reported in the literature. It is unlikely that all of the individual measurements used to derive the reported statistical parameters are appropriate for a given site and deposition problem. You should develop any input parameter distribution by using the best available data for the particular site and assessment question being addressed by the model. One of the major challenges associated with stochastic modeling is specifying the input parameter distributions. A great deal of scientific judgment is always required in this process. As a result, undue confidence should not be placed in these distributions or in the resulting output distribution (Hoffman and Miller 1983).

At this time, stochastic models are not routinely used in regulatory assessments. They are, however, finding application in other risk assessment activities (e.g., Farris et al. 1994).

## Guidelines for Selecting Models

It is clear that there are a number of models available for use in assessing the impact of radionuclides released to the atmosphere. Even nuclear licensees have some latitude in selecting models for demonstrating regulatory compliance (e.g., NRC 1977). While it is often tempting to use the same, familiar model or computer code for all assessment questions, this approach could result in a serious misapplication of an otherwise useful model. The following guidelines are designed to help you select the atmospheric transport, diffusion, and deposition models that are most appropriate for a given assessment question:

1. Carefully and completely define the assessment question or scenario that is to be addressed. This should always be the first step in the model selection process. Failure to define the assessment problem carefully enough could result in a model predicting the correct results for the wrong problem (IAEA 1989).
2. Evaluate the data that are available for modeling the particular site and facility to be considered in the assessment. No model produces results that are any better than the input data used to initiate the model. Carefully evaluate the quality and representativeness of any data that are to be used in assessment modeling.
3. Evaluate the strengths and weaknesses of various models using the available input data. It is likely that many models will be available that could be used to address the assessment problem. Not all of the models, however, are likely to be compatible with the available input data, nor do they necessarily contain all of the features important to the particular assessment. For example, a complex numerical model requiring detailed specification of the wind field as input may not be useful for an assessment problem when the only wind data are from one height above ground at an offsite location.
4. Select the simplest model that will appropriately address the specific assessment question using the available input data. There is no need to use a complex model, for example, if the results of a screening model will answer the problem being addressed. This is the basic tenant behind the U.S. Environmental Protection Agency (EPA) regulatory assessment procedures for users of small amounts of radioactivity (EPA 1989a). Also, if the results

of the assessment calculation are to be presented to members of the public, a simple model may be more understandable and believable.
5. Provide a defensible estimate of the uncertainty associated with calculations made by the model selected. You should always be able to at least bound the uncertainty associated with any model calculation. This should be done even when using standard regulatory models to ensure that the model is truly applicable to the assessment question being asked.

## *Regulatory Models*

There are a number of regulatory models used to do assessments. Different models are used for different purposes.

## AERMOD Model

AERMOD is a steady-state model developed by the EPA that includes the many recent advances in boundary layer physics. It calculates the concentration statistics at various points in the region of interest from long-term releases using data from a single meteorological station. The AERMOD system consists of two preprocessors and a dispersion module. The meteorological preprocessor is called AERMET and provides AERMOD with the physical parameters to characterize the vertical structure of the planetary boundary layer. A second preprocessor, AERMAP, provides the characterization of the terrain and generates receptor grids for AERMOD.

AERMOD uses a Gaussian treatment for horizontal and vertical dispersion in the stable boundary layer. Also, for stable conditions, AERMOD uses the Briggs (1984) plume rise equations with wind and temperature gradient at stack top and halfway to final plume rise. During unstable conditions the model uses a non-Gaussian probability density function in the vertical for dispersion calculations. The treatment of plume rise during unstable conditions has random convective velocities superimposed on the displacements.

The basic hourly meteorological observations are used to help determine friction velocity, Monin-Obukhov length, convective velocity scale, mechanical and convective mixing heights, and sensible heat flux. The surface characteristics are flexible with provision for different fetches, time of year (month), roughness length, albedo, and Bowen ratio. AERMOD can accommodate meteorological data from National Weather Service stations as well as more sophisticated tall towers with more than just a single measurement level and measurements of turbulence. Wet and dry deposition can also be included.

Air pollution sources in both urban and rural terrain can be modeled, and a large number of individual sources can be input into the model. Variable urban treatments are available through input of a city's population. Diurnal changes in the boundary layer structure are also included in AERMOD's formulation.

AERMOD also has the capability of specifying multiple receptor networks in a single run and may mix Cartesian and polar receptor networks in the same run. This would be useful in cases where the receptor domain needs to be sampled more densely in one area than in another.

For more information about the AERMOD model, see www.epa.gov/scram001/dispersion_prefrec.htm.

## CALPUFF Model

The CALPUFF modeling system includes a meteorological preprocessor called CALMET that develops hourly wind and temperature fields on a three-dimensional, gridded modeling domain. CALMET also generates two-dimensional fields of mixing height, surface characteristics, and dispersion properties. CALPOST is a postprocessing module used to produce tabulations that summarize the results of the simulations by identifying the highest and second highest 3-h average concentrations at each receptor. The dispersion module CALPUFF can also perform visibility-related modeling that involves the computation of extinction coefficients and related measures of visibility for different averaging times and locations.

As its name implies, CALPUFF is a puff model, in particular, a Lagrangian puff model that keeps track of puffs as they are carried along by the wind. As the wind direction changes from hour to hour, the puffs take on new paths following the wind changes. The basic dispersion is based on the Gaussian distribution, and the concentration is superimposed from the many puffs (up to 99 puffs per hour) as they move past a given downwind receptor.

The model can be run in a "slug" mode that is computationally more efficient since a number of successive overlapping circular puffs form the stretched-out slug. In the slug mode, the hourly averaged pollutant mass is spread evenly throughout the slug. Eventually, as downwind distance increases, a slug's lateral growth will cause the slug to resemble a puff, and the code then switches back to individual puff mode to improve computational efficiency. Thus, there is no advantage to the slug simulation beyond a certain downwind distance.

The effects of near-field buildings and far-field terrain are both accommodated with CALPUFF. Deposition and transformation processes are also handled in the code.

For more information about the CALPUFF model, see www.src.com/calpuff/calpuff1.htm and www.epa.gov/scram001/dispersion_prefrec.htm.

## CAP88 Model

The CAP88 model is a steady-state EPA model using data from a single meteorological station. It is a modified Gaussian plume model specifically tailored for estimating dispersion of radionuclides. The model can accommodate up to six emitting sources, which may consist of elevated stacks or even area sources. Plume rise calculations are done for either a momentum- or buoyancy-dominated plume. The dose estimates are provided for a circular grid extending out to 80 km around the release point. Since the code is specifically tailored for radionuclide releases, the population's spatial distribution must be specified as part of the input to the model.

A user need only supply the state name or agricultural productivity values; agricultural arrays of milk cattle, beef cattle, and agricultural crop area are then generated automatically. State-specific or user-supplied agricultural productivity values are generated to match the distances used in the population arrays supplied to the code.

Users may override the default agricultural productivity values by entering the data directly on an Agricultural Data tab form. When Alaska, Hawaii, or Washington, DC, are selected, agricultural productivity values are set to zero and must be provided by the user.

One may choose to do either "Radon-only" or "Non-Radon" runs, to conform to the format of the 1989 Clean Air Act's National Emission Standards for Hazardous Air Pollutants (NESHAPs) rulemaking (EPA 1989b). "Radon-only" assessments have only $^{222}$Rn in the source term to automatically include working-level calculations; any other source term ignores working levels.

Deposition velocities for gases are automatically set to zero, whereas the deposition velocity is $3.5 \times 10^{-2}$ m s$^{-1}$ for iodine, and $1.8 \times 10^{-3}$ m s$^{-1}$ for particulates. The scavenging coefficient is calculated as a function of the annual precipitation and enters through the meteorological data.

Organ dose and related weighting factors follow the Federal Guidance Report No. 13 (FGR 13; Eckerman et al. 1999) method, which provides the dose to 23 internal organs and the total effective dose equivalent. Cancer risk for 15 cancer sites are reported as per FGR 13.

Unfortunately, the code cannot work with several sources at different locations of the same facility; all the sources are modeled as if they were at the same point. Another disadvantage is that variations in radionuclide concentrations due to complex terrain cannot be modeled. Also, dose and risk are applicable only from chronic exposures; the code cannot be used for short-term or high-level radionuclide exposure or intake.

For more information about the CAP88 model, see www.epa.gov/radiation/assessment/CAP88/index.html.

## Conclusions

Mathematical models are an important part of any procedure designed to assess the impact of radionuclides released into the atmosphere. Nuclear licensees are provided with a significant amount of proscriptive guidance as to what models to use under what circumstances. At the same time, they have significant freedom within that guidance. Professional health physicists have a responsibility to use the guidance wisely and carefully. Probably the most important thing to remember is that all models are approximations of reality, not reality itself. The results of any model calculation should be evaluated carefully before the results are used in any decision-making process, especially if the decisions to be made are related to human health and safety.

## Problems

1. A meteorological tower is located in a rural area on a site that is best characterized as a grassy, flat surface. Under neutral atmospheric stability conditions, the wind speed at a height of 10 m on this tower is measured to be 5 m s$^{-1}$.

Based on this information, what is the estimated wind speed at a height of 150 m in the atmosphere under these conditions?

2. A small nuclear facility located in a rural area releases the radionuclide $^{85}$Kr to the atmosphere. The annual average release rate for $^{85}$Kr from this site is approximately $5 \times 10^{15}$ Bq s$^{-1}$. The building from which this release occurs is 100 m long, 100 m wide, and 20 m high. The radionuclide is released from a small stack located on the roof of the building. The stack is 2 m above the roof level. The annual average volumetric flow rate from this stack is approximately 5 m$^3$ s$^{-1}$. The operators of the facility have been asked to calculate a simple screening air concentration assuming that the wind blows 25% of the time toward the nearest potential receptor. What is that screening air concentration?

3. The operators of the facility described in problem 2 decide, after reviewing the results of problem 2, that they would like to make a more refined annual average air concentration calculation. The facility has no onsite meteorological instrumentation, but a nearby airport reports that the annual average wind speed for this area is approximately 3 m s$^{-1}$ and that, on average, the stability of the atmosphere is neutral. The facility operators believe that the nearest potential receptor lives 750 m from the facility. Assuming that the wind blows 25% of the time toward this receptor, use an appropriate Gaussian plume model from this chapter to calculate the air concentration at the specified location.

4. The operators of the site described in problem 2 decide to begin discharging their effluent from a stack 60 m tall that is located next to the building described in the problem. Using an appropriate form of the Gaussian plume model, calculate the annual average ground-level centerline air concentration at the location of the nearest potential receptor.

5. The operators of the site described in problems 2–4 realize that a sector average air concentration is probably more appropriate for an annual average air concentration than is the ground-level centerline value calculated in problem 4. Furthermore, they understand that there is some uncertainty associated with any model calculation. Using an appropriate form of the Gaussian plume model, calculate the annual average ground-level sector-averaged air concentration at the location of the nearest potential receptor and provide a quantitative estimate of the uncertainty associated with this value.

References

Bailey, D.T. 2000. *Meteorological Monitoring Guidance for Regulatory Modeling Applications*. EPA 454/R-99-005. U.S. Environmental Protection Agency, Washington, DC.

Barr, S., and W.E. Clements. 1984. "Diffusion Modeling: Principles of Application." In *Atmospheric Science and Power Production*, ed. D. Randerson. DOE/TIC-27601. U.S. Department of Energy, Office of Scientific and Technical Information, Technical Information Center, Oak Ridge, TN. Pages 584–619.

Batchelor, G.K. 1950. "The Application of the Similarity Theory of Turbulence to Atmospheric Diffusion." *Quarterly Journal of the Royal Meteorological Society* 76(328): 133–146.

Battan, L.J. 1974. *Weather*. Foundations of Earth Science Series. Prentice-Hall, Englewood Cliffs, NJ.

Brenk, H.D., J.E. Fairobent, and E.H. Markee, Jr. 1983. "Transport of Radionuclides in the Atmosphere." In *Radiological Assessment, A Textbook on Environmental Dose Analysis*, ed. J.E. Till and H.R. Meyer. NUREG/CR-3332 & ORNL-5968. U.S. Nuclear Regulatory Commission, Washington, DC.

Briggs, G.A. 1974. "Diffusion Estimation for Small Emissions." In *1973 Annual Report, Environmental Research Laboratories. Air Resources, Atmospheric Turbulence and Diffusion Laboratory, Oak Ridge, TN*. U.S. Department of Energy, Office of Scientific and Technical Information, Technical Information Center, Oak Ridge, TN. Pages 83–145.

Briggs, G.A. 1984. "Plume Rise and Buoyancy Effects." In *Atmospheric Science and Power Production*, ed. D. Randerson. DOE/TIC-27601. U.S. Department of Energy, Office of Scientific and Technical Information, Technical Information Center, Oak Ridge, TN. Pages 327–366.

Briggs, G.A., I. Van der Hoven, R.J. Engelmann, and J. Halitsky. 1968. "Processes Other than Natural Turbulence Affecting Effluent Concentrations." In *Meteorology and Atomic Energy—1968*, ed. D.H. Slade. TID-24190. U.S. Atomic Energy Commission, Washington, DC. Pages 189–255.

Cramer, H.E. 1957. "A Practical Method for Estimating the Dispersal of Atmospheric Contaminants." *Proceedings of the First National Conference on Applied Meteorology*. Section C. American Meteorological Society, Hartford, CT. Pages C33–C35.

Crawford, T.V. 1966. *A Computer Program for Calculating the Atmospheric Dispersion of Large Clouds*. UCRL-50179. University of California Lawrence Radiation Laboratory Report. University of California Lawrence Radiation Laboratory, Livermore, CA.

Crawford, T.V. 1967. "Atmospheric Diffusion of Large Clouds." In *Proceedings of the USAEC Meteorological Information Meeting, Chalk River Nuclear Laboratories, September 11–14, 1967*, ed. C.W. Mason. Atomic Energy of Canada Ltd., Chalk River, Ontario. Pages 191–214.

Crawford, T.V. 1968. *The Long Range Diffusion of the Effluent Cloud from the Phoebus 1B-IV Reactor Test of February 23, 1967*. UCRL-50418. University of California Lawrence Radiation Laboratory Report. University of California Lawrence Radiation Laboratory, Livermore, CA.

Crawford, T.V. 1969. "Atmospheric Transport, Diffusion, and Deposition of Radioactivity." In *Proceedings for the Symposium on Public Health Aspects of Peaceful Uses of Nuclear Explosives Sponsored by the Southwestern Radiological Health Laboratory, Bureau of Radiological Health, Las Vegas, Nevada, April 7–11, 1969*, ed. M.W. Carter. Bureau of Radiological Health, Las Vegas, NV. Pages 249–280.

Crawford, T.V. 1970. "Diffusion and Deposition of the Schooner Clouds." In *Engineering with Nuclear Explosives, a Symposium, Sponsored by the American Nuclear Society in Cooperation with the United States Atomic Energy Commission, Las Vegas, Nevada, January 14–16, 1970*. U.S. Atomic Energy Commission, Washington, DC. Pages 381–399.

Crawford, T.V. 1990. "Meteorological Measurements for Emergency Response." In *Meteorological Aspects of Emergency Response*, ed. M.L. Kramer and W.M. Porch. American Meteorological Society, Boston, MA. Pages 15–36.

Dickerson, M.H. 1973. *A Mass Consistent Wind Field Model for the San Francisco Bay Area*. UCRL-74265. University of California Lawrence Radiation Laboratory Report. University of California Lawrence Radiation Laboratory, Livermore, CA.

Drake, R.L., and S.M. Barrager. 1979. *Mathematical Models for Atmospheric Pollutants*. EPRI EA-1131. Electric Power Research Institute, Palo Alto, CA.

Eckerman, K.F., and J.C. Ryman. 1993. *External Exposure to Radionuclides in Air, Water, and Soil.* Federal Guidance Report No. 12. U.S. Environmental Protection Agency, Washington, DC.

Eckerman, K.F., R.W. Leggett, C.B. Nelson, J.S. Puskin, and A.C.B. Richardson. 1999. *Cancer Risk Coefficients for Environmental Exposure to Radionuclides.* Federal Guidance Report No. 13. EPA 402-R-99–001. U.S. Environmental Protection Agency, Washington, DC.

Egan, B.A. 1975. "Turbulent Diffusion in Complex Terrain." In *Lectures on Air Pollution and Environmental Impact Analysis.* American Meteorological Society, Boston, MA. Pages 112–135.

EPA (U.S. Environmental Protection Agency). 1989a. *User's Guide for the COMPLY Code (Revision 1).* EPA 520/1-89–003. U.S. Environmental Protection Agency, Washington, DC.

EPA. 1989b. "National Emission Standards for Hazardous Air Pollutants: Radionuclides." 40 CFR Part 61. Final Rule. *Federal Register* 54(240): 51654–51715.

Farris, W.T., B.A. Napier, T.A. Ikenberry, J.C. Simpson, and D.B. Shipler. 1994. *Atmospheric Pathway Dosimetry Report, 1944–1992.* PNWD-2228 HEDR. Battelle Pacific Northwest Laboratories, Richland, WA.

Fast, J.D., S. Berman, and R.P. Addis. 1991. *A Comparison of the WIND System Atmospheric Models and MATS Data.* WSRC-RP-91–1209. Westinghouse Savannah River Company Report. Westinghouse Savannah River Company, Aiken, SC.

Garratt, J.R. 1994. *The Atmospheric Boundary Layer.* Cambridge Atmospheric and Space Science Series. Cambridge University Press, New York.

Gifford, F.A., Jr. 1968. "An Outline of Theories of Diffusion in the Lower Layers of the Atmosphere." In *Meteorology and Atomic Energy–1968,* ed. D.H. Slade. TID-24190. U.S. Atomic Energy Commission, Washington, DC. Pages 65–116.

Gifford, F.A., Jr. 1975. "Atmospheric Dispersion Models for Environmental Pollution Applications." *Lectures on Air Pollution and Environmental Impact Analysis.* American Meteorological Society, Boston, MA. Pages 35–58.

Gifford, F.A., Jr. 1976. "Turbulent Diffusion-Typing Schemes: A Review." *Nuclear Safety* 17(1): 68–86.

Gifford, F.A., Jr. 1977. "Tropospheric Relative Diffusion Observations." *Journal of Applied Meteorology* 16(3): 311–313.

Golder, D. 1972. "Relations among Stability Parameters in the Surface Layer." *Boundary-Layer Meteorology* 3(1): 47–58.

Hage, K.D., G. Arnason, N.E. Bowne, P.S. Brown, W.D. Entrekin, M.A. Levitz, and J.A. Sekorski. 1966. *Particle Fallout and Dispersion in the Atmosphere, Final Report.* SC-CR-66–2031. Aerospace Nuclear Safety, Sandia Corporation, Albuquerque, NM.

Hanna, S.R., G.A. Briggs, J. Deardorff, B.A. Egan, F.A. Gifford, Jr., and F. Pasquill. 1977. "Meeting Review: AMS Workshop on Stability Classification Schemes and Sigma Curves—Summary of Recommendations." *Bulletin of the American Meteorological Society* 58: 1305–1309.

Hanna, S.R., G.A. Briggs, and R.P. Hosker, Jr. 1982. *Handbook on Atmospheric Diffusion.* DOE/TIC-11223. U.S. Department of Energy, Office of Scientific and Technical Information, Technical Information Center, Oak Ridge, TN.

Healy, J.W., and R.E. Baker. 1968. "Radioactive Cloud-Dose Calculations." In *Meteorology and Atomic Energy—1968,* ed. D.H. Slade. TID-24190. U.S. Atomic Energy Commission, Washington, DC. Pages 301–377.

Heffter, J.L., B.J.B. Stunder, and G.D. Rolph. 1990. "Long-Range Forecast Trajectories of Volcanic Ash from Redoubt Ash from Redoubt Volcano Eruptions." *Bulletin of the American Meteorology Society* 71(12): 1731–1738.

Hoffman, F.O. 1991. "The Use of Chernobyl Fallout Data to Test Model Predictions of the Transfer of $^{131}$I and $^{137}$Cs from the Atmosphere through Agricultural Food Chains." *Proceedings, Third Topical Meeting on Emergency Preparedness and Response. Chicago, Illinois, April 16–19, 1991.* American Nuclear Society, La Grange Park, IL. Pages 118–123.

Hoffman, F.O., and C.W. Miller. 1983. "Uncertainties in Environmental Radiological Assessment Models and Their Implications." In *Environmental Radioactivity. Proceedings No. 5*. National Council on Radiation Protection and Measurements, Bethesda, MD. Pages 110–138.

Hoffman, F.O., C.W. Miller, and Y.C. NG. 1984. "Uncertainties in Environmental Radiological Assessment Models and their Implications." In *Environmental Transfer to Man of Radionuclide Releases from Nuclear Installations*. Commission of the European Communities, Luxembourg.

Hoffman, F.O., B.G. Blaylock, M.L. Frank, and K.M. Thiessen. 1993. "A Risk-based Screening Approach for Prioritizing Contaminants and Exposure Pathways at Superfund Sites." *Environmental Monitoring and Assessment* 28(3): 221–237.

Holzworth, G.C. 1972. *Mixing Heights, Wind Speeds, and Potential for Urban Air Pollution Throughout the United States*. PB-207 103. U.S.Environmental Protection Agency, Research Triangle Park, NC.

Hosker, R.P., Jr. 1984. "Flow and Diffusion Near Obstacles." *Atmospheric Science and Power Production*, ed. D. Randerson. DOE/TIC-27601. U.S. Department of Energy, Office of Scientific and Technical Information, Technical Information Center, Oak Ridge, TN. Pages 241–326.

IAEA (International Atomic Energy Agency). 1989. *Evaluating the Reliability of Predictions Made Using Environmental Transfer Models*. Safety Series No. 100. International Atomic Energy Agency, Vienna, Austria.

Jensen, N.O. 1981. "A Micrometeorological Perspective on Deposition." *Health Physics* 40(6): 887–891.

Kohler, H., S.R. Peterson, and F.O. Hoffman, eds. 1991. *Multiple Model Testing Using Chernobyl Fallout Data for $^{131}$I in Forage and Milk and $^{137}$Cs in Forage, Milk, Beef, and Grain*. BIOMOVS Technical Report No. 13, Scenario A4. National Institute for Radiation Protection, Stockholm, Sweden.

Knox, J.B., and J.J. Walton. 1984. "Modeling Regional to Global Air Pollution." In *Atmospheric Science and Power Production*, ed. D. Randerson. DOE/TIC27601. U.S. Department of Energy, Office of Scientific and Technical Information, Technical Information Center, Oak Ridge, TN. Pages 621–651.

Knox, J.B., H.A. Tewes, T.V. Crawford, and T.A. Gibson. 1970. *Radioactivity Released from Underground Nuclear Detonations: Source, Transport, Diffusion and Deposition*. UCRL-50230, rev. 1. University of California Lawrence Radiation Laboratory Report. University of California Lawrence Radiation Laboratory, Livermore, CA.

Kolmogorov, A.N. 1941. "The Local Structure of Turbulence in Incompressible Viscous Fluid for Very Large Reynolds Numbers." *Comptes Rendus de l'Académie des Sciences de l'URSS* 30: 301.

Lahti, G.P., R.S. Hubner, and J.C. Golden. 1981. "Assessment of Gamma-ray Exposures Due to Finite Plumes." *Health Physics* 41(2): 319–340.

Lange, R. 1978. "A Three-Dimensional Particle-In-Cell Model for the Dispersal of Atmospheric Pollutants and its Comparison to Regional Tracer Studies." *Journal of Applied Meteorology* 17(3): 320–329.

Lewellen, W.S., R.I. Sykes, and S. Parker. 1985. *Comparison of the 1981 INEL Dispersion Data with Results from a Number of Different Models*. NUREG/CR-4159. U.S.Nuclear Regulatory Commission, Washington, DC.

Little, C.A. 1984. "Comparison of Exposure Pathways." *Models and Parameters for Environmental Radiological Assessments*, ed. C.W. Miller. DOE Critical Review Series DOE/TIC-11468. U.S. Department of Energy, Office of Scientific and Technical Information Center, Oak Ridge, TN. Pages 4–10.

Lyons, W.A. 1975. "Turbulent Diffusion and Pollutant Transport in Shoreline Environments." *Lectures on Air Pollution and Environmental Impact Analysis*. American Meteorological Society, Boston, MA. Pages 136–208.

McKenna, T.J., J.A. Martin, C.W. Miller, L.M. Hively, R.W. Sharpe, J.G. Gitter, and R.M. Watkins. 1987. *Pilot Program: NRC Severe Reactor Accident Incident Response Training Manual: Severe Reactor Accident Overview*. NUREG-1210, vol. 2. U.S. Nuclear Regulatory Commission, Washington, DC.

Miller, C.W., and L.M. Hively. 1987. "A Review of Validation Studies for the Gaussian Plume Atmospheric Dispersion Model." *Nuclear Safety* 28(4): 522–531.

Miller, C.W., and F.O. Hoffman. 1979. "Analysis of NRC Methods for Estimating the Effects of Dry Deposition in Environmental Radiological Assessments." *Nuclear Safety* 20(4): 458–467.

Miller, C.W., A.L. Sjoreen, C.L. Begovich, and O.W. Hermann. 1986. *ANEMOS: A Computer Code to Estimate Air Concentrations and Ground Deposition Rates for Atmospheric Nuclides Emitted from Multiple Operating Sources*. ORNL-5913. Martin Marietta Energy Systems. Oak Ridge National Laboratory, Oak Ridge, TN.

Monin, A.S., and A.M. Obukhov. 1954. "Basic Regularity in Turbulent Mixing in the Surface Layer of the Atmosphere." *Trudy Geofizicheskogo Instituta Akademiya Nauk SSSR. Sb. Statei* 24: 163–167.

Murphy, C.E., Jr., and J.T. Sigmon. 1990. "Dry Deposition of Sulfur and Nitrogen Oxide Gases to Forest Vegetation." In *Acidic Precipitation. Vol. 3: Sources, Deposition and Canopy Interactions*, ed. S.E. Lindberg, A.L. Page, and S.A. Norton. Springer-Verlag, New York. Pages 217–240.

NCRP (National Council on Radiation Protection and Measurements). 1984. *Radiological Assessment: Predicting the Transport, Bioaccumulation, and Uptake by Man of Radionuclides Released to the Environment*. NCRP Report No. 76. National Council on Radiation Protection and Measurements, Bethesda, MD.

NCRP. 1989. *Screening Techniques for Determining Compliance with Environmental Standards*. NCRP Commentary No. 3. National Council on Radiation Protection and Measurements, Bethesda, MD.

NCRP. 1993. *Uncertainty in NCRP Screening Models Relating to Atmospheric Transport, Deposition and Uptake by Humans*. NCRP Commentary No. 8. National Council on Radiation Protection and Measurements, Bethesda, MD.

NCRP. 1996. *Screening Models for Releases of Radionuclides to Atmosphere, Surface Water, and Ground*. NCRP Report No. 123 I and II. National Council on Radiation Protection and Measurements, Bethesda, MD.

NRC (U.S. Nuclear Regulatory Commission). 1972. *Onsite Meteorological Programs, Safety Guide 23*. U.S. Nuclear Regulatory Commission, Washington, DC.

NRC. 1977. *Methods for Estimating Atmospheric Transport and Dispersion of Gaseous Effluents in Routine Releases from Light-Water-Cooled Reactors*. Regulatory Guide 1.111. U.S. Nuclear Regulatory Commission, Washington, DC.

NRC. 1987. *U.S. Nuclear Regulatory Commission. Pilot Program: NRC Severe Reactor Accident Incident Response Training Manual*. NUREG-1210, vol. 2. U.S. Nuclear Regulatory Commission, Washington, DC.

NRC. 1992. *Response Technical Manual, RTM-92*. NUREG/BR-0150, vol. 1, rev. 2. U.S. Nuclear Regulatory Commission, Washington, DC.

Obukhov, A.M. 1941. "Energy Distribution in the Spectrum of Turbulent Flow." *Izvestiya Akadema Nauk SSSR Seria Geographia i Geofizika* 5: 453.

Ogram, G.L., and S.C. Wright. 1991. "Experimental Evaluation of Ontario's Nuclear Emergency Atmospheric Dispersion Model." *Proceedings, Third Topical Meeting on Emergency Preparedness and Response, April 16–19, 1991, Chicago, Illinois*. American Nuclear Society, La Grange Park, IL. Pages 114–117.

Overcamp, T.J. 2007. "Solutions to the Gaussian Cloud Approximation for Gamma Absorbed Dose." *Health Physics* 92(1): 78–81.

Overcamp, T.J., and R.A. Fjeld. 1983. "An Exact Solution to the Gaussian Cloud Approximation for Gamma Absorbed Dose Due to a Ground-Level Release." *Health Physics* 44(4): 367–372.

Overcamp, T.J., and R.A. Fjeld. 1987. "A Simple Approximation for Estimating Centerline Gamma Absorbed Dose Rates Due to a Continuous Gaussian Plume." *Health Physics* 53 (2): 143–146 [Errata: 56(6): 974 (June 1989)].

Overcamp, T.J., and T. Ku. 1988. "Plume Rise from Two or More Adjacent Stacks." *Atmospheric Environment* 22(4): 625–637.

Pasquill, F. 1961. *Atmospheric Diffusion*. D. Van Nostrand, London.

Pasquill, F. 1974. *Atmospheric Diffusion, Second Edition*. Ellis Horwood Series in Environmental Science. John Wiley & Sons, New York.

Pasquill, F., and F.B. Smith. 1983. *Atmospheric Diffusion*, 3rd ed. Ellis Horwood Series in Environmental Science. John Wiley & Sons, New York.

Pendergast, M.M. 1984. "Meteorological Fundamentals." In *Atmospheric Science and Power Production*, ed. D. Randerson. DOE/TIC-27601. U.S. Department of Energy, Office of Scientific and Technical Information, Technical Information Center, Oak Ridge, TN. Pages 33–79.

Pendergast, M.M., and T.V. Crawford. 1974. "Actual Standard Deviations of Vertical and Horizontal Wind Direction Compared to Estimates from Other Measurements." In *Pre-print Volume from Symposium on Atmospheric Diffusion and Air Pollution. Santa Barbara, California, September 9–13, 1974*. American Meteorological Society, Boston, MA. Pages 1–6.

Pepper, D.W. 1995. *Selecting a Numerical Package for Heat Transfer and Fluid Flow*. ASME HTC Tech Brief. American Society of Mechanical Engineers, New York.

Pepper, D.W., and A.J. Baker. 1988. "Finite Differences vs. Finite Elements." In *Handbook of Numerical Heat Transfer*. John Wiley & Sons, New York. Pages 519–577.

Pepper, D.W., and J.A. Marino. 1996. "A Set of Environmental Transport and Diffusion Models for Calculating Hazardous Releases." *Nuclear Technology* 113(2): 190–203.

Pepper, D.W., R.E. Cooper, and A.J. Baker. 1982. "A Comparison of Numerical Models for Calculating Dispersion from Accidental Releases of Pollutants." In *Atmospheric Pollution 1992*, ed. M. Benarie. Elsevier, New York. Pages 127–140.

Ramsdell, J.V., Jr., and C.J. Fosmire. 1995. *Atmospheric Dispersion Estimates in the Vicinity of Buildings*. PNL-10286. Battelle Pacific Northwest Laboratories, Richland, WA.

Ramsdell, J.V., Jr., C.A. Simonen, and K.W. Burk. 1994. *Regional Atmospheric Transport Code for Hanford Emission Tracking (RATCHET)*. PNWD-2224 HEDR. Battelle Pacific Northwest Laboratories, Richland, WA.

Randerson, D. 1984. "Atmospheric Boundary Layer." In *Atmospheric Science and Power Production*, ed. D. Randerson. DOE/TIC-27601. U.S. Department of Energy, Office of Scientific and Technical Information, Technical Information Center, Oak Ridge, TN. Pages 147–188.

Richardson, L.F. 1920. "The Supply of Energy from and to Atmospheric Eddies." *Proceedings Royal Society (London), Series A* 97: 354–373.

Richardson, L.F. 1926. "Atmospheric Diffusion Shown on a Distance-Neighbor Graph." *Proceedings Royal Society (London), Series A* 110: 709.

Sehmel, G.A. 1984. "Deposition and Resuspension." In *Atmospheric Science and Power Production*, ed. D. Randerson. DOE/TIC-27601. U.S. Department of Energy, Office of Scientific and Technical Information, Technical Information Center, Oak Ridge, TN. Pages 533–583.

Slade, D.H. 1968. *Meteorology and Atomic Energy—1968*. AEC/TID-24190. U.S. Atomic Energy Commission, National Technical Information Service, Springfield, VA.

Slinn, W.G.N. 1984. "Precipitation Scavenging." In *Atmospheric Science and Power Production*, ed. D. Randerson. DOE/TIC-27601. U.S. Department of Energy, Office of Scientific and Technical Information, Technical Information Center, Oak Ridge, TN. Pages 466–532.

Snyder, W.H., and R.E. Lawson, Jr. 1991. "Fluid Modeling Simulation of Stack-Tip Downwash for Neutrally Buoyant Plumes." *Atmospheric Environment* 25A: 2837–2850.

Snyder, W.H., R.E Lawson, Jr., R.S. Thompson, and G.C. Holzworth. 1980. *Observations of Flow around Cinder Cone Butte, Idaho*. EPA-600/7-80-150. U.S. Environmental Protection Agency, Research Triangle Park, NC.

Snyder, W.H., R.S. Thompson, R.E. Eskridge, R.E. Lawson, Jr., I.P. Castro, J.T. Lee, J.C.R. Hunt, and Y. Ogawa. 1985. "The Structure of Strongly Stratified Flow over Hills: Dividing-Streamline Concept." *Journal of Fluid Mechanics* 152: 249–288.

Sullivan, T.J., J.S. Ellis, C.S. Foster, K.T. Foster, R.L. Baskett, J.S. Nasstrom, and W.W. Schalk, III. 1993. "Atmospheric Release Advisory Capability: Real-Time Modeling of Airborne Hazardous Materials." *Bulletin of the American Meteorological Society* 74(12): 2343–2361.

Sutton, O.G. 1953. *Micrometeorology*. McGraw-Hill, New York.

Taylor, G.I. 1959. "The Present Position in the Theory of Turbulent Diffusion." In *Atmospheric Diffusion and Air Pollution, Advances in Geophysics*, vol. 6, ed. F.N. Freckle and P.A. Sheared. Academic Press, New York. Page 101.

Thuillier, R.H., and R.L. Mancuso. 1980. *Building Effects on Effluent Dispersion from Roof Vents at Nuclear Power Plants*. EPRI NP-1380. Electric Power Research Institute, Palo Alto, CA.

Turner, D.B. 1967. *Workbook of Atmospheric Dispersion Estimates*. Public Health Service Publication 999-AP-26. Robert A. Taft Sanitary Engineering Center, Cincinnati, OH.

Uliasz, M., and R.A. Pielke. 1991. "Lagrangian-Eulerian Dispersion Modeling System for Real-Time Mesoscale Applications." *Proceedings, Third Topical Meeting on Emergency Preparedness and Response. Chicago, Illinois, April 16–19, 1991*. American Nuclear Society, La Grange Park, IL. Pages 95–98.

Van der Hoven, I. 1968. "Deposition of Particles and Gases." *Meteorology and Atomic Energy—1968*, ed. D.H. Slade. TID-24190. U.S. Atomic Energy Commission, Washington, DC. Pages 202–208.

Weber, A.H., W.R. Buckner, and J.H. Weber. 1982. "Statistical Performance of Several Mesoscale Atmospheric Dispersion Models." *Journal of Applied Meteorology* 21(11): 1633–1644.

Widner, T.E., K.E. Shank, and B.H. Carson. 1991. "Improving Radioactive Release Readiness Using Scale Model Wind Tunnel Tests." *Proceedings, Third Topical Meeting on Emergency Preparedness and Response*. American Nuclear Society, La Grange Park, IL. Pages 110–113.

Wilkins, E.M. 1963. *New Applications of Atmospheric Turbulence and Diffusion Theory*. Report No. 0–71000/312–19. LTV Research Center, Ling-Temco-Vought, Dallas, TX.

# 4

# Surface Water Transport of Radionuclides

Yasuo Onishi

Radionuclides released to surface water (rivers, lakes, estuaries, coastal waters, and oceans) are passed along the aquatic food chain to humans. Potential aquatic pathways to humans include drinking water; consumption of vegetables, meat, and milk, and soil affected by the use of contaminated irrigation water; consumption of fish, shellfish, and seaweeds; external exposure though swimming and boating; and direct exposure from (and potential consumption of) contaminated shore sediment (NCRP 1996). To evaluate these aquatic contributions to human dose, we need to determine radionuclide concentrations in water and suspended bottom and shore sediments in and along the surface water.

Surface water is one component of the natural hydrology cycle and interacts with other environmental media (i.e., atmosphere, land, and subsurface water). Because many radionuclides migrate from one environmental medium (e.g., atmosphere) to others (e.g., land and surface and subsurface waters), their associated effects on humans and environments can derive from more than one environmental medium (see figure 4.1). Radionuclides come to surface water from various point and nonpoint sources through direct discharges to a receiving surface-water body, dry and wet deposition from the atmosphere, runoff and soil erosion from land surfaces, and seepage from groundwater.

Radionuclides in surface water are, in turn, a potential contamination source to groundwater and atmosphere (if volatilized). Thus, we must treat surface water as a part of the overall hydrologic system to assess radionuclide transport, fate, and its impact to biota and humans.

This chapter discusses how to estimate radionuclide concentrations in surface water (i.e., dissolved concentration in water and concentrations attached to

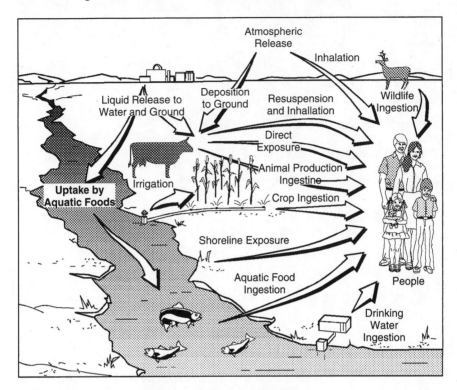

Figure 4.1  Radionuclide pathways.

suspended, bottom, and shore sediments) through both routine and accidental radionuclide releases to surface water. The types of surface water described here are rivers, estuaries, coastal waters and oceans, small lakes, and large lakes.

There are three basic models for estimating radionuclide concentrations in surface waters (IAEA 1985a): analytical models, box models, and numerical models. Analytical models solve the basic radionuclide transport and fate equations, which are usually advection–diffusion equations with significant simplifications to yield exact solutions to the governing equations. These simplifications usually include water body geometry, flow conditions, and dispersion processes that are constant over the entire water body and time; thus, their applicability and potential accuracy are lowest among these three model types. However, they are easy and quick to use and are very useful tools for preliminary assessment of radionuclide impacts. In many situations, simple models applied with conservative assumptions have proved to be adequate for establishing compliance with regulatory limits (IAEA 1985a, 2001).

The box models, sometimes called compartment models, treat an entire water body as a homogeneous (totally mixed) compartment or as a series of homogeneous compartments connected to represent the whole water body. These models often include some sediment–radionuclide interactions (Burns and Cline 1985).

The box models are used extensively for general water-quality problems in lakes (e.g., lake eutrophication) to account for complex chemical and biological reactions to determine chemical concentrations (Baca and Arnett 1976).

The numerical models are denoted as "advanced" by the International Atomic Energy Agency (IAEA 1985a). They usually transform one-, two-, and three-dimensional governing equations into finite-difference, finite-volume, or finite-element forms to allow varying water-body geometry, flow and sediment conditions, dispersion, and possible chemical and biological processes. Thus, they have a wider applicability than the other two model types (Onishi 1994a). These models are useful for cases requiring detailed or accurate estimates of radionuclide migration (e.g., remediation assessment of surface water contaminated by the Chernobyl nuclear accident).

This chapter discusses relevant transport and fate mechanisms of radionuclide transport in surface water. Some useful analytical and empirical models and some representative numerical models are presented for radionuclide transport in surface water. The Chernobyl nuclear accident is also discussed to illustrate the radionuclide aquatic pathway.

## Basic Transport and Fate Mechanisms

Radionuclides, once released to surface water, undergo complex transport and fate processes. They are controlled by transport, contaminant exchange between dissolved and particulate phases, fate processes, transformation, and contaminant transfer between surface water and other environmental media, as shown in table 4.1 (Onishi 1985).

### *Transport*

Transport is a physical process that moves a radionuclide from one location to another by movements of water, sediment, and biota in a water body.

#### Water Movement

Water movement is the principal means of radionuclide migration. The primary mechanisms are advection and dispersion, especially for radionuclides easily dissolved in water. A current in surface water can be very complex. In estuarine and marine environments, tide, wind, waves, salinity, and temperature variations affect flows. Although flows induced by wind, water density variations, and, to a lesser extent, waves are much smaller than tidal flows in many areas, the long-term migration and accumulation of radionuclides are significantly affected by these nonperiodic flows (Onishi et al. 1993b). The flow velocities and depths are either calculated by hydrodynamic codes or determined by field measurements. For further discussion, refer to studies on hydrodynamics (Chow 1959; Rouse 1961) and dispersion (Fischer et al. 1979; Bowie et al. 1985).

Table 4.1  Major mechanisms affecting radionuclide transport and fate in surface water

1. **Transport**
    i. Water movement
        a. Discharge-induced advection and dispersion
        b. Ambient flow-induced advection and dispersion
    ii. Sediment movement
        a. Sediment transport
        b. Sediment deposition to surface-water bed/bottom
        c. Sediment erosion from surface-water bed/bottom
    iii. Bioturbation within the surface-water bed/bottom

2. **Intermedia Transport**
    i. Adsorption/desorption
    ii. Precipitation/dissolution
    iii. Volatilization
    iv. Physical breakup

3. **Degradation/Decay**
    i. Radionuclide decay

4. **Transformation**
    i. Yield of daughter products

5. **Radionuclide Contributions from Other Environmental Media**
    i. Dry and wet deposition from atmosphere
    ii. Runoff and soil erosion from land surface
    iii. Seepage from/to groundwater

## Sediment Movement

Sediment movement affects the transport of radionuclides both indirectly and directly. Indirectly, sediment transport produces or changes bed sediment forms (e.g., ripples and dunes), changing the bed friction factor. The change in the bed friction factor in turn affects the flow depth and velocity (Vanoni 1975; Onishi 1994b), which affects radionuclide transport. Directly, both suspended and bed sediments adsorb many radionuclides, which then migrate with the sediment. The deposition of contaminated suspended sediment and the direct adsorption of the radionuclides by the bed sediment can build up the radionuclides in the bed, producing a long-term source of contamination through subsequent resuspension or desorption. However, radionuclide adsorption by sediments can also reduce the concentration of the dissolved contaminant, which is usually more readily taken up by aquatic biota (NCRP 1984). Because the sediment transport rate is roughly proportional to the third to fifth power of the flow velocity, the sediment transport is highly variable with time and space (Vanoni 1975). For example, storms can significantly alter sediment and radionuclide/contaminant distributions, especially those strongly affiliated with sediment (Onishi et al. 1993b). In the Irish Sea, more than 90% of plutonium discharged from the Sellafield (formerly Windscale)

Nuclear Site is associated with sediment (Hetherington 1976). Transport and accumulation are controlled by these nonperiodic flows and episodic events for long-term migration.

The effects of transport, deposition, and resuspension of sediment on the migration and accumulation of a radionuclides become important if (1) the radionuclide is hydrophobic and has a high affinity for sorption, (2) the concentration of suspended sediment (especially fine sediment, e.g., silt, clay, and organic matter) in a receiving water is high, and (3) the radionuclide is persistent, so contaminated bed sediments create a long-term source of contamination.

## Bioturbation

Movements of benthic organisms within the bed can move radionuclides deposited within bottom sediment (Onishi et al. 1993b).

### Intermedia Transfer

Intermedia transfer is a chemical process to change a phase of a radionuclide to another phase, for example, a dissolved radionuclide to a particulate or gaseous radionuclide.

## Adsorption and Desorption

Through adsorption and desorption, a dissolved radionuclide is transferred to the surface of a solid material or removed from it. The quantification of adsorption/desorption is usually expressed in terms of a nonlinear form (the Langmuir or the Freundlich isotherms) or a linear form (a distribution coefficient) (Onishi 1985). At low radionuclide concentrations (as expected to occur in the environment), the contaminant partitioning between solids and water is usually expressed using the distribution coefficient, $K_d$:

$$K_d = \frac{\text{amount of contaminant sorbed by sediment}}{\text{amount of contaminant left in solution}}.$$

Rough ranges of $K_d$ values for many radionuclides in fresh and marine waters are shown in table 4.2. Median values shown in table 4.2 are somewhat conservative (lower) to provide higher dissolved radionuclide concentrations. There is uncertainty associated with the use of $K_d$ for modeling adsorption and desorption because of the complexity of the various geochemical mechanisms (e.g., ion exchange, precipitation–mineral formation, complexation–hydrolysis, oxidation–reduction, and colloid and polymer formulations) and because of the interactions among them (Onishi et al. 1981; IAEA 1985b). Therefore, a great deal of caution must be exercised in selecting $K_d$ values.

A more rigorous approach is using geochemical models to solve reactions of aqueous chemical species and the formation of solids by adsorption/desorption (Chapman 1982; Felmy et al. 1984; Onishi 2005). However, this approach requires

Table 4.2 Estimates of distribution coefficient, $K_d$ (mL/g), for radionuclides in natural waters, with emphasis on oxidizing conditions

| Element | Fresh water Range | Fresh water Median | Salt water Range | Salt water Median |
|---|---|---|---|---|
| Am | $10^2$–$10^5$ | $5 \times 10^3$ | $10^2$–$10^5$ | $10^4$ |
| Sb | 20–500 | 50 | 20–500 | 50 |
| C | 1–100 | 10 | 10–$10^4$ | 1,000 |
| Ce | $10^2$–$10^4$ | $10^3$ | $10^3$–$10^7$ | $10^4$ |
| Cr(III) | 50–1,000 | 100 | $10^2$–$10^4$ | $10^3$ |
| Cr(VI) | 0–50 | 5 | 10–100 | 30 |
| Cs | 50–$10^5$ | 1,000 | 10–$10^4$ | 500 |
| Co | 50–$10^4$ | 100 | $10^2$–$10^5$ | $10^3$ |
| Cm | $10^2$–$10^5$ | $5 \times 10^3$ | $10^2$–$10^5$ | $10^4$ |
| Eu | $10^2$–$10^4$ | $5 \times 10^2$ | $10^3$–$10^5$ | $10^4$ |
| Fe(II) | 10–1,000 | 50 | 10–1,000 | 100 |
| Fe(III) | $10^3$–$10^4$ | $10^3$ | $10^4$–$10^5$ | $5 \times 10^4$ |
| I | 0–75 | 3 | 0–100 | 10 |
| Mn(II) | 50–$10^4$ | 300 | $10^2$–$10^4$ | $10^3$ |
| Np(IV) | $10^3$–$10^6$ | $10^5$ | $10^4$–$10^5$ | $10^4$ |
| Np(V) | 0.2–1,000 | 5 | 5–1,000 | 20 |
| $^{32}PO_4$ | 5–100 | 30 | 50–1,000 | 100 |
| Pu(IV) | $10^3$–$10^7$ | $10^5$ | $10^2$–$10^5$ | $5 \times 10^4$ |
| Pu(V) | 50–$10^3$ | 50 | 50–$10^3$ | 50 |
| Pu(VI) | $10^2$–$10^3$ | $10^2$ | $10^2$–$10^4$ | $5 \times 10^2$ |
| Pm | $10^2$–$10^3$ | $5 \times 10^2$ | $10^3$–$10^5$ | $10^4$ |
| Ra | $10^2$–$10^3$ | 100 | 10–$10^3$ | 50 |
| Ru(VI) | 10–1,000 | 100 | 10–$10^3$ | 20 |
| Sr | 8–4,000 | 100 | 6–400 | 50 |
| Tc(VII) | 0–100 | 0.5 | 0 | 0 |
| Th | $10^3$–$10^6$ | $10^5$ | $10^4$–$10^5$ | $10^4$ |
| $^3$H | 0 | 0 | 0 | 0 |
| U(IV) | $10^3$–$10^5$ | $5 \times 10^3$ | $10^3$–$10^5$ | $5 \times 10^3$ |
| U(VI) | 10~1,000 | 20 | $10^2$–$10^4$ | 50 |
| Zn | $10^2$–$10^3$ | $5 \times 10^2$ | $10^3$–$10^4$ | $5 \times 10^3$ |
| Zr | $5 \times 10^2$–$10^4$ | $10^3$ | $10^3$–$10^5$ | $10^4$ |

detailed information such as competing chemical species, geochemical equilibrium constants, and a complete mineralogical description of the sediment or soil.

## Precipitation and Dissolution

Effects of precipitation and dissolution on radionuclide transport are usually minor because levels of radionuclide concentrations in surface-water environments are very low. Some geochemical models can estimate the amounts of chemicals to be precipitated or dissolved in a given condition (Felmy et al. 1984; Felmy 1990; Onishi

2005), but thermodynamic databases needed to simulate radionuclide aqueous reactions and solid dissolution/precipitation are quite limited.

## Volatilization

Volatilization is another intermedia exchange mechanism; some radionuclides are in the form of volatilizing liquids (e.g., tritium and iodine) or dissolved gas (e.g., radon) and may volatilize through the water–air interface.

## Physical Breakup

Some radionuclides in a particulate phase must first break up physically (e.g., through weathering) before they can become chemically reactive to dissolve into water. Such is the case for $^{90}$Sr and $^{137}$Cs deposited around the Chernobyl nuclear plant in Ukraine after the accident in 1986 (Zheleznyak et al. 1992).

## *Degradation/Decay*

Degradation and decay processes are chemical and biological processes that reduce the radionuclide concentration in a water body.

## Radionuclide Decay

Unlike organic chemicals that have many degradation mechanisms, radionuclide fate mechanism is only associated with radionuclide decay. Radionuclide decay is commonly expressed in terms of the radionuclide's half-life.

## *Transformation*

Radionuclide transformation produces radionuclides in the decay chain of an originally released radionuclide. If these decay products are as toxic as, or more toxic than, the original radionuclide, these new products must be included in the assessment.

## Yield of Daughter Products

Transformation is related to degradation and decay mechanisms because contaminants of concern may themselves be products of degradation or radionuclide decay. For example, uranium buried at mill tailing sites produces very toxic radon gas as one of its daughter products before it eventually degrades to lead. The time scale of transport in surface water is usually too small for daughter products to become important. Radionuclide accumulation in a surface-water bottom, however, can have a much longer time frame than radionuclide transport in water. In some cases, contaminant degradation and decay chains may also be included in the assessments.

## Radionuclide Contributions from Other Environmental Media

Environmental intermedia exchanges of radionuclides include dry and wet deposition from the atmosphere, radionuclide influx through overland runoff and soil erosion (or transfer from surface water to land by irrigation), and exchange between surface and groundwater through seepage.

These transport and fate mechanisms are further illustrated under "Chernobyl Nuclear Accident Aquatic Assessment," below.

One must evaluate which of these mechanisms are important for a specific case under evaluation and how much accuracy is needed for environmental health and remediation assessment

## Radionuclide Transport Models

This section addresses radionuclide migration models for main surface-water bodies: rivers, estuaries, coastal waters and oceans, small lakes, and large lakes.

Radionuclide transport and fate in surface water may be expressed in the following three-dimensional advection–diffusion equation cast in Cartesian coordinates (IAEA 1985a; Onishi 1994a):

$$\frac{\partial C}{\partial t} + U \frac{\partial C}{\partial x} + V \frac{\partial C}{\partial y} + (W - W_s) \frac{\partial C}{\partial z}$$
$$= \frac{\partial}{\partial z}\left(\varepsilon_x \frac{\partial C}{\partial x}\right) + \frac{\partial}{\partial y}\left(\varepsilon_y \frac{\partial C}{\partial y}\right) + \frac{\partial}{\partial z}\left(\varepsilon_z \frac{\partial C}{\partial z}\right) - \lambda C \qquad (4.1)$$

where

$C$ = radionuclide concentration (Bq m$^{-3}$)
$T$ = time (s)
$U, V,$ and $W$ = velocity components in the x-, y-, and z-directions, respectively (m s$^{-1}$)
$W_s$ = contaminant fall velocity (m s$^{-1}$)
$x, y,$ and $z$ = longitudinal, lateral, and vertical directions (m)
$\varepsilon_x, \varepsilon_y,$ and $\varepsilon_z$ = dispersion coefficients in x-, y-, and z-directions (m), respectively (m$^2$ s$^{-1}$)
$\lambda$ = radionuclide decay (s$^{-1}$)

The equations in this section are applicable to transport and fate of radionuclides and other chemicals (e.g., heavy metals and pesticides). Equation 4.1 may be integrated over the entire width (y-direction) or depth (z-direction) to obtain laterally or vertically averaged, two-dimensional equations. Integrating the equation over the entire cross section (both y- and z-directions) yields a one-dimensional governing equation, as discussed further below.

Many numerical models solve this advection–diffusion equation with known flow characteristics (e.g., velocity, depth, width, and dispersion) and radionuclide information (e.g., release, decay, and adsorption/desorption rates). However, various simplifications (e.g., constant depth, velocity, dispersion, and straight bank and shore lines) can be imposed to obtain analytical solutions to this general governing equation. These solutions are used to determine radionuclide concentrations for surface-water screening methodologies (rivers, estuaries, coastal waters, oceans, and lakes).

## Accidental Radionuclide Releases

For an accidental radionuclide release lasting a short time (i.e., instantaneous release), equation 4.1 yields the following analytical solution, assuming $V$, $W$, and $W_s$ are zero (Sayre 1975):

$$C(x,y,z,t) = W_R f_x(x,t) f_y(y,t) f_z(z,t) \exp(-\lambda t), \tag{4.2}$$

$$f_x(x,t) = \frac{1}{2\sqrt{\pi \varepsilon_x t}} \exp\left\{-\frac{(x-Ut)^2}{4\varepsilon_x t}\right\}, \tag{4.3}$$

$$f_y(y,t) = \frac{1}{2\sqrt{\pi \varepsilon_y t}} \exp\left\{-\frac{y^2}{4\varepsilon_y t}\right\}, \tag{4.4}$$

$$f_z(z,t) = \frac{1}{2\sqrt{\pi \varepsilon_z t}} \exp\left\{-\frac{z^2}{4\varepsilon_z t}\right\}, \tag{4.5}$$

where $W_R$ is the total radionuclide release (Bq).

The functions $f_x(x,t)$, $f_y(y,t)$, and $f_z(z,t)$ are probability density functions of the normal (Gaussian) distribution with respective means of $\bar{x} = Ut$, $\bar{y} = \bar{z} = 0$, and standard deviation of $\sigma_x = 2\varepsilon_x t$, $\sigma_y = 2\varepsilon_y t$, $\sigma_z = 2\varepsilon_z t$. This solution can be extended to temporally and spatially varying release sources by the method of superposition or convolution (Sayre 1975).

As the unfiltered (sum of dissolved and sediment-sorbed) radionuclide is dispersed farther downstream, its concentration becomes more vertically and laterally uniform. This leads to $f_y(y,t)$ and $f_z(z,t)$ approaching to $1/B$ and $1/d$ ($B$ and $d$ are width and depth of the surface water). For the horizontal (longitudinal and lateral) dispersal from an instantaneous line source that is uniformly distributed over the entire water depth, equation 4.1 becomes

$$\frac{\partial C}{\partial t} + U\frac{\partial C}{\partial x} + V\frac{\partial C}{\partial y} = \varepsilon_x \frac{\partial^2 C}{\partial x^2} + \varepsilon_y \frac{\partial^2 C}{\partial y^2} - \lambda C. \tag{4.6}$$

The solution to this equation is

$$C(x,y,t) = \frac{W_R}{4\pi Bt\sqrt{\varepsilon_x \varepsilon_y}} \exp\left[-\left\{\frac{(x-Ut)^2}{4\varepsilon_x t} + \frac{y^2}{4\varepsilon_y t}\right\} - \lambda t\right]. \tag{4.7}$$

For an instantaneous source that is originally uniformly distributed over the entire cross section, the governing equation and its solution are

$$\frac{\partial C}{\partial t} + U\frac{\partial C}{\partial x} = \varepsilon_x \frac{\partial^2 C}{\partial x^2} - \lambda C, \tag{4.8}$$

$$C(x,t) = \frac{W_R}{2A\sqrt{\pi \varepsilon_x t}} \exp\left[-\left\{\frac{(x-Ut)^2}{4\varepsilon_x t} + \lambda t\right\}\right], \tag{4.9}$$

where $A$ = cross-sectional area $(B \cdot d)$ of the flow channel (m²).

### EXAMPLE 4.1

A plant along the river accidentally spilled $5 \times 10^5$ Bq of $^{137}$Cs into the river. Average river width, depth, and velocity are 40 m, 5 m, and 3 m s$^{-1}$, respectively. The longitudinal dispersion coefficient of the river is estimated to be 200 m² s$^{-1}$. A nearby city uses the river for drinking water and has its water intake station 5 km downstream from the plant. Assuming that $^{137}$Cs will be fully mixed across the river cross section by the time it reaches the city intake station and that radionuclide decay is negligible,

1. What will the $^{137}$Cs concentration be at the intake station 30 min after the accidental spill?
2. What will the $^{137}$Cs concentration be and at what time will it occur when the center of $^{137}$Cs plume patch arrives at the city intake station?

*Solution*: Because $^{137}$Cs is totally mixed across the river cross section, the concentration there changes only with the longitudinal distance and time. Use equation 4.9 to determine the concentration:

1. At $t = 30$ min $= 1{,}800$ s, $^{137}$Cs concentration, neglecting the radionuclide decay, is

$$C = \frac{500{,}000}{(40 \times 5)} \frac{1}{2\sqrt{\pi(200 \times 1{,}800)}} \exp\left\{-\frac{\{5{,}000 - (3 \times 1{,}800)\}^2}{(4 \times 200 \times 1{,}800)}\right\}$$
$$= 1.06 \text{ Bq m}^{-3}.$$

2. The center of $^{137}$Cs plume patch arrives at the intake station at time $t$ such that $x - Ut = 0$. Thus, $5{,}000 - 3t = 0$. So $t = 1{,}667$ s $= 27$ min and 47 s. At that time, the $^{137}$Cs concentration is

$$C = \frac{500{,}000}{(40 \times 5)} \frac{1}{2\sqrt{\pi(200 \times 1{,}667)}} = 1.22 \text{ Bq m}^{-3}.$$

### *Routine Long-Term Radionuclide Releases*

Because a nuclear facility likely releases radionuclides in a consistent, steady mode, we need to determine a steady-state radionuclide concentration at a given downstream location to estimate a potential dose resulting from the normal plant operation.

After the radionuclide is completely mixed vertically, the radionuclide transport equation for the steady-state radionuclide release will be reduced from the three-dimensional equation 4.1 to the following steady-state, two-dimensional, advection–diffusion equation:

$$U \frac{\partial C}{\partial x} = \varepsilon_x \frac{\partial^2 C}{\partial x^2} + \varepsilon_y \frac{\partial^2 C}{\partial y^2} - \lambda C. \tag{4.10}$$

Assuming that the flow field is infinitely wide and that the receiving surface water has no initial radionuclide content, equation 4.10 yields the unfiltered radionuclide concentration at a given location:

$$C = \frac{Q}{2\pi d \sqrt{\varepsilon_x \varepsilon_y}} \left\{ \exp\left( \frac{Ux}{2\varepsilon_x} - \frac{\lambda x}{U} \right) \right\} \cdot K_0 \left\{ \frac{U}{2\varepsilon_x} \sqrt{x^2 + \frac{\varepsilon_x}{\varepsilon_y}(y - y_0)^2} \right\}, \tag{4.11}$$

where

$Q$ = radionuclide release rate (Bq s$^{-1}$)
$y_0$ = lateral distance of the radionuclide release point measured from the riverbank (m)
$K_0\{\}$ = modified Bessel function of the second kind of the zeroth order. (Values of $K_0\{\}$ are presented in table 4.3.)

Once a radionuclide plume reaches the riverbank, its spread is restricted. In this case, instead of equation 4.11, a solution to equation 4.10 can be calculated by using the reflection or the mirror image source technique (Sayre 1975):

$$C_r = C(x, y) + \sum_{n=1}^{\infty} \left[ C\left\{ x, nB - \left( y_0 - \frac{B}{2} \right) + (-1)^n \left( y - \frac{B}{2} \right) \right\} \right.$$
$$\left. + C\left\{ x, -nB - \left( y_0 - \frac{B}{2} \right) + (-1)^n \left( y - \frac{B}{2} \right) \right\} \right] \tag{4.12}$$

where

$B$ = river width (m)
$C_r$ = radionuclide concentration affected by the riverbank (Bq m$^{-3}$)
$N$ = number of reflection cycles

There is generally no significant contribution of $C(x, y)$ from terms with $n > 4$ or 5. By using equations 4.11 and 4.12, one can obtain radionuclide concentrations at any location within a river with an arbitrary radionuclide release point ($y = y_0$).

A solution to the steady-state, one-dimensional advection–diffusion equation with the first-order decay term (equation 4.13) for a continuous release at $x = 0$ is (O'Conner and Jawler 1965):

$$U \frac{\partial C}{\partial x} = \varepsilon_x \frac{\partial^2 C}{\partial x^2} - \lambda C \tag{4.13}$$

$$C = \frac{Q}{AU\sqrt{1 + 4\lambda \varepsilon_x}} \cdot \exp\left\{ \frac{Ux}{2\varepsilon_x}\left( 1 \pm \sqrt{1 + \frac{4\lambda \varepsilon_x}{U^2}} \right) \right\}. \tag{4.14}$$

Table 4.3  Modified Bessel functions of the second kind of the zeroth order

| | | | Modified Bessel functions | | | | |
|---|---|---|---|---|---|---|---|
| $x$ | $e^x K_0(x)$ | $x$ | $e^x K_0(x)$ | $x$ | $e^x K_0(x)$ | $x$ | $e^x K_0(x)$ |
| 0.0 | | 3.8 | 0.62429 15812 | 7.6 | 0.44763 71996 | 13.0 | 0.34439 86455 |
| 0.1 | 2.68232 61023 | 3.9 | 0.61665 73147 | 7.7 | 0.44480 55636 | 13.2 | 0.34182 59943 |
| 0.2 | 2.14075 73233 | 4.0 | 0.60929 76693 | 7.8 | 0.44202 70724 | 13.4 | 0.33931 01806 |
| 0.3 | 1.85262 73007 | 4.1 | 0.60219 65064 | 7.9 | 0.43930 00819 | 13.6 | 0.33684 91405 |
| 0.4 | 1.66268 20891 | 4.2 | 0.59533 89889 | 8.0 | 0.43662 30185 | 13.8 | 0.33444 09142 |
| 0.5 | 1.52410 93857 | 4.3 | 0.58871 14486 | 8.1 | 0.43399 43754 | 14.0 | 0.33208 36383 |
| 0.6 | 1.41673 76214 | 4.4 | 0.58230 12704 | 8.2 | 0.43141 27084 | 14.2 | 0.32977 55402 |
| 0.7 | 1.33012 36562 | 4.5 | 0.57609 67897 | 8.3 | 0.42887 66329 | 14.4 | 0.32751 49332 |
| 0.8 | 1.25820 31216 | 4.6 | 0.57008 72022 | 8.4 | 0.42638 48214 | 14.6 | 0.32530 02091 |
| 0.9 | 1.19716 33803 | 4.7 | 0.56426 24840 | 8.5 | 0.42393 59993 | 14.8 | 0.32312 98364 |
| 1.0 | 1.14446 30797 | 4.8 | 0.55861 33194 | 8.6 | 0.42152 89433 | 15.0 | 0.32100 23534 |
| 1.1 | 1.09833 02828 | 4.9 | 0.55313 10397 | 8.7 | 0.41916 24781 | 15.2 | 0.31891 63655 |
| 1.2 | 1.05748 45322 | 5.0 | 0.54780 75643 | 8.8 | 0.41683 54743 | 15.4 | 0.31687 05405 |
| 1.3 | 1.02097 31613 | 5.1 | 0.54263 53519 | 8.9 | 0.41454 68462 | 15.6 | 0.31486 36051 |
| 1.4 | 0.98806 99961 | 5.2 | 0.53760 73540 | 9.0 | 0.41229 55493 | 15.8 | 0.31289 43424 |
| 1.5 | 0.95821 00533 | 5.3 | 0.53271 69744 | 9.1 | 0.41008 05783 | 16.0 | 0.31096 15880 |
| 1.6 | 0.93094 59808 | 5.4 | 0.52795 80329 | 9.2 | 0.40790 09662 | 16.2 | 0.30906 42269 |
| 1.7 | 0.90591 81386 | 5.5 | 0.52332 47316 | 9.3 | 0.40364 41245 | 16.4 | 0.30720 11919 |
| 1.8 | 0.88283 35270 | 5.6 | 0.51881 16252 | 9.5 | 0.40156 51322 | 16.6 | 0.30537 14592 |
| 1.9 | 0.86145 06168 | 5.7 | 0.51441 35938 | 9.6 | 0.39951 79693 | 16.8 | 0.30357 40487 |
| 2.0 | 0.84156 82151 | 5.8 | 0.51012 58183 | 9.7 | 0.39750 18313 | 17.0 | 0.30180 80193 |
| 2.1 | 0.82301 71525 | 5.9 | 0.50594 37583 | 9.8 | 0.39551 59416 | 17.2 | 0.30007 24678 |
| 2.2 | 0.80565 39812 | 6.0 | 0.50186 31309 | 9.9 | 0.39355 95506 | 17.4 | 0.29836 65276 |
| 2.3 | 0.78935 61312 | 6.1 | 0.49787 98929 | 10.0 | 0.39163 19344 | 17.6 | 0.29668 93657 |
| 2.4 | 0.77401 81407 | 6.2 | 0.49399 02237 | 10.2 | 0.38786 02539 | 17.8 | 0.29504 01817 |
| 2.5 | 0.75954 86903 | 6.3 | 0.49019 05093 | 10.4 | 0.38419 55846 | 18.0 | 0.29341 82062 |
| 2.6 | 0.74586 82430 | 6.4 | 0.48647 73291 | 10.6 | 0.38063 29549 | 18.2 | 0.29182 26987 |
| 2.7 | 0.73290 71515 | 6.5 | 0.48284 74413 | 10.8 | 0.37716 77125 | 18.4 | 0.29025 29472 |
| 2.8 | 0.72060 41251 | 6.6 | 0.47929 77729 | 11.0 | 0.37379 54971 | 18.6 | 0.28870 82654 |
| 2.9 | 0.70890 49774 | 6.7 | 0.47582 54066 | 11.2 | 0.37051 22156 | 18.8 | 0.28718 79933 |
| 3.0 | 0.69776 15980 | 6.8 | 0.47242 75723 | 11.4 | 0.36731 40243 | 19.0 | 0.28569 14944 |
| 3.1 | 0.68713 11010 | 6.9 | 0.46910 16370 | 11.6 | 0.36419 73076 | 19.2 | 0.28421 81554 |
| 3.2 | 0.67697 51139 | 7.0 | 0.46584 50959 | 11.8 | 0.36115 86616 | 19.4 | 0.28276 73848 |
| 3.3 | 0.66725 91831 | 7.1 | 0.46265 55657 | 12.0 | 0.35819 48784 | 19.6 | 0.28133 86117 |
| 3.4 | 0.65795 22725 | 7.2 | 0.45953 07756 | 12.2 | 0.35530 29318 | 19.8 | 0.27993 12862 |
| 3.5 | 0.64902 63377 | 7.3 | 0.45646 85618 | 12.4 | 0.35247 99643 | 20.0 | 0.27854 48766 |
| 3.6 | 0.64045 59647 | 7.4 | 0.45346 68594 | 12.6 | 0.34972 32746 | | |
| 3.7 | 0.63221 80591 | 7.5 | 0.45052 36991 | 12.8 | 0.34703 03081 | | |

The plus/minus signs in the argument of the exponential function in equation 4.14 are associated with downstream and upstream concentrations, respectively. This solution is useful in estuaries, where radionuclide concentrations must be estimated in both downstream (seaward) and upstream (landward) directions.

Radionuclide transport models for the screening (preliminary) assessment described below are based on these types of analytical solutions and can be used to estimate constant, continuous releases of radionuclides to rivers, estuaries, coastal waters, oceans, small lakes and large lakes with steady-state flow conditions. To be conservative, low-flow conditions are to be selected for the radionuclide concentrations for screening purposes. The following two steps are involved to obtain dissolved and sediment-sorbed radionuclides:

*Step 1:* Apply an appropriate screening methodology for a specific surface-water body to estimate an unfiltered (sum of dissolved and suspended-sediment sorbed) radionuclide concentration in a river, estuary, coastal water, ocean, small lake, or large lake.

*Step 2:* Calculate filtered (dissolved) radionuclide concentrations as well as those sorbed by suspended sediment, bed sediment, and shoreline/beach sediment, as described under "Sediment Effects," below.

These calculated radionuclide concentrations are then used for subsequent aquatic pathway and dose calculations. Numerical models described further below are general enough that they can be used to determine dissolved and sediment-sorbed radionuclide concentrations under both steady and time-varying flow and release conditions.

## Rivers

Rivers discussed here are nontidal rivers; thus, their flows are not affected by tides. Rivers receive radionuclides directly released to them from nuclear facilities located along the rivers, as well as radionuclides introduced to them through atmospheric fallout, groundwater seepage, and overland surface runoff.

### Basic River Characteristics

Basic river characteristics required to obtain the radionuclide concentration in a river are the river velocity, water depth, river width, diffusion/dispersion, sediment concentration, and possible sediment deposition/resuspension.

*Flow Characteristics* Most rivers experience a wide range of flows with varying width, depth, and velocity as well as sediment transport. A river (alluvial) flow is characterized by the interdependency of flow, sediment transport, bed form, and friction factor, as discussed under "Sediment Movement," above. This interdependency introduces a major difficulty for analyzing and simulating river flow and associated sediment and radionuclide transport. Various methods and formulas are available for predicting depth-discharge relationships and calculating sediment discharge in rivers (Vanoni 1975; Onishi 1994b).

In spite of this interdependency, it is generally accepted that the width, depth, velocity, and sediment transport rate increase with a river discharge as power functions for a wide range of mean annual river flow rates (Leopold et al. 1964). Based on these data, the following relationships can be derived:

$$B = 10 \, (q_w)^{0.460}, \tag{4.15}$$

$$d = 0.163 \, (q_w)^{0.447}, \tag{4.16}$$

where $q_w$ is the river flow rate ($m^3 s^{-1}$). The corresponding flow velocity, $U$ (m s$^{-1}$), may be obtained with the known discharge, depth, and width as:

$$U = \frac{q_w}{dB}. \tag{4.17}$$

We recommend selecting the 30-year-low annual flow rate for river discharge, $q_w$, to calculate conservative radionuclide concentrations (higher than actually expected). If site-specific values of river depth and flow velocity under a 30-year-low annual river flow rate are not known, the following steps may be used to obtain them (NCRP 1996):

*Step 1:* Estimate the river width, $\bar{B}$(m), which represents annual average river width from observation or from a map.

*Step 2:* Obtain the corresponding mean annual river flow rate, $\bar{q}_w$(m$^3$ s$^{-1}$), from the river width, $\bar{B}$, by using equation 4.15.

*Step 3:* Assume that the 30-year-low annual river-flow rate, $q_w$ (m$^3$ s$^{-1}$), is one-third of the mean annual river flow rate, $\bar{q}_w$, obtained in step 2.

*Step 4:* Obtain the corresponding river width, $B(m)$, from the 30-year-low annual river flow rate, $q_w$, by using equation 4.15 again.

*Step 5:* Obtain the corresponding river depth, $d(m)$, from the 30-year-low annual river flow rate, $q_w$, by using equation 4.16.

*Step 6:* Obtain the corresponding river flow velocity, $U$ (m s$^{-1}$), using equation 4.17.

Although it is desirable to use locally measured river characteristics, these values described above may be used to calculate default values of river width, depth, discharge, and velocity to obtain radionuclide concentrations based on analytical solutions to the general advection–diffusion equation 4.1.

*Dispersion* Another important transport process is diffusion/dispersion, which consists of molecular diffusion, turbulence diffusion, and dispersion. Dispersion is the largest among them and mostly results from averaging flow characteristics over space. Dispersion coefficients vary significantly from river to river or with river channel and flow conditions even in the same river. Examples of longitudinal and lateral dispersion coefficients for different rivers are shown in tables 4.4 and 4.5 (Sayre 1975; Fischer et al. 1979; Bowie et al. 1985).

The common expression of vertical dispersion coefficient in a river is

$$\varepsilon_z = 0.067 u_* d, \tag{4.18}$$

where $\varepsilon_z$ is the vertical dispersion coefficient (m$^2$ s$^{-1}$) and $u_*$ is shear velocity (m s$^{-1}$) (Sayre 1975).

Longitudinal and lateral dispersion coefficients in rivers vary by several orders of magnitude from small creeks to large rivers (e.g., the Mississippi and Columbia rivers), and many formulations are available to estimate these coefficients. The following expressions provide reasonably good estimates of the longitudinal and

Table 4.4 Examples of longitudinal dispersion coefficients in rivers

| Channel | Width, B (m) | Mean depth, $d$ (m) | Mean velocity, $U$ (m/s) | Shear velocity, $u_*$ (m/s) | Dispersion coefficient, $\varepsilon_x$ (m/s) | $\varepsilon_x/du_*$ |
|---|---|---|---|---|---|---|
| Missouri River | 200 | 2.70 | 1.55 | 0.074 | 1,500 | 7,500 |
| Sabine River | 104 | 2.04 | 0.58 | 0.05 | 315 | 3,100 |
|  | 127 | 4.75 | 0.64 | 0.08 | 670 | 1,800 |
| Yadkin River | 70 | 2.35 | 0.43 | 0.10 | 110 | 470 |
|  | 72 | 3.84 | 0.76 | 0.13 | 260 | 520 |
| Wind/Bighorn | 59 | 1.10 | 0.88 | 0.12 | 42 | 320 |
| Rivers | 69 | 2.16 | 1.55 | 0.17 | 160 | 440 |
| Nooksack River | 64 | 0.76 | 0.67 | 0.27 | 35 | 170 |
| Clinch River, | 47 | 0.85 | 0.32 | 0.067 | 14 | 235 |
| Tennessee | 60 | 2.10 | 0.94 | 0.104 | 54 | 245 |
|  | 53 | 2.10 | 0.83 | 0.107 | 47 | 210 |
| Chicago Ship Canal | 48.8 | 8.07 | 0.27 | 0.0191 | 3.0 | 20 |
| Clinch River, Virginia | 36 | 0.58 | 0.21 | 0.049 | 8.1 | 280 |
| Powell River | 34 | 0.85 | 0.15 | 0.055 | 9.5 | 200 |
| Bayou Anacoco | 26 | 0.94 | 0.34 | 0.067 | 33 | 520 |
|  | 37 | 0.91 | 0.40 | 0.067 | 39 | 640 |
| John Day River | 25 | 0.58 | 1.01 | 0.14 | 14 | 170 |
|  | 34 | 2.47 | 0.82 | 0.18 | 65 | 150 |
| Coachella Canal | 24 | 1.56 | 0.71 | 0.043 | 9.6 | 140 |
| Green-Duwamish River | 20 | 1.10 |  | 0.049 | 6.5–8.5 | 120–160 |
| Copper Creek | 16 | 0.49 | 0.27 | 0.080 | 20 | 500 |
|  | 18 | 0.85 | 0.60 | 0.100 | 21 | 250 |
|  | 16 | 0.49 | 0.26 | 0.080 | 9.5 | 245 |
|  | 19 | 0.40 | 0.16 | 0.116 | 9.9 | 220 |
| Comite River | 16 | 0.43 | 0.37 | 0.05 | 14 | 650 |
| Sacramento River |  | 4.00 | 0.53 | 0.051 | 15 | 74 |
| River Derwent |  | 0.25 | 0.38 | 0.14 | 4.6 | 131 |
| South Platte River |  | 0.46 | 0.66 | 0.069 | 16.2 | 510 |
| Yuma Mesa A Canal |  | 3.45 | 0.68 | 0.345 | 0.76 | 8.6 |

lateral dispersion coefficients over a wide range of river conditions (Sayre 1975; Fischer et al. 1979):

$$\varepsilon_x = \frac{U^2 B^2}{30\, du_*}, \quad (4.19)$$

$$\varepsilon_y = \alpha\, d\, u_*, \quad (4.20)$$

where $\varepsilon_x$ and $\varepsilon_y$ are longitudinal and lateral dispersion coefficients, respectively ($m^2\, s^{-1}$), and $\alpha$ is a proportionality coefficient (nondimensional).

Table 4.5  Examples of lateral dispersion coefficients in rivers

| Rivers | Channel description | Channel width, $B$ (m) | Mean depth, $d$ (m) | Mean velocity, $U$ (m/s) | Shear velocity, $u_*$ (m/s) | Lateral Dispersion Coefficient, $\varepsilon_y$ (m$^2$/s) | $\varepsilon_y/Du_*$ |
|---|---|---|---|---|---|---|---|
| Mackenzie River from Fort Simpson to Norman Wells | Generally straight alignment or slight curvature; numerous islands and sand bars | 1,240 | 6.7 | 1.77 | 0.152 | 0.67 | 0.66 |
| Mobile River | Mostly straight, one mild curve | 430 | 4.93 | 0.30 | 0.018 | 0.639 | 7.20 |
| Potomac River; 29 km reach below the Dickerson Power Plant | Gently meandering river with up to 60° bends | 350 | 0.73–1.74 | 0.29–0.58 | 0.033–0.051 | 0.013–0.058 | 0.52–0.65 |
| Athabasca River | Below Fort McMurray | 373 | 2.19 | 0.95 | 0.056 | 0.092 | 0.75 |
| Athabasca River | Below Athabasca | 320 | 2.05 | 0.86 | 0.079 | 0.066 | 0.41 |
| Columbia River | | 305 | 3.05 | 1.35 | 0.088 | 0.199 | 0.74 |
| Missouri River downstream of Cooper Nuclear Station, Nebraska | Reach includes one 90° and one 180° bend | 210–270 | 4.0 | 5.4 | 0.08 | 1.1 | 3.4 |
| Missouri River | Sinuous, severe bends | 234 | 3.96 | 1.98 | 0.042 | 0.549 | 3.30 |
| Missouri River near Blair, Nebraska | Meandering river | 200 | 2.7 | 1.75 | 0.074 | 0.12 | 0.6 |
| Missouri River | Two mild alternating bends | 183 | 2.66 | 1.74 | 0.073 | 0.117 | 0.60 |
| North Saskatchewan River | Below Edmonton | 213 | 1.55 | 0.58 | 0.080 | 0.030 | 0.25 |
| Bow River | At Calgary | 104 | 1.00 | 1.05 | 0.139 | 0.085 | 0.61 |
| Ijssel River | Groins on sides and gently curvature | 69.5 | 4.0 | 0.96 | 0.075 | 0.15 | 0.51 |
| Grand River | Below Kitchener | 59.2 | 0.51 | 0.35 | 0.069 | 0.009 | 0.26 |
| Lesser Slave River | Contorted meander | 43.0 | 2.53 | 0.65 | 0.049 | 0.041 | 0.33 |
| Beaver River | Near Cold Lake | 42.7 | 0.96 | 0.50 | 0.044 | 0.042 | 1.00 |
| Bernado Conveyance Channel | | 20.1 | 0.70 | 1.25 | 0.061 | 0.013 | 0.30 |
| South River | | 18.3 | 0.40 | 0.18 | 0.040 | 0.0046 | 0.29 |
| South River | Few mild bends | 18.2 | 0.40 | 0.21 | 0.040 | 0.0048 | 0.30 |
| Aristo Feeder Canal | | 18.3 | 0.67 | 0.67 | 0.062 | 0.0093 | 0.22 |

The proportionality coefficient, α, in equation 4.20 varies with a width-to-depth ratio, ranging from 0.1–0.2 for small laboratory flumes and medium-sized irrigation canals to 0.6–2.0 for the Missouri and MacKenzie rivers (Sayre 1975; Fischer et al. 1979). As a somewhat conservative value for the size of a river receiving a radionuclide release, let α = 0.6. Assuming (Fischer et al. 1979) that

$$u_* = 0.1\, U, \tag{4.21}$$

substituting equation 4.21 into equation 4.18 yields

$$\varepsilon_z = 0.0067\, Ud. \tag{4.22}$$

Similarly, substituting equation 4.21 into equations 4.19 and 4.20 yields the following expressions of longitudinal and lateral dispersion coefficients for rivers:

$$\varepsilon_x = \frac{UB^2}{3d}, \tag{4.23}$$

$$\varepsilon_y = 0.06\, dU. \tag{4.24}$$

To determine longitudinal distances to achieve complete mixing, it is assumed that complete vertical and lateral mixing is achieved when the minimum concentration is half the maximum concentration along the same vertical and lateral lines, respectively. Assuming that the radionuclide is released from one of the riverbanks at mid-depth, the longitudinal distances required to achieve this complete vertical and lateral mixing are given by $L_z$ and $L_y$, respectively:

$$L_z = 0.045\, \frac{Ud^2}{\varepsilon_z}, \tag{4.25}$$

$$L_y = 0.18\, \frac{UB^2}{\varepsilon_y}. \tag{4.26}$$

These distances were obtained solving two-dimensional advection–diffusion equations with the mirror image source technology (equation 4.12). Substituting equations 4.22 and 4.24 into equations 4.25 and 4.26 yields the following results:

$$L_z = 7d, \tag{4.27}$$

$$L_y = 3\, \frac{B^2}{d}. \tag{4.28}$$

Because almost all rivers are wider than they are deep, the distance $L_y$ is usually greater than $L_z$. Thus, the concentration in a region downstream of $L_y$ is completely mixed in both vertical and lateral directions, or over the river cross section.

### Screening River Model

These river models are for screening to determine whether a radionuclide release under consideration is acceptable or requires further detailed assessment. They estimate radionuclide concentrations in rivers for routine long-term radionuclide releases and are not intended for short-term or accidental releases. (Use methods described above under "Accidental Radionuclide Releases" for the latter case.) Because a radionuclide concentration is highest along the radionuclide plume centerline, we assume the radionuclide is released from a riverbank. If there is more than one source of radionuclide discharge, the radionuclide concentration for each discharge should be calculated separately and summed at the same receptor location to obtain the cumulative concentration from multiple releases. As discussed above, two steps are used: first, determine the unfiltered radionuclide concentration with an appropriate screening methodology for a specific surface water, and second, obtain dissolved and sediment-sorbed radionuclide concentrations.

Rivers are divided into the following three regions based on the degree of radionuclide mixing within a river cross section (see figure 4.2):

Region 1: The area where complete mixing in either vertical or lateral direction is not achieved

Region 2: The area where complete mixing is achieved in the vertical direction

Region 3: The area near the bank of the river opposite from the bank where a radionuclide release occurs

*Region 1*  If water is taken from a location before complete vertical mixing is achieved (i.e., $x \leq L_z = 7d$), the radionuclide concentration in this area is assumed not to be diluted by river water yet and is conservatively assumed to be the same as that within an effluent discharge outfall. Note that region 1 is a downstream area within seven times the river depth. The calculation of radionuclide concentration, which is an effluent radionuclide concentration, is given as

$$C = \frac{Q}{F}, \quad (4.29)$$

where $F$ is the flow rate of the liquid effluent (m³ s⁻¹).

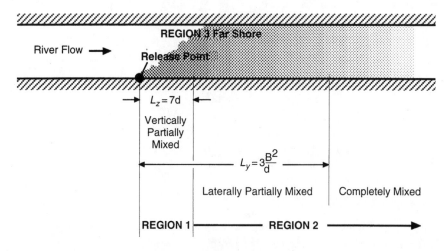

Figure 4.2   River mixing regions.

It is highly unlikely that drinking water would be taken from a location before complete vertical mixing has taken place (or worse, at the place where the radionuclide release has occurred). Under some circumstances, however, this case may be valid for the collection of some aquatic biota (e.g., fish). However, with care, one may take advantage of initial dilution within this region (Jirka et al. 1983).

*Region 2* After a radionuclide is completely mixed vertically ($x > L_z$), the radionuclide distribution is reduced from three to two dimensional. The mathematical model selected for this case is equation 4.10 and its associated solutions expressed in equations 4.11 and 4.12. A radionuclide concentration will be highest along the center of the radionuclide plume at any river cross section. To obtain a radionuclide concentration along the center of the plume, $y = y_0$ is entered into equations 4.11 and 4.12. Assuming that the radionuclides are released from a riverbank (i.e., $y_0 = 0$), as a conservative measure, equations 4.11 and 4.12 yield the following unfiltered radionuclide concentrations along the same side of the riverbank ($y = 0$) that the radionuclide is released:

$$C(x,y) = \frac{Q}{\pi d \sqrt{\varepsilon_x \varepsilon_y}} \cdot \exp\left(\frac{Ux}{2\varepsilon_x} - \frac{\lambda x}{U}\right) \cdot K_o\left(\frac{Ux}{2\varepsilon_x}\right), \quad (4.30)$$

Substituting longitudinal and lateral dispersion coefficients in equations 4.23 and 4.24 into equation 4.30, one can obtain

$$C(x,y) = \frac{Q}{0.142\pi \, dUB} \cdot \exp\left(\frac{1.5 \, dx}{B^2} - \frac{\lambda x}{U}\right) \cdot K_o\left(\frac{1.5 \, dx}{B^2}\right), \quad (4.31)$$

or it can be written as

$$C_r = C(x,y) = C_{tr} P_r, \quad (4.32)$$

where

$$C_{tr} = \frac{Q}{q_w} \exp\left(-\frac{\lambda x}{U}\right), \quad (4.33)$$

$$P_r = \frac{1}{0.142\pi} \exp(E) \cdot K_o(E), \quad (4.34)$$

$$E = \frac{1.5 \, dx}{B^2}, \quad (4.35)$$

where $q_w = dUB$ = river flow rate (m$^3$ s$^{-1}$).

Values of $P_r$ are given in figure 4.3 as a function of index $E$. Note that $C_{tr}$ in equation 4.33 is the completely mixed radionuclide concentration over a river cross section. Except for radionuclides with very short half-lives (e.g., $^{131}$I), the exponential term in equation 4.33 may be taken to be unity. The variable $P_r$ can be regarded as a correction factor for partial mixing and approaches unity as the downstream distance $x$ increases. Note that $P_r$ should be greater than or equal to unity.

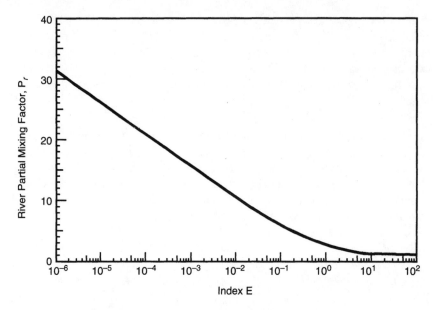

Figure 4.3  Partial mixing coefficient for rivers, $P_r$.

When the downstream distance $x$ becomes greater than $L_y$, complete lateral mixing has been achieved over the entire river cross section. For this case ($x > L_y$), $P_r$ becomes unity, and equation 4.32 is simplified to

$$C = C_{tr}. \tag{4.36}$$

*Region 3*  If water usage occurs on the opposite riverbank, the maximum radionuclide concentration on that bank is the cross-sectionally averaged concentration because a radionuclide must traverse at least half a river's width to reach this region. The unfiltered (sum of dissolved and suspended-sediment sorbed) radionuclide concentration in this region can be conservatively assigned as

$$C(x) = C_{tr}. \tag{4.37}$$

### EXAMPLE 4.2

After the Chernobyl nuclear accident, the Pripyat River in Ukraine received $5.0 \times 10^{11}$ Bq yr$^{-1}$ of $^{90}$Sr from around the Chernobyl nuclear plant through runoff water and groundwater seepage, totaling $1.0 \times 10^7$ m$^3$ yr$^{-1}$ of the effluent discharge. The Pripyat River width under a normal (mean annual) flow condition is 200 m. The nearest resident is living at the town of Chernobyl, which is 10 km downstream on the same (south) side of the river as the nuclear plant. Determine the $^{90}$Sr concentration in the river at the town of Chernobyl.

*Solution*: First determine the river characteristics from equation 4.15 with the normal river width of 200 m: $200 = 10 \, (\bar{q}_w)^{0.460}$. Thus, the mean annual river discharge $\bar{q}_w = 673$ m³ s⁻¹. The 30-year-low annual river discharge is one-third of 673 m³ s⁻¹, so $q_w = 224$ m³ s⁻¹. With $q_w$ of 224 m³ s⁻¹ and equation 4.15, the river width under the 30-year-low annual river discharge $B$ is $10(224)^{0.460} = 121$ m. Similarly, with equation 4.16, the river depth in this 30-year-low annual river discharge $d = 0.163(121)^{0.447} = 1.39$ m. From equation 4.17, the river velocity for this discharge $U = q_w/dB = 224/(1.39 \times 121) = 1.33$ m s⁻¹. The distance required to have vertical complete mixing is $L_z = 7d = 7 \times 1.39$ m $= 9.7$ m from equation 4.27. Because the town of Chernobyl is 10 km downstream from the Chernobyl nuclear plant, $x = 10{,}000 > L_z = 9.7$ m. Thus, this river location is in region 2 (see figure 4.2). The $^{90}$Sr release rate $Q = 5.0 \times 10^{11}$ Bq yr⁻¹ $= 15{,}900$ Bq s⁻¹. The half-life of $^{90}$Sr is 29.1 years, so the decay rate $\lambda = \ln 2 / (29.1 \times 365 \times 24 \times 60 \times 60) = 7.55 \times 10^{-10}$ s⁻¹. Equation 4.33 yields the cross-sectional averaged $^{90}$Sr concentration to be

$$C_{tr} = \frac{15{,}900}{224} \exp\left\{-\frac{(7.55 \times 10^{-10})(10{,}000)}{1.33}\right\} = 71.0 \text{ Bq m}^{-3}.$$

As shown in equation 4.35, index $E = 1.5 \, dx / B^2 = (1.5 \times 1.39 \times 10{,}000)/121^2 = 1.42$. The river partial mixing factor is $P_r = 1.8$ from figure 4.3 with index $E$ of 1.42. From equation 4.32, the $^{90}$Sr concentration in the Pripyat River at the town of Chernobyl is $C_r = (71.0 \times 1.8) = 128$ Bq m⁻³ $= 0.128$ Bq l⁻¹.

## Estuaries

An estuary is defined as a semienclosed coastal water body that is connected at one end to a river and at the other end to a sea. The estuary is considered to extend to the point where tide no longer exists. An estuary can have fresh or saline water, although salinity is generally much lower than the 34–35 parts per thousand of seawater. Estuaries receive radionuclides released directly from nuclear facilities located along them, as well as those coming from upstream rivers, surrounding coastal waters/oceans, and atmospheric fallout.

### Basic Estuarine Characteristics

Estuaries are natural deposits of various radionuclides and other contaminants. They are complex water bodies affected by various factors (e.g., tide, river flow, wind, and salinity). Basic estuarine characteristics required to obtain the radionuclide concentration in an estuary are the flow velocity, tide, water depth, estuary width, diffusion/dispersion, and sediment deposition.

*Flow Characteristics* The main factors controlling an estuarine flow are tides, river flow, and currents induced by water density differences, wind, and waves. Tide is an oscillatory current caused by the gravitational attractions of the moon and sun on the rotating earth. It dominates flows in most estuaries and coastal waters. Based on the tidal cycle, flows may be categorized as semidiurnal, diurnal, and mixed tides (Ippen 1966). A semidiurnal tide has 12.42 h in an ideal cycle (twice per lunar day) and two roughly equal tidal ranges. The U.S. East Coast tends to have semidiurnal tides. A diurnal tide has 24.84 h in a tidal cycle (once per lunar day) and is more common on the Gulf of Mexico. A mixed tide is a combination of these two and tends to have one large and one small tide occurring in a lunar day, as often occurs on the U.S. West Coast. Depending on how tides move through the estuary, they may consist of progressive, standing, and mixed waves affected by the shape and depth of theestuary (Weigel 1964).

Because an estuary is connected to a sea and a river, some water density variations occur. The density differences in an estuary are mostly due to the fresh, lighter (and often warmer) river water meeting saline, heavier (and usually colder) seawater and producing salinity and temperature variations. The density difference can generate both vertical and horizontal circulation within an estuary and produces a net downstream flow near the water surface and a net upstream flow near the estuarine bottom. When there is a significant density difference, an estuary is stratified. This is usually caused by a large river (e.g., the Mississippi River) coming into an estuary. If an estuary has a small river inflow, the density variation becomes insignificant, and it becomes a well-mixed estuary. Falling between these two mixing conditions is a partially mixed estuary (e.g., the Columbia River Estuary).

Wind over an estuary can push the water near its surface to the downwind direction, producing a countercurrent near the estuarine bottom (Weigel 1964). The wind also may generate standing waves, or seiches. Wind-induced waves are also potentially important in an estuary because of their enhanced effect on sediment transport (Onishi et al. 1993b).

Even through a tidal flow dominates most estuarine flows, these nonoscillating flows induced by density variations, wind, and waves have significant impacts on a long-term radionuclide transport in an estuary (Onishi et al. 1993b).

Many radionuclides are adsorbed by sediment, especially by fine silt and clay (see table 4.2). The transport, deposition, and resuspension of fine sediment are important in the evaluation of radionuclide migration in surface waters. An estuary is an effective sediment trap, receiving much of the sediment carried by a river and passing only a small amount on to the sea. The estuarine bed becomes a long-term source of pollution (Onishi 1994a).

*Dispersion* Dispersion processes in an estuary are generally more complex than are those in a river because of the additional effects of tides, density stratification, wind, and complex estuarine channel geometry and bathymetry. Unlike rivers, no general theoretical expressions cover a wide range of estuarine conditions. Table 4.6 presents examples of longitudinal dispersion coefficients in some estuaries (Ward 1974, 1976; Fischer et al. 1979).

For the vertical dispersion coefficient, equation 4.18 is still valid. However, this equation, as well as equations 4.19 and 4.20 in this case, uses the mean flow speed over a tidal cycle, $U_t$, instead of the actual flow velocity used for a river case.

Table 4.6  Longitudinal dispersion coefficients for estuaries

| Estuary | Freshwater inflow, $q_w$(m$^3$/s$^{-1}$) | Tidally averaged velocity, $U$(m/s$^{-1}$) | Longitudinal dispersion coefficient, $\varepsilon_x$(m$^2$/s$^{-1}$) |
|---|---|---|---|
| Cooper River, South Carolina | 280 | 0.076 | 900 |
| Savannah River, Georgia/South Carolina | 200 | 0.052–0.21 | 300–600 |
| Hudson River, New York | 140 | 0.011 | 450–1500 |
| Delaware River, Delaware | 70 | 0.037–0.31 | 100–1500 |
| Houston Ship Channel, Texas | 30 | 0.015 | 800 |
| Cape Fear River, North Carolina | 30 | 0.0092–0.15 | 60–300 |
| Potomac River, Virginia | 15 | 0.00092–0.0018 | 6–300 |
| Lower Raritan River, New Jersey | 4 | 0.0088–0.014 | 150 |
| South River, New Jersey | 0.7 | 0.0031 | 150 |
| Compton Creek, New Jersey | 0.3 | 0.0040–0.031 | 30 |
| Wappinger and Fishkill Creek, New York | 0.06 | 0.00031–0.0012 | 15–30 |
| East River, New York | 0 | 0 | 300 |
| San Francisco Bay | | | |
| Southern arm | | | 20–200 |
| Northern arm | | | 50–2000 |
| Rotterdam Waterway, Holland | | | 280 |
| Rio Quayas, Equador | | | 760 |
| Severn Estuary, UK | | | |
| Summer | | | 50–120 |
| Winter | | | 120–500 |
| Thames River, UK | | | |
| Low river flow | | | 50–90 |
| High river flow | | | 300 |
| Tay Estuary, UK | | | 50–150 |
| Narrows of Mercey, UK | | | 100–400 |

Thus,

$$u_* = 0.1\ U_t, \quad (4.38)$$

$$\varepsilon_z = 0.0067\ U_t d, \quad (4.39)$$

where $U_t$ is 0.32 ($|U_e| + |U_f|$). Note that $U_e$ and $U_f$ are maximum ebb and flood velocities (m s$^{-1}$), respectively, and that the tidal speed is assumed to vary sinusoidally with time.

The lateral mixing in an estuary tends to be several times larger than the corresponding lateral mixing in a nontidal river, mainly due to large lateral flows caused by irregular channel geometry and cross section, tides, density stratification, and wind (Fischer et al. 1979).

The proportionality factor, $\alpha$, in equation 4.20 may be assigned to be 3 for an estuary:

$$\varepsilon_y = 3 \, du_*. \quad (4.40)$$

Substituting equation 4.38 for equation 4.40 yields the following lateral dispersion coefficient for an estuary:

$$\varepsilon_y = 0.3 \, dU_t. \quad (4.41)$$

The longitudinal mixing in an estuary is also affected by an oscillating tidal flow. The ratio $N$ of the longitudinal dispersion coefficient in an estuary to that in a river is expressed by the ratio $M$ of the tidal period to the time scale for cross-sectional mixing (Fischer et al. 1979). Expressing the cross-sectional mixing time by $B^2/\varepsilon_y$, and $\varepsilon_y$ as expressed in equation 4.40, the time-scale ratio $M$ becomes

$$M = \frac{0.3 \, d \, U_t T_p}{B^2}, \quad (4.42)$$

where $T_p$ is a tidal period. $T_p$ will be 45,000 s for a semidiurnal tide occurring twice per day (e.g., those on the U.S. East Coast) and 90,000 s for a diurnally dominating tide occurring once per day (e.g., those occurring along the Gulf of Mexico and the U.S. West Coast):

$$M = \frac{13,500 \, dU_t}{B^2} \text{ for twice per day tide,}$$

$$M = \frac{27,000 \, dU_t}{B^2} \text{ for once per day tide.}$$

The ratio of the longitudinal dispersion coefficient in an estuary to that in a river, $N$, is obtained as a function of $M$, as shown in figure 4.4. Because the longitudinal dispersion coefficient in a river is obtained by equation 4.19, the longitudinal dispersion coefficient in an estuary is

$$\varepsilon_x = \frac{N \, U_t \, B^2}{3 \, d}. \quad (4.43)$$

Substituting equation 4.40 into equation 4.26 yields the following longitudinal distance, $L_y$, required to achieve complete lateral mixing:

$$L_y = 0.6 \frac{B^2}{d}. \quad (4.44)$$

Similarly, the longitudinal distance, $L_z$, required to achieve complete vertical mixing can be obtained by substituting equation 4.39 into equation 4.25 as

$$L_z = 7d. \quad (4.45)$$

As in the river case, vertical (or lateral) complete mixing is achieved in the estuary when the minimum concentration is half of the maximum concentration along the same vertical (or lateral) line.

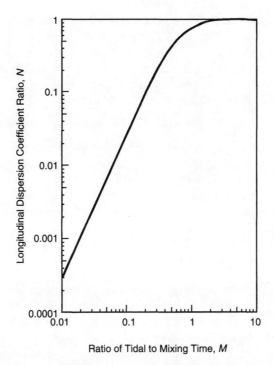

**Figure 4.4** Variation of longitudinal dispersion coefficient ratio, $N$, with the ratio of tidal period to mixing time, $M$.

## Screening Estuary Methodology

Estuarine models presented here are for screening purposes to estimate unfiltered radionuclide concentrations in estuaries for routine, long-term radionuclide releases. For short-term or accidental releases, the instantaneous release models described above under "Accidental Radionuclide Releases" may be used with the estuarine dispersion coefficients and the tidally averaged velocity. Because a radionuclide concentration is highest along the plume centerline, we assume the radionuclide is released from an estuarine bank. The water use location may be downstream or upstream from the release location.

A mathematical model selected for constant, continuous release of a radionuclide into an estuary is based on the same equation 4.10 and its solutions used for rivers. The following are differences from river flow conditions to express distinct estuarine tidal flow characteristics:

An actual flow velocity (a tidally varying velocity) is not used, but a tidally averaged flow velocity (net freshwater flow velocity) is used to advect (transport) a radionuclide for the analysis.

Vertical dispersion coefficient is still evaluated by equation 4.18 but is based on a mean flow speed over a tidal cycle.

Longitudinal and lateral dispersion coefficients are estimated based on a mean flow speed over a tidal cycle, corrected to reflect tidally varying hydrodynamics.

Water-use locations, both upstream and downstream from the radionuclide release point, are analyzed, and a radionuclide concentration in the upstream location is corrected by the tidal effect.

Estuaries are divided into the following four regions (see figure 4.5):

Region 1: The downstream and upstream areas where complete mixing in either vertical or lateral direction is not achieved

Region 2: The downstream area where complete mixing is achieved in the vertical direction

Region 3: The upstream area where complete mixing is achieved in the vertical direction

Region 4: The area near the bank of the estuary opposite from the bank where a radionuclide release occurs

*Region 1* If water usage occurs before complete vertical mixing is achieved (i.e., $x \leq L_z = 7d$), the radionuclide concentration in this area is assumed not yet diluted by estuarine water and is the same as a location within an effluent discharge outfall. Region 1 is an upstream or downstream area within seven times the estuarine depth. The calculation of radionuclide concentration is given as

$$C = \frac{Q}{F}, \qquad (4.46)$$

where $F$ is the flow rate of the liquid effluent in $m^3\ s^{-1}$. However, one may take advantage of initial mixing dilution within this region (Jirka et al. 1983).

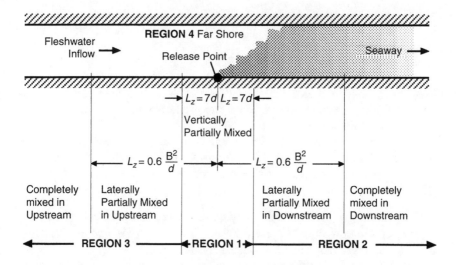

Figure 4.5  Estuarine mixing regions.

*Region 2* After a radionuclide is completely mixed vertically ($x > L_z$), the radionuclide distribution is reduced from three to two dimensional. The mathematical model selected for this region is the same as for a river (equation 4.10 and its associated solutions expressed in equations 4.11 and 4.12). Radionuclide concentration will be highest along the center of the radionuclide plume at any estuarine cross section. Assuming that the radionuclides are released from an estuarine bank (i.e., $y_0 = 0$) as a conservative measure, these equations yield the following concentrations:

$$C(x, y) = \frac{Q}{\pi d \sqrt{\varepsilon_x \varepsilon_y}} \cdot \exp\left(\frac{Ux}{2\varepsilon_x} - \frac{\lambda x}{U}\right) \cdot K_0\left(\frac{U}{2\varepsilon_x}\sqrt{x^2 + \frac{\varepsilon_x}{\varepsilon_y}(y - y_0)^2}\right). \quad (4.47)$$

Substituting longitudinal and lateral dispersion coefficients in equations 4.41 and 4.43 into equation 4.47, one can obtain the following estuarine equation:

$$C(x, y) = \frac{Q}{0.632 \, \pi \, d \, U_t \, B \sqrt{N}} \exp\left(\frac{1.5 \, dx}{N \, B^2} \frac{U_a}{U_t}\right) \cdot$$

$$K_0\left(\frac{1.5 \, d}{N \, B^2} \frac{U_a}{U_t}\sqrt{x^2 + \frac{N \, B^2}{0.9 \, d^2}(y - y_0)}\right) \exp\left(-\frac{\lambda x}{U_a}\right), \quad (4.48)$$

where the tidally averaged velocity is $U_a = q_w / Bd$.

The following radionuclide concentration can be obtained along the estuarine shore downstream from where a radionuclide is released ($y = y_0 = 0$):

$$C_{ed} = C(x, y_0) = C_{te} \, P_e, \quad (4.49)$$

where the concentration for complete mixing is given by

$$C_{te} = \frac{Q}{q_t} \exp\left[-\frac{\lambda x}{U_a}\right], \quad (4.50)$$

and $q_t = dU_t B$ = average tidal discharge.

The estuarine partial mixing factor, $P_e$, is given as

$$P_e = \frac{1}{0.32 \, \pi \sqrt{N}} \exp(E) \, K_0 \, (E), \quad (4.51)$$

where

$$E = \frac{1.5 \, dx}{N \, B^2} \frac{U_a}{U_t}. \quad (4.52)$$

Values of $P_e$, which should be greater than 1 and less than $q_t / Q$, are given in figure 4.6 as a function of the argument $E = 1.5 dx \, U_a / NB^2 U_t$ and longitudinal dispersion coefficient ratio, $N$ (see figure 4.4). Equation 4.49 is used to calculate radionuclide concentration in the area downstream from the radionuclide release point to reflect the estuarine bank effects restricting radionuclide lateral spread. Conservatively, this downstream concentration may also be used for the upstream concentration without accounting for reduced upstream mixing. Note that $C_{te}$ is a concentration of a radionuclide after complete mixing over a cross section is reached, and $P_e$ can be considered as the correction factor for estuarine partial mixing.

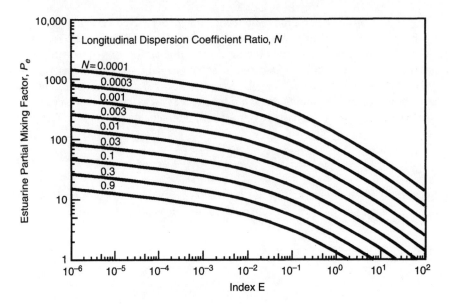

Figure 4.6  Estuarine partial mixing factor, $P_e$.

*Region 3*  To estimate upstream radionuclide concentrations more accurately, a one-dimensional equation is used to derive the necessary correction factor. As discussed under "Routine Long-Term Radionuclide Releases," above, the one-dimensional solution of radionuclide concentrations either upstream or downstream is obtained by equation 4.14. The ratio of upstream to downstream concentration is

$$\text{UCF} = \exp\left\{ \frac{U_a x}{\varepsilon_x} \sqrt{1 + \frac{4\lambda\varepsilon_x}{U_a^2}} \right\}. \tag{4.53}$$

Values of $x$ should be taken as negative to calculate the upstream correction factor, UCF, in equation 4.53. UCF is less than unity. Substituting equation 4.43 into equation 4.53 yields

$$\text{UCF} = \exp\left( \frac{3 x d\, U_a}{N B^2 U_t} \right). \tag{4.54}$$

The partial mixing of the unfiltered (sum of dissolved and suspended-sediment sorbed) radionuclide concentration, $C_{eu}$, in an upstream region before complete mixing over the entire cross section is:

$$C_{eu}(x, y) = \text{UCF} \cdot C_{ed}, \tag{4.55}$$

where $C_{ed}$ is calculated by equation 4.49.

Note that the upstream distance, $x$, must be shorter than the actual distance the radionuclide can travel during a flood tide period. The upstream travel distance, $L_u$, may be calculated by

$$L_u = 0.32\, U_f\, T_p. \tag{4.56}$$

*Region 4* If water usage occurs either upstream or downstream on the opposite estuarine bank, the maximum radionuclide concentration in that bank is the cross-sectionally averaged concentration because a radionuclide must traverse at least half the estuarine width to reach this region. The maximum unfiltered radionuclide concentration in this region can be obtained by

$$C = C_{te}, \tag{4.57}$$

where $C_{te}$ is given by equation 4.50 and the upstream water use location is within the distance, $L_u$, calculated by equation 4.56.

### EXAMPLE 4.3

Ruthenium-106 is released to an estuary along the U.S. East Coast at the rate of $3.7 \times 10^{11}$ Bq yr$^{-1}$, with the radioactive effluent discharge of $3.15 \times 10^7$ m$^3$ yr$^{-1}$. A 30-year-low annual freshwater discharge is not known, but the normal river width just before the river becomes an estuary is 66 m. The estuarine width and depth are 100 m and 2.0 m, respectively. The maximum ebb velocity is 1.0 m s$^{-1}$. The nearest water use location is 1 km upstream from the radionuclide outfall on the same side of the estuary. Determine $^{106}$Ru concentration at the water use location.

*Solution*: First determine the characteristics of the river and estuary from equation 4.15 with the normal river width of 66 m, $66 = 10(\bar{q}_w)^{0.460}$. Thus, the mean annual freshwater (river) discharge $\bar{q}_w = 60$ m$^3$ s$^{-1}$. The 30-year-low annual river discharge is one-third of 60 m$^3$ s$^{-1}$, so $q_w = 20$ m$^3$ s$^{-1}$. The freshwater river velocity (tidally averaged velocity) in the estuary $U_a = q_w/(Bd) = 20/(100 \times 2) = 0.10$ m s$^{-1}$. Since the maximum flood velocity $U_f$ is not known, assume its magnitude is the same as that of the maximum ebb velocity, $U_e$. The mean tidal velocity $U_t = 0.32(|U_e| + |U_f|) = 0.32(1.0 + 1.0) = 0.64$ m s$^{-1}$. The distance required to have complete vertical mixing is $L_z = 7d = 7 \times 2$ m $= 14$ m from equation 4.45. Since the water use location is 1 km upstream from the release point, $x = 1,000 > L_z = 14$ m. We also need to examine whether $^{106}$Ru will travel 1 km upstream during the flood tide. The U.S. East Coast generally has a semidiurnal tide, so the tidal period $T_p = 45,000$ s (or 12.5 h). Since the travel distance during the flood tide $L_t = 0.32 T_p U_f = 0.32(45,000 \times 1.0) = 14,400$ m, $^{106}$Ru can reach 1 km upstream during the flood period. This water use location is thus in region 3 (see figure 4.5).

The $^{106}$Ru release rate $Q = 3.7 \times 10^{11}$ Bq yr$^{-1}$ = 11,700 Bq s$^{-1}$. The half-life of $^{106}$Ru is 368 days, so the decay rate $\lambda = \ln 2 / (368 \times 24 \times 60 \times 60) = 2.18 \times 10^{-8}$ s$^{-1}$. Equation 4.50 yields the cross-sectionally averaged $^{106}$Ru concentration:

$$C_{tr} = \frac{11,700}{(100 \times 2 \times 0.64)} \exp\left\{-\frac{(2.18 \times 10^{-8})(1,000)}{0.10}\right\} = 91.4 \text{ Bq m}^{-3}$$

From equation 4.44, the distance required to complete mixing across the estuarine cross section $L_y = 0.6 B^2/d = (0.6 \times 100^2)/2 = 3,000$ m. Since $x = 1,000$ m $< L_y = 3,000$ m, $^{106}$Ru is not completely mixed across the estuarine cross section at the water use location. For the East Coast, the ratio of the tidal period to the mixing time, $M$, is

$$M = \frac{(13,500 \times 2 \times 0.64)}{(100)^2} = 1.72.$$

From figure 4.4, the longitudinal dispersion coefficient ratio $N = 0.85$. Thus, index $E$ is

$$E = \frac{(1.5 \times 1,000 \times 2 \times 0.1)}{(100)^2(0.64 \times 0.85)} = 0.0551.$$

From figure 4.6, with $N = 0.85$ and $E = 0.0551$, $P_e = 4.0$. From equation 4.49, the partial mixed concentration $C_{ed} = 91.4 \times 4.0 = 366$ Bq m$^{-3}$. To further calculate the concentration with upstream correction, use equation 4.54. The upstream correction factor, UCF, is

$$\text{UCF} = \exp\left[\frac{(3)(-1,000 \times 2 \times 0.10)}{(0.85)(100)^2(0.64)}\right] = 0.90.$$

From equation 4.55, the $^{106}$Ru concentration at the water use location 1 km upstream is $C_{eu} = 0.90 \times 366 = 329$ Bq m$^{-3}$.

## Coastal Waters and Oceans

Coastal waters and oceans receive radionuclides released directly from nuclear facilities located along them, as well as those entered into these water bodies from atmospheric fallout and rivers.

### Basic Coastal Water and Ocean Characteristics

Basic characteristics of costal waters and oceans required to obtain the radionuclide concentration are the current velocity, water depth, and diffusion/dispersion.

*Flow Characteristics* Flows in coastal water and oceans may consist of currents driven by density variation, wind and waves, and tide (Weigel 1964; Ippen 1966). The ocean has surface, interior, and bottom boundary layers. The ocean surface layer with its thickness of a few hundred meters is the most active area, where all these flows can occur. Similar to estuaries, the surface layer has (1) a rotary tidal flow in open coast/sea, a reverse tidal current in narrow coastal water, or hydrantic tide caused by connecting two water bodies with two different tides; (2) a flow induced by the wind stress on the water surface and resulting changes on density and flow patterns; and (3) wave-induced currents. As waves break near the shore, a long-shore current parallel to the shore and a rip current roughly perpendicular to the shore appear within a near-shore zone (Weigel 1964).

Below this surface layer, there is a deep interior region where the water temperature is almost constant and has interior deep water flows induced by density variation. At the ocean bottom, there are thin bottom boundary layer flows (Onishi et al. 1987).

Density variations generate large-scale, relatively stable ocean currents (Weigel 1964). For example, strong currents appear along western boundaries of the oceans, such as the Gulf Stream in the North Atlantic and the Kuroshio in the North Pacific. These currents may have velocities up to 1.5 m s$^{-1}$ and can shed a series of eddies (warm core rings). However, flows in most ocean areas are much slower (about 10 cm s$^{-1}$) and more variable (Weigel 1964; Ippen 1966; Onishi et al. 1987).

Figure 4.7 shows the general surface layer flow pattern off the U.S. East Coast, together with 2,800-m and 3,800-m disposal sites (formerly U.S. low-level radioactive waste ocean disposal sites on 2,800-m- and 3,800-m-deep ocean bottoms) (Ingham et al. 1977).

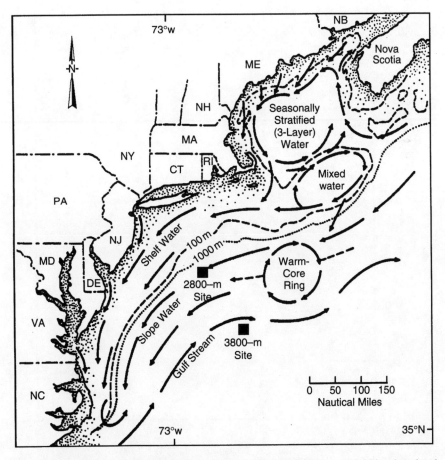

Figure 4.7 Surface current along the U.S. East Coast around 2,800-m and 3,800-m low-level radioactive waste ocean disposal sites. From Ingham et al. (1977).

*Dispersion Coefficient and Mixing* Unlike dispersion in rivers and estuaries, the dispersion in coastal and ocean environments is scale dependent. Characteristic mixing lengths in rivers and estuaries are probably related to the water depth and channel width (both with fixed sizes), while in coastal water and ocean, the characteristic mixing length is related to the distance between the contaminant release point and a receptor location; thus, it keeps increasing as the contaminant travels farther with time. This scale-dependent mixing characteristic is clearly revealed in figure 4.8, showing the dispersion coefficient increasing with 4/3 power of the distance (Okubo 1971).

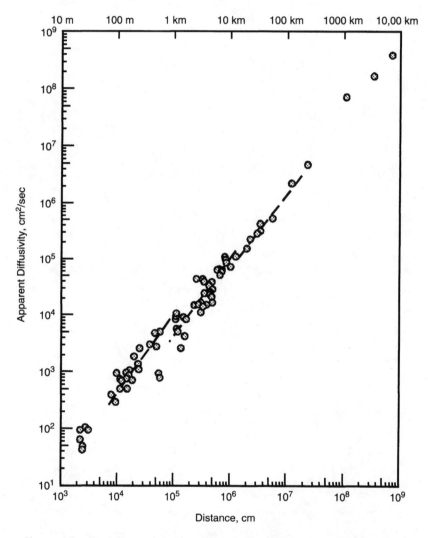

Figure 4.8  Spatially varying ocean dispersion coefficient. From Okubo (1971).

## Coastal Water Screening Model

A mathematical model for coastal water may be based on the following steady, vertically averaged, two-dimensional advection–diffusion equation:

$$U_c \frac{\partial C}{\partial x} = \varepsilon_y \frac{\partial^2 C}{\partial y^2} - \lambda C, \qquad (4.58)$$

where $U_c$ = the coastal current (m s$^{-1}$). Note that the longitudinal dispersion is not included in equation 4.58, resulting in a somewhat conservative estimate of the radionuclide concentration. Based on Okubo's (1971) dispersion expression shown in figure 4.8, the longitudinal and lateral dispersion coefficient, $\varepsilon_x$ and $\varepsilon_y$ (m$^2$ s$^{-1}$), may be expressed as (NCRP 1996)

$$\varepsilon_x = 4.66 \times 10^{-6} \left(\frac{x}{U_c}\right)^{1.34}, \qquad (4.59)$$

$$\varepsilon_y = 3.44 \times 10^{-7} \left(\frac{x}{U_c}\right)^{1.34}. \qquad (4.60)$$

If the site-specific, scale-dependent dispersion coefficient is available, it should be used instead of equations 4.59 and 4.60.

Assume that (1) the shoreline is straight along the x-axis ($y = 0$); (2) water depth, $d$ (m), is constant; and (3) the coastal current, $U_c$ (m s$^{-1}$), is constant and parallel to the shoreline (see figure 4.9). Then, the following analytical solution to equation 4.58 may be obtained for a continuous, steady radionuclide release from a discharge point that is $y_0$ (m) away from the shore ($x = 0, y = y_0$) (Sayre 1975; Edinger and Buchak 1977):

$$C(x,y) = \frac{Q}{d\sqrt{\pi\, U_c \varepsilon_y x}} \cdot \exp\left\{-\frac{U_c(y - y_0)^2}{4\,\varepsilon_y x} - \frac{\lambda x}{U}\right\}. \qquad (4.61)$$

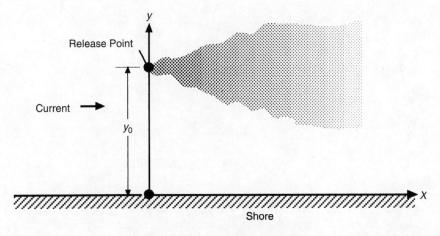

Figure 4.9  Ocean mixing area.

The unfiltered radionuclide concentration expressed in equation 4.61 includes a shoreline effect to restrict lateral dispersion of a radionuclide plume using the mirror image technique expressed in equation 4.12. If the site-specific current velocity is not known, $U_c = 0.1$ m s$^{-1}$ may be used as a default value (NCRP 1996).

Note that equations 4.58 and 4.61 can be obtained from the more general equations 4.10 and 4.11 by assuming that the longitudinal dispersion is not important, that is, imposing the following conditions (Sayre 1975):

$$\frac{\varepsilon_y}{\varepsilon_x} \left(\frac{y - y_0}{x}\right)^2 \ll 1 \qquad (4.62)$$

and

$$\frac{x U_c}{2\varepsilon_x} \gg 1. \qquad (4.63)$$

Using the scale-dependent dispersion coefficients expressed in equations 4.59 and 4.60 and $U_c = 0.1$ m s$^{-1}$, these two conditions may be rewritten as the following conditions to be satisfied:

$$7d < x \ll 8 \times 10^7 \text{ m}, \qquad (4.64)$$

$$\left|\frac{y - y_0}{x}\right| \ll 3.7. \qquad (4.65)$$

Substituting equation 4.60 into equation 4.61 yields

$$C(x,y) = \frac{962 \, U_c^{0.17} \, Q}{d x^{1.17}} \exp\left\{-\frac{7.28 \times 10^5 \, U_c^{2.34} \, (y - y_0)^2}{x^{2.34}} - \frac{\lambda x}{U_c}\right\}. \qquad (4.66)$$

Radionuclide concentrations along the plume center and shoreline can be obtained by assigning $y = y_0$ and $y = 0$, respectively, in equations 4.61 and 4.66. Thus, along the plume centerline,

$$C(x,y) = \frac{Q}{d \sqrt{\pi \varepsilon_y x}} \cdot \exp\left\{-\frac{\lambda x}{U_c}\right\}, \qquad (4.67)$$

$$C(x,y) = \frac{962 \, U_c^{0.17} \, Q}{d x^{1.17}} \exp\left\{-\frac{\lambda x}{U_c}\right\}. \qquad (4.68)$$

Along the shoreline,

$$C(x,y) = \frac{Q}{d \sqrt{\pi \varepsilon_y x}} \cdot \exp\left\{-\frac{U_c y_0^2}{4 \varepsilon_{yx}} - \frac{\lambda x}{U_c}\right\}, \qquad (4.69)$$

$$C(x,y) = \frac{962 U_c^{0.17} Q}{d x^{1.17}} \cdot \exp\left\{-\frac{7.28 \times 10^5 \, U_c^{2.34} y_0^2}{x^{2.34}} - \frac{\lambda x}{U_c}\right\}. \qquad (4.70)$$

For cases where a receptor is located within seven times the water depth, equation 4.29 may be used, assuming there is no dilution within a receiving coastal water, or one may consider initial mixing (Jirka et al. 1983).

## EXAMPLE 4.4

Plutonium-241 is released to the Pacific coastal water at a rate of $3.7 \times 10^{11}$ Bq yr$^{-1}$. The release point is 200 m off the beach. Water depth at the release point is 40 m. Determine $^{241}$Pu concentrations at the plume center and on the beach 3 km downcurrent.

*Solution*: The $^{241}$Pu release rate $Q = 3.7 \times 10^{11}$ Bq yr$^{-1}$ = 11,700 Bq s$^{-1}$. Since the half-life of $^{241}$Pu is 14.4 years, its decay rate $\lambda = \ln 2 / (14.4 \times 365 \times 24 \times 60 \times 60) = 1.53 \times 10^{-9}$ s. The velocity of the coastal current is not known here, so we will use the default coastal current velocity, $U_c = 0.1$ m s$^{-1}$.

From equation 4.68, $^{241}$Pu concentration along the plume centerline at 3 km is

$$C = \frac{(962)(0.1)^{0.17}(11,700)}{(40)(3,000^{1.17})} \exp\left\{-\frac{(1.53 \times 10^{-9})(3,000)}{0.1}\right\} = 16.3 \text{ Bq m}^{-3}.$$

The beach location is 3 km downcurrent and 200 m away from the plume center. From equations 4.64 and 4.65,

$$7(40) = 280 \text{ m} < 3{,}000 \text{ m} \ll 8 \times 10^7 \text{m},$$

$$\left|\frac{0-200}{3,000}\right| = 0.0667 \ll 3.7.$$

We can use equation 4.70 to calculate the $^{241}$Pu concentration on the beach at 3 km:

$$C = \frac{(962)(0.1)^{0.17}(11,700)}{(40)(3,000^{1.17})} \exp\left\{-\frac{(7.28 \times 10^5)(0.1^{2.34})(200^2)}{3,000^{2.34}}\right.$$
$$\left. -\frac{(1.53 \times 10^{-9})(3,000)}{0.1}\right\} = 6.15 \text{ Bq m}^{-3}.$$

### Lakes

Lakes are biologically very active water bodies. Lakes receive radionuclides released directly from nuclear facilities located along them, as well as those entering lakes from upstream rivers, groundwater seepage, overland surface runoff, and atmospheric fallout.

### Basic Lake Water Characteristics

Basic lake characteristics required to obtain the radionuclide concentration are the lake size and depth, lake velocity, diffusion/dispersion, and sediment deposition.

*Flow Characteristics*  Unique processes are responsible for radionuclide distributions and movements in lakes. The important processes are (1) lake flows, including wind-induced currents and seiches, (2) stratification and seasonal turnover, (3) sediment deposition and adsorption/desorption, (4) chemical reactions and water quality, and (5) biotic interaction (Onishi 1985). Although they may not be important for radionuclide distributions in lakes except for possibly affecting radionuclide adsorption and aqueous radionuclide reactions, chemical and biological reactions are important for the well-being of lakes (e.g., nitrogen, phosphorus, carbon, and dissolved oxygen cycles, as well as phytoplankton, zooplankton, and fish, all interacting to create a very dynamic chemical and biological environment) (Baca and Arnett 1976).

Flows are slower because lakes are confined and relatively deep. In lakes, especially large lakes, flows are mostly wind induced (Weigel 1964; NRC 1976, 1978). Large lakes have a large residence time, and their mixing is dominated by wind-induced currents. They are large enough to have nonuniform radionuclide concentrations.

The wind-induced flow velocity near the water surface is usually less than a few percent of the wind speed in large, open, deep water (Weigel 1964). However, no established way can determine whether a lake is large enough to account for the wind-induced flow on the radionuclide transport assessment. Because wind induces both surface waves and current by transmitting energy to surface water, the following guideline is provided to determine whether a lake is large or small.

Wind-induced waves and flows are expected to be a function of wind speed and duration, fetch length (length of a water body along the wind direction), and depth and width of the surface water (ACERC 1966). For the wind-induced waves to reach their final conditions, 20 km of open surface on a deep lake may be assumed sufficient to produce a steady wind-induced current of probably 1–2% of the wind speed. Furthermore, the narrow width of a water body reduces an effective fetch length by approximately the square root of the width to the length ratio (ACERC 1966). A simple guideline is to call a lake large if the lake water surface area (approximately the product of an average lake length and width) is greater than 400 $km^2$. Otherwise, the lake is considered small.

*Dispersion Coefficient and Mixing* Unlike estuaries, stratification in lakes is produced by water temperature variations caused by solar radiation. In the temperate region, water temperature near the lake surface is heated by sun during late spring, summer, and early fall. This generates an upper stratum of warm water (epilimnion) overlying colder lake bottom water (hypolimnion) separated by a thin thermocline layer. This thermal stratification severely limits radionuclide mixing between the epilimnion and hypolimnion. Various formulas of dispersion coefficients accounting for the stratification are presented in Bowie et al. (1985).

As the lake surface water cools in winter, its temperature approaches 4°C (at which fresh water is the heaviest), and it may become heavier than the water in the lake bottom, producing a vertical turnover. This seasonal turnover results in vigorous mixing of radionuclides throughout, even in large lakes (NRC 1976, 1978). The turnover may also occur in spring, depending on water temperatures in the epilimnion and hypolimnion. For a long-term (annual or longer) assessment, radionuclide distributions in small lakes may be regarded as totally uniform throughout, eliminating the effect of lake stratification on radionuclide mixing.

For large lakes, the dispersion and mixing are controlled by wind-induced flows (NRC 1978). Dispersion coefficients for large lakes are scale dependent, like those in coastal water and oceans. They are selected to be the same as those in coastal water and oceans, as indicated in equations 4.59 and 4.60.

### Small Lake Screening Model

For a small lake or reservoir, an unfiltered radionuclide concentration is assumed to be uniform within the entire impoundment (see figure 4.10). Although a lake or reservoir may be stratified at times, there are seasonal turnovers that mix water through the thermocline, as discussed above. Because the screening methodology provides a long-term (e.g., an annual) average concentration, the effect of stratification on radionuclide distribution is neglected. Under this condition, the governing equation of a radionuclide concentration in a lake is

$$\frac{dC}{dt} = \frac{(q_w + \lambda V)C}{V} + \frac{Q}{V}, \qquad (4.71)$$

where $q_w$ is the river inflow rate to a lake (m$^3$ s$^{-1}$), and $V$ is the lake volume (m$^3$).

If the 30-year-low annual river flow rate into and out from the lake is to be used, it may be estimated as follows:

*Step 1:* Estimate the river width, $\bar{B}$, under normal flow conditions.
*Step 2:* Obtain the mean annual river flow rate, $\bar{q}_w$ (m$^3$ s$^{-1}$), by equation 4.15.
*Step 3:* Calculate the 30-year-low annual flow rate, $q_w$ (m$^3$ s$^{-1}$), to be one-third of $\bar{q}_w$ (m$^3$ s$^{-1}$).

If there is atmospheric deposition of radionuclides onto the lake and its watershed, the radionuclide release rate to the lake may account for this contribution in addition to the liquid radionuclide effluent directly discharged to the lake. To account for this contribution, it is assumed that the size of the lake watershed is 100 times the lake surface area and that 2% of a radionuclide deposited onto the watershed

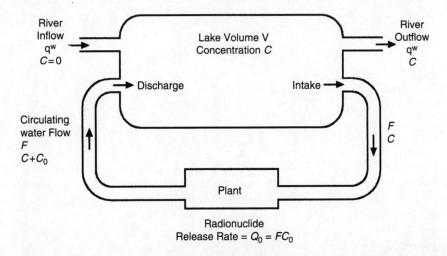

Figure 4.10 Small lake.

reaches the lake through runoff, surface soil erosion, and groundwater seepage. The combined radionuclide release rate, $Q$, is

$$Q = Q_0 + \frac{3 d_i A_l}{86,400}, \qquad (4.72)$$

where

$A_l$ = lake surface area (m$^2$)
$d_i$ = atmospheric deposition rate (m day$^{-1}$)
$Q_0$ = annual average rate of radionuclide directly discharged to a lake (Bq s$^{-1}$)

The atmospheric deposition rate, $d_i$, can be estimated as the product of the deposition velocity and radionuclide concentration in the air above the lake (see chapter 3). Assuming that at time $t = 0$, $C = 0$, the solution to equation 4.71 is

$$C = C_{lt} = \frac{Q}{q_w + \lambda V} \left[ 1 - \exp\left\{ -\left( \frac{q_w}{V} + \lambda \right) t \right\} \right]. \qquad (4.73)$$

If

$$\frac{q_w}{V} + \lambda > 10^{-8} \text{ s}^{-1}, \qquad (4.74)$$

then for 30 years of facility operation,

$$\exp\left\{ -\left( \frac{q_w}{V} + \lambda \right) t \right\} \ll 1. \qquad (4.75)$$

Under this condition, equation 4.73 can be simplified to yield the following steady-state solution to equation 4.71:

$$C_{lt} = \frac{Q}{q_w + \lambda V}. \qquad (4.76)$$

If a lake does not have river inflow and outflow, and the half-life of a radionuclide is greater than 0.693 times the duration of the radionuclide discharge, equation 4.71 may be reduced to

$$\frac{dC}{dt} = \frac{Q}{V}. \qquad (4.77)$$

The solution to equation 4.77 is

$$C_{lt} = \frac{Q}{V} t. \qquad (4.78)$$

### EXAMPLE 4.5

Cobalt-60 is released to a lake for 30 years. The lake does not have a river flowing into or out of it. A radionuclide release rate is $3.70 \times 10^{10}$ Bq yr$^{-1}$ with a radionuclide effluent discharge of $3.15 \times 10^5$ m$^3$ yr$^{-1}$. The average lake length, width, and depth are 400 m, 100 m, and 10 m, respectively. The nearest water use location is 100 m from the outfall on the same side of the lake. Determine a $^{60}$Co concentration at the water use location.

*Solution*: This lake does not have river inflow or outflow, so $q_w = 0$. Because the half-life of $^{60}$Co (5.27 years) is less than 0.693 times the total radionuclide discharge duration (30 years), use equation 4.73. The $^{60}$Co release rate $Q = 3.75 \times 10^{10}$ Bq yr$^{-1}$ = 1,170 Bq s$^{-1}$. The radionuclide decay rate $\lambda = \ln 2 / (5.27 \times 365 \times 24 \times 60 \times 60) = 4.17 \times 10^{-9}$ s$^{-1}$. The lake volume $V = 400$ m $\times$ 100 m $\times$ 10 m = 400,000 m$^3$. The total time $^{60}$Co is discharged is 30 years, so $t = 30 \times 365 \times 24 \times 60 \times 60 = 9.46 \times 10^8$ s. The totally mixed $^{60}$Co concentration in the lake is calculated from equation 4.73 as

$$C_{lt} = \frac{1,170}{(4.17 \times 10^{-9})(4.00 \times 10^5)} \left[1 - \exp\left\{-(4.17 \times 10^{-9})(9.46 \times 10^8)\right\}\right]$$

$$= 6.88 \times 10^5 \text{Bq m}^{-3}$$

The lake surface area, $S, = 400$ m $\times$ 100 m $= 40,000$ m$^2 = 0.04$ km$^2 < 400$ km$^2$. The lake is considered small, and the $^{60}$Co concentration at the water use location is $6.88 \times 10^5$ Bq m$^{-3} = 688$ Bq l$^{-1}$.

## EXAMPLE 4.6

This example is the same as example 4.5, but the lake has a river flowing in and out. The 30-year-low annual river discharge $q_w = 1$ m$^3$ s$^{-1}$. Determine $^{60}$Co concentration at the water use location in this case.

*Solution*: The totally mixed $^{60}$Co concentration in the lake is calculated by equation 4.73:

$$C_{lt} = \frac{1,170}{1 + (4.17 \times 10^{-9})(4.0 \times 10^5)}$$

$$\times \left[1 - \exp\left\{-\left(\frac{1}{4.0 \times 10^5} + 4.17 \times 10^{-9}\right)(9.46 \times 10^8)\right\}\right] = 1,170 \text{ Bq m}^{-3}$$

Because this is a small lake, the $^{60}$Co concentration at the water use location is 1,170 Bq m$^{-3}$ = 1.17 Bq l$^{-1}$. Even with a small river inflow and outflow, there is a significant concentration reduction.

## Large Lake Screening Model

Large lakes are those with a large residence time, $q_w/V$, and with flows dominated by windinduced currents (e.g., the U.S. Great Lakes). The radionuclide transport in large lakes is mainly controlled by a windinduced flow, stratification and seasonal turnover, and scale-dependent mixing, similar to the situation in coastal waters (NCRP 1996). Complete mixing over the entire lake can be achieved within a relatively short time (<1 yr) even in a large lake because of a very large dispersion coefficient that changes with distance and seasonal turnover (NRC 1976, 1978). However, near the release point, complete mixing has not been reached, and this partially mixed radionuclide concentration is calculated by using the coastal water methodology (equation 4.58). To include a contribution from buildup of background-level radionuclide concentrations due to a long-term radionuclide discharge, one can add the concentration obtained from equation 4.73 or 4.78 to the concentration calculated from equation 4.61 to obtain the unfiltered radionuclide concentration along the plume centerline in a large lake, as follows:

$$C = C_{lt} + C_{l1}, \qquad (4.79)$$

where

$$C_{l1} = C(x, y) = \frac{962 \, U_c^{0.17} \, Q}{d \, x^{1.17}} \exp\left\{-\frac{\lambda x}{U}\right\}. \qquad (4.80)$$

Radionuclide concentrations along a large lake's shoreline are calculated by

$$C = C_{lt} + C_{l2}, \qquad (4.81)$$

$$C_{l2} = C(x, y) = \frac{962 \, U_c^{0.17} \, Q}{d \, x^{1.12}} \exp\left\{\frac{-7.28 \times 10^5 \, U_c^{2.34} \, y_o^2}{x^{2.34}} - \frac{\lambda x}{U}\right\}. \qquad (4.82)$$

As in the coastal water and oceans, default velocity $U_c$ is 0.1 m s$^{-1}$.

### EXAMPLE 4.7

A nuclear plant has been discharging $^{60}$Co from the lakeshore to a lake for 30 years. The lake has river inflow and outflow. A radionuclide release rate is 3.70 × 10$^{10}$ Bq yr$^{-1}$ with a radionuclide effluent discharge of 3.15 × 10$^7$ m$^3$ yr$^{-1}$. The average lake length, width, and depth are 80 km, 80 km, and 7 m, respectively. A 30-year-low annual river discharge is not known, but a representative river width under a normal (mean annual) river discharge is 50 m. The nearest water use location is 1 km from the outfall on the same side of the lake. Determine $^{60}$Co concentration at the water use location in this case.

*Solution*: From equation 4.15 with the normal river width of 50 m, 50 = 10($\bar{q}_w$)$^{0.460}$. The mean annual freshwater (river) discharge $\bar{q}_w$ = 33 m$^3$ s$^{-1}$. The 30-year-low annual river discharge is one-third of 33 m$^3$ s$^{-1}$, so $q_w$ = 11 m$^3$ s$^{-1}$. The lake volume $V$ = (80,000 m) × (80,000 m) × (7 m) = 4.48 × 10$^{10}$ m$^3$.

The $^{60}$Co release rate $Q$ = 3.75 × 10$^{10}$ Bq yr$^{-1}$ = 1,170 Bq s$^{-1}$. The half-life of $^{60}$Co is 5.27 years. Thus, the decay rate $\lambda$ = ln 2 / (5.27 × 365 × 24 × 60 × 60) = 4.17 × 10$^{-9}$ s$^{-1}$. Equation 4.73 yields the totally mixed $^{60}$Co concentration in the lake as

$$C_{lt} = \frac{1,170}{11 + (4.17 \times 10^{-9})(4.48 \times 10^{10})}$$

$$\times \left[1 - \exp\left\{-\left(\frac{11}{4.48 \times 10^{10}} + 4.17 \times 10^{-9}\right)(9.46 \times 10^8)\right\}\right] = 5.82 \text{ Bq m}^{-3}$$

The lake surface area $S = 80 \text{ km} \times 80 \text{ km} = 6,400 \text{ km}^2 > 400 \text{ km}^2$, so the lake is considered large. The partial mixing concentration, $C_{l1}$, contribution to the total $^{60}$Co concentration is calculated by equation 4.80 as

$$C_{l1} = \frac{(962)(0.1)^{0.17}(1,170)}{(7)(1,000^{1.17})} \exp\left\{-\frac{(4.17 \times 10^{-9})(1,000)}{0.1}\right\} = 33.6 \text{ Bq m}^{-3}.$$

The total $^{60}$Co concentration at the water use location $C = C_{lt} + C_{l1} = 5.82 + 33.6 = 39.4 \text{ Bq m}^{-3} = 0.0394 \text{ Bq l}^{-1}$.

## Sediment Effects

Many radionuclides are adsorbed by suspended, bed, and shore sediments in surface waters and are divided into dissolved and sediment-sorbed phases. Due to adsorption onto sediment particles, the radionuclide concentration in the dissolved phase will be reduced, while the radionuclide concentrations on the banks and bed of the water body will be increased. Moreover, radionuclides absorbed by suspended sediment may deposit in slow-moving areas, creating a long-term source of contamination through resuspending contaminated bed sediment or desorbing radionuclides back into overlying water over many years (Hetherington 1976; Voitsekhovitch et al. 1994).

Some aquatic pathways (e.g., direct exposure from shore and dredged sediments) are associated with sediment-sorbed radionuclides, and others (e.g., shellfish) are associated with both dissolved and sediment-sorbed radionuclides, but most aquatic pathways (e.g., drinking water and consumption of irrigated food and animal products) are more closely associated with dissolved radionuclides. When surface water is used for drinking, many suspended sediments are removed by water treatment processes, so most of the radionuclide adsorbed by the sediment is removed, as well (Voitsekhovitch et al. 1994).

Screening models described previously estimate the unfiltered, or total (sum of dissolved and sediment-sorbed), radionuclide concentrations in rivers, estuaries, coastal waters and oceans, and small and large lakes. Treating total radionuclide concentrations calculated by these models as dissolved radionuclide concentrations is conservative (to estimate higher dissolved concentrations) on most occasions for continuous radionuclide release cases, except for those pathways directly related to the sediment-sorbed radionuclides.

The distribution coefficient, $K_d$, is commonly used to express the exchange of radionuclides between dissolved and sediment-sorbed phase in a final equilibrium state:

$$K_d = \frac{\text{sediment} - \text{sorbed radionuclide concentration per unit weight of sediment (Bq kg}^{-1})}{\text{dissolved radionuclide concentration per unit volume of water (Bq m}^{-3})}.$$

The dissolved (filtered) radionuclide concentration (Bq m$^{-3}$) in surface water under an equilibrium condition between the water and suspended sediment can be obtained by

$$C_w = \frac{C}{1 + K_d S_s}, \qquad (4.83)$$

where $S_s$ is a suspended sediment concentration (kg m$^{-3}$). Note that $K_d$ in this equation is expressed in m$^3$ kg$^{-1}$, rather than mL g$^{-1}$ as in table 4.2. This equation only reflects the sorption by suspended sediment. The unfiltered (sum of dissolved and sediment-sorbed) radionuclide concentration, $C$, is calculated in preceding sections for each of the surface-water bodies. This assumes that there is no deposition of contaminated suspended sediment or reduction in a radionuclide amount due to adsorption by bed and shore sediments before the radionuclide arrives at a given downcurrent location. Radionuclide decay during transport, if necessary, is already reflected in the estimation of the unfiltered radionuclide concentration obtained previously for a steady, continuous release.

Some values of $K_d$ associated with suspended sediment in fresh water and salt water are provided in table 4.2. $K_d$ values can vary by several orders of magnitude for each of the radionuclides affected by sediment types, water quality, and other conditions, and one must be careful to select an appropriate value for a specific site under consideration (Onishi et al. 1981; IAEA 1985b). Even though the bulk of adsorption occurs in a relatively short period, it still takes a finite time. Figure 4.11 shows the fraction of the total radioactivity that the suspended sediment carries.

The radionuclide concentration, $C_p$ (Bq kg$^{-1}$), adsorbed by suspended sediment may be obtained by

$$C_p = K_d \, C_w = \frac{K_d C}{1 + K_d S_s}. \tag{4.84}$$

The suspended sediment sorbed radionuclide concentration, $C_p$ (Bq kg$^{-1}$), may be converted to the unit of Bq m$^{-3}$ by

$$C_{ps} = C_p S_s. \tag{4.85}$$

Deposition of contaminated suspended sediment and adsorption of dissolved radionuclides from overlying water contaminate the bottom of the surface-water body. Many field data (Onishi et al. 1981; Zheleznyak et al. 1992) show much smaller (approximately one to three orders of magnitude smaller) $K_d$ values associated with bottom sediment than with the suspended sediment. This is largely because of relatively coarser sediment in the bottom and the large amount of bed sediment per unit volume of the bed compared with suspended sediment concentration. The apparent $K_d$ value for bottom sediment may be taken to be 1/10 of the $K_d$ values associated with suspended sediment, producing a more conservative (higher) radionuclide concentration estimate for the bed. The bottom radionuclide concentration, $C_b$ (Bq kg$^{-1}$), may be calculated to be 1/10 of the value estimated by equation 4.84:

$$C_b = \frac{(0.1) K_d C}{1 + K_d S_s}. \tag{4.86}$$

Assuming the top soil layer of 5.0 cm and bulk sediment density of 1,200 kg m$^{-3}$, the surface radiation contamination of shore/beach sediment per unit area, $C_s$ (Bq m$^{-2}$), may be estimated as

$$C_s = \frac{(0.1)(60) K_d C}{1 + K_d S_s}. \tag{4.87}$$

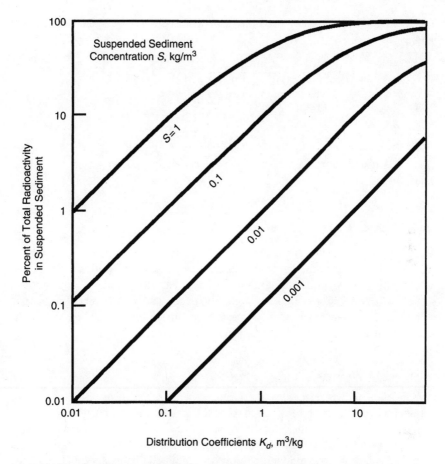

Figure 4.11  Percentage of total radioactivity in suspended sediment.

For continuous radionuclide release, suspended, bed, and shore sediments are continuously exposed to newly arriving radionuclides. Radionuclide decay is accounted for through the decay of the traveling radionuclide in water.

### EXAMPLE 4.8 (CONTINUATION OF EXAMPLE 4.1)

If the city drinking water intake plant has a filtering system to remove half of the suspended solids, what will the $^{137}$Cs concentration be at 30 min after the accidental release? Assume that the suspended sediment concentration in the river is 1.4 kg m$^{-3}$ (or 1,400 mg l$^{-1}$), and that the $^{137}$Cs distribution coefficient is 30,000 mL g$^{-1}$.

*Solution*: From equation 4.83, the dissolved $^{137}$Cs concentration is

$$C_w = \frac{1.06}{1 + (0.001)(30,000)(1.4)} = 0.0246 \, \text{Bq m}^{-3}.$$

The $^{137}$Cs concentration sorbed by the suspended sediment is obtained by equation 4.84 as

$$C_p = \frac{(0.001 \times 30,000 \times 1.06)}{1 + (0.001 \times 30,000 \times 1.4)} = 0.740 \, \text{Bq kg}^{-3}.$$

From equation 4.85,

$$C_{ps} = (0.740 \times 1.4) = 1.04 \, \text{Bq m}^{-3}.$$

The $^{137}$Cs distribution after filtering the river water is $0.0246 + (0.5 \times 1.04) = 0.545 \, \text{Bq m}^{-3}$.

## Numerical Modeling

Analytical solutions presented in preceding sections can estimate radionuclide concentrations only when flow conditions are steady and uniform. However, as discussed below under "Chernobyl Nuclear Accident Aquatic Assessment," surface-water conditions (e.g., flow velocity, depth, and width) vary significantly with time and distance. Under these varying conditions, numerical modeling must be used to determine radionuclide concentrations in surface water.

### Governing Equations

Radionuclide transport models for surface water must account for the major radionuclide transport and fate mechanisms discussed above under "Basic Transport and Fate Mechanisms." In the deterministic approach, all the mathematical expressions are framed by conservation of mass. In the three-dimensional Cartesian coordinate system with appropriate initial and boundary conditions, the transport equation expressed in Eulerian mode becomes the advection–diffusion equation:

$$\frac{\partial C}{\partial t} + U\frac{\partial C}{\partial x} + V\frac{\partial C}{\partial y} + (W - W_s)\frac{\partial C}{\partial z}$$
$$= \frac{\partial}{\partial z}\left(\varepsilon_x \frac{\partial C}{\partial x}\right) + \frac{\partial}{\partial y}\left(\varepsilon_y \frac{\partial C}{\partial y}\right) + \frac{\partial}{\partial z}\left(\varepsilon_z \frac{\partial C}{\partial z}\right) - \lambda C + S, \quad (4.88)$$

where, $S$ is a sink/source term, representing all mechanisms listed in table 4.1, except advection, dispersion, and radioactive decay.

Equation 4.88 is applicable to transport and fate of radionuclides as well as to other chemicals (e.g., heavy metals and toxic organic chemicals). The equation may be integrated over the entire width (y-direction) or depth (z-direction) to obtain laterally or vertically averaged two-dimensional equations. Integrating equation 4.88 over the entire cross section (both y- and z-directions) yields a one-dimensional governing equation, as discussed above (see equation 4.13).

Most radionuclides exist in surface water as liquid and solid phases and are transported by water and sediment (Voitsekhovitch et al. 1994). Numerical models for radionuclide transport must simulate hydrodynamics, sediment transport, and transport and fate of both dissolved and sediment-sorbed radionuclides with their interactions. Continuity and the Navier Stokes equations (including the Saint-Venant equation for river flows) are used to simulate hydrodynamics (see Chow 1959; Rouse 1961).

## Some Representative Models

Reviews of contaminant transport models (e.g., Onishi et al. 1981, 1987; Onishi 1985; NCRP 1984) indicate that many transport and fate models have been developed for contaminants (radionuclide and toxic chemicals) for rivers, estuaries, coastal waters, oceans, and lakes by including some of the major mechanisms (see table 4.1).

Many of the models predict only dissolved contaminant concentrations. Examples are one-dimensional models developed by Dailey and Harleman (1972) and Eraslan et al. (1977); two-dimensional models developed by Leendertse (1970), Webb and Morley (1973), Yotsukura and Sayer (1976), and Eraslan et al. (1977); and three-dimensional models developed by Simons (1973), Leendertse and Liu (1975), Shepard (1978), Eraslan et al. (1983), and Blumberg and Herring (1986). These models do not include the sediment–contaminant interactions (e.g., adsorption/desorption and subsequent transport, deposition, and erosion of sediment-sorbed contaminants), so are suited for cases where the contaminants have very low affinity to sediments or a surface-water body has very low sediment concentrations. However, in many instances, the concern is for situations where contaminants have high affinity to sediments, a surface-water body has high concentrations of sediments (especially fine sediments), or there is long-term contaminant accumulation in a surface-water body. These include $^{137}$Cs in the Pripyat and Dnieper rivers in Ukraine (Zheleznyak et al. 1992, 1995), the Cattaraugus and Buttermilk creeks in New York (Onishi et al. 1982b), and the Clinch River in Tennessee (Onishi et al. 1982b); and plutonium in South Mortandad Canyon in New Mexico (Onishi et al. 1982a).

For such cases, models must also include sediment–contaminant interactions. Models incorporating these interactions include compartment codes such as MARINARD (Koplik et al. 1984) and MARKA (Robinson and Marietta 1985); one-dimensional codes such as those of White and Gloyna (1969), CHNSED (Fields 1976), TODAM (Onishi et al. 1982a), Bencala's code (Bencala 1983), RIVTOX (Zheleznyak et al. 1992), and CHERIMA (Holly et al. 1990); two-dimensional codes such as SERATRA (Onishi et al. 1982b), FETRA (Onishi 1981), Lick's code (Lick 1983), and WATOX 2 (Zheleznyak et al. 1992); and three-dimensional codes such as RMA-10 (King 1982), Spaulding and Parish's code (Spaulding and Parish 1984), TEMPEST (Trent and Eyler 1994), FLESCOT (Onishi et al. 1993b), Sheng's code (Sheng 1993), and WATOX-3 (Zheleznyak et al. 1992). With known flow and sediment distributions, compartment models such as EXAMS (Burns and Cline 1985) and WASP4 (Ambrose et al. 1988) also predict transport and fate of both dissolved and sediment-sorbed contaminants. Figure 4.11 illustrates at a quick glance when sediment effects on radionuclides must be accounted for in the assessment.

All of these models use very simple approaches to handle chemical reactions such as adsorption and precipitation. Some effort has been made to improve modeling of the transport and fate of contaminants by coupling transport models with chemical models (Chapman 1982; Felmy et al. 1983; Onishi 2005). Chemical models essentially solve various chemical reactions based on mass conservation, chemical equilibrium principles, and kinetics with the aid of thermodynamics (Harvie et al. 1987). Some chemical computer codes such as MINEQL (Westall et al. 1976), EQ3/EQ6 (Wolery 1980), MINTEQ (Felmy et al. 1984), and GMIN (Felmy 1990) also calculate adsorption/desorption and precipitation/dissolution. For example, the compartment transport model, EXAMS was coupled to MINTEQ to form the model MEXAMS (Felmy et al. 1983). It was tested to calculate heavy metal transport and chemical reactions (e.g., the amounts of adsorption/desorption and precipitation/dissolution) in a lake (Onishi et al. 1983). The reactive transport code, ARIEL, couples three-dimensional hydrodynamics-mass/heat transport with the equilibrium and chemical kinetics to simulate chemical mixing and reactions (Onishi 2005). It was applied to underground radioactive waste storage tanks to simulate mixing of chemically reactive wastes in dissolved and solid phases, equilibrium and kinetic chemical reactions, and associated rheology changes (e.g., viscosity and yield and shear stresses).

The following processes are illustrated below by using the 1986 Chernobyl nuclear accident as an example:

- Surface-water transport and fate mechanisms (see table 4.1)
- Migration and accumulation of radionuclides in surface water
- Radionuclide transport modeling for rivers
- Aquatic environmental impacts
- Human health impacts through the aquatic pathways

## Chernobyl Nuclear Accident Aquatic Assessment

The Chernobyl nuclear plant is located about 100 km north of Kiev, Ukraine, along the Pripyat River, which joins the Dnieper River, discharging to the Black Sea. During 11 days following the April 26, 1986, accident, its 1,000-MW (electric) Unit 4 reactor released radionuclides into the atmosphere in an amount six times that of the Hiroshima atomic bomb (IAEA 1991; UNSCEAR 2000). The largest amounts of radionuclides released from the Chernobyl plant were the short-lived $^{133}$Xe, $^{133}$I, $^{131}$I, $^{132}$Tc, and $^{239}$Np, with half-lives of just hours to several days. The main radionuclides affecting human health are the longer lived $^{137}$Cs, $^{90}$Sr, $^{238,239,240}$Pu, besides short-lived $^{131}$I.

The shock wave of the Chernobyl explosion, the high temperature, and oxidation of the nuclear fuel (uranium dioxide) led to the formation of many hot fuel particles, more than 90% of which had an activity on the order of 40 Bq per particle (Voitsekhovitch et al. 1994). Radionuclide deposited amounts are approximately $1.8 \times 10^{18}$, $8.5 \times 10^{16}$, $1 \times 10^{16}$, and $1.1 \times 10^{14}$ Bq of $^{131}$I, $^{137}$Cs, $^{90}$Sr, and $^{238,239,240}$Pu, respectively (IAEA 1991; UNSCEAR 2000). About 120,000 people

Surface Water Transport of Radionuclides  193

were evacuated from the area within 30 km of the plant. The fallout of these airborne Chernobyl radionuclides formed one near and two distant contamination zones in southern Belarus, the western Bryansk region of Russia, and the northern and central regions of Ukraine, as shown in figure 4.12 (Onishi et al. 1993a). The near contamination zone (including a 30-km exclusion zone) around the Chernobyl nuclear plant is mostly due to the deposition of heavy hot particles (containing $^{90}$Sr, $^{137}$Cs, and plutonium), while the distant contamination zones are due to fallout of lighter condensed

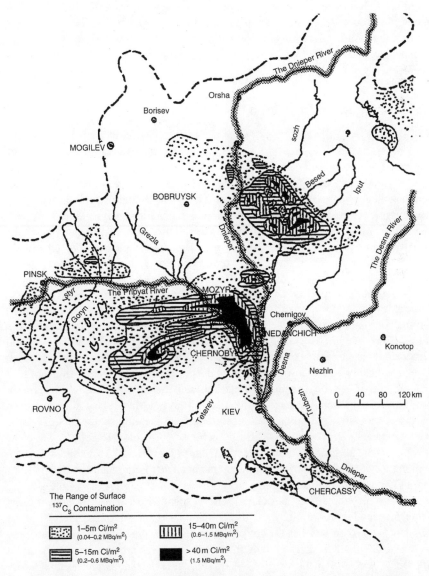

Figure 4.12   $^{137}$Cs contamination in the Pripyat and upper Dnieper river basins.

particles (mostly $^{137}$Cs). The contaminated surface soil became a long-term source of radionuclides, especially $^{90}$Sr and $^{137}$Cs, into the Pripyat and Dnieper rivers (Konoplev et al. 1992).

The health of people living in the heavily contaminated areas has been affected mostly by $^{137}$Cs, $^{90}$Sr, $^{131}$I, and plutonium, and it is expected that about 4,000 people will die from cancers and leukemia (Chernobyl Forum 2005). Although terrestrial pathways (e.g., direct irradiation from contaminated solids, inhalation of radionuclides) account for more than 90% of the health impacts, the radionuclides in the rivers reach 20 million people in Ukraine through aquatic pathways (e.g., drinking the contaminated Dnieper River water, eating fish, and consuming foods produced with Dnieper River irrigation water) (Prister et al. 1992). Assessment of the radionuclide migration into and within the rivers is important to these 20 million people, and to effectively mitigate the consequences of the Chernobyl nuclear accident.

## Radionuclide Transport in Rivers

The Dnieper and Pripyat river system is the main surface-water artery of Ukraine with six reservoirs. The Pripyat joins the Dnieper at the upstream end of the Kiev Reservoir above the city of Kiev. The Dnieper discharges into the Black Sea after a 1,000-km journey passing through six dams and reservoirs: Kiev, Kanev, Kremenchug, Dneprodzerzhinsk, Dnieprovskoe, and Kakhovka, in this sequence. The hydrologic features of these rivers and their tributaries are typical for European rivers with plain watersheds and a forest-steppe landscape. The spring snowmelts usually cause flooding of the Pripyat and Dnieper rivers near the Chernobyl area from early March to mid-May. Summer storms generate somewhat smaller floods.

Radionuclides deposited on the ground are transported to these receiving rivers through runoff and soil erosion caused by rainfall and snowmelt, infiltration into subsurface water and subsequent groundwater seepage into these rivers, and, to a lesser extent, wind resuspension and subsequent fallout. Figure 4.13 shows measured concentrations of $^{90}$Sr and $^{137}$Cs in the Pripyat River at the town of Chernobyl, which is approximately 10 km downstream from the plant along the river, for the first four years after the Chernobyl accident (Voitsekhovitch et al. 1994). The highest levels in the Pripyat River were observed in the first few days of May 1986 and were caused by direct deposition of radioactive substances from the atmosphere that exceeded a total beta activity level of 4,000 Bq l$^{-1}$ with a $^{90}$Sr level of 15 Bq l$^{-1}$. This $^{90}$Sr level exceeds the local drinking water standard of 2.0 Bq l$^{-1}$. This illustrates the importance of the fifth transport and fate mechanism, "Radionuclide Contributions from Other Environmental Media," listed in table 4.1. Almost the whole activity (99%), except iodine, was associated with suspended particles during that time. The Dnieper River reservoirs were also contaminated almost immediately after the accident by deposition of radioactive substances from the atmosphere. From May to June 1986, the beta activity level in the Pripyat River dropped sharply to about 200 Bq l$^{-1}$, with $^{90}$Sr in the range of 1–2 Bq l$^{-1}$, due to the radionuclide influx mostly caused by overland runoff and soil erosion; subsequently, the levels became even

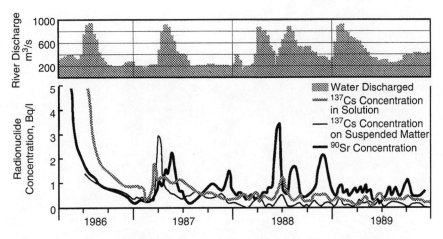

Figure 4.13 Measured Pripyat River discharge and $^{90}$Sr and $^{137}$Cs concentrations at the Town of Chernobyl. From Voitsekhovitch et al. (1994).

lower. Although $^{131}$I contributed significantly to the river contamination during the Chernobyl accident, its share of the river contamination reduced very rapidly due to its short half-life of 8.04 days, illustrating the importance of accounting for the third mechanism, "Degradation/Decay," listed in table 4.1. High $^{90}$Sr radionuclide peaks for subsequent years were associated with flooding caused by snow melting and some summer storms, bringing radionuclides deposited on watersheds, floodplains, and river bottoms into the river system. This indicates the importance of the first mechanism, "Transport," and runoff and soil erosion from the land surface listed under the fifth mechanism listed in table 4.1.

Figure 4.14 also shows a significant amount of $^{137}$Cs being carried by suspended sediment, indicating the importance of sediment movement listed under "Transport" in table 4.1. Ratios of $^{137}$Cs adsorbed by each sediment size fraction to the total sorbed $^{137}$Cs over several years are shown in figure 4.14, revealing the clear preference of $^{137}$Cs to associate with finer sediment of silt (whose diameter is between 4 μm and 62 μm) and clay (whose diameter is <4 μm). This illustrates adsorption/desorption listed under "Intermedia Transport" in table 4.1. In 2004, $^{90}$Sr and $^{137}$Cs concentrations in the Pripyat River were 0.1–0.5 Bq l$^{-1}$ and 0.03–0.1 Bq l$^{-1}$, respectively, showing (i) high peaks associated with floodings and (ii) continuous decline of the radionuclide concentrations over the years.

Many of these sediment-sorbed radionuclides were deposited in reservoirs before reaching the Black Sea. A significant amount of the radionuclides originally deposited on the watersheds reached the Pripyat and Dnieper rivers and remains in the six reservoirs of the Dnieper River. This deposition has formed long-term aquatic pathways of radionuclides to humans. To evaluate these aquatic contributions to human dose, we need to determine radionuclide concentrations in water and suspended, bottom, and shore sediments in and along the rivers.

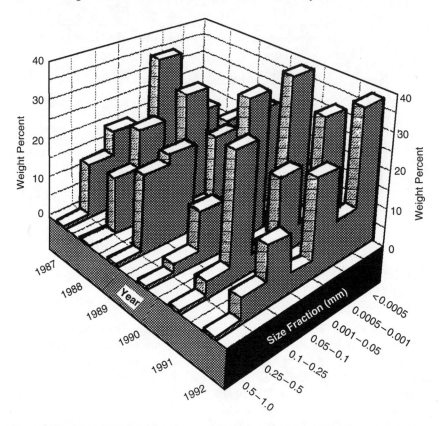

Figure 4.14 Cesium-137 distribution on suspended sediments of the Pripyat River. From Voitsekhovitch et al. (1994).

As in the Pripyat River, the highest contamination levels in the Dnieper reservoirs were registered in the initial period after the nuclear accident. Later, with the sharp decrease of the radioactive releases from the Chernobyl nuclear plant, the impact of the aerosol component became lower, and a fairly rapid sedimentation took place in the Kiev Reservoir (the first of the six reservoirs). By May 7, 1986, the total beta activity in the water in the lower part of the Kiev Reservoir was down to about 0.4 Bq $l^{-1}$. Figure 4.15 shows $^{90}$Sr concentrations at the ends of Kiev and Kanev reservoirs predicted by the TODAM code (Onishi et al. 1982a) and measured data. This figure shows the rapid decline of $^{90}$Sr concentrations shortly after the accident and its rise during the flooding periods thereafter. Most $^{90}$Sr influx to the Kiev Reservoir passes through the six reservoirs of the Dnieper River and enters the lower Dnieper area, where $^{90}$Sr concentration is near the water irrigation standard of 0.3 Bq $l^{-1}$.

Unlike $^{90}$Sr, significant portions of the $^{137}$Cs in the rivers are adsorbed by suspended and bottom sediment (figures 4.13 and 4.14). Much $^{137}$Cs is subject to sediment deposition in slower moving areas. The reduction in $^{137}$Cs concentration as it moves across the reservoirs is apparent from measured $^{90}$Sr and $^{137}$Cs concentrations. Variations of $^{90}$Sr concentration within the reservoirs and with time are

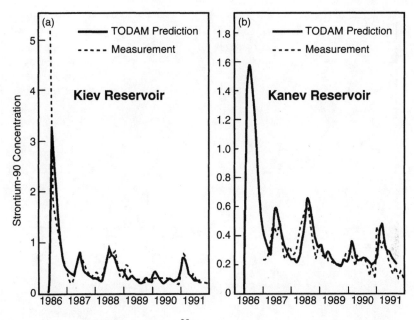

Figure 4.15 Predicted and measured $^{90}$Sr concentrations at the Kiev Reservoir (a) and the Kanev Reservoir (b).

much smaller than are those of $^{137}$Cs. For the first several postaccident years, almost 95% of the $^{137}$Cs that entered the Dnieper River was deposited at the bottoms of the Dnieper reservoirs (Voitsekhovitch et al. 1994). Moreover, Kiev Reservoir collects the majority of $^{137}$Cs deposited in the reservoirs. Figure 4.16 shows the measured $^{137}$Cs concentrations in the water and bottom sediment of the Pripyat and Dnieper rivers, as well as $^{90}$Sr concentrations in the water. The contribution of $^{137}$Cs discharged to a northeastern part of the Black Sea during this period (when annual Dnieper River flow was below the norm) was not significant.

### Aquatic Pathways and Their Radiation Dose Contributions

The six reservoirs on the Dnieper River are a main source of surface water for consumption in Ukraine, and 20 million Ukrainians are affected by aquatic pathways (see figure 4.1). The Dnieper River water is used in municipal water systems for more than eight million people along the Dnieper River reservoirs and the Crimea Republic (Voitsekhovitch et al. 1994). The Dnieper reservoirs are used intensively for commercial fishing. The dominant radionuclide found in fish is $^{137}$Cs. As expected, many of the fish in the Dnieper River show elevated levels of radionuclides. Table 4.7 presents $^{137}$Cs concentrations in various fish in Kiev Reservoir, showing that the $^{137}$Cs concentrations in most fish peaked in 1987 and have gradually declined since then (Voitsekhovitch et al. 1994). In 2001, $^{137}$Cs concentrations were down by 5–20 times, depending on the fish (e.g., fish size, and predatory or

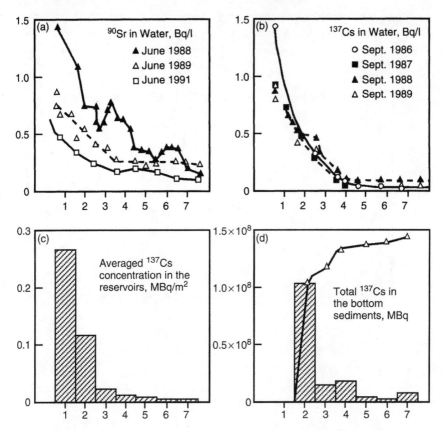

Figure 4.16 Measured $^{137}$Cs concentrations in water and bottom sediment in the Pripyat and Dnieper rivers, together with $^{90}$Sr concentration in water. On the x-axes, 1 indicates the mouth of the Pripyat River, and 2–7 indicate the six reservoirs in the Dnieper River.

not). It mostly accumulated in soft tissues, especially in muscles, liver, and kidney. On the other hand, $^{90}$Sr selectively accumulated in the bony tissues of the skeleton, scales, and fins. After the 1986 accident, there has not been a significant decrease in the fish catch in the reservoirs. The annual catch averages about 1,200 tons and 25,000 tons of commercial fish from Kiev Reservoir and all six reservoirs, respectively.

More than 1.6 million hectares are irrigated with 4.8 km$^3$ of Dnieper River water annually, including 1.13 million hectares from Kakhovka Reservoir alone. The irrigation water is used for various produce: rice, wheat, beets, corn, potatoes, and vegetables (Prister et al. 1992). After the 1986 Chernobyl accident, $^{137}$Cs and $^{90}$Sr concentrations in rice grain and straw increased by 50–100%. Almost half of the irrigated lands are used for fodder crops, which determine the contamination of meat and milk. Radionuclide concentrations of $^{137}$Cs in meat and milk where Dnieper River water is used for irrigation are 1.5–2.5 times higher than are those in the uncontaminated control regions, as shown in table 4.8. Vegetables also exhibit noticeably higher $^{137}$Cs contamination from Dnieper River irrigation water than do

Table 4.7  $^{137}$Cs concentration (Bq kg$^{-1}$ of fresh muscle weight) in fish species caught in the Kiev reservoir

| Fish species | 1987 | 1988 | 1989 | 1990 | 1991 |
|---|---|---|---|---|---|
| Perch | 1.5 – 3.0 | 1.5 – 3.4 | 1.2 – 3.4 | 0.7 – 1.7 | 0.6 – 1.3 |
| Pike | 0.8 – 2.8 | 0.9 – 1.6 | 0.8 – 1.5 | 0.6 – 1.3 | 0.5 – 1.1 |
| Tench | 0.7 – 1.7 | 0.6 – 0.9 | 0.5 –0.8 | 0.2 – 0.8 | 0.3 – 0.5 |
| Bream | 0.5 – 1.0 | 0.4 – 0.9 | 0.3 –0.7 | 0.2 – 0.5 | 0.1 – 0.3 |
| Rudd | 0.4 – 1.1 | 0.3 – 0.7 | 0.2 – 0.5 | 0.1 – 0.4 | 0.1 –0.3 |

Table 4.8  $^{137}$Cs concentrations (Bq/kg) in produce and animal products obtained from the irrigated lands in 1988 (Prister et al. 1992)

| Region, district | Alfalfa | Winter wheat | Tomato | Cabbage | Cucumber | Milk | Meat |
|---|---|---|---|---|---|---|---|
| Cherkassy, Kanivsky | 0.34 | 0.11 | 0.01 | 0.002 | 0.01 | 0.22 | 1.11 |
| Poltava, Kremenchutsky | 0.34 | 0.15 | 0.01 | 0.001 | 0.01 | 0.19 | 1.11 |
| Dniepropetrovsk, Nizhnednieprovsky | 0.26 | 0.04 | 0.01 | 0.001 | 0.01 | 0.30 | 1.48 |
| Kherson, Dnieprovsky | 0.17 | 0.11 | 0.01 | 0.001 | 0.01 | 0.30 | 0.74 |
| Crimea, Sympheropolsky | 0.34 | 0.07 | 0.01 | 0.002 | 0.01 | 0.22 | 1.11 |
| Reference districts without water use from the Dnieper River | | | | | | | |
| Kharkov, Lazovsky | 0.08 | 0.04 | 0.001 | 0.001 | 0.1 | 0.11 | — |
| Donetsk, Krasnoarmeysky | 0.08 | 0.04 | 0.001 | 0.001 | 0.1 | 0.11 | 0.37 |

those in the Kharkov and Donetsk regions, where irrigation water is not connected with the Dnieper drainage basin (Prister et al. 1992).

During the first two months after the accident, $^{131}$I in drinking water was a main contributor to exposure. It contributed $2 \times 10^{-4}$ Sv of the effective equilibrium dose to a Kiev citizen through the Dnieper River drinking water (Voitsekhovitch et al. 1994). Individual dose contribution from $^{90}$Sr, $^{89}$Sr, $^{134}$Cs, and $^{137}$Cs in treated drinking water was about $8.7 \times 10^{-5}$ Sv per year. Seventy percent of the radiation dose to the Kiev population during 1986 was due to $^{131}$I in the drinking water. However, because $^{131}$I has a very short half-life (eight days), its contribution decreased rapidly with time.

Expected individual dose from the aquatic pathways varies from 10–12 $\mu$Sv yr$^{-1}$ in the Crimea and Poltava regions to 102 $\mu$Sv yr$^{-1}$ in the Kiev region. The collective committed effective dose from aquatic pathways to Ukrainians is estimated to be 3,000–3,500 man-Sv. Strontium-90 contributes about 80% of the aquatic pathway contribution to radiation dose. Contributions from various aquatic pathways vary from regions to region. In the Kiev region, irrigation foods, drinking water, and fish contribute 18%, 43%, and 39%, respectively, among the aquatic pathways, while in the more downstream Poltava region, they are 8%, 25%, and 67%, respectively (Berkovskiy et al. 1996).

The collective dose from the aquatic pathways corresponds to about 5% of the expected total population exposure of about 55,000–70,000 man-Sv. The Chernobyl Forum (2005) estimates that about 4,000 people will die from the Chernobyl accident. Although the aquatic pathway is not a dominant factor for radiation risk compared with other contributors, it is nonetheless important, because the Pripyat and Dnieper aquatic system is the only radiation link to 20 million Ukrainians who do not live near the Chernobyl nuclear plant.

## New Chernobyl Development

The new development at the Chernobyl site is a movable, arch-shaped, hanger-like, "New Safe Confinement" (NSC) structure (National Geographic 2006). The 100-m-high, 270-m-wide, 166-m-long NSC will cover the Chernobyl shelter (sarcophagus). The NSC intends to (1) reduce or eliminate future radionuclide releases to atmosphere and groundwater, (2) provide safer working conditions for dismantling the shelter and for radioactive waste management, and (3) generate economic and social benefits. The evaluation of the NSC impacts on aquatic environment indicates that building the NSC would reduce the radionuclide flux to the subsurface and river environment, even though the Chernobyl shelter collapse without the NSC would not cause measurable harmful effects from $^{90}$Sr, $^{137}$Cs, and $^{239}$Pu on the Pripyat River environment (Onishi et al. 2003).

It should be noted that the greatest public health problem caused by the Chernobyl accident is a mental problem (e.g., helplessness and anxiety) induced by fear of radiation health effects, the lack of a vital local economy, lower living standards, high unemployment, and increased poverty (Chernobyl Forum 2005). Regardless of the remedial benefits of the NSC, its construction will provide employment and revitalize the economy; thus, it addresses the most critical Chernobyl problem: the mental health and socioeconomic issues.

The Chernobyl example illustrates what happens to rivers and aquatic food chains, and its consequences to humans, in the event of a nuclear accident. This assessment shows that aquatic pathways (e.g., drinking water, fishing, and consumption of irrigated crops, meat, and milk) contribute to human health impacts. Transport and fate assessment in surface water is important for estimating radionuclide concentrations through the aquatic food chain and for evaluating potential cleanup activities.

## Problems

1. A truck accidentally spilled 10,000 Bq of $^3$H into a river from a riverbank. There is a park 10 km downstream on the same side of the river, and some people are swimming there near the bank. The discharge, average depth, and

average width of the river are 1,000 m³ s⁻¹, 10 m, and 500 m, respectively. Tritium is fully mixed vertically but not yet fully mixed laterally when the ³H plume reaches 10 km downstream from the accidental spill site. What is the expected ³H concentration at the swimming beach 6 h after the accident?

2. A nuclear plant accidentally released $1 \times 10^6$ Bq of $^{137}$Cs through an effluent outfall located on the river bottom in the middle (center) of the river. The river velocity is 2.0 m s⁻¹. The longitudinal, lateral, and vertical dispersion coefficients are 100, 0.5, and 0.05 m² s⁻¹, respectively. A city drinking water intake is located 5 m above the river bottom, 50 m laterally from the middle of the river and 1 km downstream from the effluent outfall. When will the center of the $^{137}$Cs plume patch pass through the water intake station? What will the unfiltered (total) $^{137}$Cs concentration (sum of the dissolved and suspended sediment-sorbed $^{137}$Cs) be at the water intake station at that time? Assume that sediment-sorbed $^{137}$Cs does not settle to the river bottom and that it moves downstream at the river velocity like the dissolved $^{137}$Cs.

3. (continued from problem 2) If the city drinking water intake plant has a filtering system to remove 80% of the suspended solids, what will the $^{137}$Cs concentration in the city water be, after filtration, 8.3 min after an accidental release? Assume that the suspended sediment concentration in the river is 1,400 mg l⁻¹ and the $^{137}$Cs distribution coefficient is 30,000 mL g⁻¹.

4. Strontium-90 is continuously released to a river at a rate of $5 \times 10^{10}$ Bq yr⁻¹, with a radionuclide effluent discharge of 1 m³ s⁻¹. The 30-year-low annual discharge of the river is 1,000 m³ s⁻¹. The nearest water intake for drinking and irrigation is 10 km downstream from the radionuclide outfall on the same side of the river. Determine the $^{90}$Sr concentration at the water intake location.

5. Groundwater seepage continuously releases 1,000 Bq s⁻¹ of ³H to a river. The corresponding depth, width, and velocity of the river for the 30-year-low annual discharge are 10 m, 500 m, and 0.2 m s⁻¹, respectively. The ³H is in a dissolved form, and the radionuclide decay can be neglected. What will an expected concentration be at a municipal river water intake station 30 km downstream from the groundwater seepage location on the same side of the river?

6. Ruthenium-106 is released to an estuary along the Gulf of Mexico at the rate of $1.00 \times 10^{11}$ Bq yr⁻¹, with a radioactive effluent discharge of $1.00 \times 10^8$ m³ yr⁻¹. A 30-year-low annual freshwater discharge is not known, but the normal river width just before the river becomes an estuary is 50 m. The estuarine width and depth are 50 and 5 m, respectively. The maximum ebb and flood velocities are 1.0 and 0.80 m s⁻¹, respectively. The nearest water use location is 2 km upstream from the radionuclide outfall on the same side of the estuary. Determine $^{106}$Ru concentration at the water use location.

7. Strontium-90 is released to an estuary along the Pacific Coast at the rate of $3.7 \times 10^{11}$ Bq yr⁻¹ with a radionuclide discharge of $3.15 \times 10^7$ m³ yr⁻¹. The 30-year-low annual freshwater river discharge is 2,000 m³ s⁻¹; the average estuarine width and depth are 1,000 and 5 m, respectively. The maximum ebb and flood velocities are 1.0 and 0.7 m s⁻¹, respectively. The nearest

receptor is 2 km downstream (seaward) from the outfall on the same side of the estuary as the outfall. Determine $^{90}$Sr concentration at the receptor location.

8. Plutonium-239 is released to a coastal water at a rate of $3.7 \times 10^{11}$ Bq yr$^{-1}$ with a radionuclide effluent discharge of $3.15 \times 10^7$ m$^3$ yr$^{-1}$. The release point is 100 m off the beach. Water depth at the release point is 20 m.

   (a) A potential receptor is living 1 km downcurrent, and all the exposure pathways (e.g., swimming, fish consumption) are based on the Pu concentration 1-km downcurrent location along the plume centerline. Calculate the $^{239}$Pu concentration there.
   (b) You argue that a potential receptor is living 1 km downcurrent, but the nearest fishing site is really 5 km downcurrent. Calculate the $^{239}$Pu concentration 5 km downcurrent along the plume centerline.
   (c) You further argue that the potential receptor is not living in the water. Calculate $^{241}$Pu concentration at the beach 1 km downcurrent from the release point.

9. A city is to discharge a sewage effluent containing $^{137}$Cs into a coastal water. The sewage effluent discharge is $1.00 \times 10^8$ m$^3$ yr$^{-1}$, and the amount of radionuclide discharged through the municipal effluent is $1.00 \times 10^{11}$ Bq yr$^{-1}$. Because you are a city water facility planner, you want to minimize the cost of the sewage effluent pipeline construction. Select the shortest distance, $y_0$, from the coastal shore line to the effluent discharge point such that this discharge can still satisfy the requirement that the $^{137}$Cs concentration be less than 5 Bq m$^{-3}$ in coastal water at the nearest coastal water use location (i.e., a fishing site). This fishing site is 2 km downcurrent from the proposed release point. The sea bottom slopes down such that the water depth increases 1 m every 50 m from the shoreline (e.g., the water depth is 1 m 50 m from the shore and 10 m 500 m from the shore).

10. Iodine-129 is released to a lake that does not have any river inflow or outflow. The radionuclide release rate is $3.70 \times 10^7$ Bq yr$^{-1}$ with a radionuclide effluent discharge of $3.15 \times 10^6$ m$^3$ yr$^{-1}$. The average lake length, width, and depth are 400, 100, and 10 m, respectively. The nearest water use location is 100 m downcurrent from the outfall on the same side of the lake. Determine $^{129}$I concentration at the water use location for 30 years of release.

11. Iodine-131 is released to a lake without any river flowing into or discharging from the lake. The radionuclide release rate is $3.70 \times 10^7$ Bq yr$^{-1}$, with a radionuclide effluent discharge of $3.15 \times 10^5$ m$^3$ yr$^{-1}$. The average lake length, width, and depth are 400, 100, and 10 m, respectively. The nearest water use location is 100 m from the outfall on the same side of the lake. Determine the $^{131}$I concentration at the water use location after the 30 years of release.

12. Iodine-131 is released to a lake with river inflow and outflow. The radionuclide release rate is $3.70 \times 10^{10}$ Bq yr$^{-1}$ with a radionuclide effluent discharge of $3.15 \times 10^7$ m$^3$ yr$^{-1}$. The average lake length, width, and depth are 80 km, 80 km, and 7 m, respectively. The 30-year-low annual river discharge is not known, but a representative river width under normal (mean

annual) river discharge is 50 m. The nearest water use location is 1 km from the outfall on the same side of the lake. Determine the $^{131}$I concentration at the water use location for 30 years of this release.

13. A nuclear facility discharges $1.00 \times 10^{10}$ Bq yr$^{-1}$ of $^{90}$Sr with an effluent discharge of $1.00 \times 10^{8}$ m$^3$ yr$^{-1}$ into a lake for 30 years. The lake is 1 km long, 500 m wide, and 10 m deep. A river flows into and out of the lake, and the average river width under a normal flow is 200 m. Farmers want to use river water for irrigation before it flows into the lake. The irrigation return flow will not go back to the lake, so the amount of river water the farmers use will be lost from the river. You are a regional water use planner. How much river water (if any) can you allocate to the farmers without $^{90}$Sr concentration in the lake getting greater than 10 Bq m$^{-3}$?

References

ACERC (Army Coastal Engineering Research Center). 1966. *Shore Protection, Planning and Design*. Technical Report No. 4. U.S. Department of Army, Corps of Engineers, Washington, DC.

Ambrose, R.B., Jr., T.M. Wool, J.P. Connolly, and R.W. Schanz. 1988. *WASP4, a Hydrodynamic and Water Quality Model—Model Theory, User's Manual, and Programmer's Guide*. EPA/600/3-87/039. U.S. Environmental Protection Agency, Environmental Research Laboratory, Athens, GA.

Baca, R.G., and R.C. Arnett. 1976. *A Limnological Model for Eutrophic Lakes and Impoundments*. Battelle, Pacific Northwest Laboratories, Richland, WA.

Bencala, K.E. 1983. "Simulation of Solute Transport in a Mountain Pool-Riffle Stream with a Kinetic Mass Transfer Model for Sorption." *Water Resources Research* 14(3): 732–738.

Berkovskiy, V., G. Ratia, and O. Nasvit. 1996. "Internal Doses to Ukrainian Population Using Dnieper River Water." *Health Physics* 71(1): 37–44.

Blumberg, A.F., and H.J. Herring. 1986. *Circulation Modeling Using Curvilinear Coordinates*. Report No. 81. Dynalysis, Princeton, NJ.

Bowie, G.L., W.B. Mills, D.B. Porcella, C.L. Campbell, J.R. Pagenkopf, G.L. Rupp, K.M. Johnson, P.W.H. Chan, S.A. Gherini, and C.E. Chamberlin. 1985. *Rates, Constants, and Kinetics Formulations in Surface Water Quality Modeling*, 2nd ed. Prepared for the U.S. Environmental Protection Agency by Tetra Tech, Lafayette, CA, and Humboldt State University, Arcata, CA.

Burns, L.A., and D.M. Cline. 1985. *Exposure Analysis Modeling System—Reference Manual for EXAMS II*. EPA/600/3-85/038. U.S. Environmental Protection Agency, Environmental Research Laboratory, Athens, GA.

Chapman, B.M. 1982. "Numerical Simulation of the Transport and Speciation of Nonconservative Chemical Reactions in Rivers." *Water Resources Research* 18(1): 155–167.

Chernobyl Forum. 2005. "Chernobyl Legacy: Health, Environmental, and Socio-economic Impacts and Recommendations to the Governments of Belarus, the Russian Federation and Ukraine." International Atomic Energy Agency, Vienna, Austria.

Chow, V.T. 1959. *Open-Channel Hydraulics*. McGraw-Hill, New York.

Dailey, J.E., and D.R.F. Harleman. 1972. *Numerical Model for the Prediction of Transient Water Quality in Estuary Networks*. MIT Report No. 158. Ralph M. Parsons Laboratory, Massachusetts Institute of Technology, Cambridge, MA.

Edinger, J., and E. Buchak. 1977. *Description and Application of the LARM (Laterally-Averaged Reservoir Model)*. J.E. Edinger and Associates, Wayne, PA.

Eraslan, A.H., E.J. Akin, J.M. Barton, J.L. Bledsoe, K.E. Cross, H. Diament, D.E. Fields, S.K. Fischer, J.L. Harris, D.M. Hetrick, J.T. Holdeman, K.K. Kim, M.H. Lietzke, J.E. Park, M.R. Patterson, R.D. Sharp, and B. Thomas, Jr. 1977. *Development of a Unified Transport Approach for the Assessment of Power-Plant Impact.* ORNL/NUREG/TM89. Oak Ridge National Laboratory, Oak Ridge, TN.

Eraslan, A.H., W.L. Lin, and R.D. Sharp. 1983. *FLOWER: A Computer Code for Simulating Three-Dimensional Flow, Temperature, and Salinity Conditions in Rivers, Estuaries, and Oceans.* NUREG/CR-3172. U.S. Nuclear Regulatory Commission, Washington, DC.

Felmy, A.R. 1990. *GMIN: A Computerized Chemical Equilibrium Model Using a Constrained Minimization of the Gibbs Free Energy.* PNL-7281. Pacific Northwest National Laboratory, Richland, WA.

Felmy, A.R., S.M. Brown, Y. Onishi, R.S. Argo, and S.B. Yabusaki. 1983. *MEXAMS—the Metals Exposure Analysis Modeling System.* Battelle, Pacific Northwest Laboratories, Richland, WA.

Felmy, A.R., D.C. Girvin, and E.A. Jenne. 1984. *MINTEQ—a Computer Program for Calculating Aqueous Geochemical Equilibrium.* EPA 600/3-84-032. U.S. Environmental Protection Agency, Athens, GA.

Fields, D.E. 1976. *CHNSED: Simulation of Sediment and Trace Contaminant Transport with Sediment/Contaminant Interaction.* ORNL/NSF/EATC-19. Oak Ridge National Laboratory, Oak Ridge, TN.

Fischer, H.B., E.J. List, R.C.Y. Koh, J. Imberger, and N.H. Brooks. 1979. *Mixing in Inland and Coastal Waters.* Academic Press, New York.

Harvie, C.E., J.P. Greenberg, and J.H. Weare. 1987. "A Chemical Equilibrium Algorithm for Highly Non-ideal Multiphase Systems: Free Energy Minimization." *Geochemica et Cosmochimica Acta* 51: 1045–1057.

Hetherington, J.A. 1976. "The Behavior of Plutonium Nuclides in the Irish Sea." In *Environmental Toxicity of Aquatic Radionuclides: Models and Mechanisms*, ed. M.W. Miller and J.N. Stannard. Ann Arbor Science, Ann Arbor, MI; 81–106.

Holly, F.M., Jr., J.C. Yang, P. Schwarz, J. Schaefer, S.H. Hsu, and R. Einhelling. 1990. *CHARIMA, Numerical Simulation of Unsteady Water and Sediment Movement in Multiply Connected Networks of Mobile-Bed Channels.* IIHR Report No. 343. Iowa Institute of Hydraulic Research, University of Iowa, Iowa City, IA.

IAEA (International Atomic Energy Agency). 1985a. *Hydrological Dispersion of Radioactive Material in Relation to Nuclear Power Plant Siting.* Safety Guide 50-SG-S6. International Atomic Energy Agency, Vienna, Austria.

IAEA. 1985b. *Sediment $K_d$s and Concentration Factors for Radionuclides in the Marine Environment.* Technical Report Series No. 247. International Atomic Energy Agency, Vienna, Austria.

IAEA. 1991. *The International Chernobyl Project, Technical Report of the Advisory Committee* IAEA-ISBN-92-0-129191-4. International Atomic Energy Agency, Vienna, Austria.

IAEA. 2001. *Generic Models for Use in Assessing the Impact of Dischargings of Radioactive Substances to the Environment.* Safety Reports Series No. 19. International Atomic Energy Agency, Vienna, Austria.

Ingham, M.C., J.J. Bisagni, and D. Mizenko. 1977. "The General Physical Oceanography of Deepwater Dumpsite 106." *Baseline Report of Environmental Conditions in Deepwater Dumpsite 106.* NOAA Dumpsite Evaluation Report 77-1. NOAA National Ocean Service, Rockville, MD.

Ippen, A.T. 1966. *Estuary and Coastline Hydrodynamics.* McGraw-Hill, New York.

Jirka, G.H., A.N. Findikakis, Y. Onishi, and P.J. Ryan. 1983. "Transport of Radionuclides in Surface Waters." *Radiological Assessment—a Textbook on Environmental Dose*

*Analysis*, ed. J.E. Till and H.R. Meyer. NUREG/CR-3332, ORNL-5968. U.S. Nuclear Regulatory Commission, Washington, DC.

King, I.P. 1982. "A Three-Dimensional Finite Element Model for Stratified Flow." In *Finite Element Flow Analysis*, ed. T. Kawai. University of Tokyo Press, Tokyo, Japan; 513–527.

Konoplev, A.V., A. Rodhe, and O.V. Voitsekhovitch. 1992. "Chapter 4—Migration of Contaminant in Water Bodies." *Hydrological Aspects of Accidental Pollution of Water Bodies*. WHO No. 754. Operational Hydrology Report No. 37. World Meteorological Organization, Geneva, Switzerland; 27–48.

Koplik, C.M., M.F. Kaplan, J.Y. Nalbandian, J.H. Simonson, and P.G. Clark. 1984. *User's Guide to MARINRAD: Model for Assessing the Consequences of Release of Radioactive Material into the Oceans*. SAND83-7104. Analytic Sciences Corporation, Reading, MA.

Leendertse, J.J. 1970. "A Water Quality Simulation Model for Well Mixed Estuaries and Coastal Seas." *Principles of Computation*, vol. 1. RM-6230-RC. Rand Corporation, Santa Monica, CA.

Leendertse, J.J., and S.K. Liu. 1975. "A Three-Dimensional Model for Estuaries and Coastal Seas."*Aspects of Computation*, vol. 2. R-1784-OWRT. Rand Corporation, Santa Monica, CA.

Leopold, L.B., M.G. Norman, and J.P. Miller. 1964. *Fluvial Processes in Geomorphology*. W.H. Freeman, San Francisco, CA.

Lick, W. 1983. "Entrainment, Disposition, and Long-Term Transport of Fine-Grained Sediment." *Proceedings of an Evaluation of Effluent Dispersion and Fate Models for OSC Platforms*. U.S. Department of Interior, Minerals Management Service, Santa Barbara, CA.

*National Geographic*. 2006. "The Long Shadow of Chernobyl." April, 32–53.

NCRP (National Council of Radiation Protection and Measurement). 1984. "Radiological Assessment: Predicting the Transport, Bio-accumulation, and Uptake by Man of Radionuclides Released to the Environment." NCRP No. 76. NCRP Scientific Committee 64, Bethesda, MD.

NCRP. 1996. *Screening Models for Releases of Radionuclides to Atmosphere, Surface Water, and Ground*. NCRP Report No. 123 I. National Council of Radiation Protection and Measurement, Bethesda, MD.

NRC (Nuclear Regulatory Commission). 1976. *Regulatory Guide I-113, Estimating Aquatic Dispersion of Effluents from Accidental and Routine Reactor Releases for the Purpose of Implementing Appendix I*. Nuclear Regulatory Commission, Washington, DC.

NRC. 1978. *Liquid Pathway Generic Study: Impacts of Accidental Radionuclide Releases to the Hydrosphere from Floating and Land-Based Nuclear Power Plants*. NUREG-0440. Nuclear Regulatory Commission, Washington, DC.

O'Conner, J.D., and J.P. Jawler. 1965. *Mathematical Analysis of Estuarine Pollution*. Report No. 31a. American Institute of Chemical Engineers, Houston, TX.

Okubo, A. 1971. "Oceanic Diffusion Diagrams." *Deep Sea Research* 18: 789–802.

Onishi, Y. 1981. "Sediment and Contaminant Transport Model." *Journal of Hydraulics Division American Society of Civil Engineers* 107(HY9): 1089–1107.

Onishi, Y. 1985. "Chemical Transport and Fate in Risk Assessment." In *Principals of Health Risk Assessment*, ed. P.F. Ricci. Prentice-Hall, Englewood Cliffs, NJ; 117–154.

Onishi, Y. 1994a. "Contaminant Transport Models in Surface Waters." In *Applied Sciences*. Vol. 274: *Computer Modeling of Free Surface and Pressurized Flows*, ed. M.H. Chaudrey and L.W. Mays. NATO ASI Series E. Kluwer Academic, Dordrecht, The Netherlands; 313–341.

Onishi, Y. 1994b. "Sediment Transport Models and Their Testing." In *Applied Sciences*. Vol. 274: *Computer Modeling of Free Surface and Pressurized Flows*, ed. M.H. Chaudrey and L.W. Mays. NATO ASI Series E. Kluwer Academic, Dordrecht, The Netherlands; 281–312.

Onishi, Y. 2005. "Slurry Transport." In *Multiphase Flow Handbook*, ed. C. Crowe. CRC Press, Boca Raton, FL; 13.119–13.128.

Onishi, Y., R.J. Serne, E.M. Arnold, C.E. Cowan, and F.L. Thompson. 1981. *Critical Review: Radionuclide Transport, Sediment Transport and Water Quality Mathematical Modeling, and Radionuclide Sorption/Desorption Mechanisms*. NUREG/CR-1322, PNL-2901. Pacific Northwest National Laboratory, Richland, WA.

Onishi, Y., G. Whelan, and R.L. Skaggs. 1982a. *Development of a Multi-media Radionuclide Exposure Assessment Methodology for Low-Level Waste Management*. PNL3370. Pacific Northwest National Laboratory, Richland, WA.

Onishi, Y., S.B. Yabusaki, and C.T. Kincaid. 1982b. "Performance Testing of Sediment-Contaminant Transport Model, SERATRA." *Proceedings of the ASCE Hydraulics Conference, Applying Research to Hydraulic Practice. Jackson, Mississippi, August 17–20, 1982*. ed. P.E. Smith. American Society of Civil Engineers, New York; 623–632.

Onishi, Y., A.R. Felmy, and S.B. Yabusaki. 1983. "Three Modeling Approaches for Transport and Geochemical Interactions in Surface Water." *EOS Transactions* 64(45): 696.

Onishi, Y., L.F. Hibler, and C.R. Sherwood. 1987. *Review of Hydrodynamic and Transport Models and Data Collected Near the Mid-Atlantic Low-Level Radioactive Waste Disposal Sites*. PNL-6331. Pacific Northwest National Laboratory, Richland, WA.

Onishi, Y., G.M. Petrie, L.W. Vail, O.V. Voitsekhovitch, M.J. Zheleznyak, V.M. Shershakov, and A.V. Konoplev. 1993a. "USA/CIS Coordinating Committee and Its Hydrologic Studies." *Proceedings of an International Workshop on Hydrologic Impact in Relation to Nuclear Power Plants, 23–25 September, 1992, Paris*. IHPIV SC.93/WS/51. UNESCO (United Nations Educational, Scientific and Cultural Organization), Paris; 8–15.

Onishi, Y., H.C. Graber, and D.S. Trent. 1993b. "Preliminary Modeling of Wave-Enhanced Sediment and Contaminant Transport in New Bedford Harbor." In *Nearshore and Estuarine Cohesive Sediment Transport*, ed. A.J. Mehta. Coastal Estuarine Series, Vol. 42. American Geophysical Union, Washington, DC; 541–557.

Onishi, Y., C.R. Cole, M.J. Zheleznyak, S.L. Kivva, N. Dzjuba, and O.V. Voitsekhovitch. 2003. "Radionuclide Migration in Surface Water and Groundwater Affected by the Chernobyl New Safe Confinement." *Proceedings of the 48th Annual Meeting of Health Physics Society. San Diego, California, July 20–24, 2003*, Health Physics Society, McLean, VA.

Prister, B., N. Loschilov, L. Perepelyatnikova, and G. Perepelyatnikov. 1992. "Efficiency of Measures Aimed at Decreasing the Contamination of Agricultural Products in Areas Contaminated by the Chernobyl NPP Accident." *The Science of the Total Environment* 112(1): 79–87.

Robinson, A.R., and M.G. Marietta. 1985. *Research, Progress and the Mark A Box Model for Physical, Biological, and Chemical Transports*. SAND84-0646. Sandia National Laboratories, Albuquerque, NM.

Rouse, H. 1961. *Fluid Mechanics for Hydraulic Engineers*. Dover Publications, New York.

Sayre, W.W. 1975. "Natural Mixing Processes in Rivers." In *Environmental Impact on Rivers (River Mechanics III)*, ed. H.W. Shen. Hseigh Wen Shen, Fort Collins, CO; 6.1–6.37.

Sheng, Y.P. 1993. "Hydrodynamics, Sediment Transport and Their Effects on Phosphorus Dynamics in Lake Okeechobee." In *Nearshore and Estuarine Cohesive Sediment Transport*, ed. A.J. Mehta. American Geophysical Union, Washington, DC; 558–571.

Shepard, J.G. 1978. "A Simple Model for the Dispersion of Radioactive Wastes Dumped on the Deep-Sea Bed." *Marine Science Communications* 4(4): 293–327.

Simons, T.J. 1973. *Development of Three-Dimensional Numerical Models of the Great Lakes*. Scientific Series No. 12. Canada Center for Inland Waters, Burlington, Ontario, Canada.

Spaulding, M.L., and D. Pavish. 1984. "A Three-Dimensional Numerical Model of Particulate Transport for Coastal Water." *Continental Shelf Research* 3(1): 55–67.

Trent, D.S., and L.L. Eyler. 1994. *TEMPEST: A Computer Program for Three Dimensional Time Dependent, Computational Fluid Dynamics*. PNL-8857. Pacific Northwest National Laboratory, Richland, WA.

UNSCEAR (U.N. Scientific Committee on the Effects of Atomic Radiation). 2000. "Sources and Effects of Ionizing Radiation." In *Report to the General Assembly (with Scientific Annexes), UN Science Committee on the Effects of Atomic Radiation*, vol. 2. United Nations, New York; 451–566.

Vanoni, V.A., ed. 1975. *Sedimentation Engineering*. ASCE Manuals and Reports on Engineering Practice. American Society of Civil Engineers, New York.

Voitsekhovitch, O.V., M.J. Zheleznyak, and Y. Onishi. 1994. *Chernobyl Nuclear Accident Hydrologic Analysis and Emergency Evaluation of Radionuclide Distribution in the Dnieper River, Ukraine During the 1993 Summer Flood*. PNL9980. Pacific Northwest Laboratories, Richland, WA.

Ward, P.R.B. 1974 "Traverse Dispersion in Oscillatory Channel Flow." *Journal of the Hydraulics Division Proceedings, American Society of Civil Engineers* 100: 755–772.

Ward, P.R.B. 1976. "Measurements of Estuary Dispersion Coefficient." *Journal of the Environmental Engineering Division Proceedings, American Society of Civil Engineers* 102: 855–859.

Webb, G.A.M., and F. Morley. 1973. *A Model for the Evaluation of Deep Ocean Disposal of Radioactive Waste*. National Radiological Protection Board, Harwell, UK.

Weigel, R.L. 1964. *Oceanographical Engineering*. Prentice-Hall, Englewood Cliffs, NJ.

Westall, J.C., J.L. Zachary, and F.M. Morel. 1976. *MINEQL: A Computer Program for the Calculation of Chemical Equilibrium Composition of Aqueous Systems*. Tech. Note 18. Department of Civil Engineering, Massachusetts Institute of Technology, Cambridge, MA.

White, A., and E. F. Gloyna. 1969. *Radioactivity Transport in Water—Mathematical Simulation*. EHE-70-04. Technical Report No. 19. University of Texas, Austin, TX.

Wolery, T.J. 1980. *Chemical Modeling of Geologic Disposal of Nuclear Waste: Progress Report and a Perspective*. UCRL-52748. University of California, Lawrence Livermore National Laboratory, Livermore, CA.

Yotsukura, N., and W.W. Sayer. 1976. "Transverse Mixing in Natural Channels." *Water Resources Research* 12(4): 695–704.

Zheleznyak, M.J., R.I. Demchenko, S.L. Khursin, Y.I. Kuzmenko, P.V. Tkalich, and N.Y. Vitjuk. 1992. "Mathematical Modeling of Radionuclide Dispersion in the Pripyat-Dnieper Aquatic System after the Chernobyl Accident." *The Science of the Total Environment* 112: 89–114.

Zheleznyak, M., G. Blaylock, J. Garnier-Laplace, G. Gontier, A.V. Konoplev, A. Marinets, L. Monte, Y. Onishi, K.-L. Sjoblom, M. Tschurlovits, O. Voitsekhovitch, V. Berkosky, A. Bulgakov, T. Tkalitch, N. Tkachenko, and G. Wincler. 1995. "Modeling of Radionuclide Transfer in Rivers and Reservoirs: Validation Study within the IAEA/CEC Vamp Programme on Environmental Impacts of Radionuclide Releases." *Proceedings of 1995 IAEA Symposium*. IAEASM-339. International Atomic Energy Agency, Vienna, Austria; 335–376.

# 5

# Transport of Radionuclides in Groundwater

Richard B. Codell
James O. Duguid

The radiological effect of nuclear waste released to or disposed of into the ground depends on at least the following phenomena:

- Rate of release of radionuclides from the waste form (i.e., the source term)
- Transport of radionuclides from the source term to the eventual consumer
- Consequences of releases when they reach the eventual consumer of the contaminated water

Potential doses to humans from nuclear waste disposed to the ground could occur through the contamination of drinking water and food and from contaminated surfaces, such as flood plains and beaches. Groundwater flow is one of the likely pathways for radionuclides released from waste disposal areas. Groundwater transport is also a potential pathway for certain classes of accidental and normal releases from nuclear facilities, such as power plants, fuel reprocessing plants, and mining or milling operations.

Transport of radionuclides through the ground can be estimated by using tracers, groundwater dating, mathematical models, or by a combination of all these techniques. Chemical or radioactive tracers can be deliberately introduced to the groundwater and monitored through wells for directly determining groundwater velocity and transport. Alternatively, natural or manufactured materials present in the environment but not deliberately introduced to the groundwater can also be traced.

Groundwater dating is a technique by which the age of a region of groundwater can be estimated from the concentration of a natural or manufactured radionuclide, or nonradioactive material, that it contains. Manufactured tracers are limited to estimating the age of relatively young groundwater. For example, the radionuclides $^{14}C$, $^{36}Cl$, and $^{3}H$ released since the beginning of the nuclear era can be used to date water up to several decades (although these radionuclides are also naturally occurring). Manufactured substances, such as chlorofluorocarbons, can also be used as unique markers because these materials have been in existence only since the 1930s (Thompson et al. 1976). For longer periods, water may be dated using naturally occurring $^{14}C$ over hundreds to thousands of years. Measurements of the migration of radionuclides released from naturally occurring uranium and thorium ore bodies can be used as analogs to manufactured radioactive waste disposal situations (Prikryl et al. 1997; Smellie et al. 1997).

Using groundwater flow and transport models, along with measurements conducted in the field, provides a means to calculate the expected concentrations of radionuclides following their release to the environment. This chapter discusses current practice in groundwater flow and transport modeling as well as data requirements and possible misuses of models. A set of analytical models is presented along with illustrations of their use.

## Applications of Groundwater Models for Radionuclide Migration

Groundwater models are useful for a number of radionuclide migration problems, which are described in the following sections.

### Geologic Isolation of High-Level Waste

Actual demonstrations of the long-term behavior of a high-level waste repository system cannot be performed. Therefore, we must rely on mathematical models for performance assessments, using data collected over comparatively short periods of time to predict the long-term performance of the system. This is the only means by which the cumulative effects of changes in the properties of the repository system, the effects of various design features of the repository, and the effects of the repository on the environment can be analyzed. Performance assessment not only provides this type of analysis but also offers information that is useful in guiding research and development activities in site selection, repository design, and waste package design. Performance assessment deals with concepts that can be quantified (i.e., failure analysis and consequence assessment) (Klingsberg and Duguid 1980).

An assessment of the long-term performance of a repository analyzes the events and processes that could release radionuclides from the waste and the phenomena that might transport radionuclides to the biosphere. These phenomena may be roughly classified as those that occur in the near field, where waste and repository phenomena dominate, and those that occur in the far field, at a greater distance from the repository where natural phenomena dominate. Although these two regions are not separated by a precisely defined boundary, the distinction is useful because the physical and chemical effects of heat and radiation from the waste are limited to the near field (Klingsberg and Duguid 1980).

Different methods of analysis are therefore appropriate for the two regions. Near-field analysis studies the combined effects of heat, radiation, repository design and construction, and the waste package. Far-field analysis studies the effects of events that arise from natural phenomena and from potential human actions after the repository has been sealed. These far-field phenomena usually appear in the geosphere and the biosphere outside of the repository. Both near-field and far-field performance must be considered in determining how well the natural and the manufactured components of the disposal system meet the criteria for repository performance. A summary of the total system performance that includes all of the analyses described here can be found in the site recommendation for the potential high-level waste repository at Yucca Mountain, Nevada (U.S. Department of Energy 2002).

## Shallow Land Burial

For near-surface disposal, such as shallow land burial, the analysis of system performance bears many similarities to, and some differences from, isolation of high-level waste. The major distinctions are (a) the wastes are not heat producing, (b) the waste packages for high-level waste are generally much more robust, and (c) high-level waste is generally buried much deeper. The analysis proceeds in much the same fashion as for high-level waste and includes developing a source term through corrosion or breaching of waste containers, defining an appropriate leach rate for the waste form, developing a system release scenario, and calculating groundwater flow and radionuclide transport for use in the biosphere models and dose codes (Aikens et al. 1979).

## Uranium Mining and Milling

Several potential groundwater contamination problems are associated with the mining, milling, and waste disposal operations necessary to produce uranium fuels (NRC 1979; Shepard and Cherry 1980). The greatest waterborne contamination hazard to groundwater is the seepage from tailings ponds resulting from conventional milling procedures. The waste stream contains about 50% solids and 50% water, which is usually disposed of in tailings ponds formed behind earth or rubble dams. Tailings are sometimes reburied in the ore pits. Acid leach mills are the most prevalent type. Tailings ponds usually receive highly acidic (pH 0.5–2) water and tailings, but in some cases, tailings are first neutralized. The wastes from the tailings ponds differ most from other forms of nuclear waste because of their unique chemistry. In acidic tailings, most of the radioactive and other chemical wastes are in the dissolved state. Acidic wastes are sometimes neutralized to reduce the solubility of pollutants; however, in some cases the wastes slowly become acidic because of the oxidation of pyrite (iron sulfide). The behavior of the radioactive contaminants varies from simple to complex. Radium, a radiologically significant radioactive waste component present, has a fairly simple chemistry because it exists only in the +2 valence state. Uranium and several other radioactive elements behave in a much more complicated fashion because they may exist in several different valence states and form complex compounds (Landa 1980). The solubility of most of the contaminants is high for low-pH conditions and decreases markedly at higher pH. Neutralization by carbonates such as limestone, however, either added deliberately or encountered in the environment, can mobilize uranium in the form of soluble carbonate complexes. Uranium may also be mobilized in an oxidizing environment or by certain organic chemicals in groundwater (Shepard and Cherry 1980).

## Nuclear Power Plant Accidents

Postulated accidental releases of radioactivity to the groundwater pathway have been evaluated for a wide range of nuclear facilities for generic sites and actual reactor licensing reviews. The accidental releases considered ranged from small leaks from contaminated water streams in nuclear plants to major releases caused by a core meltdown accident (NRC 1975, 1978; Niemczyk et al. 1981).

Consideration given to nuclear power plant accident releases to groundwater differs from those for high- and low-level waste disposal or for other fuel cycle problems in several important respects. First, the risk of contamination generally would exist only for the lifetime of the plant. Administrative controls would be in effect during this period, so mitigative measures could presumably be taken should an accidental release occur. Second, the isotopes of importance in nuclear power plant liquid pathway accidents are generally those with high dose factors and half-lives of years to tens of years, notably $^3$H, $^{134}$Cs, $^{137}$Cs, $^{89}$Sr, $^{90}$Sr, and $^{106}$Ru. Unlike nuclear waste, long-lived radionuclides, actinides, and transuranics have been shown to be of much lower importance (NRC 1978). Third, for a given event, consequences of radioactive releases to the groundwater pathway typically present much smaller risks than do releases to the airborne pathway.

## Types of Groundwater Models

Performance assessment for nuclear waste disposal generally requires three types of models: (1) models that determine the portion of the radioactive source released by infiltrating water contacting the waste, (2) models in terms of measurable hydrologic parameters that predict the migration of radionuclides from the source to locations accessible to the public, and (3) models that determine the potential radiation dose using the radionuclide concentrations that reach accessible locations. This chapter discusses only the first and second types of models. However, doses that could result from radionuclides carried in groundwater and other pathways are covered in chapters 8–10.

## Groundwater Models for High-Level Waste Repositories

Although modeling of high-level waste repositories shares many things in common with other waste disposal situations, it differs from most other groundwater models in several respects. First, the models must deal with the substantial effects of heat released from spent nuclear fuel and reprocessed waste over thousands of years into the future. Second, the repositories are deep underground in media chosen primarily for their low permeability (U.S. Environmental Protection Agency 1998) or, in the unique case of the proposed Yucca Mountain repository, its relative dryness (CRWMS M&O 2000, section 1.8 and chapter 3).

Near-Field Performance

Models for assessing the performance of a high-level waste repository in the near field must take into account mechanical stresses, heat flow, chemical interactions, and radiation-induced physical-chemical processes. All of these phenomena, in addition to the properties of the host rock, affect the environment of the emplaced waste. The following sections discuss the principal types of models required for near-field analysis: heat transfer models, thermomechanical models, and models of physical and chemical interactions among the emplaced waste, the waste package, and the host rock.

*Heat Transfer Models* Thermal models based on physical laws provide an accurate portrayal of heat flow and changes in temperature. The experimental results to date suggest that predictions of temperature within a few percent of measured values can be achieved. Considering heat transfer is important because temperature gradients can be a driving force in groundwater flow and chemical changes in the engineered and natural materials. Models for fully saturated systems are generally more successful than are those for partially saturated systems. However, in recent years considerable progress has been made in developing and applying thermal models to unsaturated groundwater systems (CRWMS M&O 2000, chapter 3; Pruess 1991; Nitao 1998). Heat transfer in saturated systems generally is dominated by thermal conduction through the rock and soil, with convective heat transfer by flowing water important only in those cases where there are large, concentrated sources of heat. Partially saturated systems have the added component of heat transfer by latent heat (i.e., the evaporation and condensation of water).

*Thermomechanical Models* Thermomechanical models are based on functional relationships for heat transfer, thermal expansion, and stress and strain that are obtained from laboratory tests. Because repository rocks are inhomogeneous and may be fractured, generic functional laws are more difficult to obtain for rocks than for most other construction materials. Thermomechanical codes are currently being used to analyze the uplift and subsidence; room stability and rate of closure; hole stability and rate of closure; canister movement; pillar stability; thermomechanical effects on groundwater flow, stresses, and strains at critical locations in the rock mass; and mechanical failure of the rock mass (Science Applications, Inc. 1979).

*Thermal-Hydrologic Models* Thermal-hydrologic models are used to predict the temperature and relative humidity in the vicinity of the waste package over time (CRWMS M&O 2000, chapter 3; Pruess 1991; Nitao 1998). In the case of the proposed Yucca Mountain repository, this information and the near-field geochemical conditions are used as input to waste package and drip shield degradation models to analyze waste package failure and expected mode of release from the waste package (CRWMS M&O 2000, section 3.4). The expected mode of release from the waste package changes with the degree of waste package and drip shield failure. Release by diffusion may dominate when there are only a few penetrations of the package and the drip shield is intact. Release by advection may dominate after the drip shield and waste package have failed to the point of allowing flow through them (CRWMS M&O 2000, sections 3.6 and 3.7).

*Chemical Models* To predict the near-field behavior of a repository requires analyses of the interactions between the emplaced package components and the host rock. These interactions fall into six general categories: (1) movement of fluids in the vicinity of the waste package, (2) corrosion of the waste package materials by these fluids, (3) dissolution of the waste form by groundwater containing the added corrosion products, (4) sorption of radionuclides by the rocks and the degraded engineered components of the repository, (5) absorption of radiation emitted by the waste, and (6) alteration of chemical phases and properties in the vicinity of the waste packages. A study of these interactions predicts the kinds, amounts, and chemical state of the radionuclides available for entry into the groundwater system (Jenne 1979; CRWMS M&O 2000, sections 3.3–3.6 and chapter 5; Williams 2001; Wolery 1992; Cloke 2000).

*Models of the Source Term* The source term describes the waste at all times; it specifies the radionuclides present, the physical and chemical conditions of the waste, and the rate of release to the geosphere. The radionuclide concentrations at the time of the breach can be calculated from their original concentrations in the waste. Source term evaluation is highly site specific, depending on such factors as the chemistry of the waste, the host rock, and the groundwater. Interaction between natural and manufactured components also can play an important role in defining release mechanisms (Mohanty et al. 1997).

## Far-Field Performance

In general, the procedure for calculating the far-field effects of a repository breach is shown in figure 5.1. After the scenarios have been identified, the next step in a performance assessment is to predict their consequences. Whether the scenarios will actually occur cannot be predicted with complete certainty, but when possible, the probabilities of their occurrences are estimated. Probabilities are, however, highly uncertain for the events that have occurred in the region around a repository site only a few times in geologic history or that may not have occurred there at all. For this reason, the assessment of repository performance relies heavily on predictions of the consequences of scenarios rather than on predictions of their probabilities.

After the source term has been defined, transport through the geosphere is determined by modeling fluid flow into and away from a repository. The output from the geosphere transport codes is a prediction of the radionuclide concentrations reaching the biosphere as a function of time. As contaminants move with groundwater, they may interact with the rock or be sorbed and thus effectively retarded with respect to the water flow. The parts of the transport models that describe sorption generally

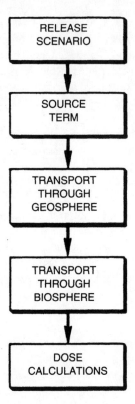

Figure 5.1   Elements of far-field risk assessment. From Klingsberg and Duguid (1980).

assume equilibrium between the contaminants in the fluid and the rock. Because models of flow through fractures tend to be specific to particular types of fracture systems, they are less universally applicable than are porous media models (Duguid and Lee 1977).

There are two basic problems for modeling material transport in fractured media. One problem is to assemble sufficient data to be able to adequately describe the hydrology of the far-field region surrounding a repository site. Determining hydraulic and transport properties of fractured rock is difficult, and a considerable amount of research remains before reliable methods will be available. The second problem is to understand the sorption process in the fractured rocks; although sorption is effective in porous rocks, it might be much less effective in fractured rocks. Much of the sorption occurs on surfaces rather than the bulk material, and the area of fracture surfaces is smaller than the area of a porous medium.

Practical modeling of flow and transport in fractured aquifers is limited, but it is becoming increasingly more feasible. Several approaches for modeling flow and transport through fractured media are currently in use: (1) equivalent continuum formulations that use porous media models with coefficients that approximate the behavior of a fractured system, (2) dual continuum models that have two interconnected compartments for the fractures and matrix, and (3) discrete fracture network models. Equivalent continuum models are most commonly used because they are well understood and available. They are most useful for well-connected fracture systems, for large scales, and for predicting the average behavior of a system. One of the major weaknesses of the equivalent continuum formulation is that it requires the assumption that the matrix and fractures in the continuum are always in equilibrium. In practice, fractures are often fast pathways through the medium and are largely disconnected from the water in the matrix of the rock (National Academy of Sciences 1996).

Dual continuum models treat the fracture and matrix as two distinct continua. However, the two continua are coupled; therefore, mass and pressure can communicate between them. Dual continuum models overcome the serious restriction of single continuum models that the fractures and the matrix must be in equilibrium. Therefore, they are capable of representing fast pathways through the medium more realistically (Updegraff et al. 1991; Bodvarsson et al. 1997; Wu et al. 2002).

Discrete fracture models attempt to represent the important flow conducting pathways as discrete interconnected planes and describe how these elements interact (National Academy of Sciences 1996). Except for a few large, obvious features, the geometry and interconnection of fractures are generally not known. Therefore, these models represent the field of interconnected fractures stochastically (i.e., randomly) from inferences about fracture orientations, spacings, and lengths based on indirect observations, such as outcrops and core samples. After generating numerous potential random sets of fractures, the results are presented in terms of the set of possible outcomes of the random fracture fields. In some cases, the sets can be refined to narrower sets based on how well the individual runs comply with other information available to the modeler, such as hydraulic test data and tracer information (Dershowitz et al. 1996).

### Groundwater Models for Shallow Land Burial of Low-Level Waste

Although difficult to analyze, the near field can be as important and complex for shallow land disposal as for deep geological disposal. Most of the action is in the disposal trench, and these interactions determine the amount of radionuclides that are available for future groundwater transport. Determining the water balance and the amount of water infiltrating is difficult, as is determining the leaching and release from chemically and physically heterogeneous materials such as low-level waste. Calculating transport of radionuclides from shallow land burial sites is complicated by the waste frequently being leached in the unsaturated zone. The movement of waste leachate to the water table must consider both flow and transport through the unsaturated zone. In the simplest case, flow and transport may be assumed to be downward in one dimension. An approximate model for release that may be suitable for low-level waste is discussed under "Source Term Models for Low-Level Waste," below.

### Groundwater Models for Mill Tailings Waste Migration

The unique aspects of modeling mill tailings wastes are the complex chemistry of the wastes and the process of neutralization, especially by rocks in the transport pathway. Concentrations of chemicals may be quite high in a mill tailings pond, which complicates the transport processes. Typical equilibrium concepts, such as the "retardation factor," will not work well for complicated, nonlinear phenomena such as precipitation when concentrations exceed solubility limits. Unsaturated flow in some of the pond settings and the transient existence of the milling operations may present special modeling problems (Shepard and Cherry 1980; NRC 1979).

## Equations for Groundwater Flow and Radioactivity Transport

The movement of radionuclides in groundwater can be described by two equations: one for the movement of water and one for the mass transport of the radionuclides. In using these equations, the movement of water in the region under consideration must be known before the transport equation can be solved. The following discussion of the equations associated with groundwater movement and transport of dissolved radioactive substances can be used only as a general guide. A person intending to use models to analyze a specific problem would usually need the aid of an experienced groundwater hydrologist and modeler. A set of highly simplified transport models and examples of their use are provided under "Analytical Solutions of the Convective-Dispersive Equations," below.

### Groundwater Flow

Groundwater moves in openings in the soil or rock (pores and fractures) under the influence of gravity, capillary forces, and friction. Figure 5.2 conceptually illustrates an experiment for determining the flow through a packed column and the relationship of flow to the properties of the medium. Under conditions of full saturation (i.e., all of the pores or fractures are filled with water), friction balances gravity, and the flux of water (volume of flow $Q$ per unit cross-sectional area $A$) is proportional to the pressure gradient $\Delta h/\Delta L$ and a frictional term, $K$, generally called hydraulic conductivity:

$$V = -K\frac{\Delta h}{\Delta L}. \tag{5.1}$$

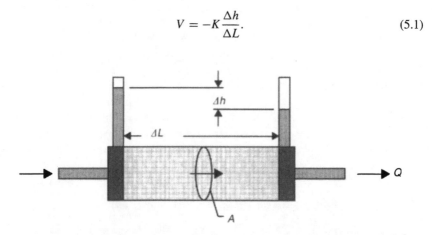

Figure 5.2  Flow in a packed column. Adapted from Brown (2003).

Equation 5.1 is known as Darcy's Law (D'Arcy 1856) and is the basis for groundwater flow equations in the saturated zone. An approximation of the flux in the major flow direction can be obtained using Darcy's Law:

$$V_x = -K\frac{dh}{dx} \cong -K\frac{\Delta h}{\Delta x}, \qquad (5.2)$$

where $dh/dx$ is the hydraulic gradient in the direction of flow. Use of this equation assumes a homogeneous isotropic medium in which the gradient is constant over the increment. The actual velocity of an inert tracer would be greater than the flux because water is moving only in the pore spaces. The pore velocity (or seepage velocity), $U$, can be approximated by dividing the flux by the effective porosity, $n_e$:

$$U = \frac{V_x}{n_e} \qquad (5.3)$$

### EXAMPLE 5.1

For saturated groundwater flow, calculate the pore velocity in fine sandstone under a gradient of 0.01. (Use arithmetic mean values in tables 5.4 and 5.5.)

The arithmetic mean hydraulic conductivity, $K$, is $3.31 \times 10^{-4}$ cm s$^{-1}$ (from table 5.5). The arithmetic mean effective porosity, $n_e$, is 0.21 (from table 5.4). Therefore,

$$U = \frac{V_x}{n_e} = \frac{-K(\Delta h/\Delta x)}{n_e}$$
$$= 1.58 \times 10^{-5} \text{ cm s}^{-1}.$$

Equation 5.2 applies only to the saturated zone. The unsaturated zone is usually regarded as the region above the regional water table where water flow is affected strongly by capillary forces. (The term "unsaturated" is in common use, but it is somewhat inaccurate because there is usually a substantial amount of water present in the soil or rock.) Equations for unsaturated flow are generally much more difficult to solve because the hydraulic conductivity term is now a function of the degree of saturation. Consequently, simple models in the unsaturated zone often assume flow to be downward and steady.

Flow in the unsaturated zone can be represented by a set of coupled equations for the movement of gas and water (Lappala 1981). When the assumptions are made that the water moves as a single phase and that no trapped air pockets exist, a single governing equation for saturated-unsaturated flow is obtained (ANS 1980):

$$\left(\frac{\theta}{n'}\alpha' + \theta\beta' + \frac{d\theta}{dh}\right)\frac{\partial h}{\partial t} = \nabla \cdot \left[\overline{\overline{K}}(h) \cdot (\nabla h + \nabla z)\right], \qquad (5.4)$$

where

$\theta$ = moisture content (dimensionless)
$n'$ = total porosity (dimensionless)
$\alpha'$ = modified coefficient of compressibility of the medium (cm$^{-1}$)
$\beta'$ = modified coefficient of compressibility of water (cm$^{-1}$)

$h$ = pressure head (cm) (i.e., the pressure expressed as the height to which a column of water would rise above the location of measurement)
$t$ = time (s)
$\overline{\overline{K}}$ = hydraulic conductivity tensor (cm s$^{-1}$)
$z$ = the elevation head (cm) (i.e., the elevation of the location of measurement above a datum, $z = 0$)
$\nabla$ = the del operator

Equation 5.4 is nonlinear because, for unsaturated flow, both hydraulic conductivity and moisture content are functions of pressure head. The evaluation of equation 5.4 in three dimensions is difficult but feasible, generally requiring numerical solutions (Reeves and Duguid 1975; Nitao 1989).

The hydraulic conductivity can account for directional properties (anisotropy) that arise in formations such as layered sediments. Hydraulic conductivity is a "tensor," which has directional properties. It is usually easier, however, to align the coordinate system of the problem being addressed to the main direction of the tensor so that only the diagonal components are necessary.

## Saturated Flow

If the medium is further assumed to be saturated, homogeneous, and isotropic (i.e., properties of the medium are not direction dependent), hydraulic conductivity becomes a scalar and equation 5.4 becomes (ANS 1980)

$$\nabla^2 H = \frac{S_s}{K} \frac{\partial H}{\partial t}, \qquad (5.5)$$

where

$H$ = total head = $h + z$ (cm) (i.e., the height to which a column of water would rise above the datum, $z = 0$)
$S_s$ = specific storage coefficient (cm$^{-1}$) = $\rho g(\alpha + n'\beta)$
$\rho$ = water density (g cm$^{-3}$)
$g$ = acceleration of gravity (cm s$^{-2}$)
$\alpha$ = coefficient of compressibility of the medium, (cm s$^2$ g$^{-1}$)
$\beta$ = coefficient of compressibility of water (cm s$^2$ g$^{-1}$)

The $S_s$ term defines the storage of water caused by the compressibility of water and the medium under conditions of confinement in an aquifer (i.e., water is confined by a low-permeability layer). For a confined aquifer of thickness, $b$, the storage coefficient and transmissivity are respectively defined as $S = S_s b$ and $T = Kb$. Equation 5.5 becomes

$$\nabla^2 H = \frac{S}{T} \frac{\partial H}{\partial t}. \qquad (5.6)$$

In simulations using equation 5.6, the boundary conditions of leakage between water-bearing units should be used where appropriate. Aquifers that are not totally confined are known as leaky aquifers and require methods that are more complicated, such as those described by Bredehoeft and Pinder (1970).

Under water table conditions, where there is no confining unit, the compressibility of water and the medium become relatively unimportant, and the storage is related to the change in the water level. For these situations, the continuity equation can be written as (ANS 1980)

$$\nabla^2 H^2 = \frac{2S_y}{K}\frac{\partial H}{\partial t}, \quad (5.7)$$

where $S_y$ is the specific yield of the aquifer (dimensionless), which is a measure of the volumetric storage in the water table aquifer for a change in the water table height. For specific yield of steady flow in either confined or unconfined aquifers, equations 5.6 and 5.7 reduce respectively to

$$\nabla^2 H = 0, \quad (5.8)$$

$$\nabla^2 H^2 = 0. \quad (5.9)$$

Equations 5.8 and 5.9 can be solved analytically for a few cases with simple geometries and uniform conditions (see "Advection Models," below), but they generally require computer techniques. A short synopsis of some of the available analytical and computer techniques for solving flow and transport equations is presented under "Methods of Solution for Groundwater Flow and Transport," below.

### Unsaturated Flow

The difficulty associated with the solution of the flow and transport in the unsaturated zones leads naturally to approximation methods. The time of travel can be estimated by assuming that the mean downward velocity, $U$, is proportional to the rate of recharge of water at the surface, $r$, and inversely proportional to the mean volumetric water content, $\theta$:

$$U = \frac{r}{\theta}. \quad (5.10)$$

The volumetric water content, $\theta$, can be conservatively assumed equal to the field capacity, which is the maximum water content where moisture can no longer be held against gravity. Field capacity is equal to the specific retention, $S_r$, which is defined as

$$S_r = n_e - n, \quad (5.11)$$

where $n_e$ is the effective porosity and $n$ is the total porosity. Representative values of $n_e$ and $n$ are tabulated under "Porosity and Effective Porosity," below. Alternatively, $\theta$ can be estimated by trial and error from hydraulic conductivity by setting $K(\theta)$ equal to $r$. In this case, $K$ is expressed as a function of water content.

### *Mass Transport*

Radioactive releases may travel in the unsaturated region (i.e., region above the water table) before entering the zone of saturation (i.e., region below the water table). However, releases can also be made directly into the zone of saturation. The predominant direction of the unsaturated flow is downward until the flow reaches the zone of saturation. Within the zone of saturation, the flow is predominantly lateral.

The most general form of the mass transport equation is for transport in saturated–unsaturated media. If local equilibrium of the dissolved species between water and rock can be assumed, the general mass transport equation can be written as (ANS 1980)

$$R_d \theta \frac{\partial c}{\partial t} - \nabla \cdot (\theta \overline{\overline{D}} \cdot \nabla c) + \nabla \cdot (\overline{V}c) + \left[ R_d \frac{\partial \theta}{\partial t} + \lambda \theta R_d \right] c = 0, \quad (5.12)$$

where

$R_d$ = retardation coefficient
$C$ = concentration of dissolved constituent (g cm$^{-3}$)
$\overline{\overline{D}}$ = dispersion tensor (cm$^2$ s$^{-1}$)
$\overline{V}$ = flux vector (cm s$^{-1}$)
$\lambda$ = radioactive decay constant (s$^{-1}$), which is

$$\lambda = \frac{\ln 2}{t_{1/2}}, \quad (5.13)$$

where $t_{1/2}$ is the radioactive half-life of the radionuclide(s). The $R_d$ term is discussed further under "Sorption, Retardation, and Colloids," below.

If the dispersion tensor is assumed to be homogeneous and isotropic and the flux is assumed to be parallel to the x-axis, equation 5.12 can be written as

$$R_d \frac{\partial c}{\partial t} - \nabla \cdot (\overline{\overline{D}} \cdot \nabla c) + \frac{\overline{V}}{n} \cdot \nabla c + \lambda R_d c = 0. \quad (5.14)$$

If fluid flux is assumed to be uniform along the x-axis, equation 5.14 becomes

$$\frac{\partial c}{\partial t} - \frac{D_x}{R_d} \frac{\partial^2 c}{\partial x^2} - \frac{D_y}{R_d} \frac{\partial^2 c}{\partial y^2} - \frac{D_z}{R_d} \frac{\partial^2 c}{\partial z^2} + \frac{U}{R_d} \frac{\partial c}{\partial x} + \lambda c = 0, \quad (5.15)$$

where $U$ is the pore velocity defined by equation 5.3 and $D_x, D_y$, and $D_z$ are the dispersion coefficients in the x-, y-, and z-directions, respectively (cm$^2$ s$^{-1}$), as described under "Diffusion and Dispersion in Porous Media," below.

The approximate rate of movement of the radionuclide is $U/R_d$. The above equations are strictly valid only for isotropic media (i.e., media whose hydraulic conductivity is uniform in all directions), but they may be applied to slightly anisotropic formations when the dispersivities are obtained from field studies.

## Chain Decay of Radionuclides

Radionuclides decay either to stable products or to another radioactive species called a *daughter* or *progeny*. In some species, several progeny may be produced before the parent species decays to a stable element. This process is particularly important for modeling actinides and transuranics. In considering this process over the transport path of radionuclides, one transport equation must be written for each original species and each progeny to yield the concentration of each radionuclide (parent and progeny) at points of interest along the flow path. In a constant one-dimensional velocity field, the general equations can be written as (Burkholder and Rosinger 1980)

$$R_{d1}\frac{\partial c_1}{\partial t} + U\frac{\partial c_1}{\partial z} = D\frac{\partial^2 c_1}{\partial z^2} - R_{d1}\lambda_1 c_1, \tag{5.16a}$$

$$R_{d2}\frac{\partial c_2}{\partial t} + U\frac{\partial c_2}{\partial z} = D\frac{\partial^2 c_2}{\partial z^2} - R_{d2}\lambda_2 c_2 + R_{d1}\lambda_1 c_1, \tag{5.16b}$$

$$R_{di}\frac{\partial c_i}{\partial t} + U\frac{\partial c_i}{\partial z} = D\frac{\partial^2 c_i}{\partial z^2} - R_{di}\lambda_i c_i + R_{di-1}\lambda_{i-1} c_{i-1}, \tag{5.16c}$$

where

$R_{di}$ = retardation factor for species $i$
$U$ = pore velocity
$c_i$ = concentration of species $i$
$D_x$ = dispersion coefficient in the direction of flow
$\lambda_i$ = decay coefficient for species $i$

Equations 5.16a–5.16c describe the material balances of the $i$th chain and all preceding members of the chain.

A relatively simple analytical formulation applies if all progeny are assumed to have equal retardation. The concentration $c_i$ of the $i$th progeny in terms of the parent concentration is

$$c_i = \frac{\lambda_i c_1}{\lambda_1} \prod_{m=1}^{i-1} \lambda_m \sum_{j=1}^{i} \frac{e^{-\lambda_j t}}{\prod_{k \neq j}^{i} (\lambda_k - \lambda_j)}. \tag{5.17}$$

For long-chain decays with $R_d$ values different for the parent and the progeny, numerical solutions are practically mandatory. Several useful closed-form solutions for chain decay with unequal sorption are available (Burkholder and Rosinger 1980; Dillon et al. 1978; Gureghian 1994). Another useful class of high-efficiency solutions relies on numerical inversion of Laplace transforms (Hodgkinson and Maul 1988; Sudicky 1989).

## Percolation of Water into the Ground

An important part of the analysis of contaminants migrating in groundwater is determining the rate of release of the contaminant at the source (e.g., leaching of low-level waste) and determining the speed of transport of the groundwater. Both of these aspects of the migration problem involve knowing the rate at which water infiltrates the ground either from a surface-water body (e.g., a river or pond) or directly from rainfall (percolation). Infiltration is most important for shallow land burial but less important for a deep repository not affected by local recharge. For example, the source of radioactive contamination at a low-level waste site may be limited by the amount of rainfall that penetrates the land surface and comes in contact with the buried waste. The flow of groundwater in the water table aquifer is related to the rate at which surface water recharges it. This rainwater infiltration rate may be an important boundary condition for shallow land burial problems (Aikens et al. 1979). Percolation of rainwater is frequently estimated by calculating the water budget for the root zone. Water enters the root zone through infiltration from rainfall and dissipates by evaporation directly from the surface, by transpiration from vegetation, and by seepage vertically to the water table. Rigorous (Gupta et al. 1978) and empirical (Thornthwaite and Mather 1957) methods of performing a water budget are in common use. These methods can be found in standard hydrology textbooks along with coefficients that apply to a variety of soil and vegetation types and climates (Chow 1964).

## Parameters for Transport and Flow Equations

The definitions of parameters used in the transport and flow equations are described below.

### Diffusion and Dispersion in Porous Media

Dispersion in equation 5.12 is actually a combination of molecular diffusion and mechanical dispersion, which are processes that irreversibly distribute dissolved constituents within porous media.

### Molecular Diffusion

Molecular diffusion results from the random movement of molecules at a very small scale. Diffusion within fluids depends on fluid properties, such as temperature, concentration, and viscosity, as well as temperature and concentration gradients. In a one-dimensional, nonflowing diffusion process, transport because of diffusion is usually related to Fick's Law:

$$\frac{\partial c}{\partial t} = \frac{\partial}{\partial x} D' \frac{\partial c}{\partial x}, \qquad (5.18)$$

where $D'$ is the effective diffusion coefficient for porous media, which typically varies from about $10^{-8}$ to $10^{-5}$ cm$^2$ s$^{-1}$. The effective diffusion coefficient, $D'$, will be lower than the molecular diffusion coefficient in a free liquid because diffusion will be inhibited by the pore structure (Evenson and Dettinger 1980).

### Dispersion

Dispersion describes the mechanical mixing of dissolved constituents by the complex flow paths the fluid must take in the porous medium. Variability of path length and velocity from the mean results in longitudinal and lateral spreading of the dissolved constituents. Laboratory investigations have shown that in porous media, longitudinal dispersion is related to the seepage velocity. For an isotropic medium, the dispersion coefficients, $D_{ij}$, can be described in terms of the longitudinal and transverse dispersivity (Scheidigger 1961):

$$\theta D_{ij} = \alpha_T V \delta_{ij} + (\alpha_L - \alpha_T) \frac{V_i V_j}{V}, \qquad (5.19)$$

where

$\theta$ = volumetric water content
$\alpha_T$ = transverse dispersivity (cm)
$V$ = magnitude of the flux (cm s$^{-1}$)
$\delta_{ij} = 1$ for $i = j$, $\delta_{ij} = 0$ for $i \neq j$ (Kronecker delta function)
$\alpha_L$ = longitudinal dispersivity (cm)
$V_i$ and $V_j$ = components of the flux (cm s$^{-1}$)

Even in small-scale laboratory experiments in uniform porous media, dispersion processes usually dominate the diffusion processes. Dispersion depends on flow, however. For very low flow rates, molecular diffusion, which is independent of flow, may dominate.

## Macrodispersion

Experiments with packed laboratory columns generally yield dispersivities having dimensions on the order of the median grain diameter, ranging from millimeters to centimeters. If the dispersivities measured in the laboratory were used in a transport model for a large aquifer, dispersion could be grossly underpredicted. At the aquifer scale, it appears that the heterogeneities in permeability, fracturing, stratification, and other properties of the medium; sampling errors; and model approximations are more important to producing dispersive behavior than mixing around individual grains and pores in the laboratory-scale experiments (Evenson and Dettinger 1980; Anderson 1979).

Field studies tend to support the hypothesis that macrodispersion is a result of heterogeneity in hydraulic conductivity. Dispersivity also apparently increases with the length scale of the experiment. There is a tendency for large dispersivities to coincide with experiments involving large distances. The Fickian analogy for dispersion, equation 5.18, does not always behave satisfactorily, and the dispersion cannot be characterized with a parameter as simple as the dispersivity. The assumptions of homogeneity of the medium break down if the heterogeneities are not random or if they are large in comparison to the scale of the aquifer being modeled (Winograde and Pearson 1976). The cases in which the simple dispersion models are likely to fail include the following (Evenson and Dettinger 1980):

- Media in which a few extensive conductivity variations dominate the transport process
- Media in which conductivity variations are abrupt and severe and tend to follow well-defined paths
- Observations that are made on a scale that is small compared to the scale of the variation

- Media that show variations in conductivity that cannot be modeled as a random field with apparently random values, spatial extents, and orientation assumed for the aquifer properties

These phenomena have been described generically as channeling. Evenson and Dettinger (1980) suggest that media in which these phenomena are likely to occur (e.g., fracture systems and karst) should not be modeled according to Fick's Law without extensive justification. Models capable of dealing with these problems are beyond the scope of this book.

### Determination of Dispersion

It is frequently the case that the only way the values of dispersion coefficients can be determined for a given site is by direct observation of either manufactured or naturally occurring tracers. Tracers that have been deliberately introduced are used in groundwater studies in single- or multiple-well pumping experiments over relatively short distances and times. Direct tracer methods have several disadvantages in groundwater studies. Because groundwater velocities are rarely large under natural conditions, undesirably long times are normally required for tracers to move significant distances through the flow system. For this reason, only small, nonrepresentative portions of the flow field can be measured. Because geological materials are typically quite heterogeneous, numerous observations are usually required to adequately monitor the passage of the tracer through the portion of the flow field under investigation. The measurements themselves may actually disturb the flow field significantly. Values of dispersivities obtained from a wide range of tracer experiments and those based on numerical models of observed groundwater solute transport cases are presented in tables 5.1 and 5.2, respectively (Anderson 1979; Evenson and Dettinger 1980). These values represent site-specific cases; they should be extrapolated to other cases only with extreme caution. Furthermore, the dispersivities reported in table 5.2 probably reflect processes, such as numerical dispersion, that are inaccuracies of the mathematical model and are not measured in nature.

### *Porosity and Effective Porosity*

The parameters porosity and effective porosity (or specific yield) are necessary to solve the flow and solute transport equations. The porosity of a soil or rock is a measure of the interstitial space relative to the space occupied by solid material, and it is expressed quantitatively as the percentage of the total volume occupied by the interstices.

The porosity of a sedimentary deposit depends chiefly on the shape and arrangement of its constituent particles, the degree of assortment of its particles, the cementation and compaction to which it has been subjected, the dissolution of mineral matter by water, and the fracturing resulting in open joints other than interstices. The porosity of many sedimentary deposits is increased by the irregular angular shapes of their grains. Porosity decreases with increases in the variety of size of

Table 5.1 Dispersivity values $\alpha_L$ and $\alpha_T$ obtained directly through measurements of tracer breakthrough curves in groundwater solute transport (Evenson and Dettinger 1980)

| Setting | $\alpha_L$ (m) | $\alpha_T$ (m) | $\Delta x^a$ (m) | $\overline{U}^b$ (m day$^{-1}$) | Method |
|---|---|---|---|---|---|
| Chalk River, Ontario alluvial aquifer | 0.034 | | | | Single-well tracer test |
| Chalk River, strata of high velocity | 0.034–0.1 | | | | Single-well |
| Alluvial aquifer | 0.5 | | | | Two-well |
| Alluvial, strata of high velocity | 0.1 | | | | Two-well |
| Lyons, France, alluvial aquifer | 0.1–0.5 | | | | Single-well |
| Lyons (full aquifer) | 5 | | | | Single-well |
| Lyons (full aquifer) | 12.0 | 3.1–14 | | 7.2 | Single-well test with resistivity |
| Lyons (full aquifer) | 8 | 0.015–1 | | 9.6 | Single-well test with resistivity |
| Lyons (full aquifer) | 5 | 0.145–14.5 | | 13 | Single-well test with resistivity |
| Lyons (full aquifer) | 7 | 0.009–1 | | 9 | Single-well test with resistivity |
| Alsace, France, alluvial sediments | 12 | 4 | | | Environmental tracer |
| Carlsbad, New Mexico, fractured dolomite | 38.1 | | 38.1 | 0.15 | Two-well tracer |
| Savannah River, South Carolina, fractured schistgneiss | 134.1 | | 538 | 0.4 | Two-well |
| Barstow, California, alluvial sediments | 15.2 | | 6.4 | | Two-well |
| Dorest, England, chalk (fractured) | 3.1 | | 8 | | Two-well |
| (intact) | 1.0 | | 8 | | Two-well |
| Berkeley, California, sand/gravel | 2–3 | | 8 | 311–1,382 | Multiwell trace test |
| Mississippi limestone | 11.6 | | | | Single-well |
| Nevada Test Site carbonate aquifer | 15 | | | | Two-well tracer |
| Pensacola, Florida, limestone | 10 | | 312 | 0.6 | Two-well |

$^a$ $\Delta x$ = distance between wells in two-well test.
$^b$ $\overline{U}$ = groundwater seepage velocity.

grains because small grains fill interstices between larger grains. Table 5.3 gives representative values of porosity for a wide range of soils and rocks.

The effective porosity is the portion of the porosity that can be considered to be available for the flow of groundwater through a porous medium. Not all of the water in the interstices of saturated rock or soil is available for flow. Part of the

Table 5.2 Dispersivity values $\alpha_L$ and $\alpha_T$ obtained by calibration of numerical transport models against observed groundwater solute transport (Evenson and Dettinger 1980)

| Setting | $\alpha_L$ (m) | $\alpha_T$ (m) | $\Delta x^a$ (m) | $\overline{U}^b$ (m day$^{-1}$) | Method$^c$ |
|---|---|---|---|---|---|
| Rocky Mountain Arsenal alluvial sediments | 30.5 | 30.5 | 305 | | Areal (moc) |
| Arkansas River Valley coalluvial sediments | 30.5 | 9.1 | 660 × 1,320 | | Areal (moc) |
| California alluvial sediments | 30.5 | 9.1 | 305 | | Areal |
| Long Island glacial deposits | 21.3 | 4.3 | Variable (50–300) | 0.4 | Areal (fe) |
| Brunswick, Georgia, limestone | 61 | 20 | Variable | | Areal (moc) |
| Snake River, Idaho, factured basalt | 91 | 136.5 | 640 | | Areal |
| Idaho, fractured basalt | 91 | 91 | 640 | | Areal (fe) |
| Hanford site, Washington fractured basalt | 30.5 | 18 | | | Areal (rw) |
| Barstow, California, alluvial deposits | 61 | 18 | 305 | | Areal (fe) |
| Roswell Basin, New Mexico, limestone | 21.3 | | | | Areal |
| Idaho Falls, lava flows and sediments | 91 | 137 | Variable | | Areal |
| Barstow, California, alluvial sediments | 61 | 0.18 | 3 × 152 | | Profile (fe) |
| Alsace, France, alluvial sediments | 15 | 1 | | | Profile |
| Florida (SE) limestone | 6.7 | 0.7 | Variable | | Profile |
| Sutter Basin, California, alluvial sediments | 80–200 | 8–20 | (2–20 km) | | 3-D (fe) |

$^a$ $\Delta x$ = grid size in program.
$^b$ $\overline{U}$ = groundwater seepage velocity.
$^c$ (fe), use of a finite element model; (moc); method of characteristics; (rw), random walk model.

water is retained in the interstices by the forces of molecular attraction or is trapped in dead-end pores. The amount of water trapped is greatest in media having small interstices. Table 5.4 gives representative values of effective porosity for a wide range of soils and rocks.

## Hydraulic Conductivity for Saturated Flow

The hydraulic conductivity, $K$, for an isotropic, homogeneous saturated medium determines the rate at which water moves through the porous medium for a given hydraulic gradient. Hydraulic conductivity depends on the properties of both the fluid and the medium and has units of velocity (cm s$^{-1}$). A measure of the hydraulic conductivity, which is a property of the porous medium alone, is the intrinsic permeability, $k$, that has units of length squared and is usually expressed in darcys (1 darcy = $9.87 \times 10^{-9}$ cm$^2$). Hydraulic conductivity, $K$, and intrinsic permeability,

Table 5.3 Typical values of porosity of aquifer materials (McWhorter and Sunada 1977)

| Aquifer material | Number of analyses | Range | Arithmetic mean |
|---|---|---|---|
| *Igneous rocks* | | | |
| Weathered granite | 8 | 0.34–0.57 | 0.45 |
| Weathered gabbro | 4 | 0.42–0.45 | 0.43 |
| Basalt | 94 | 0.03–0.35 | 0.17 |
| *Sedimentary materials* | | | |
| Sandstone | 65 | 0.14–0.49 | 0.34 |
| Siltstone | 7 | 0.21–0.41 | 0.35 |
| Sand (fine) | 245 | 0.25–0.53 | 0.43 |
| Sand (coarse) | 26 | 0.31–0.46 | 0.39 |
| Gravel (fine) | 38 | 0.25–0.38 | 0.34 |
| Gravel (coarse) | 15 | 0.24–0.36 | 0.28 |
| Silt | 281 | 0.34–0.51 | 0.45 |
| Clay | 74 | 0.34–0.57 | 0.42 |
| Limestone | 74 | 0.07–0.56 | 0.30 |
| *Metamorphic rocks* | | | |
| Schist | 18 | 0.04–0.49 | 0.38 |

Table 5.4 Typical values of effective porosity (or specific yield) of aquifer materials (McWhorter and Sunada 1977)

| Aquifer material | Number of analyses | Range | Arithmetic mean |
|---|---|---|---|
| *Sedimentary materials* | | | |
| Sandstone (fine) | 47 | 0.02–0.40 | 0.21 |
| Sandstone (medium) | 10 | 0.12–0.41 | 0.27 |
| Siltstone | 13 | 0.01–0.33 | 0.12 |
| Sand (fine) | 287 | 0.01–0.46 | 0.33 |
| Sand (medium) | 297 | 0.16–0.46 | 0.32 |
| Sand (coarse) | 143 | 0.18–0.43 | 0.30 |
| Gravel (fine) | 33 | 0.13–0.40 | 0.28 |
| Gravel (medium) | 13 | 0.17–0.44 | 0.24 |
| Gravel (coarse) | 9 | 0.13–0.25 | 0.21 |
| Silt | 299 | 0.01–0.39 | 0.20 |
| Clay | 27 | 0.01–0.18 | 0.06 |
| Limestone | 32 | ~0–0.36 | 0.14 |
| *Wind-laid materials* | | | |
| Loess | 5 | 0.14–0.22 | 0.18 |
| Eolian sand | 14 | 0.32–0.47 | 0.38 |
| Tuff | 90 | 0.02–0.47 | 0.21 |
| *Metamorphic rock* | | | |
| Schist | 11 | 0.22–0.33 | 0.26 |

$k$, are generally related by the equation

$$K = \frac{kg\rho}{\mu}, \tag{5.20}$$

where $g$ is the acceleration of gravity, $\rho$ is the density of the fluid, and $\mu$ is the viscosity of the fluid. Table 5.5 gives representative values of hydraulic conductivity in centimeters per second for a sample of common porous materials (McWhorter and Sunada 1977).

Environmental factors may affect the hydraulic conductivity of a given porous medium. For example, ion exchange on clay and colloid surfaces will cause changes in mineral volume, pore size, and shape. Changes in pressure may cause compaction of the material or may cause gases to come out of solution, which would reduce the hydraulic conductivity (Davis and De Wiest 1965).

### Sorption, Retardation, and Colloids

Sorption is an important mechanism in retarding the migration of dissolved radionuclides in groundwater. Sorption is defined to include all rock–water interactions that cause the radionuclides to migrate at a slower rate than the groundwater itself. The amount of sorption is dependent on both the chemistry of the water and of the rocks; because some of the chemical reactions are slow, it is also a function of time.

### Transport Based on Assumption of Equilibrium (Retardation Factor)

The most widespread use of sorption in groundwater modeling requires the assumption that all chemical species in the water, and the soil or rock in which it is in contact, are at equilibrium. The concentration of the species in the water then can be related to the concentration of the same species on the rock or soil by a constant, $K_d$. For

Table 5.5 Typical values of hydraulic conductivity of porous materials (McWhorter and Sunada 1977)

| Material | Number of analyses | Range (cm s$^{-1}$) | Arithmetic mean (cm s$^{-1}$) |
|---|---|---|---|
| *Igneous rocks* | | | |
| Weathered granite | 7 | $(3.3–52) \times 10^{-4}$ | $1.65 \times 10^{-3}$ |
| Weathered gabbro | 4 | $(0.5–3.8) \times 10^{-4}$ | $1.89 \times 10^{-4}$ |
| Basalt | 93 | $(0.2–4,250) \times 10^{-8}$ | $9.45 \times 10^{-6}$ |
| *Sedimentary materials* | | | |
| Sandstone (fine) | 20 | $(0.5–2,270) \times 10^{-6}$ | $3.31 \times 10^{-4}$ |
| Siltstone | 8 | $(0.1–142) \times 10^{-8}$ | $1.9 \times 10^{-7}$ |
| Sand (fine) | 159 | $(0.2–189) \times 10^{-4}$ | $2.88 \times 10^{-3}$ |
| Sand (medium) | 255 | $(0.9–567) \times 10^{-4}$ | $1.42 \times 10^{-2}$ |
| Sand (coarse) | 158 | $(0.3–6,610) \times 10^{-4}$ | $5.20 \times 10^{-2}$ |
| Gravel | 40 | $(0.3–31.2) \times 10^{-1}$ | $4.03 \times 10^{-1}$ |
| Silt | 39 | $(0.09–7,090) \times 10^{-7}$ | $2.83 \times 10^{-5}$ |
| Clay | 19 | $(0.1–47) \times 10^{-8}$ | $9 \times 10^{-8}$ |
| *Metamorphic rocks* | | | |
| Schist | 17 | $(0.002–1,130) \times 10^{-6}$ | $1.9 \times 10^{-4}$ |

these conditions, the retardation coefficient, $R_d$ (i.e., the ratio of the pore velocity to the apparent velocity of the sorbed material), can be defined as

$$R_d = \frac{n}{n_e} + \frac{\rho_b}{n_e} K_d, \qquad (5.21)$$

where

$n$ = total porosity
$n_e$ = effective porosity
$\rho_b$ = bulk density (g cm$^{-3}$)
$K_d$ = distribution coefficient (mL g$^{-1}$)

More conservatively, by assuming $n = n_e$, $R_d$ can be estimated as

$$R_d = 1 + \frac{\rho_b}{n_e} K_d. \qquad (5.22)$$

An equivalent retardation factor may be defined for fracture flow where the exposed area of the fracture is used rather than the porosity (Freeze and Cherry 1979).

### EXAMPLE 5.2

Calculate the retardation coefficient, $R_d$, for strontium in a fine sandstone with a bulk density, $\rho_b$, of 2.8 and a distribution coefficient, $K_d$, of 20 mL g$^{-1}$.

The arithmetic mean values of $n$ and $n_e$ are found from tables 5.3 and 5.4 to be 0.34 and 0.21, respectively. The retardation coefficient, $R_d$, calculated from equation 5.21 is therefore

$$R_d = \frac{0.34}{0.21} + \frac{2.8}{0.21} \times 20 = 268.3.$$

Equation 5.22 gives a slightly lower value of 267.7.

Where $K_d$ depends on concentrations or the reaction rates are slow, more complicated geochemical transport models may be necessary. Values of $K_d$ are required to calculate the travel time of key radionuclides from the source to the biosphere. The sorption coefficients are usually obtained using a standard batch test, where rocks are put in contact with groundwater in which small amounts of radionuclides have been mixed. The problem with this type of approach is that more detailed geochemical data are necessary to support the validity of the sorption measurement over the expected travel time of the radionuclides (which may be of the order of thousands of years).

To justify using simple sorption coefficients, a detailed understanding of the geochemical mechanisms of rock–water interactions must be attained. Such mechanisms as dissolution, precipitation, complexing, adsorption/desorption, phase transformations, and solubility should be understood for radionuclides of interest in the geochemical environment. The effect of heat, radiation, or high concentrations of chemicals will be particularly important close to the source of release in some situations. Much of this understanding for shorter periods of time and close to the points of release can be obtained by combining laboratory and field experiments with data from natural systems that can be used as analogs (Prikryl et al.

Table 5.6  Distribution coefficients: strontium and cesium (Isherwood 1981)

|  | $K_d$ (mL g$^{-1}$) | |
| --- | --- | --- |
| Materials | Strontium | Cesium |
| Basalt, 32–80 mesh | 16–135 | 792–9,520 |
| Basalt, 0.5–4 mm, 300 ppm TDS | 220–1,220 | 39–280 |
| Basalt, 0.5–4 mm, seawater | 1.1 | 6.5 |
| Basalt, fractured in situ measurement | 3 |  |
| Sand, quartz—pH 7.7 | 1.7–3.8 | 22–314 |
| Sands | 13–43 | 100 |
| Carbonate, >4 mm | 0.19 | 13.5 |
| Dolomite, 4,000 ppm TDS | 5–14 |  |
| Granite, >4 mm | 1.7 | 34.3 |
| Granodiorite, 100–200 mesh | 4–9 | 8–9 |
| Granodiorite, 0.5–1 mm | 11–23 | 1,030–1,810 |
| Hanford sediments | 50 | 300 |
| Tuff | 45–4,000 | 800–17,800 |
| Soils | 19–282 | 189–1,053 |
| Shaley siltstone < 4 mm | 8 | 309 |
| Shaley siltstone > 4 mm | 1.4 | 102 |
| Alluvium, 0.5–4 mm | 48–2,454 | 121–3,165 |
| Salt, > 4 mm saturated brine | 0.19 | 0.027 |

1997). However, over longer time periods or far from the points of release, the data should be obtained from studies of the natural system. Natural analogs of interest for application to radionuclide migration are hydrothermal ore deposits, intrusive magmas into generic host rocks, uranium ore bodies (Prikryl et al. 1997; Smellie et al. 1997), natural fission reactors (International Atomic Energy Agency 1975), and underground nuclear explosions (Klingsberg and Duguid 1980). In addition, the behavior of natural radionuclides and their decay products in host rock can provide the data necessary to choose conservative sorption coefficients.

Tables 5.6 and 5.7 give typical ranges of distribution coefficients, $K_d$, for significant radionuclides in an assortment of rocks and soils (Isherwood 1981). These tables illustrate some of the sensitivity of $K_d$ to several factors, such as particle size and chemistry of the water phase. Values of $K_d$ should be extrapolated to situations other than those for which they were determined only with extreme caution. Distribution coefficients can be measured in the field (Yu et al. 1993). It is also possible to estimate $K_d$ from other information. The U.S. Department of Energy code RESRAD (Yu et al. 2001) describes alternative methods for deriving the distribution coefficients. Two of those methods are described here.

1. *Concentrations measured in groundwater and soil*. If both concentration $C_s$ (Bq g$^{-1}$) on the soil particles and $C$ (Bq mL$^{-1}$) in the groundwater are available, $K_d$ can be estimated directly as the ratio of the two, that is,
$K_d$ (mL g$^{-1}$) = $C_s/C$.

Table 5.7 Distribution coefficients: thorium and uranium (Isherwood 1981)

| $K_d$ (mL g$^{-1}$) | Conditions |
|---|---|
| *Thorium* | |
| 160,000 | Silt loan, Ca-saturated clay, pH 6.5 |
| 400,000 | Montmorillonite, Ca-saturated clay, pH 6.5 |
| 160,000 | Clay soil, 5 mM Ca(NO$_3$)$_2$, pH 6.5 |
| 40–130 | Medium sand, pH 8.15 |
| 310–470 | Very fine sand, pH 8.15 |
| 270–10,000 | Silt/clay, pH 8.15 |
| 8 | Schist soil, 1 g L$^{-1}$ Th, pH 3.2 |
| 60 | Schist soil, 0.1 g L$^{-1}$ Th, pH 3.2 |
| 120 | Illite, 1 g L$^{-1}$ Th, pH 3.2 |
| 1,000 | Illite, 0.1 g L$^{-1}$ Th, pH 3.2 |
| <100,000 | Illite, 0.1 g L$^{-1}$ Th, pH >6 |
| *Uranium* | |
| 62,000 | Silt loam, U(VI), Ca-saturated, pH 6.5 |
| 4,400 | Clay soil, U(VI), 5 mM Ca(NO$_3$)$_2$, pH 6.5 |
| 300 | Clay soil, 1 ppm UO$^{2+}$, pH 5.5 |
| 2,000 | Clay soil, 1 ppm UO$^{2+}$, pH 10 |
| 270 | Clay soil, 1 ppm UO$^{2+}$, pH 12 |
| 4.5 | Dolomite, 100–325 mesh, brine, pH 6.9 |
| 2.9 | Limestone, 100–170 mesh, brine, pH 6.9 |

2. *Correlation with plant/soil concentration ratios.* This method relies on the observation from large numbers of field and laboratory samples that there is a strong correlation between the plant/solid concentration ratio and $K_d$ (Sheppard and Thibault 1990):

$$\ln K_d = a - 0.5(\ln \text{CR}), \tag{5.23}$$

where CR is the plant:soil concentration ratio, which is the ratio of the radionuclide content per unit weight of wet plant to dry soil (Yu et al. 2001). The value of the coefficient $a$ is 2.11 for sandy soil, 3.36 for loamy soil, and 4.62 for organic soil. The concentration ratio CR can be approximated by the vegetable/soil transfer factors, $B_{jv}$, presented in table 5.8.

### Transport Based on Geochemical Models

Considerable progress has been made on incorporating detailed chemistry into models of groundwater transport, especially because of the advent of powerful, inexpensive computers that can solve the difficult equations they present. Models at the simpler end of the spectrum are for one-dimensional, saturated transport, with chemical equilibrium among reacting species (Parkhurst 1995). At the complicated end of the spectrum are models for the coupled transport in at least two dimensions for the kinetic and equilibrium reactions in saturated and unsaturated flow (Yeh and Salvage 1995; Lichtner and Seth 1996). These models, while useful for research, often lack the necessary input data to be effective production tools.

Table 5.8  Vegetable/soil transfer factors ($B_{jv}$) for root uptake (Yu et al. 2001)

| Element | $B_{jv}$ | Element | $B_{jv}$ | Element | $B_{jv}$ | Element | $B_{jv}$ |
|---|---|---|---|---|---|---|---|
| H | 4.8 | In | $3.0 \times 10^{-3}$ | Cu | $1.3 \times 10^{-1}$ | W | $1.8 \times 10^{-2}$ |
| Be | $4.0 \times 10^{-3}$ | Sn | $2.5 \times 10^{-3}$ | Zn | $4.0 \times 10^{-1}$ | Ir | $3.0 \times 10^{-2}$ |
| C | 5.5 | Sb | $1.0 \times 10^{-2}$ | Ge | $4.0 \times 10^{-1}$ | Au | $1.0 \times 10^{-1}$ |
| N | 7.5 | Te | $6.0 \times 10^{-1}$ | As | $8.0 \times 10^{-2}$ | Hg | $3.8 \times 10^{-1}$ |
| F | $2.0 \times 10^{-2}$ | I | $2.0 \times 10^{-2}$ | Se | $1.0 \times 10^{-1}$ | Tl | $2.0 \times 10^{-1}$ |
| Na | $5.0 \times 10^{-2}$ | Xe | 0 | Br | $7.6 \times 10^{-1}$ | Pb | $1.0 \times 10^{-2}$ |
| Al | $4.0 \times 10^{-3}$ | Cs | $4.0 \times 10^{-2}$ | Kr | 0 | Bi | $1.0 \times 10^{-x}$ |
| P | 1.0 | Ba | $5.0 \times 10^{-3}$ | Rb | $1.3 \times 10^{-1}$ | Po | $1.0 \times 10^{-3}$ |
| S | $6.0 \times 10^{-1}$ | La | $2.5 \times 10^{-3}$ | Sr | $3.0 \times 10^{-1}$ | Rn | 0 |
| Cl | 20.0 | Ce | $2.0 \times 10^{-3}$ | Y | $2.5 \times 10^{-3}$ | Ra | $4.0 \times 10^{-2}$ |
| K | $3.0 \times 10^{-1}$ | Pr | $2.5 \times 10^{-3}$ | Zr | $1.0 \times 10^{-3}$ | Ac | $2.5 \times 10^{-3}$ |
| Ar | 0 | Nd | $2.4 \times 10^{-3}$ | Nb | $1.0 \times 10^{-2}$ | Th | $1.0 \times 10^{-3}$ |
| Ca | $5.0 \times 10^{-1}$ | Pm | $2.5 \times 10^{-3}$ | Mo | $1.3 \times 10^{-1}$ | Pa | $1.0 \times 10^{-2}$ |
| Sc | $2.0 \times 10^{-3}$ | Sm | $2.5 \times 10^{-3}$ | Tc | 5.0 | U | $2.5 \times 10^{-3}$ |
| Cr | $2.5 \times 10^{-4}$ | Eu | $2.5 \times 10^{-3}$ | Ru | $3.0 \times 10^{-2}$ | Np | $2.0 \times 10^{-2}$ |
| Mn | $3.0 \times 10^{-1}$ | Gd | $2.5 \times 10^{-3}$ | Rh | $1.3 \times 10^{-x}$ | Pu | $1.0 \times 10^{-3}$ |
| Fe | $1.0 \times 10^{-3}$ | Tb | $2.6 \times 10^{-3}$ | Pd | $1.0 \times 10^{-1}$ | Am | $1.0 \times 10^{-3}$ |
| Co | $8.0 \times 10^{-2}$ | Ho | $2.6 \times 10^{-3}$ | Ag | $1.5 \times 10^{-1}$ | Cm | $1.0 \times 10^{-3}$ |
| Ni | $5.0 \times 10^{-2}$ | Ta | $2.0 \times 10^{-2}$ | Cd | $3.0 \times 10^{-1}$ | Cf | $1.0 \times 10^{-3}$ |

## Colloid Migration

Radionuclides can also be transported as, or by, colloids rather than in the dissolved state. Colloids can exist in subsurface environments in several forms, including clay particles, iron oxyhydroxides, bacteria, humic substances, and radionuclides (Ibaraki and Sudicky 1995). They can be important in radionuclide release and transport (Avogadro and de Marsily 1984). Colloids can be generated from high-level waste, especially waste glass (Bates et al. 1992). Colloidal radionuclides are also produced in shallow land burial and may be in the form of bacteria (Schäfer et al. 1998). The colloids can be the actual radionuclides themselves (e.g., plutonium), often called true colloids, or nonradioactive particles onto which radioactive materials are reversibly or irreversibly sorbed, often called pseudocolloids (McCarthy and Zachara 1989).

Radionuclides carried as colloids will behave differently from dissolved radionuclides in several important respects (Grindrod 1993):

- They may be filtered out if they are transported through fine material, such as clay, or may coagulate because of electrostatic forces.
- They may move faster than dissolved radionuclides where there are relatively big passageways through the medium. Electrostatic forces may keep the colloidal particles flowing in the center of flow channels (Bonano and Beyeler 1985).

- Diffusion of colloids will be small because of their large size relative to dissolved molecules, limiting the diffusion of colloid-borne radionuclides into the rock matrix.

Details on models of colloid formation and transport and the importance of colloids to radiological risk assessments are beyond the scope of this book.

## Methods of Solution for Groundwater Flow and Transport

Numerous mathematical models have been developed to simulate the flow of groundwater and the transport of radioactive and chemical substances, particularly in the field of waste management. Discussion of the virtually hundreds of groundwater models is beyond the scope of this chapter, but several excellent compilations of groundwater models are available (Bachmat et al. 1978; Science Applications, Inc. 1979; Domenico and Schwartz 1990). Groundwater mathematical models can be broadly classified as either numerical or analytical. Numerical techniques are usually direct solutions of the differential equations describing water movement and solute transport, using methods such as finite differences or finite elements. These methods always require a digital computer, a large quantity of data, and an experienced modeler/hydrologist. The validity of the results from numerical models depends strongly on the quality and quantity of the input parameters. Analytical models are usually approximate or exact solutions to simplified forms of the equations for water movement and solute transport. Such models are simpler to use than numerical models and can generally be solved with the aid of a calculator, although computers are also used.

Analytical models are much more limited to simplified representations. However, they are useful for scoping the problem to determine data needs or the applicability of more detailed numerical models. Several of the more important types of numerical and analytical models are discussed in the following sections.

### Numerical Methods

The following sections broadly describe the most popular numerical methods currently in use for solving groundwater problems.

### Finite Difference

One approach that has been applied to solving groundwater equations involves finite-difference approximations. To apply these approximations, the region under consideration is usually divided into a regular orthogonal grid in one, two, or three dimensions. The intersections of the grid are called nodal points, and they represent the position at which the solution for unknown values, such as hydraulic head, is obtained. When difference equations are written for all nodes and boundary conditions are applied, a system of algebraic equations can be solved for the variables at each node for each time increment (Faust and Mercer 1980).

## Finite Element

The finite-element method is a numerical method where a region is divided into subregions, called elements, whose shapes are determined by a set of points called nodes (similar to the finite-difference grid). The first step is to derive an integral representation of the partial differential equations. This is commonly done by using the method of weighted residuals or the variational method. The next step is to approximate the dependent variables (head or concentration) in terms of interpolation functions, called basis functions. After the basis functions are specified and the elements defined, the integral relationship must be expressed for each element as a function of the coordinates of all nodal points of the element. Then the values of the integrals are calculated for each element. The values for all elements are combined and boundary conditions applied to yield a system of first-order linear differential equations in time (Faust and Mercer 1980).

## Method of Characteristics

This method is used in convection-dominated transport problems where finite-difference and finite-element approaches suffer from numerical dispersion or solutions that oscillate. The approach is not to solve the transport equations directly but to solve an equivalent system of ordinary differential equations that are obtained by rewriting the transport equation using the fluid particles as a reference point. This is accomplished numerically by introducing a set of moving points (reference particles) that can be traced within the stationary coordinates of a finite-difference grid block and that are allowed to move a distance proportional to the velocity and elapsed time. The moving particles simulate the convective transport because concentration is a function of spreading or convecting of the particles. After the convective effects are known, the remaining parts of the transport equation are solved using finite-difference approximations (Faust and Mercer 1980).

## Random Walk Method

The random walk method is similar in many ways to the method of characteristics. In this approach, a particle-tracing advection model is used to simulate advection. At each advection time step, the particles are dispersed by being displaced a random distance in a random direction. Concentrations are calculated by counting the resulting number of elements in each cell and comparing this to the initial conditions. This solution technique is based on the realization that a normal probability distribution is a solution to Fick's Law of diffusion. This method of transport modeling is easily implemented, and it provides simulations with accuracy that is limited only by the number of particles that can be traced. The main disadvantage of this method and the method of characteristics is the difficulty and expense of keeping track of large numbers of particles (Evenson and Dettinger 1980).

## Flow Network Models

The numerical simulation by finite differences or finite elements of groundwater flow and solute transport problems in two and three dimensions can be costly in terms of computational resources. Flow network models, such as the network flow and transport model NEFTRAN2 (Olague et al. 1991), can be used to describe two- or three-dimensional fields in a much more efficient way by a network of interconnecting one-dimensional flow segments. Fluid discharge and velocity are determined by requiring conservation of mass at the segment junctions. Radionuclide migration from the points of release is calculated by assuming that transport occurs along a single one-dimensional path having a length equal to the total migration path length. The network model is particularly useful when it is used in conjunction with a more complicated two- or three-dimensional model to first define the flow and concentration field for a particular example. The network model is first matched to the results of the complicated model. The matched network model may then be used for further computations with a much smaller commitment of computational resources than the original model for further runs and sensitivity experiments.

## Advection Models

There are groundwater solute modeling situations where the phenomenon of dispersion is only a minor factor in defining the transport of contaminants in groundwater. For example, the flux of contaminant entering a river that is recharged from a contaminated aquifer is much less sensitive to dispersion than the concentration in a particular well. In the former case, the contaminated groundwater would enter over a wide area, which would tend to smear out the effect of dispersion. For similar reasons, the transport from nonpoint sources of contamination, such as large low-level radioactive waste landfills, would diminish the sensitivity of modeled results to dispersion (Nelson 1978).

Javandel et al. (1984) describes RESSQ, a computer program for modeling the transport of contaminants in a two-dimensional plane. Point or areal sources of contamination, pumping wells, injection wells, and background flow fields can be defined for steady-state conditions. The paths of contaminant particles then can be traced from the source to the points of use. The user first identifies simple flow components of the system, such as uniform regional flow, point sources representing recharge and discharge wells, and circles representing area sources. The velocity fields of the components are then superimposed to generate an overall flow field. After the velocity field is defined, the path and time of tracer particles from their points of entry to the points at which they are withdrawn or leave the system boundaries are tracked by numerically integrating their velocities with respect to time at each point in the field. This technique will calculate the arrival time and concentration of particles, which can then be used to represent concentrations of radionuclides.

## Analytic Elements

The analytic element approach bears some similarity to the advection methods described in the preceding section, but it is considerably more flexible and powerful.

Analytic functions are used to describe point sinks and line sinks that represent wells and rivers. It is also possible to define analytic functions that bound regions with different hydrologic properties. The analytic functions are superimposed onto the regional flow field. Only the boundary conditions of the domain need to be discretized, unlike finite-element or finite-difference solutions. Some of the implementations of the method allow parameters of the analytic functions to be left as unknowns and to be solved from the other constraints imposed on the system. Because of its high efficiency, the analytic element method has been used to describe multilayer aquifers and large regional systems that might be unwieldy for solutions that are more conventional (Strack 1989; Haitjema 1995).

## Analytical Solutions of the Convective-Dispersive Equations

Analytical groundwater transport models can be used for certain types of analyses where available data do not warrant a more complicated study. Such models are useful for scoping the transport problem and may frequently be adequate for regulatory needs if model coefficients are chosen conservatively. A series of simple analytical models is presented below. The models are developed here for the limiting case of unidirectional saturated transport of a single dissolved substance with one-, two-, and three-dimensional dispersion in an isotropic aquifer, with no disturbances to the flow from either the source of the radionuclides or the users of water (e.g., wells). While computer solutions are available (e.g., Codell et al. 1982), in their simplest forms, they may be used with the aid of only a calculator.

As discussed above under "Mass Transport," the differential equation describing concentration would be

$$\frac{\partial c}{\partial t} + \frac{U}{R_d}\frac{\partial c}{\partial x} = \frac{D_x}{R_d}\frac{\partial^2 c}{\partial x^2} + \frac{D_y}{R_d}\frac{\partial^2 c}{\partial y^2} + \frac{D_z}{R_d}\frac{\partial^2 c}{\partial z^2} - \lambda c, \qquad (5.24)$$

where

- $c$ = concentration in the liquid phase (Bq L$^{-1}$)
- $D_x, D_y,$ and $D_z$ = dispersion coefficients in the x-, y-, and z-directions, respectively (cm$^2$ yr$^{-1}$)
- $\lambda$ = decay coefficient (yr$^{-1}$)
- $U$ = x component groundwater pore velocity (m yr$^{-1}$)
- $R_d$ = retardation coefficient (dimensionless)

The dispersion coefficient can be approximated from equation 5.19. In this case, $V_2 = V_3 = 0$, $V = U$, and $\theta$ can be approximated for saturated flow by the effective porosity, $n_e$. Also, since $U = V/n_e$,

$$D_x = \alpha_x U,$$
$$D_y = \alpha_y U, \qquad (5.25)$$
$$D_z = \alpha_z U,$$

where $\alpha_x, \alpha_y,$ and $\alpha_z$ are the dispersivities in the x-, y- and z-directions, respectively.

## Point Concentration Model

The first model developed is used for calculating the concentration in the aquifer at some point downgradient of a release (e.g., water supply well). Equation 5.24 is solved in terms of Green's functions (Wylie 1975):

$$c_i = \frac{1}{n_e R_d} X(x,t) Y(y,t) Z(z,t), \tag{5.26}$$

where

$c_i$ = concentration at any point in space for an instantaneous 1 Bq release
$n_e$ = effective porosity of the medium
$X, Y$, and $Z$ = Green's functions in the x-, y-, and z-directions, respectively

Equation 5.26 is developed in the following cases for a variety of boundary and source configurations:

1. For the case of a point source at $(0, 0, z_s)$ in an aquifer of infinite lateral $(x, y)$ extent and depth $b$, as illustrated in figure 5.3:

$$c_i = \frac{1}{n_e R_d} X_1 Y_1 Z_1, \tag{5.27}$$

where

$$X_1 = \frac{1}{\sqrt{4\pi D_x t/R_d}} \exp\left[-\frac{(x - Ut/R_d)^2}{4D_x t/R_d} - \lambda t\right], \tag{5.28}$$

$$Y_1 = \frac{1}{\sqrt{4\pi D_y t/R_d}} \exp\left(-\frac{y^2}{4D_y t/R_d}\right), \tag{5.29}$$

$$Z_1 = \frac{1}{b}\left[1 + 2\sum_{m=1}^{\infty} \exp\left(-\frac{m^2 \pi^2 D_z t}{b^2 R_d}\right) \cos m\pi \frac{z_s}{b} \cos m\pi \frac{z}{b}\right]. \tag{5.30}$$

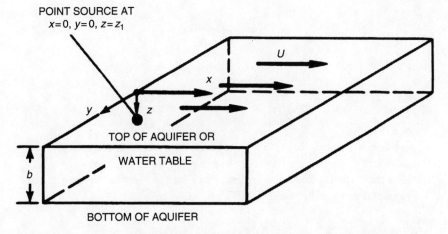

Figure 5.3  Idealized groundwater system for point concentration model, point source.

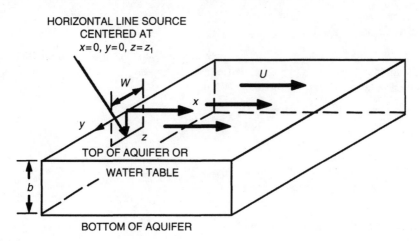

Figure 5.4 Idealized groundwater system for point concentration model, horizontal line source.

2. For the vertically averaged concentration in case 1 above (equivalent to a vertical line source of length $b$):

$$c_i = \frac{1}{n_e R_d} X_1 Y_1 Z_2, \qquad (5.31)$$

where

$$Z_2 = \frac{1}{b}. \qquad (5.32)$$

3. For a horizontal line source of length $w$ centered at $(0, 0, z_s)$, as illustrated in figure 5.4:

$$c_i = \frac{1}{n_e R_d} X_1 Y_2 Z_1, \qquad (5.33)$$

where

$$Y_2 = \frac{1}{2w}\left[\text{erf}\frac{(w/2+y)}{\sqrt{4D_y t/R_d}} + \text{erf}\frac{(w/2-y)}{\sqrt{4D_y t/R_d}}\right], \qquad (5.34)$$

and erf is the error function. Tables of the error function are available in standard mathematical texts (Abramowitz and Stegun 1972).

4. For the vertically averaged concentration in case 3 above (equivalent to an area source of width $w$ and depth $b$):

$$c_i = \frac{1}{n_e R_d} X_1 Y_2 Z_2. \qquad (5.35)$$

Transport of Radionuclides in Groundwater  239

5. For a point source at $(0, 0, z_s)$ in an aquifer of infinite lateral extent and depth:

$$c_i = \frac{1}{n_e R_d} X_1 Y_1 Z_3, \qquad (5.36)$$

where

$$Z_3 = \frac{1}{\sqrt{4\pi D_z t/R_d}} \left\{ \exp\left[-\frac{(z-z_s)^2}{4D_z t/R_d}\right] + \exp\left[-\frac{(z+z_s)^2}{4D_z t/R_d}\right] \right\}. \qquad (5.37)$$

6. For a horizontal line source of width $w$ centered at $(0, 0, z_s)$ in an aquifer of infinite lateral extent and depth:

$$c_i = \frac{1}{n_e R_d} X_1 Y_2 Z_3. \qquad (5.38)$$

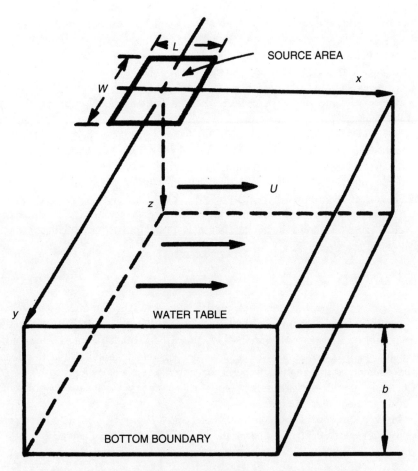

Figure 5.5  Vertically averaged groundwater dispersion model.

7. The vertically averaged concentration for a horizontal area source of length $L$ and width $w$ centered at (0, 0, 0) in an aquifer of constant depth $b$, as illustrated in figure 5.5:

$$c_i = \frac{1}{n_e R_d} X_2 Y_2 Z_2, \tag{5.39}$$

where

$$X_2 = \frac{1}{2L} \left[ \text{erf} \frac{\left(x + \frac{L}{2}\right) - \frac{ut}{R_d}}{\sqrt{4 D_x t / R_d}} - \text{erf} \frac{\left(x - \frac{L}{2}\right) - \frac{ut}{R_d}}{\sqrt{4 D_x t / R_d}} \right] \exp(-\lambda t). \tag{5.40}$$

**EXAMPLE 5.3 CONCENTRATION IN AN AQUIFER OF LIMITED THICKNESS**

One megabecquerel of a radioactive pollutant leaks quickly into a water table aquifer through a highly permeable ground cover over a square surface area 50 m on a side. The pollutant has a half-life of 30 years. A well tracer test indicates that the groundwater is moving in the direction of two wells at a speed, $U$, of 1.5 m day$^{-1}$ and that the dispersivities, $\alpha_x$ and $\alpha_y$, are 20 m and 10 m, respectively. The saturated thickness of the water table aquifer, $b$, is 50 m and has an effective porosity, $n_e$, of 0.2. The pollutant has been determined to have a retardation coefficient, $R_d$, of 20 in the aquifer. Calculate the concentration of the pollutant in wells with downgradient coordinates that are (a) $x = 200$ m, $y = 0$ m, and (b) $x = 400$ m, $y = 50$ m with respect to the center of the source area. The wells are screened over the entire depth of the aquifer.

*Solution*: Case 7 above applies to this example because the source is a horizontal area type and the wells are screened over the total depth, which would vertically average the concentration. Equation 5.39 is therefore evaluated with Green's functions: $X_2$ determined by equation 5.40, $Y_2$ determined by equation 5.34, and $Z_2$ determined by equation 5.32. The dispersion coefficients $D_x$ and $D_y$ are calculated by equation 5.25 to be 30 m$^2$ yr$^{-1}$ and 15 m$^2$ yr$^{-1}$, respectively. Figure 5.6 shows the concentration as a function of time calculated for the two wells.

## Flux Model

As depicted in figure 5.7, the flux model is used to calculate the discharge rate of a radionuclide entering a surface-water body that has intercepted the aquifer containing the transported material. It is assumed that all material entering the aquifer eventually enters the surface water except for the part lost through radioactive decay. The assumptions that apply to the point concentration model also apply to this model. The model provides only the rate of input to the surface water at an average distance $x$ downgradient from the surface.

In the unidirectional flow field assumed, the flux, $F$ (Bq yr$^{-1}$), of material crossing an area $dA = dy\, dz$ perpendicular to the x-axis is described by the equation

$$\frac{dF}{dA} = \left( Uc - D_x \frac{\partial c}{\partial x} \right) n_e, \tag{5.41}$$

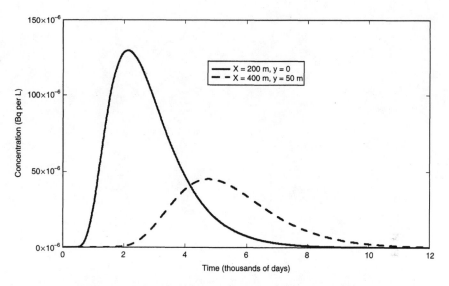

**Figure 5.6** Concentration in downgradient wells for example 5.3.

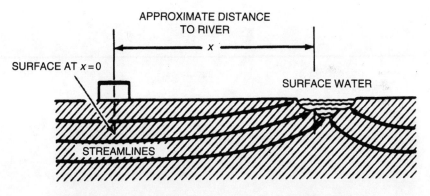

**Figure 5.7** Groundwater/surface-water interface, flux model.

where $c$ is the concentration in the dissolved phase. The total flux across the plane would be

$$F = n_e \int\limits_0^b \int\limits_{-\infty}^{\infty} \left( Uc - D_x \frac{\partial c}{\partial x} \right) dy\, dx. \tag{5.42}$$

*Source Released from a Vertical Plane* As described by equation 5.27, if $c_i$ is the concentration from an instantaneous release of 1 Bq at $x = 0$ and time $t = 0$, then the resulting flux at distance $x$ downgradient would be

$$F_i = \frac{\left(x + \frac{Ut}{R_d}\right)}{4\sqrt{D_x \pi t^3/R_d}} \exp\left[\frac{\left(x - \frac{Ut}{R_d}\right)^2}{4D_x t/R_d} - \lambda t\right]. \quad (5.43)$$

*Horizontal Area Source* For conditions expressed by equation 5.39, the corresponding flux would be

$$F_i = \frac{1}{2L\sqrt{\pi D_x t/R_d}} \left\{ \frac{u}{R_d} \sqrt{\pi D_x t/R_d} \left[\text{erf}(z_1) - \text{erf}(z_2)\right] - \frac{D_x}{R_d} \right. \quad (5.44)$$

$$\left. \times \left[\exp(-z_1^2) - \exp(-z_2^2)\right] \right\} \times \exp(-\lambda t),$$

where

$$z_1 = \frac{x - \frac{u}{R_d}t + \frac{L}{2}}{\sqrt{4D_x t/R_d}}, \quad z_2 = \frac{x - \frac{u}{R_d}t - \frac{L}{2}}{\sqrt{4D_x t/R_d}}. \quad (5.45)$$

### EXAMPLE 5.4

For the same conditions in example 5.3, calculate the flux of the pollutant into a river intercepting the flow, which is a distance $x$ of 2,000 m downgradient from the center of the source.

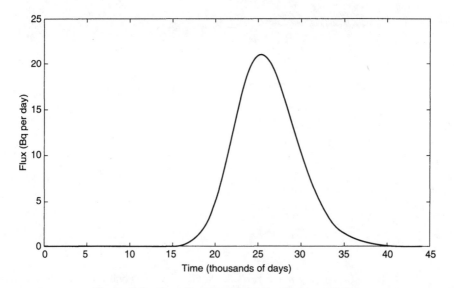

Figure 5.8 Flux of pollutant into river for example 5.4.

Transport of Radionuclides in Groundwater   243

*Solution*: Equation 5.44 applies in this case. Figure 5.8 shows the flux to the river as a function of time.

### Generalization of Instantaneous Models

Equations above under "Analytical Solutions of the Convective-Dispersive Equations" have been formulated only in terms of instantaneous releases. They can be generalized for arbitrary releases by using the convolution integral:

$$\theta = \int_0^t f(\tau)\,\theta_i(t-\tau)\,d\tau, \qquad (5.46)$$

where

$\theta$ = solution at time $t$ for the arbitrary release
$\theta_i(t-\tau)$ = solution at time $(t-\tau)$ for an instantaneous release at $(t-\tau) = 0$
$f(\tau)$ = source release rate at time $\tau$ (Bq yr$^{-1}$)

Certain analytical solutions can be found to equation 5.46 for simple source release rate functions. For example, Wilson and Miller (1978) develop the solution to equation 5.46 for continuous release in terms of a definite integral known as the "well function." Most useful solutions to equation 5.46 use numerical integration, generally involving a computer. Several special precautions must be taken, however, to preserve computational accuracy because the terms within the integral of equation 5.46 can be very nearly zero over most of the integration range. Computer programs solving the equations in this section are described in Codell et al. (1982).

### Superposition of Solutions

The equations described in this section all have the property of linearity, which means that solutions can be superimposed on each other. Complicated areas can be solved by representing the source area as a series of point sources and linearly summing the solutions. The same applies for solutions different in time. For example, the solution for two instantaneous point sources is simply the sum of the two individual solutions.

### *Simplified Analytical Methods for Minimum Dilution*

This section develops simplified forms of the equations presented under "Analytical Solutions of the Convective-Dispersive Equations," above. These equations can be used to calculate the minimum dilutions (i.e., maximum concentration) of volume $V_r$ of a substance instantaneously released from a point source into an aquifer.

*Concentration at Downgradient Wells in Confined Aquifers for an Instantaneous Point Source at the Surface*   For a confined aquifer with a release at the surface, concentrations close to the point of release would be unaffected by the bottom boundary and well mixed vertically far from the point of release. Between these extremes, there is a region where the concentration is in between the unmixed and vertically mixed conditions. The degree of vertical mixing can be characterized in a confined aquifer of constant thickness and uniform transport properties by the factor

$$\phi = \frac{b^2}{\alpha_z x}, \tag{5.47}$$

where

$b$ = thickness of the aquifer (m)
$\alpha_z$ = vertical dispersivity (m)
$x$ = distance downgradient of the release

The factor $\phi$ can be used to characterize the aquifer into three approximate regions:

- If $\phi < 3.3$, the release may be considered to be within 10% of being vertically mixed in the aquifer.
- If $\phi > 12$, the release may be considered to be within 10% of being unaffected by the vertical boundaries of the aquifer.
- If $3.3 < \phi < 12$, the release is neither completely mixed nor unaffected by the boundaries.

Different methods apply to each of the three regions.

*Vertically Mixed Region* ($\phi < 3.3$) For an instantaneous release at $x = 0$, the minimum dilution corrected for decay directly downgradient of a source would be

$$D_L = R_d 4\pi n_e \frac{\sqrt{\alpha_x \alpha_y}\, xb}{V_T} \exp(\lambda t), \tag{5.48}$$

where

$D_L$ = minimum dilution = $c_0/c$
$R_d$ = retardation coefficient (dimensionless)
$n_e$ = effective porosity (dimensionless)
$\alpha_x, \alpha_y$ = dispersivities (m) in the indicated direction
$V_T$ = volume of liquid source term (m$^3$)
$x$ = distance downgradient (m)
$b$ = aquifer thickness (m)
$\lambda$ = decay constant = $\ln 2 / t_{1/2}$
$t$ = travel time (yr)

The travel time $t$ can be approximated as

$$t = \frac{x}{U} R_d, \tag{5.49}$$

where $U$ is the pore velocity defined by equation 5.3.

*Unmixed Region ($\phi > 12$)*  For an instantaneous release at $x = 0$ on the surface of the aquifer, the minimum dilution of the surface of the aquifer directly downgradient from the source would be

$$D_L = \frac{n_e R_d (4\pi x)^{3/2} \sqrt{\alpha_x \alpha_y \alpha_z}}{2 V_T} \exp(\lambda t). \tag{5.50}$$

*Intermediate Region ($3.3 < \phi < 12$)*  For an instantaneous release at $x = 0$ on the surface of an aquifer, the minimum dilution on the surface of the aquifer directly downgradient from the source would be

$$D_L = \frac{R_d 4\pi n_e \sqrt{\alpha_x \alpha_y}\, xb}{V_T F(\phi)} \exp(\lambda t), \tag{5.51}$$

where

$$F(\phi) = 1 + 2 \sum_{n=1}^{\infty} \exp\left(\frac{-n^2 \pi^2}{\phi}\right), \tag{5.52}$$

and the other terms are as defined above. The function $F(\phi)$ is conveniently plotted in figure 5.9. For small values of $\phi$, $F$ approaches 1.0, which yields the vertically mixed case. For large values of $\phi$, the slope of $F$ is 1/2, and the semi-infinite vertical case prevails.

*Groundwater/Surface-Water Interface, Instantaneous Source*  For an instantaneous release to the groundwater at $x = 0$, the minimum dilution in an intercepting river, corrected for decay, would be

$$D_L = \frac{2 R_d Q \sqrt{\pi \alpha_x x}}{U V_T} \exp(\lambda t), \tag{5.53}$$

where

$Q$ = flow rate of river (m$^3$ yr$^{-1}$)
$\alpha_x$ = longitudinal dispersivity of the aquifer (m)
$U$ = pore velocity of groundwater (m yr$^{-1}$)
$V_T$ = volume of release (e.g., tank volume, m$^3$)

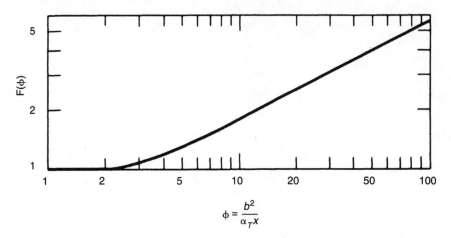

Figure 5.9  Mixing factor for confined aquifers.

## Models for Population Doses

If interdictive or mitigative methods to restrict water use are not taken into account, population dose from contaminated groundwater is proportional to the time-averaged concentration. Relatively simple equations can be used for estimating average concentration in groundwater or in surface-water supplies contaminated by groundwater. The two equations presented below are useful for population dose estimates.

*Quantity of Released Radioactivity Crossing a Vertical Plane*  In the case of groundwater flow to an intercepting river, the total quantity, $M$ (Bq), of the dissolved substance entering the river would be

$$M = \int_0^\infty F \, dt, \tag{5.54}$$

where $F$ is the flux defined for either an instantaneous point or vertical plane source by equation 5.40 or a horizontal area source by equation 5.44. This equation can be integrated graphically or numerically and, in some cases, may have an analytical solution.

If dispersion is relatively small (e.g., $\alpha_L \ll L$), the following approximation may be used:

$$M = M_0 e^{-\lambda t}, \tag{5.55}$$

where $M_0$ is the quantity of radioactivity released instantaneously from the source (Bq), $t$ is the travel time (yr), and $\lambda$ is the decay coefficient (yr$^{-1}$).

If the substance is being released from the source at a rate proportional to the quantity remaining (e.g., a leaching source term):

$$M = M_0 \frac{\lambda'}{\lambda' + \lambda} e^{-\lambda t}, \tag{5.56}$$

where $\lambda'$ is the fractional release rate from the source (yr$^{-1}$), and $M_0$ is the initial quantity of material in the source term (Bq).

*Direct Groundwater Use* This model calculates the quantity of a radionuclide ingested by a population using the contaminated groundwater. Groundwater use is considered to be spatially continuous instead of being from discrete well points. The total amount of the released radionuclide ingested by the population is

$$I = \int_0^\infty \int_{-\infty}^\infty \int_{-\infty}^\infty cQ_g \, dx \, dy \, dt, \tag{5.57}$$

where

$I$ = ultimate amount of radionuclide ingested from the release
$c$ = groundwater concentration (Bq L$^{-1}$)
$Q_g$ = groundwater withdrawal rate for drinking water purposes (m$^3$ yr$^{-1}$ km$^{-2}$)

If all use is restricted beyond downgradient distance $L$ from the release point, equation 5.57 may be integrated in closed form to give

$$I = \frac{\frac{M_0 Q_g}{2n_e R_d b} \exp\left\{\frac{UL}{2D_x} - \left[\frac{R_d L^2 (\lambda+\gamma)}{D_x}\right]^{1/2}\right\}}{\sqrt{\lambda+\gamma}\left(\sqrt{\lambda+\gamma} - \sqrt{\gamma}\right)}, \tag{5.58}$$

where

$$\gamma = \frac{U^2}{4 R_d D_x}. \tag{5.59}$$

$M_0$ is the total quantity of the radionuclide discharged to the point source, and the other terms are as previously defined.

If use of the groundwater is restricted between two downgradient distances, $L_1$ and $L_2$, the quantity of radionuclide ingested would be defined as

$$I = I(L_1) - I(L_2), \tag{5.60}$$

where $I(L_1)$ and $I(L_2)$ are evaluations of equation 5.58 for $L_1$ and $L_2$, respectively.

### EXAMPLE 5.5

This hypothetical example demonstrates the use of several of the simpler analytical models described under "Simplified Analytical Methods for Minimum Dilution" and "Models for Population Doses," above. Leakage into the ground rapidly empties a 10-m$^3$ tank containing 4 MBq L$^{-1}$ of $^3$H and 200 Bq L$^{-1}$ of $^{90}$Sr. The site is 15 m above the mean river level and 1,000 m upgradient from a river that has a representative low flow of 100 m$^3$ s$^{-1}$, and it is the sink for all surficial groundwater in the area. Two shallow wells are located 100 m and 700 m directly downgradient from the site of the spill. Groundwater exists in a homogeneous alluvial sand layer 30 m thick under water table conditions. Dispersivities for the sand have been determined in the near field from tracer tests to be 0.15 m for $\alpha_y$ and $\alpha_z$ and 0.5 m for $\alpha_x$. The bulk density, $\rho_b$, of the sand is 2.6 g cm$^{-3}$. Its total porosity, $n$, is 0.4, and its effective porosity, $n_e$, is 0.25. The permeability, $K$, is 0.02 cm s$^{-1}$. Distribution coefficients, $K_d$, for the sand have been determined to be 0 and 2.0 mL g$^{-1}$ for dilute solutions of $^3$H and $^{90}$Sr, respectively. From the above information, calculate the following:

(a) The maximum concentrations of the radioactive components in the river
(b) The maximum concentrations of the components in the near well
(c) The maximum concentrations of the components in the far well
(d) The total quantity of each radionuclide escaping to the river

*Solution*: (a) If it is assumed that the source is released over a short time period, you may use equation 5.53 for instantaneous releases to calculate the maximum river concentrations of $^3$H and $^{90}$Sr. First, determine the pore velocity, $U$, from equation 5.3 and the effective porosity, $n_e$:

$$U = \frac{V_x}{n_e} = -\frac{K\frac{\Delta H}{\Delta x}}{n_e}.$$

The gradient is

$$\frac{\Delta H}{\Delta x} = \frac{-50\,\text{m}}{3000\,\text{m}} = -0.0167,$$

$$U = \frac{-2 \times 10^{-2}\,\text{cm s}^{-1} \times -0.0167}{0.25} \times \frac{86{,}400\,\text{s day}^{-1}}{100\,\text{cm m}^{-1}} = 1.037\,\text{m day}^{-1}.$$

The retardation coefficients for $^3$H and $^{90}$Sr can be determined from equation 5.22:

$$^3\text{H} \quad R_d = 1 + \frac{2.6}{0.4} \times 0.0 = 1$$

$$^{90}\text{Sr} \quad R_d = 1 + \frac{2.6}{0.4} \times 2.0 = 14$$

The travel times for the two components are calculated by equation 5.49:

$$^3\text{H} \quad t = \frac{xR_d}{U} = \frac{1000\,\text{m} \times 1}{1.037\,\text{m day}^{-1}} \times \frac{\text{yr}}{365\,\text{day}} = 2.64\,\text{yr}$$

$$^{90}\text{Sr} \quad t = \frac{1000\,\text{m} \times 14}{1.037\,\text{m day}^{-1}} \times \frac{\text{yr}}{365\,\text{day}} = 37\,\text{yr}$$

The half-lives of $^3$H and $^{90}$Sr are 12.3 yr and 29 yr, respectively. The decay-corrected minimum dilutions in the river are found by applying equation 5.53:

$$^3\text{H} \quad D_L = \frac{2 \times 1.0 \times 100\,\text{m}^3\,\text{s}^{-1} \times \sqrt{\pi \times 0.5\,\text{m} \times 1000\,\text{m}}}{1.037\,\text{m day}^{-1} \times 10\,\text{m} \times \frac{\text{day}}{86{,}400\,\text{s}}}$$

$$\times \exp\left[\frac{\ln 2}{12.3\,\text{yr}} \times 2.64\,\text{yr}\right] = 7.66 \times 10^7,$$

$$^{90}\text{Sr} \quad D_L = \frac{2 \times 14 \times 100\,\text{m}^3\,\text{s}^{-1} \times \sqrt{\pi \times 0.5\,\text{m} \times 1000\,\text{m}}}{1.037\,\text{m}\,\text{day}^{-1} \times 10\,\text{m} \times \dfrac{\text{day}}{86,400\,\text{s}}}$$

$$\times \exp\left[\frac{\ln 2}{29\,\text{yr}} \times 37\,\text{yr}\right] = 9.24 \times 10^8.$$

The peak concentrations in the river are determined by dividing the tank concentrations by the dilution factors:

$$c(^3\text{H}) = \frac{4\,\text{MBq}\,\text{L}^{-1}}{7.66 \times 10^7} = 0.052\,\text{Bq}\,\text{L}^{-1},$$

$$c(^{90}\text{Sr}) = \frac{200\,\text{Bq}\,\text{L}^{-1}}{9.24 \times 10^8} = 2.2 \times 10^{-7}\,\text{Bq}\,\text{L}^{-1}.$$

(b) To calculate the maximum concentrations in a well 100 m downgradient, first determine whether the thickness of the aquifer would affect the result by calculating the factor $\phi$ from equation 5.47:

$$\phi = \frac{b^2}{\alpha_z x} = \frac{(30\,\text{m})^2}{0.15\,\text{m} \times 100\,\text{m}} = 60$$

Therefore, in this region, the release will be relatively unaffected by the thickness of the aquifer, and equation 5.50 applies. The travel times are estimated using the retardation factors and pore velocity calculated above:

$$^3\text{H} \quad t = \frac{100\,\text{m} \times 1}{1.037\,\text{m}\,\text{day}^{-1}} \times \frac{\text{yr}}{365\,\text{day}} = 0.26\,\text{yr}$$

$$^{90}\text{Sr} \quad t = \frac{100\,\text{m} \times 14}{1.037\,\text{m}\,\text{day}^{-1}} \times \frac{\text{yr}}{365\,\text{day}} = 3.7\,\text{yr}$$

Applying equation 5.50,

$$^3\text{H}\,D_L = 0.25 \times 1 \times (4\pi \times 100\,\text{m})^{3/2} \frac{\sqrt{0.5\,\text{m} \times 0.15\,\text{m} \times 0.15\,\text{m}}}{2 \times 10\,\text{m}^3}$$

$$\times \exp\left(\frac{\ln 2}{12.3\,\text{yr}} \times 0.264\,\text{yr}\right) = 59.$$

$$^{90}\text{Sr}\,D_L = 0.25 \times 14 \times (4\pi \times 100\,\text{m})^{3/2} \frac{\sqrt{0.5\,\text{m} \times 0.15\,\text{m} \times 0.15\,\text{m}}}{2 \times 10\,\text{m}^3}$$

$$\times \exp\left(\frac{\ln 2}{29\,\text{yr}} \times 3.7\,\text{yr}\right) = 903.$$

The peak well concentrations are therefore 0.068 MBq L$^{-1}$ for $^3$H and 0.22 Bq L$^{-1}$ for $^{90}$Sr.

(c) To calculate the maximum concentrations in a well 700 m downgradient, calculate $\phi$ for this region from equation 5.47:

$$\phi = \frac{(30\,\text{m})^2}{0.15\,\text{m} \times 700\,\text{m}} = 8.57.$$

Therefore, this well is in the intermediate region, and equation 5.51 applies. The factor $F(\phi)$ can be read from figure 5.9 to be about 1.7. Travel times for each component calculated from equation 5.49 are

$$^{3}\text{H} \quad t = \frac{700 \text{ m} \times 1}{1.037 \text{ m day}^{-1}} \times \frac{\text{yr}}{365 \text{ day}} = 1.85 \text{ yr},$$

$$^{90}\text{Sr} \quad t = \frac{700 \text{ m} \times 14}{1.037 \text{ m day}^{-1}} \times \frac{\text{yr}}{365 \text{ day}} = 25.9 \text{ yr}.$$

Applying equation 5.51,

$$^{3}\text{H} \quad D_L = \frac{1 \times 4\pi \times 0.25\sqrt{0.15 \text{ m} \times 0.15 \text{ m}} \times 700 \text{ m} \times 30 \text{ m}}{10 \text{ m}^3 \times 1.7}$$

$$\times \exp\left[\frac{\ln 2}{12.3 \text{ yr}} \times 1.85 \text{ yr}\right] = 2583,$$

$$^{90}\text{Sr} \quad D_L = \frac{14 \times 4\pi \times 0.25\sqrt{0.15 \text{ m} \times 0.15 \text{ m}} \times 700 \text{ m} \times 30 \text{ m}}{10 \text{ m}^3 \times 1.7}$$

$$\times \exp\left[\frac{\ln 2}{29 \text{ yr}} \times 25.9 \text{ yr}\right] = 6.05 \times 10^4.$$

The peak well concentrations are 0.0015 MBq L$^{-1}$ for $^{3}$H and 0.0031 Bq L$^{-1}$ for $^{90}$Sr.

(d) To calculate quantity $Q$ of each radionuclide eventually reaching the river, apply equation 5.55, because $\alpha_x \ll L$ (i.e., 0.5 m vs. 1,000 m). Travel times are estimated above in (a). The quantity of each radionuclide initially in the tank is the concentration multiplied by the volume. Therefore,

$$^{3}\text{H} \quad Q = 4 \text{ MBq L}^{-1} \times 10 \text{ m}^3 \times 1000 \text{ L m}^{-3} \times \exp\left(\frac{-\ln 2}{12.3 \text{ yr}} \times 2.64 \text{ yr}\right)$$

$$= 3.45 \times 10^4 \text{ MBq},$$

$$^{90}\text{Sr} \quad Q = 200 \text{ Bq L}^{-1} \times 10 \text{ m}^3 \times 1000 \text{ L m}^{-3} \times \exp\left(\frac{-\ln 2}{29 \text{ yr}} \times 37 \text{ yr}\right)$$

$$= 8.26 \times 10^5 \text{ Bq}.$$

## Source Term Models for Low-Level Waste

A simplified source term model useful for shallow land burial has been developed to estimate the release of radionuclides from contaminated soil on the basis of infiltration rate, retardation, and radioactive decay (Baes and Sharp 1983).

The concentration of the radionuclide in the soil is

$$C_t^s = C_0^s \exp(-\lambda_R t), \tag{5.61}$$

where

$C_t^s$ = soil concentration at time $t$ (Bq m$^{-3}$)
$C_0^s$ = initial soil concentration (Bq m$^{-3}$)
$\lambda_R$ = decay coefficient for all first-order removal processes (yr$^{-1}$)

$$\lambda_R = \lambda_i + \lambda_h + \lambda_L, \qquad (5.62)$$

where

$\lambda_i$ = radioactive decay coefficient (yr$^{-1}$)
$\lambda_h$ = plant uptake and harvesting (yr$^{-1}$)
$\lambda_L$ = leaching constant (yr$^{-1}$)

The rate at which the radionuclides enter the groundwater depends on the quantity of radionuclide present in the source term and the rate of release from the waste form per unit quantity. The present methodology relates the rate of removal of radionuclides from the soil to the rate of water infiltration, soil bulk density, porosity, and the distribution coefficient of the solute. The leach rate, $\lambda_L$, is defined as

$$\lambda_L = \frac{I}{R_d H n}, \qquad (5.63)$$

where $I$ is the groundwater infiltration rate (m yr$^{-1}$), $R_d$ is the retardation coefficient (dimensionless), $H$ is the waste thickness, and $n$ is the total soil porosity (dimensionless). The retardation coefficient is as defined in equation 5.22.

Soil partition coefficients presented in table 5.9 are derived from the values suggested in Kennedy and Strenge (1992). The infiltration rate is chosen to be 0.18 m yr$^{-1}$, which is based on the high end of the range of infiltration rates determined for low-level radioactive waste sites in the U.S. Southeast (Oztunali et al. 1981). Default values for porosity of 0.3 and waste thickness, $H$, of 0.5 m are also chosen for the present model. The leach rates of all progeny in a chain are assumed to be the same as the first member of the chain.

## Model Validation and Calibration

In general, a model is a set of equations and validation consists of comparing the solution of these equations with field-measured data. High-quality field data on contaminant or radionuclide transport in groundwater are scarce. Collecting data necessary for detailed modeling efforts is extremely costly because the aquifer in which the transport is taking place can be measured only indirectly from wells. Several well-known validation efforts are discussed in Robertson et al. (1974), Pinder (1973), Wilson and Miller (1978), Evenson and Dettinger (1980), Anderson (1979), Isherwood (1981), and Nuclear Energy Agency (1995). Although the agreement of a numerical solution with an analytical solution of the same equations is an important step in assuring that the model is correct, it shows only that the numerical techniques work as expected.

Table 5.9  Soil Partition Coefficients (Kennedy and Strenge 1992)

| Element | $K_d$ (mL g$^{-1}$) | Element | $K_d$ (mL g$^{-1}$) | Element | $K_d$ (mL g$^{-1}$) |
|---|---|---|---|---|---|
| Ac | 420 | H | 0 | Ra | 500 |
| Ag | 90 | Hg | 19 | Rb | 52 |
| Am | 1900 | Ho | 240 | Re | 14 |
| As | 110 | I | 1 | Rh | 52 |
| u | 30 | In | 390 | Rn | 0 |
| Ba | 52 | Ir | 91 | Ru | 55 |
| Be | 240 | K | 18 | S | 14 |
| Bi | 120 | Kr | 0 | Sb | 45 |
| Br | 14 | La | 1200 | Sc | 310 |
| C | 6.7 | Mn | 50 | Se | 140 |
| Ca | 8.9 | Mo | 10 | Sm | 240 |
| Cd | 40 | Na | 76 | Sn | 130 |
| Ce | 500 | Nb | 160 | Sr | 15 |
| Cf | 510 | Nd | 240 | Tb | 240 |
| Cl | 1.7 | Ni | 400 | Tc | 0.1 |
| m | 4000 | Np | 5 | Te | 140 |
| Co | 60 | Os | 190 | Th | 3200 |
| Cr | 30 | P | 8.9 | Tl | 390 |
| Cs | 270 | Pa | 510 | U | 15 |
| Cu | 30 | Pb | 270 | W | 100 |
| Eu | 240 | Pd | 52 | Xe | 0 |
| F | 87 | Pm | 240 | Y | 190 |
| Fe | 160 | Po | 150 | Zn | 200 |
| Gd | 240 | Pr | 240 | Zr | 580 |
|   |   | Pu | 550 |   |   |

Models are calibrated for a specific problem by defining an initial set of parameter estimates (field or laboratory measured), running the model for the problem, and comparing the results with observed values. If the comparison is poor, the parameter estimates are modified, the model is rerun, and results are again compared with observed data. This process is continued until the desired level of agreement between observed data and the simulation is obtained (Mercer and Faust 1980). Modifying boundary conditions and parameters is subjective and requires a considerable amount of knowledge of the region being simulated and experience on the part of the modeler. The boundary conditions and parameters used in the final simulation must still be in agreement with the knowledge and understanding of the geology and hydrology of the site. Through this process, the equations of groundwater flow in porous media have been well tested and verified. The equations of flow through fractured media have been verified through calibration of unsaturated zone flow models at Yucca Mountain, Nevada (Bechtel SAIC Company 2002; Bodvarsson et al. 1997; Wu et al. 2002). Transport models have been applied to fractured media in both the saturated and unsaturated zone at Yucca Mountain (Robinson et al. 1997; Zyvoloski et al. 1996).

Konikow and Bredehoeft (1992) point out that groundwater models cannot be validated because they are scientific hypotheses that can only be tested and falsified. Case histories have shown that calibrated models sometimes produce a nonunique solution that can lead to highly erroneous predictions of future states, even for the systems for which the models have been calibrated (Konikow 1986).

Where enough data on hydraulic heads and variation of head over long time periods are available, the inverse of the equations for head can be solved for the spatial distribution of permeability. The solution of the inverse problem (Neuman 1973; Carrera and Neuman 1986) is useful because it yields a spatial distribution of parameters that are consistent with the hydrology of the site under consideration. In this process, field-measured parameters are useful for comparison with computer-generated parameters to ensure that the generated values are realistic. For the transport equations, the inverses are not unique unless there can be other constraints on the system (Kunstmann et al. 1997).

## Misuse of Models

We end this chapter with some cautions about applying models to groundwater flow and transport estimation. The three most common misuses of models are overkill, misinterpretation, and blind faith in model results. *Overkill* is defined as using a more sophisticated model than is appropriate for the available data or the level of result desired. The temptation to apply the most sophisticated computational tool to a problem is difficult to resist. A question that often arises is when three-dimensional models should be used as opposed to two-dimensional or one-dimensional models. Including flow in the third dimension is recommended only in thick aquifers or if permeability changes drastically across the thickness of the aquifers. Including the third dimension requires substantially more data than do one- and two-dimensional models. For example, saturated–unsaturated flow through a shallow land burial site is truly a three-dimensional problem. However, the data are seldom available to consider more than one dimension above the water table. In many cases, sophisticated models are used too early in the analysis of a problem. Begin with the simplest model appropriate to the problem and progress toward more sophisticated models until you achieve the desired level of results. In transport problems, the flow modeling should be completed and checked against the understanding of site hydrology before applying a transport model.

*Misinterpretations* usually arise because inappropriate boundary conditions were selected or the history of a site has been misread. Under either of these conditions, the simulated data will not match the hydrologic history of the site.

Perhaps the worst misuse of a model is *blind faith in model results*, leading to unwarranted confidence in the prediction. Simulated results that contradict hydrologic intuition almost always arise from making a mistake in data entry, using a computer code that contains an error, or applying a model to a problem for which it was not designed. The latter case can occur in applying an analytical solution that was obtained using boundary conditions that are different from those to which the solution is being applied (Mercer and Faust 1980).

## Problems

1. Saturated groundwater flow: Determine the seepage velocity for an average coarse gravel for a groundwater gradient of 0.002.
2. Unsaturated flow: A water balance shows that 10 cm yr$^{-1}$ of water infiltrates the ground and recharges the water table. Calculate the average downward velocity of the water in the unsaturated zone, which is a fine sand.
3. Retardation coefficient: Calculate the retardation coefficient for cesium in an average sand with a bulk density of 2.8 g cm$^{-3}$ and $K_d$ of 50 mL g$^{-1}$.
4. Groundwater concentration: Calculate the concentration as a function of time 1,500 m directly downgradient of a 1-MBq instantaneous point source at the surface in an infinitely deep aquifer. The seepage velocity is 1 m day$^{-1}$. The $x, y,$ and $z$ dispersivities are 50 m, 20 m, and 1 m, respectively. The effective porosity is 0.2. The retardation coefficient is 20. The half-life of the substance is 1,000 yr.
5. Dilution in groundwater: For the same conditions as problem 4, consider that the 1 MBq was dissolved in 1,000 L of water. Calculate the minimum dilution in the well using equations 5.47–5.52. Compare results to the peak determined in problem 4, and explain any differences.
6. Dilution in river: For the same conditions as problem 4, calculate the minimum dilution in an intercepting river having an average flow rate of 10 m$^3$ s$^{-1}$.
7. Population dose for an average individual: Waste is being discharged to an aquifer and ingested by downgradient users at a rate of 0.1 m$^3$ day$^{-1}$ km$^{-2}$. All users are more than 5,000 m and less than 10,000 m downgradient. The properties of the radionuclide and the aquifer are $U = 1$ m day$^{-1}, \alpha_x = 100$ m, $n_e = 0.2, R_d = 3, b = 100$ m, and $t_{1/2} = 30$ yr. Calculate the becquerels ingested for each becquerel released.

### Notes

This chapter is the opinion solely of the authors and does not represent the official positions of either the U.S. Nuclear Regulatory Commission or the U.S. Department of Energy.

### References

Abramowitz, M., and I. Stegun, eds. 1972. *Handbook of Mathematical Functions.* Applied Mathematics Series 55. National Bureau of Standards. U.S. Government Printing Office, Washington, DC.

Aikens, A.E., R.E. Berlin, J. Clancy, and O.I. Oztunali. 1979. *Generic Methodology for Assessment of Radiation Doses from Groundwater Migration of Radionuclides in LWR Wastes in Shallow Land Burial Trenches.* Atomic Industrial Forum, Washington, DC.

Anderson, M.P. 1979. "Using Models to Simulate the Movement of Contaminants through Groundwater Flow Systems." In *CRC Critical Reviews in Environmental Control.* Chemical Rubber Corporation, Boca Raton, FL; pp. 97–156.

ANS (American Nuclear Society). 1980. *Evaluation of Radionuclide Transport in Ground Water for Nuclear Power Sites.* Report 2.17. La Grange Park, IL.

Avogadro, A., and G. de Marsily. 1984. "The Role of Colloids in Nuclear Waste Disposal." In *Scientific Basis for Nuclear Waste Management VII*, ed. G.L. McVay. Material Research Society Symposium Proceedings, vol. 26. Elsevier, New York; pp. 495–505.

Bachmat, Y., B. Andrews, D. Holtz, and S. Sebastian. 1978. *Utilization of Numerical Groundwater Models for Water Resources Management*. EPA 600/08–78/012. U.S. Environmental Protection Agency, Ada, OK.

Baes, C.F., III, and R.D. Sharp. 1983. "A Proposal for Estimation of Soil Leaching or Leaching Constant for Use in Assessment Models." *Journal of Environmental Quality* 12(1): 17–28.

Bates, J.K., J.P. Bradley, A. Teetsov, C.R. Bradley, and M. Buchholtz ten Brink. 1992. "Colloid Formation during Waste Form Reaction: Implications for Nuclear Waste Disposal." *Science* 256(5057): 649–651.

Bechtel SAIC Company. 2002. *Validation Test Report (VTR) for iTOUGH2 V5.0*. 10003-VTR-5.0–00. Bechtel SAIC Company, Las Vegas, NV.

Bodvarsson, G.S., T.M. Bandurraga, and Y.S. Wu, eds. 1997. *The Site-Scale Unsaturated Zone Model of Yucca Mountain, Nevada, for the Viability Assessment*. LBNL-40376. Lawrence Berkeley National Laboratory, Berkeley, CA.

Bonano, E.J., and W.E. Beyeler. 1985. "Transport and Capture of Colloidal Particles in Single Fractures." In *Material Research Society Symposium Proceedings*, vol. 44. Materials Research Society, Warrendale, PA; pp. 385–392.

Bredehoeft, J.D., and G.F. Pinder. 1970. "Digital Analysis of Areal Flow in Multiaquifer Groundwater Systems: A Quasi Three-Dimensional Model." *Water Resources Research* 6(3): 883–888.

Brown, G.O. 2003. *Henry Darcy and His Law*. Oklahoma State University, Stillwater, OK. Available online at biosystems.okstate.edu/darcy/references.htm.

Burkholder, H.C., and E.L.J. Rosinger. 1980. "A Model for the Transport of Radionuclides and Their Decay Products through Geologic Media." *Nuclear Technology* 49(1): 150–158.

Carrera, J., and S.P. Neuman. 1986. "Estimation of Aquifer Parameters under Transient and Steady State Conditions." *Water Resources Research* 22(2): 199–210.

Chow, V.T., ed. 1964. *Handbook of Applied Hydrology*. New York: McGraw-Hill.

Cloke, P.L. 2000. *Data Qualification Report for Thermodynamic Data File, Data0.ymp.R0 for Geochemical Code, EQ3/6*. TDR-EBS-MD-000012 REV 00. Civilian Radioactive Waste Management System Management and Operating Contractor, U.S. Department of Energy, Yucca Mountain Project, Las Vegas, NV.

Codell, R.B., K.T. Key, and G. Whalen. 1982. *A Collection of Mathematical Models for Dispersion in Surface and Ground Water*. NUREG-0868. U.S. Nuclear Regulatory Commission, Washington, DC.

CRWMS M&O (Civilian Radioactive Waste Management System Management and Operating Contractor). 2000. *Total System Performance Assessment for the Site Recommendation*. TDR-WIS-PA-000001 REV 00 ICN 01. Civilian Radioactive Waste Management System Management and Operating Contractor, U.S. Department of Energy, Yucca Mountain Project, Las Vegas, NV.

D'Arcy, H. 1856. *Les Fontaines Publiques de la Ville de Dijon*. Victor Dalmont, Paris.

Davis, S.M., and R.M.M. De Wiest. 1965. *Hydrogeology*. John Wiley & Sons, New York.

Dershowitz, W.S., G. Lee, J. Geier, T. Foxford, P. LaPointe, and A. Thomas. 1996. *FRACMAN—Interactive Discrete Fracture Data Analysis, Geometric Modeling, and Exploration Simulation*. Golder Associates, Redmond, WA.

Dillon, R.T., R.B. Lantz, and S.B. Pahwa. 1978. *Risk Methodology for Geologic Disposal of Radioactive Waste: The Sandia Waste Isolation Flow and Transport (SWIFT) Model.* SAND 78–1267. Sandia National Laboratories, Albuquerque, NM.

Domenico, P.A., and F.W. Schwartz. 1990. *Physical and Chemical Hydrogeology.* John Wiley & Sons, New York.

Duguid, J., and R.C.Y. Lee. 1977. "Flow in Fractured Porous Media." *Water Resources Research* 13(3): 558–566.

Evenson, D.E., and M.D. Dettinger. 1980. *Dispersive Processes in Models of Regional Radionuclide Migration.* University of California, Lawrence Livermore Laboratory, Livermore, CA.

Faust, C.R., and J.W. Mercer. 1980. "Ground Water Modeling: Numerical Models." *Ground Water* 18(4): 395–409.

Freeze, R.A., and J. Cherry. 1979. *Groundwater.* Prentice-Hall, Englewood Cliffs, NJ.

Grindrod, P. 1993. "The Impact of Colloids on the Migration and Dispersal of Radionuclides within Fractured Rock." *Journal of Contaminant Hydrology* 13: 167–181.

Gupta, S.K., K. Tanji, D. Nielsen, J. Biggar, C. Simmons, and J. MacIntyre. 1978. *Field Simulation of Soil-Water Movement with Crop Water Extraction.* Water Science and Engineering Paper 4013. Department Land, Air, and Water Resources, University of California, Davis, CA.

Gureghian, A.B. 1994. *FRAC-SSI: Far-Field Transport of Radionuclide Decay Chains in a Fractured Rock.* Center for Nuclear Waste Regulatory Analyses, San Antonio, TX.

Haitjema, H. 1995. *Analytic Element Modeling of Groundwater Flow.* Academic Press, San Diego, CA.

Hodgkinson, D.P., and P.R. Maul. 1988. "1-D Modeling of Radionuclide Migration Through Permeable and Fractured Rock for Arbitrary Length Decay Chains Using Numerical Inversion of Laplace Transforms." *Annals of Nuclear Energy* 15(4): 179–189.

Ibaraki, M., and E.A. Sudicky. 1995. "Colloid-Facilitated Contaminant Transport in Discretely Fractured Porous Media." *Water Resources Research* 31(12): 2945–2960.

International Atomic Energy Agency. 1975. "The Oklo Phenomenon." In *Proceedings of an IAEA-CEA Symposium. Libreville, June 23–27.* STI/PUB/405. International Atomic Energy Agency, Vienna, Austria.

Isherwood, D. 1981. *Geoscience Data Base Handbook for Modeling a Nuclear Waste Repository.* NUREG/CR-0912, Vols. 1 and 2. U.S. Nuclear Regulatory Commission, Washington, DC.

Javandel, I., C. Doughty, and C. Tsang. 1984. *Groundwater Transport: Handbook of Mathematical Models.* Water Resources Monograph Series 10. American Geophysical Union, Washington, DC.

Jenne, E.A., ed. 1979. *Chemical Modeling in Aqueous Systems.* American Chemical Society Symposium Series 93. American Chemical Society, Washington, DC.

Kennedy, W.E., and D.L. Strenge. 1992. *Residual Radioactive Contamination from Decommissioning—Technical Basis for Translating Contamination Levels to Annual Total Effect Dose Equivalent.* NUREG/CR-5512 (PNL-7994). Prepared by Pacific Northwest Laboratory for the Nuclear Regulatory Commission, Washington, DC.

Klingsberg, C., and J. Duguid. 1980. *Status of Technology for Isolating High-Level Radioactive Waste in Geologic Repositories.* U.S. Department of Energy, Technical Information Center, Oak Ridge, TN.

Konikow, L.F. 1986. "Predictive Accuracy of a Ground-Water Model—Lessons from a Postaudit." *Ground Water* 24(2): 173–184.

Konikow, L.F., and J.D. Bredehoeft. 1992. "Ground-Water Models Cannot Be Validated." *Advances in Water Resources* 15(1): 75–83.

Kunstmann, H., W. Kinzelbach, P. Marschall, and G. Li. 1977. "Joint Inversion of Tracer Tests Using Reversed Flow Fields." *Journal of Contaminant Hydrology* 26(1–4): 215–226.

Landa, E. 1980. *Isolation of Uranium Mill Tailings and the Component Radionuclides from the Biosphere, Some Earth Science Perspectives*. USGS Circular 814. U.S. Geological Survey, Reston, VA.

Lappala, E.G. 1981. *Modeling of Water and Solute Transport Under Variably Saturated Conditions: State of the Art, Modeling and Low-Level Waste Management: An Interagency Workshop*, ed. C.A. Little and L.E. Stratton. National Technical Information Service, Springfield, VA.

Lichtner, P.C., and M.S. Seth. 1996. *User's Manual for MULTIFLO: Part II—MULTIFLO 1.0 and GEM 1.0 Multicomponent Multiphase Reactive Transport Model*. CNWRA96–010. Center for Nuclear Waste Regulatory Analyses, San Antonio, TX.

McCarthy, J.F., and J.M. Zachara. 1989. "Subsurface Transport of Contaminants." *Environmental Science and Technology* 23(5): 496–502.

McWhorter, D.B., and D.K. Sunada. 1977. *Ground-Water Hydrology and Hydraulics*. Water Resources Publications, Fort Collins, CO.

Mercer, J.W., and C.R. Faust. 1980. "Ground-Water Modeling: An Overview." *Ground Water* 18(2): 108–115.

Mohanty, S., G.A. Cragnolino, T. Ahn, D.S. Dunn, P.C. Lichtner, R.D. Manteufel, and N. Sridhar. 1997. *Engineered Barrier System Performance Assessment Code: EBSPAC Version 1.1, Technical Description and User's Manual*. CNWRA 97–006. Center for Nuclear Waste Regulatory Analyses, San Antonio, TX.

National Academy of Sciences. 1996. *Rock Fractures and Fluid Flow: Contemporary Understanding and Applications*. National Academies Press, Washington, D.C.

Nelson, R.W. 1978. "Evaluating the Environment Consequences of Groundwater Contamination, Parts 1, 2, 3, and 4." *Water Resources Research* 14(3): 409–450.

Neuman, S. P. 1973. "Calibration of Distributed Parameter Groundwater Flow Models Viewed as a Multiple-Objective Decision Process Under Uncertainty." *Water Resources Research* 9(4): 1006–1021.

Niemczyk, S.J., K. Adams, W.B. Murfin, L.T. Ritchle, E.W. Eppel, and J.D. Johnson. 1981. *The Consequences from Liquid Pathways After a Reactor Meltdown Accident*. NUREG/CR-1598. U.S. Nuclear Regulatory Commission, Washington, DC.

Nitao, J.J. 1989. *V-TOUGH—an Enhanced Version of the TOUGH Code for the Thermal and Hydrologic Solution of Large-Scale Problems in Nuclear Waste Management*. UCID-21954. Lawrence Livermore National Laboratory, Livermore, CA.

Nitao, J.J. 1998. *Reference Manual for the NUFT Flow and Transport Code*, version 2.0. UCRL-MA-130651. Lawrence Livermore National Laboratory, Livermore, CA.

NRC (U.S. Nuclear Regulatory Commission). 1975. *Reactor Safety Study, an Assessment of Accident Risks in U.S. Commercial Nuclear Power Plants*. WASH-1400. U.S. Nuclear Regulatory Commission, Washington, DC.

NRC. 1978. *Liquid Pathway Generic Study*. NUREG-0440. U.S. Nuclear Regulatory Commission, Washington, DC.

NRC. 1979. *Draft Generic Environmental Impact Statement on Uranium Milling*. NUREG-0511. Collective vols. 1 and 2. U.S. Nuclear Regulatory Commission, Washington, DC.

Nuclear Energy Agency. 1995. "GEOVAL 94: Validation through Model Testing." In *Proceedings of an NEA/SKI Symposium. Paris, France, October 11–14*. Organization for Economic Cooperation and Development, Nuclear Energy Agency, Paris, France.

Olague, N.E., D.E. Longsine, H.E. Campbell, and C.D. Leigh. 1991. *User's Manual for the NEFTRAN II Computer Code*. NUREG/CR-5618. U.S. Nuclear Regulatory Commission, Washington, DC.

Oztunali, O.I, G.C. Re, P.M. Moskowitz, E.D. Picazo, and C.J. Pitt. 1981. *Data Base for Radioactive Waste Management*. NUREG/CR-1759. vol. 3, prepared by Dames and Moore, White Plains, NY, for Office of Nuclear Material Safety and Safeguards, U.S. Nuclear Regulatory Commission, Washington, DC.

Parkhurst, D.L. 1995. *User's Guide to PHREEQC—a Computer Program for Speciation, Reaction-Path, Advective-Transport and Inverse Geochemical Calculations*. WRI 95–4227. U.S. Geological Survey. Denver, CO.

Pinder, G.F. 1973. "A Galerkin Finite Element Simulation of Groundwater Contamination on Long Island, New York." *Water Resources Research* 9(6): 1657–1669.

Prikryl, J.D., D.A. Pickett, W.M. Murphy, and E.C. Pearcy. 1997. "Migration Behavior of Naturally Occurring Radionuclides at the Nopal I Uranium Deposit, Chihuahua, Mexico." *Journal of Contaminant Hydrology* 26(1): 61–69.

Pruess, K. 1991. *TOUGH2—a General-Purpose Numerical Simulator for Multiphase Fluid and Heat Flow*. Technical Report LBL-29400. Lawrence Berkeley National Laboratory, Berkeley, CA.

Reeves, M., and J. Duguid. 1975. *Water Movement Through Saturated-Unsaturated Porous Media: A Finite Element Galerkin Model*. ORNL-4927. Union Carbide Corporation, Nuclear Division, Oak Ridge National Laboratory, Oak Ridge, TN.

Robertson, J.B., R. Schoen, and J.T. Barraclough. 1974. *The Influence of Liquid Waste Disposal on the Geochemistry of Water at the National Reactor Testing Station*. Open File Report IDO-22053. U.S. Geological Survey, Water Resources Division, Idaho Falls, ID.

Robinson, B.A., A.V. Wolfsberg, H.S. Viswanathan, G.Y. Bussod, and C.W. Gable. 1997. *The Site-Scale Unsaturated Zone Transport Model of Yucca Mountain*. YMP Milestone SP25BM3. Los Alamos National Laboratory, Los Alamos, NM.

Schäfer, A., P. Ustohal, H. Harms, F. Stauffer, T. Dracos, and A. Zehnder. 1998. "Transport of Bacteria in Unsaturated Porous Media." *Journal of Contaminant Hydrology* 33(1–2): 149–169.

Scheidigger, A. E. 1961. "General Theory of Dispersion in Porous Media." *Journal of Geophysical Research* 66(18): 3273–3278.

Science Applications, Inc. 1979. *Tabulation of Waste Isolation Computer Models*. ONWI-78. Battelle Project Management Division, Office of Nuclear Waste Isolation, Columbus, OH.

Shepherd, T.A., and J.A. Cherry. 1980. "Contaminant Migration in Seepage from Uranium Mill Tailings Impoundments—an Overview." In *Uranium Mill Tailings Management, Proceedings of the Third Symposium. Civil Engineering Department, Colorado State University, Fort Collins, CO, November 24–25*. Colorado State University, Fort Collins, CO; pp. 299–331.

Sheppard, M.I., and D.H. Thibault. 1990. "Default Soil Solid/Liquid Partition Coefficients, $K_d$s, for Four Major Soil Types: A Compendium." *Health Physics* 59(4): 471–482.

Smellie, J.A.T., F. Karlsson, and W.R. Alexander. 1997. "Natural Analogue Studies: Present Status and Performance Assessment Implications." *Journal of Contaminant Hydrology* 26(1): 3–17.

Strack, O.D. 1989. *Groundwater Mechanics*. Englewood Cliffs, NJ: Prentice Hall.

Sudicky, E.A. 1989. "The Laplace Transform Galerkin Technique: A Time Continuous Finite Element Theory and Application to Mass Transport in Groundwater." *Water Resources Research* 25(8): 1833–1846.

Thompson, G.M., J.M. Hayes, and S.N. Davis. 1976. "Fluorocarbon Tracers in Hydrology." *Geophysical Research Letters* 1: 177–180.

Thornthwaite, C.W., and J. Mather. 1957. "Instructions and Tables for Computing Potential Evapotranspiration and the Water Balance." *Publications in Climatology.* Laboratory of Climatology, Clifton, NJ. Available online at www.udel.edu/Geography/CCR/PIC.html.

Updegraff, C.D., C.E. Lee, and D.P. Gallegos. 1991. *DCM3D: A Dual-Continuum Three Dimensional Groundwater Flow Code for Unsaturated, Fractured Porous Media.* NUREG/CR-5536. U.S. Nuclear Regulatory Commission, Washington, DC.

U.S. Department of Energy. 2002. *Yucca Mountain Science and Engineering Report.* DOE/RW-0539, rev. 1. U.S. Department of Energy, Office of Civilian Radioactive Waste Management, Washington, DC.

U.S. Environmental Protection Agency. 1998. "40 CFR 194 Criteria for the Certification and Recertification of the WIPP's Compliance with the Disposal Regulations: Certification Decision, Final Rule." *Federal Register* 63(95): 27354–27406.

Williams, N.H. 2001, December 11. Letter to J.R. Summerson, Department of Energy/YMSCO. Subject: Contract No. DE-AC08–01RW12101—Total System Performance Assessment—Analyses for Disposal of Commercial and DOE Waste Inventories at Yucca Mountain—Input to Final Environmental Impact Statement and Site Suitability Evaluation, REV 00 ICN 02. RWA:cs-1204010670, with enclosure. Bechtel SAIC Company, Las Vegas, NV.

Wilson, J.L., and P.J. Miller. 1978. "Two Dimensional Plume in Uniform Groundwater Flow." *Journal of Hydraulics Division, ASCE* 104(HY4): 503–514.

Winograde, I.J. and F.J. Pearson. 1976. "Major Carbon 14 Anomaly in a Regional Carbonate Aquifer: Possible Evidence of Megascale Channelling, South Central Great Plains." *Water Resources Research* 12(6): 1125.

Wolery, T.J. 1992. *EQ3/6, a Software Package for Geochemical Modeling of Aqueous Systems: Package Overview and Installation Guide*, version 7.0. UCRL-MA-110662 PT I. Lawrence Livermore National Laboratory, Livermore, CA.

Wu, Y.S., L. Pan, W. Zhang, and G.S. Bodvarsson. 2002. "Characterization of Flow and Transport Processes within the Unsaturated Zone of Yucca Mountain, Nevada, Under Current and Future Climates." *Journal of Contaminant Hydrology* 54(3–4): 215–247.

Wylie, C.R. 1975. *Advanced Engineering Mathematics.* McGraw Hill, New York.

Yeh, G.T., and K.M. Salvage. 1995. *Users' Manual for HYDROGEOCHEM 2.0: A Coupled Model of Hydrologic Transport and Mixed Geochemical Kinetic/Equilibrium Reactions in Saturated-Unsaturated Media.* The Pennsylvania State University, State College, PA.

Yu, C., C. Loureiro, J.J. Cheng, L.G. Jones, Y.Y. Wang, Y.P. Chia, and E. Faillace. 1993. *Data Collection Handbook to Support Modeling of the Impacts of Radioactive Material in Soil.* ANL/EAIS-8. Argonne National Laboratory, Argonne, IL.

Yu, C., A.J. Zielen, J.J. Cheng, D.J. LePoire, E. Gnanapragasam, S. Kamboj, J. Arnish, A. Wallo III, W.A. Williams, and H. Peterson. 2001. *User's Manual for RESRAD*, version 6. ANL/EAD-4. Argonne National Laboratory, Argonne, IL.

Zyvoloski, A.G., B.A. Robinson, Z.A. Dash, and L.L. Trease. 1996. *Models and Methods Summary for the FEHM Application.* LA-UR-94-3787 Rev 01. Los Alamos National Laboratory, Los Alamos, NM.

# 6

# Terrestrial Food Chain Pathways: Concepts and Models

F. Ward Whicker
Arthur S. Rood

The ability to predict the accumulation of contaminants in human and other biological tissues is often a key step in evaluating dose and risk resulting from environmental releases. This statement applies equally to radioactive and nonradioactive contaminants. It is true whether we are referring to dose and risk to exposed human beings, or to impacts on the plants and animals with which we share the environment. Although this chapter focuses on the movement of radionuclides through the environment to human receptors, it will be apparent that accumulation processes in plant and animal tissues are inherently involved, and the approaches and methodologies can also be applied for assessing ecological risks from a wide variety of contaminants.

This chapter deals almost exclusively with the terrestrial environment where most humans live and derive sustenance. Aquatic environments, both freshwater and marine, may be as important to human existence as land; therefore, chapter 7 is devoted to them. The terrestrial transport processes are not only numerous and complex, but each piece of landscape also differs from most others, at least to some degree. Variations in soil characteristics, climate, and topography can produce extreme differences in vegetation and animal communities. Variations in agricultural management practices, such as tillage, fertilizer application, and irrigation, also produce conditions that alter contaminant behavior. These sorts of variations alone and in combination can cause qualitative and quantitative differences in radionuclide transport, accumulation, and dose.

This chapter may be viewed within the context of pathways by which radionuclides can be transported through the ecosystems of the environment, ultimately resulting in a dose to humans. As conceptually illustrated in figure 6.1,

# Terrestrial Food Chain Pathways

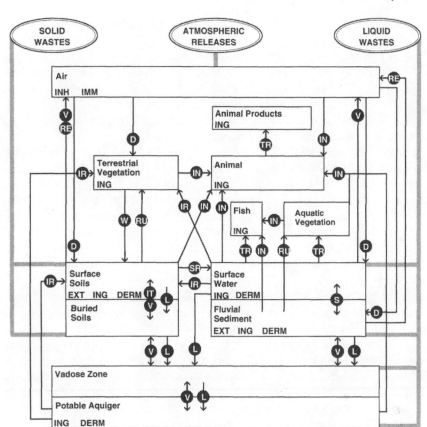

Figure 6.1 Transport pathways by which radionuclides and other contaminants can move through aquatic and terrestrial ecosystems and, ultimately, to humans.

radionuclides can follow many different pathways before they might be ingested or inhaled or externally expose human tissues. This chapter explores those pathways. For example, radionuclides that are leached from soil via infiltrating waters may eventually end up in potable groundwater or water used to irrigate crops. The relative importance of the various pathways differs with circumstances. The pathways of minor importance can often be identified so that more effort can be devoted to evaluating the more important routes. Identifying the more important routes is useful since all assessments have time and resource constraints.

The aquatic pathways in figure 6.1 are addressed in chapter 7, but some functional linkages do exist between aquatic and terrestrial ecosystems. Some of those linkages, such as irrigation and runoff, are mentioned in this chapter.

In general, this chapter addresses the environmental transport of radionuclides introduced or concentrated in the environment by human activity. Such radionuclides can be naturally occurring (e.g., uranium, thorium, and radium) or anthropogenic (e.g., plutonium). For example, uranium mining and milling concentrate and distribute naturally occurring uranium isotopes and their decay products into quantities that are not observed in nature. Although the models and methods presented here are designed to address impacts from radionuclides introduced or concentrated by human activity, they apply equally well to naturally occurring radionuclides in concentrations observed in nature. figure 6.2 shows how physical and biological transport processes must be evaluated to perform an environmental dose and risk assessment.

Before getting into the technical details of this chapter, the question of level of detail should be considered. For example, it would be tempting to cover a great deal of theory and experimental evidence for each of the processes and transport pathways discussed. Such an approach, however, would quickly expand this chapter into a book. It would also be somewhat difficult to read, digest, and apply to practical problems. Therefore, we treat the subject in a manner that is immediately useful in practice. Some of the underlying mathematical concepts are developed to provide an understanding of why the practical predictive equations normally used in assessments look as they do. This chapter further allows the student to construct models that address site-specific concerns that may not be addressed using generic models. In an additional attempt to make this chapter useful, we include some tables of typical parameter values that are often used in practice. Despite this quest for simplicity and usefulness, we are forced to offer caveats and uncertainties, as well as numerous references that might be consulted for further information and to support the generalities offered.

## Conceptual Model of the Terrestrial Environment

Before translating our understanding of the terrestrial environment into mathematical formulations for predictive uses, it is helpful to briefly review how land ecosystems function. The normal functional processes of ecosystems primarily drive the movement of most radioactive and nonradioactive contaminants. Radionuclides, once released to the environment, are ordinarily present in very tiny mass concentrations. Because of this, they tend to behave as tracers, in the sense that their presence does not alter the normal movement and transfer of substances such as air, water, nutrients, or biological fluids. In essence, the tracers simply track movements of materials that are present in much larger mass quantities.

Terrestrial Food Chain Pathways 263

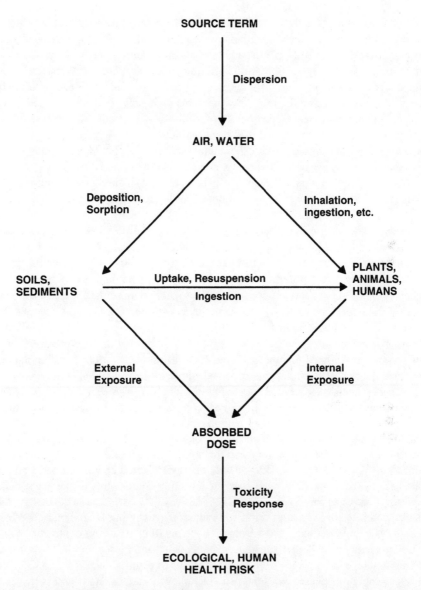

Figure 6.2 Typical sequence of events leading to an ecological or human health risk from environmental contaminants.

The question of which substance a particular tracer will tend to follow is governed by the physical and chemical properties of the tracer and the media in which it is found. For example, if the tracer is $^{85}$Kr, a noble gas, its movements will track those of other nonreactive gases. If released to the atmosphere, $^{85}$Kr will disperse in the air stream and essentially stay there until it physically decays.

If the tracer is $^{137}$Cs, the portion of it that is not strongly adsorbed to clay particles will generally follow its chemical analogue, potassium, in biogeochemical cycles. It will move through soils, plants, and animals in a manner qualitatively similar to that of potassium. Since potassium is physiologically regulated and cesium is not, there would be quantitative differences between the two. Another example might be $^{239}$Pu, which tends to be relatively insoluble in the environment and binds very strongly to soil particles. Once bound to soil, the movement of this radionuclide is governed largely by movement of the soil particles. Natural processes (e.g., wind and water erosion, animal activities) or human activities (e.g., tillage and other forms of mechanical disturbance) are usually responsible for soil movement.

As described in greater detail elsewhere (e.g., Whicker and Schultz 1982; Whicker 1983), the transport of radionuclides is governed by properties of the radionuclide, the characteristics and behavior of organisms, and attributes of the ecosystem of concern. Here, it seems appropriate to briefly describe the more general functions of terrestrial ecosystems, particularly those that accomplish the transport of nutrients or other materials from place to place, which in this context refers to components of ecosystems, such as air, soil, water, plants, and animals. These components and their subcomponents, such as soil horizons, plant species, and animal species, form the structure of a particular ecosystem. Connections between the components are made up of functional processes that govern energy flow, nutrient transport, and material transport. We like to think of the ecosystem as an assemblage of components, linked together by functional processes, to form a system having characteristics and attributes that supersede those of the individual components.

A generalized ecosystem is illustrated in figure 6.3. The basic components are conceptualized with boxes and functional connections with arrows. The structural components include plants or, in ecological jargon, "primary producers," which capture energy from the sun and, through photosynthesis, incorporate carbon dioxide from the atmosphere and water from the soil to produce organic compounds and release oxygen. This conversion of light energy to stored chemical energy forms the base of the food chain, which drives all biotic components of the ecosystem. Energy transformations in ecosystems are one-way flows, with the sun providing the input, and heat loss via respiration and radiation accounting for the output. While energy is being continually lost from ecosystems (thus requiring continual input), nutrient elements and other substances are recycled and reused within ecosystems. For example, carbon cycles from the atmosphere to plants and animals, to soil, and back to the atmosphere. Nutrients, such as phosphorus, calcium, potassium, and sodium, cycle from soil to plants, to animals, and back to soil via the decomposition of organic detritus. Mineral nutrients may be locked up for a long time in deep sediments and rocks, but various geological forces continually release a fraction to the living environment, or biosphere. Extraterrestrial dust deposits continuously on the earth, providing another input to the biospheric nutrient pool; however, the rates of these elemental inputs and losses (e.g., to deep sediments) are small compared to the rates at which elements are recycled.

The basic ideas of material transport in ecosystems are quite familiar to most people because we are surrounded by obvious examples, unless our sphere of observation is limited to the indoors or to highly urban settings. Understanding the details

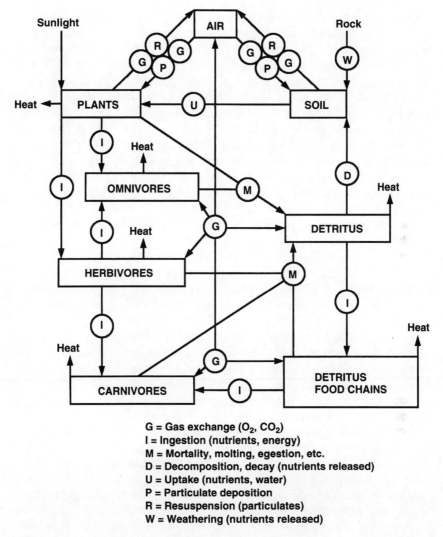

Figure 6.3 Conceptual model of generalized ecosystem showing structural components and functional processes.

of these processes, despite many decades of research, is much more difficult. Many details remain only partially understood, and quantitative prediction is a process fraught with uncertainty. Figure 6.3 and the following discussion provide the basic conceptual framework to understand how ecosystems function as a system. After the framework is established, we can begin to add the details that will enable us to develop quantitative simulation models for material and radionuclide transport.

Air is a highly mobile, dynamic medium that transports gases and aerosols over great distances. Processes important to ecosystem function are the exchanges of oxygen, carbon dioxide, and nitrogen between the atmosphere and plants and animals, including microorganisms in the soil. Atmospheric motions also cause the resuspension of small particulates from soil and plant surfaces and transport the material some distance before it is deposited back to soil or vegetation. Radioactive materials released to the atmosphere behave like gases or aerosols, depending on their form and physical properties. The process of deposition from air to soil or vegetation is accomplished by several mechanisms, including gravitational settling, surface impaction, electrostatic attraction, and washout or rainout in conjunction with precipitation. These processes result in certain mineral nutrient transfers. In fact, this is the primary means by which certain organisms, such as lichens, obtain nutrients. Another example is the deposition of salts from sea air over coastal landscapes. The atmosphere also provides a direct route of gas and particulate exchange with animals through inhalation and expiration.

The soil functions as the primary reservoir of nutrients in most terrestrial ecosystems. It plays a crucial role in supporting plants, both physically and chemically. It also hosts the microorganisms responsible for the decomposition of dead organic material, or detritus. Decomposition releases nutrients that can reenter the food chain through root uptake. Soil itself is a chemically dynamic medium, with minerals, organic compounds, and gases continually entering and leaving. It can also be physically dynamic, for example, through wind and water erosion. In well-developed natural ecosystems, the soil is normally protected to varying degrees from erosion by established vegetation cover. Tillage is a common practice in agricultural ecosystems that mixes the soil vertically, sometimes leaving it more vulnerable to erosion. The structure and productivity of the plant community are partially controlled by the properties of a particular soil. Such properties include texture (i.e., content of sand, silt, and clay), which affects the water-holding capacity and the content of mineral nutrients and organic matter. In addition to serving as the main nutrient reservoir, soil is the primary storage medium for most contaminants, especially those that are particle reactive. Soil is generally not the primary storage medium for inert gases or highly volatile contaminants, or for substances that tend not to react strongly with soil particles. Direct or inadvertent ingestion of soil is a potential transport pathway for soil-bound radionuclides to animals and humans. Infiltrating water, coupled with dissolution or colloid transport, provides a mechanism for removal of radionuclides from soil to deeper strata and eventually to groundwater systems.

Plants form the energy and nutrient base of the food chain. They draw their nutrients from the soil and the atmosphere. The rate at which energy can be stored as plant biomass for a given area is called "primary productivity." Primary productivity is controlled by the amount of sunlight, the availability of water, the temperature, and the quality of the soil. The quantity of animal biomass that can be maintained in a given area is directly related to the available primary productivity. The plants box in figure 6.3 is highly simplified because many kinds of plants exist in the same ecosystem at the same time. In natural ecosystems, we often refer to plant communities, which are assemblages of a number (sometimes a very large number) of species that coexist. Such assemblages coevolve in a manner that tends to maximize the stability of the ecosystem. Different plant species can have competitive as well as mutualistic relationships. In contrast, agricultural ecosystems are usually managed to maximize productivity of a single species. Such systems have been termed "monocultures." Agricultural ecosystems typically require a great deal of human manipulation and expenditures of energy (e.g., tractors and fuel). Plant biomass is also a dynamic medium. When conditions permit, growth occurs through primary production. Loss of plant biomass occurs continually through consumption, respiration, mechanical damage (e.g., trampling), and natural senescence.

Animals exist in all ecosystems. Their presence is made possible by the energy and nutrients available in plant biomass. Animals that eat plants are called herbivores; those capable of eating other animals are called carnivores; and those that eat both plant and animal tissues are called omnivores. Animals that are small enough to live in or on host animals, often deriving food from their fluids or other tissues, are called parasites. There are many examples of animal food chains where numerous species are involved in a linear sequence from an herbivore to a top predator—an animal that is not preyed upon by any other animal. Food chains are not strictly linear sequences; they are more often termed "food webs" because of alternative food sources at a particular feeding (trophic) level. Most herbivores can feed on a variety of plant species, and most carnivores rely on more than one kind of prey. There is often a distinct selective disadvantage to specialization in food habits. Not all food chains or webs originate with the consumption of green vegetation. Detritus food chains originate from consumption of dead organic matter, which in turn is generated by both plant and animal tissue. Plant and animal populations continually increase their biomass through reproduction and growth and yet simultaneously lose biomass through mortality, respiration, and other processes. Individual animals can move, through muscle power, from place to place. Some species move only a few meters; others, such as birds, can move hundreds or thousands of miles. Plants move by other mechanisms, such as seed dispersal.

All living things, both plant and animal, produce dead organic matter—detritus—the majority of which accumulates in the upper layers of soil. As with plant and animal biomass, detritus is continually produced and lost. The rate of detritus loss depends on the activity of microorganisms that break this material down into its basic chemical constituents. The activity of these tiny decomposers is enhanced with moisture and warm temperatures and depends on the amount of detritus available. The net change in the pool of organic detritus depends on which is larger, the rate of input or the rate of loss. Dead organic matter is certainly important to soil fertility, and the decomposition of this material is crucial to the release of nutrients for reuse by plants and animals. In agricultural ecosystems, soils become nutrient depleted because biomass is harvested and removed from the land. The only way to mitigate this loss is to apply commercial fertilizers or mulch that is rich in organic matter and inorganic nutrients. Again, this requires significant human intervention.

Energy is required for animals to accomplish critical functions, such as movement, feeding, reproduction, and growth. A large fraction of this energy flow is lost from the ecosystem as heat. Because of this energy loss, the structure of the ecosystem has certain limitations in terms of the amount of biomass that can be supported at various trophic levels. As a general rule, about 10% of the biomass production at a given trophic level can be supported at the next level. Because of this, animal biomass production is normally much smaller than plant production, and predator biomass is significantly reduced from its supporting prey biomass. These relationships have a fundamental influence on the partitioning of radionuclides and other contaminants in ecosystems and within specific biological components.

## Strategies for Evaluating Food Chain Transport

There are various ways to measure and evaluate the transport of radionuclides through the various steps in the food chain. Historically, actual measurements of concentrations in soil, plants, animals, and people were undertaken to obtain the needed information. More recently, our knowledge has increased to the point that models, both statistical and mechanistic, can be used effectively for the same purpose. This section reviews some of the specific approaches and discusses the strengths and weaknesses of each.

### Predictive Approaches

Three basic approaches are used to estimate the degree of radionuclide concentration in environmental media, such as air, water, soil, plants, and animals; foods, such as meat, milk, and vegetables; and human tissues. These approaches include direct measurements, statistical models, and mechanistic models. Each offers certain advantages, yet all have limitations. Our goals should always be to come as near to finding the "truth" as possible, to improve our understanding of real phenomena, and to obtain results that are credible and useful for decision making. This often requires the use of more than one predictive approach.

Direct Measurements

The direct measurement approach is usually preferable to model predictions for current or after-the-fact situations where representative environmental media or biological tissues can be sampled adequately and the radionuclide content determined accurately. This approach avoids the problems of uncertainty in model structure and lack of knowledge about parameter values. The direct measurement approach tends to carry more credibility, especially with researchers familiar with the inherent weaknesses of predictive models. Of course, direct measurements have their own shortcomings. For example, inferences made from sample measurements are subject to both sampling and measurement errors. Ordinarily, the first problem (taking truly representative samples) is more serious than is the second problem. With proper sampling, however, estimates of central tendency and dispersion (variability) can be obtained. Direct measurements of a particular quantitative end point, for example, dose from $^{131}$I to the human thyroid, may not be possible or feasible. In such a case, perhaps the best that can be done is to make measurements of $^{131}$I in the thyroid with an external detector and calculate the dose, or measure the radionuclide in foods and then calculate the intake and dose. In these cases, a combination of the direct measurement and modeling approaches might be the most useful. Such a combined approach is often simpler and more reliable than if the entire process, from source term to dose, is modeled. Obviously, the direct measurement approach is not useful for evaluating future events or historical releases, unless the radionuclides of interest are long-lived and still measurable.

Another significant limitation of direct measurements is that it is often not feasible to obtain measurements of all pertinent contaminants at all locations and times of interest. For example, in an emergency response situation, it may not be feasible to perform comprehensive sampling that will answer the assessment question. In practice, a combined approach is applied, using mechanistic models coupled with sampling to validate or calibrate the model.

## Statistical Models

Statistical models usually rely on empirical relationships that have been established among various measurements and on quantities presumed to influence the measurements. This approach may be appropriate where mathematical or statistical relationships can be demonstrated with an adequate degree of statistical rigor and where the sampling upon which the relationships were based was representative of and applicable to the assessment question. An example of this type of model is the statistically significant relationship between readily extractable potassium levels in soil (K) and the plant: soil concentration ratio (CR) for $^{137}$Cs, which is shown in figure 6.4. In this case, $CR = 40.9\,K^{-0.95}$ represents a data-based statistical model for the uptake of $^{137}$Cs by vegetable crops on the exposed lake bed of Par Pond at the U.S. Department of Energy's Savannah River Site in South Carolina

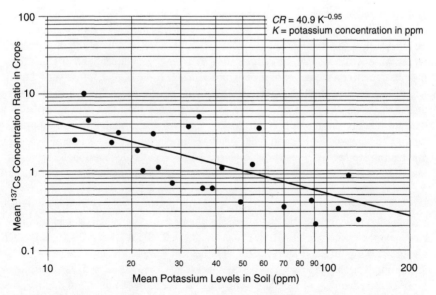

Figure 6.4 Effects of soil potassium on the concentration ratio for $^{137}$Cs in vegetable crops, Par Pond, Savannah River Site.

(Seel et al. 1995). It is likely that this relationship has powerful predictive value for the same vegetable crops in the same location. A reasonable question is how well the relationship applies to other locations with differing soil types and perhaps to different crops or varied agricultural practices. The likely answer is not very well. With sufficient data, one might be able to include additional independent variables in this model to improve its range of applicability. Powerful techniques are available for describing multivariate relationships, and these relationships may have a high degree of predictability. Some statistical models provide little insight into the processes giving rise to an observed relationship. Other models may provide important clues about the importance of specific processes, simply from the variables required in the model and the strength of their relationship to the predicted quantity. Clearly, both statistical models and direct measurements require sufficient measurements of radionuclide concentrations in the exposure media, which may limit their utility if exposure media concentrations are limited or are not available.

## Mechanistic Models

Mechanistic models represent an attempt to synthesize experimental data and knowledge about individual processes to produce a predictive tool. Because of their flexibility to address past, present, and future situations, mechanistic models are probably the most widely used models in risk assessments. In this approach, individual processes can be independently investigated and formulated mathematically, and parameters can be estimated. Some of the simpler models may require only a few parameters, and those parameters may cover more than one process. The finer the detail and the more processes that are explicitly accounted for, the more complex the model is said to be. Complex models may be able to simulate a wide variety of situations because special conditions, which are dealt with according to the structure and parameter values in the model, can be built into the model. Complex models may be inherently easier to test than simpler models because of their ability to simulate a wider variety of conditions and circumstances (e.g., time dependence). Thus, these models can offer a greater choice of actual data sets for making comparisons. On the negative side, complex, highly mechanistic models are more difficult to construct, and their predictive performance may not necessarily be better than that of much simpler models. Ample opportunity exists for complex models to wrongly predict simply because of the large number of uncertain variables. Extensive peer review and testing are needed to establish the credibility of mechanistic models. Nevertheless, with sufficient effort, a reasonable degree of credibility can be established, and a great deal can be learned by going through the process.

Mechanistic models of generally low complexity have been adopted by regulatory agencies for prospective and ongoing assessments of nuclear and industrial facilities. Some examples include RESRAD (Yu et al. 2001), CAP88PC (EPA 2006), MILDOS (Argonne National Laboratory 1998), and the Soil Screening Guidance (EPA 1996). These models are typically used to demonstrate compliance with regulations or address the adequacy of cleanup. With careful selection of model parameters, one may be able to demonstrate that a dose standard will not be exceeded; however, these models were not designed to predict the true dose in the present or future.

## Choosing a Predictive Approach

Choosing a predictive approach can be difficult. Certainly, the assessment question and the type and quantity of data available are primary determinants. Other determinants might be the expertise, time, and resources available to address the problem. Questions that could be asked include how much uncertainty is acceptable, and what are the economic, regulatory, or potential health impacts of a bad decision resulting from a faulty assessment. Typically, the more important the assessment, the more likely it is to be critically reviewed. Therefore, more effort is required to earn acceptance. In many cases, a combination of the three approaches can be useful. The result is strengthened when different approaches lead to the same conclusion.

## Model Attributes

The mechanistic modeling approach has generally become mainstream over the last two or three decades in the radionuclide transport field, although the degree of complexity and the assessment questions to be answered by the models have varied greatly. Models have ranged from very simple, steady-state formulations to highly complex, time-dependent (dynamic), multicompartment computer codes. Some models were developed to demonstrate compliance with regulatory standards, while others were intended to provide realistic estimates of dose. Compliance models tend to be conservative in that they err on the side of maximizing impacts and doses, while realistic models are constructed to simulate reality as accurately as possible. Models that produce single-output estimates are termed deterministic models. Increasingly, emphasis has been placed on providing best estimates, along with measures of uncertainty. This has led to stochastic models that propagate uncertainty in structure or parameter values through to the output. Such models can be implemented using readily available computer software. Stochastic (randomized error propagation) models, implemented using readily available computer software, can meet this requirement.

Virtually all the mechanistic models for radionuclide transport implicitly or explicitly incorporate expressions for both inputs and outputs from a compartment. figure 6.5 illustrates this fundamental concept. The environmental compartment

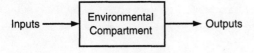

- High input rate increases the amout in the compartment.

- High output rate decreases the amout in the compartment.

Figure 6.5  Diagram of a basic compartment with inputs and outputs.

might be one of the boxes in figure 6.3, or it might represent a particular biological tissue or organ. The input might be represented as a transfer rate into the compartment (amount per unit time), with a particular process, such as deposition, inhalation, or ingestion, accounting for this transfer. Outputs from the compartment, which carry the same dimensional units as the inputs, represent processes that remove the radionuclide from the compartment, such as weathering loss from plants, excretion from animals, and radioactive decay. Sometimes, more than one process is involved in each of the input and output rates. The conceptual model shown in figure 6.5 can be formulated mathematically with a simple differential equation. The question of whether the content of the compartment is increasing or decreasing over time depends on the relative magnitudes of the inputs and outputs. When inputs equal outputs, the content of the compartment will achieve a constant value, and it is said to be in steady state. The steady-state content of a radionuclide in a compartment will tend to be comparatively high when the input rate is relatively high and the output rate constant is relatively low.

There is a class of models that assume steady-state conditions have been reached, and so time dependencies are not present in the formulations. These models are computationally simple because different compartments in ecological or physiological systems maintain a constant relationship to one another under steady-state or equilibrium conditions. Although steady-state conditions seldom exist in the real world, at least for very long, such conditions are often assumed in assessments for regulatory compliance or in certain situations involving constant, chronic releases. The assumptions necessary for equilibrium models are generally fewer than for dynamic models, and computations typically can be done by calculator or even by hand.

In contrast, dynamic models are more general because steady-state conditions do not have to be assumed. These models are more useful for describing acute releases such as those resulting from accidents, and chronic routine releases that vary temporally in magnitude. Dynamic models can handle time-varying input and output rates, sometimes with analytical expressions and other times using numerical algorithms performed by computer. Explicit, time-dependent formulations, although often useful, are limited in terms of the number of linked compartments that can be handled and the mathematical nature of varying input and output rates. Often, less effort is required to implement a numerical algorithm than to code a complex analytical solution. Numerical solutions are conveniently implemented using commercial software such as MatLab[1] and Mathematica[2] or using numerical algorithms in traditional programming languages (Press et al. 1992). Dynamic models often are more easily tested than are steady-state models because specific conditions can be accounted for, and a larger array of real data sets can be used for comparison.

The approach in this chapter is to apply some of the fundamental mathematics of kinetic systems to develop a series of compartment models. These range from very simple equilibrium models to dynamic models. The models start with single compartments and progress to multicompartment systems. In addition, we emphasize the application of kinetic-based models to practical situations involving food chain transport.

## Mechanistic Models: The Mathematical Foundations for Single Compartments

This section presents fundamental concepts of tracer kinetics that apply directly to predicting radionuclide transport and accumulation in single-compartment systems. Mathematical equations that employ intake and loss functions to estimate accumulation in specific ecological or physiological compartments are developed. This section begins with a discussion of basic concepts and terminology used in tracer kinetics, before proceeding to the mathematics of specific compartment systems.

### Concepts and Terminology of Tracer Kinetics

*System*: A system may be considered an aggregate of parts that interact in such a way as to lend particular properties to the whole. Systems may be visualized at the level of the cell, organism, and ecosystem in the biological realm. This discussion concentrates primarily on physiological and ecological systems and the individual components of these systems.

*Compartment*: A compartment can be thought of as a space, usually having defined boundaries, within which materials are free to mix, thereby achieving somewhat homogeneous concentrations throughout. A compartment that can exchange materials with the outside is considered an open compartment; otherwise, it is said to be a closed compartment. In an open compartment, the rate at which tracer substances mix within the space should be considerably more rapid than the rates of entry or loss. If this is not the case, then it may not be appropriate to consider the space as a compartment, at least in a pure mathematical context. Some examples of compartments discussed in this chapter include water, soil, plant tissue, blood, thyroid gland, bone, and muscle. Clearly, these compartments may not always fit the true definition of having complete and rapid mixing. Therefore, compartment models are only mathematical approximations of reality. Judgment, peer review, and testing must be used in developing compartment models.

*Steady State*: A compartment is considered in steady state when substances involved in normal processes are entering and leaving it at the same rate. In this condition, the content of the compartment is constant through time. Compartments in steady state are also said to be in equilibrium. Few natural compartments are in perfect steady state because of dynamic processes, such as growth and ecological change. Steady state is often assumed for compartments as long as their volumes and rates of input and loss do not change appreciably during the time frame of interest. The mathematics of non-steady-state systems is not treated here.

*Tracee*: The kinetics of a compartment system is usually governed by the flow of a fluid medium and often by the behavior of a natural substance within the medium. A common example of a fluid medium is water, and particular elements, compounds, or solid aggregates in the medium may represent the natural substances that control the behavior of a trace substance. The physical and chemical properties of the trace substance determine which natural substance is traced. The substance traced is termed the tracee. The tracee for tritium, if in the form of $^3HOH$, is the water with which it has an opportunity to mix. The tracee for $^{131}$I is stable iodine in the same chemical form.

*Tracer*: Radionuclides, and most other pollutants, are normally present in atom concentrations that are very low in comparison to concentrations of their tracees. In this situation, their kinetic behavior is governed by that of the appropriate tracee. A substance present in trace concentrations that is carried through normal processes by a tracee is called a tracer. If the behavior of the tracer is independent of its concentration, then it is considered to be within the "tracer range." If this range were to be exceeded, then the mass properties of the tracer would affect the kinetics of the tracee, and the system would behave abnormally. This situation would be rare, especially for radionuclides; thus, it is not dealt with here.

*Transport Pathway*: This term refers to a route by which material passes into or from a particular compartment or a mechanism producing the passage between two specific compartments. A transport pathway involves specific physical structures and represents a certain functional process. Common examples of transport pathways include deposition, resuspension, sorption, desorption, ingestion, inhalation, and excretion.

*Conceptual Model*: Conceptual models of compartment systems are symbolic representations that use diagrams to illustrate system structure and functional relationships. Figures 6.1 and 6.3 provide some examples of conceptual models. These are often referred to as box-and-arrow diagrams. Developing, understanding, and reviewing mathematical models are assisted considerably by first constructing a conceptual model to provide a framework for visualizing the components and functional processes embodied in the mathematical model.

*Mathematical Model*: A mathematical model is a mathematical equation, set of equations, or a computer code that embodies equations, algorithms, and parameter values that may be used to perform computations. Mathematical models represent human understanding of real systems in quantitative terms, and they also represent an attempt to provide a simulation of reality. The degree to which a model succeeds in simulating reality is dependent on how well the model is structured and parameterized. Mathematical models may provide single estimates (deterministic models) or a distribution of estimates (stochastic or probabilistic models).

*First-Order Systems and Rate Constants*: The models dealt with in this chapter are primarily treated as first-order compartment systems. A first-order process is one in which a rate of transfer is assumed to be proportional to the quantity of material available for transfer. For example, if $q$ represents the quantity of a tracer in a compartment, the rate of loss from the compartment $dq/dt$ might be formulated as the differential equation

$$\frac{dq}{dt} = -kq, \qquad (6.1)$$

where $k$, the first-order rate constant, represents the fraction of $q$ that is lost per unit time. Ordinarily, $k$ is assumed to remain constant with time, and because $q$ decreases with time, the loss rate also declines with time. In this case, the loss rate and $q$ decline according to an exponential function; hence, such systems are sometimes referred to as exponential loss systems. A practical example of the conceptual model formulated by equation 6.1 is radioactive decay, a well-known and highly predictable first-order process.

*Half-time*: Half-time for a first-order compartment represents the time required for half of a specific substance to be lost from the compartment. This is analogous to the term "half-life" for the radioactive decay process. The half-time ($T_{1/2}$) for the system described by equation 6.1 is shown to be related to $k$:

$$T_{1/2} = \frac{\ln 2}{k}. \tag{6.2}$$

Each radionuclide has a well-defined decay constant and, thus, half-life. Most other physical loss processes and most biological loss processes, however, do not have such well-defined rate constants and half-times, even though many estimates can be found in the literature. The terms "half-time" and "half-life" are commonly used because the concept is easy to visualize, and specific values are generally easier to remember than are those of rate constants.

## Single-Compartment, First-Order Loss Systems

A single-compartment system considers a single mixing pool with various rates of input and first-order or exponential loss. This section reviews the most common of these systems.

### Source and Sink Compartments

The conceptual model for a source compartment exhibiting first-order loss is as follows:

$$\boxed{q} \xrightarrow{k}.$$

The differential equation that describes this model is $dq/dt = -kq$ (equation 6.1), where $q$ is the tracer quantity in the compartment, $k$ is the loss rate constant (with units of $t^{-1}$), and $t$ is time (with such units as days, hours, or seconds). The differential equation may be converted to an ordinary algebraic expression by integration and the assignment of a boundary condition, in this case letting $q = q(0)$ at $t = 0$. The resulting algebraic equation is the familiar form

$$q = q(0)e^{-kt}. \tag{6.3}$$

### EXAMPLE 6.1

A familiar example of this model is radioactive decay. Suppose we have a sealed ampoule containing 100 Bq of $^{131}$I on March 2, 1993. The only loss mechanism is radioactive decay. The half-life of $^{131}$I is eight days; therefore, the decay constant is ln 2/8 days or 0.693/8 days = 0.0866 day$^{-1}$. This value implies a loss rate of about

8.7% per day. With these facts in hand, we can estimate the quantity of $^{131}$I in the ampoule at any time $t$ from

$$q = 100 \, \text{Bq} \cdot e^{-0.0866t}.$$

On March 20, for instance, $t = 20 - 2$ or 18 days, and $q$ is 21 Bq.

A common situation is one where there is more than one loss mechanism operating on the compartment. For example, we might have the following system:

$$\xleftarrow{k_2} \boxed{q} \xrightarrow{k_1}.$$

In this case, the appropriate model is

$$q = q(0)e^{-k_{\mathit{eff}}t}, \tag{6.4}$$

where $k_{\mathit{eff}}$, called the effective loss rate constant, is simply $k_1 + k_2$. For conceptual models of this form, any number of rate constants resulting in loss from a single compartment may be summed to form an effective loss rate constant. The units for each rate constant must be the same before the numerical sum is calculated.

### EXAMPLE 6.2

Instead of a sealed ampoule, suppose that the 100 Bq of $^{131}$I in example 6.1 is in the thyroid gland of a deer. Now, radioactive decay and biological loss mechanisms are operating. If the biological half-time were estimated to be 15 days, for instance, the biological loss rate constant, $k_b$, would be ln 2/15 days, or 0.0462 day$^{-1}$. In example 6.1, we determined that the physical decay constant, $k_p$, for $^{131}$I was 0.0866 day$^{-1}$. Therefore, the effective loss rate constant for the deer thyroid may be estimated from $0.0866 + 0.0462$ or 0.1328 day$^{-1}$. Now, if the elapsed time were 18 days, as before, the thyroid would contain only 9.2 Bq of $^{131}$I instead of the 21 Bq of $^{131}$I in the sealed ampoule.

It is rare when dealing with ecological or biological systems to find compartments that cannot lose radioactive materials by one or more mechanisms in addition to radioactive decay. Biological compartments lose tracers by processes such as weathering, senescence, excretion, molting, and leaching. Physical compartments lose tracers by such processes as leaching, resuspension, surface outflow, and weathering. Quantifying rate constants appropriate to specific compartments in particular circumstances is necessary for most assessment models, and it is one of the more difficult tasks. Part of the difficulty lies in determining whether a rate constant is appropriate (i.e., whether the process is first order) and part lies in estimating parameters. Ideally, these questions should be answered by proper experimentation, but risk assessors are often forced by lack of time and resources to use published values and to justify their applicability.

A sink compartment is a space that may or may not receive materials but that has no significant loss mechanisms operating on it. The conceptual model for a sink with a constant rate of input ($R$) might look like this:

$$R \longrightarrow \boxed{q \qquad\qquad}.$$

Depending on the form of the input function, the mathematical description will vary. For instance, if the rate of input ($R$) is constant, and $q = q(0)$ at $t = 0$, then

$$q = Rt + q(0). \tag{6.5}$$

If the rate of input happens to decline exponentially, for example, according to

$$R(t) = R(0)e^{-kt},$$

then

$$q = \frac{R(0)}{k}\left(1 - e^{-kt}\right) + q(0). \tag{6.6}$$

Of course, many other forms are possible, again depending on the input function. If we are dealing with a radionuclide as the contaminant of concern, then the sink compartment is never a true sink because radioactive decay always occurs at some known rate; thus, there is a loss mechanism. If we are dealing with a very long-lived radionuclide and the time domain of interest is relatively short, it may be possible to ignore the decay loss without losing much computational accuracy. One way that this type of modeling becomes something of an art is in knowing when simplifications can be made that will have minimal impact on the final result.

### Single Compartments with Constant Input Rates

Perhaps the most commonly encountered models in the field of radionuclide transport originate with the idea of a single compartment having first-order loss processes and a constant input rate. The conceptual model for this situation is

$$R \longrightarrow \boxed{q \qquad\qquad} \longrightarrow k.$$

where $R$ is the rate of entry of the tracer into the compartment, $q$ is the quantity of tracer in the compartment, and $k$ is the first-order loss rate constant. This conceptual model may be translated to mathematical terms by writing the differential equation:

$$\frac{dq}{dt} = R - kq. \tag{6.7}$$

Simply put, this equation says that the rate of change in $q$ is the difference between the rate of input and the rate of loss. $R$ is the rate of input with units of quantity per

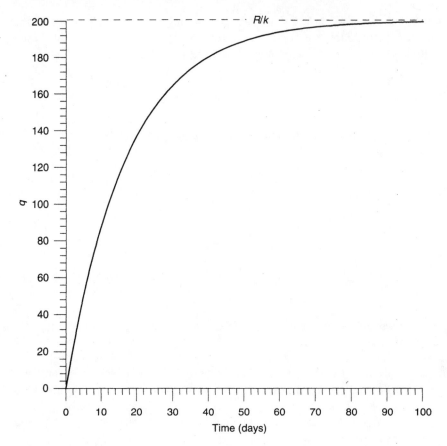

**Figure 6.6** Buildup of a tracer in a compartment receiving constant input and exhibiting first-order loss.

time. The product $kq$ is the loss rate and has the same units as $R$ because $q$ represents quantity and $k$ is the fraction of that quantity lost per time. Note that differential equations for first-order compartment systems are relatively easy to formulate; they are simply the rates of input minus the rates of loss. Equation 6.7 may be translated to the more useful algebraic form by integration. This can be accomplished by several methods (Whicker and Schultz 1982). Setting the boundary condition that $q = q(0)$ at $t = 0$, it can be shown that

$$q = \frac{R}{k}\left(1 - e^{-kt}\right) + q(0)e^{-kt}. \tag{6.8}$$

If $q(0) = 0$, then the second term drops out and a plot of $q$ versus $t$ will asymptotically approach $R/k$ (figure 6.6). Note that under chronic input conditions, $q$ will eventually stabilize at a numerical value of $R/k$, which is sometimes also called $q(eq)$ to indicate equilibrium or steady-state conditions. The rate at which equilibrium is established is governed by the expression $(1 - e^{-kt})$. Hence, the fraction of equilibrium attained depends only on the values of $k$ and $t$. The numerical value of $q(\text{eq})$ depends on both $R$ and $k$.

Equation 6.8 is a time-dependent or dynamic model because $t$ is one of the independent variables. For regulatory compliance and other purposes, such as simple screening calculations, equilibrium is often assumed. This assumption has the advantages of being conservative as well as computationally simple. For instance, the equilibrium version of the single-compartment model with a constant input and first-order loss is simply

$$q = \frac{R}{k}. \tag{6.9}$$

The application of equations 6.8 and 6.9 to practical problems of radionuclide accumulation in physical or biological compartments is usually straightforward and involves the formulation of $R$ and $k$ in meaningful terms. For example, if we are trying to estimate the rate of transfer of a contaminant from air to vegetation and if it can be reasonably assumed that the rate is constant, $R$ can be estimated from

$$R = XV_d, \tag{6.10}$$

where $X$ is the average air concentration (e.g., Bq m$^{-3}$), and $V_d$ is a term called the deposition velocity with units of distance per time (e.g., m s$^{-1}$). Note that the product of $XV_d$ has the units of flux (Bq m$^{-2}$ s$^{-1}$). The effective loss rate constant for vegetation ($k_{\it{eff}}$) is calculated from the sum of the physical decay constant ($k_p$) and the weathering loss rate constant ($k_w$). Applying these terms to equation 6.9, the equilibrium quantity on vegetation becomes

$$q = \frac{XV_d}{(k_p + k_w)}. \tag{6.11}$$

Here, the units work out to be Bq m$^{-2}$ for the quantity ($q$). Often, we are interested more in the concentration of contaminant on the vegetation (in Bq kg$^{-1}$) than in the above units for $q$. This requires the division of $q$ by the vegetation biomass ($M$) in units of kg m$^{-2}$. Therefore, the equilibrium concentration of contaminant on the vegetation ($C_{\text{veg}}$) is

$$C_{\text{veg}} = \frac{XV_d}{M(k_p + k_w)}. \tag{6.12}$$

The corresponding time-dependent expression for the concentration on vegetation, $C_{\text{veg}}(t)$, using equation 6.8, is

$$C_{\text{veg}}(t) = \frac{XV_d}{M(k_p + k_w)} \left[ 1 - e^{-(k_p+k_w)t} \right] + C_{\text{veg}}(0) e^{-(k_p+k_w)t}. \tag{6.13}$$

### EXAMPLE 6.3

Suppose that a $^{137}$Cs air concentration of 1 Bq m$^{-3}$ is maintained over a pasture having a mean biomass of 0.3 kg m$^{-2}$. If the deposition velocity is 200 m day$^{-1}$, and the weathering half-time is 14 days ($k_w = \ln 2/14$ days $= 0.0495$ day$^{-1}$), we can calculate the equilibrium concentration on vegetation:

$$C_{veg} = \frac{(1\,\text{Bq m}^{-3})(200\,\text{m d}^{-1})}{(0.0495\,\text{d}^{-1})(0.3\,\text{kg m}^{-2})} = 1.35 \times 10^4 \text{ Bq kg}^{-1}.$$

Note that in this example, $k_{eff}$ is essentially the same value as $k_w$ because $k_p$ is a very small number.

This estimate, as well as the same approach, can be used to estimate the equilibrium concentration of $^{137}$Cs in beef raised on this pasture. As with the vegetation example, the first step is to estimate the rate of input to muscle tissue resulting from the ingestion of contaminated vegetation. This may be estimated from the product of the ingestion rate, the concentration in vegetation, and the fraction of $^{137}$Cs ingested that goes to the muscle compartment. For example, if a cow eats 10 kg of vegetation per day, and 40% of the ingested $^{137}$Cs goes to muscle tissue, then

$$R = (1.35 \times 10^4 \text{ Bq kg}^{-1})(10\,\text{kg d}^{-1})(0.4) = 5.39 \times 10^4 \text{ Bq d}^{-1}.$$

The effective loss rate constant may be estimated from the sum of the biological loss rate constant and the physical decay constant. For instance, if the biological half-time is 20 days, $k_b$ would be 0.0347 day$^{-1}$. As with the vegetation example above, the physical decay constant is trivial, so the $k_{eff}$ is essentially the same value. Now, assuming the mass of muscle in a beef animal is 200 kg, we can calculate the equilibrium concentration in muscle tissue:

$$C_{meat} = \frac{5.39 \times 10^4 \text{ Bq d}^{-1}}{(0.0347\,\text{d}^{-1})(200\,\text{kg})} = 7.77 \times 10^3 \text{ Bq kg}^{-1}.$$

Example 6.3 illustrates some of the logic and mathematical basis for simple equilibrium food chain models. The good news here is the simplicity; the bad news is the fact that other transport pathways, such as root uptake of $^{137}$Cs from its buildup in the soil over time, and inhalation, have not been considered. Again, the intent of this chapter is to start with simple concepts and building blocks, to refine them one step at a time, and ultimately, to connect them to form systems that provide a reasonable simulation of all important pathways.

## Single Compartments with Time-Dependent Input Rates

In most real-world situations, contaminant input rates to compartments are more likely to vary with time than to be constant. If the time dependence is somewhat chaotic or unpredictable, then simple mathematical relationships will not suffice to describe the dynamic behavior of a compartment. There are a few situations, however, where the time-dependent input rate is reasonably well defined by a mathematical equation and, furthermore, where a dynamic solution to the compartmental content also exists. Below, we examine several such cases.

The conceptual model for single, first-order loss compartments receiving time-dependent input rates may be illustrated as

$$R(t) \longrightarrow \boxed{Q} \quad k \longrightarrow,$$

where $R(t)$ is a mathematical function with time as the independent variable.

*Model for Exponentially Declining Input Rate* One of the most common cases is the exponentially declining input rate:

$$R(t) = R(0)e^{-\lambda t}.$$

The differential equation for the single, first-order loss compartment with an exponentially declining input is

$$\frac{dq}{dt} = R(0)e^{-\lambda t} - kq, \tag{6.14}$$

where $R(0)$ is the initial rate of input, and $\lambda$ is the rate constant (having units of $t^{-1}$) that describes the rate at which the input rate changes. Integration of equation 6.14 with the boundary condition that $q = q(0)$ at $t = 0$ yields $q$ as a function of time:

$$q = \frac{R(0)}{k - \lambda}\left(e^{-\lambda t} - e^{-kt}\right) + q(0)e^{-kt}. \tag{6.15}$$

### EXAMPLE 6.4

Suppose that an acute deposition event occurred over a pasture, leaving an average initial $^{131}$I concentration on vegetation of 50 Bq kg$^{-1}$. If the weathering half-time is 14 days and the physical half-life is 8 days, the effective loss rate constant from vegetation is easily calculated (equations 6.2 and 6.4) to be 0.136 day$^{-1}$ (this corresponds to an effective half-time of 5.1 days). If we were interested in the content of $^{131}$I in the thyroid of a deer continually grazing on the contaminated pasture, the time-dependent intake rate could be estimated from

$$R(t) = r C_{\text{veg}}(0) a e^{-\lambda t},$$

where $r$ is the vegetation ingestion rate, $C_{\text{veg}}(0)$ is the initial concentration on vegetation, $a$ is the assimilation fraction to the thyroid, and $\lambda$ is 0.136 day$^{-1}$ as calculated above. Based on a study in Gist and Whicker (1971), $r$ was estimated to be about 1.5 kg day$^{-1}$, and $a$, approximately 0.8. In the same study, the biological half-time of $^{131}$I in the deer thyroid was found to be about 6.5 days; thus, $k_{\textit{eff}}$ can be estimated (equations 6.2 and 6.4) to be 0.193 day$^{-1}$. Applying these values to equation 6.15 with the assumption that $q(0) = 0$, we obtain an expression for the estimated $^{131}$I burden ($q$) as a function of time:

$$q = \left[\frac{(50\,\text{Bq}\,\text{kg}^{-1})(1.5\,\text{kg}\,\text{d}^{-1})(0.8)}{0.193\,\text{d}^{-1} - 0.136\,\text{d}^{-1}}\right]\left(e^{-0.136t} - e^{-0.193t}\right)$$

$$= 1{,}050\,\text{Bq}\,\left(e^{-0.136t} - e^{-0.193t}\right).$$

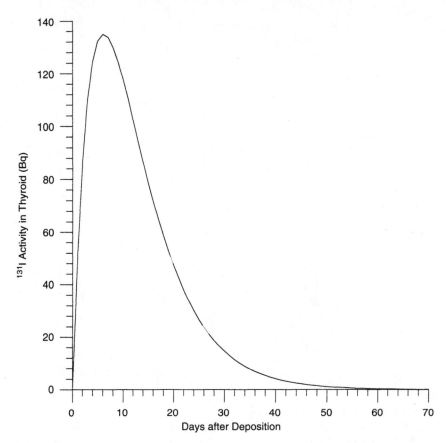

Figure 6.7 Time-dependent content of $^{131}$I in a deer thyroid after a fallout deposition event.

A plot of this function is provided in figure 6.7. Note that this function goes through a maximum value of ∼135 Bq at $t = 6.1$ days.

It is interesting to note that equation 6.15 becomes undefined if $k = \lambda$. This is unlikely for real applications, but if the situation does arise, the problem can be handled by setting $\lambda = k$ and rewriting the differential equation

$$\frac{dq}{dt} = R(0)e^{-kt} - kq. \tag{6.16}$$

Integration of equation 6.16 yields

$$q = R(0)te^{-kt} + q(0)e^{-kt}. \tag{6.17}$$

Four additional single-compartment, first-order loss models involving time-dependent inputs are provided below.

*Model for Two-Component, Exponentially Decreasing Input*

$$R(t) = R(0)(ae^{-\lambda_1 t} + be^{-\lambda_2 t}) \longrightarrow \boxed{q} \xrightarrow{k}$$

Solution:

$$q = R(0)\left[\frac{a}{k-\lambda_1}\left(e^{-\lambda_1 t} - e^{-kt}\right) + \frac{b}{k-\lambda_2}\left(e^{-\lambda_2 t} - e^{-kt}\right)\right] + q(0)e^{-kt} \quad (6.18)$$

*Model for Exponential Approach to Equilibrium Input*

$$R(t) = R(eq)(1 - e^{-\lambda t}) \longrightarrow \boxed{q} \xrightarrow{k}$$

Solution:

$$q = \frac{R(eq)}{k}\left(1 - e^{-kt}\right) - \frac{R(eq)}{k-\lambda}\left(e^{-\lambda t} - e^{-kt}\right) + q(0)e^{-kt} \quad (6.19)$$

*Model for Linear Slope Input*

$$R(t) = at + R(0) \longrightarrow \boxed{q} \xrightarrow{k}$$

Solution:

$$q = \frac{a}{k^2}\left(e^{-kt} + kt - 1\right) + \frac{R(0)}{k}\left(1 - e^{-kt}\right) + q(0)e^{-kt} \quad (6.20)$$

Note that this model has a limited domain of validity if the input slope is negative (i.e., $R$ cannot become negative).

*Model for Sine Function Input*

$$R(t) = a \sin bt + \overline{R} \longrightarrow \boxed{q} \xrightarrow{k}$$

Solution:

$$q = \frac{a}{k^2+b^2}\left(k \sin bt - b \cos bt + be^{-kt}\right) + \frac{\overline{R}}{k}\left(1 - e^{-kt}\right) + q(0)e^{-kt} \quad (6.21)$$

The sine function input model and modifications of it can be useful for cases where inputs undergo diurnal or seasonal cycles. An interesting application of this function was made by Peters (1994) for aquatic turtles in a $^{137}$Cs-contaminated environment.

## Single-Compartment, Non–First-Order Loss Systems

Occasionally one encounters a situation where loss from a single compartment is not first order. Examples might include loss functions that can be represented as the sum of exponential terms or a power function. Uses of the convolution integral and Borel's theorem to solve such situations are described below.

### The Convolution Integral

The convolution integral is sometimes a useful tool to solve single-compartment systems subject to a variety of input and loss functions. It can be particularly useful in certain cases where the loss function is something other than first order. The basic forms of the convolution integral as applied to single compartments are

$$q(t) = \int_0^t R(T)L(t-T)dT \tag{6.22}$$

and

$$q(t) = \int_0^t R(t-T)L(T)dT, \tag{6.23}$$

where $R(T)$ represents a time-dependent input rate function and $L(T)$ represents a time-dependent loss function. The input function may be of the forms illustrated in the preceding section, with the same units. The loss function, $L(T)$, is the fraction of any single quantity of material in the compartment that remains after some initial time, namely, $q(t) / q(0)$. Because this function is a ratio of similar quantities, it is unitless. For example, the loss function for a compartment exhibiting first-order loss would be $e^{-kt}$. Note that equation 6.22 and equation 6.23 have two time scales, with $t$ representing the basic time scale and $T$ representing the scale on which the loss of any introduced tracer bolus is timed (Whicker and Schultz 1982). Should $q(0)$ be a nonzero value, then $q(0)L(T)$ must be added to the solution.

### EXAMPLE 6.5

Suppose we have an input function $R(0)e^{-\lambda t}$ and a loss function $a_1 e^{-k_1 t} + a_2 e^{-k_2 t}$. Using these functions in either equation 6.22 or 6.23, the following expression is obtained:

$$q(t) = \frac{a_1 R(0)}{k_1 - \lambda}\left(e^{-\lambda t} - e^{-k_1 t}\right) + \frac{a_2 R(0)}{k_2 - \lambda}\left(e^{-\lambda t} - e^{-k_2 t}\right).$$

This equation assumes that $q(0) = 0$.

A very useful special case involving the convolution integral is that where the input ($R$) is constant in time. In this case, it may be shown that

$$q(t) = R \int_0^t L(t)dt. \tag{6.24}$$

For cases where $L(t)$ is a power function, the mathematics require the integral to be evaluated from 1 to $t$:

$$q(t) = R \int_1^t L(t)dt. \tag{6.25}$$

Applying equations 6.24 and 6.25 to the case of a constant input rate $(R)$, it may be shown, for example, that if

$$L(t) = \sum_{i=1}^{n} \left( a_i e^{-\lambda_i t} \right), \tag{6.26}$$

then

$$q(t) = R \sum_{i=1}^{n} \left[ \frac{a_i}{\lambda_i} \left( 1 - e^{-\lambda_i t} \right) \right]. \tag{6.27}$$

In the case of the power function, if

$$L(t) = t^p, \tag{6.28}$$

where $p$ is negative, but $\neq -1$, then

$$q(t) = \frac{R}{p+1} \left( t^{p+1} - 1 \right). \tag{6.29}$$

Equation 6.29 was fit to a radionuclide uptake curve by growing barn swallows at the Idaho National Laboratory (Millard et al. 1990). The fitted curve and actual data points are illustrated in figure 6.8.

### Borel's Theorem

For those familiar with the use of Laplace transforms, Borel's theorem is a handy tool for solving a number of single-compartment systems where the input and loss functions are known. The basic application of Borel's theorem to single-compartment systems is

$$q(t) = \mathscr{L}^{-1} \left[ \mathscr{L}R(t) \cdot \mathscr{L}L(t) \right], \tag{6.30}$$

where $\mathscr{L}R(t)$ and $\mathscr{L}L(t)$ represent the Laplace transforms of the time-dependent input rate and loss functions, respectively, and $\mathscr{L}^{-1}$ represents the inverse transformation back to the time-dependent function, $q(t)$. Laplace transforms of many functions can be looked up in most mathematical handbooks. The procedure is to transform the input and loss functions, multiply them, and, if possible, algebraically manipulate the product into a form amenable to inverse transformation. The inverse, if it exists, is the expression for $q(t)$. If $q(0)$ is a nonzero value, then $q(0)L(t)$ must be added to the solution. Several applications of Laplace transforms and Borel's theorem are provided in Whicker and Schultz (1982).

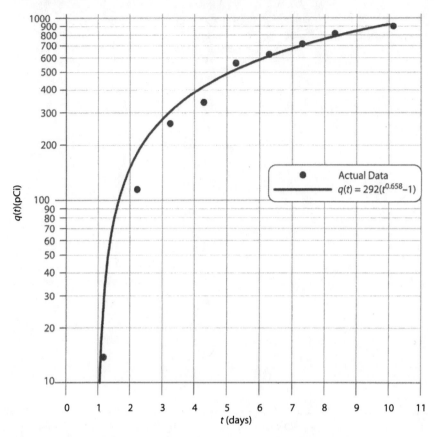

**Figure 6.8** Fit of the power function model to data on buildup of radioactivity in growing barn swallows at the Idaho National Engineering and Environmental Laboratory (redrawn from Millard et al. 1990).

### Derivation of Rate Constants Involving Fluid Flow Compartments

Many applications involve determining the radionuclide inventory in a compartment where there is flow into and out of the compartment. Examples include ponds, soil compartments with infiltration, buildings, and atmospheric mixing cells. If first-order processes are assumed to govern the behavior of the system, then the rate constant ($k$) that describes the loss from the compartment can be described by

$$k = \frac{F}{V}, \qquad (6.31)$$

where $F$ is the steady-state flow rate in and out of the system (m$^3$ s$^{-1}$), and $V$ is the volume of the compartment (m$^3$). In instances of unidirectional flow, the above equation can be reduced to

$$k = \frac{F}{V} = \frac{uA}{AL} = \frac{u}{L},$$

where

$A$ = cross section area of the compartment perpendicular to flow (m$^2$)
$u$ = fluid velocity (m s$^{-1}$)
$L$ = length of the compartment parallel to flow (m)

### EXAMPLE 6.6

A 100-m$^2$ area of the soil surface emits a $^{222}$Rn flux of 1 Bq m$^{-2}$ s$^{-1}$. What is the air concentration in a 1.5-m-high compartment directly above the radon source assuming a wind speed of 2 m s$^{-1}$?

*Solution*: The loss rate constant from the compartment is

$$k = \frac{F}{V} = \frac{2\,\text{m s}^{-1}(10\,\text{m} \times 1.5\,\text{m})}{(10\,\text{m} \times 1.5\,\text{m})\,10\,\text{m}} = \frac{2\,\text{m s}^{-1}}{10\,\text{m}} = 0.2\text{s}^{-1}.$$

The $^{222}$Rn decay rate constant is 0.693/3.82 day (day/86,400 s) = 2.1 × 10$^{-6}$ s$^{-1}$. Since the decay rate constant is insignificant compared to the removal rate constant, we can ignore decay and write the differential equation as

$$\frac{dq}{dt} = R - kq; \quad q(t) = \frac{R}{k}\left(1 - e^{-kt}\right);$$

$$t \longrightarrow \infty, q(eq) = \frac{R}{k} = \frac{(1\,\text{Bq m}^{-2}\,\text{s}^{-1}) \times 100\,\text{m}^2}{0.2\,\text{s}^{-1}} = 500\,\text{Bq}.$$

The concentration is determined by dividing the radon inventory by the volume of the compartment (150 m$^3$) = 3.33 Bq m$^{-3}$.

### Numeric Solutions

Numerical solutions are approximations to the true solution for differential equations. Desktop and laptop computers provide a practical option for solving differential equations using numerical solutions, and for this reason, they are commonly implemented. The Runge-Kutta method is perhaps the most efficient and robust numerical solver for environmental pathway models, as well as many other applications. However, crude numerical algorithms often provide an adequate solution to a problem because of the uncertainty inherent in modeling the environmental transport of radionuclides. Commercial software products such as Mathematica and MatLab have built-in advanced ordinary differential equation solvers. Algorithms in conventional programming languages such as FORTRAN 77, FORTRAN 90, and C++ are available on the Internet (cm.bell-labs.com/netlib/master/index.html) or in published texts such as *Numerical Recipes* (Press et al. 1992).

### Individual Transport Processes: Concepts and Mathematical Formulations

This section reviews different types of transport processes and provides details on the concepts and mathematical formulations that are specific to various transfers within ecological or agricultural systems. The formulations described are relatively simple and generally considered "mainstream"; however, they are not the only formulations that have been used in practice.

## Types of Processes

Before describing the concepts underlying individual transport processes and their mathematical formulations, it is important to recognize that such processes are, in reality, of three distinct types: continuous, discrete, and stochastic. These processes are briefly discussed below.

### Continuous Processes

Continuous processes occur essentially without interruption, at some average rate. The rate of most processes in nature, however, is seldom constant through time. One process rate that is continuous, and essentially constant, is radioactive decay of a large population of unstable atoms. Of course, viewed at the scale of atoms, radioactive decay is a series of discrete events. If a reasonably large population of radioactive atoms is involved, however, the measured decay rate constant is extremely consistent and unaffected by external variables, such as temperature, sunlight, or pressure. A decay rate constant can be applied to a specific radionuclide with a high degree of confidence. Many rate processes, even though they may fluctuate somewhat, are sufficiently predictable that they may be simulated as constant, continuous processes. In other cases, continuous rate processes require that time dependence be incorporated into their formulation. An example is seasonal change in the metabolic rates of ectothermic animals that cause significant cycles in intake and loss rates of nutrients and contaminants (Peters 1994).

### Discrete Processes

Discrete processes are events that occur periodically and cause an abrupt change in pathways or rates of material flow. Such discrete events are commonly encountered in managed agricultural ecosystems. For example, tillage, harvest, and diet changes in livestock are management decisions that cause sudden alterations in pathways and rates of material transport. Accurate modeling of agricultural systems, therefore, may require the scheduling of discrete events in the equations or algorithms. Discrete events also occur in natural ecosystems. Some of these could result from human activities, such as burning, logging, and stream diversion or impounding, or from natural events such as migrations, leaf fall, or reproductive cycles.

### Stochastic Processes

Events, sometimes very significant or dramatic, that occur periodically yet with rather unpredictable timing are said to be stochastic. Stochastic events are often processes that occur in nature, but the timing and magnitude of occurrence appears to be random and, therefore, not accurately predictable. The random nature of stochastic events distinguishes them from discrete events, of which timing and intensity are predictable. Examples of stochastic processes include intense rainfall events, high winds, lightning, earthquakes, fires, and certain biological phenomena, such as chance movements, mating, and invasions. Historical records of stochastic events can be used to develop probability statements for recurrence of future events in the same area. The probabilities will vary over time and space. Because some stochastic processes affect the transport and distribution of radionuclides in the environment, models that are more realistic often attempt to incorporate probabilistic treatment of such processes. An example might be rainfall, which affects processes such as wet deposition, resuspension, erosion, and plant growth. A dynamic model with sufficient historical data could simulate both the occurrence and magnitude of rainfall events, which in turn could be coupled to various transport processes in the model. Stochastic events may be incorporated into long-term averages that may then be used in simpler models. The predictive capability of these simple models is limited, however, to long-term averages of the predicted quantity.

## Deposition from Air to Soil and Vegetation

Deposition from air is a function of atmospheric conditions (wind speed and intensity of turbulence), the roughness height of the surface, particle size and density, type and intensity of precipitation, and, in some cases, chemical form. The amount of activity deposited from the air is depleted from the airborne mass. In this section, we review deposition process, focusing on empirical data. Models are introduced that describe the amount of activity that initially deposits on vegetation and soil and is retained on vegetation over time.

Contamination of terrestrial environments is most frequently caused by the deposition of materials suspended in the atmosphere onto the earth's surface. Airborne contaminants can originate from a diverse number of sources and can be carried for some distance—even global-scale distances—by the air stream. The contaminants can be in the form of reactive or inert gases and particulates, and the latter especially may exist in various sizes and shapes. The physical size and form and the chemical properties of a particular contaminant affect its behavior in the air stream, including its length of residence. The transport of airborne contaminants to the earth's surface is accomplished by several processes, including gravitational settling, dry deposition, and wet deposition. The following sections discuss these processes and the more specific mechanisms within each process.

### Gravitational Settling

The deposition rate of particles that are large enough to be affected more by the force of gravity than by atmospheric turbulence can be predicted from physical principles. As a general rule, deposition rates of particles having an effective diameter $> 20 \,\mu m$ and a density of at least $1.0 \text{ g cm}^{-3}$ can be estimated from Stokes's law (Whicker and Schultz 1982). Where Stokes's law is valid, the terminal fall velocity, $V_g$, may be estimated from

$$V_g = \frac{2\rho_s g r^2}{9\nu \rho_m}, \tag{6.32}$$

where

$\rho_s$ = particle density (g cm$^{-3}$)
$g$ = gravitational constant (cm s$^{-2}$)
$r$ = particle radius (cm)
$\nu$ = kinematic viscosity of the atmosphere (cm$^2$ s$^{-1}$)
$\rho_m$ = density of the atmosphere (g cm$^{-3}$)

McDonald (1960) addresses the theory, use, and typical results of Stokes's law and related phenomena.

### Dry Deposition

Small aerosol particles (those <20 μm) and reactive gases tend to deposit on vegetation or other surfaces at rates greater than those that might be predicted from gravitational settling. Other mechanisms, including surface impaction, electrostatic attraction, and chemical interaction, can become far more important than gravity for very small particles and reactive gases. Collectively, these mechanisms, which operate in the absence of precipitation, give rise to dry deposition. The most common way of expressing dry deposition is through the use of the deposition velocity, $V_d$. The deposition velocity is the ratio of the deposition rate ($\psi$ in Bq m$^{-2}$ s$^{-1}$) to the near surface air concentration ($X$ in Bq m$^{-3}$):

$$V_d = \frac{\psi}{X}. \tag{6.33}$$

Note that $V_d$ has the units of velocity (m s$^{-1}$), hence the term "deposition velocity." Experimentally, the deposition velocity is determined from measurements of $\psi$ and $X$ in various media, for a variety of contaminants and environments. To estimate the rate of deposition of an aerosol from the atmosphere to the soil or to vegetation, the near-surface air concentration (either measured or predicted from an atmospheric dispersion model) is multiplied by the appropriate deposition velocity:

$$\psi = X V_d. \tag{6.34}$$

The biggest challenge for a particular application is to find and be able to justify the appropriate value(s) for $V_d$.

Table 6.1 lists a small sampling of experimentally determined values for $V_d$. The model PATHWAY (Whicker and Kirchner 1987) employed a most likely value of 0.2 cm s$^{-1}$ for radionuclides in particulate form based on a review of the literature. Based upon expert elicitations conducted by the U.S. Nuclear Regulatory Commission and the Commission of European Communities, dry deposition velocities are expected to range between 0.1 and 1.0 cm s$^{-1}$ for particulate aerosols and from 0.6 to 1.6 cm s$^{-1}$ for elemental iodine (NRC/CEC 1994).

Table 6.1 Some early deposition velocities measured by field experimentation, compiled by Whicker and Schultz (1982)

| Substance | Surface | Remarks | $V_d$ (cm s$^{-1}$) |
|---|---|---|---|
| $^{131}$I | Grass | 200–610 g grass m$^{-2}$ | 1.1–3.7 |
| $^{131}$I | Clover leaves | | 0.5–1.3 |
| $^{131}$I | Paper | | 0.3–2.0 |
| $^{131}$I | Grass | 153–246 g grass m$^{-2}$ | 0.6–1.0 |
| $^{131}$I | Soil | | 0.4–0.8 |
| $^{131}$I | Snow | | 0.2 |
| $^{131}$I | Sticky paper | | 0.2–0.6 |
| $^{131}$I | Grass | | 1.2–2.1 |
| $^{131}$I | Soil | | 0.5–1.4 |
| $^{131}$I | Sticky paper | | 0.1–1.5 |
| $^{131}$I | Water | | 1.4–2.3 |
| $^{137}$Cs | Water | <10 μm particles | 0.9 |
| | Soil | <10 μm particles | 0.04 |
| | Grass | <10 μm particles | 0.2 |
| $^{103}$Ru | Water | <10 μm particles | 2.3 |
| | Soil | <10 μm particles | 0.4 |
| | Grass | <10 μm particles | 0.6 |
| $^{95}$Zr-Nb | Water | <10 μm particles | 5.7 |
| | Soil | <10 μm particles | 2.9 |
| $SO_2$ | Short grass | | 0.55 |
| | Medium grass | | 0.77–1.19 |
| | Soil | | 1.1 |
| | Water | | 0.46 |
| | Coniferous forest | | <2 |
| Ozone | Soil | | 0.84–1.76 |
| | Grass | | 0.55 |
| | Sand | | 0.14 |
| $^{238}$Pu | Bean leaves | Exposure chamber, 0.5–1.8 μm particles | 0.003–0.02 |
| Pollens | Soil | Open field | 2–15 |
| $^{141}$Ce | Deciduous tree leaves | Calculated from fallout data | 0.3–0.8 |
| $^7$Be | Seawater | | 0.6–5.3 |
| $^{134}$Cs, $^{141}$Ce | Grass | Submicron aerosols | 0.02–0.07 |
| | Sagebrush | Submicron aerosols | 0.15–0.18 |

In estimating deposition of radioiodine from releases at the Hanford Site, which was part of the U.S. nuclear weapons development complex during World War II, scientists used a resistance model to estimate radioiodine deposition from three general forms: organic slightly reactive gases (CH$_3$I), inorganic reactive gases (I$_2$), and iodine attached to aerosol particles (Ramsdell et al. 1994). Typical deposition velocities for these three forms are $1 \times 10^{-3}$ cm s$^{-1}$ for CH$_3$I, 1.0 cm s$^{-1}$ for I$_2$, and 0.1 cm s$^{-1}$ for particles (Voillequé and Keller 1981; Risk Assessment Corporation 1998). The resistance model was used in the Regional Atmospheric Transport Code for Hanford Emission Tracking (RATCHET) to provide deposition velocities for the three iodine forms. In the resistance model used by RATCHET, deposition velocities are a function of wind speed, atmospheric stability, and surface roughness (Ramsdell et al. 1994). The resistance model in RATCHET was adapted from the MESOPUFF II atmospheric transport model (Scire et al. 1984), and it is given by

$$V_d = (r_a + r_s + r_t)^{-1}, \qquad (6.35)$$

where

$r_a$ = aerodynamic resistance (s m$^{-1}$)
$r_s$ = surface-layer resistance (s m$^{-1}$)
$r_t$ = transfer resistance (s m$^{-1}$)

Aerodynamic resistance is a function of the wind speed, atmospheric stability, and surface roughness. Surface-layer resistance is a function of the wind speed and surface roughness. Transfer resistance is a function of the characteristics of the depositing material and surface type.

The transfer resistance in RATCHET is used to place an upper bound on the deposition velocity and can be approximated by taking the inverse of the maximum observed deposition velocity for the given form of iodine. In general, deposition velocity increases with wind speed, atmospheric instability, and surface roughness. The resistance model incorporates variability in the deposition velocity with changing wind speed, atmospheric stability, and surface roughness, and thereby provides a distribution of deposition velocities that are incorporated into a model simulation. For example, using the resistance model incorporated into the CALPUFF air dispersion model (Scire et al. 2000) for a particle with negligible gravitational settling and a 5-cm roughness height, the dry deposition velocity varies from about 0.01 cm s$^{-1}$ for extremely stable conditions and calm winds (<1 m s$^{-1}$) to about 8 cm s$^{-1}$ for extremely unstable conditions and higher wind speeds (>14 m s$^{-1}$). These extremes do not often occur in nature, and under more typical atmospheric conditions (e.g., neutral conditions and a wind speed of ~3 m s$^{-1}$), the deposition velocity calculated with the resistance model is about 1 cm s$^{-1}$. Another typical approach is to treat $V_d$ as one of the stochastic variables having a distribution of possible values. If large particles (>10 μm) are involved, then the gravitational settling term equation 6.32 is added to $V_d$ in equation 6.35.

A more complete discussion of the resistance model can be found in Seinfeld and Pandis (1998). The resistance model has been incorporated into the U.S. Environmental Protection Agency recommended atmospheric dispersion models (EPA 1992, 2004; Scire et al. 2000).

### Wet Deposition

Precipitation in the form of rainfall and snowfall provides important mechanisms for deposition of aerosol particles and reactive gases. During precipitation episodes, wet deposition will likely dominate over the dry deposition phenomenon. The general process of wet deposition has also been called precipitation scavenging (Slade 1968). Precipitation scavenging has been subdivided into two major categories: in-cloud scavenging, termed rainout, and below-cloud scavenging, termed washout.

The process of washout is more effective for particles >1 μm than is rainout, and the efficiency of the washout process increases with particle size (Slade 1968). The deposition rate from washout ($R_{wo}$, in Bq m$^{-2}$ s$^{-1}$) can be described by

$$R_{wo} = \lambda \int_0^z X(z)dz, \quad (6.36)$$

where $\lambda$ is the washout coefficient with units of per second and $X(z)$ is the air concentration as a function of height above the ground surface ($z$). If the temporal as well height dependence of $X$ were known, then the total, time-integrated deposition ($S$, in Bq m$^{-2}$) could be estimated by

$$S = \lambda \int_0^t \int_0^z X(z,t)dz\,dt. \quad (6.37)$$

Values for the washout coefficient, according to experts, range from 0.01 to 10 s$^{-1}$, depending on rainfall intensity and particle size or chemical form (NRC/CEC 1994). These equations may be of more theoretical and conceptual value than practical utility because of the effort required to measure $X(z, t)$. The equations could more easily be applied if $X(z, t)$ were predicted from an atmospheric dispersion model.

While washout results primarily from the impaction of falling raindrops on supramicrometer particles below a cloud, rainout operates within the cloud and upon submicrometer and larger particles. The mechanisms of rainout are not thoroughly understood, but they are believed to involve electrical effects, nucleation, diffusiophoresis, and Brownian motion (Slade 1968). These mechanisms result in the condensation of water around particles; when the droplet has grown to a sufficient size, it will fall under the influence of gravity. Rainout may be modeled using a rainout coefficient, $\psi$, which has units per time. For example, a parcel of in-cloud air with a radionuclide concentration, $X$, could be depleted over time according to

$$X = X(0)e^{-\psi t}. \quad (6.38)$$

Values of $\psi$ between $10^{-4}$ and $5 \times 10^{-3}$ s$^{-1}$ have been estimated from field measurements of cosmogenic radionuclides by Perkins et al. (1970). The approach suggested by equation 6.38 could be used to estimate deposition of in-cloud contaminants if the functions $X(t, z)$ and cloud dimensions were known. The data requirements to treat both rainout and washout are generally difficult to satisfy unless the data are predictable from atmospheric dispersion models.

Depending on particle size, the fraction of a contaminant removed from the atmosphere by a rainfall event of 0.33 mm in 10 min would be expected by experts to range from approximately 0.01 to 0.4 (NRC/CEC 1994).

A more practical method of predicting wet deposition is to use the volumetric washout factor, $W_v$, which is defined as

$$W_v = \frac{C_{rain}}{C_{air}}, \quad (6.39)$$

where $C_{rain}$ and $C_{air}$ are the measured contaminant concentrations in rain and air, respectively. These quantities require surface-level measurements, but these are practical to obtain. Because numerous factors could affect $W_v$, time- and site-specific measurements are desirable. Assuming that appropriate values of $W_v$ are available, the deposition rate may be estimated from

$$R_d = C_{air} W_v R_{rain}, \qquad (6.40)$$

where $R_{rain}$ is the rainfall rate in m s$^{-1}$. In this case, the product $W_v R_{rain}$, having the units m s$^{-1}$, can be thought of as a "wet deposition velocity," or $V_{d(wet)}$. Likewise, the total deposition, $S$, might be predicted from the total rainfall:

$$S = C_{air} W_v \int_0^t R_{rain}(t) dt. \qquad (6.41)$$

Equation 6.41 assumes that $C_{air}$ and $W_v$ are constant throughout the rainfall episode, which of course may not be the actual case. Equations 6.40 and 6.41 deal only with deposition on the earth's surface. The deposition on vegetation specifically requires knowing either the fraction of the total deposit intercepted and retained by vegetation, or another empirical factor. One such factor described by Hoffman et al. (1992) is the vegetation/rain concentration ratio, $CR_{veg/rain}$:

$$CR_{veg/rain} = \frac{C_{veg}}{C_{rain}}, \qquad (6.42)$$

where the ratio was found to be relatively constant for $^{131}$I at about 2.62 L kg$^{-1}$.

## Soil–Vegetation Partitioning of Deposition

In a number of practical assessments, the total quantity of radionuclide deposition is known from after-the-fact measurements or is estimated from models. In many cases, the fraction of the deposit that was intercepted and initially retained by the vegetation needs to be estimated. The early work of Chamberlain (1970) laid the foundation for a model that predicts the fractional interception of atmospheric deposits on vegetation using the vegetative biomass and an empirical constant. The fraction of a deposit allocated to the bare soil surface, $f_s$, is estimated from

$$f_s = e^{-\alpha B}, \qquad (6.43)$$

where $\alpha$ is the foliar interception constant in m$^2$ kg$^{-1}$ and $B$ is the vegetation biomass in units of kg m$^{-2}$. The foliar interception constant depends on particle size, vegetation characteristics, and other variables, so site- and situation-specific measurements are important. Measurements of $B$ are straightforward, so uncertainty in this parameter can be determined relatively easily. The fraction of a deposit allocated to vegetation, $f_v$, is assumed to be $1 - f_s$, or

$$f_v = 1 - e^{-\alpha B}. \qquad (6.44)$$

Table 6.2  Typical biomass values for various terrestrial ecosystems (Whittaker, 1970)

| Ecosystem | Biomass per unit area (dry kg m$^{-2}$) | |
|---|---|---|
| | Mean | Range |
| Agricultural land | 1 | 0.2–12 |
| Swamp and marsh | 12 | 3–50 |
| Tropical forest | 45 | 6–80 |
| Temperate forest | 30 | 6–200 |
| Boreal forest | 20 | 6–40 |
| Shrubland | 6 | 2–20 |
| Savannah | 4 | 0.2–15 |
| Temperate grassland | 1.5 | 0.2–5 |
| Tundra and alpine | 0.6 | 0.1–3 |
| Desert scrub | 0.7 | 0.1–4 |
| Extreme desert | 0.02 | 0–0.2 |

This equation indicates that as the vegetation biomass increases, the fraction of an atmospheric deposit intercepted by it approaches 1.0 asymptotically. While this equation has been shown to be useful for grasses and certain other plant forms, it may not always apply, particularly for shrub species in arid regions that maintain spacing to ensure adequate water and nutrient availability. Values of $B$ can range from a few g m$^{-2}$ to well over 1 kg m$^{-2}$, depending on species, season, and degree of maturation. A few typical values for $B$ are provided in table 6.2.

The foliar interception constant, $\alpha$, has been determined from a variety of field measurements. Modal values used in the PATHWAY model were 0.39 m$^2$ kg$^{-1}$ for fallout of particles between 129 and 416 km from the Nevada Test Site, where weapons testing occurred, and 2.8 m$^2$ kg$^{-1}$ for global fallout of submicrometer particles associated with atmospheric weapons testing (Whicker and Kirchner 1987). It is believed that a major reason for these differences is the effect of particle size on $\alpha$. Simon (1990) provides a good review of this topic. Based on the PATHWAY values for $\alpha$ and a typical value of 0.3 kg m$^{-2}$ for the biomass of semiarid region pasture grasses, $f_v$ would be estimated to be 0.11 for regional Nevada Test Site fallout and 0.57 for global fallout. An additional review of foliar interception constants yielded values ranging from 0.23 to 9.6 m$^2$ kg$^{-1}$ (IAEA 1994).

Hoffman et al. (1992) studied foliar interception of $^{131}$I, $^7$Be, and three sizes of insoluble microspheres under simulated rainfall. The ratio, $f_v/B$, varied from about 0.05 to 7 m$^2$ kg$^{-1}$. In all cases, this ratio was an inverse function of rainfall amount.

### Transport from Soil to Vegetation

Radionuclides in the soil can be transported to vegetation by three primary mechanisms:

- Resuspension from the soil surface
- Root uptake from the deeper soil layers
- Rain splash

The relative importance of these mechanisms depends on the vertical distribution of the radionuclide in the soil; soil properties; the physical-chemical properties of the radionuclide; the nature, condition, and growth rate of the vegetation; and other factors that include human, animal, natural, and mechanical disturbance of soil. Under most terrestrial scenarios, radioactive isotopes of most elements tend to accumulate, at least ultimately, in the soil more than in any other ecosystem compartment. Therefore, soil typically becomes the primary reservoir of contamination. From soil, common mechanisms of transport to animals and people involve transport through vegetation.

## Suspension and Resuspension

Suspension and resuspension are processes that result in the lifting of small (usually <50 μm) particles from the ground surface into the air stream by wind forces (Toy et al. 2002). Once suspended in the air stream, and if these particles are small enough (typically <20 μm), they can move considerable distances by the wind and ultimately be redeposited on soil and vegetation. If radionuclides are attached to these smaller soil particles, then suspension transports these radionuclides from soil to air where they represent a potential inhalation hazard to animals and humans (Whicker and Schultz 1982). Suspension processes are important for surface soil (e.g., <1 cm in depth), although contamination may extend to some depth below the surface. The distribution of radionuclides in the soil column depends on the mobility of the radionuclide, mechanical disturbances, and the processes that resulted in soil contamination (i.e., deposition from airborne plumes, liquid spills, etc.).

Resuspension, as the name implies, is the reemission of radionuclides from soil into the air stream that were initially deposited on the surface from airborne plumes. Resuspension is often applied to relatively large areas of uniform contamination because deposition from airborne plumes (except in the case of episodic release events) typically results in a substantial area of contamination that generally deceases with distance from the source. Some resuspension functions contain a temporal component by which the resuspension flux decreases with time from the initial deposit (Linsley 1978; Sehmel 1980; Smith et al. 1982).

Suspension rates, on the other hand, may show seasonal variations depending on soil moisture, wind speed, and vegetation cover (Whicker et al. 2006). In practical application, however, suspension rates are generally considered constant with time provided the source mass is not depleted appreciably over the time frame of interest (Gillette and Chen 2001).

*Suspension Models* Mechanistic suspension models simulate the effects of wind erosion on an erodible soil surface. Two basic processes are modeled: saltation (horizontal flux) and suspension (vertical flux). With regard to long-distance particle transport, on a scale of tens to hundreds of meters, we are primarily interested in the latter, but suspension is also a result of the saltation process (Bagnold 1941; Toy et al. 2002). Saltation refers to relatively large particles (80–1,000 μm) that are lifted by the wind but are too heavy to be held aloft. Once a saltating particle is lifted, it quickly falls back to earth, and its impact results in the release of finer particles. These fine particles are suspended in the air stream and can be transported considerable distances.

Gillette and colleagues studied suspended soil aerosols above eroding agriculture fields (Gillette et al. 1972, 1974; Gillette 1974a, 1974b; Gillette and Goodwin 1974) and developed a model for suspension flux that recognized the dependency of suspension flux on the saltation flux. He noted that suspension flux increases more rapidly than saltation flux with increasing wind velocity. The magnitude of suspension flux (or vertical flux) is dependent on wind speed and the particle size distribution in soil. The size distribution of suspended airborne particles is also dependent on size distribution of the originating soil. Travis (1974) used these relationships and developed a combined saltation–suspension flux model for describing suspension fluxes for particles <20 μm, as summarized in Smith et al. (1982):

$$F_v = F_h \left( \frac{C_v}{C_h U_{*t}^3} \right) \left[ \left( \frac{U'_*}{U_{*t}} \right)^{P/3} - 1 \right], \tag{6.45}$$

where

$F_v$ = vertical flux (g cm$^{-2}$ s$^{-1}$)
$F_h$ (g m$^{-1}$ s$^{-1}$) = horizontal (saltation) flux
$U_{*t}$ = threshold friction velocity (cm s$^{-1}$)
$U'_*$ = friction velocity during wind erosion (cm s$^{-1}$)
$C_h$ (1 × 10$^{-6}$ g s$^2$ cm$^{-4}$) = empirical constant of proportionality
$C_v$ (2 × 10$^{-10}$ g cm$^{-2}$ s$^{-1}$) = empirical constant of proportionality
$P$ = mass percentage of particles < 20 μm

The friction velocity is related to the surface roughness height ($z_0$), which is the height above the ground surface where the wind speed is zero. The threshold friction velocity is the friction velocity at which erosion begins. This model was implemented in the MILDOS code (Strenge and Bander 1981) for assessment of radiation doses from fugitive dust emissions from uranium mill tailings piles. The suspension rates of particles <20 μm are given for six wind speed classes (0.67, 2.5, 4.5, 6.9, 9.6, and 12.5 m s$^{-1}$). The suspension flux of the first two wind speed classes is zero. The next four wind speed classes have suspension fluxes of 3.9 × 10$^{-7}$, 9.7 × 10$^{-6}$, 5.7 × 10$^{-5}$, and 2.1 × 10$^{-4}$ g m$^{-2}$ s$^{-1}$.

Cowherd et al. (1985) developed a relatively simple but practical model based on the work of Gillette (1981) and others to estimate annual average emission of particulate matter with aerodynamic diameters <10 μm (PM$_{10}$) from surfaces with an unlimited reservoir of eroding particles. The model is an adaptation of the power function introduced by Gillette (1981) and is a function of the mean annual wind speed, the threshold wind speed, fraction of vegetative cover, and the soil particle size mode. The emission of PM$_{10}$ is given by

$$\varphi = 0.036 \left(1 - f_V\right) \left(\frac{U_m}{U_t}\right)^3 F(x), \tag{6.46}$$

where

$\varphi$ = particulate emission factor (g of $PM_{10}$ $m^{-2}$ $hr^{-1}$)
$f_V$ = fraction of vegetative cover (unitless)
$U_m$ = mean annual wind speed at 7 m (m $s^{-1}$)
$U_t$ = equivalent threshold wind speed at 7 m (m $s^{-1}$)
$F(x)$ = function dependent on $U_t/U_m$ in Cowherd et al. (1985) (unitless)
$x = 0.886(U_t/U_m)$ (EPA 1996)
0.036 = empirical constant (g of $PM_{10}$ $m^{-2}$ $hr^{-1}$)

The function $F(x)$ is presented graphically in Cowherd et al. (1985). For use in a spreadsheet or computer code, the function was fitted piecewise in Cowherd for $x \geq 2$ and in Weber et al. (1999) for $x < 2$:

$$F(x) = 1.91207 - 0.0278085x + 0.48113x^2 - 1.09871x^3 + 0.335341x^4$$

for $x < 2$

and

$$F(x) = 0.18\,(8x^3 + 12x)\,e^{-x^2} \quad \text{for } x \geq 2. \tag{6.47}$$

The threshold friction velocity at the surface is related to the mode of the soil aggregate particle size distribution. Cowherd et al. (1985) includes methods for determining the threshold friction velocity for aggregate particle size modes of 0.01–100 mm. The threshold friction velocity for this range of aggregate particle size modes is given by

$$u_* = 65.5315 p^{0.417673}, \tag{6.48}$$

where

$u_*$ = threshold friction velocity at the surface (cm $s^{-1}$)
$p$ = soil aggregate particle size mode (mm)

The threshold friction velocity is corrected for nonerodible elements. A nonerodible element can be large particles, a rock outcropping, pavement, or other obstacle and is quantified in terms of the parameter Lc. The parameter Lc is defined as the fraction of area as viewed from directly overhead that is occupied by nonerodible elements. The overhead area is then corrected for the equivalent frontal area. For example, if a spherical nonerodible element is half embedded in the soil, then the equivalent frontal area is one-half the overhead area (Cowherd et al. 1985). Cowherd et al. provides a graph relating Lc to the ratio of the corrected and uncorrected $u_*$ values (figure 6.9).

The corrected threshold friction velocity is extrapolated up to the height of the wind speed measurement by the diabatic wind profile for neutral conditions and is given in Cowherd et al. (1985) as

$$U_t = \frac{u_*}{0.4} \ln\left(\frac{z}{z_0}\right), \tag{6.49}$$

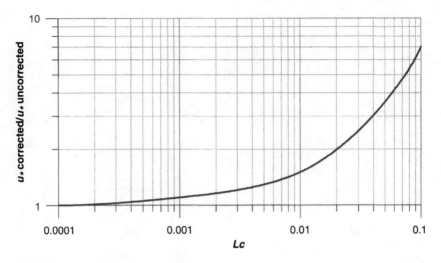

**Figure 6.9** Ratio of corrected and uncorrected threshold friction velocity ($u^*$) as a function of Lc (adapted from Cowherd et al. 1985).

where

$z_0$ = roughness height of the erodible surface (m)
$z$ = wind speed measurement height (m)

Roughness heights are provided in figure 6.10.

The particulate emission factor provides the annual average mass flux of $PM_{10}$ from a soil surface with an unlimited reservoir of erodible particles. The air concentration at some point distant from the contaminated soil site can be calculated using an air dispersion model. For example, the following equation may be used to estimate annual average concentrations of radionuclides attached to soil $PM_{10}$:

$$C_a = \frac{X}{Q} \varphi \, C_{ss} \text{ER} \cdot \text{CF}, \qquad (6.50)$$

where

$X/Q$ = annual average dispersion factor (g m$^{-3}$ of $PM_{10}$ per g m$^{-2}$ s$^{-1}$)
$\varphi$ = particulate emission factor (g of $PM_{10}$ m$^{-2}$ s$^{-1}$)
$C_{ss}$ = radionuclide concentration in soil (Bq kg$^{-1}$)
$C_a$ = radionuclide concentration in air (Bq m$^{-3}$)
CF = conversion factor from kg to g (0.001 kg g$^{-1}$)
ER = enrichment ratio (g of soil / g of contaminated airborne particles)

The annual average dispersion factor, $X/Q$, is dependent on the size of the source. Therefore, the particulate emission factor approach is best applied to sources that are limited in areal extent.

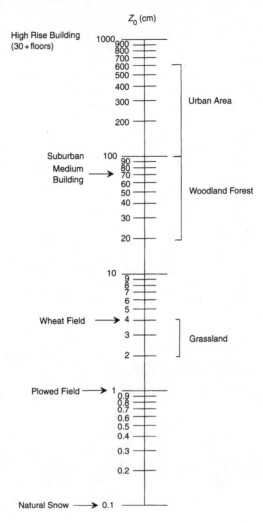

Figure 6.10  Roughness heights for various surfaces. Adapted from Cowherd et al. (1985).

*Resuspension*  Wind, raindrop splash, and other mechanical disturbances are the major causes of resuspension. The forces causing resuspension and the factors modifying the process are numerous and complex. For example, atmospheric variables that affect the rate of resuspension include air speed, turbulence, density, and viscosity. Factors that modify resuspension potential include soil texture, cohesiveness, moisture content, density and other local site features such as plant cover, ground surface roughness, and topography (Toy et al. 2002). In time and space, measured rates of resuspension have varied over 10 orders of magnitude (Sehmel 1980), making selection of a single number difficult.

Several approaches to the problem of estimating resuspension exist. One of the most straightforward is the mass loading approach. In this case, the concentration of a contaminant in air, $C_{\text{air}}$, is estimated from

$$C_{\text{air}} = M_d C_{ss} \, \text{ER}, \tag{6.51}$$

where

$M_d$ = dust load in air or mass loading factor (g m$^{-3}$)
$C_{ss}$ = concentration of contaminant in the soil surface (Bq g$^{-1}$)
ER = enrichment factor

The enrichment factor is the ratio of the activity density of the airborne respirable particles (Bq g$^{-1}$) to the activity density in the underlying soil (Bq g$^{-1}$). Enrichment factors can range from 1 to almost 5 (Healy 1980) for undisturbed soil, depending on the mechanism by which the soil was originally contaminated. Soil disturbances such as grading, wildfires, and even soil thawing in the springtime can increase the ER value by a factor of about 3–6 (Shinn 1992). The dust load in air is relatively easy to measure with standard air sampling equipment and appropriate filters. The term $C_{ss}$ is the concentration measured or predicted in the soil surface layer that is subject to resuspension. Ordinarily, this layer is considered to be about 0.1–0.3 cm in depth. Values for $M_d$ have ranged from an average of about 40 µg m$^{-3}$ in rural locations to more than 10,000 µg m$^{-3}$ above bare fields during high winds (Hinds 1982). Healy (1980) reports mass loading factors in urban areas to range from 33 to 254 µg m$^{-3}$ and 9 to 79 µg m$^{-3}$ in rural areas. Anspaugh et al. (1975) chose an average value of 100 µg m$^{-3}$ for radiological assessment purposes. This approach inherently assumes that the soil where $C_{ss}$ is measured is representative of that suspended in the air stream. This assumption is most likely valid when the contaminant is fairly evenly distributed over a large area; it is much less likely to be valid for small or heterogeneous patches of surface contamination. In this latter case, the radionuclide concentration in the aerosol is likely to be somewhat less than the concentration in the locally contaminated soil because of dilution with less contaminated aerosol that originated elsewhere.

The most common approach to estimating resuspension is the use of the resuspension factor. The resuspension factor, $F_r$, is defined as

$$F_r = \frac{C_{\text{air}}}{S}, \tag{6.52}$$

where $C_{\text{air}}$ and $S$, the surface deposition quantity in Bq m$^{-2}$, are estimated at the same location. Note that $F_r$ has units of m$^{-1}$. The concentration of a contaminant in air, therefore, may be estimated from the surface deposition and $F_r$ by

$$C_{\text{air}} = F_r S. \tag{6.53}$$

In addition to the variables mentioned above, the resuspension factor declines with the age of the deposited contaminant. For fresh deposits on undisturbed soils with a reasonable amount of vegetation cover, $F_r$ might be expected to range from about $10^{-6}$ to $10^{-4}$ m$^{-1}$. For aged deposits, however, $F_r$ can be expected to more or less stabilize in the range of about $10^{-10}$ to $10^{-8}$ m$^{-1}$, provided the soil is not mechanically disturbed (Sehmel 1980). The decline through time in the resuspension of a deposit can be explained by the percolation of contaminants into the soil profile and by concurrent natural deposition of dust and organic material. The downward percolation of particle-reactive contaminants can be accomplished by physical processes, such as soil cracking, frost heaving, colloid transport with water infiltration, and dissolved-phase transport. Biological activities such as animal burrowing and root penetration can also cause downward and upward migration of soil contaminants. Anspaugh et al. (1975) provides data and a model of the time dependence of the resuspension factor, while Oksza-Chocimowski (1977) provides some generalized models on the topic. Garger et al. (1997) provide an excellent environmental data set for resuspension measurements that covers an eight-year period in the Ukraine following the Chernobyl accident. Good general reviews of resuspension factors are provided by Linsley (1978) and Sehmel (1980).

The general form of the time-dependent resuspension factor is given by (Smith et al. 1982)

$$F_r(t) = F_r(0)e^{-\lambda_r f(t)} + F_r(\infty), \qquad (6.54)$$

where

$\lambda_r$ = weathering rate constant describing the decrease in resuspension over time (day$^{-1}$)
$F_r(0)$ = initial resuspension factor (m$^{-1}$)
$F_r(\infty)$ = long-term resuspension factor (m$^{-1}$)
$f(t)$ = time function

For deposition from uranium milling facilities, the U.S. Nuclear Regulatory Commission (NRC 1982) uses a value for $\lambda_r$ of 5.06 yr$^{-1}$, which is equivalent to a 50-day half-life, and $f(t) = t$. Other values for $\lambda_r$ include an estimate by Linsley (1978) of 0.01 day$^{-1}$ (3.65 yr$^{-1}$) and 0.6769 yr$^{-1}$ that was used in the Reactor Safety Study (NRC 1975).

The air concentration from resuspension from continuous deposition of particles from an anthropogenic source (i.e., emissions from a stack) over a period of time, $t$, can be found by integration of the time-dependent resuspension function and the steady-state deposition flux from the anthropogenic source. The U.S. Nuclear Regulatory Commission (NRC 1982) uses an expression that also includes the effective loss rate from the soil layer that is subject to resuspension. The air concentration from resuspension is given by

$$C_{\text{air}}(t) = \psi 10^{-5} \left\{ \frac{1 - \exp[-(K+\lambda_r)(t-a)]}{K + \lambda_r} + 10^{-9} \delta(t) \frac{\exp[-K(t-a)] - \exp(-Kt)}{K} \right\}, \qquad (6.55)$$

where

$\psi$ = annual deposition flux (Bq m$^{-2}$ yr$^{-1}$)
$K$ = effective removal rate constant from soil (yr$^{-1}$)
$a = t - 1.82$ for $t > 1.82$ yr, and 0 for $t \leq 1.82$ yr
$\delta(t)$ = zero for $t \leq 1.82$ yr and 1 for $t > 1.82$ yr
$t$ = time from the start of operations (yr)

# ERRATA

Page 302, Equation 6.55

Equation 6.55 should read

$$C_{air}(t) = y\, 10^{-5} \left\{ \frac{1 - \exp[-(K + \lambda_{r_r})(t-a)]}{K + \lambda_{r_r}} + 10^{-4}\, \delta(t)\, \frac{\exp[K(t-a)] - \exp(Kt)}{K} \right\}$$

OR

$$C_{air}(t) = y\left\{ 10^{-5}\left( \frac{1 - \exp[-(K + \lambda_{r_r})(t-a)]}{K + \lambda_{r_r}} \right) + 10^{-9}\, \delta(t)\, \frac{\exp[K(t-a)] - \exp(Kt)}{K} \right\}$$

The effective removal rate constant typically includes leaching and radioactive decay. Note that equation 6.55 assumes $F_r(0) = 10^{-5}$ m$^{-1}$ and $F_r(\infty) = 10^{-9}$ m$^{-1}$. If the deposition flux from the anthropogenic source varies with time, then the air concentration from resuspension can be found by convolution:

$$C_{\text{air}}(t) = \int_0^t \psi(\tau) L(t-\tau) F_r(t-\tau) d\tau, \quad (6.56)$$

where $L(t - \tau)$ is a loss function of the radionuclide from soil and $t - \tau$ is the age of the deposit. Equation 6.56 was implemented in a model for environmental exposure to plutonium from airborne releases at the Rocky Flats plant (Rood et al. 2002). Both the resuspension factor and the mass loading factor essentially partition radionuclides between the soil and air for radionuclides present on the soil surface.

Mechanical disturbance can have a substantial impact on resuspension and, depending on the nature of the disturbance, can have suspended radionuclides that are deeper in the soil column as compared to wind-driven resuspension. Sehmel (1980) showed that measured resuspension factors for mechanical disturbances could be roughly two to three orders of magnitude greater than could those for wind-driven resuspension.

The uncertainty in the resuspension factor is very high because of the many processes that affect it. Short-term variations can range up to 10 orders of magnitude because of the episodic nature of wind erosion and its strong dependence on numerous variables. Garger et al. (1997) showed, however, that temporal averaging can reduce the uncertainty to perhaps an order of magnitude but advise the use of site-specific information where possible. Temporal averaging can be entirely appropriate for dose assessments involving long-lived radionuclides and relatively stable land-use patterns (i.e., agriculture, industrial, or residential).

After we have an estimate of $C_{\text{air}}$, from equation 6.54, 6.55, or 6.56, the deposition velocity parameter can be used to predict the deposition rate on vegetation. One useful formulation to estimate the rate of transfer from the soil surface to vegetation through resuspension is

$$R_{\text{res}} = SF_r V_d, \quad (6.57)$$

where $R_{\text{res}}$ is the soil-to-vegetation resuspension rate in Bq m$^{-2}$ s$^{-1}$, and the other terms are as previously defined. Applying this formulation to a specific problem requires considerable care in selecting values for $F_r$ and $V_d$ to assure that they adequately represent the specific case in question.

## Root Uptake

The process of root uptake involves the transport of contaminants from the soil to the plant through active uptake by plant roots. Only those contaminants located within the rooting zone that are in a physical/chemical form that renders them available for absorption by the roots are subject to this process. The degree of root uptake depends on many factors, including the specific contaminant and its chemical form, the physical and chemical properties of the soil, and the species, tissue, and physiological state of the plant. The classical formulation for root uptake uses the equilibrium concentration ratio, CR, which is simply the ratio of the plant concentration to the concentration in the soil that the roots can access:

$$\mathrm{CR} = \frac{C_{\mathrm{veg}}}{C_{\mathrm{soil}}}, \qquad (6.58)$$

where $C_{\mathrm{veg}}$ is the concentration of radionuclide in the vegetation part of interest (Bq kg$^{-1}$ dry mass) and $C_{\mathrm{soil}}$ is the average concentration in the associated soil (Bq kg$^{-1}$ dry mass).

Concentration ratio values are determined from empirical measurements in field or laboratory settings. In assessment applications, if $C_{\mathrm{soil}}$ is predicted or measured, and appropriate values of the equilibrium concentration ratio are available, then $C_{\mathrm{veg}}$ can be estimated from the product of these terms. It is usually necessary to know from the design of the experiments that gave rise to a particular equilibrium concentration ratio, whether the value represents strictly root uptake, or whether it also represents some combination of resuspension or airborne transport and root uptake. Field-determined values usually represent the latter. Ordinarily, there is substantial uncertainty involved in applying equilibrium concentration ratios because of the factors mentioned above, unless appropriate site-specific measurements are available. Additionally, equilibrium concentration ratios may sometimes be reported in the fresh or wet-weight equivalent. That is,

$$\mathrm{CR}_{\mathrm{wet}} = \mathrm{CR}_{\mathrm{dry}} \left( \frac{W_{\mathrm{dry}}}{W_{\mathrm{wet}}} \right), \qquad (6.59)$$

where

$\mathrm{CR}_{\mathrm{wet}}$ = wet weight concentration ratio
$\mathrm{CR}_{\mathrm{dry}}$ = dry weight concentration ratio
$W_{\mathrm{dry}}$ = dry weight of the crop (kg)
$W_{\mathrm{wet}}$ = wet weight of the crop (kg)

The wet-weight conversion is typically applied to fruits, vegetables, and grains since dietary intakes for humans are reported in terms of wet-weight ingestion. Dry weight is typically used for animal forage because animal consumption rates of forage are in terms of dry weight. It is important to know whether the equilibrium concentration ratios are dry or wet weight and to adjust dietary intakes or the equilibrium concentration ratios accordingly. The water content of crops may range up to 90–95% of the total weight. table 6.3 presents representative dry-weight to wet-weight ratios for various crops and animal forage. Some equilibrium concentration ratios that have been used in generic assessments are provided in table 6.4.

Some of the more complex, dynamic transport models that require rate process formulations use various modifications of equation 6.58. For example, the model PATHWAY (Whicker and Kirchner 1987) estimates the rate of root uptake ($R_{\mathrm{up}}$, in

Table 6.3 Ratio of dry weight to wet weight for food crops consumed by humans and animal forage, from Baes et al. (1984) except as noted

| Food type | Crop | $W_{dry}/W_{wet}$ | Food type | Crop | $W_{dry}/W_{wet}$ |
|---|---|---|---|---|---|
| Leafy vegetables | Asparagus | 0.070 | | Grapefruit | 0.11 |
| | Lettuce[a] | 0.050 | | Orange | 0.13 |
| | Spinach[a] | 0.083 | | Peach | 0.13 |
| | Broccoli[a] | 0.11 | | Strawberry | 0.10 |
| Root vegetables | Potato | 0.22 | | Cantaloupe | 0.060 |
| | Sweet potato | 0.32 | | Watermelon | 0.080 |
| | Carrot | 0.12 | | Lemon | 0.11 |
| | Onion | 0.13 | Grains | Barley | 0.89 |
| Fruits | Apple | 0.16 | | Wheat | 0.88 |
| | Cherry | 0.17 | | Corn | 0.90 |
| | Pear | 0.17 | Forage[a] | Alfalfa | 0.23 |
| | Plum | 0.54 | | Clover | 0.20 |
| | Cucumber | 0.039 | | Grass | 0.18 |
| | Eggplant | 0.073 | | Silage | 0.24 |
| | Squash | 0.082 | Miscellaneous | Pea | 0.26 |
| | Tomato | 0.059 | | Peanut | 0.92 |

[a] Source: NRC (1983).

units of Bq m$^{-2}$ day$^{-1}$) from

$$R_{up} = C_{soil} \, CR \left(\frac{dB}{dt}\right), \quad (6.60)$$

where $dB/dt$ is the time derivative (rate of change) of the plant biomass in units of kg m$^{-2}$ day$^{-1}$. This formulation is intuitive because it implies that the rate of uptake is proportional to the rate of plant growth. Root uptake can only occur when the plant is growing. This relationship also preserves the ratio of concentrations in the plant and soil, even though this may not always be true in actual field situations. Plant biomass and its derivative can be measured at relevant points in time, or they may be modeled using formulations such as

$$\frac{dB}{dt} = k_g f_l f_t B \left(\frac{B_{max} - B}{B_{max}}\right) - D_A F_V f_r, \quad (6.61)$$

where

$k_g$ = growth rate constant (day$^{-1}$)
$f_l, f_t$ = growth rate modifiers for light and temperature, respectively (unitless, range 0–1)
$B$ = current plant biomass (kg m$^{-2}$ dry weight)
$B_{max}$ = maximum plant biomass (kg m$^{-2}$ dry weight)
$D_A$ = grazing animal density (number m$^{-2}$)
$F_V$ = ingestion rate of grazing animals (kg day$^{-1}$ animal$^{-1}$)
$f_r$ = fraction of ingestion rate contributed by field grazing

Table 6.4 Typical plant/soil concentration ratios for selected elements and crops, adapted from IAEA (1994) unless otherwise noted

| Element | Crop | Concentration ratio (dry mass basis) | |
|---|---|---|---|
| | | Expected | Range (95%) |
| Americium | Cereal grains | $2.2 \times 10^{-5}$ | $1.5 \times 10^{-7}$ to 0.7 |
| | Fruits, tubers | 0.001 | $1 \times 10^{-5}$ to 0.2 |
| | Grass | 0.001 | 0.0005 to 0.2 |
| Cesium | Cereal grains | 0.02 | 0.001 to 1 |
| | Fruits, tubers | 0.16 | 0.001 to 5 |
| | Grass | 0.3 | 0.01 to 6 |
| Copper | Root vegetables, fruits, and grains | $0.25^a$ | |
| | Leafy vegetables | $0.4^a$ | |
| Chlorine | Root vegetables, fruits, and grains | $70^a$ | |
| | Leafy vegetables | $70^a$ | |
| Iodine | Leafy vegetables | $0.16^a$ | |
| Lead | Cereal grains | 0.005 | 0.0004 to 0.05 |
| | Fruits, tubers | 0.014 | 0.0001 to 0.2 |
| | Grass | 0.001 | |
| Neptunium | Cereal grains | 0.0027 | $2.3 \times 10^{-5}$ to 0.083 |
| | Tubers (potato) | 0.0067 | $7.1 \times 10^{-4}$ to 0.14 |
| | Grass | 0.069 | |
| Nickel | Cereal grains | 0.03 | 0.003 to 0.3 |
| Plutonium | Cereal grains | $8.6 \times 10^{-6}$ | $3 \times 10^{-7}$ to 0.4 |
| | Fruits, tubers | $7.2 \times 10^{-4}$ | $1 \times 10^{-6}$ to 0.06 |
| | Grass | $3.4 \times 10^{-4}$ | $5 \times 10^{-6}$ to 0.6 |
| Radium | Cereal grains | 0.001 | 0.0002 to 0.006 |
| | Fruits, tubers | 0.016 | 0.0002 to 1 |
| | Grass | 0.08 | 0.01 to 0.4 |
| Strontium | Cereal grains | 0.15 | 0.002 to 1.4 |
| | Fruits, tubers | 0.86 | 0.002 to 14 |
| | Grass | 1 | 0.03 to 8 |
| Technetium | Cereal grains | 0.7 | 0.07 to 4 |
| | Fruits, tubers | 10 | 2 to 7000 |
| | Grass | 76 | 10 to 760 |
| Thorium | Cereal grains | $3 \times 10^{-5}$ | $3 \times 10^{-6}$ to 0.0009 |
| | Fruits, tubers | 0.006 | $5 \times 10^{-6}$ to 0.4 |
| | Grass | 0.01 | 0.001 to 0.1 |
| Uranium | Cereal grains | 0.001 | |
| | Fruits, tubers | 0.01 | 0.0008 to 0.14 |
| | Grass | 0.02 | 0.002 to 0.2 |
| Zinc | Cereal grains | $0.90^a$ | 0.29 to 4.8 |
| | Tubers (potato) | $0.90^a$ | 0.12 to 1.1 |
| | Grass | 0.99 | 0.33 to 3.0 |
| Zircon | Not specified | 0.001 | |

$^a$ Source: Baes et al. (1984).

Whicker and Kirchner (1987) provide information on the use of equation 6.61, some typical parameter values, and their justification. Certainly, all these parameters are modified by local site conditions and by the plant and animal species involved.

Equation 6.61 also carries the implicit assumption that water is not limiting plant growth. This may be a reasonable assumption for irrigated land but not for rangeland.

Another term that relates to soil-to-plant transfers is the translocation ratio. This can be expressed as the concentration of a radionuclide in an edible tissue (in becquerels per unit mass) divided by the deposition on the ground (in becquerels per unit area). This kind of expression can have units of $m^2\ kg^{-1}$. In using this parameter, as well as the equilibrium concentration ratio, it is critical to specify whether the root uptake process alone is represented, or whether a combination of both root uptake and aerial transport is involved. This is important for most radionuclides, particularly soon after a contaminating event. It is also more important in arid or semiarid locales where resuspension is dominant. The translocation ratio may also be expressed as the concentration in one tissue divided by that in another of the same plant. This term, as well as many others discussed in this chapter, is described in Thiessen et al. (1999), and measured values are summarized by the International Atomic Energy Agency (IAEA 1994).

## Transport from Vegetation to Soil

Contaminants associated with vegetation are eventually transported, either directly or indirectly, to the soil surface. The primary mechanisms involved depend on whether the contaminants are on the surfaces of the foliage or incorporated within the plant tissues. The dominant mechanism for loss of surficial deposits is called weathering; the main way in which tissue-bound material is lost is senescence (dying back) of the foliage. These mechanisms are discussed below. Radioactive decay is another loss mechanism requiring consideration. Another mechanism that can apply to soluble contaminants in plant tissues is leaching by rainfall, but this is usually a secondary pathway and is not treated further here. Another phenomenon not specifically treated here is the incorporation of long-lived contaminants in woody tissues, which may remain for many years or decades, depending on the persistence of the wood itself.

### Weathering

Weathering refers to the mechanical loss of contaminants from the surfaces of foliage that were likely contaminated by aerial deposition or resuspension. Most commonly, the weathering phenomenon is described as a first-order process; thus, a weathering rate constant ($k_w$, with units of per time) is used in modeling the process. For example, if weathering were the only significant loss process, the concentration of a radionuclide on vegetation, $C_{\text{veg}}$, could be estimated from

$$C_{\text{veg}} = C_{\text{veg}}(0)e^{-k_w t}, \qquad (6.62)$$

where $C_{\text{veg}}(0)$ is the initial concentration. If radioactive decay or other loss mechanisms of significance are identified, the appropriate rate constants may be summed with $k_w$ to form an effective loss rate constant (see equation 6.4). As illustrated in equation 6.13, another utility of the weathering rate constant is to estimate $C_{\text{veg}}$ from aerial deposition. In models that are more complex, it is often necessary to express a rate of transfer of a contaminant from vegetation to the soil surface. For example, the rate of material transfer by weathering, $R_w$, could be described as

$$R_w = k_w C_{veg} B, \tag{6.63}$$

where $R_w$ would have units of Bq m$^{-2}$ day$^{-1}$ and $B$ is the vegetation biomass in kg m$^{-2}$. Typical values used for the weathering half-time, $T_w$, are approximately 14 days for most particulate-form radionuclides and 10 days for semivolatile radionuclides, such as $^{131}$I (IAEA 1994). In some instances, the weathering of surficial deposits is more accurately described by a two- (or more) component exponential function. In such cases, the mathematical expressions illustrated in equations 6.26 and 6.27 may be useful. The weathering half-time for lichens can be much longer than that for herbaceous vegetation. For example, values on the order of 10 years or longer have been measured for $^{137}$Cs in arctic lichens, which are very important in food chains of certain arctic people such as Eskimos in Alaska and Canada and Laplanders in northern Scandinavia and Siberia (Liden and Gustafsson 1967; Hanson and Eberhardt 1969).

## Senescence

Senescence is a natural phenomenon dramatically illustrated by the seasonal color change observed in the leaves of deciduous trees and shrubs, grasses, and other forms of herbaceous vegetation. As plant tissues undergo the transition from a green, actively photosynthesizing tissue to a dying tissue that usually becomes brown, tan, yellow, orange, or red in color, the foliage becomes increasingly susceptible to mechanical breakage, detachment from the host plant, and transport to the soil surface. Once accumulated on the soil surface, such dead foliage becomes organic litter that is subject to further mechanical breakdown, and when moisture and temperature are adequate, the material is subject to microbial decomposition, or decay. The latter process forms humus and releases nutrients and contaminants to the soil surface. The senescence process, therefore, starts the chain of events that leads to a transfer of contaminants associated with plant foliage to the soil surface. A senescence rate constant, $k_s$, may be used in modeling formulations in much the same way as $k_w$; however, because senescence is seasonally dependent and operates to its full extent only during a portion of the year, this needs to be considered in modeling applications. Furthermore, while weathering applies only to surficial deposits, senescence applies to both surficial and tissue-bound material. An example application of $k_s$ in a rate equation for the transfer of contamination from vegetation to soil is

$$R_{sen} = k_s C_{veg} (B - B_{min}), \tag{6.64}$$

where

$R_{sen}$ = senescence rate (Bq m$^{-2}$ day$^{-1}$)
$B_{min}$ = minimum overwinter biomass

This formulation was used in the PATHWAY model (Whicker and Kirchner 1987), but the process was only allowed to operate from August 1 to December

31, the primary period of plant senescence in the western United States. Note from this equation that the rate of senescence is proportional to the amount of biomass available above the overwintering value.

Values of $k_s$ may be estimated from the literature or from observations of biomass changes during the senescence season. The value 0.05 day$^{-1}$ was used in the model PATHWAY for vegetable crops and pasture grasses. A value of 0.15 day$^{-1}$ was assigned to alfalfa. Values used for $B_{min}$ ranged from 0.015 to 0.08 kg m$^{-2}$ depending on the species.

## *Transport within the Soil Column*

Numerous natural and anthropogenic processes alter the vertical distribution of contaminants within the soil column. It is important to understand these processes if they are to be modeled appropriately. In some situations for certain contaminants, the vertical distribution may not affect a dose or risk assessment significantly. In many cases, however, the depth distribution has a profound influence on the transport pathways to plants, animals, and people. For example, a given amount of $^{239}$Pu on the ground surface will have a larger health impact than the same quantity evenly distributed in the upper 15 cm of soil. This is because inhalation of contaminated soil particles is a more effective dose pathway than is external exposure or ingestion of contaminated food crops grown on the soil. Only $^{239}$Pu on or very near the ground surface is available for resuspension and subsequent inhalation (figure 6.11). This section describes two natural physical/chemical processes that produce a downward migration of contaminants into the soil: percolation and leaching. Other natural phenomena, such as frost heaving and biologically

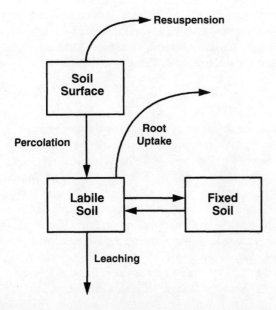

Figure 6.11  Diagram showing how the rate of percolation can affect the dominant pathways of contaminant transport.

induced material movements, are described briefly. This section also discusses the effects of plowing agricultural fields, referred to as tillage, which is a common anthropogenic process that causes vertical mixing of the soil and its associated contaminants.

## Percolation

Percolation refers to the downward migration of particles, and contaminants attached to those particles, through the soil profile. Ordinarily, particles of a few micrometers or less tend to move downward at more rapid rates than do larger particles. This movement is fundamentally influenced by gravity. One type of movement occurs when dry particles fall into open pore spaces, cracks, or vertical tunnels left after the decomposition of roots or the movements of animals that live in the soil. Another type of movement, called eluviation, occurs when particles move down through the soil profile under the influence of infiltrating water. Such particles may be in the colloidal size range. Because most contaminants adhere preferentially to the smaller soil particles, these mechanisms can account for much of the percolation or eluviation process. Colloid-sized particles (those with diameters in the range of $0.001–1\mu$ m) generally account for most of the contaminants in surface soil via percolation and eluviation processes because contaminants adhere preferentially to smaller soil particles.

## Leaching

Leaching involves the movement of dissolved materials down through the soil under the influence of water infiltration. This process is not much different from colloidal transport, except that the dissolved contaminants, if reactive, may undergo almost continuous sorption–desorption with solid surfaces as the water moves through the soil or geological matrix. Some contaminants do not readily sorb to surfaces, so they tend to move at greater effective rates than those materials that tend to sorb strongly. The tendency of a dissolved substance to sorb to a given solid material may be measured by the distribution coefficient, or $K_d$, usually defined as

$$K_d = \frac{C_{\text{solid}}}{C_{\text{liquid}}}, \tag{6.65}$$

where $C_{\text{solid}}$ is the concentration in the solid phase and $C_{\text{liquid}}$ is the concentration in the liquid phase. Ordinarily, the $K_d$ is determined experimentally under specific conditions and is measured after equilibrium has been established between the solid and liquid phases. The definition of equilibrium in this case is the condition where the rates of sorption to and desorption from the solid are equal; therefore, the ratio of concentrations in the solid and liquid phases is constant. In actual practice, some radionuclides, such as radiocesium, reach an apparent equilibrium fairly rapidly but then continue to more fully equilibrate over a longer time period (Stephens et al. 1998). In addition, while some $K_d$ values are measured by sorption experiments, others are determined by desorption, and the resulting $K_d$ values may differ considerably between the two methods. Because of these variables, as well as others, published $K_d$ values should be critically evaluated before they are applied in models (Jenne 1998). $K_d$ values are often used in groundwater models because of their relationship to the retardation factor, which is the ratio of the velocity of groundwater to that of a contaminant (Jenne 1998).

Many estimates of the rate of downward movement of contaminants in soil are empirically determined and do not necessarily distinguish between percolation and leaching mechanisms. One estimate of the rate of movement from the soil surface to deeper soil layers was made possible from indirect inference. Anspaugh et al. (1975) measured a temporal decline in the rate of resuspension, which for the early time period was described as having a 35-day half-time. Because resuspension depends directly on the quantity of contamination at the soil surface, it could be argued that the 35-day half-time was actually a measure of the rate at which the material was transported to deeper soil layers, where resuspension did not operate. In this case, a rough estimate of the percolation rate constant depleting the soil surface might be ln 2/35 days, or 0.02 day$^{-1}$. Estimates of radionuclide transport from the root zone layer in soil to deeper layers, which effectively precludes further root uptake, were formulated as leaching rate constants by Schreckhise (1980). Values in the general range of $10^{-6}$ day$^{-1}$ to $10^{-5}$ day$^{-1}$ were determined. These values are similar to a generic leaching rate constant of approximately $1.6 \times 10^{-5}$ day$^{-1}$ proposed by Coughtrey et al. (1985).

One commonly used formation of a leaching rate constant can be derived from a simplified formulation for dissolved-phase unsaturated transport. The rate constant is derived from the fundamental relationship given previously in equation 6.31 for compartments with steady-state fluid flow. First, we must determine the dissolved-phase concentration of radionuclides in the pore water. Consider a unit volume of soil, $V$, having a moisture content (the fraction of the total unit volume occupied by water) of $\theta$ and a bulk density of $\rho$. The total mass in the system can be written as

$$M = (C_w \theta + C_s \rho) V, \qquad (6.66)$$

where

$C_w$ = dissolved-phase concentration in pore water (mass m$^{-3}$)
$C_s$ = sorbed-phase concentration (mass of radionuclide/mass of soil)

The relationship between the dissolved and sorbed phase is described by the distribution coefficient. Equation 6.66 can be rewritten as

$$M = (C_w \theta + K_d C_w \rho) V. \qquad (6.67)$$

Solving for $C_w$ gives

$$C_w = \frac{M}{(\theta + K_d \rho) V} = \frac{M}{\theta (1 + K_d \rho/\theta) V}. \qquad (6.68)$$

The term $(1 + K_d \rho/\theta)$ is the retardation factor, $R_d$. The fraction of total radionuclide mass in the soil compartment that is in the dissolved phase is given by $1/R_d$. Note that when $K_d = 0, R_d = 1$.

Now consider the mass balance equation for a radionuclide in a soil compartment with one-dimensional fluid flow through the compartment. The mass balance equation for the compartment can be written as

$$R_d \theta A T \frac{dC_w}{dt} = -A C_w v, \tag{6.69}$$

where

$v$ = Darcy flux through the soil compartment, which is equivalent to infiltration (m s$^{-1}$)
$A$ = cross-sectional area of the compartment perpendicular to the direction of fluid flow (m$^2$)
$T$ = thickness of the compartment (m)

The product of $C_w$, $R_d$, $\theta$, $A$, and $T$ is equivalent to the total mass, $M$, of a radionuclide in the compartment given in equation 6.68 where $V = AT$. Substitution of $C_w$ from equation 6.68 into the right-hand side of equation 6.69 gives

$$\frac{dM}{dt} = -A \frac{M}{\theta(1 + K_d \rho/\theta)V} v = -M \left( \frac{v}{\theta R_d T} \right). \tag{6.70}$$

The term on the right side of equation 6.70 in parentheses is equivalent to the leach rate constant as given in Baes and Sharp (1983). Note that the term $v/\theta T$ is equivalent to equation 6.31. The difference between equations 6.70 and 6.31 is that only the radionuclide mass in the dissolved phase is subject to leaching. Therefore, $v/\theta T$ is multiplied by $1/R_d$, the fraction of the total radionuclide mass in the soil compartment that is in the dissolved phase.

The moisture content can be determined with measurements; however, numerous measurements over the course of a year should be taken because moisture content will vary (especially near the surface) depending on season and recent precipitation. In assessment modeling, long-term average moisture contents often are not available, and moisture retention curves are used to estimate moisture content. One such simple scheme is implemented in the RESRAD code (Yu et al. 2001) using a relationship given in Clapp and Hornberger (1978):

$$\theta = \theta_{sat} \left( \frac{v}{K_{sat}} \right)^{1/(2b+3)}, \tag{6.71}$$

where

$R_s$ = saturation ratio (dimensionless)
$K_{sat}$ = saturated hydraulic conductivity (m yr$^{-1}$)
$b$ = soil-specific exponential parameter (dimensionless)
$\theta_{sat}$ = saturated moisture content (i.e., all pore space is filled with water)

Representative values for $\theta_{sat}$, $K_{sat}$, and $b$ are provided in table 6.5. Note that saturated hydraulic conductivity is highest for sand and decreases over two orders of magnitude for clay.

The leach rate constant concept can be applied to multiple soil layers in series and thereby provide soil and pore water concentrations in deeper soil layers and

Table 6.5  Representative values for saturated hydraulic conductivity, saturated moisture content, and soil-specific exponential parameter[a]

| Soil type | Saturated hydraulic conductivity (m yr$^{-1}$) | Saturated moisture content | Soil-specific exponential parameter |
|---|---|---|---|
| Sand | $5.55 \times 10^3$ | 0.395 | 4.05 |
| Loamy sand | $4.93 \times 10^3$ | 0.410 | 4.38 |
| Sandy loam | $1.09 \times 10^3$ | 0.435 | 4.90 |
| Silty loam | $2.27 \times 10^2$ | 0.485 | 5.30 |
| Loam | $2.19 \times 10^2$ | 0.451 | 5.39 |
| Sandy clay loam | $1.99 \times 10^1$ | 0.420 | 7.12 |
| Silty clay loam | $5.36 \times 10^1$ | 0.477 | 7.75 |
| Clay loam | $7.73 \times 10^1$ | 0.476 | 8.52 |
| Sandy clay | $6.84 \times 10^1$ | 0.426 | 10.40 |
| Silty clay | $3.26 \times 10^1$ | 0.492 | 10.40 |
| Clay | $4.05 \times 10^1$ | 0.482 | 11.40 |

[a] Data from Clapp and Hornberger (1978) as presented in Yu et al. (2001).

contaminant fluxes to an aquifer. For example, the differential equations for a two-compartment system where the first compartment lies on top of the second compartment may be written as

$$\frac{dQ_1}{dt} = -(k_1 + \lambda) \, Q_1,$$
$$\frac{dQ_2}{dt} = k_1 Q_1 - (k_2 + \lambda) \, Q_2, \tag{6.72}$$

where

$Q_1$ = radionuclide inventory in the first compartment (Bq)
$Q_2$ = radionuclide inventory in the second compartment (Bq)
$k_1$ and $k_2$ = leach rate constants for the first and second compartments, respectively (yr$^{-1}$)
$\lambda$ = decay rate constant (yr$^{-1}$)

The general equation written in terms of the radionuclide mass in each soil layer for a decay chain series may be written as (Rood 2004):

$$\frac{dQ_{i,j}}{dt} = \left(\frac{v_{i-1}}{\theta_{i-1} R_{d_{i-1,j}} T_{i-1}}\right) Q_{i-1,j} - \left(\frac{v_i}{\theta_i R_{d_{i,j}} T_i}\right) Q_{i,j} - \lambda_j Q_{i,j}$$
$$+ \mathrm{BR}_j \lambda_{j-1} Q_{i,j-1} + S_{i,j} \qquad j \neq 1, i \neq 1,$$

$$\frac{dQ_{i,j}}{dt} = \left(\frac{v_{i-1}}{\theta_{i-1} R_{d_{i-1,j}} T_{i-1}}\right) Q_{i-1,j} - \left(\frac{v_i}{\theta_i R_{d_{i,j}} T_i}\right) \tag{6.73a}$$
$$\times Q_{i,j} - \lambda_j Q_{i,j} + S_{i,j} \qquad j = 1, i \neq 1,$$

$$\frac{dQ_{i,j}}{dt} = -\left(\frac{v_i}{\theta_i R_{d_{i,j}} T_i}\right) Q_{i,j} - \lambda_j Q_{i,j} + BR_j \lambda_{j-1} Q_{i,j-1} + S_{i,j} \qquad j \neq 1, i = 1,$$

$$\frac{dQ_{i,j}}{dt} = -\left(\frac{v_i}{\theta_i R_{d_{i,j}} T_i}\right) Q_{i,j} - \lambda_j Q_{i,j} + S_{i,j} \qquad j = 1, i = 1 \qquad (6.73b)$$

where

$Q_{i,j}$ = number of atoms in soil layer $i$ of radionuclide $j$ in the decay series
$BR_j$ = fraction of radionuclide $j-1$ that decays to radionuclide $j$
$\lambda_j$ = decay rate constant of radionuclide $j$ (s$^{-1}$)
$S_{i,j}$ = source term into soil layer $i$ for radionuclide $j$ (atoms s$^{-1}$)

Note that for a radionuclide decay series, $Q$ must be given in the number of atoms and not activity. If there is no decay series, then $Q$ can be calculated in terms of activity, and the radionuclide ingrowth terms of equation 6.73, $BR_j$, $\lambda_{j-1}$, $Q_{i,j-1}$) are omitted. Equation 6.73 is a relatively simple formulation for vertical transport of radionuclides in the soil. It is intended, however, only as an assessment tool. Unsaturated flow and solute transport is a complex, evolving science and beyond the scope of this chapter.

Values for $K_d$ are extremely variable depending on factors such as soil characteristics, the radionuclide and its chemical form, pH, and redox conditions. Some default values for selected radionuclides and four general soil types, compiled from extensive literature reviews, were developed by Sheppard and Thibault (1990). Selected geometric mean values are provided in table 6.6. In most cases, the uncertainty ranges span roughly two orders of magnitude.

In general, there is a great deal of room for further contributions to the development of formulations and parameter values for processes causing downward movement of contaminants into the soil. In particular, there seems to be a need to account explicitly for percolation, eluviation, and biologically mediated transport, in addition to the more standardized procedure of considering only solution phase transport.

### EXAMPLE 6.7

A proposed uranium mill is planning to operate for 50 years. Atmospheric dispersion modeling of the proposed facility estimated deposition rates of $^{238}$U and $^{234}$U of 2,360 Bq m$^{-2}$ yr$^{-1}$ and 1,180 Bq m$^{-2}$ yr$^{-1}$ for $^{230}$Th, $^{226}$Ra, and $^{210}$Pb at a point outside the facility boundary. What are the radionuclide soil concentrations at the surface and at depth 50 years, 100 years, and 1,000 years from the startup of the mill?

*Solution*: For this problem, equation 6.73 is used to compute radionuclide inventories in surface and subsurface soils. We simplify the problem by considering the principal $^{238}$U decay chain members and ignoring the short-lived isotopes of thorium, protactinium, bismuth, lead, polonium, and radon. We also assume no radon is released to the air, a branching ratio of 1 for all radionuclides in the decay series, and leaching as the only means of transport. The soil column is discretized into four

Table 6.6  Geometric mean $K_d$ values (mLg$^{-1}$) for selected elements in various soil types (Sheppard and Thibault 1990)

| Element | Sand | Loam | Clay | Organic |
| --- | --- | --- | --- | --- |
| Americium | 1,900 | 9,600 | 8,400 | 112,000 |
| Beryllium | 250 | 800 | 1,300 | 3,000 |
| Cadmium | 80 | 40 | 560 | 800 |
| Cerium | 500 | 8,100 | 20,000 | 3,300 |
| Curium | 4,000 | 18,000 | 6,000 | 6,000 |
| Cobalt | 60 | 1,300 | 550 | 1,000 |
| Chromium | 70 | 30 | 1,500 | 270 |
| Cesium | 280 | 4,600 | 1,900 | 270 |
| Iron | 220 | 800 | 165 | 600 |
| Iodine | 1 | 5 | 1 | 25 |
| Manganese | 50 | 750 | 180 | 150 |
| Niobium | 160 | 550 | 900 | 2,000 |
| Neptunium | 5 | 25 | 55 | 1,200 |
| Lead | 270 | 16,000 | 550 | 22,000 |
| Polonium | 150 | 400 | 3,000 | 7,300 |
| Plutonium | 550 | 1,200 | 5,100 | 1,900 |
| Radium | 500 | 36,000 | 9,100 | 2,400 |
| Ruthenium | 55 | 1,000 | 800 | 66,000 |
| Strontium | 15 | 20 | 110 | 150 |
| Technetium | 0.1 | 0.1 | 1 | 1 |
| Thorium | 3,200 | 3,300 | 5,800 | 89,000 |
| Uranium | 35 | 15 | 1,600 | 410 |
| Zinc | 200 | 1,300 | 2,400 | 1,600 |
| Zirconium | 600 | 2,200 | 3,300 | 7,300 |

compartments, each having a thickness of 7.5 cm. The site in question is estimated to have an annual net infiltration rate of 15 cm yr$^{-1}$, and the soil is reported to be primarily sand (bulk density of 1.5 g cm$^{-3}$). Therefore, the moisture retention parameters from table 6.5 for sand are used to determine the moisture content, and the $K_d$ values for sand reported in table 6.6 are used and assumed to be the same for all compartments.

We first calculate the moisture content using the moisture retention data in table 6.5 for sand:

$$\theta = 0.395 \left( \frac{0.15 \text{ m yr}^{-1}}{5{,}550 \text{ m yr}^{-1}} \right)^{1/(2(4.05)+3)} = 0.153.$$

Equation 6.73 can be rewritten in terms of a leaching rate constant, as shown in equation 6.70. For uranium isotopes, the leach rate constant is

$$k = \frac{I}{T \theta R_d} = \frac{15 \text{ cm yr}^{-1}}{(7.5 \text{ cm})(0.153)\left(1 + \frac{35 \text{ mL g}^{-1} \, 1.5 \text{ g cm}^{-3}}{0.153}\right)} = 0.03798 \text{ yr}^{-1}.$$

The differential equations describing $^{238}$U mass balance in soil compartments 1 and 2 are

$$\frac{dQ_{1,1}}{dt} = \left(2,360 \text{ Bq yr}^{-1}\text{m}^{-2}\right) \text{CF}_1 - \left(0.03798 \text{ yr}^{-1} + \lambda_1\right) Q_{1,1},$$

$$\frac{dQ_{2,1}}{dt} = \left(0.03798 \text{ yr}^{-1}\right) Q_{1,1} - \left(0.03798 \text{ yr}^{-1} + \lambda_1\right) Q_{2,1},$$

where

$\lambda_1$ = decay rate constant for $^{238}$U (s$^{-1}$)
$Q$ = number of atoms in the compartment per square meter
$\text{CF}_1$ = conversion factor from activity to number of atoms and given by $1/\lambda_1$

For $^{234}$U, the mass balance equations in soil layers 1 and 2 are

$$\frac{dQ_{1,2}}{dt} = \left(2,360 \text{ Bq yr}^{-1}\text{m}^{-2}\right) \text{CF}_2 - \left(0.03798 \text{ yr}^{-1} + \lambda_2\right) Q_{1,2} + \lambda_1 Q_{1,1},$$

$$\frac{dQ_{2,2}}{dt} = \left(0.03798 \text{ yr}^{-1}\right) Q_{1,2} + \lambda_1 Q_{2,1} - \left(0.03798 \text{ yr}^{-1} + \lambda_2\right) Q_{1,2}.$$

A similar set of equations is used to describe the mass balance of the remaining radionuclides in all four soil compartments. Equation 6.73 is best solved numerically using published algorithms or simulation software discussed earlier in this chapter (although analytical solutions do exist, e.g., Birchall 1986). In this example, the system of coupled differential equations was solved using a fourth-order Runge-Kutta routine adapted from Press et al. (1992).

Figure 6.12 shows the depth distribution of key radionuclides at 50, 100, and 1,000 years. Radionuclide inventories were converted to concentrations (in Bq g$^{-1}$) by converting the radionuclide inventory in atoms to becquerels and dividing by the mass of soil per square meter in the soil compartment [(0.075 m)(1.5 × 10$^6$ g m$^{-3}$) = 1.125 × 10$^5$ g m$^{-2}$]. The concentration profiles at 50 years show a sharp decrease in concentration with depth for $^{230}$Th, $^{226}$Ra, and $^{210}$Pb, but the profile is more gradual for $^{238}$U because uranium is more readily leached, as indicted by its lower $K_d$ value. At 100 years, leaching results in substantial depletion of uranium from the soil column, while the profile for the other radionuclides has not changed substantially from that at 50 years. At 1,000 years, most of the uranium has been leached from the soil column, and $^{226}$Ra and $^{210}$Pb exhibit depletion in the surface compartment. Thorium-230 continues to show high concentrations in the surface compartment relative to concentrations at depth. Lead-210 appears to have nearly the same activity concentration profile as $^{226}$Ra, despite the fact that the $K_d$ value for lead is about half that of radium. While $^{210}$Pb is more mobile than $^{226}$Ra (because of its lower $K_d$ value), its half-life (22.3 yr) is relatively short compared to $^{226}$Ra (1,600 yr), and it therefore decays before it can travel a significant distance from its parent.

The results of this exercise are important for estimating doses from resuspension, external exposure, and the food chain because all three exposure pathways depend on the radionuclide distribution in soils.

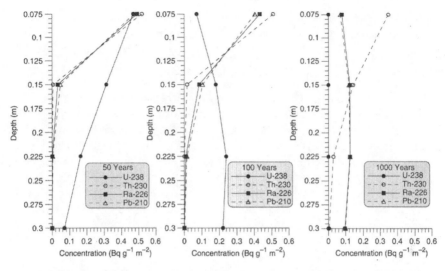

Figure 6.12 Soil profiles at 50, 100, and 1,000 years from atmospheric deposition of the $^{238}$U decay series from a hypothetical uranium milling operation.

## Other Natural Processes Producing Vertical Migration in Soil

Several other natural processes can produce vertical movements of contaminants in soil. These include the physical processes of soil cracking (Whicker and Ibrahim 2006) and frost heaving, and biological processes including burrowing of insects and small mammals, vertical movements of invertebrates such as earthworms, and root penetration and decomposition. Although there is considerable evidence for such phenomena, the relative magnitude of these processes is not well quantified. For example, during dry periods at Rocky Flats in Colorado, the clay-dominated soils will form cracks, which may exceed 1 cm in width at the surface and extend to depths of up to 20 cm or more, depending on the duration of drought. Particles can fall into such cracks, carrying contaminants with them (Whicker and Ibrahim 2006). Higley (1994) demonstrated this process in the laboratory. Also, the freezing of damp soils can produce subsurface ice crystals. As the ice crystals grow, they can push subsurface particles to the surface, which results in vertical mixing over the upper centimeter or so of the soil profile. Earthworms burrow up and down over the upper 40 cm or so of the soil profile, all the while ingesting soil particles and depositing fecal material. This process transports $^{239}$Pu both downward and upward in the soil (Litaor et al. 1994).

In the Rocky Flats environment, the burrowing activities of pocket gophers have also been found to result in significant mixing of $^{239}$Pu in the upper 15 cm of soil (Winsor and Whicker 1980). It is not difficult to imagine root penetration moving soil particles or root uptake followed by senescence and decomposition as causing some

### Tillage

The agricultural process of tillage or plowing is nearly as old as agriculture itself. It is well known that tillage can mix a recent surface deposit throughout the vertical extent of the plow layer. In fact, this technique was used to reduce radionuclide concentrations in the surface layer both at Chernobyl and near Rocky Flats. Radionuclides in the surface layer are mixed with clean soil brought to the surface by the plow. Tilling therefore dilutes radionuclides in the surface layer with clean soil, resulting in a reduced source term for resuspension. This process is of particular value for radionuclides such as $^{239}$Pu, which have a high inhalation dose factor but a low ingestion dose factor. In constructing transport models for agricultural ecosystems, tillage events can be scheduled to occur on certain dates and the plow depth can be specified. In the PATHWAY model, the plow layer was assumed to be 25 cm, and the event was assumed to mix radionuclides evenly within this layer (Whicker and Kirchner 1987). Other assumptions could be more appropriate in different situations.

### *Transport from Vegetation to Animals*

Ingestion of vegetation is a primary pathway for radionuclide transport to herbivorous animals. The consumption of fresh pasture or rangeland vegetation is often the overwhelmingly dominant pathway for the transfer of short-lived radionuclides, such as $^{131}$I, to the thyroid of herbivores and to dairy products, such as milk and cottage cheese. When exposed vegetation is consumed directly, surficial and tissue-bound contaminants enter the gut, and depending on the chemical/physical nature of the material, some fraction will be absorbed into the blood and transferred to various tissues. The rate of ingestion to a specific animal, $R_{\text{ing}}$, may be formulated as

$$R_{\text{ing}} = a \sum_i r_i C_{\text{veg}_i}, \tag{6.74}$$

where

$a$ = fraction of an ingested quantity that is transferred to a particular organ or tissue
$r_i$ = rate of ingestion of species or item $i$
$C_{\text{veg}_i}$ = average concentration of the radionuclide in species or item $i$

The units for $R_{\text{ing}}$ are quantity (e.g., becquerels) per time. This equation allows for a complex diet made up of more than one item. For animals that graze on natural rangeland, there may not be sufficient information to consider each dietary item separately, so it may be necessary to focus on the predominant forage species or to construct a weighted average (see example 6.3). The necessary information is often available to construct diets for penned or fed animals. For example, figure 6.13 shows a typical dietary regime for a dairy cow in the western United States.

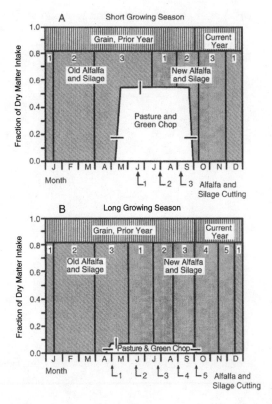

Figure 6.13 Typical dairy cow dietary regime for the western United States.

Equation 6.74 implies that the assimilation fraction, $a$, is independent of the type of feed. This may be the case in some instances, but not in others. For example, it has been shown that $^{137}$Cs incorporated in grain is assimilated more readily than that associated with foliage (Johnson et al. 1968). Values used for $a$ are widely scattered in the literature, but some general information is tabulated by the International Commission on Radiological Protection (ICRP 1960). A brief review for assimilation fractions in mammals by Kitchings et al. (1976) yielded ranges of 0.25–0.85 for cesium, 0.5–1.0 for strontium, 0.03–0.4 for iodine, 0.05–0.25 for cobalt, and 1.0 for hydrogen. The estimation of ingestion rates, $r_i$, can be approached using a body mass and metabolic demand approach (Whicker and Schultz 1982), or it can be estimated from direct measurements under controlled conditions. The latter may need modification if inferences to free-ranging animals are attempted from experiments with captive animals. A review of dry matter intake rates for food-producing animals (referred to later as $F_v$) is provided by the International Atomic Energy Agency (IAEA 1994), and table 6.7 summarizes selected values.

Equation 6.74 can be used to estimate $R$ in equation 6.8, and with knowledge of the biological loss rate constant, the time-dependent quantity of a radionuclide in animal tissue can be calculated. example 6.3 above determines the equilibrium

Table 6.7 Dry matter and water intake rates of domestic animals (IAEA 1994)

|  | Dry matter | | |
| --- | --- | --- | --- |
| Domestic animals | Expected (kg day$^{-1}$) | Range (kg day$^{-1}$) | Water (L day$^{-1}$) |
| Dairy cows | 16.1 | 10–25 | 50–100 |
| Beef cattle (500 kg) | 7.2 | 5–10 | 20–60 |
| Calves (160 kg) | 1.9 | 1.5–3.5 | 5–15 |
| Dairy goats | 1.3 | 1.0–3.5 | 5–10 |
| Dairy sheep | 1.3 | 1.0–2.5 | 5–8 |
| Lambs (50 kg) | 1.1 | 0.5–2.0 | 3–5 |
| Pigs (110 kg) | 2.4 | 2.0–3.0 | 6–10 |
| Laying hens | 0.1 | 0.07–0.15 | 0.1–0.3 |
| Chickens | 0.07 | 0.05–0.15 | 0.1–0.3 |

concentration of $^{137}$Cs in beef muscle tissue. For most practical applications, the input–loss modeling approach just illustrated is more cumbersome than the transfer coefficient approach, covered under "Transfers to Animal-Derived Human Food Products," below. The transfer coefficient approach uses empirically measured factors that inherently embody intake and loss processes, but these processes do not require explicit treatment. The transfer coefficient, however, only estimates equilibrium conditions. The concepts and models illustrated in this chapter are required for the construction of dynamic models, which do not need to assume equilibrium conditions (see example 6.4, where an acute deposition event leading to a time-dependent ingestion rate of $^{131}$I requires the use of a dynamic model).

## Transport from Soil to Animals

The transfer to animals of contaminants bound to small particles of soil or organic detritus within the upper few millimeters of the soil profile can occur by two primary processes: ingestion and inhalation. These processes and their formulations are described below.

### Ingestion

Animals ingest soil, both advertently and inadvertently. They directly ingest it from the soil surface, they lick their hair or skin, and they ingest surficial deposits on vegetation and particles suspended in water. Typical values for $F_s$, the soil ingestion rate, are around 500, 100, and 10 g day$^{-1}$ for cattle, sheep, and poultry, respectively (see literature cited in Whicker and Kirchner 1987). People, particularly young children, also intentionally or inadvertently ingest soil, and estimates range from 30 mg day$^{-1}$ for adults to 100 mg day$^{-1}$ for children (LaGoy 1987). This is sometimes called pica. These values are often derived from measuring titanium in soil and fecal material. Titanium is essentially insoluble and serves as a good tracer for soil particles. The rate of soil surface contaminant ingestion, $R_{soil}$, may be estimated from

$$R_{soil} = aF_sC_{ss}, \qquad (6.75)$$

where $a$ is the fraction of the ingested quantity that goes to a particular organ or tissue, and $C_{ss}$ is the radionuclide concentration on the soil surface. This surface layer is normally considered to be 1–3 mm thick (Pinder et al. 1979). In cases where the soil ingestion pathway might be significant, equations 6.74 and 6.75 may be summed to describe the total intake from vegetation and soil. In most circumstances, intake via soil is considerably less than that from vegetation. This question is addressed by National Council on Radiation Protection and Measurements (NCRP 1993).

## Inhalation

In the case of relatively insoluble radionuclides with high inhalation dose coefficients, such as $^{239}$Pu, the inhalation of contaminated resuspended soil particles may be a more important exposure pathway than ingestion. This is particularly true if most of the soil contamination is at the surface and if conditions are favorable for resuspension. As described in equations 6.54, 6.55, and 6.56, the air concentration resulting from resuspension, $C_{air}$, may be estimated either from mass loading or from a resuspension factor. The air concentration resulting from suspension can be estimated from the formulations shown in equation 6.50. If $C_{air}$ is reasonably well known, the rate of lung deposition via inhalation, $R_{inh}$, in amount per time, is

$$R_{inh} = C_{air}TR_bf_d, \qquad (6.76)$$

where

$T$ = tidal volume of the lung (m$^3$)
$R_b$ = breathing rate in breaths per unit time
$f_d$ = fraction of the inhaled material that is deposited in a specific region of the lung

Some material deposited in the upper respiratory passages may be coughed up and swallowed; material deposited in deeper lung tissue may be absorbed or otherwise cleared at some rate depending on particle size, solubility, and so forth. The location of deposition, as well as the relative amount, is particle size dependent. The quantity of material deposited in the lung that is absorbed into the body and the fate of it once absorbed is covered in detail in chapter 9.

## *Transfers to Animal-Derived Human Food Products*

As mentioned above, the equilibrium concentrations of radionuclides in common food products, such as meat, milk, and eggs, may be estimated from empirically derived parameters called transfer coefficients. These parameters are determined from experiments in which a radionuclide is fed daily to an animal and the product

Table 6.8 Expected values for transfer coefficients (day L$^{-1}$ for milk products, day kg$^{-1}$ for meat) for selected elements in various animal food products (IAEA 1994)[a]

| Element | Cow milk | Goat milk | Beef | Lamb | Pork | Poultry | Eggs |
|---|---|---|---|---|---|---|---|
| Technetium | $2 \times 10^{-5}$ | $1 \times 10^{-2}$ | $1 \times 10^{-5}$ | | $2 \times 10^{-4}$ | $3 \times 10^{-2}$ | 3 |
| Manganese | $3 \times 10^{-5}$ | | $5 \times 10^{-4}$ | $6 \times 10^{-3}$ | $4 \times 10^{-3}$ | $5 \times 10^{-2}$ | $6 \times 10^{-2}$ |
| Iron | $3 \times 10^{-5}$ | | $2 \times 10^{-2}$ | $7 \times 10^{-2}$ | $3 \times 10^{-2}$ | 1 | 1 |
| Strontium | $3 \times 10^{-3}$ | $3 \times 10^{-2}$ | $8 \times 10^{-3}$ | $3 \times 10^{-1}$ | $4 \times 10^{-2}$ | $8 \times 10^{-2}$ | $2 \times 10^{-1}$ |
| Iodine | $1 \times 10^{-2}$ | $4 \times 10^{-1}$ | $4 \times 10^{-2}$ | $3 \times 10^{-2}$ | $3 \times 10^{-3}$ | $1 \times 10^{-2}$ | 3 |
| Cesium | $8 \times 10^{-3}$ | $1 \times 10^{-1}$ | $5 \times 10^{-2}$ | $5 \times 10^{-1}$ | $2 \times 10^{-1}$ | 10 | $4 \times 10^{-1}$ |
| Barium | $5 \times 10^{-4}$ | $5 \times 10^{-3}$ | $2 \times 10^{-4}$ | | | $9 \times 10^{-3}$ | $9 \times 10^{-1}$ |
| Cerium | $3 \times 10^{-5}$ | | $2 \times 10^{-5}$ | $2 \times 10^{-4}$ | $1 \times 10^{-4}$ | $2 \times 10^{-3}$ | $9 \times 10^{-5}$ |
| Uranium | $4 \times 10^{-4}$ | | $3 \times 10^{-4}$ | | $6 \times 10^{-2}$ | 1 | 1 |
| Plutonium | $1 \times 10^{-6}$ | $1 \times 10^{-5}$ | $1 \times 10^{-5}$ | $3 \times 10^{-3}$ | $8 \times 10^{-5}$ | $3 \times 10^{-3}$ | $5 \times 10^{-4}$ |
| Americium | $2 \times 10^{-6}$ | $1 \times 10^{-5}$ | $4 \times 10^{-5}$ | $4 \times 10^{-3}$ | $2 \times 10^{-4}$ | $6 \times 10^{-3}$ | $4 \times 10^{-3}$ |
| Cobalt | $7 \times 10^{-5}$ | | $1 \times 10^{-4}$ | $6 \times 10^{-2}$ | $1 \times 10^{-3}$ | 2 | $1 \times 10^{-1}$ |
| Zirconium | $6 \times 10^{-7}$ | $6 \times 10^{-6}$ | $1 \times 10^{-6}$ | | | $6 \times 10^{-5}$ | $2 \times 10^{-4}$ |

[a] See original source for other data and ranges of values.

of interest is measured after it has reached equilibrium with the daily intake. For example, the transfer coefficient, TC, is defined as

$$\text{TC} = \frac{C_{\text{prod}}(\text{eq})}{R}, \tag{6.77}$$

where $C_{\text{prod}}(\text{eq})$ is the measured concentration (Bq kg$^{-1}$) in the product of interest at equilibrium and $R$ is the radionuclide ingestion rate (Bq day$^{-1}$), in this case the rate of entry into the mouth. Repeated measurements of $C_{\text{prod}}$ may be needed to assure that equilibrium has been reached. The time required for equilibrium varies depending on the radionuclide and its kinetic behavior in the tissues of interest. The transfer coefficient can be related fundamentally (and mathematically) to the assimilation fraction going to the product of interest, as well as to the loss rate of material from the product. Obviously, the value of the transfer coefficient varies greatly with the radionuclide and its physical/chemical form, the animal product, and other factors. For example, some default values for TC$_{\text{milk}}$ are $8 \times 10^{-3}$, $1 \times 10^{-2}$, and $1 \times 10^{-7}$ day l$^{-1}$ for cesium, iodine, and plutonium, respectively (NCRP 1989). Some default TC$_{\text{beef}}$ values are $3 \times 10^{-2}$, $1 \times 10^{-2}$, and $1 \times 10^{-5}$ day kg$^{-1}$ for the same three elements, respectively (NCRP 1989). table 6.8 provides a more complete review of transfer coefficient values based on a review by the International Atomic Energy Agency (IAEA 1994).

Applying the transfer coefficient to predict radionuclide concentrations in food products is quite simple. For example, the concentration of $^{137}$Cs in milk could be estimated from

$$C_{\text{milk}} = C_{\text{diet}} F_v T C_{\text{milk}}, \tag{6.78}$$

where $C_{\text{diet}}$ is the average concentration of $^{137}$Cs in the diet, and $F_v$ is the total ingestion rate of dietary items in kg day$^{-1}$. The generic value of $8 \times 10^{-3}$ day l$^{-1}$ for TC$_{\text{milk}}$ could be used. Equation 6.78 could also be used to calculate the concentration in beef, where $C_{\text{diet}}$ and $F_v$ would be the same, but TC$_{\text{beef}}$ would be used instead of TC$_{\text{milk}}$. The generic value of $3 \times 10^{-2}$ day kg$^{-1}$ could be used for TC$_{\text{beef}}$.

## Ingestion Pathways to Humans

In principle, formulation of the ingestion pathway to human beings is the same as that for animals, and the use of equation 6.74 or modifications of it might be appropriate. However, additional questions and complications normally must be dealt with in the case of humans. For example, the intake rates of the various dietary items depend on the season of the year, geographic location, lifestyle, age, and sex. Dietary habits have changed over the last several decades, so dose reconstructions also involve food habit reconstructions. The human diet is incredibly complex, at least in the more developed parts of the world, with literally hundreds of different items possible. A major complication is determining the geographic source of the food. Foods may be grown and consumed in a local area of interest, but increasingly, foods consumed in one area may have been produced in many other, sometimes distant, areas. Many food items are prepared and stored before consumption. These practices can alter the concentrations of shorter lived and/or leachable radionuclides in the food items.

Although various summary statistics are available (e.g., EPA 1989), a major challenge in estimating intake and dose to people from ingestion pathways is in getting appropriate and accurate information on food habits, food distribution patterns, storage times, and preparation methods. Because intake can vary dramatically with factors such as age, sex, and lifestyle, the information needs to be specific to each category of individual. We can attempt to deal with some of the complexity of category-specific ingestion with equations of the general form

$$R_{\text{ing}} = \sum_i a_i r_i C_{\text{food}_i} f_{c_i} f_{r_i} e^{-\lambda t_i}, \qquad (6.79)$$

where

$R_{\text{ing}}$ = rate of radionuclide ingestion to a specific organ or tissue (Bq day$^{-1}$)
$a_i$ = fraction of ingested radionuclide in item $i$ that goes to a specific organ or tissue
$r_i$ = rate of ingestion of food (or nonfood) item $i$ (kg day$^{-1}$)
$C_{\text{food}_i}$ = average concentration of radionuclide in food (and nonfood) item $i$ (Bq kg$^{-1}$)
$f_{c_i}$ = fraction of ingested food item $i$ that was exposed to radionuclide contamination
$f_{r_i}$ = fraction of original radionuclide concentration remaining in food item $i$ after preparation
$\lambda$ = physical decay constant of radionuclide (day$^{-1}$)
$t_i$ = storage time of food (or nonfood) item $i$ prior to consumption (day)

The rate of radionuclide ingestion is likely to be a time-dependent quantity because of temporal changes in $C_{\text{food}_i}$ and because other factors in equation 6.79

may vary. The committed effective dose (in sieverts) to an organ, tissue, or effective whole body from $t$ days of ingestion would be

$$D = F_d \int_0^t R_{\text{ing}}(t) dt, \tag{6.80}$$

where $F_d$ is the dose coefficient in Sv Bq$^{-1}$, and $R_{\text{ing}}(t)$ represents the time-dependent ingestion rate. The dose coefficient for ingestion represents the 50-year committed dose per becquerel ingested, and it is radionuclide and organ specific. The most commonly used tables are provided in units of Sv Bq$^{-1}$ (EPA 1988). The concepts embodied in the dose coefficients are treated in detail in chapter 9.

The list of foods to be included in equation 6.79 should also consider nonfood items, such as water and other beverages, and especially in the case of children, soil (EPA 1989). As long as the respective concentrations and ingestion rates are known or estimated, they can be added to the terms to be summed for total ingestion. Only in a few instances is water likely to be a significant pathway of intake. Well water and treated water are seldom contaminated significantly with anthropogenic radionuclides, although there are cases where well water contains high concentrations of naturally occurring radionuclides such as $^{226}$Ra or $^{238}$U. The importance of soil ingestion varies with circumstances (EPA 1989; NCRP 1993).

## Dynamic Multicompartment Models: Putting It All Together

The individual transport processes just reviewed can be used in single-compartment models, as illustrated with a few selected examples given earlier in this chapter. The same transport processes, and corresponding rate equations, can be used in formulating dynamic (time-dependent) multicompartment models. Dynamic multicompartment models are usually more accurate representations of the actual complexity in ecosystems, in the sense that they account for functional linkages and material feedback between individual compartments, as well as changing rates of transport throughout the system. When compartments are linked, or coupled, any temporal change in one will affect the rates of flow into other compartments. Natural and agricultural ecosystems are linked systems with regard to energy, nutrient, and contaminant flows. Of course, humans depend on these ecosystems for food and other needs, so people can be conceptualized as being coupled to them in both a real and a modeling sense. The basic mathematical scheme typically used to describe multicompartment systems starts by writing, for each environmental compartment, $h$, a differential equation of the general form

$$\frac{dq_h}{dt} = \sum_{i=1}^n R_{\text{in}_i} - \sum_{j=1}^m R_{\text{out}_j}, \tag{6.81}$$

where

$q_h$ = quantity of radionuclide in compartment $h$ (Bq m$^{-2}$)
$R_{\text{in}_i}$ = the $i$th rate of inflow to compartment $h$
$R_{\text{out}_j}$ = the $j$th rate of outflow from compartment $h$

In a coupled system, some of the $R_{in_i}$ values will be functions of the quantities ($q$) in other compartments, and some of the $R_{out_j}$ values become inputs to other compartments. A system of $h$ compartments will be described by a system of $h$ differential equations. Because these equations will have mutual dependencies, they are coupled and must be solved simultaneously as a set of equations.

Because a differential equation gives only the net rate of change in a compartment, it must be integrated to allow computation of the actual content in the compartment at any point in time. For two-, three-, or sometimes four-compartment systems, the differential equations can sometimes be solved by analytical means. A common way of doing this is to convert the differential equations into algebraic forms by Laplace transformation (Whicker and Schultz 1982). Then, the transformed equations sometimes can be solved by regular algebraic manipulation or by matrix operations and the use of Cramer's Rule (Whicker and Schultz 1982). The success of this process depends on the form of the various rate equations and the general model structure. Many systems of equations do not have explicit analytical solutions. For this reason, as well as the fact that systems of five or more highly linked compartments typically are unlikely to have analytical solutions, it is usually necessary and practical to use numerical methods to solve the sets of differential equations describing the model. For example, PATHWAY and similar dynamic system models rely on numerical algorithms, such as the fourth-order Runge-Kutta, Adams-Bashforth, and so on, implemented in computer language for their solution (Whicker and Kirchner 1987; Carnahan et al. 1969). Contemporary model developers can be assisted considerably by the availability of commercial software (e.g., UCALC,[3] STELLA,[4] MatLab,[5] Mathematica[6]) designed to numerically solve sets of differential equations, or can develop their own code using published algorithms (Press et al. 1992).

Models written as sets of differential equations that employ numerical solution techniques offer several distinct advantages over those that can be written as explicit algebraic equations. For example, the former models can handle complex inputs, such as a series of acute "spikes," or data files containing virtually any time-dependent input pattern imaginable, while the latter are unlikely to have such a capability. In addition, the differential equation-based model can easily handle parameters that are time dependent and can handle discrete changes in the various input values, such as scheduled harvest dates. Because of the enormous flexibility of differential equation-based models, they can be more testable than algebraic models due to the complex of variables that affect model predictions. Such variables, if known, can be relatively simple to account for in the differential equation-based models.

The solution of the set of equations depicted in equation 6.81 provides estimates of the time-dependent inventories in the various system compartments. This involves being able to track the amounts of contamination in all parts of the system at any point in time. The computed contaminant inventories (usually in Bq m$^{-2}$) may then be converted to concentrations through division by the appropriate compartment masses per unit area. Concentrations of radionuclides in animal forage can then be used to estimate concentrations in human food products such as meat, milk, eggs, and so forth, using equation 6.77 or its variants. Contaminant concentrations thus estimated for vegetables and animal-derived foods can then be used to compute time-dependent intake rates to humans. Such intake rates can be modified as appropriate using the concepts in equation 6.78. Resulting functions can be integrated (usually by numerical techniques) and converted to estimates of dose, as shown in equation 6.80.

A relatively simplified conceptual version of the PATHWAY model (Whicker and Kirchner 1987), called "PATH," is shown in figure 6.14. The corresponding differential equations describing PATH are shown below:

$$\frac{dQSS}{dt} = \text{Dep} \cdot e^{-\alpha B} + (\text{KW} \cdot \text{QVS}) + [\text{KS} \cdot (\text{QVS} + \text{QVI})]$$
$$- [(\text{RF} \cdot V) + \text{KR} + \text{KP} + \text{KPER}] \cdot \text{QSS}$$

$$\frac{dQVS}{dt} = \text{Dep} \cdot (1 - e^{-\alpha B}) + [(\text{RF} \cdot V + \text{KR}) \cdot \text{QSS}]$$
$$- [(\text{KW} + \text{KP} + \text{KA}B + \text{KS}) \cdot \text{QVS}]$$

$$\frac{dQVI}{dt} = \text{KA}B \cdot \text{QVS} + \left[\text{CR} \cdot DB \cdot \frac{\text{QLS}}{(\text{XR} \cdot \text{PS})}\right] - [(\text{KP} + \text{KS}) \cdot \text{QVI}]$$

$$\frac{dQLS}{dt} = \text{KPER} \cdot \text{QSS} - \left[\text{CR} \cdot DB \cdot \frac{\text{QLS}}{(\text{XR} \cdot \text{PS})}\right] - [(\text{KP} + \text{KL}) \cdot \text{QLS}],$$

where QSS, QVS, QVI, and QLS are the "state variables" (figure 6.14), Dep is the deposition quantity (Bq m$^{-2}$), and other parameter definitions, units, and numerical values (for illustration) are listed in table 6.9. The above differential equations were coded and solved using the UCALC software. In this model structure, the soil and vegetation compartments are the only "state variables" computed from differential equations. The animal components are assumed to be in equilibrium with radionuclide concentrations in their food. This is generally warranted in terrestrial ecosystems because animal or human components store less than that 1% of the total radionuclide inventory in the system. This does not mean that the *concentrations* in animal or human tissues and the resulting doses are necessarily trivial.

The inventories in the soil and vegetation compartments can be used to drive straightforward algebraic computations for animals and humans. The soil is broken into the surface layer and deeper layers because these two layers are involved in quite different transport processes. Similarly, the vegetation is broken into two compartments because of their respective connections to different processes (see figure 6.14).

One possible implementation of the PATH model is illustrated for a hypothetical scenario involving an acute deposition of 1.0 Bq $^{134}$Cs per m$^2$ of land surface. The question is what is the estimated effective dose from $^{134}$Cs to a reference person (ICRP 1975) who is consuming each day, for 1,000 days, meat, milk, and leafy vegetables produced on the contaminated land. The person is also assumed to be inhaling the material being resuspended from the land surface. For this illustration, external exposure from $^{134}$Cs photons will be neglected, although it could be calculated from the amount in soil without very much difficulty. A corresponding set of parameter values chosen for illustration of this problem is provided in table 6.9.

# PATH: A SIMPLIFIED VERSION OF PATHWAY

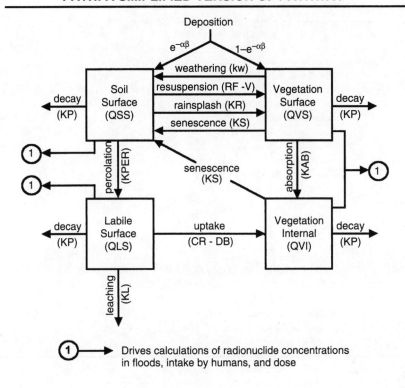

## PATHWAY: DIFFERENTIAL EQUATIONS FOR STATE VARIABLES

Soil Surface:

$$D{:}QSS = DEP \cdot EXP(-ALPHA \cdot B) + KW \cdot QVS + KS \cdot (QVS + QVI)$$
$$-(RF \cdot V + KR + KP + KPER) \cdot QSS$$

Vegetation Surface:

$$D{:}QVS = DEP \cdot (1 - EXP(-ALPHA \cdot B)) + (RF \cdot V + KR) \cdot QSS$$
$$-(KW + KP + KAB + KS) \cdot QVS$$

Vegetation Internal:

$$D{:}QVI = KAB \cdot QVS + CR \cdot DB \cdot QLS/(XR \cdot PS) - (KP + KS) \cdot QVI$$

Labile Soil:

$$D{:}QLS = KPER \cdot QSS - CR \cdot DB \cdot QLS/(XR \cdot PS) - (KP + KL) \cdot QLS$$

Figure 6.14 Conceptual diagram of a simplified version of PATHWAY and the set of differential equations that describe the system.

Table 6.9  Parameter values chosen to illustrate the model in figure 6.14

| Parameter | Description (symbol used in chapter 6) | Value (units) |
|---|---|---|
| ALPHA | Foliar interception constant ($\alpha$) | 0.39 (m² kg⁻¹) |
| B | Plant biomass ($B$) | 0.3 (dry kg m⁻²) |
| DB | Time derivative of plant biomass ($dB/dt$) | 0.036 (dry kg m⁻² d⁻¹) |
| KW | Weathering rate constant ($k_w$) | 0.0495 (d⁻¹) |
| KS | Senescence rate constant ($k_s$) | 0.035 (d⁻¹) |
| RF | Resuspension factor ($F_r$) | 10⁻⁵ (m⁻¹) |
| V | Deposition velocity ($V_d$) | 173 (m d⁻¹) |
| KR | Rainsplash rate constant ($k_{rsp}$) | 8.6 × 10⁻⁴ (d⁻¹) |
| KPER | Percolation rate constant ($k_{per}$) | 0.0198 (d⁻¹) |
| KP | Physical decay constant ($\lambda$) | 9.4 × 10⁻⁴ (d⁻¹) |
| KAB | Absorption rate constant ($k_{ab}$) | 5.3 × 10⁻³ (d⁻¹) |
| CR | Plant/soil concentration ratio (CR) | 0.01 |
| XR | Depth of rooting zone | 0.25 (m) |
| PS | Dry soil bulk density ($\rho$) | 1.46 × 10³ (kg m⁻³) |
| KL | Leaching rate constant ($k_{leach}$) | 6.6 × 10⁻⁶ (d⁻¹) |
| CRV | Cow's rate of vegetation intake ($F_v$) | 17 (dry kg d⁻¹) |
| HRV | Human's rate of leafy vegetable intake ($r_i$) | 0.01 (dry kg d⁻¹) |
| HRMK | Human's rate of milk intake ($r_i$) | 0.36 (l d⁻¹) |
| HRMT | Human's rate of meat intake ($r_i$) | 0.277 (kg d⁻¹) |
| TCMK | Transfer coefficient to milk (TC$_{milk}$) | 0.008 (d L⁻¹) |
| TCMT | Transfer coefficient to meat (TC$_{meat}$) | 0.05 (d kg⁻¹) |
| BR | Breathing rate, human ($TR_b$) | 20 (m³ d⁻¹) |
| FDING | Dose conversion factor for ingestion ($F_d$) | 1.98 × 10⁻⁸ (Sv Bq⁻¹) |
| FDINH | Dose conversion factor for inhalation ($F_d$) | 1.25 × 10⁻⁸ (Sv Bq⁻¹) |

Many of the parameter values are taken from the PATHWAY code (Whicker and Kirchner 1987). These parameter values were used in the set of four differential equations shown to represent the conceptual model in figure 6.14, and a numerical solution was employed to solve the equations on daily time steps. The quantities of $^{134}$Cs in each of the four compartments are shown in figure 6.15 as a function of time following an acute deposition event.

Figure 6.15 shows that, in accordance with the structural model and parameter values for $\alpha$ and $B$, the initial values of the soil surface (QSS) and vegetation surface (QVS) are 0.89 and 0.11 Bq m⁻², respectively, which can be calculated from equations 6.43 and 6.44. The other compartments (boxes) initially contain no $^{134}$Cs. The inventory of $^{134}$Cs in the surface soil compartment declines exponentially for nearly 500 days due to various losses, the main loss mechanism being percolation into the labile soil pool, which occurs at the rate of about 2% per day (KPER = 0.0198 day⁻¹). After about 100 days, the dominant reservoir for $^{134}$Cs becomes the labile soil, the layer from which root uptake can occur. After about 700 days, the labile soil (QLS) and the surface soil (QSS) achieve equilibrium with one another because the rates of transport from one to the other become essentially

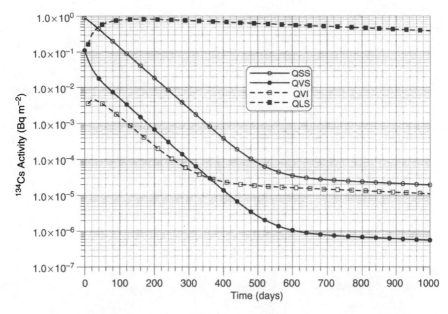

Figure 6.15 Simulated inventories of $^{134}$Cs in the major compartments of PATH following the acute deposition of 1 Bq m$^{-2}$ versus time (days) after deposition. The compartments are the soil surface (QSS), the labile soil (QLS), the vegetation surface (QVS), and the vegetation internal tissues (QVI).

balanced. A slow and nearly equal rate of decline occurs for both QLS and QSS due to physical decay of the $^{134}$Cs, which has a half-life of 2.05 years, and a smaller loss occurs from the system due to leaching.

The vegetation surface compartment (QVS) tends to follow the trend in the QSS compartment. This is because QSS is its only source of $^{134}$Cs once the initial deposition has been lost through weathering and senescence. The vegetation internal compartment (QVI) is receiving $^{134}$Cs from two sources, with absorption from QVS predominating for about the first 300 days, but with root uptake from QLS dominating after about 450 days. The trend pattern for QVI illustrates this rather clearly. To simulate ingestion of vegetation by cattle or by humans, the concentration of $^{134}$Cs in vegetable matter (CVEG, in Bq kg$^{-1}$) was computed on daily time steps by UCALC from the quantity (QVS + QVI) / $B$. The assumption was made that leafy vegetables would contain $^{134}$Cs concentrations similar to that of cattle forage.

The next step was to estimate the $^{134}$Cs intake rates by the reference person from both ingestion and inhalation. The intake rate from ingestion of leafy vegetables (IRV) was computed on daily steps by CVEG · HRV, where HRV is the rate of leafy vegetable intake (table 6.9). The intake rate from ingestion of milk (IRMK) was computed from CVEG · CRV · TCMK · HRMK, where the latter three terms are the cow's forage intake rate, the transfer coefficient to milk, and the person's intake rate of milk, respectively (see table 6.9). Similarly, the intake rate from meat consumption (IRMT) was calculated from CVEG · CRV · TCMT · HRMT, where

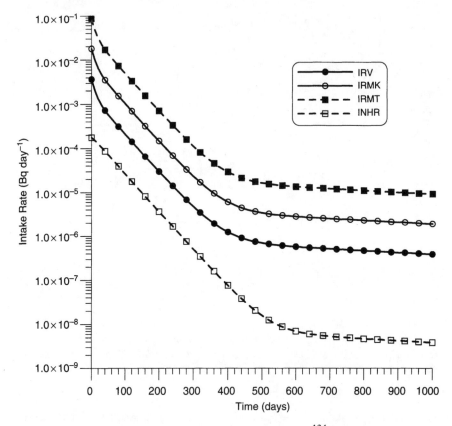

Figure 6.16 Simulated intake rates by a reference person of $^{134}$Cs following the acute deposition of 1 Bq m$^{-2}$ versus time (days) after deposition. The ingestion rates of leafy vegetables (IRV), milk (IRMK), and meat (IRMT) are shown, as well as the inhalation rate (INHR).

the latter two terms are the transfer coefficient to meat and the person's intake rate of meat, respectively (table 6.9). Finally, the inhalation rate for $^{134}$Cs (INHR) was estimated from the product QSS · RF · BR, where the latter two terms are the resuspension factor and the person's breathing rate, respectively (table 6.9). The results of these daily intake computations versus time are illustrated in figure 6.16.

The $^{134}$Cs intake rates all decline from their initial values by about three orders of magnitude within the first year after the initial deposition (figure 6.16). This is due to the decline of QSS, which drives plant contamination processes at the surface such as resuspension and rain splash. The main process driving the decline in QSS is percolation of contamination into the labile soil. After about 500 days, the rates of decline in the ingestion rates decrease considerably. At this point, the primary pathway driving the levels of $^{134}$Cs in food products is root uptake from QLS, and the primary mechanism causing the slow decline is physical decay of $^{134}$Cs. The curve for the inhalation rate is one to several orders of magnitude less than the food product ingestion rates, and it is a direct reflection of QSS, since this is the only compartment subject to the resuspension process.

In order to carry out a dose calculation, the ingestion and inhalation rates were numerically integrated using the expression, for example, for inhalation, of $IINH = IINH + (INHR \cdot DT)$ where IINH is the time-integrated inhalation quantity (Bq), INHR is the inhalation rate (Bq day$^{-1}$), and DT is the time step (1 day). A corresponding algorithm was used for integrating the food product ingestion rates, as well as the total ingestion rate. The cumulative integrated intakes are shown as a function of time in figure 6.17. The cumulative totals do not change much after 200 days. For example, in the case of total ingestion (leafy vegetables, milk, and meat summed), the 200-day value was 2.82 Bq, while the 1,000-day value was 2.88 Bq. The 1,000-day value for inhalation was only 0.0097 Bq. The 1,000-day value of 2.88 Bq, multiplied by an effective dose coefficient for ingestion of $1.98 \times 10^{-8}$ Sv Bq$^{-1}$ (EPA 1988), gives an effective dose of $5.7 \times 10^{-8}$ Sv resulting from the 1 Bq $^{134}$Cs deposited per square meter. The 1,000-day value for inhalation of 0.0097 Bq, multiplied by the corresponding dose coefficient for inhalation of $1.25 \times 10^{-8}$ Sv Bq$^{-1}$ (EPA 1988), yields an effective inhalation dose of $1.21 \times 10^{-10}$ Sv. The ingestion dose in this case exceeds the inhalation dose by a factor of 470, so inhalation was negligible in this case.

Although the preceding example ignored several other pathways, processes, and mostly minor complications, it provides a general idea of how the individual processes described earlier in this chapter can be integrated into a more realistic model of the whole system. Clearly, even a model of modest complexity such as this requires more than a few algebraic equations and parameters. Fortunately, computer software can greatly simplify the mechanics of solving differential equations and producing graphical or numerical output. Model building tends to be an iterative process, where the quality of the product reflects the time and effort spent on it. An important outcome of the model-building process is the understanding of the model and its limitations as a tool. A real disadvantage of using "off-the-shelf" models can be the considerable difficulty of knowing precisely what they are doing and why. It is often much easier to defend in a credible manner a tool that one has had a hand in constructing.

## Conclusions

In conclusion, the development of concepts and models for pathway analysis of environmental contaminants is probably as much an art as it is a science. Although models, by necessity, must rely on a base of knowledge and information gained through scientific efforts and practical experience, the manner in which such knowledge and information are synthesized and incorporated into mathematical formulations is very much dependent on the modeler. Alternative approaches can be equally useful or equally wrong. Complexity is not always preferable to simplicity. For some applications, two or more parameters can be "lumped" into a single parameter, with the net result possibly being just as adequate for the purpose at

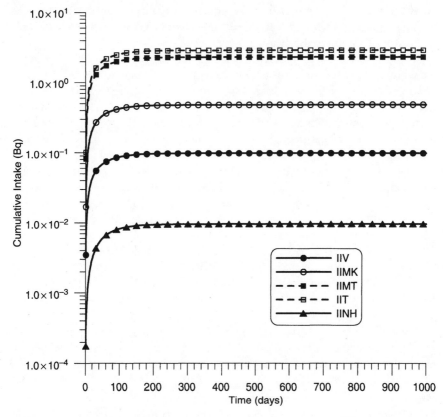

**Figure 6.17** Simulated cumulative intakes by a reference person of $^{134}$Cs following the acute deposition of 1 Bq m$^{-2}$ versus time (days) following deposition. The time-integrated total intakes are shown for leafy vegetables (IIV), milk (IIMK), meat (IIMT), the sum for ingested items (IIT), and the quantity inhaled (IINH).

hand. In addition to alternative model structures that the modeler may employ, the selection of parameter values is also subject to an almost infinite number of choices (see, e.g., models and parameter values in Simmonds et al. 1995). One of the most important jobs of the modeler is to make intelligent selections of parameter values, which should be based on good scientific knowledge, and which should be subject to critical peer review and, ultimately, tested against real, independent data. Building a model is one thing; successfully defending it during critical review is quite another matter. A recent discussion on the interface between the state of the science and models of radionuclide transport in the environment is available in Whicker et al. 1999). In addition to reviewing the state of the science of radioecology, Whicker et al. show how models are dependent on the science and how models, in turn, can be used to improve the science.

Despite all the caveats about the large effort normally required to construct, parameterize, and defend pathway analysis models, the power of modern desktop computers and the availability of software for solving sets of differential equations make model building a practical and stimulating exercise for many professionals. Model building can have many purposes, but one of the more rewarding is that of serving as a tool to organize, summarize, and test one's state of knowledge about the transport of radionuclides or other trace substances in the environment. When it comes to using models as a serious dose assessment tool, however, it is important for the modeler and users of the model output to be critical and ever skeptical. The credibility of any model needs to be earned through critical review and testing.

## Problems

1. A surface soil has a present-day $^{137}$Cs concentration of 0.05 Bq m$^{-2}$. If leaching and other removal processes are inconsequential, what is the concentration 50 years from the present?
2. If 1 MBq of $^{241}$Am is disposed in a landfill, what is that activity of $^{237}$Np after 500 years? (Hint: Look up the decay chain for $^{241}$Am.)
3. The 1-h average air concentration of $^{131}$I after an accidental release is 100 MBq m$^{-3}$. Assuming a deposition velocity of 0.02 m s$^{-1}$ for iodine and a weathering half-time of 14 days, how much $^{131}$I deposits on pasture grass during the 1-h averaging period?
4. Assuming the 1-h average concentration in problem 3 is maintained for 24 h, how much $^{131}$I remains on the pasture grass at the end of the 24-h period? State all assumptions used in the calculations.
5. If deposition ceases after 24 h in problem 4, how much $^{131}$I remains on the pasture grass five days after the start of the release?
6. Some cropland is contaminated with 1 MBq m$^{-2}$ of $^{90}$Sr down to a depth of 15 cm. Assuming no tillage for the next 50 years, a net infiltration rate of 50 cm yr$^{-1}$, a bulk density of 1.5 g cm$^{-3}$, and a loam soil, what is the $^{90}$Sr concentration (in MBq m$^{-2}$) in the top 15 cm of soil 50 years from the present?
7. Suppose that a $^{90}$Sr (half-life = 28.79 yr) air concentration of 10 Bq m$^{-3}$ is maintained over a pasture having a mean biomass of 0.3 kg m$^{-2}$. If the deposition velocity is 0.002 m s$^{-1}$ and the weathering half-time is 14 days, what is the equilibrium concentration *on* vegetation (i.e., ignore root uptake)?
8. Suppose that a $^{85}$Sr (half-life = 64.84 days) air concentration of 100 Bq m$^{-3}$ is maintained over a pasture having a same mean biomass, deposition velocity, and weathering half-time as in problem 7 (0.3 kg m$^{-2}$, 0.002 m s$^{-1}$, and 14 day, respectively). What is the equilibrium concentration in beef flesh raised on this pasture? Assume 10 kg day$^{-1}$ (dry) ingestion rate and ignore root uptake and soil ingestion.
9. What is the equilibrium concentration of $^{85}$Sr in milk from a cow grazing on the pasture described in problem 8?
10. Compute the suspension flux above a soil region that is contaminated at the surface with 200 MBq m$^{-2}$ of $^{226}$Ra. Assume a mean wind speed of 3 m s$^{-1}$, roughness height ($z_o$) of 5 cm, fraction of vegetative cover ($f_v$) of 0.4,

particle size mode ($p$) of 0.8 mm, Lc = 0.0001, a surface soil density of 1.2 g cm$^{-3}$, and thickness of contamination of 3 cm.

11. Using the particulate suspension flux determined in problem 10, compute the concentration in air and on vegetation at a location having a $X/Q$ value for an area source of $1 \times 10^{-4}$ s m$^{-1}$. Assume a deposition velocity of 0.001 m s$^{-1}$, biomass of 0.2 kg m$^{-2}$ (dry), a foliar interception constant ($\alpha$) of 1 m$^2$ kg$^{-1}$, and a weathering half-time of 14 days.

12. An accidental release from a facility results in $^{137}$Cs contamination across a region that is farmed. Assuming 100 MBq m$^{-2}$ of $^{137}$Cs were deposited on soil, compute the $^{137}$Cs concentration from root uptake in a wheat crop that is grown on the contaminated land immediately after the accident. Assume that the soil is tilled to a depth of 0.3 m and a bulk density of 1.5 g cm$^{-3}$.

13. Using the data in problem 12, compute the $^{137}$Cs concentration from root uptake in the wheat crop 15 years and 30 years after the accident. Assume a net annual infiltration rate of 0.7 m yr$^{-1}$, a sandy loam soil, and a $K_d$ of 100 mL g$^{-1}$.

14. Using the results of problem 13, determine the effective half-time of $^{137}$Cs in the agriculture soil. Is leaching an important process in reducing $^{137}$Cs concentrations in the wheat crop?

Notes

1. MathWorks, www.mathworks.com/products/product_listing/index.html.
2. Wolfram Research, www.wolfram.com.
3. Available from Dr. Thomas B. Kirchner, Carlsbad Environmental Monitoring and Research Center, New Mexico State University, 1400 University Drive, Carlsbad, NM 88220–3575 (phone 505–234–5504).
4. Available from High Performance Systems, Inc., 45 Lyme Road, Suite 200, Hanover, NH 03755–1221 (phone 603–643–9636).
5. Available from MathWorks, 3 Apple Hill Drive, Natick, MA 01760–2098 (phone 508–647–7000).
6. Available from Wolfram Research Inc., 100 Trade Center Drive, Champaign, IL 61820–7237 (phone 217–398–7000).

References

Anspaugh, L.R., J.H. Shinn, P.L. Phelps, and N.C. Kennedy. 1975. "Resuspension and Redistribution of Plutonium in Soils." *Health Physics* 29: 571–582.
Argonne National Laboratory. 1998. *MILDOS-AREA User's Guide*. Environmental Assessment Division, Argonne National Laboratory, Argonne, IL.
Baes, C.F., and R.D. Sharp. 1983. "A Proposal for Estimation of Soil Leaching and Leaching Constants for Use in Assessment Models." *Journal of Environmental Quality* 12: 17–28.
Baes, C.F., R.D. Sharp, A.L. Sjoreen, and R.W. Shor. 1984. *A Review and Analysis of Parameters for Assessing Transport of Environmentally Released Radionuclides through Agriculture*. ORNL-5786. Oak Ridge National Laboratory, Oak Ridge, TN.
Bagnold, R.A. 1941. *The Physics of Blown Sand and Desert Dunes*. Methuen & Co., London.
Birchall, A. 1986. "A Microcomputer Algorithm for Solving Compartmental Models Involving Radionuclide Transformations." *Health Physics* 50(3): 389–397.

Carnahan, B., H.A. Luther, and J.O. Wilkes. 1969. *Applied Numerical Methods*. John Wiley & Sons, New York.

Chamberlain, A.C. 1970. "Interception and Retention of Radioactive Aerosols by Vegetation." *Atmospheric Environment* 4(1): 57–78.

Clapp, R.B., and G.M. Hornberger. 1978. "Empirical Equations for Some Soil Hydrologic Properties." *Water Resources Research* 14: 601–604.

Coughtrey, P.J., D. Jackson, and M.C. Thorne. 1985. *Radionuclide Distribution and Transport in Terrestrial and Aquatic Ecosystems*, vol. 6. A.A. Balkema, Rotterdam.

Cowherd, C., Jr., G.E. Muleski, P.J. Englehart, and D.A. Gillette. 1985. *Rapid Assessment of Exposure to Particulate Emissions from Surface Contamination Sites*. EPA/600/8-85/002. Office of Health and Environmental Assessment, Office of Research and Development, U.S. Environmental Protection Agency, Washington, DC.

EPA (U.S. Environmental Protection Agency). 1988. *Limiting Values of Radionuclide Intake and Air Concentration and Dose Conversion Factors for Inhalation, Submersion, and Ingestion*. Federal Guidance Report No. 11, EPA-520/1-88-020. Office of Radiation Programs, Washington, DC.

EPA. 1989. *Exposure Factors Handbook*. EPA/600/8-89/043. Exposure Assessment Group, Office of Health and Environmental Assessment, U.S. Environmental Protection Agency, Washington, DC.

EPA. 1992. *User's Guide for the Industrial Source Complex (ISC) Dispersion Models*. Vol. 1: *User's Instructions*. EPA-450/4-92-008a. U.S. Environmental Protection Agency, Research Triangle Park, NC.

EPA. 1996. *Soil Screening Guidance: Technical Background Document*. EPA/540/R95/128 U.S. Environmental Protection Agency, Office of Solid Waste and Emergency Response, Washington, DC.

EPA. 2004. *User's Guide to the AMS/EPA Regulatory Model AERMOD*. EPA-454/B-03-001. U.S. Environmental Protection Agency, Office of Air Quality and Planning, Research Triangle Park, NC.

EPA. 2006. *CAP88-PC Version 3.0 User's Guide*. U.S. Environmental Protection Agency, Office of Radiation and Indoor Air, Washington, DC.

Garger, E.K., F.O. Hoffman, and K.M. Thiessen. 1997. "Uncertainty of the Long-Term Resuspension Factor." *Atmospheric Environment* 31: 1647–1656.

Gillette, D.A. 1974a. "On the Production of Soil and Wind Erosion Aerosols Having Potential for Long Range Transport." *Journal de Recherches Atmospheriques* 8(3/4): 735–744.

Gillette, D.A. 1974b. "Production of Fine Dust by Wind Erosion of Soil: Effect of Wind and Soil Texture." In *Atmosphere-Surface Exchange of Particulate and Gaseous Pollutants. ERDA Symposium Series, Richland, Washington, Sept 4–6*. Energy Research and Development Administration, Washington, DC; pp. 591–943.

Gillette, D.A. 1981. "Production of Dust that May Be Carried Great Distances." In *Desert Dust: Origin, Characteristics, and Effects on Man*, ed. Troy Pewe. Geological Society of America Special Paper 186. Geologic Society of America, Boulder, CO; pp. 11–26.

Gillette, D.A., and W. Chen. 2001. "Particle Production and Aeolian Transport from a Supply-Limited Source Area in the Chihuahuan Desert, New Mexico." *Journal of Geophysical Research* 106(D6): 5267–5278.

Gillette, D.A., and P.A. Goodwin. 1974. "Microscale Transport of Sand Sized Soil Aggregates Eroded by Wind." *Journal of Geophysical Research* 79(27): 4068–4075.

Gillette, D.A., I.H. Blifford, Jr., and C.R. Fenster, 1972. "Measurements of Aerosols Size Distributions and Vertical Fluxes of Aerosols on Land Surface Subject to Wind Erosion." *Journal of Applied Meteorology* 11: 977–987.

Gillette, D.A., I.H. Blifford, Jr., and D.W. Fryrear. 1974. "The Influence of Wind Velocity on the Size Distribution of Aerosols Generated by Wind Erosion of Soils." *Journal of Geophysical Research* 79(27): 4068–4075.

Gist, C.S., and F.W. Whicker. 1971. "Radioiodine Uptake and Retention by the Mule Deer Thyroid." *Journal of Wildlife Management* 35: 461–468.

Hanson, W.C., and L.L. Eberhardt. 1969. "Effective Half-Times of Radionuclides in Alaskan Lichens and Eskimos." In *Symposium on Radioecology*, ed. D.J. Nelson and F.C. Evans. USAEC Report CONF-670503. National Technical Information Service, Springfield, VA; p. 627.

Healy, J.W. 1980. "Review of Resuspension Models." In *Transuranic Elements in the Environment*, ed. Wayne C. Hanson. DOE/TIC-22800. U.S. Department of Energy, Office of Health and Environmental Research, Washington, DC.

Higley, K.A. 1994. *Vertical Movement of Actinide-Contaminated Soil Particles*. Ph.D. dissertation, Department of Radiological Health Sciences, Colorado State University, Fort Collins, CO.

Hinds, W.C. 1982. *Aerosol Technology: Properties, Behavior, and Measurement of Airborne Particles*. John Wiley & Sons, New York.

Hoffman, F.O., K.M. Thiessen, M.L. Frank, and B.G. Blaylock. 1992. "Quantification of the Interception and Initial Retention of Radioactive Contaminants Deposited on Pasture Grass by Simulated Rain." *Atmospheric Environment* 26A: 3313–3321.

IAEA (International Atomic Energy Agency). 1994. *Handbook of Parameter Values for the Prediction of Radionuclide Transfer in Temperate Environments*. Technical Report Series 364. International Atomic Energy Agency, Vienna.

ICRP (International Commission on Radiological Protection). 1960. "Report of Committee 11 on Permissible Dose for Internal Radiation (1959)." *Health Physics* 3: 1–380.

ICRP. 1975. *Reference Man: Anatomical, Physiological, and Metabolic Characteristics*. ICRP Publication 23. International Commission on Radiological Protection, Stockholm, Sweden.

Jenne, E.A. 1998. *Adsorption of Metals by Geomedia*. Academic Press, New York, chapter 1.

Johnson, J.E., G.M. Ward, E. Firestone, and K.L. Knox. 1968. "Metabolism of Radioactive Cesium and Potassium by Dairy Cattle as Influenced by High and Low Forage Diets." *Journal of Nutrition* 94(3): 282–288.

Kitchings, T., D. DiGregorio, and P. Van Voris. 1976. "A Review of the Ecological Parameters of Radionuclide Turnover in Vertebrate Food Chains." In *Radioecology and Energy Resources*, ed. C.E. Cushing. Dowden, Hutchinson & Ross, Stroudsburg, PA; pp. 301–313.

LaGoy, P.K. 1987. "Estimated Soil Ingestion Rates for Use in Risk Assessment." *Risk Analysis* 7(3): 355–359.

Liden, K., and M. Gustafsson. 1967. "Relationships and Seasonal Variation of $^{137}$Cs in Lichen, Reindeer and Man in Northern Sweden 1961–1965." In *Radioecological Concentration Processes*, ed. B. Aberg and F.P. Hungate. Pergamon Press, New York; p. 193.

Linsley, G.S. 1978. *Resuspension of the Transuranic Elements—A Review of Existing Data*. NRPB-R75. U.K. National Radiological Protection Board, Chilton, U.K.

Litaor, M.I., M.L. Thompson, G.R. Barth, and P.C. Molzer. 1994. "Plutonium-239+240 and Americium-241 in Soils East of Rocky Flats, Colorado." *Journal of Environmental Quality* 23: 1231–1239.

McDonald, J.E. 1960. "An Aid to Computation of Terminal Fall Velocities of Spheres." *Journal of Meterology* 17(4): 463.

Millard, J.B., F.W. Whicker, and O.D. Markham. 1990. "Radionuclide Uptake and Growth of Barn Swallows Nesting by Radioactive Leaching Ponds." *Health Physics* 58(4): 429–439.

NCRP (National Council on Radiation and Measurements). 1989. *Screening Techniques for Determining Compliance with Environmental Standards.* NCRP Commentary No. 3. National Council on Radiation and Measurements, Bethesda, MD.

NCRP. 1993. *Uncertainty in NCRP Screening Models Relating to Atmospheric Transport, Deposition, and Uptake by Humans.* NCRP Commentary 8. National Council on Radiation and Measurements, Bethesda, MD.

NRC (U.S. Nuclear Regulatory Commission). 1975. *Reactor Safety Study—an Assessment of Accident Risks in Commercial Nuclear Power Plants.* WASH-1400 NUREG-75/014. U.S. Nuclear Regulatory Commission, Washington, DC; appendix VI.

NRC. 1982. *Regulatory Guide 3.51.* U.S. Nuclear Regulatory Commission, Office of Nuclear Research, Washington, DC.

NRC. 1983. *Radiological Assessment: A Textbook on Environmental Dose Analysis.* NUREG/CR-3332 (ORNL-5968). Prepared by Oak Ridge National Laboratory. U.S. Nuclear Regulatory Commission, Washington, DC.

NRC/CEC (U.S. Nuclear Regulatory Commission/Commission of European Communities). 1994. *Probabilistic Accident Consequence Uncertainty Analysis: Dispersion and Deposition Uncertainty Assessment.* NUREG/CR-6244/EUR 15855EN/SAND94-1453, vol. 1. National Technical Information Service, Springfield, VA.

Oksza-Chocimowski, G.V. 1977. *Generalized Model of the Time-Dependent Weathering Half-Life of the Resuspension Factor.* U.S. EPA Report ORP/LV-77-4. U.S. Environmental Protection Agency, Las Vegas, NV.

Perkins, R.W., C.W. Thomas, J.A. Young, and B.C. Scott. 1970. "In-Cloud Scavenging Analysis from Cosmogenic Radionuclide Measurements." In *Precipitation Scavenging*, ed. R.J. Engleman and W.G.N. Slinn. AEC Symposium Series 22. U.S. Atomic Energy Commission, Washington, DC.

Peters, E.L. 1994. *Environmental Influences on the Kinetics of Group IA Elements in Ectotherms and Their Use in Estimating Field Metabolic Rates.* Ph.D. dissertation. Colorado State University, Fort Collins, CO.

Pinder, J.E., III, M.H. Smith, A.L. Boni, J.C. Corey, and J.H. Horton. 1979. "Plutonium Inventories in Two Old-Field Ecosystems in the Vicinity of a Nuclear Fuel Reprocessing Facility." *Ecology* 60: 1141–1150.

Press, W.H., S.A. Teukolsky, W.T. Vettering, and B.P. Flannery. 1992. *Numerical Recipes: The Art of Scientific Computing.* Cambridge University Press, New York.

Ramsdell, J.V., C.A. Simonen, and K.W. Burke. 1994. *Regional Atmospheric Transport Code for Hanford Emission Tracking (RATCHET).* PNWD-2224-HEDR. Pacific Northwest Laboratories, Richland, WA.

Risk Assessment Corporation. 1998. *Centers for Disease Control and Prevention Technical Workshop Report: Issues Related To Estimating Doses Due to I-131 Releases to the Atmosphere from the Hanford Site.* 2-CDC-Task Order-1998-Final. Risk Assessment Corporation, Neeses, SC.

Rood, A.S. 2004. "A Mixing Cell Model for Assessment of Contaminant Transport in the Unsaturated Zone under Steady-State and Transient Flow Conditions." *Environmental Engineering Science* 21(6): 661–677.

Rood, A.S., H.A. Grogan, and J.E. Till. 2002. "A Model for a Comprehensive Evaluation of Plutonium Released to the Air from the Rocky Flats Plant, 1953–1989." *Health Physics* 82(2): 182–212.

Schreckhise, R.G. 1980. *Simulation of the Long Term Accumulation of Radiocontaminants in Crop Plants.* PNL-2636/UC-70. Battelle Pacific Northwest Laboratory, Richland, WA.

Scire, J.S., F.W. Lurmann, A. Bass, and S.R. Hanna. 1984. *User's Guide to the MESOPUFFII Model and Related Processor Programs*. EPA-600/8–84–013. Environmental Sciences Research Laboratory, U.S. Environmental Protection Agency, Research Triangle Park, NC.

Scire, J.S., D.G. Strimaitis, and R.J. Yamartino. 2000. *A User's Guide for the CALPUFF Dispersion Model Version 5*. Earth Tech Inc., Concord, MA.

Seel, J.F., Whicker, F.W., and Adriano, D.C. 1995. "Uptake of $^{137}$Cs in Vegetable Crops Grown on a Contaminated Lakebed." *Health Physics* 68(6): 793–799.

Sehmel, G.A. 1980. "Particle Resuspension: A Review." *Environment International* 4: 107–127.

Seinfeld, J.H., and S.N. Pandis. 1998. *Atmospheric Chemistry and Physics*. John Wiley & Sons, New York.

Sheppard, M.I., and D.H. Thibault. 1990. "Default Soil Solid/Liquid Partition Coefficients, $K_d$s for Four Major Soil Types: A Compendium." *Health Physics* 59(4): 471–482.

Shinn, J.H. 1992. "Enhancement Factors for Resuspended Aerosol Radioactivity: Effects of Topsoil Disturbance." In *Proceedings of the Fifth International Conference on Precipitation Scavenging and Atmosphere-Surface Exchange Processes*, ed. S.E Schwartz and W.G.N. Slinn. Hemisphere Publishing Corporation, Washington, DC; pp. 1183–1193.

Simmonds, J.R., G. Lawson, and A. Mayall. 1995. *Methodology for Assessing the Radiological Consequences of Routine Releases of Radionuclides to the Environment*. Report EUR 15760 EN. Directorate-General, Environment, Nuclear Safety and Civil Protection, European Commission, Luxembourg.

Simon, S.L. 1990. "An Analysis of Vegetation Interception Data Pertaining to Close-In Weapons Test Fallout." *Health Physics* 59(5): 619–626.

Slade, D.H., ed. 1968. *Meteorology and Atomic Energy 1968*. TID-24190. U.S. Atomic Energy Commission, Washington, DC.

Smith, W.J., III, F.W. Whicker, and H.R. Meyer. 1982. "Review and Categorization of Saltation, Suspension, and Resuspension Models." *Nuclear Safety* 23(6): 685–699.

Stephens, J.A., F.W. Whicker, and S.A. Ibrahim. 1998. "Sorption of Cs and Sr to Profundal Sediments of a Savannah River Site Reservoir." *Journal of Environmental Radioactivity* 38(3): 293–315.

Strenge, D.L., and T.J. Bander. 1981. *MILDOS—a Computer Program for Calculating Environmental Radiation Doses from Uranium Recovery Operations*. NUREG CR-2011, PNL-3767. Pacific Northwest Laboratory, Richland, WA.

Thiessen, K.M., M.C. Thorne, P.R. Maul, G. Prohl, and H.S. Wheater. 1999. "Modeling Radionuclide Distribution and Transport in the Environment." *Environmental Pollution* 100: 151–177.

Toy, T.J., G.R. Foster, and K.G. Renard. 2002. *Soil Erosion: Processes, Prediction, Measurement, and Control*. John Wiley & Sons, New York.

Travis, J.R. 1974. "A Model for Predicting the Redistribution of Particulate Contaminants from Soil Surfaces." In *Atmosphere-Surface Exchange of Particulate and Gaseous Pollutants. ERDA Symposium Series. Richland, WA, Sept 4–6, 1974*. CONF-740921, NTIS 1976. Energy Research and Development Administration, Washington, DC; pp. 591–943.

Voillequé, P.G., and J.H. Keller. 1981. "Air-to-Vegetation Transport of $^{131}$I as Hypoiodous Acid (HOI)." *Health Physics* 40(1): 91–94.

Weber, J.M., A.S. Rood, and H.R. Meyer. 1999. *Development of the Rocky Flats Plant 903 Area Plutonium Source Term*. 8CDPHE-RFP-1998-Final (Rev. 1). Risk Assessment Corporation, Neeses, SC.

Whicker, F.W. 1983. "Radionuclide Transport Processes in Terrestrial Ecosystems." *Radiation Research* 94(1): 135–150.

Whicker, F.W., and T.B. Kirchner. 1987. "PATHWAY: A Dynamic Food-Chain Model to Predict Radionuclide Ingestion after Fallout Deposition." *Health Physics* 52(6): 717–737.

Whicker, F.W., and V. Schultz. 1982. *Radioecology: Nuclear Energy and the Environment.* CRC Press, Boca Raton, FL.

Whicker, F.W., G. Shaw, G. Voigt, and E. Holm. 1999. "Radioactive Contamination: State of the Science and its Application to Predictive Models." *Environmental Pollution* 100(1): 133–149.

Whicker, J.J., J.E. Pinder, III, D.D. Breshears, C.F. Eberhart, 2006. "From Dust to Dose: Effects of Forest Disturbance on Increased Inhalation Exposure. *Science of the Total Environment* 368: 519–530.

Whicker, R.D., and S.A. Ibrahim. 2006. "Vertical Migration of $^{134}$Cs Bearing Particles in Arid Soils: Implications for Plutonium Redistribution." *Journal of Environmental Radioactivity* 88(2):171–188.

Whittaker, R.H. 1970. *Communities and Ecosystems.* Macmillan, London.

Winsor, T.F., and F.W. Whicker. 1980. "Pocket Gophers and Redistribution of Plutonium in Soil." *Health Physics* 39(2): 257–262.

Yu, C., A.J. Zielen, J.J. Cheng, D.J. LePoire, E. Gnanapragasam, S. Kamboj, J. Arnish, A. Wallo, III, W.A. Williams, and H. Peterson. 2001. *User's Manual for RESRAD Version 6.* ANL/EAD-4. Argonne National Laboratory, Environmental Assessment Division, Argonne, IL.

# 7

# Aquatic Food Chain Pathways

Steven M. Bartell
Ying Feng

Radionuclides and other toxic chemicals often accumulate in the surface waters and sediments of aquatic ecosystems. Airborne contaminants deposited to the surface of lakes, rivers, estuaries, and coastal oceans combine with inputs from accidental spills, routine releases from permitted point sources, and nonpoint-source runoff from urban, agricultural, and residential landscapes to generate complex source terms that become the subjects of comprehensive human health and ecological risk analyses. Complex and dynamic physical, chemical, and biological processes define pathways that determine the distribution and concentration of contaminants in diverse assemblages of aquatic organisms, as well as in the sediments and suspended particulate matter that constitute the structure of aquatic ecosystems. The ultimate fate of many chemical contaminants, including selected radionuclides, is closely tied to the hydrologic cycle. It is therefore not surprising that substantial emphasis has been placed on the development of methods, models, and data for characterizing radionuclide transport, distribution, and bioaccumulation in aquatic ecosystems.

The accumulation of radionuclides may pose direct risks to aquatic organisms (Higley et al. 2003). In addition, terrestrial organisms that live nearby or feed at the water's edge might drink contaminated water, be exposed to sediments, or consume aquatic organisms contaminated with radionuclides. Importantly, the human consumption of fish and shellfish provides a pathway of exposure and potential health risks from ionizing radiation. This chapter describes methods for assessing the movement of radionuclides along pathways of transport and accumulation in rivers, reservoirs, lakes, estuaries, and coastal oceans. The materials presented here focus primarily on the accumulation of radionuclides by aquatic organisms. Subsequent consumption of aquatic resources by humans is briefly described in relation to human health risk assessment.

Where possible, materials that describe the transport and accumulation of radionuclides along aquatic pathways have been selected from previous human health and ecological risk assessments.

The discussion of aquatic pathways in this chapter makes use of the kinds of results produced by mathematical models that describe the advective and diffusive transport of radionuclides in aquatic systems. These transport models do not receive detailed treatment here; we instead refer the reader to chapter 4 in this textbook. However, this chapter does examine the implications of spatial and temporal variability in radionuclide exposures on bioaccumulation through the application of a two-dimensional transport model developed for the Chernobyl cooling pond.

## Aquatic Ecosystem Classification

Aquatic ecosystems are diverse in structure and function. The pathways of radionuclide transport and bioaccumulation described in this chapter refer primarily to four classes of aquatic systems: (1) streams and rivers, (2) impounded reservoirs and lakes, (3) estuaries, and (4) coastal oceans. Radionuclide contamination of the open ocean is not addressed in relation to human health and ecological risk assessment because humans are not routinely exposed directly or indirectly to such contamination. The open ocean is typically less ecologically productive when compared to estuaries and coastal oceans; ecological risks posed by radionuclides in the open ocean have correspondingly received less attention. Finally, the capacity for dilution of radionuclides (and other contaminants) considerably reduces exposure concentrations in the open oceans.

Reasons for the above classification of aquatic systems include fundamental differences in hydrodynamics and physical mixing characteristics, unique food web structures, and differences in water chemistry (e.g., freshwater vs. saltwater environments). Each of these factors can influence radionuclide transport along aquatic pathways and subsequent bioaccumulation by aquatic organisms. For example, advective transport is a dominant process that can rapidly mix and dilute contaminants introduced to streams and rivers. In contrast, the thermal stratification of water commonly observed in temperate-zone lakes and reservoirs can concentrate contaminants during the significant portions of the ecologically productive months. As a result, some aquatic populations might be exposed to unusually high radionuclide concentrations, while other populations might be thermally isolated from exposure during this period. Impounded reservoirs provide complex physical mixing characteristics, with upper regions providing a riverlike environment and regions nearer the dam typically more lakelike in their mixing patterns. These two areas are separated by a zone of transition from riverlike to lakelike conditions. The dominant tidal mixing features of many estuaries can create periodic exposure regimes unique to these systems. The effects of tidal mixing processes in establishing spatial-temporal patterns of exposure can be further complicated by the differences in water chemistry between the overlying fresh water and deeper saline waters that define a moving saltwater wedge that distinguishes tidal estuaries from other aquatic systems.

These four classes of aquatic ecosystems also differ in their food web structures and ecological production dynamics. For example, ecological production in streams and rivers is often regulated by the external (allochthonous) inputs of particulate organic matter (e.g., leaves, partially decomposed vegetation, animal wastes). These systems are commonly heterotrophic, with the processing of particulate organic matter by insects and other invertebrates and bacterial decomposition providing the energetic basis that supports higher order consumer populations (e.g., fish). This trophic dynamic pattern contrasts with the predominantly autotrophic nature of lakes, where photosynthesis by planktonic algae and other aquatic vegetation provides the energetic basis for ecological production. Impounded reservoirs can be mainly heterotrophic in their upstream regions, while primarily autotrophic in the lakelike region nearer the dam. Depending on the particular system, tidal estuaries might be net importers or exporters of organic materials in relation to the adjacent coastal oceans. The freshwater river portion of the estuary might be mainly heterotrophic in nature, while primary production by phytoplankton can dominate in the lower reaches. The particular species assemblages that make up the food webs in these aquatic ecosystems differ markedly, with the major differences occurring between the freshwater and marine environments. Such differences in species composition are important in influencing the pattern of radionuclide movement throughout the food webs, as well as in determining pathways of human exposures. For example, marine benthic invertebrates (e.g., clams, oysters, shrimp, and lobsters) are more commonly consumed by humans than are benthic invertebrates from freshwater systems (e.g., freshwater mussels and crayfish).

## Conceptual Model for an Aquatic Environment

Aquatic ecosystems are exceedingly diverse in form and function. Detailed and individualized descriptions of streams, larger rivers, lakes, reservoirs, tidal estuaries, coral reefs, coastal oceans, and the open sea are well beyond the intended scope of this chapter. Nevertheless, many of the structural and functional relationships common to these diverse aquatic systems are illustrated in the conceptual model of an aquatic food web developed for the Clinch River in Tennessee (figure 7.1; Cook et al. 1999). Radionuclides can be introduced to the river through inputs directly to water. Direct inputs might result from atmospheric deposition or from point-source additions to surface water. Radionuclides can also enter aquatic systems as allochthonous inputs adsorbed to inorganic or organic particulate matter that settles to the sediments and becomes part of the settled detritus pool. This food web also illustrates the potential importance of direct exposure of aquatic populations to radionuclides dissolved (or complexed) in water or sorbed to bed sediments and suspended sediments. Populations of aquatic plants and animals can be directly exposed to ionizing radiation from radionuclides that are dissolved in water, or sorbed to suspended particulate matter or to settled detritus and sediments. In addition to direct exposure, these populations can also be exposed to radiation

## Aquatic Food Chain Pathways 343

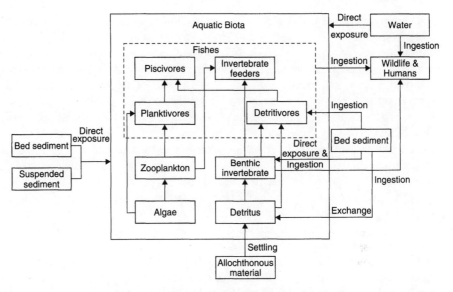

Figure 7.1 Conceptual model of the Clinch River, Tennessee, food web. Unlabeled arrows denote food web transfers. Adapted from Cook et al. (1999).

by ingesting contaminated food or prey. Sediments and settled detritus can also be ingested by benthic invertebrates.

Radionuclides move throughout the aquatic food web through the pathways of trophic transfer illustrated in figure 7.1. The conceptual model illustrates two quite different kinds of aquatic food webs. The aquatic biota on the left side of figure 7.1 describe a typical autotrophic food chain where primary production in the form of planktonic algae is consumed by zooplankton, which are in turn eaten by planktivorous fish that are prey for piscivorous fish. This conceptual model also shows the potential for heterotrophic transfer from contaminated detritus through benthic invertebrates to fishes that feed on detritus or fishes that feed on benthic invertebrates (the right side of the conceptual model). From a risk assessment perspective, the bed sediments and allochthonous inputs designate potential sources of radionuclide contamination that might be remediated.

### Physicochemical Processes

The transport and bioavailability of radionuclides (and other chemicals) are influenced by the dynamic and complex chemical environments specific to these aquatic ecosystems (Stuum and Morgan 1970). Water temperature can influence the uptake and accumulation of radionuclides and other chemicals by aquatic organisms because many of the biological processes involved in accumulation are dependent on temperature. For example, rates of biological elimination of contaminants increase with increasing temperatures. Rates of feeding by fish and other aquatic organisms also increase with increasing water temperatures; thus, accumulation of ingested contaminants increases correspondingly with warming water temperatures.

Seasonal changes in the dissolved carbon dioxide chemistry of fresh water presents a variable complexation and redox environment that can influence the chemical speciation of radionuclides. Spatial and temporal variations in the concentration of particulate and dissolved carbon (i.e., bicarbonates, carbonates, and hydroxyl groups) can determine the relative amounts of freely dissolved versus organically and inorganically bound concentrations of radionuclides. Changes in pH and redox conditions can similarly affect the relative distribution of radionuclides between dissolved and particulate fractions. These processes that provide for complex chemical speciation are further complicated by seasonal changes in ecological production that can add organic carbon (i.e., allochthonous inputs) to aquatic ecosystems. The particular chemistry of salt water can provide for even more complicated partitioning of radionuclides between dissolved and sorbed fractions. Such partitioning can determine the relative importance of direct pathways versus food web pathways of radionuclide exposure for populations of aquatic plants and animals. Finally, the chemical characteristics of individual radionuclides also influence their transport and ultimate distribution in aquatic ecosystems. For example, several nuclides have a propensity to be bound to particulate matter (e.g., $^{137}$Cs, $^{90}$Sr), while others are more commonly found in solution (e.g., $^{3}$H).

## Radionuclide Uptake and Concentration Factors

Assessing human health and ecological risks posed by radionuclides in aquatic ecosystems requires quantification of exposures and subsequent uptake and accumulation by populations of aquatic biota that might ultimately be consumed by humans. The most straightforward approach to quantifying exposure and accumulation would be to collect the necessary samples of water, sediments, and biota and directly measure the concentrations of radionuclides of concern in these samples. The measurement of exposure and bioaccumulation would be limited, in theory, only by the availability of resources for sample collection and data development. In some circumstances, direct measurement is not possible. For example, in reconstructing exposures and risks from historical releases of radioactivity, it might no longer be possible to obtain the necessary samples; short-lived nuclides would have long since decayed below practical detection limits, and biota of concern in possible historical exposure pathways to humans would no longer be available. On the other hand, a risk assessment might be conducted to estimate the possible impacts of radioactive waste management alternatives on risks associated with future exposures. In this case, the necessary data are not available, and future risks will have to be estimated using projected values of exposures and bioaccumulation. Mathematical descriptions of exposure, accumulation, and dose will necessarily play an important role in reconstructing historical risks or forecasting future risks.

The following sections review the mathematical development of expressions for the accumulation and concentration of radionuclides by aquatic biota. The material follows closely from the derivation and uses the same notation presented by Peterson (1983). For greater detail, the interested reader should review the material on aquatic pathways as presented originally in Peterson (1983). One purpose of this chapter is to outline the mathematical foundation for the derivation of the bioconcentration factor (BCF) and indicate some of the basic simplifications and assumptions inherent in its derivation and application.

The time-varying concentration of a chemical in an aquatic organism can be described by the following expression:

$$\frac{dC}{dt} = \frac{I_w}{m} C_w - rC, \tag{7.1}$$

where

$C$ = chemical concentration in the organism (e.g., Bq kg$^{-1}$)
$C_w$ = chemical concentration in water (e.g., Bq L$^{-1}$)
$I_w$ = uptake rate of water by an aquatic organism (e.g., L kg$^{-1}$ t$^{-1}$), where $t$ is the unit of time
$m$ = mass of the aquatic organism (kg)
$r$ = biological elimination rate of the chemical by the organism ($t^{-1}$)

The analytical solution of equation 7.1 specifies the concentration of a chemical in the organism at any time $t$ as

$$C(t) = \frac{I_w C_w}{mr}[1 - \exp(-rt)]. \tag{7.2}$$

One important implication of equation 7.2 is that, after a sufficiently long time, the concentration of a chemical in the organism will approach a constant or equilibrium value of

$$C_{equil} = \lim_{t \to \infty} C(t) = \frac{I_w C_w}{mr}. \tag{7.3}$$

If it is assumed that the concentration of the chemical in water is also constant or at equilibrium, then it is possible to estimate the accumulation of a chemical in an organism simply from measuring the chemical concentration in water and knowing the ratio of the chemical in water to the chemical in the organism. The ratio of the concentration of the chemical in the organism to that in water is

$$\frac{C_{equil}}{C_w} = \frac{I_w}{mr}. \tag{7.4}$$

The ratio defined by equation 7.4 has been called the concentration factor (CF) and is described as

$$CF = \frac{\text{equilibrium concentration in organism}}{\text{concentration in water}}. \tag{7.5}$$

Note that the CF pertains to the direct accumulation of a chemical from water. This definition of the CF (i.e., Peterson 1983) is analogous to the BCF used to describe the accumulation of nonradioactive chemical contaminants (Bartell et al. 1998). Neither the CF nor the BCF addresses the accumulation of chemical contaminants from ingestion of contaminated food or prey. In the assessment of nonradioactive chemicals, the bioaccumulation factor includes the uptake and accumulation of chemical contaminants from food web pathways; a corresponding biota-to-sediment accumulation factor defines the equilibrium accumulation of chemicals taken up from contaminated sediments (Bartell et al. 1998).

The preceding derivation (equation 7.1) applies more correctly to radionuclides by adding a term for radioactive decay ($\lambda$) in addition to biological elimination and replacing $r$ by $r + \lambda$. Equation 7.2 can then be reformulated as

$$C_f(t) = \frac{I_w C_w}{m(r+\lambda)}[1 - \exp-(r+\lambda)t], \tag{7.6}$$

and the corresponding definition of the CF (i.e., equation 7.4) is redefined as

$$\mathrm{CF}^* = \frac{I_w}{m(r+\lambda)}. \tag{7.7}$$

The concentration of the radionuclide in the organism accumulated directly from water can be expressed as

$$C_f(t) = \mathrm{CF}^* C_w, \tag{7.8}$$

where the revised CF now considers both biological elimination and radioactive decay.

The concept of half-life in pathway analysis was introduced in chapter 6. Following Peterson (1983), the effective half-life ($T_e$) can be defined as

$$T_e = \frac{T_{\mathrm{bio}} \cdot T_{\permil}}{T_{\mathrm{bio}} + T_{\permil}} \quad \text{or} \quad T_e = \frac{\ln 2}{\lambda + r}, \tag{7.9}$$

where $T_{\mathrm{bio}}$ is the rate of biological elimination of radionuclide by the organism ($t^{-1}$) and $T_{\permil}$ is rate of radionuclide decay ($t^{-1}$).

Thus, the decrease in the concentration of a radionuclide in an organism that has come to equilibrium with the concentrations in its surrounding environment can be estimated from knowledge of the rates of biological elimination and radioactive decay.

In the practice of assessing risks, BCFs might be available for stable isotopes of radionuclides of concern, but not for the radionuclide itself. Under equilibrium conditions, the CF for the radionuclide (i.e., equation 7.7) is slightly smaller than the corresponding CF for the stable element (e.g., equation 7.4), providing that both elements are in the same chemical form. This can be demonstrated by taking the ratio of the two CFs:

$$\frac{\mathrm{CF}}{\mathrm{CF}^*} = \frac{r}{r+\lambda} = \frac{1}{1+\lambda/r} \quad \text{and} \quad \mathrm{CF}^* = \frac{\mathrm{CF}}{1+\lambda/r} = \frac{\mathrm{CF}}{1+T_{\mathrm{bio}}/T_{\permil}}. \tag{7.10}$$

Using BCFs for stable isotopes will tend to overestimate bioaccumulation, which might suffice for screening-level assessments that are intentionally biased toward overestimating exposure, accumulation, and risk.

Using the CF approach to characterize the importance of exposure pathways in aquatic ecosystems assumes that the equilibrium concentration of an element accumulated by an organism is directly proportional to the concentration of that element dissolved in water. However, aquatic organisms exhibit the physiological ability to regulate the concentration of elements in their tissues. Therefore, aquatic organisms inhabiting waters with high concentrations of elements (e.g., calcium) do not accumulate the elements without limit. Similarly, organisms placed under conditions of low element concentrations still acquire sufficient concentrations to maintain themselves or grow. The regulated constant composition can be described by

$$(x/m)_{\text{equil}} = \frac{I_w C_w}{mr} = \xi, \qquad (7.11)$$

where

$x$ = amount of the stable element (e.g., moles) in an organ of $m$ grams
$I_w$ = intake rate (L g$^{-1}$t$^{-1}$)
$C_w$ = concentration of the element in water (e.g., moles L$^{-1}$)
$r$ = biological elimination rate constant ($t^{-1}$)
$\xi$ = size of the stable element pool (moles g$^{-1}$)

From the definition of the CF specified in equation 7.5, the CF is correctly given by

$$\text{CF} = \frac{(x/m)}{C_w} = \frac{\xi}{C_w}. \qquad (7.12)$$

Equation 7.12 can be plotted as a straight line with a slope = 1.0 on logarithmic scales:

$$\ln \text{CF} = \ln \xi - \ln C_w \qquad (7.13)$$

These simple plots can be constructed to compare the relative tendencies for elements to bioaccumulate (Peterson 1983).

The discussion thus far has considered the accumulation of individual radioactive elements under the assumption that the accumulation of one compound is essentially independent of the accumulation of other radionuclides (or chemicals). However, when two chemically analogous elements are present, they can compete for uptake and retention:

$$\xi = \frac{X_a}{m} + \frac{X_b}{m} = \frac{I_{wa} C_a}{mr_a} + \frac{I_{wb} C_b}{mr_b} = \frac{I_{wb}}{mr_b}[C_b + (\text{OR})_w C_a], \qquad (7.14)$$

where OR is the observed ratio, which is equivalent to $\text{OR}_w = I_{wa}/I_{wb} \cdot r_a/r_b$. The term outside of the brackets in equation 7.14 ($I_{wb}/mr_b$) can be seen to be the CF for element $b$ (see equation 7.4); therefore,

$$\text{CF}_b = \frac{\xi}{C_b + (\text{OR})_w C_a}. \qquad (7.15)$$

An equivalent expression for element $a$ can be derived from equation 7.14 and the observed ratio as defined above:

$$\text{CF}_a = \frac{(\text{OR})_w \xi}{C_b + (\text{OR})_w C_a}. \qquad (7.16)$$

The concepts and formulations in the form of equations 7.14 and 7.16 can be used to characterize or quantify the interaction between two chemically similar elements with regard to bioconcentration.

## Examples of Bioconcentration Factors

Peterson (1983) collated many of the previously published values of radionuclide BCFs for freshwater and marine plants, crustaceans, mollusks, and fish. Our purpose is not to repeat these extensive tables of BCFs, which in many instances remain useful in assessing exposure, dose, and risk. We instead refer the reader to Peterson (1983) for specific values and the sources of their derivation.

There have been additions to these earlier BCFs since the 1983 publication; much of the more recent work has focused on radionuclide accumulation by freshwater and marine fish. Table 7.1 lists more recently determined BCFs for selected radionuclides for fish. These values largely reflect the work of Poston and Klopfer (1988), the International Atomic Energy Agency (IAEA 1985), and the National Council on Radiation Protection and Measurements (NCRP 1996).

The derivation of CFs or BCFs relies on the assumption of equilibrium in the concentrations of radionuclides in the water and in the organism of interest. Departures from equilibrium either in the concentration of radionuclides in water or in the concentration in biota will introduce bias and imprecision into the estimation of a BCF. The fundamental nature of aquatic ecological systems suggests that these systems seldom, if ever, present equilibrium conditions in relation to the assumptions inherent in the derivation, estimation, and application of CFs to estimate bioaccumulation of radionuclides (or other chemicals). Physical and hydrodynamic forces that circulate water and distribute chemicals, combined with spatial and temporal variations in the chemical and biological processes that transform chemicals, result in far-from-equilibrium conditions in relation to bioaccumulation. However, in the absence of detailed scale-dependent kinetic information in relation to these dynamic physical, chemical, and biological environments, the concentration or BCFs might provide a useful approximation of the concentration of chemicals by aquatic organisms. The summary of spatial and temporal variations in the concentrations of several nuclides in Clinch River, Tennessee, water and fish described by Widner et al. (1999) provides a good example of the deviations from equilibrium conditions and their impacts on the corresponding BCF estimates (see table 7.2).

The BCFs listed in table 7.2 show variation in space and time. Estimated BCFs varied for different river miles in the Clinch River, and also from year to year. The year-to-year variability was summarized by calculating the geometric mean and geometric standard deviation for $^{137}$Cs, $^{90}$Sr, and $^{60}$Co for miles 2–5 and for miles 10–15 in the Clinch River using the yearly values from table 7.2. The results of this analysis are summarized in table 7.3. The summary suggests that $^{137}$Cs is more highly bioconcentrated than $^{60}$Co and $^{90}$Sr. All three nuclides indicated higher BCFs for Clinch River miles 10–15 than for miles 2–5. The year-to-year variability, expressed as the geometric standard deviation divided by the geometric mean, showed that the bioconcentration of $^{137}$Cs was less variable in time. $^{90}$Sr and $^{60}$Co were similar in their temporal variability and considerably more variable than $^{137}$Cs.

Widner et al. (1999) addressed the uncertainties in using BCFs to estimate radionuclide accumulation by fish by specifying the BCFs as log-triangular distributions. These distributions were developed to characterize uncertainties on the mean value of the BCFs that resulted from the representativeness of local data,

Table 7.1 Bioconcentration factors (L kg$^{-1}$) for selected elements for freshwater and marine fish (adapted from Poston and Klopfer [1988])

| Element/nuclide | Freshwater fish | Marine fish |
|---|---|---|
| Americium | 5[a] (30)[d] | 5[a] (10, 50) |
| | 25[b] | 25[b] |
| | 250[c] | 250[c] |
| Antimony | 200 (100) | 1,000 (1,000; 400) |
| Barium | 200 (4) | 10 (10, 50) |
| Cerium | 500 (30) | 100 (10, 50) |
| Cesium | 15,000/[K][a] | 100 (50, 100) |
| | (2,000) | |
| | 5,000/[K][e] | |
| Cobalt | 30 (300) | 50 (100; 1,000) |
| | | 10,000[f] |
| Curium | 5[a] (30) | 5[a] (10, 50) |
| | 25[b] | 25[b] |
| | 250[c] | 250[c] |
| Iodine | 200 (40) | 10 (10, 10) |
| | 500 | |
| Iron | 250[g] (200) | 3,000 (3,000; 3,000) |
| | 2,000[h] | |
| Manganese | 0.32/[Mn] (500) or 400[i] | 400 (500, 400) |
| Molybdenum | 10 (10) | 40 (10) |
| Neptunium | 5[a] (30) | 5[a] (10, 10) |
| | 25[b] | 25[b] |
| | 250[c] | 250[c] |
| Nickel | 100 (100) | 670 (500; 1,000) |
| Niobium | 200 (300) | 1[j] (100, 30) |
| Phosphorus | 3,000 (50,000) | 1,286 (30,000) |
| Plutonium | 5[a] (30) | 5[a] (1, 40) |
| | 25[b] | 25[b] |
| | 250[c] | 250[c] |
| Radium | 70 (50) | 50 (100, 500) |
| Ruthenium | 100 (10) | 2 (1, 2) |
| Scandium | 100 (100) | 750 (2; 1,000) |
| Sodium | 100 (20) | 0.07 (10, 0.10) |
| Strontium | 5.18-1.21 ln [Ca] | 5 (1, 2) |
| | 50[k] (60) | |
| Technetium | 15 (20) | 30 (10, 30) |
| Thorium | 100 (100) | 600 (10,000; 600) |
| Uranium | 50 (10) | 1 (1, 50) |
| Zinc | 2,500 (1,000) | 5,000 (2,000; 1,000) |
| Zirconium | 200 (300) | 1[j] (100, 20) |

[a] Piscivorous fish.
[b] Planktivorous fish.
[c] Bottom-feeding fish.
[d] First value in parentheses is from NCRP (1996), second value is from IAEA (1985).
[e] Non-piscivorous fish.
[f] Salmonids.
[g] Predatory fish.
[h] Omnivorous and planktivorous fish.
[i] Use 400 when [Mn] is unknown.
[j] For muscle only; IAEA (1985) values include contribution from bone.
[k] Generic value of 50 used when [Ca] is unknown.

local water conditions, ranges of possible BCFs (e.g., IAEA 1994), and incomplete understanding and quantification of the nonequilibrium dynamics of radionuclides in the Clinch River, Tennessee (see table 7.4).

Table 7.2 Summary of observed bioconcentration factors (BCFs), based on reported measurements of radionuclides in Clinch River fish and water (modified from Widner et al. 1999)

| Radionuclide and year | Clinch River miles 2–5[a] | | | Clinch River miles 10–15[b] | | |
|---|---|---|---|---|---|---|
| | Average concentrations | | Observed | Average concentrations | | Observed |
| | Fish ($Bq\ kg^{-1}$) | Water ($Bq\ L^{-1}$) | BCF ($L\ kg^{-1}$) | Fish ($Bq\ kg^{-1}$) | Water ($Bq\ L^{-1}$) | BCF ($L\ kg^{-1}$) |
| $^{137}Cs$ | | | | | | |
| 1960 | 18 | 0.085 | 210 | 30 | 0.094 | 320 |
| 1960–1962 | | | | 27 | 0.048 | 560 |
| 1963 | | | | 17 | 0.094 | 180 |
| May 1962–April 1963 | | | | 18 | 0.029 | 600 |
| 1978 | 8.7 | 0.026 | 330 | 3.2 | 0.043 | 73 |
| 1979 | 3.2 | 0.00074 | 4,300 | 26 | 0.00074 | 36,000 |
| 1980 | 3.1 | 0.0027 | 1,100 | 4.4 | 0.0030 | 1,500 |
| 1981 | 3.2 | 0.0039 | 830 | 3.3 | 0.0043 | 770 |
| 1982 | 2.1 | 0.0023 | 94 | 3.5 | 0.025 | 140 |
| 1983 | 2.9 | 0.0094 | 310 | 4.8 | 0.010 | 470 |
| 1985 | 2.0 | 0.0057 | 350 | 2.3 | 0.0063 | 370 |
| $^{90}Sr$ | | | | | | |
| 1960 | 24 | 0.35 | 68 | 29 | 0.39 | 75 |
| 1960–1962 | | | | 14 | 0.23 | 58 |
| 1963 | | | | 1.9 | 0.12 | 16 |
| 1978 | 0.15 | 0.0037 | 39 | 0.19 | 0.0041 | 46 |
| 1979 | 0.17 | 0.012 | 14 | 0.57 | 0.015 | 38 |
| 1980 | 0.38 | 0.025 | 15 | 0.36 | 0.028 | 13 |
| 1981 | 0.34 | 0.048 | 7.0 | 0.30 | 0.053 | 5.6 |
| 1982 | 0.41 | 0.061 | 6.8 | 0.33 | 0.067 | 4.9 |
| 1983 | 0.58 | 0.067 | 8.6 | 0.39 | 0.074 | 5.3 |
| 1984 | 0.24 | 0.037 | 6.6 | 0.58 | 0.041 | 14 |
| 1985 | 0.54 | 0.064 | 8.5 | 0.43 | 0.070 | 6.1 |
| $^{106}Ru$ | | | | | | |
| 1960 | 9.9 | 14 | 0.73 | 3.9 | 15 | 0.26 |
| 1960–1962 | | | | 4.8[c] | 11 | 0.44 |
| $^{60}Co$ | | | | | | |
| 1960 | 2.9 | 0.20 | 15 | 6.2 | 0.22 | 29 |
| 1960–1962 | | | | 2.6[c] | 0.17 | 16 |
| 1978 | 0.18 | 0.0041 | 43 | 0.31 | 0.0059 | 52 |
| 1979 | 0.16 | 0.0015 | 100 | 0.53 | 0.0019 | 280 |
| 1980 | 0.10 | 0.0071 | 14 | 0.18 | 0.0078 | 23 |
| 1981 | 0.14 | 0.0032 | 44 | 0.11 | 0.0035 | 32 |
| 1982 | 0.040 | 0.020 | 2.0 | 0.10 | 0.022 | 4.5 |
| 1985 | 0.11 | 0.0047 | 23 | 0.16 | 0.0052 | 30 |

[a] Fish were caught between Clinch River mile (CRM) 2 and CRM 5; water concentrations were for CRM 4.5 or adjusted to CRM 4.5 based on reported measurements at CRM 14.5.
[b] Fish were caught between CRM 10 and CRM 15; water concentrations were reported for CRM 14.5 or adjusted to CRM 14.5 based on reported measurements at CRM 4.5.
[c] Fish were caught between CRM 4.5 and CRM 19.1.

Table 7.3  Summary of spatial and temporal variability in BCF values in the Clinch River, Tennessee (based on Table 7.2)

|  | Clinch River miles 2–5 | Clinch River miles 10–15 |
|---|---|---|
| $^{137}Cs$ | | |
| Geometric mean | 479.0 | 551.0 |
| Geometric standard deviation | 3.25 | 5.06 |
| GSD/GM[a] | 0.0068 | 0.0092 |
| $^{90}Sr$ | | |
| Geometric mean | 13.27 | 16.16 |
| Geometric standard deviation | 2.30 | 2.80 |
| GSD/GM | 0.173 | 0.173 |
| $^{60}Co$ | | |
| Geometric mean | 21.04 | 30.09 |
| Geometric standard deviation | 3.48 | 3.18 |
| GSD/GM | 0.165 | 0.106 |

[a] GSD = geometric standard deviation; GM = geometric mean.

Table 7.4  Distributions used to estimate radionuclides in fish flesh for the Clinch River, Tennessee (adapted from Widner et al. [1999])

| Radionuclide | Mean[a] | Minimum | Maximum |
|---|---|---|---|
| $^{137}Cs$ | 600 | 120 | 3,000 |
| $^{90}Sr$ | 10 | 1 | 100 |
| $^{106}Ru$ | 1 | 0.1 | 10 |
| $^{60}Co$ | 30 | 6 | 150 |

[a] Distributions defined as log triangular.

## *Bioconcentration Factors in Screening-Level Risk Estimations*

Values of BCFs are used in screening-level assessments of potential human health risks (NCRP 1996). The objective of this type of screening assessment is to bias risk estimation by intentionally selecting parameter values to overestimate exposure. Radionuclides that appear to pose minimal health risks as the result of these pessimistic assumptions concerning exposure might be removed or given lower priority in subsequent, more detailed (and costly) assessment. Radionuclides that appear to pose unacceptable risks using this biased approach can be examined further using more realistic assumptions or, if possible, through direct measures of exposure.

An example of using BCFs in screening-level risk estimation is illustrated by the following general equation that describes risk posed by the consumption of contaminated fish (Widner et al. 1999):

$$\text{SLRE}_{\text{fish}} = (C_w \text{BCF}) \cdot U_{\text{fish}} \cdot F_{\text{cf}} \cdot \text{EF} \cdot \text{ED} \cdot \text{SF}_{\text{ing}}, \quad (7.17)$$

where

$\text{SLRE}_{\text{fish}}$ = screening-level risk estimate for ingestion of fish
$C_w$ = concentration of radionuclide in surface water (Bq L$^{-1}$)
BCF = bioconcentration factor (L kg$^{-1}$)
$U_{\text{fish}}$ = average daily consumption of fish, e.g., 0.03 (kg days$^{-1}$)
$F_{\text{cf}}$ = fraction of fish consumed that is contaminated, e.g., 0.80 (unitless)
EF = exposure frequency, e.g., 365 (days yr$^{-1}$)
ED = exposure duration, e.g., 30 (yr)
$\text{SF}_{\text{ing}}$ = oral slope factor (risk Bq$^{-1}$)

The values indicated above for several of the exposure parameters ($U_{\text{fish}}$, $F_{\text{cf}}$, EF, ED) were combined with the values of BCF and $\text{SF}_{\text{ing}}$ to estimate annual excess cancer risks posed by consuming fish that were exposed to concentrations ($C_w$) of selected radionuclides potentially released to the Clinch River, Tennessee, in 1956 (Widner et al. 1999). The greatest risks were posed by $^{137}$Cs and $^{32}$P. These radionuclides also exhibited the highest BCF values in this analysis. Lesser risks resulted for $^{90}$Sr and $^{60}$Co, while the remaining radionuclides suggested minimal risks. Risk posed by all the examined radionuclides associated with the ingestion of contaminated fish summed to $8.52 \times 10^{-5}$ for 1956. Again, it must be emphasized that these are annual estimates and that the original assessment integrated annual risk estimates from 1944 to 1991 to develop a more comprehensive screening-level assessment than is summarized in table 7.5. The 1956 data were selected for this example because the greatest releases to the Clinch River apparently occurred during the middle to late 1950s (Widner et al. 1999).

This example demonstrates one use of the BCF values in human health risk assessment; analogous exposures can be developed for birds, mammals, and other wildlife in the assessment of potential ecological risks posed by radionuclides. In these assessments, the uncertainties associated with the derivation of the BCF values can dramatically influence the results of the assessment. This example focuses on the bioconcentration of radionuclides by fish. However, BCF values for organisms representative of lower trophic levels (e.g., planktonic algae, planktonic animals, and organisms that inhabit the sediments) might also become important in the assessment, particularly if fish derive significant amounts of contamination from these and other contaminated food sources.

## Bioaccumulation Factors in Estimating Exposure

In addition to screening-level assessments, BCFs are also important in estimating more realistic human and nonhuman exposures to radionuclides and other chemical contaminants. The objective of the following calculations is to replace the biased estimation characteristic of the screening-level risk assessment with estimates of risk that are as accurate as possible, given limitations in available data. The annual average intake of a radionuclide can be expressed as

$$\text{Intake} = (C_w \text{BCF}) \cdot N_{\text{fish}} \cdot P_{\text{fish}} \cdot F_{\text{cf}} \cdot \text{EF} \cdot \text{ED} \cdot F_{\text{R,fish}}, \tag{7.18}$$

where

Intake = amount of radioactivity ingested (Bq)
$C_w$ = concentration of radionuclide in surface water (Bq L$^{-1}$)
BCF = bioconcentration factor (L kg$^{-1}$)
$N_{fish}$ = number of fish meals (fish meals days$^{-1}$)
$P_{fish}$ = size of fish meal portions (kg fish meal$^{-1}$)
$F_{cf}$ = fraction of fish consumed that is contaminated (unitless)
EF = exposure frequency (days yr$^{-1}$)
ED = exposure duration (yr)
$F_{R,fish}$ = fraction of radionuclide remaining in fish after processing (unitless)

Compared to the screening-level calculation, the generic exposure equation above adds detail in the form of the number of fish meals consumed, the size of each meal, and the impact of fish processing (e.g., boiling, frying) on the residual radioactivity in the meal. Clearly, the exposure estimate depends directly on the value of the BCF used in the calculation. The exposure assessment replaces the upper value estimates of the BCFs used in screening with values more representative of the expected value (e.g., mean, median, and mode).

Table 7.5 Example screening-level risk estimates (SLREs) for ingestion of contaminated fish (adapted from Widner et al. [1999])

| Radionuclide | Oral slope factor[a] (risk Bq$^{-1}$) | Screening-level BCF[b] (L kg$^{-1}$) | Estimated concentration in water (1956)[c] (Bq L$^{-1}$) | 1956 SLRE$_{fish}$ (risk) |
|---|---|---|---|---|
| $^{137}$Cs | 8.5 × 10$^{-10}$ | 2,000 | 4.40 | 6.55 × 10$^{-05}$ |
| $^{106}$Ru | 9.3 × 10$^{-10}$ | 10 | 0.75 | 6.11 × 10$^{-08}$ |
| $^{90}$Sr | 1.5 × 10$^{-09}$ | 60 | 2.60 | 2.05 × 10$^{-06}$ |
| $^{60}$Co | 5.1 × 10$^{-10}$ | 300 | 1.20 | 1.61 × 10$^{-06}$ |
| $^{144}$Ce | 8.0 × 10$^{-10}$ | 30 | 1.50 | 3.15 × 10$^{-07}$ |
| $^{95}$Zr | 1.1 × 10$^{-10}$ | 300 | 0.31 | 8.96 × 10$^{-08}$ |
| $^{95}$Nb | 6.1 × 10$^{-11}$ | 300 | 0.39 | 6.25 × 10$^{-08}$ |
| $^{235}$U | 1.3 × 10$^{-09}$ | 10 | 0.0073 | 8.31 × 10$^{-10}$ |
| $^{238}$U | 1.2 × 10$^{-09}$ | 10 | 0.0073 | 7.67 × 10$^{-10}$ |
| $^{240}$Pu | 8.5 × 10$^{-09}$ | 30 | 0.0073 | 1.63 × 10$^{-08}$ |
| $^{232}$Th | 8.9 × 10$^{-10}$ | 100 | 0.0073 | 5.69 × 10$^{-09}$ |
| $^{241}$Am | 8.9 × 10$^{-10}$ | 30 | 0.0073 | 1.71 × 10$^{-09}$ |
| $^{154}$Eu | 2.5 × 10$^{-10}$ | 50 | 3.60 | 3.94 × 10$^{-07}$ |
| $^{147}$Pm | 3.8 × 10$^{-11}$ | 30 | 3.60 | 3.60 × 10$^{-08}$ |
| $^{140}$Ba | 3.2 × 10$^{-10}$ | 5 | 0.097 | 1.36 × 10$^{-09}$ |
| $^{32}$P | 1.6 × 10$^{-10}$ | 10,000 | 0.97 | 1.36 × 10$^{-05}$ |
| $^{91}$Y | 3.6 × 10$^{-10}$ | 30 | 3.60 | 3.41 × 10$^{-07}$ |
| $^{143}$Pr | 1.8 × 10$^{-10}$ | 30 | 3.60 | 1.70 × 10$^{-07}$ |
| $^{147}$Nd | 1.6 × 10$^{-10}$ | 30 | 3.60 | 1.51 × 10$^{-07}$ |

[a] Values from 1995 HEAST tables (see www.epa.gov/radiation/heast/).
[b] Values from IAEA (1994).
[c] Values from Table 3.B.5 in Widner et al. (1999).

## Bioaccumulation under Nonequilibrium Conditions: The Chernobyl Cooling Pond Example

Radioecological exposure assessments in aquatic systems involve trade-offs between detailed models of water circulation, contaminant transport, and ecological interactions and the frequently encountered simplifying assumptions of well-mixed and steady-state systems (Breck 1980). As discussed in this chapter, among the simplified approaches, the BCF is the most widely used approach to characterize the accumulation of radionuclides by aquatic organisms (Williams 1960; Friend 1963; Harvey 1964; Fujita et al. 1966; Garder and Skulberg 1966; Fleishman 1973). The BCF may be useful for screening-level analysis (NCRP 1991; Evans 1988; IAEA 1993). However, significant spatial and temporal variability in environmental conditions and the variability of ecological processes complicate the estimation of exposures in natural systems. Importantly, the effects of such variability are not addressed in estimation of the BCF.

Ecological exposure assessments and estimates of bioaccumulation should be integrated with ecosystem science. Under nonequilibrium conditions, ecologically realistic assessments of exposure and subsequent accumulation of radionuclides should consider, for example, density-dependent predator–prey relationships in a multidimensional physical and chemical environment, population dynamics, changes in the distribution and abundance of population biomass, and the variability in ingestion and metabolic process rates among organisms within populations.

Spatial and temporal heterogeneity is an integral part of natural systems (Mackay and Paterson 1984), and such heterogeneity complicates exposure modeling. Biological populations are often distributed unevenly in space, that is, a clumped distribution (Patalas 1969; Lewis 1979). The effects of spatial and temporal variations of contaminant concentrations on bioaccumulation have been examined (e.g., Bartell and Brenkert 1991; Breck et al. 1988).

The following example developed for the Chernobyl cooling pond systematically examines the implications of these kinds of variability on the estimation of bioaccumulation of $^{137}$Cs. Reactor No. 4 at the Chernobyl nuclear power plant exploded on April 26, 1986. The total release of fission products from the damaged unit was estimated as $1.85 \times 10^{18}$ Bq, including approximately $8.5 \times 10^{16}$ Bq of $^{137}$Cs (IAEA 1986). The analysis focused on radioactive contaminant transport during the period of May–September 1986, when there was no pumped intake or release of water from the Chernobyl power plant. The major forcing function for water circulation during this period was wind. After September 1986, the unit returned to operation, and water intake–discharge activities resumed.

Water is supplied to the Chernobyl nuclear power plant from a cooling pond that is located southeast of the plant site, as shown in figure 7.2 (BIOMOVS II 1996). The cooling pond has a total area of approximately 22 km$^2$ and a volume of approximately $1.5 \times 10^8$ m$^3$ (see table 7.6). There are a number of deep areas in the cooling pond with depths greater than 10 m; about 28% of the total pond volume is contained in these deeper areas.

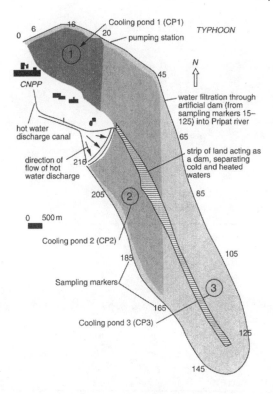

Figure 7.2 Schematic illustration of the Chernobyl nuclear power plant (CNPP) cooling pond (Feng 1995).

Table 7.6  Geometry information of the Chernobyl cooling pond

| | |
|---|---|
| Volume (m$^3$) | $1.5 \times 10^8$ |
| Area (km$^2$) | 22.0 |
| Length (km) | 11.4 |
| Width, average (km) | 2.0 |
| Depth, maximum (m) | 20.0 |
| Depth, average (m) | 6.6 |

### Initial $^{137}$Cs Contamination in the Chernobyl Cooling Pond Water

The contamination in the cooling pond water was mainly from atmospheric fallout from April through May 1986. The initial deposition pattern defined three areas in the cooling pond (CP1–CP3; see figure 7.2) that received different amounts of contamination (BIOMOVS II 1996; Kryshev 1992). The concentration of $^{137}$Cs in water was highest in the northwest part of the cooling pond water and lowest in the southern part of the pond (table 7.7).

Table 7.7  Initial deposition of $^{137}$Cs in the Chernobyl cooling pond water[a]

|  | CP1[b] | CP2[b] | CP3[b] |
|---|---|---|---|
| Water volume (m$^3$) | $13 \times 10^6$ | $76 \times 10^6$ | $62 \times 10^6$ |
| $^{137}$Cs concentration in kBq m$^{-3}$ or ($\mu$Ci m$^{-1}$) | 830[b] ($2.24 \times 10^{-5}$) | 600 ($1.62 \times 10^{-5}$) | 45 ($1.2 \times 10^{-6}$) |

[a]Deposition concentrations decay corrected to May 30, 1986 (see figure 7.2), based on Kryshev (1992).
[b]See figure 7.2 for regions of cooling pond.

Table 7.8  Annual averaged biomass and biomass density in the Chernobyl cooling pond

| Population | Biomass (kg) | Biomass density (kg m$^{-3}$) |
|---|---|---|
| Phytoplankton | $1,000 \times 10^3$ | $1,000 \times 10^3$ |
| Zooplankton | $270 \times 10^3$ | $270 \times 10^3$ |
| Nonpredatory fish | $50 \times 10^3$ | $50 \times 10^3$ |
| Predatory fish | $10 \times 10^3$ | $10 \times 10^3$ |

### The Chernobyl Cooling Pond Ecosystem

The Chernobyl cooling pond is a habitat for more than 500 species of algae and more than 20 species of fish. The forage, or non-predatory fish include carp, silver bream, bream, silver carp, roach, and goldfish. The predatory fish include pike-perch, pike, perch, and asp (BIOMOVS II 1996). Each year, there are about 2,100,000 kg of total biomass production (BIOMOVS II 1996) in this cooling pond. Annual averaged biomass for phytoplankton, zooplankton, forage fish, and predatory fish have been estimated, as shown in table 7.8, and the cooling pond has been classified as slightly to moderately polluted with organic matter and heavily polluted by nitrogen (Kryshev 1992).

The exposure assessment for the Chernobyl aquatic ecosystem addressed the following issues relevant to bioaccumulation under nonequilibrium conditions:

- Radionuclide transport and distribution in water are controlled by water circulation influenced by wind, temperature change, inlet and outlet flow, and bathymetry (Bedford and Abdelrhman 1987; Connolly 1985; Thibodeaux 1979; Csanady 1970; Schwab and Bennett 1987).
- The contaminants and aquatic organisms are heterogeneously distributed (DeAngelis 1978; Lampert 1989; Steele 1974), which can result in significant variations in exposures (Smith 1992).
- Ecological processes are dynamic in nature (Odum 1971; Carpenter 1988). Correspondingly, radionuclide uptake and accumulation vary in time and space.

To address these issues, the Chernobyl cooling pond case study examined the following questions through a series of simulations:

1. What is the influence of hydrodynamics on modeled $^{137}$Cs transport and distribution?
2. How is bioaccumulation affected by the spatial and temporal heterogeneity in the distribution of $^{137}$Cs in the environment?
3. How might variability in the distribution and abundance of population biomass affect estimates of bioaccumulation?
4. How does spatial and temporal variability in biological process rates influence estimates of bioaccumulation?
5. How do different food web structures magnify or reduce the exposure to predators in a heterogeneous radioactive environment?

These questions concerning bioaccumulation were addressed through the following modeling activities. First, a two-dimensional hydrodynamic transport model was developed to calculate the concentrations of $^{137}$Cs in the Chernobyl cooling pond (Feng 1995). The estimated distributions of $^{137}$Cs were used to investigate the impacts of spatial and temporal heterogeneity of radioactivity on estimates of bioaccumulation. Second, the two-dimensional dynamic transport modeling framework was expanded to describe the $^{137}$Cs uptake and accumulation processes for a simple aquatic food chain. Third, modeled changes in abundance and distribution of population biomass were integrated into the exposure model to investigate the effects of this source of variability on simulated bioaccumulation of $^{137}$Cs by phytoplankton, zooplankton, forage fish, and piscivorous fish. Fourth, the hydrodynamic transport model was combined with a population dynamics model to simulate the distributions of phytoplankton and zooplankton biomass distributions. Biomass changes due to growth, respiration, and grazing were incorporated into estimates of $^{137}$Cs bioaccumulation. Fifth, spatial- and temporal-dependent ingestion rates were incorporated into the Chernobyl food web model.

## Chernobyl Cooling Pond Model Structure

The formulation of the model structure was based on the principle of conservation of mass for functionally defined populations in a dynamic two-dimensional space. Phytoplankton, zooplankton, forage fish, and predatory fish defined a simple aquatic food chain in the model of the Chernobyl cooling pond.

Three radionuclide concentrations were modeled: (1) $^{137}$Cs concentrations (Bq m$^{-3}$) in water, $C_w(x, y, z, t)$; (2) specific radionuclide concentrations in populations, $C_i(x, y, z, t)$ with the unit of Bq kg$^{-1}$; and (3) radionuclide volume concentration (Bq m$^{-3}$) expressed as the product of population biomass density, $m_i(x, y, z, t)$ (kg m$^{-3}$), and specific radionuclide concentration, $C_i(x, y, z, t)$ (Bq kg$^{-1}$). This representation made it possible to model population exposures in specific parcels of water in which $C_w(x, y, z, t)$ was calculated using a hydrodynamic transport model.

The volume concentration equation describes the net flux of radionuclides into the biota as the sum of the uptake and loss rates:

$$\frac{D[m_i(\vec{r},t)\,C_i(\vec{r},t)]}{Dt} = k_{ui}\,m_i(\vec{r},t)\,C_w(\vec{r},t) + k_{fi}\,m_i(\vec{r},t)\,C_{i-1}(\vec{r},t)$$
$$- k_{bi}\,m_i(\vec{r},t)\,C_i(\vec{r},t) - \lambda\,m_i(\vec{r},t)\,C_i(\vec{r},t).$$

$$i = 1, 4, \quad (7.19)$$

where

$r$ = location $(x, y, z)$ in a coordinate space
$i$ = population index (1 for phytoplankton, 2 for zooplankton, 3 for forage or non-predatory fish, and 4 for predatory fish)
$C_w(r, t)$ = radionuclide concentration (Bq m$^{-3}$) in water
$C_i(r, t)$ = radionuclide concentration (Bq kg$^{-1}$) in a population $i$
$C_{i-1}(r, t)$ = radionuclide concentration (Bq kg$^{-1}$) in a population $i-1$
$m_i(r, t)$ = biomass density (kg m$^{-3}$) of population at $i$
$k_{ui}$ = radionuclide uptake rate (m$^3$ kg$^{-1}$ day) from the water for population $i$
$k_{bi}$ = radionuclide biological elimination rate (1 day$^{-1}$) for population $i$
$k_{fi}$ = ingestion rate (kg prey kg predator$^{-1}$ day$^{-1}$) of prey for population $i$
$\lambda$ = radioactive decay constant (1 day$^{-1}$)

The first term on the right side of equation 7.19 describes the radionuclide uptake from water (i.e., bioconcentration). The second term accounts for the bioaccumulation of radionuclides from food. The third and fourth terms express radionuclide loss from the biota due to biological elimination and radioactive decay.

The advection term involved two assumptions: (1) that fish stayed within their home range (or volume), which was specified as 200 m × 200 m × depth (Breck et al. 1988), and (2) that both phytoplankton and zooplankton accumulated and eliminated radionuclides within this defined volume. The rational for the second assumption was that the radionuclides have a comparatively fast radiobiological turnover rate.

Equation 7.19 was expanded to directly express the $^{137}$Cs concentration in population $i$ as a function of $^{137}$Cs in water, food (i.e., phytoplankton), or prey biomass:

$$C_i(\vec{r},t)\frac{\partial m_i(\vec{r},t)}{\partial t} + m_i(\vec{r},t)\frac{\partial C_i(\vec{r},t)}{\partial t} = k_{ui}\,m_i(\vec{r},t)\,C_w(\vec{r},t) + k_{fi}\,m_i(\vec{r},t)\,C_{i-1}(\vec{r},t)$$
$$- k_{bi}\,m_i(\vec{r},t)\,C_i(\vec{r},t) - \lambda\,m_i(\vec{r},t)\,C_i(\vec{r},t).$$

$$(7.20)$$

Rearranging equation 7.20 gives

$$m(\vec{r},t)\frac{\partial C_i(\vec{r},t)}{\partial t} = k_{ui}\,m_i(\vec{r},t)\,C_w(\vec{r},t) + k_{fi}\,m_i(\vec{r},t)\,C_{i-1}(\vec{r},t)$$
$$- k_{bi}\,m_i(\vec{r},t)\,C_i(\vec{r},t) - \lambda\,m_i(\vec{r},t)\,C_i(\vec{r},t) - C_i(\vec{r},t)\frac{\partial m_i(\vec{r},t)}{\partial t}.$$

$$(7.21)$$

Dividing both sides by $m_i(\vec{r}, t)$ results in

$$\frac{\partial C_i(\vec{r},t)}{\partial t} = k_{ui}\, C_w(\vec{r},t) + k_{fi}\, C_{i-1}(\vec{r},t)\, C_i(\vec{r},t) \left[ \frac{\partial m_i(\vec{r},t)}{\partial t} \frac{1}{m_i(\vec{r},t)} + k_{bi} + \lambda \right]. \quad (7.22)$$

Equation 7.22 expresses the specific concentration in populations as a function of population biomass change $\partial m_i(\vec{r},t)/\partial t\; 1/m_i(\vec{r},t)$ and allows for the influence of spatial and temporally varying population dynamics to be included into the modeling of radionuclide uptake and accumulation.

The continuous exposure model (equation 7.22) was approximated by performing calculations for every spatial element with the location index $j$ in a spatial grid representation of the Chernobyl cooling pond (Feng 1995):

$$\frac{dC_{ij}(t)}{dt} = k_{ui}\, C_{wj}(t) + k_{fi}\, C_{(i-1)j}(t)\, C_{ij}(t) \left[ \frac{dm_{ij}(t)}{dt} \frac{1}{m_{ij}(t)} + k_{bi} + \lambda \right], \quad (7.23)$$

where

$C_{ij}(t)$ = concentrations in population $i$ at location $j$
$C_{wj}(t)$ = concentration in water at location $j$, which is calculated using the hydrodynamic transport model
$C_{(i-1)j}(t)$ = concentrations in prey at location $j$
$m_{ij}(t)$ = biomass of four food chain populations at location $j$

If spatial effects are ignored, equation 7.23 becomes

$$\frac{dC_i(t)}{dt} = k_{ui}\, C_w(t) + k_{fi}\, C_{i-1}(t)\, k_{bi}\, C_i(t) \left[ \frac{dm_i(t)}{dt} \frac{1}{m_i(t)} C_i(t) - \lambda\, C_i(t) \right], \quad (7.24)$$

which describes bioaccumulation in a homogeneous environment. If population biomass changes are insignificant $[dm_{ij}(t)/dt/m_{ij}(t) = 0]$, one obtains

$$\frac{dC_i(t)}{dt} = k_{ui}\, C_w(t) + k_{fi}\, C_{i-1}(t)\, k_{bi}\, C_i(t) - \lambda\, C_i(t). \quad (7.25)$$

Further, if a steady-state condition $[dC_i(t)/dt = 0]$ is assumed, the steady-state solution of equation 7.25 is

$$C_i = \frac{k_{ui}\, C_w + k_{fi}\, C_{i-1}}{k_{bi} + \lambda}$$

$$= \frac{k_{ui}}{k_{bi} + \lambda} C_w + \frac{k_{fi}}{k_{bi} + \lambda} C_{i-1}, \quad (7.26)$$

which defines the bioaccumulation factor discussed earlier in this chapter. The first term on the right-hand side of equation 7.26 defines the accumulation of a radionuclide from water (i.e., the CF, analysis to equation 7.7). The second term describes the accumulation from ingestion of contaminated prey. That is, $k_{ui}/(k_{bi} + \lambda)$ defines the BCF, and $k_{fi}/(k_{bi} + \lambda)$ is the bioaccumulation factor.

## Food Web Structure

In the simple food chain structure, the flow of radionuclides is from phytoplankton through zooplankton to forage fish and then to piscivorous fish. The radionuclide transfer for each food chain component can be stated as follows:

Phytoplankton

$$\frac{dC_{1j}(t)}{dt} = k_{u1} C_{wj}(t) - C_{1j}(t) \left[\frac{dm_{1j}(t)}{dt}\frac{1}{m_{1j}(t)} + k_{b1} + \lambda\right] \quad (7.27)$$

Zooplankton

$$\frac{dC_{2j}(t)}{dt} = k_{u2} C_{wj}(t) + k_{f2} C_{1j}(t) - C_{2j}(t) \left[\frac{dm_{2j}(t)}{dt}\frac{1}{m_{2j}(t)} + k_{b2} + \lambda\right] \quad (7.28)$$

Forage fish

$$\frac{dC_{3j}(t)}{dt} = k_{u3} C_{wj}(t) + k_{f3} C_{2j}(t) - C_{3j}(t) \left[\frac{dm_{3j}(t)}{dt}\frac{1}{m_{3j}(t)} + k_{b3} + \lambda\right] \quad (7.29)$$

Piscivorous fish

$$\frac{dC_{4j}(t)}{dt} = k_{u4} C_{wj}(t) + k_{f4} C_{3j}(t) - C_{4j}(t) \left[\frac{dm_{4j}(t)}{dt}\frac{1}{m_{4j}(t)} + k_{b4} + \lambda\right] \quad (7.30)$$

where $C_{1j}(t)$, $C_{2j}(t)$, $C_{3j}(t)$, and $C_{4j}(t)$ are radionuclide concentrations in phytoplankton, zooplankton, forage fish, and piscivorous fish in the $j$th spatial element, and $m_{1j}(t)$, $m_{2j}(t)$, $m_{3j}(t)$, and $m_{4j}(t)$ denote biomass values for these four populations.

To include effects of feeding interactions on the bioaccumulation in predatory fish, food preference factors ($f_i$) were added to equation 7.28 to give

$$\frac{dC_{4j}(t)}{dt} = k_{u4j} C_{wj}(t) + \sum f_i k_{f4i} C_{ij}(t) - C_{4j}(t) \left[\frac{dm_{4j}(t)}{dt}\frac{1}{m_{4j}(t)} + k_{b4} + \lambda\right], \quad (7.31)$$

where

$\sum f_i k_{f4i} C_{ij}(t)$ = total radionuclide uptake from the predator's consumption of contaminated prey ($i = 1, 2, 3$)

$i$ = prey, which include phytoplankton ($i = 1$), zooplankton ($i = 2$), and forage fish ($i = 3$)

$f_i$ = fraction of predator's prey that consists of populations with radionuclide concentration of $C_{ij}(t)$

## Population Dynamics and Biomass Distributions

Equation 7.22 shows that the change in biomass $\partial m_i(\vec{r}, t)/\partial t \; 1/m_i(\vec{r}, t)$ is an integral part of the radionuclide accumulation process. The effect of biomass variation on the accumulation of radionuclides in a two-dimensional space can be examined by varying $m_{ij}$ in equation 7.23. The mechanisms influencing biomass variation are discussed in the following sections for each of the four trophic components.

*Phytoplankton* Phytoplankton growth is influenced by daily variations in incident light, water temperature, nonpredatory mortality, and grazing (Mitsch et al. 1981), which can be expressed as

$$\frac{dm_1}{dt} = m_1(P_1 - M_1 - G_1), \qquad (7.32)$$

where

$m_1$ = biomass of phytoplankton (kg m$^{-3}$)
$P_1$ = photosynthetic-based phytoplankton growth rate (days$^{-1}$)
$M_1$ = mortality rate (days$^{-1}$) accounting for sinking, respiration, and nonpredatory loss
$G_1$ = zooplankton grazing rate on phytoplankton (days$^{-1}$)

In nature, these rates vary with environmental conditions and the physiological status of organisms in the population. The net growth rate, $\theta_1$ (days$^{-1}$), that determines phytoplankton population dynamics is

$$\theta_1 = P_1 - M_1 - G_1. \qquad (7.33)$$

Phytoplankton distributions also depend on water circulation patterns (Riley 1976; Pinel-Alloul et al. 1988; George and Edwards 1973; Smith et al. 1976; Taylor 1984; and Taylor and Taylor 1977).

Phytoplankton population dynamics were integrated into the two-dimensional transport model as

$$\frac{\partial m_1}{\partial t} + v_x \frac{\partial m_1}{\partial x} + v_y \frac{\partial m_1}{\partial y} - D_x \frac{\partial^2 m_1}{\partial x^2} - D_y \frac{\partial^2 m_1}{\partial y^2} + m_1 \theta_1 = 0, \qquad (7.34)$$

where

$m_1$ = phytoplankton biomass (kg m$^{-3}$)
$v_x$ and $v_y$ = flow velocity components (m s$^{-1}$) in x- and y-directions
$D_x$ and $D_y$ = turbulent diffusion coefficients (m$^2$ s$^{-1}$) in x- and y-directions
$\theta_1$ = net rate of change of phytoplankton biomass (days$^{-1}$) due to physiological and grazing processes (i.e., equation 7.33)

Solving equation 7.34 produced temporal and spatial distributions of phytoplankton biomass.

*Zooplankton* Population growth dynamics of zooplankton were formulated as

$$\frac{dm_2}{dt} = m_2(C_2 - M_2 - F_2), \qquad (7.35)$$

where

$m_2$ = zooplankton biomass (kg m$^{-3}$)
$C_2$ = zooplankton growth rate (days$^{-1}$)
$M_2$ = combined nonpredatory mortality rate (days$^{-1}$)
$F_2$ = loss rate due to consumption by forage fish (days$^{-1}$)

Therefore, the net change rate of zooplankton biomass ($\theta_2$) was described by

$$\theta_2 = C_2 - M_2 - F_2. \tag{7.36}$$

The same physical model (equation 7.34) used for phytoplankton transport was also used for zooplankton transport and growth using the net change rate for $\theta_2$.

*Fish* The population dynamics for forage fish and predatory fish were expressed separately as

Forage fish

$$\frac{dm_3}{dt} = m_3(C_3 - M_3 - F_3) \tag{7.37}$$

Predatory fish

$$\frac{dm_4}{dt} = m_4(C_4 - M_4) \tag{7.38}$$

where parameters $C_i$, $M_i$, and $F_i$ have the same meanings as those defined in equation 7.35.

The impact of environmental factors on fish distributions is complicated by fish migration, random movement, and home range movements (Harden Jones 1988; Salis and Flowers 1969). For the simulations of fish distributions in a two-dimensional environment, two scenarios were examined: (1) fish homogeneously distributed in water and (2) fish heterogeneously distributed in water.

## Spatial and Temporal Radionuclide Ingestion Rates

The specific rate at which prey are consumed by a predator is influenced by temperature and is proportional to the biomass of prey; thus, radionuclide ingestion also varies in relation to these factors, which were formulated in the models as (e.g., Bartell et al. 1992)

$$F_{fi} = aC_{mi}h(T)\frac{f_{i-1}m_{i-1}}{\sum f_{i-1}m_{i-1} + m_i}, \tag{7.39}$$

where

$F_{fi}$ = feeding rate of predator (days$^{-1}$)
$m_{i-1}$ = prey biomass (kg m$^{-3}$)
$m_i$ = predator biomass (kg m$^{-3}$)
$f_{i-1}$ = preference of the predator for the prey (unitless)
$h(T)$ = temperature dependence of feeding (unitless)
$C_{mi}$ = maximum feeding rate of predator $i$ (days$^{-1}$)
$a$ = assimilation efficiency of the prey by the predator (unitless)

The radionuclide ingestion rate ($k_{fi}$) was calculated as $k_{fi} = E_{fi}F_{fi}$ (Harvey 1964), where $E_{fi}$ is the assimilation of radionuclide from contaminated prey. The temperature and prey density-dependent varying ingestion rates were included in the

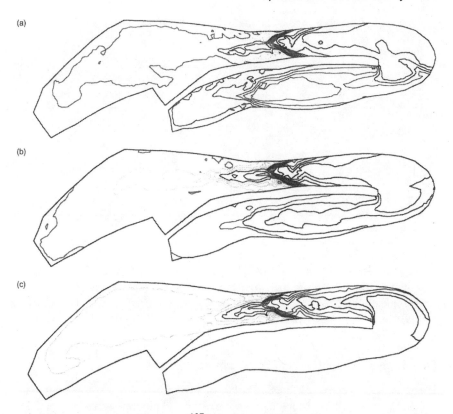

**Figure 7.3** Simulated distribution of $^{137}$Cs in the Chernobyl cooling pond from wind-driven circulation: (a) after 32 days, (b) after 56 days, and (c) after 120 days.

exposure model to investigate their influence on $^{137}$Cs accumulation by zooplankton and fish populations in the Chernobyl cooling pond. The ingestion rate ($k_{fi}$) of these consumers was represented as a function of location and time by

$$k_{fi}(x,y,t) = E_{fi}\, aC_m\, h(T) \frac{f_{i-1}\, m_{i-1}(x,y,t)}{\sum f_{i-1}\, m_{i-1}(x,y,t) + m_i(x,y,t)} \qquad i = 2,3,4. \qquad (7.40)$$

## Radionuclide Transport and Distribution

Concentrations of $^{137}$Cs in water in two-dimensional space were generated for 120 model days. Figure 7.3a–c shows the pattern of $^{137}$Cs transport after 32, 56, and 120 days, respectively. The initial concentrations were slightly dispersed after day 1 as the result of convection. The three initial differentially contaminated regions remained, and the maximum concentration was 830 kBq m$^{-3}$ in the upper north end of the cooling pond. However, the $^{137}$Cs along the central line of the cooling pond began to follow the pattern of simulated water currents toward the northwest region of the pond by day 8. The longer term, wind-driven circulation and turbulent diffusion produced highly variable $^{137}$Cs concentrations after days 16, 24, 32, 40, 56, 72, and 88. The initial concentration of 830 kBq m$^{-3}$ gradually decreased to

Table 7.9 Categorization of simulation cases based on spatial and temporal details and on spatial and temporal-dependent ecological functions

| Cases | Temporal | Spatial | Population biomass | Radionuclide ingestion rates | Multi-prey food web structure |
|---|---|---|---|---|---|
| 1 | | | | | |
| 2 | Yes | | | | |
| 3 | Yes | | Yes | | |
| 4 | Yes | Yes | | | |
| 5 | Yes | Yes | Yes | | |
| 6 | Yes | Yes | Yes | Yes | |
| 7 | Yes | Yes | Yes | Yes | Yes |

600 kBq m$^{-3}$. After 88 days, the distribution of $^{137}$Cs became more homogeneous within two distinct regions in the cooling pond that were determined mainly by the bathymetry of the pond and the simulated physical circulation. The pond was not yet completely mixed after 120 simulated days; the upper region of the pond had 540 kBq m$^{-3}$ of $^{137}$Cs in water; the model predicted 300 kBq m$^{-3}$ in the lower region of the pond.

## Case Studies in Exposure and Bioaccumulation

The principal objective of the Chernobyl modeling exercise was to examine the effects of nonequilibrium conditions on exposure and bioaccumulation of radionuclides. Seven simulations were performed with and without spatial and temporal variability to compare model results in relation to this objective (see table 7.9).

### Case 1: Homogeneous and Steady-State Exposures

The pondwide averaged concentration of $^{137}$Cs in water ($C_w$) of $4 \times 10^5$ Bq m$^{-3}$, uptake rates from water ($k_{ui}$), ingestion rates from food ($k_{fi}$), and biological elimination rates ($k_{bi}$) were used to simulate homogeneous and steady-state exposures for phytoplankton, zooplankton, and fish (see table 7.10). Radionuclide ingestion rates were calculated using equation 7.38, and average initial biomass values were used. Temperature was assumed constant at 20°C. A single-prey food chain structure was used; the preference coefficients ($f_i$) were defined as 1.0. The assimilation efficiency for $^{137}$Cs was defined as 0.2 for zooplankton, 0.35 for forage fish, and 0.6 for piscivorous fish (Kolehmainen 1972).

The simulated values of bioaccumulation resulting from the case 1 homogeneous and steady-state conditions are listed in table 7.11. These values serve as the frame of reference for comparing the results of cases 2–7.

### Case 2: Homogeneous and Dynamic Radioactive Environment

Case 2 examined $^{137}$Cs exposures for the simple food chain under time-dependent and spatially homogeneous conditions. The concentrations of $^{137}$Cs in water, $C_w(t)$,

Table 7.10  Parameter values for bioaccumulation calculations

| Populations | Uptake from water[a] ($k_{ui}$) | Maximum feeding rate[b] ($C_{mi}$) | Radionuclide ingestion[c] ($k_{fi}$) | Radiobiological elimination[c] ($k_{bi}$) |
|---|---|---|---|---|
| Phytoplankton | 0.1 | | | 0.69 |
| Zooplankton | 0.008 | 0.6 | 0.07 | 0.14 |
| Prey fish | 0.004 | 0.25 | 0.06 | 0.0069 |
| Predatory fish | 0.004 | 0.25 | 0.06 | 0.0069 |

[a] Sombre et al. (1993).
[b] Bartell et al. (1992).
[c] King (1964), Hasanen et al. (1966).

Table 7.11  $^{137}$Cs concentrations calculated for four populations under a homogeneous and steady-state condition (case 1)

| Food chain population | Concentration (Bq kg$^{-1}$) |
|---|---|
| Phytoplankton | 57,965 |
| Zooplankton | 54,161 |
| Forage fish | 689,424 |
| Piscivorous fish | 8,125,000 |

were a function of time. The changes were small during the simulated 120 days produced by a low sedimentation rate (0.0001 day$^{-1}$) and a long decay half-life of $^{137}$Cs.

Equation 7.24 was solved using the modeled time-dependent $^{137}$Cs values of $C_w(t)$. Bioaccumulation by phytoplankton and zooplankton quickly approached equilibrium (see figure 7.4) because (1) the $^{137}$Cs concentrations in water did not vary substantially and (2) the radiobiological turnover rates for modeled plankton were high. However, the fish gradually accumulated $^{137}$Cs from water and prey because of their comparatively slower elimination rates for this nuclide. Fish accumulated more $^{137}$Cs than did phytoplankton or zooplankton after 120 simulated days. Importantly, the results illustrated in figure 7.4 show that the bioaccumulation of $^{137}$Cs by phytoplankton and zooplankton had reached an approximate steady-state equilibrium during the course of the 120-day simulation. However, the accumulations of $^{137}$Cs by the forage and predatory fish continued to increase and showed no evidence of reaching equilibrium during this time period. BCFs might have been reasonably estimated and applied to the plankton; however, development and application of BCFs to estimate accumulation by fish under these circumstances would clearly violate the equilibrium assumption underlying the BCF approach to exposure assessment.

### Case 3: Homogeneous and Dynamic Radioactive Environment with Dynamic Population Biomass

Case 3 extends case 2 conditions by including population dynamics as defined by the $dm_i(t)/dt\, m_i(t)^{-1}$ terms in equation 7.22. Net biomass change rates ($\theta_i$) were

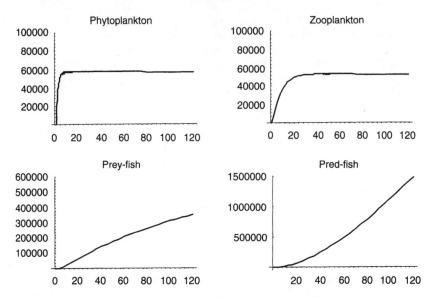

Figure 7.4 Simulated bioaccumulation of $^{132}$Cs by Chernobyl cooling pond aquatic populations under conditions of spatially homogeneous and temporally dynamic exposures (case 2).

defined as 0.003, 0.002, 0.001, and 0.001 (days$^{-1}$) for phytoplankton, zooplankton, forage fish, and piscivorous fish, respectively (Canale et al. 1976). Physical mixing impacts on biomass distributions were not addressed in case 3.

The resulting changes in biomass over the 120-day case 3 simulation were too small to produce substantially different patterns of bioaccumulation compared to case 2 (see figure 7.4). The case 3 results convey essentially the same meaning as case 2 with regard to the estimation or application of BCFs for plankton and fish under these simulated patterns of population dynamics.

### Case 4: Heterogeneous and Dynamic Radioactive Environment

A heterogeneous radioactive environment was simulated using equations 7.23 and 7.25–7.28. The parameter values in these equations were the same as for cases 1–3. Biomass variability was omitted from case 4 by defining the $dm_{ij}(t)/dt\ m_{ij}(t)^{-1}$ terms as zero. The equations were solved at selected locations (elements 15, 204, 422, 804, 1010, and 1247) in the Chernobyl cooling pond.

Results for the six locations showed that the patterns of $^{137}$Cs bioaccumulation by phytoplankton and zooplankton were largely determined by the spatial heterogeneity of $^{137}$Cs concentrations in water at these locations. For example, the temporal changes of the accumulated concentrations in grid element 15 differed from elements 422, 804, and 1010, where, for example, the accumulated $^{137}$Cs in phytoplankton and zooplankton in element 15 was 400% greater than in element 804 during the early period of the simulation and about 250% greater after 120 days of simulation. Similarly, $^{137}$Cs accumulated by piscivorous fish in element 15 was higher than in element 804 because of the higher $^{137}$Cs concentrations in water and forage fish.

## Case 5: Heterogeneous and Dynamic Radioactive Environment with Varying Biomass

Case 5 extends the case 4 conditions by retaining the spatial- and temporal-dependent biomass change terms $[dm_{ij}(t)/dt/m_{ij}(t)]$ in calculating exposure and bioaccumulation. In case 5, initial plankton biomass was distributed differently in four regions of the cooling pond. The plankton were transported and redistributed by the physical mixing model while they grew. It was assumed that the fish were distributed in four regions at different initial densities and that the fish remained at these initial locations.

Water circulation strongly affected the early biomass values. However, as the simulation progressed through time, the population growth dynamics dominated the plankton biomass changes. In case 5, population growth determined the fish biomass changes; however, the low growth rates (0.001 day$^{-1}$) for both forage fish and piscivorous fish produced relatively small changes in simulated fish biomass. The resulting accumulated concentrations of $^{137}$Cs are shown in figure 7.5. These results demonstrated that the combination of spatial-temporal variability in exposure and population dynamics produced nonequilibrium patterns of bioaccumulation for phytoplankton, zooplankton, forage fish, and piscivorous fish. The steady-state equilibrium results obtained for $^{137}$Cs bioaccumulation by phytoplankton and zooplankton in cases 2 and 3 were no longer evident. The use of BCFs to characterize

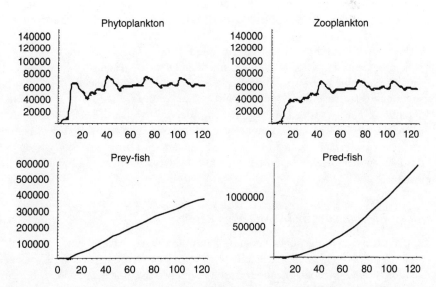

Figure 7.5 Simulated bioaccumulation of $^{137}$Cs by Chernobyl cooling pond aquatic populations in pond element 422 under conditions of spatially and temporally variable exposures (case 5).

exposure under these simulated conditions would provide inaccurate estimates of bioaccumulation with subsequent impacts on estimates of dose and risk.

### Case 6: Dynamic Exposures and Variations in Feeding Rates

In cases 1–5, the feeding rates for zooplankton, forage fish, and piscivorous fish were assumed constant and independent of prey abundance. Realistically, the feeding rates vary spatially and temporally due in part to variations in water temperature and food availability. Therefore, the feeding rate for each consumer population was varied temporally and spatially in case 6. This case also used the spatially and temporally varying $^{137}$Cs concentrations in water that were produced by the Chernobyl cooling pond transport model. The concentrations of $^{137}$Cs accumulated by zooplankton, forage fish, and piscivorous fish were calculated using the variable feeding rates.

Results from the case 6 simulation for six locations in the cooling pond were nearly identical to those presented in figure 7.5. Thus, the magnitude of variability introduced in the form of variable feeding rates was not sufficient to alter the case 5 pattern of bioaccumulation of $^{137}$Cs.

### Case 7: Dynamic Exposures and Multiple Prey

Case 7 is based on case 6 conditions but introduces a different food web structure. Cases 1–6 used a simple food chain to describe the Chernobyl cooling pond trophic structure. To examine the influence on bioaccumulation of fish feeding on different prey, a multiple prey food web structure was developed for case 7. The previously piscivorous fish was redefined as an omnivore with an assigned diet consisting of 50% phytoplankton, 20% zooplankton, and 30% forage fish. All other conditions were the same as in case 6.

The simulations for the six selected locations (figure 7.6) demonstrated that spatial heterogeneity in radioactive environment and the use of multiple food sources resulted in as much as a 250% difference in the $^{137}$Cs accumulated by predatory fish compared to case 6. The highest bioaccumulation resulted for grid locations 15 and 204. The lowest accumulation was observed for fish in location 1247. At the end of the 120-day period, none of the locations showed fish that had come into steady-state equilibrium in $^{137}$Cs accumulation. There was nearly a fourfold difference in accumulation between fish at spatial element 1247 and those at element 15.

### *Discussion of the Chernobyl Modeling Results*

The Chernobyl cooling pond modeling study investigated the effects of spatial, temporal, and ecological heterogeneity of radioactive environments on $^{137}$Cs accumulation by aquatic biota. The model results demonstrated that the temporal change in phytoplankton $^{137}$Cs concentration at different locations reflected the spatial and temporal change of the radionuclide concentration in water: the highest concentration occurred, not surprisingly, in the highest contaminated region in the modeled

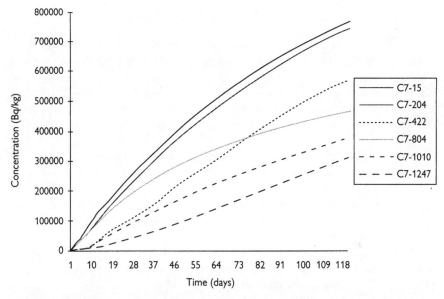

**Figure 7.6** Simulated time-dependent bioaccumulation of $^{137}$Cs by omnivorous fish exposed to contaminated water and multiple contaminated prey populations in different spatial elements of the Chernobyl cooling pond.

pond. Even though the $^{137}$Cs concentration in phytoplankton approached an approximate equilibrium state at each location, these concentrations differed from the equilibrium concentration associated with homogeneous and steady-state conditions (case 1). Thus, the use of equilibrium approaches (i.e., BCFs) can either underestimate or overestimate radionuclide bioaccumulation in phytoplankton under nonequilibrium conditions.

Zooplankton accumulated radionuclides directly from water and by consuming contaminated phytoplankton. Therefore, the accumulation patterns in different locations through time were influenced by both the variations of $^{137}$Cs concentrations in water and $^{137}$Cs accumulated in phytoplankton. Similar to the phytoplankton, the concentrations in zooplankton at six locations approached an equilibrium state toward the end of the 120-day simulation. However, the magnitudes of bioaccumulation by zooplankton differed from the calculated bioaccumulation factors (case 1) because of spatial variation of modeled $^{137}$Cs concentrations in water and food.

The implications of spatially and temporally varying exposure were more important for estimating $^{137}$Cs accumulated by fish. The concentrations in fish did not reach equilibrium during the simulated 120 days for any of the modeled scenarios. Importantly, the values of bioaccumulation differed markedly from the equilibrium concentration associated with homogeneous and steady-state conditions (case 1). This pattern resulted from the slow radiobiological processes of fish compared to the plankton and variation of $^{137}$Cs concentration in water. Therefore, using bioaccumulation factors based on equilibrium assumptions might overestimate the accumulated concentrations in fish by four- to sixfold in the Chernobyl cooling pond example.

## Temporally and Spatially Dependent Ecological Factors

Results from case 5 differed little from case 4 for phytoplankton and zooplankton at each location in the cooling pond because the calculated biomass change terms $[dm_{ij}(t)/dt/m_{ij}(t)]$ were much smaller than the $^{137}$Cs radiobiological turnover rates ($k_{bi}$) in these simulations. The radiobiological elimination rates dominated the loss of $^{137}$Cs from plankton. There was very little difference between the bioaccumulation patterns for fish using constant or variable feeding rates due to an abundance of prey.

It was anticipated that a different food web structure would have impacts on $^{137}$Cs accumulation by fish. In case 7, the concentrations accumulated in the omnivorous fish were much lower than results obtained using a simple food chain structure. The concentrations accumulated in fish feeding on multiple prey were almost four times lower than concentrations calculated using a single-prey food chain. This occurred because 50% of the fish diet was assumed to be phytoplankton, which had much lower $^{137}$Cs concentration than did forage fish (the sole prey item in cases 1–6). These model results underscore the importance of considering diet preferences of predators in assessing bioaccumulation.

Table 7.12 summarizes the relative effects among seven cases in spatial element 15 after 120 days. Element 15 is located in the most contaminated region of the Chernobyl cooling pond. The modeled concentrations of $^{137}$Cs for the homogeneous and steady-state condition (case 1) are used as the reference values. Values of the other cases are normalized to case 1 concentrations. The ratios in table 7.12 indicate that the implications of these effects on $^{137}$Cs accumulation varied among populations and differed substantially in magnitude. For example, concentration under an equilibrium condition (case 1) was almost 400% higher than under other conditions for predatory fish after 120 days; however, the concentration was 30% lower for plankton.

Table 7.12 indicates that the food web played a significant role in $^{137}$Cs bioaccumulation by predatory fish. For example, the concentration of $^{137}$Cs in predatory fish (case 7) was only 10% of equilibrium $^{137}$Cs concentration. Thus, the adequate determination of a diet for the predator population can be very important for estimating radionuclide accumulation.

Table 7.12  Comparison of concentrations of $^{137}$Cs among seven cases in spatial element 15 after 120 days to give a quantitative estimate of the relative effects (ratios normalized to case 1 values)

| Case | Phytoplankton | Zooplankton | Prey fish | Predatory fish |
|---|---|---|---|---|
| 1 | 1 | 1 | 1 | 1 |
| 2 | 0.99 | 0.99 | 0.73 | 0.27 |
| 3 | 0.99 | 0.99 | 0.72 | 0.26 |
| 4 | 1.38 | 1.35 | 0.78 | 0.31 |
| 5 | 1.38 | 1.34 | 0.69 | 0.26 |
| 6 |  | 1.35 | 0.70 | 0.26 |
| 7 |  |  |  | 0.1 |

Adding biomass change or variable ingestion rates in the exposure model had less impact than including temporal and spatial radioactive environments or changing food web structure. However, this result should not be generalized for all physical environments, species, and radionuclides. Growth rates may have considerable impact on the exposure assessment during the accelerated growth phase of biota or for radionuclides with low radiobiological turnover rates.

The Chernobyl modeling results suggested the following general conclusions:

1. The influence of hydrodynamics on radioactive material transport and distribution in the cooling pond was significant. The bathymetry of the cooling pond strongly influenced wind-driven patterns of water circulation.
2. Spatial and temporal heterogeneity of radioactive environments influenced modeled estimates of radionuclide bioaccumulation. The conventional BCFs can overestimate radionuclide bioaccumulation in fish and underestimate bioaccumulation in plankton.
3. Impacts of changes in abundance and distribution of biomass on the bioaccumulation varied among different populations in space and time. Although the impacts were not as dramatic as those associated with other ecological factors (i.e., food web structure), such impacts can be influential when biomass changes are significant or when the radiobiological turnover is comparatively slow.
4. The influences of water circulation and plankton population growth upon the plankton biomass change and distribution were roughly equivalent in the Chernobyl cooling pond.
5. Food web structure had a major impact on $^{137}$Cs bioaccumulation estimates. The impacts varied temporally and spatially. The determination of a diet of biota may be necessary to estimate bioaccumulation realistically.

## Problems

1. Describe the equilibrium CF approach for estimating ionizing radiation exposure (dose) to aquatic organisms. Discuss the strengths and limitations of this equilibrium approach for estimating exposure (dose).
2. List sources of environmental, biological, and ecological variability in estimating exposure and accumulation of ionizing radiation by aquatic plants and animals. Discuss the potential reduction of these sources of variability in relation to the opportunity to collect additional data.
3. Describe the potential impacts that climate change has on the relative rates and importance associated with each of the exposure pathways illustrated in figure 7.1. How might your response vary as a function of regional location (e.g., temperate zones, tropics)?
4. As a hypothetical scenario, presume that an accidental release of $^{137}$Cs has occurred in the headwater region of the Patuxent River and estuary. Evaluate, in qualitative terms, the applicability of the Chernobyl cooling pond exposure model for assessing exposure and dose to aquatic biota in the upper freshwater regions and lower estuarine regions of the Patuxent. What, if any, modifications might you make to the model to better characterize exposures in the

Patuxent? List critical data needed to support such modifications. How might your response to this problem change if the release was $^3$H instead of $^{137}$Cs?
5. Presume in response to problem 4 that a Patuxent River model has been implemented to estimate exposures in relation to the hypothetical release. As a water resource manager, you understand the uncertain nature of model calculations. Discuss your interpretation of modeled exposures and outline an approach for using uncertain model results in a framework for managing risks posed to exposed biota and to humans who might ingest contaminated biota.

## Acknowledgment

We gratefully acknowledge the support of the Consortium for Risk Evaluation with Stakeholder Participation (CRESP) through a contract between Rutgers University and the Cadmus Group, Inc., in the preparation of this chapter. B.G. Blaylock graciously supplied several key references discussed in this presentation. The views expressed in this chapter are solely those of the authors.

### References

Bartell, S.M., and A.L. Brenkert. 1991. "A Spatial-Temporal Model of Nitrogen Dynamics in a Deciduous Forest Watershed." In *Quantitative Methods in Landscape Ecology*, Ecological Studies Series, Vol. 82, ed. M.G. Turner, and R.H. Gardner. Springer U.S., New York. Pages 379–386.

Bartell, S.M., R.H. Gardner, and R.V. O'Neill. 1992. *Ecological Risk Estimation*. Lewis Publishers Inc., Chelsea, MI.

Bartell, S.M., J.S. Lakind, J.A. Moore, and P. Anderson. 1998. "Bioaccumulation of Hydrophobic Organic Chemicals by Aquatic Organisms: A Workshop Summary." *International Journal of Environment and Pollution* 9(1): 3–25.

Bedford, K.W., and M. Abdelrhman. 1987. "Analytical and Experimental Studies of the Benthic Boundary Layer and Their Applicability to near Bottom Transport in Lake Erie." *Journal of Great Lakes Research* 13: 628–648.

BIOMOVS II. 1996. *Assessment of the Consequences of the Radioactive Contamination of Aquatic Media and Biota*. Report No. 10. Swedish Radiation Protection Institute, Stockholm, Sweden.

Breck, J.E. 1980. "Relationships among Models for Acute Toxic Effects: Applications to Fluctuating Concentrations." *Environmental. Toxicology and Chemistry* 7: 775–778.

Breck, J.E., D.L. DeAngelis, W.V. Winkle, and S.W. Christensen. 1988. "Potential Importance of Spatial and Temporal Heterogeneity in pH, Al, and Ca in Allowing Survival of a Fish Population: A Model Demonstration." *Ecological Modeling* 41: 1–16.

Canale, R.P., P.L Freedman, P.L. Auwer, and J.J. Sygo. 1976. *Saginaw Bay Limnological Data*. Michigan Sea Grant Program Technical Report 54. MICHU-SG-76-207. University of Michigan, Ann Arbor, MI.

Carpenter, S.R. 1988. *Complex Interactions in Lake Communities*. Springer-Verlag, New York.

Connolly, J.P. 1985. "Predicting Single-Species Toxicity in Natural Water Systems." *Environmental Toxicology and Chemistry* 4: 573–582.

Cook, R.B., G.W. Suter II, and E.R. Sain. 1999. "Ecological Risk Assessment in a Large River Reservoir: 1. Introduction and Background." *Environmental Toxicology and Chemistry* 18: 581–588.

Csanady, G.T. 1970. "Dispersal of Effluent in the Great Lakes." *Water Research* 4: 79–114.

DeAngelis, D.L. 1978. *A Model for Movement and Distribution of Fish in a Body of Water.* ORNL/TM-6310. Oak Ridge National Laboratory, Oak Ridge, TN.

Evans, S. 1988. *Accumulation of Chernobyl-Related Cs-137 by Fish Populations in the Biotest Basin, Northern Baltic Sea.* STUDSVIK/NP-88/113. Studsvik Nuclear AB, Nyköping, Sweden.

Feng, Y. 1995. *Transport of Aquatic Contaminant and Assessment of Radioecological Exposure with Spatial and Temporal Effects.* Doctoral dissertation, University of Tennessee, Knoxville, TN. Page 207.

Fleishman, D.J. 1973. "Accumulation of Artificial Radionuclides in Fresh-Water Fish." In *Radioecology*. John Wiley & Sons, New York. Pages 347–369.

Friend, A.G. 1963. "The Aqueous Behavior of Strontium-85, Cesium-137, Zinc-65, and Cobalt-60 as Determined by Laboratory Type Studies." In *Transport of Radionuclides in Freshwater Systems*, ed. B.H. Kornegay, W.A. Vaughan, D.K. Jamison, and J.M. Morgan, Jr. Report TID-7664. U.S. Atomic Energy Commission, Washington, DC. Pages 43–60.

Fujita, M., J. Iwamoto, and M. Kondo. 1966. "Comparative Metabolism of Cesium and Potassium in Mammals: Interspecies Correlation between Body Weight and Equilibrium Level." *Health Physics* 12(9): 1237–1247.

Garder, K., and O. Skulberg. 1966. "Experimental Investigation on the Accumulation of Radioisotopes by Freshwater Biota." *Acta Hydrobiologica* 62: 50–69.

George, D.G., and R.W. Edwards. 1973. "Daphnia Distribution within Langmuir Circulation." *Limnology and Oceanography* 18: 790–800.

Harden Jones, F.R. 1988. *Fish Migration*. Edward Arnold Ltd., London.

Harvey, R.S. 1964. "Uptake of Radionuclides by Fresh Water Algae and Fish." *Health Physics* 10(4): 243–247.

Hasanen, E., S. Kolehmainen, and J.K. Miettinen. 1966. "Biological Half-time of Cs-137 in Three Species of Freshwater Fish: Perch, Roach, and Rainbow Trout." In *Radioecological Concentration Processes. Proceedings of a Symposium held in Stockholm 25–29 April, 1966*, ed. B. Aberg, and F.P. Hungate. Pergamon, New York. Pages 921–924.

Higley, K.A., S.L. Domotor, E.J. Antonio, and D.C. Kocher. 2003. "Derivation of a Screening Methodology for Evaluating Dose to Aquatic and Terrestrial Biota." *Journal of Environmental Radioactivity* 66(1–2): 41–59.

IAEA (International Atomic Energy Agency). 1985. *Sediment Kds and Concentration Factors for Radionuclides in the Marine Environment.* IAEA Technical Report No. 247. International Atomic Energy Agency, Vienna, Austria.

IAEA. 1986. "The Accident at the Chernobyl NPP and Its Consequences." Information Compiled for the IAEA Experts' Meeting, Vienna, Austria, August 25–29. International Atomic Energy Agency, Vienna, Austria.

IAEA. 1993. *International Meeting on Assessment of Actual and Potential Consequences of Dumping of Radioactive Waste into Arctic Seas, Oslo, 1–5 February 1993.* International Atomic Energy Agency, Vienna, Austria.

IAEA. 1994. *Handbook of Parameter Values for the Prediction of Radionuclide Transfer in Temperate Environments.* IAEA Technical Report No. 364. International Atomic Energy Agency, Vienna, Austria.

King, S.F. 1964. "Uptake and Transfer of Cesium-137 by Chlamydomonas, Daphnia, and Bluegill Fingerlings." *Ecology* 45(4): 852–859.

Kolehmainen, S.E. 1972. "The Balances of $^{137}$Cs, Stable Cesium and Potassium of Bluegill (Lepomis Macrochirus Raf.) and Other Fish in White Oak Lake." *Health Physics* 23(3): 301–315.

Kryshev, I.I. 1992. Personal communication. Subject: Radioecological Consequences of the Chernobyl Accident. Moscow, Russia.

Lampert, W.H. 1989. "The Adaptive Significance of Vertical Diel Migration in Zooplankton." *Functional Ecology* 3: 21–27.

Lewis, W.M., Jr. 1979. *Zooplankton Community Analysis—Studies on a Tropical System*. Springer-Verlag, New York.

Mackay, D., and S. Paterson. 1984. "Spatial Concentration Distributions." *Environment Science and Technology* 18(7): 207A–214A.

Mitsch, W.J., R.W. Bosserman, and J.M. Klopatek, ed. 1981. *Energy and Ecological Modeling—Developments in Environmental Modelling*, Elsevier, Amsterdam.

NCRP (National Council on Radiation Protection and Measurements). 1991. *Effects of Ionizing Radiation on Aquatic Organisms*. NCRP Report No. 109. National Council on Radiation Protection and Measurements, Bethesda, MD.

NCRP. 1996. *Screening Models for Releases of Radionuclides to Atmosphere, Surface Water, and Ground*. NCRP Report No. 123 I. National Council on Radiation Protection and Measurements, Bethesda, MD.

Odum, E.P. 1971. *Fundamentals of Ecology*. W.B. Saunders, Philadelphia, PA.

Patalas, K. 1969. "Composition and Horizontal Distribution of Crustacean Plankton in Lake Ontario." *Journal of the Fisheries Research Board of Canada* 26(8): 2135–2164.

Peterson, H.T., Jr. 1983. "Terrestrial and Aquatic Food Chain Pathways." In *Radiological Assessment: A Textbook on Environmental Dose Analysis*, ed. J.E. Till and H.R. Meyer. NUREG/CR-3332. U.S. Nuclear Regulatory Commission, Washington, DC.

Pinel-Alloul, B., J.A. Downing, M. Perusse, and G. Codin-Blumer. 1988. "Spatial Heterogeneity in Freshwater Zooplankton: Variation with Body Size, Depth, and Scale." *Ecological Society of America* 69(5): 1393–1400.

Poston, T.M., and D.C. Klopfer. 1988. "Concentration Factors Used in the Assessment of Radiation Dose to Consumers of Fish: A Review of 27 Radionuclides." *Health Physics* 55(5): 751–766.

Riley, G.A. 1976. "A Model of Plankton Patchiness." *Limnology and Oceanography* 21: 873–880.

Salis, S.B., and J.M. Flowers. 1969. "Toward a Generalized Model of Fish Migration." *Transaction of America Fishery Society* 3: 582–588.

Schwab, D.J., and J.R. Bennett. 1987. "Lagrangian Comparison of Objectively Analyzed and Dynamically Modeled Circulation Patterns in Lake Erie." *Journal of Great Lake Research* 13(4): 515–529.

Smith, I.R. 1992. *Hydroclimate—the Influence of Water Movement on Freshwater Ecology*. Elsevier, London.

Smith, L.R., C.B. Miller, and R.L. Holton. 1976. "Small-Scale Horizontal Distribution of Coastal Copepods." *Journal of Experimental Marine Biology and Ecology* 23: 241–253.

Sombré, L., Y. Thiry, and C. Myttenaere. 1993. "The Radiocesium Transfer in a Freshwater Foodchain (water-algae-mussels-fish-fish)." *Journal of Radioecology* 1: 21–27.

Steele, J.H. 1974. "Spatial Heterogeneity and Population Stability." *Nature* 248: 83.

Stuum, W., and J.J. Morgan. 1970. *Aquatic Chemistry*. Wiley-Interscience, New York. Page 583.

Taylor, L.R. 1984. "Assessing and Interpreting the Spatial Distributions of Insect Populations." *Annual Review of Entomology* 29: 321–357.

Taylor, L.R., and R.A.J. Taylor. 1977. "Aggregation, Migration, and Population Mechanics." *Nature* (London) 265: 415–421.

Thibodeaux, L.J. 1979. *Chemodynamics: Environmental Movement of Chemicals in Air, Water, and Soil*. Wiley-Interscience, New York.

Widner, T.E., J.S. Gouge, A.I. Apostoaei, B.G. Blaylock, B. Caldwell, S. Flack, F.O. Hoffman, C.J. Lewis, S.K. Nair, E.W. Reed, K.M. Thiessen, and B.A. Thomas. 1999. *Radionuclides Released to the Clinch River from White Oak Creek on the Oak Ridge Reservation—an Assessment of Historical Quantities Released, Off-Site Radiation Doses, and Health Risks*. Task 4 Report, Tennessee Department of Health. Oak Ridge Health Studies. SENES Oak Ridge Inc., Oak Ridge, TN.

Williams, L.G. 1960. "Uptake of Cs-137 by Cells and Detritus of Euglena and Chorella." *Limnology and Oceanography* 5: 301–311.

# 8

# Site Conceptual Exposure Models

James R. Rocco
Elisabeth A. Stetar
Lesley Hay Wilson

The site conceptual exposure model (SCEM) provides the framework for evaluating current and potential future risks associated with exposures to radionuclides in the environment. These risks can be estimated based on the concentrations that are present or anticipated to be present in the future in environmental media (e.g., air, soil, or water) at the locations (e.g., a residence or geographical area) where people currently or are anticipated in the future to live, work, and recreate. Exposure to radionuclides is the result of a unique combination of daily activities and the physical characteristics of environmental media and radionuclides. Each activity, along with the related characteristics, composes a pathway for exposure. The compilation of exposure pathways applicable to a person or group of people describes an exposure scenario.

The SCEM provides a profile of the behavior and activities through which people are exposed and a compilation of the current and anticipated future potential exposure pathways, exposure scenarios, and the associated exposure factors considered in their development (e.g., breathing rates, ingestion rates) that are needed to quantify risk. The SCEM guides the analysis of potential risk and provides the mechanism to organize and analyze the exposure scenarios and pathways.

The SCEM described in this chapter is associated with human health exposures. Ecological impacts may also be an important consideration during a risk analysis. Although this chapter focuses on radionuclides, potential risks associated with organic and inorganic chemicals may also be important to a particular risk analysis. The concepts and approach for developing an SCEM discussed in this chapter also apply to ecological impacts and to organic and inorganic chemicals. The details and differences for developing a SCEM for ecological impacts or organic and inorganic chemicals are not included in this chapter.

The development of a SCEM consists of four major components:

1. Defining the boundary of the area in which risks are to be analyzed and understanding the concerns and activities of the people present or anticipated to be present in the evaluation area
2. Identifying exposure pathways based on the daily activities of those people
3. Describing exposure scenarios for those people
4. Selecting the values for the exposure factors that will be used to evaluate the risks associated with these pathways and scenarios

## Evaluation Area

An essential first step in developing the SCEM is to establish the boundaries of the area to be considered in the risk analysis. The evaluation area should include the known or potential sources of radionuclides and the area of current and future potential distribution of radionuclides in the environmental media (e.g., air, soil, or water). The presence of radionuclides in environmental media can be due to many different circumstances, but in general, it is due to current or historical activities associated with the handling or use of radionuclides. Radionuclides can be present directly at the source of the radionuclide or at a location remote from the source due to transport of the radionuclides in environmental media or disposal of materials at locations remote from the original source. Identifying the evaluation area requires an understanding of the following:

- Location and composition of sources of radionuclides
- Characteristics of radionuclides present in environmental media
- Likely distribution of radionuclides in the area surrounding the source
- Mechanisms of transport for radionuclides from the source
- Population distribution and activities in the area surrounding the source

## Interested Party Input

Input from people familiar with the source of the radionuclides and from those who live, work, or recreate in the area surrounding a source of radionuclides provides the foundation for developing the SCEM and identifying the relevant evaluation area, exposure areas, exposure pathways, and exposure scenarios to be included in the SCEM. The process of gathering information from interested parties should be ongoing with new or additional information used to update and refine the SCEM as necessary.

Discussions with interested parties through individual or group meetings are an effective means of gathering information about the activities and locations through which exposures can potentially occur. Information about activities that may be unique to an area (e.g., gardening, hunting and fishing, or other recreational activities) can be obtained through these interactions, as well as information on the locations, frequency, and time spent in various activities. Interested parties can provide insight into potential sources of radionuclides and their potential presence in the evaluation area. In addition, interested parties can provide information related to specific concerns they have about potential exposures to radionuclides. The activity patterns and exposure concerns identified through these interactions can provide the basis for identifying the exposure pathways and exposure scenarios to be addressed in the SCEM.

## Exposure Pathways

Exposure pathways describe the mechanisms or paths through which a person may be exposed to radionuclides in a single environmental medium. In order for a person to be exposed to a radionuclide, the radionuclide must be present in an environmental medium (e.g., soil, air, or water) that the person would likely be in contact with while going about normal activities. For example, a jogger would not come into contact with subsurface soils and therefore would not be exposed to a radionuclide in this medium through inhalation or ingestion. However, if the radionuclide is a gamma emitter, the jogger could receive an external radiation dose. For most people, there will be exposures to multiple radionuclides in multiple environmental media through multiple exposure mechanisms (e.g., inhalation, ingestion, or external radiation). The development of exposure pathways provides a means to document each type of exposure associated with each type of activity for each location of a person.

An exposure pathway incorporates the following components:

*Sources and source areas* (e.g., stack emissions, disposal areas) that are
    identified within the evaluation area
*Radionuclides* that are associated with the sources and source areas
*Exposure areas* that describe the locations or areas where a person is likely to
    come in contact with environmental media containing a radionuclide
*Potentially exposed persons* who are likely to come in contact with
    environmental media containing a radionuclide
*Behaviors and activities of potentially exposed persons* that describe the daily
    activities of a person at home, at work, or while recreating
*Exposure media* that include air, soil, water, sediment, and other natural
    materials that a person may come in contact with during daily activities
*Routes or mechanisms of exposure* that describe the way in which a person
    comes in contact with a radionuclide (e.g., inhalation, ingestion, or
    external radiation)
*Transport media* (e.g., air, surface water) and *mechanisms* (e.g., dispersion,
    diffusion) for a radionuclide to travel or be transported by one or more
    environmental media from sources or source areas to environmental media
    in an exposure area
*Transfer mechanisms* (e.g., water-to-air volatilization, root uptake) for a
    radionuclide to transfer from one environmental medium to another
    environmental medium at a source or source area or in an exposure area

Figure 8.1 illustrates the components of an exposure pathway.

Site Conceptual Exposure Models    379

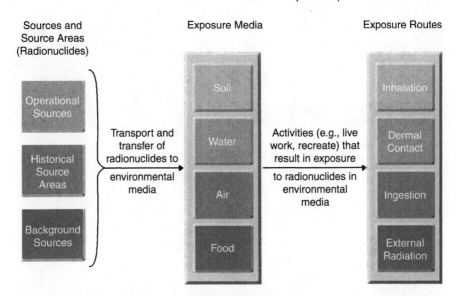

Figure 8.1  Components of an exposure pathway (adapted from an illustration by Melissa Gonzales, Ph.D., used with permission).

Risks are calculated based on concentrations of radionuclides present in environmental media in an exposure area. In the simplest form of an exposure pathway, a person is exposed to radionuclides in environmental media at the source or source area and information concerning concentrations of radionuclides in those environmental media is available. This simple form of an exposure pathway is illustrated in figure 8.2.

In many cases, radionuclides in environmental media can be transported away from the source by natural movement in the environment (e.g., groundwater movement, airborne movement of particles). This can result in a much larger area where a person may be exposed to radionuclides. For these cases, an exposure pathway incorporates additional information related to sources, source areas, transport mechanisms, and transfer mechanisms to support estimates and predictions of potential concentrations of radionuclides. Information concerning concentrations of radionuclides in environmental media may be limited or not available for exposure areas away from the source or source area. If sources and source areas are not located in the exposure area, potential future concentrations or variations in concentrations of radionuclides in environmental media should be considered.

## Sources and Source Areas

A *source* is the activity, process, or event (e.g., air stack, wastewater discharge) from which the radionuclides in environmental media originate. Sources associated with current and future activities at an industrial facility or other operations that handle radionuclides are typically governed by state and federal regulations that limit and monitor the handling and discharge of radionuclides. *Source areas* are

**380**  Radiological Risk Assessment and Environmental Analysis

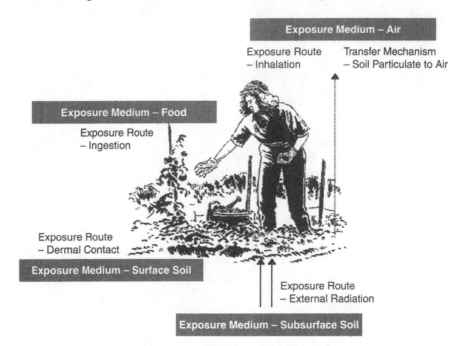

Figure 8.2   Simple exposure pathway.

areas where radionuclides were initially introduced into environmental media (e.g., liquid discharge to soil, landfills) and present an ongoing source of radionuclides through cross-media transfer and transport mechanisms, as well as direct contact by a person. In many cases, the original sources of radionuclides may no longer be present, and source areas become the primary focus of an analysis.

Both sources and source areas should be evaluated to determine the presence of radionuclides and the potential for cross-media transfer and transport of radionuclides away from the actual location of the sources or source area. In addition, characterization data associated with sources and source areas provide an understanding of the radionuclides present in the environmental media and support the selection of radionuclides for the evaluation.

## *Radionuclides*

Radionuclides can initially be identified through knowledge of the activities or processes that are the potential source of radionuclides in environmental media. Based on this knowledge, sample collection and analysis can be conducted to determine the presence and distribution of radionuclides in environmental media. In many cases, the assessment of radionuclides in environmental media will have been conducted as part of a comprehensive investigation; however, the identification of exposure pathways and exposure areas may require additional evaluation for the presence and concentration of radionuclides at identified exposure areas.

The characteristics of specific radionuclides will play a role in the identification of potential exposure pathways. Physical characteristics (e.g., solubility, half-life) can influence the applicability of various exposure pathways to particular radionuclides. For example, a radionuclide that is not readily soluble in water would not be expected to be significant as a dissolved component of a water-related exposure pathway. In addition, a short-lived radionuclide may decay before it is transported from the source to a potential exposure location or transferred from one environmental exposure medium to another. Understanding the characteristics of the radionuclides is important in identifying environmental media and exposure pathways and in calculating cumulative effects.

Additionally, some radionuclides (e.g., $^{226}$Ra, $^{40}$K) occur naturally in the environment. For radionuclides that may naturally occur, the risk analysis should include an understanding of the background concentrations of these radionuclides and the relationship between background concentrations and measured concentrations in the environmental media.

## Exposure Areas

An exposure area is an area or a number of areas in which a person may be exposed to radionuclides in environmental media. It can be unique for each environmental medium and for each behavior or activity. An exposure area can be a very large area (e.g., an area within a city boundary), a much smaller area (e.g., a building, residential property, or recreation area), or a specific location (e.g., a drinking water faucet). It can be at a source or in a source area. Exposure areas should be identified based on the known or potential presence and distribution of radionuclides and the likelihood that a person could be present in an area where these radionuclides are present or potentially present. Identification of the exposure area should also consider the potential future presence of radionuclides due to continued transport in environmental media.

The presence of a radionuclide in environmental media is dependent on many factors, including the quantity released, the duration of the release, the characteristics of the radionuclide at the source, and the characteristics of the transport media. Radionuclides can be present in environmental media from

- a release directly to environmental media (e.g., liquid discharge to the ground surface or surface water, land filling of wastes, or emissions from a stack);
- transfer from one environmental medium to another (e.g., suspension from soil to air, leaching in soil to groundwater, partitioning from sediments to surface water, or uptake from soil to plants) at the source area or at an exposure area; or
- transport from the original source or source area (e.g., by air, surface water, storm water, or groundwater) to an area removed from the original source or source area.

Actual measurement data from environmental media and information concerning potential transfer and transport mechanisms provide an initial understanding of the potential distribution of radionuclides in the area surrounding a source or source area and the locations of potential exposure areas. The quantity and distribution of actual measurement data may be limited, and the potential size of the area surrounding the source may make the collection of sufficient measurement data impractical. Therefore, the results of data interpolation, cross-media transfer calculations, and fate and transport modeling may be needed to determine potential distribution of radionuclides and to identify potential exposure areas.

Defining an exposure area requires an understanding of the daily activities of the people who currently, or may in the future, work, live, or recreate in the vicinity of the sources or source areas. This information can be gained through interviews or discussions with people who are familiar with the area and with the activities in those areas. Considering the distribution of radionuclides and the activities of people in the evaluation area, an exposure area can be defined as one of the following:

- An area where a person's activities may occur (e.g., residential property, work location)
- A point, such as the sampling location for a potable water system
- A radius around a point where a person's activities may occur (e.g., residential neighborhood)
- A set distance from an area where a person's activities may occur (e.g., hiking trail)

The size, shape, and location of an exposure area can be governed by types of behaviors or activities (e.g., residential activities on a person's property), type of environmental media (e.g., airborne materials may extend to a neighborhood or community), and the location of sources and source areas (e.g., direct exposure may occur at a source area).

## *Potentially Exposed Persons*

Potentially exposed persons are the people who are likely to come in contact with radionuclides in environmental media in the exposure areas. The identification of potentially exposed persons should consider both current and potential future land use (e.g., residential, commercial, industrial, recreational, agricultural). The description of a potentially exposed person provides information (e.g., gender, age) that can be used to identify the exposure parameters used in the risk analysis. In some cases, a more sensitive group (e.g., child, pregnant woman) may be selected as the focus for a risk analysis. For example, where incidental soil ingestion is an important consideration for the risk analysis, a preschool child (e.g., newborn to 6 years of age, based on the U.S. Environmental Protection Agency [EPA] age group for ingestion of soil [EPA 1989a, 1991]) may be a significant potentially exposed person. Depending on the location of the exposure area and the activities anticipated in that area, the potential for exposure may vary by age group. In that case, other age groups may be appropriate to consider. For example, an exposure area adjacent to a grammar school or middle school may warrant the evaluation of a school-age child. Some cases may warrant a distinction between males and females, depending on the radionuclide and specific activities identified.

## Behaviors and Activities

The behaviors and activities associated with each potentially exposed person will vary. Many behaviors and activities will be similar but distinguished by exposure factors (e.g., rate, occurrences, and duration). Behaviors and activities are often divided into groups that are defined by a major activity or land use (e.g., residential, commercial, industrial, recreational, or agricultural). In addition, construction and utility workers are often addressed as a separate group because of the unique nature of their activities.

Within each of these groups, activities can vary. For example, a commercial worker could spend the entire workday in an office or could travel from site to site. In addition, a potentially exposed person could be involved in several groups. An adult, for example, could spend time at home involved in residential activities, time at work involved in commercial or industrial activities, and time away from home engaged in recreational activities. A child may spend time at home and at a school or a daycare center. The potential for exposure would vary depending on the activity and the location. For example, a resident may spend time at home indoors cleaning, sleeping, or watching television, and time at home outdoors gardening or doing yardwork. It is important, therefore, to develop an understanding of the potential activities a person may be involved in, the location where those activities may occur, and their frequency.

## Exposure Media

Exposure media include air, water, soil, sediments, and foods that could contain concentrations of a radionuclide and that people could come in contact with as a result of their behaviors and activities. Common exposure media include the following:

Indoor air: air in a residence or other occupied building or structure
Outdoor air: ambient air outside of a residence or other occupied building or structure
Groundwater: saturated zones in the subsurface
Surface water: rivers, lakes, streams, springs
Storm water: periodic surface water runoff associated with rain events
Sediments: particles deposited in rivers, lakes, streams
Surface soil: soil that a person can come in contact with (e.g., gardening, yardwork, walking, playing)
Subsurface soil: soil below surface soil
Game meat and fish
Garden crops

## Exposure Routes

An exposure route describes a way in which a potentially exposed person comes in contact with a radionuclide. Exposure routes include the following:

- Ingestion: Radionuclides can be ingested and then absorbed into the body through the gastrointestinal tract. Ingestion of radionuclides can occur when

environmental media containing radionuclides are consumed, either intentionally (e.g., potable water, meats, fruits, and vegetables) or unintentionally (e.g., soil on a person's hands or water while swimming or showering).
- Inhalation: Radionuclides can be inhaled in ambient air containing radionuclides in the form of gases, vapors, or small particles. Radionuclides enter the body by absorption through the lungs, or by ingestion in the case of larger particles that are subsequently cleared by the respiratory system to the gastrointestinal tract (EPA 1989b).
- External radiation: A person can be exposed to ionizing radiation emitted from environmental media (e.g., soil, sediments) containing radionuclides located outside the body.
- Immersion: A person can be externally exposed to radionuclides when they are present in ambient air (e.g., indoor air, outdoor air).
- Submersion: A person can be externally exposed to radionuclides while swimming or bathing in water (e.g., surface water, potable water).

### EXAMPLE 8.1

What are the potential exposure routes that might be considered in the development of exposure pathways for an adult hiking or jogging in an area where radionuclides are present in shallow soil (i.e., 0–3 ft below the ground surface)?

*Solution*: Hiking or jogging could result in the suspension of soil as dust suspended in air. In addition, the hiker or jogger would have external exposure to radiation emitted from the soil. Exercise (e.g., jogging or walking) will result in heavier breathing and sweating, which could enhance the intake or contact with the dust or soil. With this in mind, the following exposure routes should be considered:

- Incidental ingestion of soil as particles suspended in air or soil on hands or arms that may come in contact with the mouth
- Inhalation of soil as particles suspended in air
- External radiation

## *Transport Mechanisms*

A transport mechanism is the means by which environmental media containing radionuclides move from one location to another. Transport mechanisms can play a prominent role in the estimation of future risk. They can also be used to estimate concentrations of radionuclides in environmental media in potential exposure areas remote from the sources or source areas where measurement data are not available for all environmental media. Some examples of transport mechanisms are

- soil particles suspended in air,
- radionuclides moving with groundwater as dissolved species or as a nonaqueous phase (e.g., colloid transport), and
- soil particles suspended in surface water.

## Transfer Mechanisms

A transfer mechanism moves a radionuclide from one environmental medium to another in the same area as the result of contact between the environmental media. A transfer mechanism is dependent on the characteristics of a radionuclide (e.g., volatility, solubility) and, in some cases, on the characteristics of the environmental medium (e.g., type of plant or fish species). Transfer mechanisms can be used to estimate concentrations of radionuclides in environmental media in exposure areas where measurement data are not available for all environmental media. Some examples of transfer mechanisms are

- deposition of soil or dust particles on soil or on garden crops,
- storm water deposition of suspended particles along a river or stream, and
- bioaccumulation of radionuclides in fish from surface water.

## Exposure Scenarios

An exposure scenario is a compilation of exposure pathways representing the daily activities and behaviors through which a person is potentially exposed to radionuclides. The specific exposure pathways selected for inclusion in a scenario depend on the behaviors and activities of the person described by the exposure scenario. In many cases, exposure scenarios are defined based on a particular activity (e.g., working, recreating) or land use (e.g., residential, commercial). Activity-specific or land-use-specific exposure scenarios are not uncommon in developing risk analyses for regulatory compliance or for addressing remedial action. The goal of a risk analysis, however, should be to provide realistic estimates of the potential risks to persons who live, work, and recreate in an evaluation area. These persons may live in one location, work in another location, and recreate in one or more other locations. It is likely that the magnitude of the exposure will vary at each location depending on the time spent in the area, the activity, and the concentrations of radionuclides in environmental media at the location. In some cases, exposure may occur in only one area or only at a specific time during the day. Therefore, in order to determine risk for persons living in the evaluation area, the SCEM should represent the multiple activities and multiple exposure locations through which persons are potentially exposed.

In some cases, it may not be possible to develop a single set of exposure scenarios that will reflect the concerns of all interested parties and the wide variety of exposure-related behaviors and activities of the persons who live, work, and recreate in the evaluation area. For this reason, the SCEM should provide multiple scenarios that are realistic for potentially exposed persons in the evaluation area.

### EXAMPLE 8.2

Radionuclides are known to exist in groundwater underlying a residential development in a rural area where private drinking water wells are common. Describe an exposure scenario for a resident living in this development.

*Solution*: Based on the information provided, the environmental medium to be addressed is groundwater. Because the information provided indicates that groundwater is used as a source of potable water, exposures are likely for adults and children living in this area. Activities and behaviors associated with uses of potable water include drinking water, washing, and bathing. In addition, a home garden may be present that is irrigated using the potable water source. These activities and behaviors could result in exposure routes of ingestion of water and homegrown produce, inhalation, and submersion.

Based on the information discussed above, the exposure pathways can be identified for the behaviors and activities of potentially exposed persons. Examples of exposure pathways are

- ingestion of groundwater as potable water,
- submersion while using groundwater as potable water for showering or bathing, and
- eating homegrown vegetables that are irrigated with groundwater.

## Exposure Factors

The evaluation of the potential risks associated with an exposure scenario and its corresponding exposure pathways requires the use of parameter values known as *exposure factors*, which represent the behaviors and characteristics that affect exposure to radionuclides in environmental media. They quantify the frequency and duration of exposures and the rate at which environmental media that contain radionuclides are taken into the body.

Exposure factors can vary from person to person and by age group and gender. Frequently, a mean value can be used for the exposure factors to represent the various persons within a population or grouping. Age- or gender-specific exposure factors may be used depending on (1) the amount of variation observed among subgroups within the population, (2) the importance of an exposure factor value to the overall risk estimate, and (3) the amount of data available to obtain a representative exposure value for a specific population subgroup. Common exposure factors include the following:

Exposure frequency: the regularity or rate with which the exposure takes place, such as days or events per year

Exposure time: the time per day or event over which an activity occurs

Exposure duration: the time over which the exposure is expected to occur

Body weight: the weight of a typical person, which differs based on age and gender

Inhalation rate: the rate at which a person breathes while performing a specific activity, which varies based on age and the type of activity in which a person is engaged

Ingestion rate:
- Drinking-water ingestion rate represents the amount of tap water a person consumes, including beverages (e.g., mixed juice and prepared coffee or tea) (EPA 1997).
- Food ingestion rate represents the daily amounts of vegetables, fruit, meat, milk, and other foods considered in the risk analysis. A food consumption fraction or affected food fraction can be applied to adjust these rates to account for the portion that is ingested from homegrown sources or commercial sources.
- Soil ingestion rate represents the amount of soil or dirt that individuals accidentally ingest. Some level of incidental soil ingestion is associated with routine, daily activities. The rate of soil ingestion is expected to increase with certain outdoor activities, such as work that involves excavation of soils, or recreational activities such as gardening and hiking. Soil ingestion occurs primarily outdoors, although a small amount may occur indoors from the ingestion of household dust, a small fraction of which is soil (EPA 1997).
- Surface water ingestion rate represents the amount of incidental ingestion during swimming or during contact with surface water while hiking or participating in other outdoor recreational activities (e.g., if people were to rinse their hands and faces in the surface water).

Skin-surface area: the area of skin on a person's body that is in contact with soil or water during an activity (e.g., swimming, showering, or bathing)

Shielding factor: the decrease in external exposure rates that results from the attenuation of radiation in shielding materials (e.g., a layer of clean soil above a source area in soil reduces the external exposure rate at the ground surface, and gamma radiation levels inside a building are less than those outdoors because of attenuation by the building materials)

The exposure factor values can be found in a number of sources, such as the EPA *Exposure Factors Handbook* (EPA 1997) and the *Child-Specific Exposure Factors Handbook* (EPA 2002). Additional sources include International Commission on Radiological Protection (2003) and Yu et al. (2000). The references generally rely on reviews of the open literature to develop an understanding of the potential values for the exposure factors for the general population. The studies range from large population studies (e.g., for body weight and expected lifetime) to smaller studies (for food ingestion factors and activity factors, e.g., time spent indoors vs. outdoors).

## Problems

Concentrations of radionuclides have been measured in both surface and subsurface soil adjacent to an industrial facility. The source of the radionuclides has been identified as a former process water discharge to the ground surface. Drainage for the area is toward a rural residential area approximately one-half mile downgradient of the facility. The groundwater gradient is known to be consistent with the surface gradient. Based on the limited information provided, identify potential responses for the following:

1. Exposure areas
2. Transfer mechanisms
3. Transport mechanisms
4. Exposure media
5. Potential exposed persons
6. Activities and behaviors
7. Exposure routes
8. Five potentially complete exposure pathways

## Acknowledgments

The concepts and methods discussed in this chapter are based on work products of Sage Risk Solutions, LLC, and were applied as part of the Risk Analysis, Communication, Evaluation, and Reduction project at Los Alamos National Laboratory (see www.racernm.com) and presented in a peer reviewed report titled "Site Conceptual Exposure Model," by J.R. Rocco, E.A. Stetar, and L. Hay Wilson (2005).

References

EPA (U.S. Environmental Protection Agency). 1989a. *Guidance for Soil Ingestion.* Interim Final. OSWER Directive 9850.4. U.S. Environmental Protection Agency, Office of Solid Waste and Emergency Response, Washington, DC.

EPA. 1989b. *Risk Assessment Guidance for Superfund.* Vol. 1: *Human Health Evaluation Manual (Part A).* Interim Final. EPA/540/1–89/002. U.S. Environmental Protection Agency, Office of Solid Waste and Emergency Response, Washington, DC.

EPA. 1991. *Risk Assessment Guidance for Superfund.* Vol. 1: *Human Health Evaluation Manual, Supplemental Guidance, Standard Default Exposure Factors.* Interim Final. OSWER Directive 9285.6–03. U.S. Environmental Protection Agency, Office of Solid Waste and Emergency Response, Washington, DC.

EPA. 1997. *Exposure Factors Handbook.* EPA/600/P-95/002Fa. U.S. Environmental Protection Agency, Office of Research and Development, National Center for Environmental Assessment, Washington, DC.

EPA. 2002. *Child-Specific Exposure Factors Handbook.* Interim Final Report. EPA-600-P-00–002B. U.S. Environmental Protection Agency, Office of Research and Development, National Center for Environmental Assessment, Washington, DC.

International Commission on Radiological Protection. 2003. "Basic Anatomical and Physiological Data for Use in Radiological Protection: Reference Values." ICRP Publication 89. *Annals of the ICRP* 32(3–4).

Yu, C., D. LePoire, E. Gnanapragasam, J. Arnish, S. Kamboj, B.M. Biwer, J.J. Cheng, A. Zielen, and S.Y. Chen. 2000. *Development of Probabilistic RESRAD 6.0 and RESRAD-Build 3.0 Computer Codes.* NUREG/CR-6697. U.S. Nuclear Regulatory Commission, Washington, DC.

# 9

# Internal Dosimetry

John W. Poston, Sr.
John R. Ford

## External versus Internal Exposure

The term "internal dosimetry" has always held a certain mystery about it that has confused and confounded many health physicists for a long time. It is an unfortunate term that, historically, was intended to serve in contrast to the term "external dosimetry." Generally, "external dosimetry" means the measurement of radiation exposure (or dose) due to sources located outside the body. These radiation sources, which are usually located in well-defined positions in an area, have the ability to penetrate the body, depositing energy in organs and tissues, and, potentially, causing harm to the persons being irradiated. Persons entering an external radiation field are exposed while in the area, and the exposure ceases when the individual leaves the area.

In general, radiation workers (and health physicists) are more comfortable dealing with external sources of radiation exposure than with sources that may constitute an internal hazard. If work involving radiation exposure is to be done in an area, the health physicist can make a radiation survey with portable instruments and obtain an estimate of the anticipated external exposure from this survey. "Hot spots" or other areas with high exposure rates can be identified and posted, and areas in which the doses are low also can be identified. Workers can wait in these low dose areas when they are not needed for a particular operation, and stay times for the work can be calculated, if necessary.

Most workers have become comfortable working in controlled areas in which the radiation presents only an external hazard. In addition, workers are aware of the usual methods of controlling exposure to external radiation sources by using time, distance, and shielding, and the work can be carried out with these methods in mind. These techniques are taught in all general employee training courses for radiation workers. Use of stay times, low dose-rate waiting areas, and temporary shielding are examples of how these techniques are practiced in keeping routine exposures as low as is reasonably achievable (ALARA).

For some high-risk work, it may be necessary for the health physicist responsible for monitoring the work to control exposure of workers as closely as possible. This is usually accomplished through continuous monitoring of the work as it progresses. Nevertheless, the point remains the same: radiation workers enter radiation and high radiation areas hundreds of times per day and usually give little thought to this type of exposure.

All workers entering the radiation area are required to wear personnel monitoring devices that will give reasonably accurate estimates of the doses received. Many dosimeters can be used for situations in which the radiation field varies significantly over the total body, or they can be used to monitor certain parts of the body (e.g., extremities, lens of the eye, or gonads). Some of these dosimeters can be evaluated almost immediately after the workers exit the area to provide estimates of the whole-body dose and the dose to important parts of the body. These data can be used to evaluate the effectiveness of the radiation survey, the control measures taken to keep exposures as low as is reasonably achievable, and future job assignments of the individual workers.

If the potential for internal exposure is present, then the situation facing the health physicist, and the attitudes of the workers, is entirely different. It is still possible to perform prework surveys of the area to determine the potential hazard; however, these surveys are a different type. In this case, the health physicist must make measurements of airborne radioactivity concentrations. This requires drawing a known amount of air through a filter (or other collection device) and analyzing the sample to determine total (gross) activity or to determine the radionuclides present in the sample and their activities. In addition, surface contamination surveys may be made to determine the potential for resuspension of radioactive materials into the air. However, surface contamination levels cannot be related directly to air concentrations due to the many factors influencing resuspension. Nevertheless, although the work environment can be defined with some degree of confidence, exposure of the workers to radioactive materials, which may be deposited internally, is not so easy to predict.

Once the work has begun, use of the control methods time, distance, and shielding does not play an important role in preventing an internal exposure (although limiting the time in the area would certainly reduce the probability of an internal exposure). In addition, if an internal exposure should occur, the radiation protection staff and the individual worker have little control over the time the material remains in the body. Distance is no longer a protective technique, nor is the use of temporary shielding. The radioactive material is inside the body, where almost all of the radiation energy emitted in the decay of the radionuclide will be absorbed in tissues of the body.

Because the material will remain in the body based on the metabolism of the particular element, the exposure continues after the individual leaves the airborne radioactivity area. This fact is probably one of the most important in terms of understanding workers' attitudes toward potential internal exposures. Not only does exposure of the worker continue, but the worker also takes the material out of the facility at the end of the working day. The worker is concerned with the potential that the internal deposition could be a source of radiation exposure for his or her spouse and children.

Monitoring during the work is difficult and not very effective in predicting the accumulated exposure as the work proceeds. Local or area air monitoring systems may not give a true indication of the concentrations of radioactive material in the breathing zones of individual workers. Personal air samplers may be used to obtain breathing zone samples, but low flow rates, high failure rates, and worker acceptance can be a problem. In addition, care must be taken to ensure that these samplers do not interfere with the worker's tasks such that the actual exposure period (to external or internal radiation sources) is extended.

At the completion of the work, no direct measurement techniques are available to evaluate the exposure. Samples of mucus from the nasal passages may be taken (either through nasal swabs or by blowing of the nose) to determine if radioactive material has deposited in the upper nasal passages. However, these samples must be obtained soon after exposure because the effectiveness of such samples in indicating an internal exposure is limited to a very short period after exposure (usually 15–30 min). Bioassay is the only other method available to evaluate an internal exposure. This term is used to include both the measurement of radioactive material in the body and the measurement of radioactive material excreted from the body. However, bioassay is clearly an after-the-fact evaluation technique.

Therefore, the most effective method of controlling exposure to internally deposited radioactive material is to prevent the exposure. There are many methods to accomplish this goal, and all of these are in use at most nuclear facilities. Simple restrictions such as prohibiting smoking, eating, and drinking in specified areas are intended to prevent the inadvertent intake of radioactive materials. On a larger scale, the first line of defense is containment of the radioactive material so that it cannot become airborne. Good housekeeping plays an important, but often overlooked, role in preventing material from becoming airborne and keeping exposures to such material as low as is reasonably achievable. Second, engineering controls (i.e., the design of equipment to move, exchange, filter, and clean air) are effective in keeping airborne concentrations low in most areas. When these methods are not effective, respiratory protective devices and/or protective clothing must be used to prevent (or to limit) exposure to airborne radioactive material.

If an internal exposure should occur, then internal dosimetry (the historical name) is not really the process that is followed. The term "dosimetry" literally means "dose measurement," which is not possible when the radioactive material is inside the body. However, in keeping with the traditional usage, we use the term "internal dosimetry" in this discussion, defined specifically as follows:

*Internal Dosimetry* is a process of measurement and calculation that results in an estimate of the dose to tissues of the body due to an intake of radioactive material.

The term "measurement" could apply to the measurement of the concentration of a radioactive material in air. However, this approach requires an assumption of the breathing rate to determine the intake of radioactive material. Once the intake is obtained, calculations (based on assumed metabolism of the material) are performed, and a dose estimate is obtained. The term "measurement" also applies to bioassay techniques in which the quantity (activity) of radioactive material in the body is measured by using very sensitive radiation detectors located outside the body (called direct bioassay). It also applies to the measurement of the concentration of radioactive materials excreted from the body, usually in the urine or feces (called indirect bioassay). Again, these data are combined with a mathematical model derived to explain the uptake, deposition, movement, metabolism, retention, and excretion of the particular element in the human body. This combination results in a series of calculations that produce an estimate of the dose equivalent of radioactive material.

The accuracy of the estimate must be considered in the total evaluation of an internal exposure. The accuracy depends on the measurement technique and the models selected. In some cases, the errors may approach a factor of 3 or more. The models used in these calculations represent a reference adult human. However, the individual exposed may not be well represented by this hypothetical individual. Age, gender, state of health, diet, and other factors play a role in the actual manner in which a material taken into the body may be metabolized.

From this short discussion, it should be clear that internal dosimetry really means "internal dose assessment," because there is no actual measurement of energy deposition.

Assessing internal exposures from environmental releases of radionuclides is even more complicated than that for occupational exposure. The work environment is reasonably well controlled and monitored. Data obtained from these monitoring systems can be used in any evaluation. For materials released to the environment, the transport, deposition, retention, and redistribution of radioactive materials must be considered. The intake pathways still include inhalation, but the oral pathway may become very important because of the intake of radioactive materials incorporated in foodstuffs. Thus, there is a need to the assess intakes of radionuclides through a number of pathways. In addition, such assessments must include a wide range of ages of potentially exposed individuals.

The discussions that follow will focus initially on the assessment of occupational exposure. This approach was selected because it allows a systematic discussion of the assumptions, models, and approaches taken in internal dosimetry. Below, we further broaden the discussion to include individuals of many ages and summarize some of the newest techniques that have been introduced to allow such considerations.

## *Internal Dose Control*

The current concept of controlling internal exposure to radioactive materials in most nuclear facilities is based on limiting the concentrations of these materials in air and attempting to prevent oral intake. In earlier recommendations, radionuclides were controlled by establishing maximum allowable concentrations in both air and water. These concepts were used for many years but have now been discarded.

Inhalation and ingestion of radioactive material are considered the most likely pathways of entry into the body in the work situation. Usually, no consideration is given in the internal dosimetry regulations to accidents, such as intakes through wounds (injection) or absorption of radioactive material through the intact skin. Control of exposure to radionuclides that could be internally deposited is based on the establishment of concentration limits for each radionuclide of concern in the work environment. The development of this systematic set of concentration limits is based on a four-step process that forms the basis for the establishment of all radiation protection standards in current use. These steps form a logic pattern for the establishment of internal exposure protection standards regardless of the terms used to define the concepts embodied in a particular system. These steps are as follows:

1. Establish limits for occupational radiation exposure, based on careful review of available biological data, which should not be exceeded.
2. Calculate the allowable amount (activity) of each radionuclide (and its daughters) that can be in the body at any time without exceeding the dose limits established in step 1.
3. Establish possible routes of entry into the body for each element (or radionuclide) and derive an allowable intake rate of the material that will satisfy both steps 1 and 2.
4. Based on the physiological parameters established for the routes of entry (e.g., inhalation and ingestion), calculate the allowable concentrations of the radionuclide in air that will satisfy steps 1, 2, and 3.

These steps resulted in a terminology that became well known in health physics, but the logical basis for the establishment of these terms has been lost. Recommendations of the International Commission on Radiological Protection (ICRP) and the National Council on Radiation Protection and Measurements (NCRP) published in 1959 gave each of these logical steps a name. Step 1 represented the establishment of the maximum permissible dose equivalent (MPDE). In step 2, the maximum permissible body burden (MPBB) was established, and with appropriate considerations, the maximum permissible organ burden could be derived. Step 3 called for establishment of the allowable intake rates of material. In the ICRP/NCRP formulation, this step was never assigned an official name. The final step, and the one that was used daily to control exposure to radioactive materials for many years, was the establishment of the maximum permissible concentration (MPC) values for a particular radionuclide.

New terms have been introduced to replace those above, but the fact remains that these four simple steps form the basis for all internal exposure recommendations and regulations in use throughout the world.

Note also that an internal dose assessment would proceed from step 4 to step 1 of the logic pattern. That is, a dose assessment would begin with a measurement of a radionuclide concentration (i.e., air sampling, direct or indirect bioassay) and a series of assumptions, ultimately allowing a calculation of the dose from an intake of radioactive material.

## Regulatory Requirements

Current U.S. guidance and regulations related to the control of internally deposited radionuclides are based on recommendations of the ICRP and the NCRP published between 1977 and 1988. Other publications on specific groups of radionuclides (e.g., the alkaline earths and the actinides) have been issued, but these have had little effect on the federal regulations. The ICRP recommendations on radiation protection provided in ICRP Publication 26 (ICRP 1977) and a completely revised internal dosimetry scheme in Publication 30 and its supplement (ICRP 1979a, 1979b) have been incorporated into the practice of radiation protection in this country. In addition, it should be recognized that in 1990, the ICRP adopted a revised set of recommendations and has published revised models for a number of systems, including the respiratory system. These documents have led to the recalculation of the limiting values for controlling internal exposure. However, the fact remains that these new recommendations have not been incorporated into federal or state regulations in the United States.

The following sections will introduce the recommendations and internal dosimetry techniques published by the ICRP beginning in 1977. The ICRP recommendations and techniques have been reviewed by the NCRP and the Nuclear Regulatory Commission. The current federal standards for protection against radiation (10 CFR 20, 2007) incorporate most of the new information. The initial discussion below covers the recommendations in ICRP Publication 26 (ICRP 1977) and the recommendations specific to internal dosimetry in ICRP Publication 30 (ICRP 1979a, 1979b).

## ICRP Publication 26 Techniques

The recommendations in Publication 26 were the first real pronouncements of the ICRP since the early 1960s. Recommendations were made in a number of areas, including protection from external and internal radiation sources, exposures of population groups, exposure of pregnant women, and planned special exposures. Our discussion focuses primarily on those recommendations and techniques having an impact on methods to be used for internal dosimetry.

First, the ICRP restated the objectives of radiation protection that formed the basis for the new dose limitation system. These objectives are as follows:

- No practice shall be adopted unless it provides a net positive benefit.
- All exposures shall be kept as low as reasonably achievable (ALARA), economic and social factors being taken into account.
- The dose equivalent to individuals shall not exceed the limits recommended for the appropriate circumstances by the commission's commitment to the benefit–risk philosophy and the ALARA philosophy adopted much later. In all situations, the commission warns that its recommended limits should not be exceeded.

The commission also introduced several new concepts that must be defined and explained if the recommendations are to be understood and used effectively. The

Table 9.1  Estimates of threshold doses for clinically detrimental nonstochastic effects in various tissues of adult patients exposed to conventionally fractionated therapeutic X- or γ-irradiation (adapted from Upton 1985)

| Organ | Injury at 5 years | Dose causing effect in 1-5% of patients (Gy) | Dose Causing effect in 25–50% of Patients (Gy) | Irradiation field (area) |
|---|---|---|---|---|
| Liver | Liver failure, ascites | 35 | 45 | Whole |
| Kidney | Nephrosclerosis | 23 | 28 | Whole |
| Urinary bladder | Ulcer, contracture | 60 | 80 | Whole |
| Testes | Permanent sterilization | 5–15 | 20 | Whole |
| Ovaries | Permanent sterilization | 2–3 | 6–12 | Whole |
| CNS (brain) | Necrosis | 50 | < 60 | Whole |
| Lens of the eye | Cataract | 5 | 12 | Whole |
| Bone marrow | Hypoplasia | 2 | 5.5 | Whole |

most significant change in the new recommendations is the introduction of the terms stochastic and nonstochastic effects of radiation. These terms are defined below:

*Stochastic effects* are those for which the probability of an effect occurring, rather than its severity, is regarded as a function of dose without threshold. *Nonstochastic effects* are those for which the severity of the effect varies with the dose, and for which a threshold may therefore occur.

Some somatic effects of radiation are considered stochastic. The most important of these effects is carcinogenesis, and it is considered the chief somatic risk at low doses. For this reason, cancer is the main concern in radiation protection. The commission states that, for the dose range involved in radiation protection, hereditary effects of radiation also are considered stochastic in nature.

It is very tempting to substitute the words "linear" and "threshold" effects of radiation for the terms defined above; however, beware of such temptation. The definition of stochastic effects is a little different from the concept of a linear dose–response curve because of the word "probability." In addition, it will be clear later that the ICRP includes certain effects of radiation in the two categories that may not have been included previously. For example, nonstochastic effects include cataracts of the lens of the eyes, nonmalignant damage to the skin, cell depletion in the bone marrow causing certain blood deficiencies, and gonadal cell damage leading to impairment of fertility. Data that are more recent have shown that a number of nonstochastic effects must also be considered. For example, table 9.1 lists effects of radiation that are assumed nonstochastic.

The goals of radiation protection are to prevent the detrimental nonstochastic effects of radiation exposure and to limit the probability of stochastic effects to levels deemed acceptable. Prevention of nonstochastic effects can be achieved if the dose-equivalent limits are selected such that a threshold is never reached. The ICRP goal was to select a level such that a threshold would not be reached even if the exposure lasted for an entire working lifetime. Many authors have referred to the nonstochastic limit as a capping dose. The limitation on stochastic effects was selected based on a consideration of the benefit–risk relation and keeping the ALARA philosophy always in mind.

The commission also formally defined the term "committed dose equivalent," even though a similar term was in common use in the nuclear industry (i.e., dose commitment).

*Committed dose equivalent* is the dose equivalent to a given organ or tissue that will be accumulated over 50 years, representing a working lifetime, following a single intake of radioactive material into the body.

Mathematically, the committed dose equivalent, $H_{T,50}$, is defined by

$$H_{T,50} = \int_{t_0}^{t_0+50y} H(t)dt, \tag{9.1}$$

where $H(t)$ is the relevant dose-equivalent rate to tissue $T$, and $t_0$ is the time of intake. The ICRP states this quantity may be considered a special case of the dose-equivalent commitment, but this distinction is not particularly important to the discussion of internal dosimetry. It is important to note that the definition applies to a single organ or tissue and to a single intake of radioactive material. This is a major change between the old ICRP formulation and that contained in ICRP Publication 30 (ICRP 1979a, 1979b).

The ICRP recommended a limit for stochastic effects based on total risk of all tissues irradiated. A single limit is set for uniform irradiation of the whole body, and a dose limitation system is established to minimize the total risk from irradiation of the whole body. In addition, no single tissue should be irradiated in excess of the dose limit set to prevent nonstochastic damage. Thus, for stochastic effects of radiation, the dose limitation system is based on the principle that the risk should be equal whether the whole body is being irradiated uniformly or whether there is nonuniform irradiation. For internal exposure, the ICRP concludes that this condition will be met if

$$\sum_T W_T H_{T,50} = H_{WB,L}, \tag{9.2}$$

where

$W_T$ = a weighting factor representing the proportion of the total stochastic risk resulting from irradiation of tissue ($T$), when the whole body is irradiated uniformly

$H_{T,50}$ = the committed dose equivalent to tissue $T$

$H_{WB,L}$ = the recommended annual limit for uniform irradiation of the whole body

At its 1978 meeting, the ICRP decided to call the quantity on the left-hand side of equation 9.2 the "committed effective dose equivalent"; that is,

$$H_{E,50} = \sum_T W_T H_{T,50}. \quad (9.3)$$

The conditions stated above will be met, in the opinion of the ICRP, if the limits on occupational radiation exposure are the following:

- Stochastic effects = 0.05 Sv yr$^{-1}$ for uniform irradiation of the whole body (i.e., 5 rem yr$^{-1}$).
- Nonstochastic effects = 0.5 Sv yr$^{-1}$ (50 rem yr$^{-1}$) to all tissues except the lens of the eye. The lens was finally assigned a limit of 0.15 Sv yr$^{-1}$ (15 rem yr$^{-1}$), but some of the early publications contained the value of 0.30 Sv yr$^{-1}$ (30 rem yr$^{-1}$). This change in the limit for the lens was hidden in some of the corrections later issued by the commission. More correctly, the limit for the committed effective dose equivalent is 0.05 Sv yr$^{-1}$, and the limit for the committed dose equivalent to individual organs (except the lens) is 0.5 Sv yr$^{-1}$.

## Tissues at Risk

The internal dosimetry system described in ICRP Publication 2 (ICRP 1959) introduced the concept of a critical organ, and all of the calculations of MPBB and MPC were made under the assumption that there was a need to control the dose-equivalent rate to this critical organ to the limit set by the ICRP (i.e., the MPDE). The new ICRP concept takes into account the total risk that can be attributed to the exposure of all tissues irradiated. If this concept is to be implemented, it is necessary to specify the organs and tissues of the body that should be considered at risk and to establish some measure of this risk.

The most visible change in the new ICRP recommendations is that the critical organ concept, for the most part, has been discarded. This change was necessary because the concept of a single critical organ did not fit into the scheme of specifying a committed effective dose equivalent ($H_{E,50}$) relative to a uniform whole-body irradiation. The new scheme requires that the committed effective dose equivalent be the sum of the weighted committed dose equivalent ($W_T H_{50,T}$) to each organ in the body, each with a specific sensitivity to radiation effects. This sensitivity is given by the weighting factor shown in equations 9.2 and 9.3 above.

The discussion that follows, on tissues at risk, was used to derive weighting factors needed to calculate the committed effective dose equivalent. The derivation of risk factors is based on an average risk to a particular tissue from irradiation. No consideration is given to the effects of age- or sex-dependent differences. The tissues considered and the derived risk factors are based on the following:

- Review of the susceptibility of the tissue to radiation damage
- Review of the seriousness of this radiation-induced damage
- Consideration of the extent to which this damage is treatable.

In addition, only the likelihood of inducing fatal malignant disease, nonstochastic changes, or substantial genetic defects is considered.

*Gonads* Irradiation of the gonads can cause three different effects, each of which must be considered in establishing a risk factor. First, there is the probability of tumor induction. However, the gonads appear to have a low sensitivity to radiation, and no carcinogenic effects have been documented. For this reason, the ICRP did not consider this effect important. Impairment of fertility also is a possible effect, but such an effect is clearly age dependent. Again, the ICRP did not consider this an important radiation effect. The major effect considered for irradiation of the gonads was the production of hereditary effects over the two subsequent generations. Based on an evaluation of hereditary effects over the first two generations from irradiation of either parent, the risk appears to be about $1 \times 10^{-2}$ Sv$^{-1}$ ($1 \times 10^{-4}$ rem$^{-1}$). For radiation protection purposes, the ICRP chose a value of $4 \times 10^{-3}$ Sv$^{-1}$ ($4 \times 10^{-5}$ rem$^{-1}$). This value was obtained by considering the proportion of exposures that were likely to be genetically significant. The ICRP concluded that the genetic risk must be less than the mortality risk from fatal cancers. Thus, the risk estimate was reduced by the ratio of the mean reproductive life to the total life expectancy (i.e., about 0.40).

*Bone* ICRP Publication 11 (ICRP 1968) identified the radiosensitive cells in bone as the endosteal and epithelial cells on the bone surface (BS). These sensitive cells are assumed to lie within 10 µm of the BS and are distributed uniformly throughout the skeleton (on all surfaces). The primary radiation-induced effect in these cells is cancer; however, the bone seems to be much less sensitive to radiation than other organs and tissues. For this reason, the risk factor assigned by the ICRP was $5.0 \times 10^{-4}$ Sv$^{-1}$ ($5.0 \times 10^{-6}$ rem$^{-1}$).

*Red Bone Marrow* Irradiation of the red bone marrow (RBM) is clearly linked with the induction of leukemia. Other blood-forming tissues are thought to play only a minor role in leukemia induction. For radiation protection purposes, the risk factor for leukemia was taken to be $2.0 \times 10^{-3}$ Sv$^{-1}$ ($2.0 \times 10^{-5}$ rem$^{-1}$).

*Lung* For the lung, the major radiation-induced effect is lung cancer. Evidence examined by the ICRP indicated that the risk of cancer was of the same order of magnitude as for the development of leukemia. Therefore, the risk factor assigned for lung cancer was $2.0 \times 10^{-3}$ Sv$^{-1}$ ($2.0 \times 10^{-5}$ rem$^{-1}$). In addition to considering the threat of lung cancer from radiation exposure, the ICRP again dismissed the thought that a "hot particle" in the lung would present a higher risk than that for material distributed uniformly in the lung.

*Thyroid Gland* The thyroid gland has a high sensitivity to cancer induction due to radiation exposure. In fact, it seems to be higher than that for the induction of leukemia. However, mortality from these thyroid cancers is quite low, primarily due to the success of treatment (e.g., in the United States, thyroid cancer is almost 100% survivable). The risk factor assigned to the thyroid gland was $5.0 \times 10^{-4}$ $Sv^{-1}$ ($5.0 \times 10^{-6}$ $rem^{-1}$).

*Breast* During reproductive life, the female breast may be one of the most radiosensitive tissues in the human body. For radiation protection purposes, the ICRP has assigned a risk factor of $2.5 \times 10^{-3}$ $Sv^{-1}$ ($2.5 \times 10^{-5}$ $rem^{-1}$). The breast is of primary concern for external exposures, but for some radionuclides, which distribute uniformly in the body, and for exposure by submersion in a cloud of radioactive noble gas, the breast must be included in the calculation.

*All Other Tissues* There is evidence that radiation is carcinogenic in many other organs and tissues of the body. That is, radiation is considered a general carcinogen. However, sufficient data were not available to the ICRP to allow the assignment of individual risk factors. Nevertheless, there were sufficient data to conclude that the risk factor for all other tissues was lower than those factors specified above. Based on this review, the ICRP assigned a combined risk factor for all the remaining unspecified tissues of $5.0 \times 10^{-3}$ $Sv^{-1}$ ($5.0 \times 10^{-5}$ $rem^{-1}$). The ICRP assumed that no single tissue was responsible for more than one-fifth of this value. Below, in the discussion of the ICRP calculations, these unspecified tissues are called the "remainder." However, this simple designation of the remainder is confusing to apply in some dose calculations.

### ICRP Publication 30 Techniques

ICRP Publication 30 (ICRP 1979a) introduced an entirely new dosimetry scheme for calculating the committed dose equivalent due to the intake of radionuclides in the body. The scheme was based on the concepts defined above. In the first part of ICRP Publication 30, these basic concepts were reviewed and then the dosimetry scheme was discussed in detail. Although some of the following discussion was initially introduced in ICRP Publication 26 (ICRP 1977), it seems much more appropriate to consider it in the context of the ICRP Publication 30 formulation for internal dosimetry.

### Determination of the Tissue Weighting Factors

At this point, the discussion has traced the progress of the ICRP toward deriving the tissue weighting factors needed in equation 9.3. The susceptible tissues have been identified, and a risk coefficient has been assigned to each tissue based on the available biological evidence. The next step is to calculate the individual weighting factors. This is accomplished by taking the ratio of the individual risk for a tissue to the sum of all the risk coefficients. In other words, the weighting factor is given by

$$W_T = \frac{\text{Risk coefficient for tissue } (T)}{\text{Sum of all the risk coefficients}}. \tag{9.4}$$

This calculation is summarized in table 9.2, where the tissues at risk, the radiation effects, the risk coefficients, and the weighting factors are given. The remainder

Table 9.2  Calculation of weighting factors of stochastic risks

| Tissue (T) | Radiation effect | Risk coefficient ($Sv^{-1}$) | $W_T$ |
|---|---|---|---|
| Gonads | Hereditary effects | $0.4 \times 10^{-2}$ | 0.25 |
| Breast | Cancer | $2.5 \times 10^{-3}$ | 0.15 |
| Red bone marrow | Leukemia | $2.0 \times 10^{-3}$ | 0.12 |
| Lungs | Cancer | $2.0 \times 10^{-3}$ | 0.12 |
| Thyroid gland | Cancer | $5.0 \times 10^{-4}$ | 0.03 |
| Cells on bone surfaces | Cancer | $5.0 \times 10^{-4}$ | 0.03 |
| Remainder[a] | Cancer | $5.0 \times 10^{-3}$ | 0.3 |
| Total risk | | $1.65 \times 10^{-2}$ $Sv^{-1}$ | |

[a] Assigned to any five organs or tissues not designated above. See text for a more detailed explanation.

category is assigned to the five tissues, other than those in table 9.2, that receive the highest committed dose equivalents. A weighting factor of 0.06 is assigned to each of these five tissues. If the gastrointestinal (GI) tract is irradiated, each section of the tract is considered a separate tissue.

The remainder, according to the ICRP, consists of those organs or tissues that are not mentioned in the

- metabolic model for the element,
- gastrointestinal tract model, or
- table of weighting factors (table 9.2).

As stated above, the weighting factor assigned to any single organ cannot exceed 0.06, and no more than five organs may be considered when applying these factors. A complication arises when the GI tract or organs mentioned in the metabolic model are irradiated to a significant extent. The ICRP has introduced a scheme to account for this situation, but the scheme is not discussed here.

### Secondary and Derived Limits

The ICRP has defined new terms and recommended new limits for use in radiation protection. Use of the words "maximum permissible" has been discontinued due to the misinterpretation of the intent of the concept and the misuse of the limits recommended by the ICRP. To meet the basic ICRP limits on radiation exposure of workers, intakes ($I$) of radioactive material in any one year must be limited to satisfy both of the following conditions:

$$I_S \sum_T W_T H_{T,50} \leq 0.05 \text{ Sv y}^{-1} \tag{9.5}$$

and

$$I_N H_{T,50} \leq 0.5 \text{ Sv y}^{-1}. \tag{9.6}$$

Note that equation 9.5 applies a limit to stochastic effects, whereas equation 9.6 limits nonstochastic effects in a single organ or tissue from the intake of radioactive materials. The ICRP also emphasizes that it is sufficient to limit the intake of radioactive materials in any one year to the recommended limits and that there is no need to specify a limit on the rate of intake.

A secondary limit has been defined to meet the basic conditions for occupational exposure stated in equations 9.5 and 9.6. This limit is called the annual limit on intake (ALI) and is defined as follows:

*Annual limit on intake* is the activity of a radionuclide that, if taken in alone, would irradiate a person, represented by Reference Man, to the limit set by the ICRP for each year of occupational exposure.

More specifically, the ALI is the greatest value of the annual intake $I$ that satisfies both the following inequalities:

$$I_S \sum_T W_T \, (H_{T,50} \text{ per unit intake}) \leq 0.05 \text{ Sv y}^{-1} \tag{9.7}$$

and

$$I_N \, (H_{T,50} \text{ per unit intake}) \leq 0.5 \text{ Sv y}^{-1}, \tag{9.8}$$

where $I$ (in Bq yr$^{-1}$) is the annual intake of the specified radionuclide either by ingestion or inhalation. The other parameters are identified above.

Note in equation 9.8 that the committed dose equivalent for a single tissue is used. The organ or tissue to be used in this calculation should be the organ with the largest committed dose equivalent selected from the organs with weighting factors less than 0.1. This selection criterion ensures that both dose limits will be satisfied and that, regardless of which limit controls the ALI, neither will be exceeded.

In ICRP Publication 2 (ICRP 1959), exposure to radionuclides was controlled by applying recommended maximum permissible concentrations (MPCs) in air or water to the specific exposure situation. Even though the basic recommendation of the commission under the new formulation is based on the ALI, the commission chose to include another quantity for convenience in controlling exposure to airborne radionuclides. This quantity, called the derived air concentration (DAC), is defined as follows:

*Derived air concentration* is that concentration of a radionuclide in air, which if breathed by Reference Man for a working year of 2,000 h under conditions of light activity, would result in the ALI by inhalation.

That is,

$$\text{DAC} = \frac{\text{ALI}}{(2{,}000 \text{ h y}^{-1}) \, (1.2 \text{ m}^3 \text{ h}^{-1})}, \tag{9.9}$$

or

$$\text{DAC} = \frac{\text{ALI}}{2.4 \times 10^{-3}} \text{ Bq m}^{-3}. \tag{9.10}$$

In equation 9.9, the 2,000 h is obtained from the assumption of a 40-hour work week for 50 weeks per year, and 1.2 m³ h⁻¹ is the volume of air that is assumed to be breathed by Reference Man under conditions of light activity.

## Other Definitions

Several other definitions are necessary before proceeding with a detailed discussion of the ICRP internal dosimetry scheme. Remember that the ICRP discarded the critical organ concept because it did not fit the new scheme. However, there was still a need to call the organs and tissues of the body by some name. The ICRP has introduced the terms source tissue and target tissue to describe these tissues. These are defined as follows:

*Source tissue* ($S$) is a tissue (which may be a body organ) that contains a significant amount of a radionuclide following intake of the radionuclide into the body. *Target tissue* ($T$) is a tissue (which may be a body organ) in which radiation energy is absorbed.

It should be clear that each source tissue also is a target tissue because some, if not all, of the radiation emitted by the radionuclide in the source tissue will be absorbed in the source tissue. It also should be clear that a target tissue is not necessarily a source tissue. This is true because some tissues may lie a significant distance from the source tissue but be irradiated because the radionuclide may emit X-rays or gamma rays that are able to travel large distances in tissue.

Two special cases also should be mentioned. These involve situations in which the target tissue surrounds the source tissue and vice versa. An example of the first can be found in the GI tract. Here it is assumed that radioactive material is uniformly distributed in the contents of the tract (source tissue) and the target tissue is represented by the wall of the GI tract (which does not take up the material). The opposite is found in the skeleton, where it is assumed that radioactive material is deposited in the skeleton and the target tissues are the RBM and BS cells.

Before discussing the details of the calculation of committed dose equivalent and the effective dose equivalent, two more definitions should be introduced. These definitions are important in discussing the bone model used in the new calculations. The definitions are given below:

*Volume seekers* are radionuclides that tend to be distributed throughout the volume of mineral bone. *Surface seekers* are radionuclides that tend to remain preferentially on bone surfaces.

Examples of volume seekers include $^{226}$Ra and $^{90}$Sr, whereas surface seekers are represented by $^{239}$Pu and $^{241}$Am. These definitions apply to the manner in which certain radioactive materials distribute in the skeleton and have a significant impact on the way the dose calculation proceeds.

## *Calculation of the Committed Dose Equivalent*

The committed dose equivalent is defined above as the total dose equivalent to an organ or tissue over the 50 years after intake of a radioactive material. The dose equivalent (or the committed dose equivalent) is proportional to the product of the total number of nuclear transformations occurring in the source tissue over the time period of interest and the energy absorbed per gram of target tissue per nuclear transformation of the radionuclide, modified by the appropriate quality factor. In other words,

| committed dose equivalent | = | total number of nuclear transformations in the source tissue over the period of interest | × | energy absorbed per gram of target tissue per transformation modified by the quality factor. |
|---|---|---|---|---|

In the new symbolism used by the ICRP, this word equation becomes

$$H_{T,50} = k\, U_S\, \text{SEE}\,(T \leftarrow S), \tag{9.11}$$

where $U_s$ is the total number of spontaneous nuclear transformations of a radionuclide in the source organ ($S$) over a period of 50 years after intake, and SEE is the specific effective energy imparted per gram of target tissue from a transformation occurring in a source tissue.

The total number of transformations in the source organ is obtained by integrating (or summing) over time an equation that describes the way material is retained in the organ. This equation includes loss by radioactive decay as well as loss through biological elimination. In early discussions of internal dose, the integrated activity was often called the cumulated activity and had units of microcurie-days. In the new ICRP formulation, $U_s$ has units of transformations per becquerel.

The specific effective energy is obtained from a consideration of the radiological characteristics of the nuclide deposited in the organ. These parameters, except for two, may be obtained from a review of the decay scheme of the particular radionuclide. The equation for SEE is

$$\text{SEE}\,(T \leftarrow S) = \sum_i \frac{Y_i\, E_i\, \text{AF}\,(T \leftarrow S)_i\, Q_i}{M_T}, \tag{9.12}$$

where

$Y_i$ = yield of the radiations of type $i$ per transformation of the radionuclide $j$
$E_i$ = average or unique energy of radiation $i$ in units of MeV
$\text{AF}(T \leftarrow S)_i$ = the fraction of energy absorbed in target organ $T$ per emission of radiation $i$ in source organ $S$
$Q_i$ = appropriate quality factor for radiation of type $i$
$M_T$ = mass of the target organ in units of grams

The factor $\text{AF}(T \leftarrow S)$ is called the absorbed fraction of energy and is the ratio of the energy absorbed in a target organ ($T$) to the total energy emitted by the radionuclide in the source organ ($S$). The symbols $S$ and $T$ and the arrow are to remind the user of this relationship.

Before proceeding further with the development of the ICRP equations, it is necessary to ensure that the concept of the absorbed fraction of energy is understood.

For alpha and beta radiation, all energy is assumed to be absorbed in the organ containing the radionuclide. In this case, the absorbed fraction in the source organ is equal to 1.0 (i.e., when the source irradiates itself, $S = T$). The absorbed fraction in all other target organs is assumed to be 0 (i.e., $S \neq T$). There are two exceptions to this general rule for alpha and beta radiation. These are special situations in which the source is the bone or the contents of the GI tract and the targets are BS cells and the RBM or those cells in the walls of the GI tract. These situations are discussed in more detail below in the discussion of the specific models for bone dosimetry and the GI tract.

It should be clear that the absorbed fraction concept really is important in situations where the radionuclide has a significant penetrating radiation component (i.e., X-ray or gamma-ray emission). In the new ICRP internal dose formulation, the absorbed fraction concept allows the source organ to irradiate other target organs that may be located some distance away. The British have given this concept the nickname "cross-fire."

One more complication should be mentioned before getting back to the calculation of committed dose equivalent. The ICRP has not published data that give the photon absorbed fractions of energy for radionuclides that can be used for internal dose calculations. Instead, the ICRP published data on the specific absorbed fractions (SAFs). The SAF is defined as the absorbed fraction divided by the mass of the target organ. In other words,

$$\text{SAF} = \frac{\text{AF}(T \leftarrow S)}{M_T}. \tag{9.13}$$

Note that both of the parameters on the right-hand side of equation 9.13 are included in the specification of SEE in equation 9.12. To complicate the matter further, the ICRP did not publish the data on SAFs for individual radionuclides but chose to publish the data for 12 monoenergetic photon sources with energies in the range 0.01 to 4.0 MeV. In addition, the data were not published in ICRP Publication 30 (ICRP 1979a), but were actually published as an appendix to ICRP Publication 23 (ICRP 1975) on Reference Man. Ironically, this document was published two years before Publication 26 (ICRP 1977) and four years before Publication 30.

Now, we return to the equation for committed dose equivalent and evaluate the constant in equation 9.11. The committed dose equivalent has units of sieverts per unit intake of activity. Therefore, the quantities on the right-hand side of the equation must be multiplied by a constant to bring both sides into agreement. The total number of spontaneous nuclear transformations has units of transformations per becquerel, and the specific effective energy has units of MeV $g^{-1}$ per transformation. In addition, remember that the quality factor is hidden in the calculation of SEE. Therefore, to bring both sides into agreement, it is only necessary to multiply by 1,000 g kg$^{-1}$ and $1.6 \times 10^{-13}$ J MeV$^{-1}$. Equation 9.11 then becomes

$$H_{T,50}(T \leftarrow S) = 1.6 \times 10^{-10} \, U_S \, \text{SEE}(T \leftarrow S). \tag{9.14}$$

The subscript "$T$,50" on the committed dose equivalent is intended as a reminder that the calculation of the committed dose equivalent is for a particular target organ and the time period of concern is 50 years. We elaborate on this procedure in example 9.1 below, which illustrates the techniques.

### EXAMPLE 9.1

Consider a very simple problem to illustrate the techniques and concepts introduced in this section. In this problem, the ALI and the DAC for a unit intake of $^{90}$Sr is calculated. In reality, this example duplicates some of the ICRP calculations while accepting many other values taken directly from ICRP documents.

Assume, for the purpose of this example, that the route of entry is by inhalation and that the material is soluble. (Even these designations have been changed in the new ICRP formulation, as discussed below, but these descriptions are acceptable for this example.) The steps to be taken to guide each step of the calculation are as follows:

1. Review the decay scheme data on $^{90}$Sr.
2. Review the metabolic data available on strontium.
3. List the source and target organs.
4. Assemble the data needed to calculate the committed dose equivalent for each target organ.
5. Calculate the committed dose equivalent for each target organ.
6. Calculate the weighted committed dose equivalent for each organ.
7. Calculate the committed effective dose equivalent.
8. Calculate the annual intakes for stochastic and nonstochastic effects.
9. Select the ALI.
10. Calculate the DAC.

At this point in the discussion of the new ICRP methodologies, the exercise is used to verify that the data in the ICRP documents are correct. That is, the calculation will begin at the fourth step above by tabulating data on the total number of transformations occurring in each source organ and the appropriate specific effective energy values for each target organ. These data are readily available in the supplement to ICRP Publication 30 (ICRP 1979b).

*Step 1. Review the Decay Scheme* The decay scheme for $^{90}$Sr shows that this radionuclide decays by beta emission to $^{90}$Y, which is also radioactive, decaying by beta emission to stable $^{90}$Zr (see figure 9.1, which was taken from ICRP Publication 30 [ICRP 1979a]). The half-life of $^{90}$Sr is 29.12 years, and $^{90}$Y has a half-life of 64.0 h. The calculation must include a consideration of both of these radionuclides since each transformation of $^{90}$Sr will produce an atom of $^{90}$Y, which will also transform. This consideration is reflected in the data for SEE and $U_s$.

*Step 2. Review the Metabolic Data* The metabolic data on strontium are given in ICRP 30 (ICRP 1979a). The important information here is that isotopes of strontium with half-lives greater than 15 days are assumed to be distributed uniformly throughout the volume of mineral bone (i.e., since $^{90}$Sr has a half-life greater than 15 days, this radionuclide should be considered a volume seeker).

$^{90}_{38}Sr\,(29.12y) \xrightarrow{\beta^-} {}^{90}_{39}Y\,(64.0h) \xrightarrow{\beta^-} {}^{90}_{40}Zr\,(stable)$

Figure 9.1  Simplified decay scheme for $^{90}$Sr.

The ICRP calculations assume the intake of pure material (i.e., only $^{90}$Sr), and considerations of the $^{90}$Y daughter include only those atoms of the daughter produced after the $^{90}$Sr is taken into the body. In addition, since $^{90}$Y is produced on an atom-by-atom basis as the atoms of $^{90}$Sr transform, it is assumed that the body metabolizes the daughter radionuclide in the same manner as the parent radionuclide.

*Step 3. List the Source and Target Organs*   At this point, the source and target organs can be established. These are listed below:

| Source Organs | Target Organs |
| --- | --- |
| Lungs | Lungs |
| Upper large intestine (ULI) contents | ULI wall |
| Lower large intestine (LLI) contents | LLI wall |
| Trabecular bone | Red bone marrow |
| Cortical bone | Bone surfaces |

The lungs and sections of the GI tract are always source organs because of the inhalation pathway. Bone is a source organ because the metabolic data indicated that strontium was assumed to be distributed uniformly in the volume of the bone. The source organs for bone listed above (i.e., trabecular and cortical bone) are the designations used in the bone dosimetry model. At this point, the fact that there are two kinds of bone and that radioactive material is distributed in both must be accepted without justification. Likewise, the bone model has two target organs: the RBM and BS cells.

*Step 4. Assemble the Data for the Calculation*   The data required for the calculation are shown in table 9.3. These data are taken directly from supplemental data published by the ICRP.

*Step 5. Calculate the Committed Dose Equivalents*   The products of the constant $1.6 \times 10^{-10}$, the SEE, and the $U_s$ values are shown in table 9.4. Because the radionuclides under consideration are both beta emitters, it is only necessary to consider the situation in which the source organ irradiates itself ($S = T$). That is, it is only necessary to consider the situation in which the lung irradiates the lung, and so forth. However, there are several exceptions to this in the example problem. First, notice that the contents of the GI tract irradiate the GI tract wall. Notice also that the trabecular bone irradiates both the RBM and BS. In addition, the BS is irradiated by material deposited in the cortical bone. Thus, for the BS, the contributions from each source organ (i.e., cortical bone and trabecular bone) must be summed to obtain the total committed dose equivalent for this organ.

Table 9.3  Information required for $^{90}$Sr dose calculation

| $(T \leftarrow S)$ | $^{90}$Sr | | $^{90}$Y | |
|---|---|---|---|---|
| | $U_S$ | SEE | $U_S$ | SEE |
| Lungs ← lungs | 1.9E+4$^a$ | 2.0E-4 | 3.4E+3 | 9.3E-4 |
| RBM ← TB | 7.9E+6 | 4.6E-5 | 7.9E+6 | 2.2E-4 |
| BS ← TB | 7.9E+6 | 4.1E-5 | 7.9E+6 | 1.9E-4 |
| BS ← CB | 1.9E+7 | 2.4E-5 | 1.9E+47 | 1.2E-4 |
| ULI$_w$ ← ULI$_c$ | 5.2E+3 | 4.4E-4 | 8.6E+3 | 2.1E-3 |
| LLI$_w$ ← ULI$_c$ | 9.3E+3 | 7.2E-4 | 3.1 E+3 | 3.5E-3 |

RBM, red bone marrow; TB, trabecular bone; BS, bone surfaces; CB, cortical bone; ULI$_w$, upper large intestine wall; ULI$_c$, upper large intestine contents; LLI$_w$, lower large intestine wall; LLI$_c$, lower large intestine contents.
$^a$Read 1.9E+4 as $1.9 \times 10^4$.

Table 9.4  Calculation of the committed dose equivalent and the weighted committed dose equivalent

| Tissue $(T)$ | $H_{T,50}$ (Sv Bq$^{-1}$) | $w_T$ | $w_T H_{T,50}$ (Sv Bq$^{-1}$) |
|---|---|---|---|
| Lungs | 1.1 E-9 | 0.12 | 1.3E-10 |
| Rbm | 3.3 E-7 | 0.12 | 4.0 E-8 |
| Bone surface | 7.3 E-7 | 0.03 | 2.2E-8 |
| ULI wall | 3.3 E-9 | 0.06 | 2.0E-10 |
| LLI wall | 2.8 E-9 | 0.06 | 1.7E-10 |
| | | | $H_{E,50} = 6.3E - 8$ Sv Bq$^{-1}$ |

*Step 6. Calculate the Weighted Committed Dose Equivalents* The weighted committed dose equivalent for each organ is obtained by multiplying the committed dose equivalent for each source organ by the weighting factors derived above and listed in table 9.2. For those organs not mentioned specifically in the table, a weighting factor of 0.06 is used for each.

*Step 7. Calculate the Committed Effective Dose Equivalent* The committed effective dose equivalent, $H_{E,50}$, is the sum of all the weighted committed dose-equivalent values calculated for the target organs. This value is shown at the bottom of table 9.4.

*Step 8. Calculate the Annual Intakes* The effective intakes are calculated for both the stochastic and nonstochastic limits set by the ICRP. First, consider the stochastic limit

$$I_S \leq \frac{0.05 \text{ Sv y}^{-1}}{\sum_T W_T \, (H_{T,50} \text{ per unit intake}) \text{ Sv Bq}^{-1}}. \tag{9.15}$$

**408** Radiological Risk Assessment and Environmental Analysis

Using the data from table 9.3,

$$I_S \leq \frac{0.05 \text{ Sv y}^{-1}}{6.3 \times 10^{-8} \text{ Sv Bq}^{-1}}. \tag{9.16}$$

Therefore, $I_s$ (stochastic) $\leq 8 \times 10^5$ Bq yr$^{-1}$.

For the nonstochastic limit, the committed dose equivalent for a single organ is used in the calculation. The organ selected for use is taken from a list of several organs that include the thyroid gland, the BS, and any organ in the remainder (these are organs with tissue weighting factors $< 0.10$). The procedure is to select the largest value of any of the committed dose equivalents for the organs mentioned above and to use this value in the denominator of the equation. In this example, the organ selected has to be the BS, $7.3 \times 10^{-7}$ Sv Bq$^{-1}$. Now we can continue with the calculation of the annual limit using the ICRP limit for nonstochastic effects,

$$I_N \leq \frac{0.5 \text{ Sv y}^{-1}}{(H_{T,50} \text{ per unit intake}) \text{ Sv Bq}^{-1}}. \tag{9.17}$$

Thus, for this example,

$$I_N \leq \frac{0.5 \text{ Sv y}^{-1}}{7.3 \times 10^{-7} \text{ Sv Bq}^{-1}}. \tag{9.18}$$

Therefore, $I_N$ (nonstochastic) $\leq 7 \times 10^5$ Bq yr$^{-1}$.

*Step 9. Select the Annual Limit on Intake* The ICRP has defined the ALI as the largest value of $I$ that satisfies both inequalities given above. What this really means is that the ALI is the smallest numerical value of $I$ calculated above using the stochastic and nonstochastic limits. In this case, the smallest value of $I$ is that obtained in the calculation using the nonstochastic limit, $7 \times 10^5$ Bq yr$^{-1}$. Note that, in this example, the values have been rounded to one significant figure. This serves as a reminder from the ICRP of the accuracy of the calculation.

Therefore, ALI $= 7 \times 10^5$ Bq yr$^{-1}$.

*Step 10. Calculate the Derived Air Concentration* The DAC is the ALI divided by the amount of air breathed by Reference Man in a working year:

$$\text{DAC} = \frac{\text{ALI Bq y}^{-1}}{(2{,}000 \text{ h y}^{-1})(1.2 \text{ m}^3 \text{ h}^{-1})}. \tag{9.19}$$

So, for this example,

$$\text{DAC} = \frac{7 \times 10^5 \text{ Bq y}^{-1}}{(2{,}000 \text{ h y}^{-1})(1.2 \text{ m}^3 \text{ h}^{-1})}. \tag{9.20}$$

Therefore, DAC $= 300$ Bq m$^{-3}$.

Thus, in this example, the controlling ALI is $7 \times 10^5$ Bq yr$^{-1}$ based on the 0.5 Sv yr$^{-1}$ limit to the BS. The calculated DAC is 300 Bq m$^{-3}$. These values agree with those calculated by the ICRP for $^{90}$Sr.

## Dosimetric Models Used in the ICRP 30 Calculations

In this section, we concentrate on the models and assumptions used by the ICRP in its internal dose calculations (ICRP 1979a). It is important that the parameters used in these calculations be understood so that a full appreciation of the ICRP assumptions can be achieved. The models appear to be complex, but most are relatively simple. The equations used to describe the intake, deposition, distribution, and retention of the material in the body can be reduced to a set of simple calculations. Typically, these equations are used to obtain the total number of spontaneous nuclear transformations that occur in the source organ or tissue over the time period of interest.

It is also important to become familiar with these models so an understanding of the limitations of the ICRP recommendations can be obtained. In some cases, the usefulness of a model may be limited severely even though it appears to be improved. In other situations, there is still a lack of sufficient anatomical and physiological information, and the proposed "new" model actually may contain a restatement of many of the earlier assumptions.

## Model of the Respiratory System

The dosimetric model for the respiratory system is divided into two parts: the deposition model and the retention model. However, this division is introduced simply to assist in the discussion of the model and does not represent the actual respiratory system, either anatomically or physiologically.

*Inhalation-Deposition Model* The respiratory system is divided into three regions, rather than the single region used in ICRP Publication 2 (ICRP 1959). Material inhaled by a radiation worker is assumed to be deposited in the nasopharyngeal region (N-P), the tracheobronchial region (T-B), and the pulmonary region (P). The fraction of the inhaled material deposited in these regions is determined by a set of deposition fractions that are a function of the particle size of the inhaled material. The model is based on specifying the particle size in terms of the activity median aerodynamic diameter (AMAD). The AMAD is the diameter of a unit sphere with the same settling velocity in air as that of an aerosol particle whose activity is the median for the entire aerosol. This distinction is lost on most users, so it is best to think of the AMAD as a specification of the particle diameter. Regardless of the actual diameter of the particle, it has the same characteristics as a particle that is a perfect sphere of the specified diameter (e.g., 1 $\mu$m diameter).

The deposition parameters as a function of AMAD are shown in figure 9.2. These parameters specify the fraction of the inhaled material assumed to be deposited in each of the three regions of the respiratory system model. From a practical standpoint, these data fit a general idea of the deposition as a function of particle size.

First, consider the nasopharyngeal region. This region begins at the tip of the nose and runs through the nasal passages to the epiglottis. The epiglottis is a flexible cartilage that protects the trachea during swallowing. During inhalation, large particles are usually deposited preferentially in this region. In a manner of speaking, that is

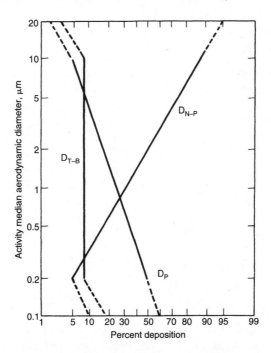

Figure 9.2 Deposition of dust in the respiratory system. The percentage of activity or mass of an aerosol that is deposited in the nasopharyngeal (N-P), tracheobronchial (T-B), and pulmonary (P) regions is given in relation to the activity median aerodynamic diameter (AMAD) of the aerosol distribution. The model is intended for use with aerosol distributions with AMADs between 0.2 and 10 μm and with geometric standard deviations of less than 4.5. Provisional estimates of deposition further extending the size range are given by the dashed lines. For an unusual distribution with an AMAD of greater than 20 μm, complete deposition in the N-P region can be assumed. The model does not apply to aerosols with AMADs of less than 0.1 μm. Adapted from ICRP (1979b).

the purpose of nasal hairs and mucus in the nasal passages. As the particle size of the inhaled material decreases, the fraction of the inhaled material deposited in the N-P region also decreases. In other words, the N-P region is very effective in removing large particles from the inhaled air. This is illustrated in figure 9.2, which shows the fraction of 10 μm particles deposited in the N-P region to be 0.87 (i.e., 87%), whereas the fraction for smaller particles (e.g., 0.2 μm particles) deposited in the N-P region is only 0.05 (5%). This assumption reenacts a common situation many have experienced when working in a dusty environment. Often the deposition of dust in the N-P region will stimulate the mucous membranes, leading to a runny nose and, sometimes, causing the worker to sneeze. The knowledge that there is a high deposition of large particles in the N-P region is also the basis for the use of nasal swabs and nose blows in the evaluation of real or suspected inhalation exposures.

The next region of the respiratory system model, the tracheobronchial region, actually extends into the lung and down to the terminal bronchioles. The trachea can be thought of as a pipe that begins to branch as it enters the lung. A typical diameter of this pipe is about 20 mm. The first branch (or "bifurcation") results in two smaller pipes, with each pipe entering the lung. These two pipes are called the right and left main bronchus (each about 8 mm in diameter). This branching continues as the bronchial tubes (called bronchi) become smaller and smaller at each bifurcation until, at the terminal bronchioles, the diameter is only about 0.6 mm.

In this region, deposition appears to be independent of particle size. In figure 9.2, the fraction deposited in the T-B region is assumed to be a constant at 0.08 (8%) for particle sizes between 0.2 and 10 μm. The distinguishing feature of the T-B region is that it is coated with a thin sheet of mucus and lined with hairlike structures called cilia. The cilia beat rhythmically in an upward motion that causes a net upward movement of the mucus sheet. This movement serves to clear particles deposited in the T-B region into the GI tract by swallowing. In other words, particles become trapped in this mucus sheet and must move with it as the cilia move the mucus up the trachea. Upon reaching the region of the epiglottis, the material and the mucus that contains it are swallowed. Thus, all exposures by inhalation also result in an exposure of the GI tract. In some cases, material may be cleared through the mouth (expectorated), but this is not a common loss pathway.

The N-P and T-B regions of the respiratory system make up what is called the "anatomical dead space" of the lung. These regions are given this name because no gaseous exchange occurs between the inhaled air and the blood. In other words, no oxygen crosses the lung into the blood and no carbon dioxide crosses into the exhaled air. However, as described below, the ICRP assumes that some of the solid material deposited in each of these regions can be absorbed in the blood. This absorption will depend on the solubility of the material inhaled.

The last major region in the deposition model is the pulmonary region. This region is sometimes called the "deep lung" and is of concern in dose calculations. The pulmonary region consists of many structures, including the respiratory bronchioles, alveolar ducts, atria, alveoli, and alveolar sacs. This region is considered the functional area of the lung. That is, gaseous exchange between the inhaled air and the blood occurs in the pulmonary region. In the pulmonary region, the deposition pattern is just the opposite of that in the N-P region. That is, as the particle size decreases, the fraction deposited in the pulmonary region increases. For 10 μm particles, the deposition is only 0.05 (5%), but the deposition fraction increases to 0.50 (50%) for particles with a diameter of 0.2 μm.

All of the new calculations published by the ICRP assume a standard particle size of 1 μm AMAD. Therefore, the fractions of the inhaled material deposited in the N-P, T-B, and P regions are 0.30, 0.08, and 0.25, respectively. Notice that the fractions deposited in all three regions represent only 0.63 (63%) of the total. Thus, this new model assumes that 0.37 (37%) of the inhaled material is exhaled.

*Inhalation-Retention Model* The basic retention model is shown in figure 9.3. Material deposited in the three regions of the respiratory system is cleared to the body fluids, the GI tract, or the pulmonary lymph nodes. A compartment model is used to explain retention and clearance from the respiratory system. This model takes the three regions of the system and breaks each region into two or more compartments. These compartments represent a convenient way of expressing the retention or movement of material from each major region of the lung. Remember that these compartments do not necessarily represent an actual anatomical or physiological structure of the system. In some cases, two compartments may

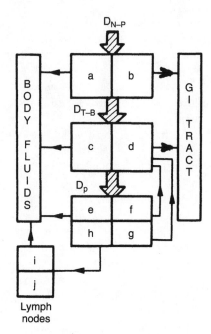

Figure 9.3 Model used to describe clearance from the respiratory system. Values for the removal half-times and compartment fractions are given in tables 9.5, 9.6, and 9.7. Adapted from ICRP (1979b).

be established to reflect differences in clearance times from a single region. Notice also that the retention model includes two compartments representing the pulmonary lymph nodes. This region does not play a role in the deposition of material during inhalation. However, the lymph nodes play an important role in the clearance of insoluble material from the deep lung.

The letters a–j in figure 9.3 indicate the various absorption and translocation processes associated with clearance from the various compartments. These processes are discussed briefly below:

a. This compartment represents rapid uptake of material deposited in the N-P region directly into the systemic body fluids by absorption. Although not precise, it is convenient to think of this compartment as the circulating blood and to consider the blood as the mechanism by which material is transported within the body.
b. Material is cleared also from the N-P region by mucus transport directly into the GI tract primarily by swallowing.
c. This compartment represents rapid absorption of material deposited in the T-B region into the systemic body fluids. The mechanism is similar to that in compartment a.
d. This compartment is analogous to compartment b and represents mucociliary clearance of material deposited in the T-B region into the GI tract.
e. As with compartments a and c, this compartment corresponds to direct translocation of material deposited in the P region into the body fluids.

f. In this compartment, it is assumed that material undergoes relatively rapid clearance that depends on mucociliary clearance as well as the possibility of material being removed by macrophages. A macrophage is a large cell whose primary function is to engulf foreign material and remove it from the lung. According to this model, the cleared material goes to the GI tract through the T-B region (i.e., compartment d).
g. There appears to be a secondary clearance process that occurs in the pulmonary region. This process is much slower than that associated with compartment f. It is important to remember that there are apparently two separate processes occurring and that there is a clear distinction in the times required for these processes to clear material from the pulmonary region.
h. The lymphatic system may play an important role in clearance of deposited material from the pulmonary region. There is a slow clearance of material into the lymph node compartments. Generally, this process is important only for insoluble material. However, as discussed below, some small fraction of the material is assumed to leave the P region and pass into the lymph nodes regardless of the solubility of the deposited material.
i. This compartment represents a secondary route of introduction of material into the body fluids. This may be strongly dependent on the ability of the material to penetrate the lymph tissue and reach the blood. It is not clear whether the mechanism represents partial or complete dissolution of the particles in the lymph system or some other less well understood mechanism.
j. Some material may reside in the lymph nodes for an infinite time. This compartment is added to reflect that fact.

Notice that compartments a, c, and e represent processes in which the deposited material is absorbed directly into the body fluids. Also, notice that compartments b, d, f, and g represent clearance mechanisms in which the inhaled material is physically transported out of a compartment and deposited in another one. Transportation to another compartment is assumed before the material can be absorbed into the body fluids. Clearance half-times associated with these transport compartments may change as the solubility of the inhaled material changes.

The ICRP has placed chemical compounds of inhaled material into three classes, D, W, and Y, depending upon the rate of clearance from the pulmonary compartment. These three pulmonary clearance classes are an indication of the solubility of the inhaled material:

D = Material cleared from the pulmonary region with a biological half-time of less than 10 days (D = days)
W = Material cleared from the pulmonary region with a biological half-time between 10 and 100 days (W = weeks)
Y = Material cleared from the pulmonary region with a biological half-time of more than 100 days (Y = years)

Fortunately for the user of these dose calculation techniques, the ICRP has placed all possible chemical compounds of the individual elements into one of these three

Table 9.5  Parameters for the ICRP respiratory model, class D aerosol

| Region | Compartment | Clearance half-time (days) | Compartment fraction |
|---|---|---|---|
| N-P | a | 0.01 | 0.5 |
|  | b | 0.01 | 0.5 |
| T-B | c | 0.01 | 0.95 |
|  | d | 0.2 | 0.05 |
| P | e | 0.5 | 0.8 |
|  | f | — | — |
|  | g | — | — |
|  | h | 0.5 | 0.2 |
| L (lymph nodes) | i | 0.5 | 1.0 |
|  | j | — | — |

Table 9.6  Parameters for the ICRP respiratory model, class W aerosol

| Region | Compartment | Clearance half-time (days) | Compartment fraction |
|---|---|---|---|
| N-P | a | 0.01 | 0.1 |
|  | b | 0.4 | 0.9 |
| T-B | c | 0.01 | 0.5 |
|  | d | 0.2 | 0.5 |
| P | e | 50 | 0.15 |
|  | f | 1 | 0.4 |
|  | g | 50 | 0.4 |
|  | h | 50 | 0.05 |
| L (lymph nodes) | i | 50 | 1.0 |
|  | j | — | — |

classes. Therefore, all the user is required to do is establish the chemical compounds inhaled or ingested and then use the classification assigned by the ICRP. These are given in what is called the metabolic data for each element. In addition, the ICRP has specified the parameters (clearance half-times and compartment fractions) associated with the retention model for each of these pulmonary clearance classes. These parameters are given in tables 9.5, 9.6, and 9.7 for classes D, W, and Y, respectively.

A short discussion of tables 9.5–9.7 will point out the significance of the pulmonary clearance classes and the parameters associated with each class. Table 9.5 shows the parameters for class D aerosols. Remember that these materials clear rapidly from the P region (<10 days), and therefore, this class includes compounds that are readily soluble. The parameters indicate that material clears quickly

Table 9.7 Parameters for the ICRP respiratory model, class Y aerosol

| Region | Compartment | Clearance half-time (days) | Compartment Fraction |
|---|---|---|---|
| N-P | a | 0.01 | 0.01 |
|  | b | 0.4 | 0.99 |
| T-B | c | 0.01 | 0.01 |
|  | d | 0.2 | 0.99 |
| P | e | 500 | 0.05 |
|  | f | 1 | 0.4 |
|  | g | 500 | 0.4 |
|  | h | 500 | 0.15 |
| L (lymph nodes) | i | 1,000 | 0.9 |
|  | j | ∞ | 0.1 |

(0.01 days = 14.4 min) from the N-P region, with 50% of the material being cleared through each compartment. In other words, 50% of the material deposited in the N-P region is absorbed into the body fluids (from compartment a) and the other 50% is swallowed and enters the GI tract (from compartment b). In some ways, this model is consistent with what has been learned about nasal swabs as an indication of potential intake—these samples must be taken soon after exposure if they are to be meaningful.

In the T-B region, 95% of the inhaled material is absorbed into the body fluids with a 0.01-day half-time, while the remainder is cleared to the GI tract with a 0.2-day half-time. Parameters for the P region indicate that the secondary clearance mechanisms that deposit material in the GI tract (compartments f and g) do not play an important role for very soluble material. Notice that 80% of the inhaled material is absorbed in the body fluids with a 0.5-day half-time, while the remaining 20% goes to the pulmonary lymph nodes. However, there is no holdup or retention of this soluble material in the lymph nodes, as the model indicates that all of the material is transferred from the lymph nodes to the body fluids. In some ways, this can be thought of as a pass-through route to the blood.

For class W aerosols (see table 9.6), the fraction cleared by each pathway changes dramatically. Remember that these compounds are only moderately soluble. Notice that only 10% of the material in the N-P region is absorbed in the body fluids, whereas the remainder of the material is deposited in the GI tract. The fraction cleared from the T-B region through each compartment is equal, and in the P region, a majority of the material actually enters the GI tract (add the fractions for compartments f and g). Pay close attention to the clearance half-times assumed for each compartment. Even though class W covers materials that clear from the pulmonary region with half-times between 10 and 100 days, a constant clearance half-time of 50 days is used for all class W materials deposited in the P and L regions. (The pulmonary clearance half-time for class D compounds was assumed to be 0.5 days.)

In table 9.7, for class Y aerosols (very insoluble), it is assumed that 99% of the material deposited in the N-P region is cleared to the GI tract. This assumption also holds for the T-B region. One important assumption included for this class of materials is the transfer of materials to the lymph nodes. For class Y, 10% of the material entering the lymph nodes is assumed to remain there with an infinite half-time. The fraction cleared to the body fluids (from compartment i) also has a different clearance half-time than the clearance half-time for compartment h. This indicates the influence of a mechanism that is acting to hold up all the material for a finite time period. Clearance class Y is the only class in which the lymph nodes serve any purpose other than passing the material through into the blood. A constant pulmonary clearance half-time of 500 days for most compartments is used for all compounds in this class.

In all three tables, no material is truly insoluble. That is, even for pulmonary clearance class Y, some material is allowed to reach the body fluids (see the fractions for compartments a, c, and e). In the old ICRP Publication 2 model (ICRP 1959), if a compound was assumed to be insoluble, no material reached the blood, and therefore, no material was deposited in organs or tissues of the body. This is an important change in the new ICRP model.

In all the ICRP calculations, particle size distributions may exist, and the standard calculations may need to be adjusted for differences between the measured and assumed AMAD. The ICRP has provided a single formula that can be used to calculate the ratio of the committed dose equivalent to tissue $T$ for the actual particle size to the committed dose equivalent for the 1 μm AMAD:

$$\frac{H_{T,50}(\text{AMAD})}{H_{T,50}(1\,\mu m)} = f(N-P)\frac{D_{N-P}(\text{AMAD})}{D_{N-P}(1\,\mu m)} + f(T-B)\frac{D_{T-B}(\text{AMAD})}{D_{N-P}(1\,\mu m)}$$
$$+ f(P)\frac{D_P(\text{AMAD})}{D_P(1\,\mu m)}, \tag{9.21}$$

where

$H_{T,50}(1\,\mu m)$ = committed dose equivalent to tissue $T$ for the standard particle size

$H_{T,50}(\text{AMAD})$ = committed dose equivalent to tissue $T$ for the new particle size

$D_{N-P}, D_{T-B}, D_P$ = fractions of material deposited in the nasopharyngeal, tracheobronchial, and pulmonary region, respectively, as a function of particle size

$f(N-P), f(T-B), f(P)$ = the fractions of the committed dose equivalent due to deposition in each of the three respiratory system regions

The deposition fractions by particle size are given in table 9.8. The parameters represented by the fractions are not given in the text of ICRP Publication 30 (ICRP 1979a) and can only be found in the supplemental material (ICRP 1979b). This information is useful only if the actual AMAD has been determined and if it is very different from the assumed particle size. However, there are situations in which a change in particle size can significantly influence the calculated committed dose equivalent and thus the ALI and the DAC.

Table 9.8 Deposition fractions as a function of aerosol AMAD

| AMAD (μm) | N-P | T-B | P |
|---|---|---|---|
| 0.2 | 0.05 | 0.08 | 0.50 |
| 0.3 | 0.088 | 0.08 | 0.43 |
| 0.4 | 0.13 | 0.08 | 0.39 |
| 0.5 | 0.16 | 0.08 | 0.35 |
| 0.6 | 0.19 | 0.08 | 0.32 |
| 0.7 | 0.23 | 0.08 | 0.30 |
| 0.9 | 0.26 | 0.08 | 0.28 |
| 1.0 | 0.30 | 0.08 | 0.25 |
| 2.0 | 0.50 | 0.08 | 0.17 |
| 3.0 | 0.61 | 0.08 | 0.13 |
| 4.0 | 0.69 | 0.08 | 0.10 |
| 5.0 | 0.74 | 0.08 | 0.088 |
| 6.0 | 0.78 | 0.08 | 0.076 |
| 7.0 | 0.81 | 0.08 | 0.067 |
| 8.0 | 0.84 | 0.08 | 0.060 |
| 9.0 | 0.86 | 0.08 | 0.055 |
| 10.0 | 0.87 | 0.08 | 0.050 |

## Model of the Gastrointestinal Tract

The GI tract model used in the internal dose calculations is very similar to the previous model in Publication 2 (ICRP 1959). The GI tract is assumed to consist of four compartments, as shown in figure 9.4: stomach (ST), small intestine (SI), upper large intestine (ULI), and lower large intestine (LLI). The most notable difference in this revised model (Publication 30, ICRP 1979a) and the previous GI tract model (ICRP 1959) is the addition of walls. That is, in the new model (see table 9.9), the mass of the wall of each individual section is specified. The mass of the contents of each section has also been reduced compared with the older model, and the time sequence in each section is specified as the mean residence time in the section. In this revised model, the mean residence times through the ULI and LLI have been increased, and the mean residence time of material in the entire GI tract is assumed to be 42 h (i.e., 42/24 days).

For each section of the GI tract in this model, the contents are the source organ in the calculation, and the wall of the particular section is the target organ. This designation is in keeping with the introduction of the concept of source and target organs given above. If the GI tract contents should contain a photon emitter, one section of the tract may irradiate all the other sections and even other organs in the body. The other organs in the body may contain photon emitters, and these may serve as source organs for the calculation of committed dose-equivalent adsorbed dose to the sections of the GI tract.

This new model appears to be much more sophisticated than the earlier one. However, in some cases, the current knowledge of the GI tract is no better than that reflected in the earlier model. Thus, the model is still very crude, and dose

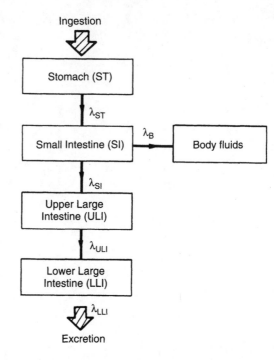

Figure 9.4  Model for the gastrointestinal tract. Parameter values are listed in table 9.9. Adapted from ICRP (1979b).

Table 9.9  Parameters for the ICRP gastrointestinal tract model (adapted from ICRP 1979b)

| Section of GI tract | Mass of wall (g) | Mass of contents (g) | Mean residence time (days) | $\lambda$ (days$^{-1}$) |
| --- | --- | --- | --- | --- |
| Stomach (ST) | 150 | 250 | 1/24 | 24 |
| Small intestine (SI) | 640 | 400 | 4/24 | 6 |
| Upper large intestine (ULI) | 210 | 220 | 13/24 | 1.8 |
| Lower large intestine (LLI) | 160 | 135 | 24/24 | 1 |

calculations depend on many simplifying assumptions (some of which were used in the earlier model). Fortunately, the mean residence times in the GI tract are still very short regardless of the solubility of the compound, and the simplifying assumptions still provide very conservative estimates of the doses to the walls of the GI tract.

As with the previous model, it is convenient to assume that the GI tract is a pipe and material entering one end of the pipe exits at the other end as excreta. The model allows material to cross into the body fluids only at the small intestine. The fraction that crosses into the blood is specified by the parameter $f_1$, as in the old ICRP Publication 2 formulation (ICRP 1959). One major difference in the new GI model is the assumption that no material is truly insoluble. In other words, all materials, regardless of the pulmonary clearance class or estimated solubility in the GI tract, are assumed to cross the small intestine wall to some extent. The fraction crossing into the body fluids may not be large, but some small amount is assumed to be absorbed for all compounds. For example, for insoluble plutonium (i.e., plutonium oxide), $f_1$ is assumed to be $1 \times 10^{-5}$. This is a very small fraction of the total inhaled activity cleared to the GI tract. In the earlier model, however, it was assumed that $f_1$ was zero for insoluble plutonium oxide and for all other insoluble compounds.

One additional important assumption is that the movement of material through the tract is governed by first-order linear kinetics. This assumption means that a set of differential equations can be written to describe the movement of material through and between the sections of the tract. This assumption is important when it is necessary to calculate the time integral of activity in each section for use in the dose calculation (i.e., the values of $U_s$).

## Dosimetric Model for Bone

The model for bone appears to be very complicated, but it is not. The ICRP has made it as simple as possible by assigning constant values of the absorbed fraction of energy for each type of radiation to the various targets in bone and by defining two major regions of interest for dose calculations. These regions were defined above as the RBM and BS cells.

Cells at risk in the skeleton (bone) are the hematopoietic stem cells close to the BS. The hematopoietic stem cells in adults (i.e., 20 years of age) are assumed to be uniformly distributed in the RBM located in trabecular bone. The dose equivalent to these cells is calculated as the average over the tissue that fills the cavities within trabecular bone (i.e., the tissue in the marrow cavities in the bone). The stem cells serve as the source of the red and white cells and are at risk because of the potential for the induction of leukemia.

The osteogenic cells are assumed to lie very close to the BS, usually within 10 μm. These cells, found in both the trabecular bone and the cortical bone of the adult, are involved in the resorption and laying down of new bone (cell remodeling). Cells on the BS are thought to be at risk because of the potential induction of bone cancer.

There are two target tissues and two source tissues in the bone. Cortical and trabecular bone have been mentioned as the source tissues. There may be additional source tissues because the possibility exists of irradiation of the target tissues by source tissues containing photon emitters located outside the bone (i.e., in other organs). Contributions to the dose from these source tissues always must be included in the complete dose calculation.

The ICRP makes another simplifying assumption regarding the location of sources and targets: that RBM does not exist within cortical bone. Cortical bone is the hard bone that makes up the shafts of the long bones and the outer surfaces of all bones. This material is very dense (density of about 2 g cm$^{-3}$). One main feature of cortical bone is the Haversian canals. These canals are located throughout the cortical bone and actually contain small blood vessels. The diameters of these canals range from 22 to 110 μm (ICRP 1975). Thus, even though the cortical bone is thought to be a hard or compact material, it is really permeated by a system of

Table 9.10  Recommended absorbed fractions for dosimetry of radionuclides in bone

| Source organ | Target organ | Class of radionuclide ||||| 
| | | α Emitter || β Emitter |||
| | | Uniform in volume | On bone surfaces | Uniform in volume | On bone surfaces ||
| | | | | | $\overline{E}_\beta \geq 0.2\,\text{MeV}$ | $\overline{E}_\beta < 0.2\,\text{MeV}$ |
|---|---|---|---|---|---|---|
| Trabecular bone | Bone surfaces | 0.025 | 0.25 | 0.025 | 0.025 | 0.25 |
| Cortical bone | Bone surfaces | 0.01 | 0.25 | 0.015 | 0.015 | 0.25 |
| Trabecular bone | Red bone marrow | 0.05 | 0.5 | 0.35 | 0.5 | 0.5 |
| Cortical bone | Red bone marrow | 0.0 | 0.0 | 0.0 | 0.0 | 0.0 |

blood vessels running through bony canals. Cells lying on the inner surface of these canals are assumed to be at risk and are included in the category "cells on bone surfaces."

Trabecular bone is the name given to the tiny pieces of bone that are found inside the shafts of the long bones or inside flat bones such as the skull, pelvis, and ribs. This bone forms the network inside the cortical bone that represents the spaces in which the RBM resides. One point that causes a great deal of confusion is that the ICRP assumes that cells lying on the surfaces of the trabecular bone (in the marrow spaces) also are included in the category "cells on bone surfaces." Thus, the target and source tissue assignments in the bone model are as follows:

| (Target ← Source) | (T ← S) Designation |
| Bone surfaces ← cortical bone | (BS ← CB) |
| Bone surfaces ← trabecular bone | (BS ← TB) |
| Red bone marrow ← trabecular bone | (RBM ← TB) |
| Red bone marrow ← cortical bone | (RBM ← CB) |

To simplify calculations using the bone model, the ICRP has assumed constant values for the absorbed fractions of energy for alpha- and beta-emitting radionuclides. The absorbed fractions are divided further according to surface and volume seekers. For beta-emitting radionuclides on BS, another division is made based on the radiation energy. The absorbed fractions of energy used in the bone model are shown in table 9.10.

The NCRP has been relatively critical of certain aspects of the bone model. These criticisms revolve around the assignment of radioelements to two classes, volume or surface seekers, and the designation of target tissues in the model (i.e., RBM and cells on BS). First, the NCRP points out that "it is well recognized that volume seekers are far from uniformly distributed, and it is unlikely that surface seekers are uniformly distributed on all surfaces, and that they remain there indefinitely" (NCRP 1985). However, the NCRP does acknowledge that the bone model possesses a degree of conservatism because of these designations.

In a similar fashion, the NCRP takes the ICRP to task regarding the assumption that the entire RBM is at risk. Again, data indicate that the sensitive cells (at least in mice) lie at approximately 120 μm from the bone surface. Thus, the bone model may overestimate the dose to the marrow by a relatively large factor. Since the radionuclides of concern are those alpha and beta emitters that deposit in bone, the use of this model may significantly overestimate the committed dose equivalent and, thus, produce unnecessarily restrictive ALI and DAC values for these radionuclides.

Finally, the NCRP was concerned that the ICRP reports "give no indication that epidemiologic experience with radium contributed to the development of the limits for bone-seeking elements" (NCRP 1985). It is well known that the original ICRP 2 model was based solely on data from radium dial painters (ICRP 1959). However, with the exception of this criticism, the NCRP proposes no alternative model.

## Submersion in a Radioactive Cloud

In Publication 30 (ICRP 1979a), the ICRP also revised the model for submersion in a radioactive cloud of noble or inert gas. Major improvements include the incorporation of newer calculations of the absorbed dose due to photon emitters in the cloud, a revision of the data on beta emitters in an infinite cloud, and calculations for what are called "room-size" clouds. The smaller clouds of radioactive gas are intended to simulate the more practical situation in which a radioactive gas escapes (and is confined) in a room or a working area. The few pages of explanation on the model included in ICRP Publication 30 are spent justifying the assumption that external exposure to the cloud is of primary importance and that the intake of the noble gas and the dose to the body while the gas is held in the lungs during inspiration are negligible.

For example, the ICRP assumed submersion in a cloud of noble gas of unit concentration (1 Bq m$^{-3}$). Three dose-equivalent rates of interest were defined, and the relative magnitude of each was evaluated:

$H(E)$ = dose-equivalent rate to any tissue from external radiation
$H(A)$ = dose-equivalent rate from internal irradiation by absorbed gas
$H(L)$ = dose-equivalent rate to the lung from contained gas

After developing an equation to estimate each of the above dose-equivalent rates and reviewing the range of possible values of the parameters in each equation, the ICRP took ratios of the rates to indicate their relative significance:

$$H(E) \geq 200 H(A) \tag{9.22}$$

and

$$H(E) \geq 130 H(L). \tag{9.23}$$

Based on these results, the ICRP decided to consider exposure to a cloud of noble gas only as an external exposure situation.

An exception to this general conclusion for submersion in a noble gas cloud was exposure to tritium. For tritium, the dose-equivalent rate to the lung ranged between 60 and 150 times the dose-equivalent rate from absorbed gas. In addition, it was assumed for tritium that $H(E) = 0$. The ICRP did not consider exposure to radon or thoron in ICRP Publication 30 (ICRP 1979a) because a separate document, ICRP Publication 32 (ICRP 1981), was in preparation to address these radionuclides.

Submersion in a cloud of radioactive gas does not represent an internal exposure situation. However, the ICRP included the discussion because it is necessary to calculate DAC for these gases.

## Recent Recommendations

The ICRP continues to update and revise recommendations as new data and methodologies become available. At the time this chapter was written, the latest update was ICRP Publication 98, *Radiation Safety Aspects of Brachytherapy for Prostate Cancer Using Permanently Implanted Sources* (ICRP 2006). The following discussion touches on recent publications that have some bearing on the internal dose assessment of environmental exposures to radionuclides. Although most of the recommendations set forth in these volumes have not yet been adopted by U.S. regulatory bodies, some of the models and biokinetic data have been incorporated in the calculation of cancer risk coefficients detailed in the U.S. Environmental Protection Agency's Federal Guidance Report 13 (Eckerman et al. 1999). Several draft recommendations were available for public comment at the time of this writing, in particular, *The Concept and Use of Reference Animals and Plants for the Purposes of Environmental Protection* and *Assessing Dose of the Representative Individual for the Purpose of Radiation Protection of the Public*, which bear directly on assessment of internal sources caused by releases into the environment.

### ICRP Publication 60

The latest general recommendations from the ICRP were published in 1991. In ICRP Publication 60 (ICRP 1991b), the ICRP revised and expanded the recommendations in ICRP Publication 26 (ICRP 1977). While the basic conceptual framework for radiation protection is largely unchanged, there are significant changes in the recommended dose limits and the basic dosimetric quantities. In addition, a draft document is available that is the latest update in this series of recommendations.

Table 9.11  Radiation weighting factors (adapted from ICRP 1991a, table 1)

| Type | $w_R$ |
|---|---|
| Photons—all energies | 1 |
| Electrons and muons – all energies[a] | 1 |
| Neutrons | |
| &lt;10 keV | 5 |
| 10 keV to 100 keV | 10 |
| &gt; 100 keV to 2 MeV | 20 |
| &gt; 2 MeV to 20 MeV | 10 |
| &gt; 20 MeV | 5 |
| Protons[b] | 5 |
| Alpha particles, heavy nuclei, fission fragments | 20 |

[a] Excludes Auger electrons emitted from nuclei bound to DNA.
[b] Protons other than recoil protons with energies greater than 2 MeV.

## Dosimetric Quantities

The first major departure from previous recommendations was the introduction of the *radiation weighting factor*, $w_R$. This value replaces the quality factor ($Q$), although the values of $w_R$ closely follow the values of $Q$ (see table 9.11). Where values of $Q$ were related to the linear energy transfer of the radiation, $w_R$ is assigned the value of unity for photons of all energies. For other radiations, the value is related to the observed relative biological effectiveness of that radiation in inducing stochastic effects at low doses compared to either X-rays or gamma rays. For internal sources, the energy of the emitted radiation determines the $w_R$; otherwise, it is determined by the energy of the radiation incident on the body (ICRP 1991a). This has been a source of some controversy because the energy of the radiation being absorbed is likely to be different from the energy that determines the radiation weighting factor (Simmons and Watt 1999). More complex methods for dealing with the radiation weighting factor as a continuous distribution in energy are detailed in annex A of ICRP Publication 60 (ICRP 1991b) and in ICRP Publication 92 (ICRP 2003). The microdosimetric treatment of the special case of Auger electron emitters bound to DNA is discussed in annex B (ICRP 1991b).

With the addition of the radiation weighting factor, the commission took the opportunity to simplify the names of many of the dosimetric quantities while making it clear that the new quantities are based on new values. The weighted absorbed dose averaged over an organ or tissue is now termed the *equivalent dose*, $H_T$. The equivalent dose is defined for a given tissue, $T$, by the expression

$$H_T = \sum_R w_R D_{T,R}, \tag{9.24}$$

where $D_{T,R}$ is the absorbed dose averaged over the tissue or organ, $T$, due to radiation of type $R$. This is similar to the earlier formulation of the dose equivalent. It should be stressed that when the radiation field consists of a mixture of different types and energies of radiation that have different values of $w_R$, then the absorbed dose must be subdivided into blocks of dose each with its own value of $w_R$ (ICRP 1991a). Remember that the absorbed dose in grays (Gy) is distinguished from the equivalent dose by the use of the special unit sievert (Sv), a quantity that is also equal to one joule per kilogram.

## EXAMPLE 9.2

A person has been exposed to a combination of radionuclides. It has been calculated that the absorbed dose to the liver due to alpha particles is 0.08 mGy ($D_{liver,alpha}$) and the absorbed dose from gamma rays is 0.05 mGy. What is the equivalent dose in the liver?

Given: $D_{liver,alpha} = 0.08$ mGy, and $D_{liver,gamma} = 0.05$ mGy. From table 9.11, $w_{alpha} = 20$, and $w_{gamma} = 1$. Therefore,

$$H_{liver} = \sum_R w_R D_{T,R} = (20 \times 0.08 \text{ mGy})_{alpha} + (1 \times 0.05 \text{ mGy})_{gamma}$$

$$= 1.65 \text{ mSv}.$$

The new term for effective dose equivalent has been simplified to *effective dose*, $E$. This still represents the sum of the weighted equivalent doses in all of the tissues and organs of the body, and it is now defined by the expression

$$E = \sum_T w_T H_T, \quad (9.25)$$

where $w_T$ is the tissue weighting factor for tissue or organ $T$ (compare with equation 9.3) and the appropriate unit is the sievert (ICRP 1991a).

Although the tissue weighting factor terminology has been retained, several critical changes have been made (compare table 9.12 with table 9.2). The values in table 9.12 are based on a reference population of equal numbers of both sexes and a range of ages so that they may be used to calculate effective doses for workers, or the whole population. The original tissue weighting factors were based on the risk of fatal cancer and severe hereditary effects. Now these factors are based on a more broadly defined concept of detriment that includes the probability of fatal cancer and the weighted probabilities of attributable nonfatal cancer, severe hereditary effects, impaired life span, and life span lost. Colon, stomach, urinary bladder, liver, esophagus, and skin have been added to the list of organs to be considered. For colon, the tissue weighting factor should be applied to the mass average of the equivalent dose in the wall of the upper and lower large intestine. As it turns out, the relative masses are largely age independent, so the equivalent dose in the colon is as follows (ICRP 1995b):

$$H_{Colon} = 0.57 H_{ULI} + 0.43 H_{LLI}. \quad (9.26)$$

The values of the tissue weighting factor are normalized to one so that a uniform equivalent dose in the whole body will equal the effective dose. In addition, the

Table 9.12  Tissue weighting factors (adapted from ICRP 1991a, table 2)

| Tissue/organ | $w_T$ |
|---|---|
| Gonads | 0.20 |
| Colon | 0.12 |
| Lung | 0.12 |
| Red bone marrow | 0.12 |
| Stomach | 0.12 |
| Bladder | 0.05 |
| Breast | 0.05 |
| Esophagus | 0.05 |
| Liver | 0.05 |
| Thyroid | 0.05 |
| Skin | 0.01 |
| Bone surfaces | 0.01 |
| Remainder[a] | 0.05 |

[a] See table 9.13.

organs of the remainder have been specified to include the following: adrenals, brain, extrathoracic airways, colon (now replaces ULI included in ICRP Publication 60 [ICRP 1991b]), kidney, muscle, pancreas, small intestine, spleen, thymus, and uterus (ICRP 1994a, 1995c).

In the treatment of the remainder tissues, a mass-weighted mean dose is used. Table 9.13 lists the masses of these particular organs. The following expression defines the equivalent dose for the remainder:

$$H_{\text{Remainder}} = \frac{\sum_{T=1}^{T=10} m_T H_T}{\sum_{T=1}^{T=10} m_T} \quad \text{if } H_T \leq H_{\max}, \quad (9.27)$$

where $m_T$ is the mass of an organ or tissue in grams (see table 9.13). In cases where one of the tissues of the remainder receives an equivalent dose ($H_{T(\text{PR})}$) greater than the highest equivalent dose in any of the 12 tissues/organs with an individual weighting factor ($H_{\max}$), the recommendation is to calculate the equivalent dose as given in the following expression:

$$H_{\text{Remainder}} = 0.5 \left[ \frac{\sum_{T=1}^{T=9} m_T H_T}{\sum_{T=1}^{T=9} m_T} + H_{T'} \right] \quad \text{for } H_{T'} > H_{\max}; \quad (9.28)$$

where $T \neq T'$. (9.29)

The mass-weighted average equivalent dose of nine organs of the remainder is combined with the equivalent dose in the organ that receives the highest equivalent dose. By calculating the equivalent dose in the remainder with this method, the tissue weighting factor is essentially split evenly between the most heavily exposed

Table 9.13 Masses of adult organs that belong to the remainder (adapted from ICRP 1995a, table 9)

| Organ | Mass (g)[a] |
|---|---|
| Adrenals | 14 |
| Brain | 1,400 |
| Extrathoracic airways | 15 |
| Kidneys | 310 |
| Muscle | 28,000 |
| Pancreas | 100 |
| Small intestine | 640 |
| Spleen | 180 |
| Thymus | 20 |
| Uterus | 80 |

[a] Total mass of the remainder tissues is equal to 30,759 g.

organ and the rest of the organs of the remainder (ICRP 1995c). The mass of the tissues must be adjusted appropriately for different age groups.

### EXAMPLE 9.3

After an intake, a person is determined to have the following equivalent doses: esophagus, 0.5 mSv; stomach, 0.7 mSv; small intestine, 0.8 mSv; and colon, 3 mSv. What is the effective dose to this individual?

Given:  From table 9.12:
$H_{Eso} = 0.5$ mSv   $w_{Eso} = 0.05$
$H_{ST} = 0.7$ mSv   $w_{ST} = 0.12$
$H_{Colon} = 3$ mSv   $w_{Colon} = 0.12$

The small intestine is a member of the remainder, since $H_{SI} = 0.8$ mSv is not greater than 3 mSv ($H_{max}$). Then,

$$H_{Remainder} = (m_{SI} \times H_{SI})/m_{Remainder}$$
$$= (640\,g \times 0.8\,mSv)/30,759\,g = 0.017\,mSv,$$

so

$$E = \sum_T w_T H_T = (0.05 \times 0.5\,mSv)_{Eso} + (0.12 \times 0.7\,mSv)_{ST}$$
$$+ (0.12 \times 3\,mSv)_{Colon} + (0.05 \times 0.02\,mSv)_{Remainder}$$
$$= 0.47\,mSv.$$

Of the other dozen or so dosimetric quantities mentioned in ICRP Publication 60 (ICRP 1991b), two are of particular concern for internal dose assessment purposes.

The *committed equivalent dose* is the time-integrated equivalent dose rate in a tissue from time $t_0$, the age at the time of intake, to age $t$, and is calculated by the following expression:

$$H_T(t - t_0) = \int_{t_0}^{t} \dot{H}_T(t, t_0) dt, \quad (9.30)$$

where

$$\dot{H}_T(t, t_0) = c \sum_S \sum_j q_{S,j}(t, t_0) \cdot \text{SEE}(T \leftarrow S; t)_j, \quad (9.31)$$

with $c$ (sometimes $C$) a conversion factor to get the appropriate units, and $q_{S,j}(t, t_0)$ representing the activity of a given radionuclide $j$ present at age $t$ after an intake at age $t_0$ in organ or tissue $S$. The term $\text{SEE}(T \leftarrow S; t)_j$ is the specific effective energy deposited in the target organ $T$ (ICRP 1990). For occupational exposures, the integration time is 50 years, and for the dose coefficients provided by the ICRP, the intake is assumed to have occurred at age 20. This also holds true for environmental exposures of adults. For children, the age at the time of intake is considered, and the integration is carried out to age 70. Since the tables consider only a single nuclide at a time, the equation is often simplified as follows:

$$H_T(\tau) = \sum_S U_S(\tau) \cdot \text{SEE}(T \leftarrow S; t), \quad (9.32)$$

where $\tau$ is the integration time after an intake in years. If this is not specified, it should be taken to be 50 years for an adult or to age 70 years for a minor. So, for occupational exposures, we obtain the expression

$$H_T(50) = \sum_S U_S(50) \cdot \text{SEE}(T \leftarrow S; t). \quad (9.33)$$

On the right side of the equation, $U_S(50)$ represents the number of nuclear transformations in 50 years in a source region, $S$. This is equivalent to the earlier term, $U_S$, of ICRP Publication 30 (see equation 9.14) and has the units of Bq · s. Of course, this term would be calculated using more recent biokinetic models. Similarly, the specific effective energy, $\text{SEE}(T \leftarrow S)$, is the equivalent dose in the target per transformation in the source region and is now expressed as

$$\text{SEE}(T \leftarrow S) = \sum_R \frac{Y_R E_R w_R \text{AF}(T \leftarrow S; t)_R}{m_T(t)}. \quad (9.34)$$

Primarily the radiation weighting factor has been substituted for $Q$ in the earlier form of the expression, and age dependence is now included (compare with equation 9.12). In order for the units to work out as Sv (Bq · s)$^{-1}$, the yield of radiation is in the units of (Bq · s)$^{-1}$, the energy of the radiation ($E_R$) must be in joules, and the mass of the tissue of interest ($m_T$) in kilograms (ICRP 1995c). In many ICRP publications, dose conversion coefficients are provided for the committed equivalent doses in the principal tissues for a unit intake of radionuclide. The term $h_T(\tau)$ is used to denote a dose conversion coefficient (ICRP 1995a).

For the committed equivalent dose in the remainder, the same methodology is applied:

$$\dot{H}_{\text{Remainder}}(t, t_0) = \frac{\sum_{T=1}^{T=10} m_T(t) \cdot \dot{H}_T(t, t_0)}{\sum_{T=1}^{T=10} m_T(t)} \quad \text{if } H_T \leq H_{\max} \quad (9.35)$$

and

$$\dot{H}_{\text{Remainder}}(t, t_0) = 0.5 \cdot \left[ \frac{\sum_{T=1}^{T=9} m_T \dot{H}_T(t, t_0)}{\sum_{T=1}^{T=9} m_T} + \dot{H}_{T'}(t, t_0) \right]$$

$$\text{for } H_{T'} > H_{\max}; T \neq T'. \quad (9.36)$$

The counterpart to the effective dose for internal exposures is the committed effective dose, which is defined as follows:

$$E(t - t_0) = \sum_{T=1}^{T=12} w_t H_T(t - t_0) + w_{\text{Remainder}} \cdot H_{\text{Remainder}}(t - t_0). \quad (9.37)$$

In the case of the tabulated values, this can again be simplified to

$$E(\tau) = \sum_T w_T H_T(\tau) = \sum_T w_T \sum_S U_S(\tau) \cdot \text{SEE}(T \leftarrow S; t). \quad (9.38)$$

The expression for the 50-year committed effective dose for workers is given as

$$E(50) = \sum_T w_T H_T(50) = \sum_T w_T \sum_S U_S(50) \cdot \text{SEE}(T \leftarrow S; t). \quad (9.39)$$

In a number of the latest publications, $e(50)$ is the dose coefficient that represents the occupational committed effective dose over 50 years due to 1 Bq of intake of a particular radionuclide (ICRP 1995c).

For environmental exposures, the committed effective dose is computed to age 70, which is designated in the publications as $e(70)$. With these new definitions, the commission has recommended that in cases in which previous dosimetric quantities need to be combined with the present quantities, no effort should be made to correct old internal dose values. The committed effective dose equivalent and the committed effective dose should be summed directly. The same applies for all the other new quantities and their ICRP Publication 30 counterparts (ICRP 1991a).

As mentioned above in the description of the tissue weighting factors, the ICRP has broadened the concept of detriment to represent both the probability of an occurrence of a harmful health effect and an assessment of the severity of that effect, which may include impairment and shortened life span. Consequently, the commission has

Table 9.14  Nominal probability coefficients (adapted from ICRP 1995a, table 9)

| Exposed group | Detriment ($10^{-2}$ Sv$^{-1}$) | | | |
|---|---|---|---|---|
| | Fatal cancer | Nonfatal cancer | Severe hereditary effects | Total |
| Adult workers | 4.0 | 0.8 | 0.8 | 5.6 |
| General population | 5.0 | 1.0 | 1.3 | 7.3 |

reassessed the probabilities of harmful effects. The most recent values for the nominal probability coefficients per sievert of effective dose are listed in table 9.14. These are useful for estimating harmful effects in cases where the equivalent dose is close to uniform such that the effective dose is representative of the exposure. In cases where the equivalent doses are not uniform, these probabilities may under- or overestimate the actual risk, so the commission has provided a listing of nominal probability coefficients for individual tissues for workers or for a whole population to be used with the appropriate equivalent dose. However, the commission stated that the risk probability obtained using the whole-body or individual tissue nominal probability coefficients would not be significantly different because the individual tissue coefficients are rough approximations. For this reason, the individual tissue coefficients are not provided in this chapter.

### EXAMPLE 9.4

What is the additional risk of fatal cancer to a member of the general population if a practice adds 1 mSv to the annual dose of the public?

Given: Detriment is equal to 0.05 Sv$^{-1}$; effective dose is 1 mSv, or 0.001 Sv; therefore,

Additional probability of fatal cancer $= 0.05$ Sv$^{-1} \times 0.001$ Sv $= 0.00005$.

## Dose Limits

The ICRP reemphasizes the importance of the three principles: the justification of a practice, the optimization of protection, and the use of individual dose and risk limits. In a departure from previous recommendations to base limits on comparisons to other safe occupations, the ICRP has analyzed recent data and models to formulate limits that they regard as justifiable. In its view, an extended occupational exposure leading to an effective dose of 1 Sv over a working life is only just tolerable, and they have set the limits accordingly (see table 9.15). With the reduction of a lifetime occupational effective dose from approximately 2.5 Sv to 1 Sv, the commission has recommended a comparable reduction in the recommended dose to the public. There remains some flexibility in these recommendations in that the annual limits of effective dose are averaged over a five-year period. Therefore, an occupationally exposed worker may have an annual exposure of up to 50 mSv in a single year, but the average over five years may not exceed 20 mSv per year. Similarly, with the

Table 9.15  Annual dose limits (adapted from ICRP 1991a, table 6)

| Application | Occupational limit | Limit for the public |
|---|---|---|
| Effective dose | 20 mSv[a] | 1 mSv |
| Equivalent dose in: | | |
| Hands and feet | 500 mSv | Not applicable |
| Lens of the eye | 150 mSv | 15 mSv |
| Skin | 500 mSv | 50 mSv |

[a] Averaged over defined periods of five years.

public, a higher exposure may be allowed in special circumstances as long as the five-year average does not exceed the limit of 1 mSv per year. For pregnant workers, an additional limit of 2 mSv to the surface of the women's abdomen should be applied once a pregnancy has been declared (ICRP 1991a).

The ICRP has also included limits for equivalent dose in three specific parts of the body that may not be sufficiently protected by the effective dose limit (see table 9.15). The effective dose limit will protect the skin from stochastic effects but may not protect the skin from deterministic effects of localized exposures. They recommend a limit of 500 mSv averaged over any 1 cm$^2$ of skin, regardless of the area exposed. This same limit is applied to the hands and feet. For the lens of the eye, the ICRP continues to recommend an annual equivalent dose limit of 150 mSv to reduce the risk of cataracts. The limits to the public for these areas of the body were based on an arbitrary reduction by a factor of 10 (ICRP 1991a).

Changing the annual dose limit also affects a number of secondary limits, in particular, the ALI. The ALI (Bq) is now defined as the annual average effective dose limit (0.02 Sv) divided by the committed effective dose coefficient (Sv Bq$^{-1}$) (ICRP 1995c)

$$\text{ALI} = \frac{0.02}{e(50)}. \tag{9.40}$$

The changes recommended by the ICRP in Publication 60 (ICRP 1991b) are wide ranging and significantly different from the recommendations in ICRP Publication 26 (ICRP 1977). Reductions of the annual limits for occupational workers and the public have led to a revision and introduction of new age-related anatomical and biokinetic models used to estimate dose due to intakes of radionuclides to ensure that these limits are not exceeded. The new models and the dose coefficients for ingestion and inhalation of radionuclides that are derived from these models are discussed in the next section.

### Age-Dependent Doses to the Public (ICRP Publications 56, 67, 69, 71, and 72)

The calculation of the equivalent dose can be laborious and often requires the use of computer software to obtain the required values. Fortunately, the ICRP has provided tabulated values for age-dependent dose conversion coefficients for ingestion or inhalation for a large number of radionuclides in a five-part series of publications. As the ICRP recommendations have evolved, they have necessitated the revision of intake dose coefficients for the whole range of radionuclides. As more information becomes available, new biokinetic models are developed or modified, which can significantly affect the dose due to an intake. In order to make sense of this potentially confusing mass of information, we describe these changes chronologically, examining the effect on the recommended dose coefficients for the intake of $^{90}$Sr.

For these publications, the dose conversion coefficients were calculated using a system of compartmental models with removal and transfer rates of each compartment dependent on the biokinetics of each radionuclide. The overall system is depicted in figure 9.5. Currently, the GI tract continues to be modeled as discussed above. The metabolic models for the transfer compartment and other tissues are scattered throughout various ICRP publications. ICRP Publication 72 (ICRP 1996, table 2) summarizes where the current models for each element can be found. If no age-dependent organ retention data were available, the biokinetic data for adults were adopted for children. In cases where there were no biokinetic data for humans, animal data were used (ICRP 1994b). For some of these publications, the bone model has a few slight modifications. ICRP Publication 66 (ICRP 1994a) introduced a new respiratory model, which was used in some of the subsequent reports. These changes are briefly described further below. In most cases where a

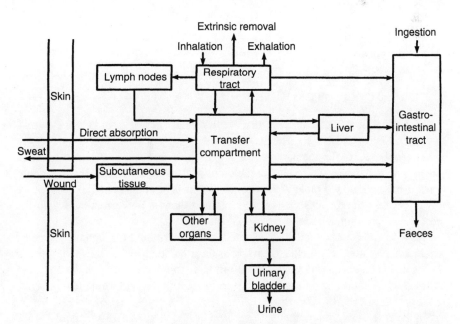

Figure 9.5 General overview of the compartmental models showing the routes of intake, transfer between major compartments, and excretion (ICRP 1998, figure 1).

radionuclide transforms to one or more radioactive daughter nuclides, the daughters are considered to have the biokinetics of the parent. Therefore, yttrium is treated biokinetically like its parent, strontium. Some exceptions have been noted for decay products of lead, radium, tellurium, thorium, uranium, and iodine. The dose conversion coefficients for a radionuclide include the contribution by the daughters in a decay chain. The reported dose conversion coefficients in all of these publications are for acute intakes; for chronic intakes, the committed effective dose should be determined for each year and then summed over the period of intake in order to take into account the effects of growth over an extended period of intake (ICRP 1995a).

## ICRP Publication 56

In this publication (ICRP 1990), the first attempt was made in determining age-dependent dose coefficients for the general population. The dose coefficients were obtained by combining specific effective energy values, obtained with a set of anthropomorphic phantoms devised by Cristy (1980) and Cristy and Eckerman (1987), with new age-adjusted values for the fraction absorbed into the blood, $f_1$, and using the existing lung and GI-tract models from ICRP Publication 30 (ICRP 1979a). The phantoms represented individuals at 3 months, 1 year, 5 years, 10 years, 15 years, and 20 years (Adult-Reference Man). In those cases of an intake occurring at an age other than the one for which the calculations were carried out or in cases of prolonged exposure, the ICRP recommends using the following age groupings:

| | |
|---|---|
| 0–12 months of age: | 3 months |
| 1–2 years of age: | 1 year |
| 2–7 years of age: | 5 years |
| 7–12 years of age: | 10 years |
| 12–17 years of age: | 15 years |
| Older than 17 years of age: | Adult |

In the rest of this chapter, we focus only on the ages of 3 months (newborn), 5 years, and 20 years (adult). The masses of the organs and tissues for these ages are given in table 9.16. For the purposes of determining the specific absorbed fraction, the thymus is used as a surrogate for the esophagus, which explains why the esophagus is not included in table 9.16. The esophagus is not considered to be a source organ because the transit time is very rapid and there is no current model for the absorption of radionuclides by the esophagus.

The systemic circulation is considered to be distributed throughout the body, excluding the contents of the GI tract, urinary bladder, and gall bladder. For a radionuclide that is considered to be uniformly distributed throughout the body, the mass given as total body mass in table 9.16 is used rather than whole-body masses used previously (ICRP 1994b).

In the case of strontium, the most exposed tissues were the BS and the RBM for all age groups. The simple model deposits fractions of the activity from the blood onto trabecular or cortical BS or into soft tissues. The removal rate of activity from the surfaces to the bone volume was taken to be 0.1 day$^{-1}$ for all ages. The fraction

Table 9.16  Organ and tissue masses (g) for selected ages adapted from ICRP 1995a, table 11)

| Organ | Three months | Five years | Adult |
| --- | --- | --- | --- |
| Adrenals | 5.83 | 5.27 | 14.0 |
| Brain | 352 | 1260 | 1,400 |
| Breasts | 0.107 | 1.51 | 360 |
| Cortical bone volume[a] | 0 | 875 | 4,000 |
| Endosteal tissue (bone surfaces) | 15.0 | 37.0 | 120 |
| Extrathoracic airways | 1.30 | 4.30 | 15.5 |
| Kidneys | 22.9 | 116 | 310 |
| Liver | 121 | 584 | 1,800 |
| Lower large intestine contents[a] | 6.98 | 36.6 | 135 |
| Lower large intestine wall | 7.96 | 41.4 | 160 |
| Lungs (thoracic airways) | 50.6 | 290 | 1000 |
| Muscle | 760 | 5,000 | 28,000 |
| Ovaries | 0.328 | 1.73 | 11.0 |
| Pancreas | 2.80 | 23.6 | 100 |
| Red bone marrow | 47 | 320 | 1500 |
| Skin | 118 | 538 | 2,600 |
| Small intestine contents[a] | 20.3 | 106 | 400 |
| Small intestine wall | 32.6 | 169 | 640 |
| Spleen | 9.11 | 48.3 | 180 |
| Stomach contents | 10.6 | 75.1 | 250 |
| Stomach wall | 6.41 | 49.1 | 150 |
| Testes | 0.843 | 1.63 | 35.0 |
| Thymus | 11.3 | 29.6 | 20.0 |
| Thyroid | 1.29 | 3.45 | 20.0 |
| Trabecular bone volume[a] | 140 | 219 | 1,000 |
| Upper large intestine contents[a] | 11.2 | 57.9 | 220 |
| Upper large intestine wall | 10.5 | 55.2 | 210 |
| Urinary bladder contents[b] | 10.4 | 67.6 | 120 |
| Urinary bladder wall | 2.28 | 14.5 | 45 |
| Uterus | 3.85 | 2.70 | 80.0 |
| Total body mass[b] | 3,536 | 19,458 | 68,831 |

[a]Bone volume and organ content masses are taken from Eckerman et al. (1999), table B.1.
[b]Total mass minus the masses of organ contents.

crossing the small intestine, $f_1$, was taken to be 0.6 for 3 months and 0.3 for 5 years and the adult, based on human and animal data. Similarly, the fraction of strontium deposited on the BS was highest at 3 months, with a deposition fraction of 0.61, which decreased to 0.29 for 5 years and 0.15 for an adult. Removal rates in the bone were 3.00, 0.56, and 0.03 (per year) for the cortical bone. Activity that was removed from the bone went directly to excretion, and recycling of activity was not considered. Dose coefficients for intake at selected ages to 70 years are given in table 9.17. The difference in dose equivalent and effective dose equivalent seen for the 3-month-old child compared to the other two ages is due to higher uptake and deposition. The 5-year-old does not significantly differ from the adult because the difference in deposition of activity is made up for by the accelerated rate of clearance in this model.

Table 9.17  Dose coefficients (Sv Bq$^{-1}$) for $^{90}$Sr (adapted from ICRP 1990, tables 3-2 and B-3)

|  | Three months | Five years | Adult (20 years) |
|---|---|---|---|
| Dose equivalent in: | | | |
| *Ingestion* | | | |
| Bone surfaces | $1.0 \times 10^{-6}$ | $3.9 \times 10^{-7}$ | $3.8 \times 10^{-7}$ |
| Red bone marrow | $7.1 \times 10^{-7}$ | $1.7 \times 10^{-7}$ | $1.8 \times 10^{-7}$ |
| Lung | $1.1 \times 10^{-8}$ | $3.8 \times 10^{-9}$ | $1.4 \times 10^{-9}$ |
| Upper large intestine | $3.7 \times 10^{-8}$ | $1.9 \times 10^{-8}$ | $5.4 \times 10^{-9}$ |
| Lower large intestine | $1.3 \times 10^{-7}$ | $7.1 \times 10^{-8}$ | $1.9 \times 10^{-8}$ |
| Effective dose equivalent | $1.3 \times 10^{-7}$ | $4.1 \times 10^{-8}$ | $3.5 \times 10^{-8}$ |
| *Inhalation (Class D)* | | | |
| Bone surfaces | $9.5 \times 10^{-7}$ | $6.7 \times 10^{-7}$ | $6.7 \times 10^{-7}$ |
| Red bone marrow | $6.8 \times 10^{-7}$ | $3.0 \times 10^{-7}$ | $3.1 \times 10^{-7}$ |
| Lung | $2.3 \times 10^{-8}$ | $1.0 \times 10^{-8}$ | $3.5 \times 10^{-8}$ |
| Upper large intestine | $1.4 \times 10^{-8}$ | $9.0 \times 10^{-9}$ | $3.0 \times 10^{-9}$ |
| Lower large intestine | $2.8 \times 10^{-8}$ | $1.7 \times 10^{-8}$ | $5.2 \times 10^{-9}$ |
| Effective dose equivalent | $1.2 \times 10^{-7}$ | $6.3 \times 10^{-8}$ | $6.0 \times 10^{-8}$ |

## ICRP Publication 67

In this publication (ICRP 1994b), the commission revised the ingestion dose coefficients for the radionuclides in ICRP Publication 56 (ICRP 1990) using the tissue-weighting factors introduced in ICRP Publication 60 (ICRP 1991b). The inclusion of the urinary bladder in the listing of tissues with specific tissue weighting factors (see table 9.17) necessitated the formulation of a bladder model. The model uses a fixed volume for the bladder contents that represents the average volume between bladder voidings. In keeping with the use of first-order kinetics to represent transfer rates of metabolic models, the bladder elimination rates are taken to be twice the number of voidings. The rate that activity enters the bladder is based on the elimination from the body tissues and the urine; fecal excretion ratios are detailed in the biokinetics models for each radionuclide. ICRP Publication 67 also introduced new age-specific biokinetic models for the alkaline earth elements (see figure 9.6) and lead, as well as a similar set of models for plutonium, americium, and neptunium.

The general compartmental model for the alkaline earth metals and lead now includes recycling of activity and additional compartments to better simulate retention (see figure 9.6). Again, the ICRP relies on modeling transfers between compartments with simple first-order kinetics. For this set of bone-seeking element models, parameters were originally developed for a 100-day-old infant and for 1-, 5-, 10-, 15-, and 25-year-old individuals. These ages were selected based on the growth of bone, which is considered to be essentially static after 25 years of age. The 100-day-old infant has been assigned to the 3-month age bracket and the 25-year-old to the adult age bracket for the purposes of dose conversion coefficients in this and

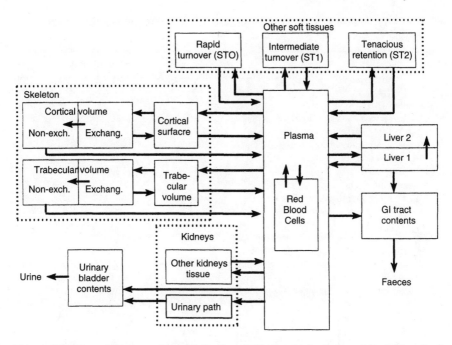

Figure 9.6 The compartmental model for the alkaline earth elements and lead. Not all of the compartments are used in the strontium model (ICRP 1998, figure 8).

later publications. The model was originally for radium but has been adapted for strontium. Strontium behaves more like calcium, and the rate of removal of strontium from plasma is less than that of radium. So strontium was assigned a rate of 15 day$^{-1}$, the same as for calcium (ICRP 1995a). Strontium is more likely to be excreted in the urine, so the ratio of urinary to fecal excretion was taken to be 3.3 rather than 1 (for calcium). The fractional loss of radionuclide to the excretion (urine and feces) is 0.15, about twice the loss rate of calcium. Age-specific parameters for this model are provided in table 9.18 (ICRP 1994b).

The bone model has been modified to include an exchangeable and nonexchangeble compartment in each bone volume. Age-related changes are found in the volumes and masses of the bone compartments and the fraction of endosteal tissue associated with the RBM. This fraction is 1 for newborns, 0.65 for 5-year-olds, and 0.5 for adults. The absorbed fractions for different forms of emitted radiation are still based on the ICRP Publication 30 values (ICRP 1994b).

The dose conversion coefficients for the ingestion of 1 Bq of $^{90}$Sr based on this new revised model and new tissue weighting factors are given in table 9.19. Comparison with Table 9.17 reveals that there is not a great difference in the committed equivalent doses or committed effective doses for 5-year-olds and adults. However, there is a marked increase in the doses for newborns that is primarily attributable to greater uptake from the plasma to the various bone compartments. The higher fraction of endosteal tissue associated with the RBM also contributes to this increase.

Table 9.18  Age-dependent parameters for strontium model (adapted from ICRP 1994b, table A-3)

|  | Three months | Five years | Adult (25 years) |
|---|---|---|---|
| $f_1$ | 0.6 | 0.3 | 0.2 |
| Transfer rates (day$^{-1}$) | | | |
| Plasma→urinary bladder contents | 0.202 | 0.488 | 0.606 |
| Plasma→upper large intestine contents | 7.26 | 17.43 | 21.79 |
| Plasma→trabecular bone surface | 10.5 | 6.22 | 9.72 |
| Plasma→cortical bone surface | 42.0 | 21.78 | 7.78 |
| Plasma→liver 1 | 0.117 | 0.280 | 0.350 |
| Plasma→soft tissue 0 (rapid turnover) | 7.56 | 18.14 | 22.68 |
| Plasma→soft tissue 1 (intermediate turnover) | 2.33 | 5.60 | 7.00 |
| Plasma→soft tissue 2 (tenacious retention) | 0.0233 | 0.0560 | 0.070 |
| Bone surface→plasma | 0.578 | 0.578 | 0.578 |
| Bone surface→exchangeable bone volume | 0.116 | 0.116 | 0.116 |
| Exchangeable bone volume→bone surface | 0.0185 | 0.0185 | 0.0185 |
| Exchangeable→nonexchangeable bone volume | 0.0046 | 0.0046 | 0.0046 |
| Nonexchangeable trabecular volume→plasma | 0.00822 | 0.00181 | 0.000493 |
| Nonexchangeable cortical volume→plasma | 0.00822 | 0.00153 | 0.0000821 |
| Liver 1→plasma | 0.0139 | 0.0139 | 0.0139 |
| Soft tissue 0→plasma | 2.52 | 6.05 | 7.56 |
| Soft tissue 1→plasma | 0.693 | 0.693 | 0.693 |
| Soft tissue 2→plasma | 0.00038 | 0.00038 | 0.00038 |
| Urinary bladder elimination rate | 40 | 12 | 12 |

Table 9.19  Ingestion dose coefficients (Sv/Bq) for $^{90}$Sr (data from ICRP 1994b, table 5-2)

|  | Three months | Five years | Adult (25 years) |
|---|---|---|---|
| Equivalent dose in: | | | |
| Bone surfaces | $2.3 \times 10^{-6}$ | $6.4 \times 10^{-7}$ | $4.1 \times 10^{-7}$ |
| Red bone marrow | $1.5 \times 10^{-6}$ | $2.7 \times 10^{-7}$ | $1.8 \times 10^{-7}$ |
| Lung | $1.2 \times 10^{-8}$ | $2.9 \times 10^{-9}$ | $6.6 \times 10^{-10}$ |
| Upper large intestine | $6.0 \times 10^{-8}$ | $2.2 \times 10^{-8}$ | $5.9 \times 10^{-9}$ |
| Lower large intestine | $1.9 \times 10^{-7}$ | $7.5 \times 10^{-8}$ | $2.2 \times 10^{-8}$ |
| Effective dose | $2.3 \times 10^{-7}$ | $4.7 \times 10^{-8}$ | $2.8 \times 10^{-8}$ |

Changes in tissue weighting factors and the new methods for dealing with the tissues of the remainder do not have a significant impact for strontium. This is also true for most other radionuclides.

### EXAMPLE 9.5

For a unit intake of $^{90}$Sr, find the committed effective dose for an adult using the values for committed equivalent doses as given in table 9.19.

**Given:**
$H_{BS}(50) = 4.1 \times 10^{-7}$ Sv
$H_{RBM}(50) = 1.8 \times 10^{-7}$ Sv
$H_{Lung}(50) = 6.6 \times 10^{-10}$ Sv
$H_{ULI}(50) = 5.9 \times 10^{-9}$ Sv

**From table 9.12:**
$w_{BS} = 0.01$
$w_{RBM} = 0.12$
$w_{Lung} = 0.12$
$H_{LLI}(50) = 2.2 \times 10^{-8}$.

Remember that $H_{Colon}(50) = 0.57 H_{ULI}(50) + 0.43 H_{LLI}(50)$ (see equation 9.26). Therefore,

$$H_{Colon}(50) = 3.36 \times 10^{-9} + 9.46 \times 10^{-9}$$
$$= 1.28 \times 10^{-8} \text{ Sv},$$

and $w_{Colon} = 0.12$. Hence,

$$E(50) = (0.01)(4.1 \times 10^{-7} \text{Sv})_{BS} + (0.12)(1.8 \times 10^{-7} \text{Sv})_{RBM} + (0.12)(6.6$$
$$\times 10^{-10} \text{Sv})_{Lung} + (0.12)(1.28 \times 10^{-8} \text{Sv})_{Colon}$$
$$= 2.73 \times 10^{-8} \text{ Sv},$$

which is very close to the tabulated value of $2.8 \times 10^{-8}$. The fact that not all of the tissues were included in this sample calculation accounts for the difference (see table 9.12).

### EXAMPLE 9.6

Estimate the committed equivalent dose in the RBM and the committed effective dose after ingestion of 8,000 Bq of $^{90}$Sr by an adult.

$$H_{RBM}(50) = 1.8 \times 10^{-7} \text{ Sv Bq}^{-1}(8,000 \text{ Bq})$$
$$= 1.44 \times 10^{-3} \text{ Sv} = 1.44 \text{ mSv}$$
$$E(50) = 2.8 \times 10^{-8} \text{ Sv Bq}^{-1}(8,000 \text{ Bq})$$
$$= 2.24 \times 10^{-4} \text{ Sv} = 0.224 \text{ mSv}.$$

## ICRP Publication 69

Part 3 of this series (ICRP 1995b) provided dose coefficients for some nuclides of iron, selenium, antimony, thorium, and uranium. There are several modifications to the biokinetic models for these elements, particularly iron, thorium (using a compartmental model resembling plutonium), and uranium (using a compartmental model like that for radium and strontium).

## ICRP Publication 71

This volume (ICRP 1995a) revises the inhalation dose coefficients in light of the lung model introduced in Publication 66 (ICRP 1994a). The new model bears little

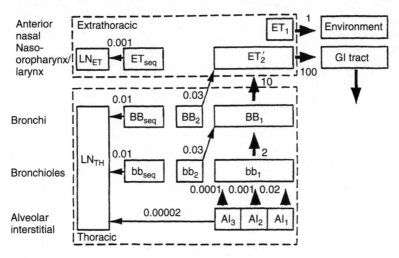

Figure 9.7 The compartments of the ICRP Publication 66 lung model and the clearance pathways. The clearance rates are depicted next to the pathway and given in units of days$^{-1}$ (ICRP 1998, figure 4).

or no resemblance to the Publication 30 (ICRP 1979a) lung model in its approach to the deposition, absorption, and clearance of radionuclides in a comprehensive manner that takes into account the effects of sex, age, and the disease state of an exposed individual. The major difference between the models is that instead of determining an average dose to the lungs (ICRP 1979a), doses are computed for each region using the masses of the putative target cells in the region and then weighted according to the estimated radiosensitivity of the target cells in the region. Figure 9.7 depicts the compartments of the ICRP Publication 66 model showing the clearance pathways and transfer rates between compartments (compare with figure 9.3). The model divides the respiratory system into five regions rather than the three used previously (see left side of figure 9.7). These are the anterior nasal passages ($ET_1$), the naso-oropharynx/larynx ($ET_2$), bronchi (BB), bronchioles (bb), and the alveolar interstitial (AI) regions. These regions are further grouped into two main regions, the extrathoracic airways and the thoracic airways (lungs), delineated by dashed boxes in the figure. Each region has several compartments to represent different rates of deposition, retention, or sequestration. The compartments marked "LN" represent the lymph nodes associated with the major airway regions (ICRP 1994a).

Deposition is age and sex dependent. This is dealt with by treating the compartments as a series of filters that successively remove respired particles (i.e., they are deposited on inhalation as well as exhalation). Particles are defined not only by their AMAD, but also by their activity mean thermodynamic diameters (AMTD). The intake of radionuclides depends on the amount of air respired, which in turn depends on the level of activity and size of the individual in question (ICRP 1994a). Table 9.20 gives some reference values for the level of activity, breathing frequency

Table 9.20  Ventilation parameters for the lung model (adapted from ICRP 1995a, tables 4 and 6)

| Activity level | Three months | | | Five Years | | | Adult (male) | | |
|---|---|---|---|---|---|---|---|---|---|
| | Time (h day$^{-1}$) | $f_R$ (min$^{-1}$) | Volume (m$^3$) | Time (h day$^{-1}$) | $f_R$ (min$^{-1}$) | Volume (m$^3$) | Time (h day$^{-1}$) | $f_R$ (min$^{-1}$) | Volume (m$^3$) |
| Sleeping | 17 | 38 | 1.53 | 12 | 23 | 2.88 | 8 | 12 | 3.60 |
| Sitting | | | | 4 | 25 | 1.28 | 6 | 12 | 3.24 |
| Light exercise | 7 | 48 | 1.33 | 8 | 39 | 4.56 | 9.75 | 20 | 14.63 |
| Heavy exercise | | | | | | | 0.25 | 26 | 0.75 |
| Total volume | | | 2.86 | | | 8.72 | | | 22.2 |

Table 9.21  Reference deposition fractions (percent of inhaled activity; adapted from ICRP 1995a, table 7; ICRP 1994a, table 7)

| Region | Environmental[a] | | | Occupational[b] worker |
|---|---|---|---|---|
| | Three months | Five years | Adult (male) | |
| ET$_1$ | 20.97 | 17.39 | 4.89 | 33.9 |
| ET$_2$ | 27.20 | 22.32 | 18.97 | 33.9 |
| BB | 1.04 | 1.03 | 1.29 | 1.8 |
| bb | 2.05 | 1.85 | 1.95 | 1.1 |
| AI | 8.56 | 9.85 | 11.48 | 5.3 |
| Total | 59.82 | 52.44 | 48.58 | 82.0 |

[a] Particles are assumed to have a density of 3.0 gcm$^{-3}$, shape factor of 1.5, 0.69 μm AMTD, and 1 μm AMAD. The deposition fractions are volume-weighted averages for the activity levels detailed in table 9.20.
[b] Particles are assumed to be 5 μm AMAD, 3.5 μm AMTD, density 3.0 gcm$^{-3}$, shape factor 1.5; fraction breathed through nose is 1; activity level is taken to be 31% sitting and 69% light exercise with a mean ventilation rate of 1.2 m$^3$ h$^{-1}$.

($f_R$), and total volume air for selected ages used in the consideration of environmental exposures. The deposition fractions obtained with the lung model using the ventilation parameters from table 9.20 and other age-specific anatomical data are given in table 9.21. These are for reference individuals performing typical activities while being exposed environmentally to standard size particles (1 μm AMAD and 0.69 μm AMTD). The deposition fractions for a worker breathing the new standard size of particles for occupational exposure (5 μm AMAD and 3.5 μm AMTD) have been included for comparison.

Activity can be cleared extrinsically from the anterior nasal passages, mechanically cleared from the airways to the GI tract or lymph nodes, or absorbed into the bloodstream. Mechanical clearance of particles is assumed to be the same for all types of material, so the transfer rates given in figure 9.7 apply to all materials. Absorption, on the other hand, is dependent on the physical and chemical properties of the particle. Particles are classed as one of the following:

Type F = fast absorption of material from the respiratory tissues to the body fluids

Type M = moderate absorption of material from the respiratory tissues to the body fluids

Type S = slow absorption of material from the respiratory tissues to the body fluids

These three types are comparable to the D, W, and Y classes of ICRP Publication 30, but they only refer to the absorption characteristics of a material, not the overall clearance rates.

Absorption and mechanical clearance are considered to be simultaneous and competing processes, and the time-dependent treatment of these processes is fairly involved. See ICRP Publication 66 (ICRP 1994a) for the details, including the full treatment of deposition and the effects of aging and disease processes on the retention of radionuclides. For the sake of brevity, we describe the general behavior of these absorption types to give a general idea of the amount of material that reaches the transfer compartment. Any material deposited in $ET_1$ is considered to be cleared by nose blowing regardless of the material type. For type F compounds, all of the material deposited in the BB, bb, and AI compartments and half of the material deposited in $ET_2$ are absorbed with an approximate half-time of 10 min. The rest of the material in $ET_2$ is transported to the GI tract. For compounds of type M, only 10% of the material deposited in BB and bb compartments is absorbed, with a half-time of 10 min. The rest is absorbed with a half-time of 140 days. Five percent of the material deposited in $ET_2$ will be cleared at the rapid rate, while the rest reaches the gut. About 70% of the material deposited in the AI compartments will ultimately reach the body fluids. Finally, most type S particles (99.9%) are absorbed with a half-time of 7,000 days, so very little is absorbed from material deposited in the $ET_2$, BB, or bb compartments. About 10% of the material in the AI compartments will reach the body fluids (ICRP 1995a).

Vapors and gases are treated separately and have their own classes. SR-0 gases are insoluble and nonreactive with negligible deposition in the respiratory system, and at present, there are no such gases reported by the ICRP. SR-1 is a gas or vapor that is either soluble or reactive and may be deposited throughout the respiratory system. An example would be radioactive carbon dioxide. SR-2 gases are highly soluble or reactive (e.g., tritium gas), and these are treated as 100% deposition in $ET_2$, with immediate absorption to the transfer compartment (ICRP 1995a).

In order to partition the detriment due to exposures of the tissues of the respiratory tract, the ICRP has divided the respiratory system into two main regions, extrathoracic and thoracic. The equivalent dose to the extrathoracic region is given by the expression

$$H_{ET} = H_{ET_1} A_{ET_1} + H_{ET_2} A_{ET_2} + H_{LN_{ET}}, \qquad (9.41)$$

and for the thoracic region (lungs in all ICRP tables),

$$H_{TH} = H_{BB} A_{BB} + H_{bb} A_{bb} + H_{AI} A_{AI} + H_{LN_{TH}} A_{LN_{TH}}, \qquad (9.42)$$

where $H_i$ is the equivalent dose to a particular region and $A_i$ (see table 9.22) is the weighting factor for the radiosensitivity of that region (ICRP 1995a).

Continuing the use of strontium as the example radionuclide, the $f_1$ values for different absorption types of strontium compounds are detailed in table 9.23. Strontium carbonate and simple ionic compounds of chloride, phosphate, and sulfate are considered to be type F. A default value of type M is suggested for compounds where specific information is lacking. Strontium titanate and strontium in an aluminosilicate matrix are considered to have type S clearance characteristics.

In table 9.24, the inhalation dose coefficients are listed for selected tissues for the three types of strontium compounds obtained using the latest lung model. Comparison with the values for a class D compound of strontium from table 9.17 to the type F values demonstrates the effect of the new lung model on dose estimates. The committed equivalent dose to the lung is much less, while more activity reaches the bone compartments, resulting in higher committed equivalent doses to the BS and a higher committed effective dose. Some of the activity that would have been previously ascribed as contributing to lung dose now ends up in the extrathoracic airways, where a committed equivalent dose of $1.7 \times 10^{-9}$ Sv per unit intake is the result (for an adult). This difference is particularly marked in the adult and less so in the newborn. This is due in part to the fact that although a larger percentage of

Table 9.22 Weighting factors for the partition of detriment among respiratory tissues (adapted from ICRP 1994a, table 31)

| Tissue | $A_i$ |
|---|---|
| *Extrathoracic region* | |
| *(a remainder tissue)* | |
| $ET_1$ | 0.001 |
| $ET_2$ | 1 |
| $LN_{ET}$ | 0.001 |
| | |
| *Thoracic region* | |
| *(lung)* | |
| BB | 0.333 |
| bb | 0.333 |
| AI | 0.333 |
| $LN_{TH}$ | 0.001 |

Table 9.23 Values of $f_1$ for inhaled strontium compounds for selected ages (adapted from ICRP 1995a, table 5.10.1.)

| Absorption type | Three months | Five years | Adult |
|---|---|---|---|
| F | 0.6 | 0.4 | 0.3 |
| M | 0.2 | 0.1 | 0.1 |
| S | 0.02 | 0.01 | 0.01 |

Table 9.24 Inhalation dose coefficients (Sv Bq$^{-1}$) for $^{90}$Sr (adapted from ICRP 1995a, tables 5.10.3a, b, c)

| | Three months | Five years | Adult (25 years) |
|---|---|---|---|
| **Equivalent dose in** | | | |
| *Type F* | | | |
| Bone surfaces | $1.3 \times 10^{-6}$ | $4.5 \times 10^{-7}$ | $3.7 \times 10^{-7}$ |
| Red bone marrow | $8.6 \times 10^{-7}$ | $1.9 \times 10^{-7}$ | $1.6 \times 10^{-7}$ |
| Lung | $6.7 \times 10^{-9}$ | $2.1 \times 10^{-9}$ | $6.1 \times 10^{-10}$ |
| Colon | $3.7 \times 10^{-8}$ | $1.2 \times 10^{-8}$ | $3.2 \times 10^{-9}$ |
| Effective dose | $1.3 \times 10^{-7}$ | $3.1 \times 10^{-8}$ | $2.4 \times 10^{-8}$ |
| *Type M* | | | |
| Bone surfaces | $4.8 \times 10^{-7}$ | $1.8 \times 10^{-7}$ | $1.6 \times 10^{-7}$ |
| Red bone marrow | $3.1 \times 10^{-7}$ | $7.7 \times 10^{-8}$ | $7.0 \times 10^{-8}$ |
| Lung | $8.2 \times 10^{-7}$ | $4.3 \times 10^{-7}$ | $2.1 \times 10^{-7}$ |
| Colon | $6.0 \times 10^{-8}$ | $2.0 \times 10^{-8}$ | $5.2 \times 10^{-9}$ |
| Effective dose | $1.5 \times 10^{-7}$ | $6.5 \times 10^{-8}$ | $3.6 \times 10^{-8}$ |
| *Type S* | | | |
| Bone surfaces | $4.5 \times 10^{-8}$ | $2.7 \times 10^{-8}$ | $8.4 \times 10^{-9}$ |
| Red bone marrow | $2.6 \times 10^{-8}$ | $1.0 \times 10^{-8}$ | $3.1 \times 10^{-7}$ |
| Lung | $3.4 \times 10^{-6}$ | $2.2 \times 10^{-6}$ | $1.3 \times 10^{-6}$ |
| Colon | $7.5 \times 10^{-8}$ | $2.5 \times 10^{-8}$ | $7.2 \times 10^{-9}$ |
| Effective dose | $4.2 \times 10^{-7}$ | $2.7 \times 10^{-7}$ | $1.6 \times 10^{-7}$ |

the activity is deposited in a newborn, most of it is deposited in the extrathoracic region, where it either is removed extrinsically or goes directly to the GI tract.

## ICRP Publication 72

This publication (ICRP 1996) is currently the last of the series and, for persons interested in evaluating potential exposures of members of the public, is probably the most useful. It provides tables for the default absorption types and information on where to locate the metabolic data for all of the radionuclides that have been considered by the ICRP to date. In addition, it provides the dose conversion coefficients for ingestion or inhalation of all the radionuclides and inhalation dose coefficients for some gases. This publication uses the default values for environmental exposures for members of the public for the five age groups described above. These tables provide dose conversion coefficients for a parent and its daughter radionuclides. One of the main simplifying assumptions carried through all of the ICRP publications is that an exposure consists of a single acute intake of a single, pure radionuclide. In practice, there are mixtures of daughters and parents, each of which must be treated separately to arrive at a good estimate of dose due to an exposure. The committed effective dose to age 70 is given for $^{90}$Sr and $^{90}$Y in table 9.25. The values for

Table 9.25 Effective dose conversion coefficients, $e(70)(\text{Sv Bq}^{-1})$, for $^{90}$Sr and $^{90}$Y (adapted from ICRP 1996, tables A.1 and A.2)

|  | Three months | Five years | Adult |
|---|---|---|---|
| *Ingestion* | | | |
| $^{90}$Sr | $2.3 \times 10^{-7}$ | $4.7 \times 10^{-8}$ | $2.8 \times 10^{-8}$ |
| $^{90}$Y | $3.1 \times 10^{-8}$ | $1.0 \times 10^{-8}$ | $2.7 \times 10^{-9}$ |
| *Inhalation, type F* | | | |
| $^{90}$Sr | $1.3 \times 10^{-7}$ | $3.1 \times 10^{-8}$ | $2.4 \times 10^{-8}$ |
| *Inhalation, type M* | | | |
| $^{90}$Sr | $1.5 \times 10^{-7}$ | $6.5 \times 10^{-8}$ | $3.6 \times 10^{-8}$ |
| $^{90}$Y | $4.0 \times 10^{-9}$ | $1.4 \times 10^{-9}$ | |
| *Inhalation, type S* | | | |
| $^{90}$Sr | $4.2 \times 10^{-7}$ | $2.7 \times 10^{-7}$ | $1.6 \times 10^{-7}$ |
| $^{90}$Y | $1.3 \times 10^{-8}$ | $4.9 \times 10^{-9}$ | $1.5 \times 10^{-9}$ |

strontium are the same as those found in tables 9.19 and 9.24, whereas the values of yttrium had not been given previously.

### EXAMPLE 9.7

A busload of first-graders is exposed to radioactive dust composed primarily of strontium. You determine that they have inhaled approximately 3,000 Bq of $^{90}$Sr and 250 Bq of $^{90}$Y. What is their estimated committed effective dose from this exposure? To what percentage of their annual limit have they been exposed?

Given: 5–7 year old children, so use tabulated values for 5 years of age. With no idea of the compound, use absorption type M:

$$e(70) = 6.5 \times 10^{-8} \text{ Sv Bq}^{-1} \text{ for } ^{90}\text{Sr and } 4.0 \times 10^{-9} \text{ Sv Bq}^{-1} \text{ for}^{90}Y$$

$$E(70) = (3000\,\text{Bq})(6.5 \times 10^{-8} \text{ SvBq}^{-1})_{\text{Sr}} + (250\,\text{Bq})(4.0 \times 10^{-9} \text{ SvBq}^{-1})_Y$$

$$= 0.000195 \text{ Sv} + 0.000001 \text{ Sv} = 0.196\,\text{mSv},$$

or 19.6% of their annual limit.

### ICRP Publication 89

The information originally compiled in ICRP Publication 23 has been supplemented by the additional anatomical and physiological information compiled in

ICRP Publication 89 (ICRP 2002). Of particular interest are the values for other populations that vary in ethnicity or national origin.

## ICRP Publications 88 and 95

Finally, with the publication of dose conversion factors for exposures to the embryo or fetus (Publication 88, ICRP 2001), the ICRP has extended the ability for us to assess doses from conception to old age. Conversion factors for 31 elements (the same elements examined in previous age-dependent dose publications) are provided with gestation-time–dependent biokinetic models for assessing doses to the embryo or fetus as the result of existing maternal body burdens at the time of conception or due to occupational/environmental exposures during the course of the pregnancy. In Publication 95 (ICRP 2004), the dose to a newborn from radionuclides concentrated in a mother's milk is examined, and a set of age- and time-dependent conversion factors are provided for a number of intake scenarios.

## Summary

The ICRP approach to internal dose assessment, summarized in this chapter, evolved over many years. Beginning in 1959, the approach focused on calculating the dose rate to a single, critical organ and controlling this dose rate within the maximum permissible dose equivalent. The models and equations used were extremely simple but were thought to provide the required degree of protection. Our discussions in this chapter began with the first complete revision of the ICRP recommendations published in 1977 and those associated with internal dose assessment published in 1979. This risk-based approach to radiation safety required the formulation of a system that allowed the expression of intakes of radioactive material through a concept of an effective dose equivalent (or, more precisely, a committed effective dose equivalent) so that the risk associated with the irradiation of many tissues in the body could be expressed properly. These recommendations and the calculated internal dose limits used the latest information available for an occupationally exposed, reference adult male, including an improved respiratory system model, a new GI tract model, a new bone model, and many other models. ICRP guidance was provided to apply these results to other situations (e.g., members of the public). In the 1990s, the ICRP began to focus on age-dependent approaches that could be used for populations of almost any age. In addition, new models for major organ systems and for the metabolism of several important elements were established based on new information. These led to recalculation of the derived limits and consideration of exposures over a wide range of ages. These developments also have been summarized here. Current activities of the ICRP appear to include development of improved metabolic models for selected elements, continued improvement or development of models of organ systems, refined dosimetric models, and more emphasis on age-dependent dose assessment. Although occupational exposure is still an important part of the ICRP's recommendations, it is anticipated that more effort will continue to be exerted on age-dependent dose assessment.

## Problems

1. For a unit intake of $^{90}$Sr, find the committed effective dose for a 3-year-old using the values for committed equivalent doses as given in table 9.19. Compare with the tabulated value.
2. Estimate the committed equivalent dose in the lung and the committed effective dose after ingestion of 37,000 Bq of $^{90}$Sr by a newborn.
3. Estimate the committed equivalent dose in the RBM and the committed effective dose after ingestion of 18,500 Bq of $^{90}$Sr by a 7-year-old.
4. An adult is exposed to a total activity of 5,000 Bq of $^{90}$Sr (type F). If you ignore clearance, what would be the equivalent dose in each of the regions of the respiratory tract? What is the equivalent dose to the lungs and the extrathoracic airways?
5. From example 9.7, what would be the estimated committed effective dose of a driver who inhaled twice the activity of the children? What percentage of the annual limit has the driver been exposed to?

References

Cristy, M. 1980. *Mathematical Phantoms Representing Children of Various Ages for Use in Estimates of Internal Dose.* NUREG/CR-1159 (also Oak Ridge National Laboratory, ORNL/NUREG/TM-367). U.S. Nuclear Regulatory Commission, Oak Ridge, TN.

Cristy, M., and K.F. Eckerman. 1987. *Specific Absorbed Fractions of Energy at Various Ages from Internal Photon Sources.* ORNL/TM-8381/V1-7. Oak Ridge National Laboratory, Oak Ridge, TN.

Eckerman, K.F., R.W. Leggett, C.B. Nelson, J.S. Pushkin, and A.C.B. Richardson. 1999. *Cancer Risk Coefficients for Environmental Exposure to Radionuclides.* Federal Guidance Report 13. EPA 402-R-99–001. U.S. Environmental Protection Agency, Washington, DC.

ICRP (International Commission on Radiological Protection). 1959. *Report of Committee II on Permissible Dose for Internal Radiation.* ICRP Publication 2. Pergamon Press, Oxford.

ICRP. 1968. *Task Group on Radiosensitivity of Tissues in Bone.* ICRP Publication 11. Pergamon Press, Oxford.

ICRP. 1975. *Report of the Task Group on Reference Man.* ICRP Publication 23. Pergamon Press, Oxford.

ICRP. 1977. *Recommendations of the International Commission on Radiological Protection.* ICRP Publication 26. Pergamon Press, Oxford.

ICRP. 1979a. "Limits for Intakes of Radionuclides by Workers." ICRP Publication 30, Part 1. *Annals of the ICRP* 2(3/4).

ICRP. 1979b. "Limits for Intakes of Radionuclides by Workers." ICRP Publication 30, Supplement to Part 1. *Annals of the ICRP* 3(1/4).

ICRP. 1981. "Limits for Inhalation of Radon by Workers." ICRP Publication 32. *Annals of the ICRP* 6(1).

ICRP. 1990. "Age-Dependent Doses to the Members of the Public from Intake of Radionuclides: Part 1." ICRP Publication 56. *Annals of the ICRP* 20(2).

ICRP. 1991a. *1990 Recommendations of the International Commission on Radiological Protection, Users' Edition*. Pergamon Press, Oxford.

ICRP. 1991b. "1990 Recommendations of the International Commission on Radiological Protection." ICRP Publication 60. *Annals of the ICRP* 21(1–3).

ICRP. 1994a. "Human Respiratory Tract Model for Radiological Protection." ICRP Publication 66. *Annals of the ICRP* 24(1–3).

ICRP. 1994b. "Age-Dependent Doses to the Members of the Public from Intake of Radionuclides: Part 2 Ingestion Dose Coefficients." ICRP Publication 67. *Annals of the ICRP* 23(3/4).

ICRP. 1995a. "Age-Dependent Doses to the Members of the Public from Intake of Radionuclides: Part 4 Inhalation Dose Coefficients." ICRP Publication 71. *Annals of the ICRP* 25(3/4).

ICRP. 1995b. "Age-Dependent Doses to Members of the Public from Intake of Radionuclides: Part 3 Ingestion Dose Coefficients." ICRP Publication 69. *Annals of the ICRP* 25(1).

ICRP. 1995c. "Dose Coefficients for Intakes of Radionuclides by Workers." ICRP Publication 68. *Annals of the ICRP* 24(4).

ICRP. 1996. "Age-Dependent Doses to the Members of the Public from Intake of Radionuclides Part 5, Compilation of Ingestion and Inhalation Coefficients." ICRP Publication 72. *Annals of the ICRP* 26(1).

ICRP. 1998. "Individual Monitoring for Internal Exposure of Workers." ICRP Publication 78. *Annals of the ICRP* 27(3/4).

ICRP. 2001. "Doses to the Embryo and Fetus from Intakes of Radionuclides by the Mother." ICRP Publication 88. *Annals of the ICRP* 31(1–3).

ICRP. 2002. "Basic Anatomical and Physiological Data for Use in Radiological Protection: Reference Values." ICRP Publication 89. *Annals of the ICRP* 32(3–4).

ICRP. 2003. *Relative Biological Effectiveness (RBE), Quality Factor (Q), and Radiation Weighting Factor (WR)*. ICRP Publication 92. Pergamon Press, Oxford.

ICRP. 2004. "Doses to Infants from Ingestion of Radionuclides in Mothers' Milk." ICRP Publication 95. *Annals of the ICRP* 34(3/4).

ICRP. 2006. *Radiation Safety Aspects of Brachytherapy for Prostate Cancer Using Permanently Implanted Sources*. ICRP Publication 98. Pergamon Press, Oxford.

NCRP (National Council on Radiation Protection and Measurements). 1985. *General Concepts for the Dosimetry of Internally Deposited Radionuclides*. NCRP Report No. 84. National Council on Radiation Protection and Measurements, Bethesda, MD.

Simmons, J.A., and D.E. Watt. 1999. *Radiation Protection Dosimetry: A Radical Reappraisal*. Medical Physics Publishing, Madison, WI.

Upton, A.C. 1985. "Nonstochastic Effects of Ionizing Radiation." *Proceedings of the Twentieth Annual Meeting of the NCRP, April 4–5, 1984*. No. 6. National Council on Radiation Protection and Measurements, Bethesda, MD.

# 10

# External Dosimetry

David. C. Kocher

In this chapter, the term "external dosimetry" refers to methods of estimating radiation doses to tissues of the human body due to exposure to radionuclides or other radiation sources located outside the body. This chapter considers approaches to estimating external doses to individuals or populations due to exposure to radionuclides dispersed in air, water, and soil that normally should be suitable for use in environmental radiological assessments of routine or accidental releases.

The primary purpose of this chapter is to discuss compilations of data, in the form of external dose rates per unit concentration of radionuclides in the environment, that can be used to estimate external dose in a straightforward manner. These discussions are intended to be helpful to practitioners of environmental radiological assessments who may not be knowledgeable about methods of calculating external dose. Simple corrections to these data to account for factors not considered in developing the data are also discussed. The complex calculations used to obtain external dosimetry data that are suitable for use in many environmental radiological assessments are not discussed in any detail, although some features of the calculations are described to provide a basic understanding of the data. This chapter also discusses an approach to calculation of external dose rates, called the point-kernel method, that often can be used when compilations of external dosimetry data are not appropriate.

Calculations of external dose are generally performed for *reference* individuals of a particular age, gender, and ethnic group (see chapter 8). Most of the external dosimetry data discussed in this chapter apply to a reference adult (Cristy and Eckerman 1987) with locations, shapes and sizes, masses, and elemental compositions of body organs and tissues essentially as prescribed by the International Commission on Radiological Protection (ICRP 1975).

Given the anatomical representation of a reference individual, the dose from external exposure to radionuclides in the environment generally depends on the following:

- Concentrations of particular radionuclides in the environment as a function of time and distance from the receptor location
- Energies and intensities of penetrating radiations emitted by each radionuclide
- Transmission of the emitted radiations from the source region through the different media (e.g., air, water, or soil) between the source and receptor locations
- Transmission of incident radiations through the body of an exposed individual, resulting in doses to particular organs and tissues

Concentrations of radionuclides in the environment used as input to calculations of external dose are assumed to be known quantities. In any calculation of external dose, the dose rate at any time from exposure to a particular radionuclide is proportional to the concentration at that time. Concentrations of radionuclides in environmental media of concern may be obtained from environmental measurements or the types of environmental transport models discussed in chapters 3–5. In general, a determination of radionuclide concentrations in the environment as a function of time and location requires detailed consideration of the rate and manner of radionuclide release, the movement of radionuclides in the environment, and radioactive decay, including the production and decay of any radioactive decay products.

Penetrating radiations of concern in assessments of external dose include photons with energies above a few electron volts (eV) and electrons with energies above about 70 keV. Photons with energies less than a few electron volts do not cause ionization in passing through matter, and electrons with energies less than about 70 keV cannot penetrate the outer dead layer of skin. Neutrons may be of concern in rare cases, but external exposure to neutrons is not considered in this chapter. Nonpenetrating radiation, such as lower energy alpha particles and electrons, is not of concern in external dosimetry. For any radionuclide, the energies and intensities of emitted photons and electrons can be obtained from available compilations (Kocher 1981a; ICRP 1983).

Transmission of emitted radiations from the source to the receptor location, transmission of incident radiations through the body of an exposed individual, and resulting doses in organs or tissues are calculated using complex methods (Eckerman and Ryman 1993). These methods are not of immediate concern to practitioners of environmental radiological assessments and are not discussed in any detail in this chapter.

## Dose Coefficients for External Exposure

This section discusses compilations of data, referred to as external dose coefficients, that often can be used in estimating external dose due to radionuclides dispersed in the environment.

## Definition of External Dose Coefficient

Most environmental radiological assessments are concerned with estimating doses due to release of radionuclides into air, water, or surface soil. For these release pathways, the most important pathways of external exposure include the following:

- Submersion in a contaminated atmospheric cloud
- Immersion in contaminated water
- Exposure to a contaminated ground surface
- Exposure to a volume of contaminated surface soil

In many assessments, the source region in which radionuclides are dispersed can be assumed to be effectively infinite or semi-infinite in extent (e.g., a semi-infinite atmospheric cloud, an infinite ground plane), and concentrations of radionuclides can be assumed to be essentially uniform throughout the source region. Based on these simplifying assumptions, the external dose-equivalent rate[1] in tissue T at time $t$ can be written as

$$dH_T(t)/dt = \chi(t) \cdot h_T, \tag{10.1}$$

where $\chi$ is the radionuclide concentration in the source region and $h$ is the external dose coefficient for the radionuclide in the specified source region.

On the basis of equation 10.1, an external dose coefficient, $h$, in tissue T (also called an external dose conversion factor or an external dose-rate conversion factor) is defined as follows:

An external dose coefficient in tissue T, $h_T$, is the dose-equivalent rate per unit concentration of a radionuclide in a specified source region.

It cannot be overemphasized that an external dose coefficient gives a dose *rate*, not a total dose over time. An external dose coefficient also may be interpreted as the dose per disintegration for a unit concentration of a radionuclide in the source region. The dose received over an exposure time $\tau$ is given by the time integral of the external dose rate:

$$H_T(\tau) = \int_\tau [dH_T(t)/dt] \, dt. \tag{10.2}$$

This time integral usually is not the same as the dose rate at a specified time (e.g., beginning of a year) multiplied by the exposure time (e.g., one year) when radionuclide concentrations in the source region are changing with time due, for example, to radioactive decay or other processes that increase or decrease concentrations in the source region.

## Compilation of External Dose Coefficients

Even with the simplifying assumptions of an infinite or semi-infinite source region and uniform concentrations of radionuclides throughout the source region, calculations of external dose coefficients can be a complex computational problem.

Calculations have been performed in a variety of ways by a number of investigators, and as a result, several compilations of external dose coefficients for radionuclides in the environment have been published.

The current state of the art in calculations of external dose coefficients for radionuclides that are distributed in the air, in water, on the ground surface, or in surface soil is represented by methods and results given in the U.S. Environmental Protection Agency's Federal Guidance Report No. 12 (Eckerman and Ryman 1993). These external dose coefficients are intended to replace previous compilations that were widely used in environmental radiological assessments, including, for example, those developed for the U.S. Nuclear Regulatory Commission (NRC 1977; Kocher 1981b, 1983) and the U.S. Department of Energy (Kocher and Eckerman 1988).

External dose coefficients in Federal Guidance Report No. 12 (Eckerman and Ryman 1993) were calculated using complex and sophisticated radiation transport models and Monte Carlo methods, and they represent an entirely new set of calculations for all of the exposure pathways listed above. For the first time, calculations for all exposure pathways take into account the proper energy and angular distributions of photons incident on the body surface. In previous compilations of dose coefficients for ground-surface exposure (NRC 1977; Kocher 1981b, 1983; Kocher and Eckerman 1988) that were calculated using the point-kernel method discussed further below, in the last section of this chapter, energy and angular distributions of photons incident on the body surface were assumed to be the same as the corresponding distributions for submersion in an atmospheric cloud, even though this assumption was known to be incorrect (Beck and de Planque 1968). Furthermore, some of the earlier compilations (Kocher 1981b, 1983; Kocher and Eckerman 1988) contain an error that resulted in underestimates of dose coefficients by about 10–30%, depending on the particular organ (Kocher and Eckerman 1988). These errors have been corrected in calculations in Federal Guidance Report No. 12. Finally, dose coefficients in the federal guidance report have considerably lower statistical uncertainties than previous values.

Other investigators have used Monte Carlo methods to estimate external dose rates from exposure to a contaminated ground surface or exposure to a volume of contaminated surface soil (Jacob and Paretzke 1986; Jacob et al. 1986, 1988a, 1988b; Chen 1991). However, dose coefficients calculated by Eckerman and Ryman (1993) are generally appropriate for use in environmental radiological assessments, in part because they represent current federal guidance. In addition, Federal Guidance Report No. 12 gives dose coefficients for a large number of radionuclides and for the most important external exposure pathways of concern in environmental radiological assessments.

## *Description of Dose Coefficients in Current Federal Guidance*

Federal Guidance Report No. 12 (Eckerman and Ryman 1993) gives external dose coefficients for the following exposure pathways:

- Submersion in a semi-infinite atmospheric cloud
- Immersion in contaminated water

External Dosimetry    451

- Exposure to a contaminated ground surface
- Exposure to a volume of surface soil that is contaminated to depths of 1 cm, 5 cm, and 15 cm, and to an infinite depth

For the last three pathways, the source region is assumed to be infinite in extent. Except for immersion in contaminated water, an exposed individual is assumed to be standing on the ground. Radionuclide distributions are assumed to be uniform in each source region.

Several dose coefficients are given for each exposure pathway in the federal guidance report. Dose coefficients that are generally useful for purposes of environmental radiological assessments are the effective dose-equivalent rates per unit concentration of radionuclides in each source region. The effective dose equivalent, denoted by $H_E$, was first defined in ICRP Publication 26 (ICRP 1977) as a weighted sum of dose equivalents to several organs or tissues, with the weighting factors intended to be proportional to the stochastic risk per unit dose in each organ or tissue (see chapter 9). Dose coefficients for each organ included in calculations of the effective dose equivalent (i.e., gonads, breast, lung, red marrow, bone surfaces, thyroid, and remainder) also are given in Federal Guidance Report No. 12 (Eckerman and Ryman 1993).

The dose coefficients described above apply only to exposure to photons because the dose to the skin is not included in the effective dose equivalent (ICRP 1977) and because electrons emitted by radionuclides in the environment are not sufficiently energetic to irradiate other organs or tissues below the body surface. Federal Guidance Report No. 12 (Eckerman and Ryman 1993) also gives dose coefficients for the skin from exposure to photons and electrons combined.[2] For ground-surface exposure and exposure to a volume of surface soil, dose coefficients for the skin from exposure to electrons apply at a height of 1 m above ground. Due to scattering and absorption of electrons in air between source and receptor locations, dose coefficients at greater heights would be lower and values at lesser heights would be higher.

In addition to calculating dose coefficients from external exposure to photons based on state-of-the-art methods, results in Federal Guidance Report No. 12 (Eckerman and Ryman 1993) differ from those in earlier compilations by including the contributions from bremsstrahlung, which is the continuous spectrum of photons produced when electrons emitted by radionuclides are decelerated by scattering in a medium. The dose from bremsstrahlung is particularly important for radionuclides that are pure beta emitters (e.g., $^{90}$Sr and $^{90}$Y). In those cases, dose coefficients for exposure to photons only are zero in earlier compilations. In practice, however, external doses from exposure to pure beta-emitting radionuclides usually will be unimportant in environmental radiological assessments compared with doses from inhalation or ingestion pathways.

In using tabulated dose coefficients in Federal Guidance Report No. 12 (Eckerman and Ryman 1993), it is important to recognize that values for each radionuclide do *not* include possible contributions from radioactive decay products that might be present in the source region. Rather, dose coefficients for any radioactive decay products are listed separately. In many cases (e.g., decay of $^{137}$Cs to

$^{137m}$Ba and the decay chain with $^{226}$Ra as its parent), all decay products are shorter lived than the parent, and the half-lives of the decay products are sufficiently short that the decay products normally can be assumed to be in activity equilibrium with the parent in the environment. In all such cases, the external dose from exposure to a radionuclide and its decay products can be calculated using the dose coefficients for the parent and its decay products and known branching fractions in the decay chain. For example, for a given concentration of $^{137}$Cs, the decay product $^{137m}$Ba normally would be in activity equilibrium, and the external effective dose-equivalent rate would be estimated using equation 10.1 as

$$dH_E/dt = \chi(^{137}\text{Cs})\{h_E(^{137}\text{Cs}) + [0.946 \cdot h_E(^{137m}\text{Ba})]\}, \tag{10.3}$$

where 0.946 is the branching fraction for production of $^{137m}$Ba in the decay of $^{137}$Cs.

A user of external dose coefficients must be cognizant of those radionuclides with radiologically significant decay products. The needed information on decay chains and branching fractions is given, for example, by Kocher (1981a) and Eckerman and Ryman (1993). When activity equilibrium in a decay chain cannot be assumed, activities of all members as a function of time can be calculated using the well-known Bateman equations (Evans 1955; Skrable et al. 1974).

## Applicability of Dose Coefficients

Dose coefficients for external exposure to radionuclides in the environment presented in Federal Guidance Report No. 12 (Eckerman and Ryman 1993) are based on the assumptions that the source region is effectively infinite or semi-infinite in extent and the concentrations of radionuclides are uniform throughout the source region. These idealized conditions are never achieved, even for naturally occurring radionuclides in the environment. However, for widely dispersed sources in air, water, or soil, use of these dose coefficients often results in reasonably realistic estimates of external dose.

For exposure to photons, the assumption of an infinite or semi-infinite and uniformly contaminated source region should provide reasonably realistic estimates of external dose whenever the concentrations of radionuclides are approximately uniform within a distance of a few photon mean free paths from a receptor location of interest. The photon mean free path is the reciprocal of the attenuation coefficient in the absorbing medium between the source and receptor locations. For air submersion and ground-surface exposure, the absorbing medium of concern is air. Water is the absorbing medium for water immersion. For photon energies that occur in radioactive decay, which usually are a few mega-electron volts or less, the mean free path is less than about 300 m in air and less than about 0.3 m in water (Schleien et al. 1998). Therefore, use of dose coefficients for air submersion and ground-surface exposure should provide reasonably realistic estimates of external dose whenever the concentrations of radionuclides are approximately uniform over a distance of about 1 km from the receptor location; the required distance is only about 1 m for water immersion. The required distance is intermediate for exposure to a volume of surface soil and depends on the depth of the source region because two absorbing media (soil, which is similar to water, and air) are involved.

The condition of applicability of dose coefficients for photons stated above should almost always be achieved for water immersion. It often will be achieved for the other exposure pathways when the sources are widely dispersed, as occurs with naturally occurring sources or with releases to the atmosphere when receptor locations are sufficiently far from the point of release. In addition, when the purpose of a dose assessment is to demonstrate compliance with a specified dose criterion, use of tabulated dose coefficients may still be appropriate even when the source region is not effectively infinite or semi-infinite in extent and the concentrations of radionuclides are not uniform, especially when conservative overestimates of dose would be provided. An example of such a situation is exposure along the centerline of a finite atmospheric plume at locations close to a ground-level release.

There are, however, important exposure situations for which use of external dose coefficients in Federal Guidance Report No. 12 (Eckerman and Ryman 1993) would not provide reasonably realistic estimates of dose and would not result in overestimates of dose. An example of such a situation is exposure to an elevated atmospheric cloud that has not reached the ground surface. Since the external dose rate at ground level is estimated based on the concentrations of radionuclides in air at the receptor location, the predicted dose rate would be zero, but this could be a significant underestimate of the actual dose rate from exposure to an elevated cloud. Whenever use of dose coefficients for exposure to photons is inappropriate (e.g., for small source regions or elevated atmospheric plumes), external dose rates can be estimated using the point-kernel method discussed in the last section of this chapter.

For electrons emitted in radioactive decay, which normally have energies of a few mega-electron volts or less, the range in air is less than 40 m (NAS/NRC 1964) and the range in water is less than 4 cm (Schleien et al. 1998). Therefore, use of tabulated dose coefficients for electrons based on the assumptions of an infinite or semi-infinite and uniformly contaminated source region should almost always be appropriate in environmental radiological assessments.

## *Effective Dose Coefficients*

In its current recommendations in ICRP Publication 60 (ICRP 1991), the ICRP has replaced the effective dose equivalent with the effective dose, which also is a weighted sum of doses to specified organs or tissues. The effective dose differs from the effective dose equivalent by including the dose to several additional organs or tissues, including the skin, and incorporating a revised set of tissue weighting factors (see chapter 9).

Differences between the effective dose equivalent and the effective dose from external exposure have been investigated by Zankl et al. (1992) and summarized in ICRP (1996). For the irradiation geometries studied, the effective dose equivalent is greater than the effective dose at photon energies above about 15 keV. For the plane-parallel or isotropic photon fields that often are used to represent environmental exposures, the difference is as large as 55% at energies near 20 keV but is less than 10% at energies above 100 keV. In most environmental radiological assessments, the dose from external exposure should be significant compared with the dose from inhalation or ingestion pathways only when radionuclides emit significant intensities of photons with energies above about 100 keV. Therefore, the difference between the effective dose equivalent and effective dose from external exposure usually should not be significant.

Given the near-equality of effective dose equivalents and effective doses for external exposure to photons at energies of primary concern, estimates of effective dose coefficients can be obtained from effective dose equivalents given in Federal Guidance Report No. 12 (Eckerman and Ryman 1993) by adding the dose coefficients for skin multiplied by 0.01. The factor 0.01 represents the tissue weighting factor for skin in the effective dose currently recommended by the ICRP (1991); skin was not included in previous recommendations on calculating effective dose equivalents (ICRP 1977). Thus, in contrast to effective dose equivalents, effective dose coefficients estimated in this way would represent the dose from exposure to photons and electrons. Given the low tissue-weighting factor of 0.01 for the skin, contributions to the effective dose from exposure of the skin generally are important only when the dose from exposure to electrons is much higher than the dose to any organ from exposure to photons. This situation occurs, for example, in cases of immersion in an atmospheric cloud containing $^{85}$Kr (Eckerman and Ryman 1993).

### Dose Coefficients for Other Age Groups

External dose coefficients in Federal Guidance Report No. 12 (Eckerman and Ryman 1993) were calculated for a reference adult. However, the exposed population of concern in environmental radiological assessments usually consists of younger age groups as well as adults. Therefore, in some cases (e.g., for purposes of dose reconstruction of past releases), it could be appropriate to take into account the age dependence of external dose.

The ICRP (1996) developed data on the age dependence of the effective dose. For an isotropic distribution of photons incident on the body surface, which is an appropriate assumption for air submersion and water immersion, the effective dose to younger age groups is somewhat higher than the effective dose to adults, and the difference increases with decreasing age. For example, at photon energies above 100 keV, the dose to an infant is about 30% greater than the dose to an adult, but the difference is only a few percent at age 15. For a plane-parallel photon field and rotational geometry, which can be used to represent exposure to a contaminated ground surface or a volume of surface soil, differences between the effective dose to adults and other age groups are less than 10% at photon energies above 100 keV. Given that external doses from radionuclides that emit only lower energy photons should be insignificant, all such differences should be unimportant in environmental radiological assessments.

### Corrections to Dose Coefficients for Photons

This section discusses simple correction factors that can be applied to estimates of external dose that are obtained using the dose coefficients described above. These corrections apply only to the dose coefficients for photons, and they are intended to provide estimates of dose that are more realistic by taking into account the effects of indoor residence, ground roughness, exposure during boating activities, and exposure to contaminated shorelines.

### Shielding during Indoor Residence

Dose coefficients for air submersion, ground-surface exposure, and exposure to a volume of surface soil described above are calculated by assuming that exposed individuals are located outdoors. Dose assessments that are intended to be more realistic for these exposure pathways should take into account the shielding effect of building structures in reducing external dose during the substantial period of time that individuals normally spend indoors. The fraction of the time that an individual spends indoors normally should be at least 0.5 (NRC 1977) and could be substantially higher for an average member of the public.

In assessing radiological impacts of routine releases from nuclear power reactors, the NRC (1977) recommended that an average shielding factor of 0.5 be used to estimate external dose during the time of indoor residence to account for the presence of building structures. However, the shielding factor should be higher for air immersion than for ground-surface exposure. For example, Burson and Profio (1977) recommended shielding factors for air immersion of 0.6 in a masonry house and 0.9 in a wood-frame house but lower shielding factors for ground-surface exposure of 0.2 and 0.4, respectively. Recommended shielding factors generally apply to radionuclides with emitted photon energies above a few hundred kilo-electron volts, and they should provide conservative overestimates of dose during indoor residence at lower photon energies (Kocher 1980).

The shielding factors during indoor residence presented above are representative of values in living areas of single-family homes. In some cases, it may be appropriate to assume residence in basements of homes or in schools, apartment houses, or office buildings, where the effects of shielding should be greater. For air submersion, representative shielding factors for higher energy photons range from about 0.6 in the basement of a wood-frame house to 0.2 or less in a large office or industrial building. For ground-surface exposure, representative values range from 0.1 in the basement of a home to 0.005 in the basement of a large, multistory structure (Burson and Profio 1977).

### Effects of Ground Roughness

Dose coefficients for ground-surface exposure described above are calculated by assuming that the source region is a smooth plane. A more realistic dose assessment for this exposure pathway would take into account the effects of normal ground roughness (i.e., the shielding provided by terrain irregularities and surface vegetation) in reducing external dose. A dose reduction factor for ground roughness would be applied in addition to the shielding factor during indoor residence discussed above.

Dose reduction factors to represent the shielding effects of ground roughness have been estimated assuming a spectrum of mostly high-energy photons that is typical of fallout following a nuclear reactor accident (Burson and Profio 1977). Recommended dose reduction factors range from nearly 1.0 in paved areas to about 0.5 in a deeply plowed field. A representative average value appears to be about 0.7. The average dose reduction factor for ground roughness should provide conservative overestimates of external doses from exposure to radionuclides that emit lower energy photons.

For ground-surface exposure, an alternative approach to estimating a dose reduction factor for ground roughness is to assume that the concentrations of radionuclides per unit area deposited on the ground surface are distributed uniformly to a depth of 1 cm. For radionuclides that emit high-energy photons, this approach gives a dose reduction factor that is consistent with the representative average value of 0.7 described above. The approach of distributing a surface deposition of activity over a depth of 1 cm in soil has the advantage that the dependence of the dose reduction factor on photon energy is taken into account.

A dose reduction factor for ground roughness generally is not needed in estimating external dose from exposure to a volume of surface soil, especially when sources are assumed to be distributed to a depth of several centimeters or more. In these cases, shielding provided by the soil, which is taken into account in the calculations, is more important than any additional shielding provided by terrain irregularities and surface vegetation.

### Exposure during Boating Activities

Dose coefficients for water immersion described above are calculated by assuming immersion in an infinite source region and are appropriate for exposure while swimming. External exposure to contaminated water also may occur during boating activities, and the external dose in this case should be less than the dose while immersed in water.

A dose reduction factor of 0.5 during boating activities normally should be a reasonable value that is unlikely to result in underestimates of actual doses. This value is based on the consideration that the source region is effectively semi-infinite in extent when an individual is located essentially at the boundary of the source region, and it assumes that the shielding provided by the typically thin hull of a boat is negligible. The latter assumption should provide conservative overestimates of dose for radionuclides that emit lower energy photons.

### Exposure to Contaminated Shorelines

Dose coefficients for ground-surface exposure described above are appropriate for large areas that are contaminated, for example, by deposition of radionuclides from the atmosphere or by irrigation with contaminated water. Ground-surface exposure also may occur along shorelines that are contaminated by deposition of radionuclides in water or sediment. The external dose from exposure to a contaminated shoreline often should be less than the dose from exposure to a large-area source, due to the narrower width of the shoreline.

The NRC (1977) developed recommended dose reduction factors for exposure to different types of shorelines, including values of 0.1 on the bank of a discharge canal, 0.2 on a river shoreline, 0.3 on a lake shore, 0.5 on an ocean shore, and 1.0 on the shore of a tidal basin. Calculations that use the point-kernel method described in the last section of this chapter provide estimates of dose reduction factors for any width of contaminated shoreline and for various energies of emitted photons (Apostoaei et al. 2000).

Dose reduction factors for contaminated shorelines described above apply to sources on the surface. Their use may result in underestimates of external dose if radionuclides are distributed with depth below the surface because most of the dose in such cases is due to sources close to the receptor location (i.e., the dose reduction factor should be closer to 1.0).

## Point-Kernel Method

For some exposure situations, use of the external dose coefficients for photons described in this chapter may be inappropriate in estimating dose rates from exposure to radionuclides in the environment. This is especially the case when the source region is relatively small (e.g., a waste pile) or the concentrations of radionuclides are highly nonuniform within a distance of a few photon mean free paths from the receptor location.

The main purpose of this section is to describe a technique, called the point-kernel method, for estimating external dose rates from exposure to arbitrary distributions of photon emitters. The point-kernel method often can be applied to exposure situations for which use of external dose coefficients for photons is not appropriate. An important example mentioned above is exposure to an elevated atmospheric plume that has not reached the ground surface. The point-kernel method also can be used in estimating dose rates from electrons.

### Description of the Point-Kernel Method

The point-kernel method is based on the concept that the dose rate from any source, either finite or infinite, is the sum of the dose rates from an infinite number of point sources that comprise the source region. Thus, the dose rate from external exposure to any source is estimated by integrating the dose rate from a point source over the extent of the source region, taking into account the distribution of sources throughout the source region.

The dose rate from a point source that emits radiations isotropically is expressed in terms of a quantity called the *specific absorbed fraction* (Loevinger and Berman 1968; Berger 1968). The specific absorbed fraction, $\Phi$, depends on the radiation type (e.g., photons or electrons) and the medium in which the emitted radiations are absorbed, and it is defined as the fraction of the emitted energy $E$ that is absorbed per unit mass of material at a distance $r$ from an isotropic point source. Units of the specific absorbed fraction are reciprocal mass (e.g., $kg^{-1}$).

Given the definition of the specific absorbed fraction described above, the absorbed dose rate, $dD/dt$, at distance $r$ from a point source with activity $A$ at time $t$ is given by Berger (1968) as

$$dD(r, E, t)/dt = k \cdot A(t) \cdot E \cdot \Phi(r, E), \tag{10.4}$$

where the constant $k$ depends on the units of absorbed dose, activity, and energy. For example, to calculate the absorbed dose rate in Gy s$^{-1}$ given the source activity in Bq, energy of the emitted radiation in MeV, and specific absorbed fraction in kg$^{-1}$, $k$ is equal to $1.6 \times 10^{-13}$ kg-Gy MeV$^{-1}$, based on the definition of the Gy as 1 J kg$^{-1}$ and the relationship that 1 MeV = $1.6 \times 10^{-13}$ J.

Given the dose rate from a point source in equation 10.4, the dose rate from any finite or infinite source is calculated as

$$dD(E, t)/dt = k \cdot E \cdot \int_\sigma \chi(\vec{r}, t) \Phi(\vec{r}, E) d\vec{r}, \tag{10.5}$$

where

$\vec{r}$ = vector coordinates of a point source in the source region, with the receptor location assumed for convenience to be at the origin of coordinates

$\chi$ = activity concentration as a function of location in the source region at time $t$

$\sigma$ = source region over which the dose rate from a point source is integrated (see Figure 10.1)

The absorbed dose rate in equation 10.5 applies to a monoenergetic source of energy $E$. For a spectrum of radiations in the decay of a radionuclide, the dose rate is the sum of the dose rates from the individual radiations, or

$$dD(t)/dt = k \cdot \sum f_i E_i \int_\sigma \chi(\vec{r}, t) \Phi(\vec{r}, E_i) d\vec{r}, \tag{10.6}$$

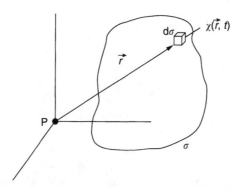

Figure 10.1 Coordinate system for calculation of external dose rate from arbitrary distribution of radioactive sources. The term $\chi(\vec{r}, t)$ is the concentration of a radionuclide in an element of the source region $d\sigma$ at vector distance $\vec{r}$ from the receptor location at the origin (P) at time $t$.

where $f_i$ is the intensity of the $i$th radiation in number per disintegration of the radionuclide and $E_i$ is its energy in MeV.

## Point-Kernel Method for Photons

The point-kernel method described above has been applied most often in estimating the dose rate from distributed sources of photon emitters. For photons ($\gamma$), the specific absorbed fraction at a distance $r$ from an isotropic point source is given by (Berger 1968)

$$\Phi_\gamma(r, E_\gamma) = (1/4\pi r^2) \cdot \mu_{en}/\rho \cdot B_{en}(\mu r) \cdot e^{-\mu r}, \qquad (10.7)$$

where

$\mu_{en}/\rho$ = mass energy-absorption coefficient (e.g., in units of m$^2$ kg$^{-1}$) in the absorbing medium between the source and receptor locations (e.g., air) at the emitted photon energy $E_\gamma$

$\mu$ = linear attenuation coefficient (e.g., in units of m$^{-1}$) in the absorbing medium at energy $E_\gamma$

$B_{en}$ = energy-absorption buildup factor (dimensionless) in the absorbing medium, which also depends on the photon energy

In equation 10.7, the term $1/4\pi r^2$ gives the reduction in dose rate due only to the effect of distance from a point source (i.e., in the absence of an absorbing medium between the source and receptor locations), $e^{-\mu r}$ describes the reduction in the number of photons at a receptor location as a function of distance from the source due to single scattering events in the absorbing medium, $\mu_{en}/\rho$ describes energy absorption of unscattered photons at the receptor location, and the energy-absorption buildup factor, $B_{en}$, accounts for the presence of scattered photons that scatter back to the receptor location and is defined as the ratio of the absorbed energy from all photons to the absorbed energy from unscattered photons only. Thus, $B_{en}$ is greater than 1.0.

Buildup factors in different absorbing media as a function of photon energy are calculated using complex methods that are beyond the scope of this discussion. Analytical approximations of the buildup factor have been developed to facilitate the integration in equation 10.5 for different source geometries (Trubey 1966).

## Applications of the Point-Kernel Method for Photons

The point-kernel method for photons represented in equations 10.5 and 10.7 has been solved analytically or numerically for several distributed source geometries with uniform concentrations of radionuclides. These include a line, infinite plane, disk, rectangular area or volume, cylindrical surface or volume, semi-infinite volume, infinite slab, spherical volume, and truncated right-circular cone volume (Blizzard et al. 1968). Solutions for these source geometries have been implemented in computer codes that can be used in environmental radiological assessments (Negin and Worku 1992). An example is provided by recent calculations of dose reduction factors for exposure to contaminated shorelines and exposure at the center of a disk or rectangular area source (Apostoaei et al. 2000).

The point-kernel method was used in calculating dose coefficients for ground-surface exposure in previous compilations (Kocher 1981b, 1983; Kocher and Eckerman 1988), and the method also was used in early calculations of dose coefficients for monoenergetic sources distributed with depth in surface soil (Kocher and Sjoreen 1985). As noted above, these dose coefficients do not take into account the proper energy and angular distributions of photons in air above the source region. Indeed, a general shortcoming of the point-kernel method is that it does not provide information on these distributions, which are important in estimating doses to organs in an individual standing on the ground (Eckerman and Ryman 1993).

For releases of radionuclides to the atmosphere, concentrations in air and on the ground surface as a function of location and time often are estimated using a Gaussian plume model (see chapter 3). When distributions of radionuclides in the air and on the ground are obtained from this model, equations 10.5 and 10.7 are not amenable to analytical (closed-form) solutions for any functional approximations of the photon buildup factor in air, unless a Gaussian distribution of sources is itself approximated by another distribution. However, because the Gaussian plume model of atmospheric transport is so widely used, numerical and graphical solutions based on the point-kernel method have received considerable attention in the literature.

It is not the purpose of this chapter to cite or review the many papers and reports on estimating external dose rates from exposure to an atmospheric plume that is described by the Gaussian model. However, any user of results found in the literature should be aware of several points.

First, published calculations of dose rates from exposure to a Gaussian atmospheric plume are strictly applicable only to radionuclides that do not deposit on the ground surface (i.e., noble gases) because the effects of wet and dry deposition on plume depletion over time generally are not taken into account. Thus, these calculations should provide conservative overestimates of dose rates from exposure to contaminated air. The degree of conservatism usually is small near the source and increases with increasing distance from the source.

Second, most calculations estimate dose rates only in the sector in which the plume is assumed to be traveling (see chapter 3), but the dose rate in other sectors is ignored, mainly because dose rates in other sectors are considerably lower due to the greater distance from the source. However, a few calculations do provide dose rates in all sectors due to plume travel in a given sector.

Finally, the dose rate from a Gaussian distribution of sources deposited on the ground surface has received far less attention than the dose rate from a Gaussian atmospheric plume. As noted above, however, use of dose coefficients for an infinite and uniformly contaminated plane (or volume) source in estimating dose rates is reasonable unless the width of the deposited plume is substantially less than about 1 km. Furthermore, if dose coefficients are applied to radionuclide concentrations on the ground along the centerline of the plume, conservative overestimates of dose rates are obtained.

In many applications of the point-kernel method, the end point of the calculations is an estimate of exposure (e.g., in roentgen, R) or absorbed dose in air at a receptor location. These results can be converted to effective dose equivalents or effective doses to an individual by using conversion factors given by the ICRP (1987, 1996) or by using the effective dose per unit air kerma (K), the kinetic energy released per unit mass (ICRP 1996), which is essentially the same as the absorbed dose. For immersion in an infinite or semi-infinite source region (air or water), conversion factors for an isotropic field are appropriate. For ground-surface exposure or exposure to a volume of soil, conversion factors for a plane-parallel photon field and rotational geometry should be reasonable. As a rule of thumb that is reasonably accurate for high-energy photons and is conservative at photon energies less than 100 keV, an exposure of 1 R is approximately equal to an effective dose or effective dose equivalent of 0.007 Sv, and an absorbed dose in air of 1 Gy is approximately equal to an effective dose or effective dose equivalent of 0.8 Sv. These conversion factors apply to adults; dose coefficients for other age groups are discussed above.

### Point-Kernel Method for Electrons

The point-kernel method described above also can be applied to electrons. For example, electron point kernels calculated by Berger (1973, 1974) were used to obtain dose coefficients for exposure of the skin to electrons in the compilations of Kocher (1981b, 1983), Kocher and Eckerman (1988), and Eckerman and Ryman (1993).

Given the short ranges of electrons emitted by radionuclides in air, water, or soil, dose coefficients for infinite or semi-infinite and uniformly contaminated sources should almost always be appropriate in estimating electron dose rates from exposure to radionuclides in the environment. Therefore, calculations for finite sources based on the point-kernel method rarely would be needed.

## Problems

1. A nuclear facility releases $^{85}$Kr, $^{89}$Sr, $^{93}$Y, $^{131}$I, $^{133}$Xe, and $^{137}$Cs such that a constant concentration of each radionuclide in air at a receptor location of 100 Bq m$^{-3}$ is maintained. Assume that $^{137m}$Ba is in activity equilibrium with its parent $^{137}$Cs in air and calculate the following external dose rates due to immersion in the atmospheric cloud at ground level: (a) the effective dose rate from exposure to photons only and (b) the effective dose rate from exposure to photons and electrons. What is the importance of external exposure of the skin to electrons for each radionuclide? Are radioactive decay products of $^{93}$Y and $^{131}$I important contributors to external dose, and if not, why not?

2. By assuming that radionuclides considered in problem 1 that are not noble gases are deposited on the ground surface with a deposition velocity ($v_d$) of 1 cm s$^{-1}$, and that radioactive decay is the only mechanism for removal of

radionuclides from the ground surface, derive an equation for the dose rate due to ground-surface exposure and calculate the following:

(a) The time after deposition begins when the total external dose from ground-surface exposure exceeds the total external dose from immersion in the atmospheric cloud, again assuming that a constant concentration of each radionuclide in air is maintained

(b) The total external dose in the first year from air immersion and ground-surface exposure [hint: The external dose rate due to exposure to each radionuclide that deposits on the ground surface as a function of time can be obtained by solving a differential equation]

3. Repeat the calculations in problem 2 by assuming a deposition velocity of 0.01 cm s$^{-1}$. What can you infer about the relative importance of external dose due to air immersion and ground-surface exposure for the two different deposition velocities?

4. For exposure to photons only, calculate the external dose rates and total external doses in problems 1 and 2 by assuming that an exposed individual resides at the receptor location 50% of the time and is located indoors in a building that provides a shielding factor of 0.7 for 80% of the time spent at that location, and that ground roughness and terrain irregularities reduce dose rates from ground-surface exposure by a factor of 0.7.

5. Using equation 10.7, derive an equation for the external dose rate at a distance $x$ above a uniformly contaminated infinite plane source by assuming no absorbing material between the source and receptor locations. Repeat the derivation by assuming an absorbing material in which the buildup factor is described by the Berger form (Trubey 1966) given by $B_{en}(\mu x) = 1 + C\mu x(e^{D\mu x})$, where $C$ and $D$ are energy-dependent coefficients that do not depend on distance from a source. Does the second derivation have an analytical (closed-form) solution?

Notes

1. The quantity "dose equivalent" is now called "equivalent dose" by the ICRP (1991).

2. Dose coefficients for the skin from exposure to electrons were calculated using methods described by Kocher and Eckerman (1988). Values for ground-surface exposure given in appendix A.3 of Kocher and Eckerman (1988), however, are in error. For this exposure pathway, correct values of external dose coefficients for electrons were used in Federal Guidance Report No. 12 (Eckerman and Ryman 1993). Correct values also were used in previous compilations of external dose coefficients for ground-surface exposure (Kocher 1981b, 1983).

References

Apostoaei, A.I., S.K. Nair, B.A. Thomas, C.J. Lewis, F.O. Hoffman, and K.M. Thiessen. 2000. "External Exposure to Radionuclides Accumulated in Shoreline Sediments with an Application to the Lower Clinch River." *Health Physics* 78(6): 700–710.

Beck, H., and G. de Planque. 1968. *The Radiation Field in Air Due to Distributed Gamma-Ray Sources in the Ground.* HASL-195. Health and Safety Laboratory, U.S. Atomic Energy Commission, New York.

Berger, M.J. 1968. "Energy Deposition in Water by Photons from Point Isotropic Sources." MIRD Pamphlet No. 2. *Journal of Nuclear Medicine* Supplement 1: 15–25.

Berger, M.J. 1973. *Improved Point Kernels for Electron and Beta-Ray Dosimetry.* NBSIR 73–107. National Bureau of Standards, U.S. Department of Commerce, Washington, DC.

Berger, M.J. 1974. "Beta-Ray Dose in Tissue-Equivalent Material Immersed in a Radioactive Cloud." *Health Physics* 26(1): 1–12.

Blizzard, E.P., A. Foderaro, N.G. Goussev, and E.E. Kovalev. 1968. "Extended Radiation Sources (Point Kernel Integrations)." In *Engineering Compendium on Radiation Shielding*, Vol. 1, ed. R.G. Jaeger. Springer-Verlag, New York.

Burson, Z.G., and A.E. Profio. 1977. "Structure Shielding in Reactor Accidents." *Health Physics* 33(4): 287–299.

Chen, S.Y. 1991. "Calculation of Effective Dose-Equivalent Responses for External Exposure from Residual Photon Emitters in Soil." *Health Physics* 60(3): 411–426.

Cristy, M., and K.F. Eckerman. 1987. *Specific Absorbed Fractions of Energy at Various Ages from Internal Photon Sources. I. Methods.* ORNL/TM-8381/V1. Oak Ridge National Laboratory, Oak Ridge, TN.

Eckerman, K.F., and J.C. Ryman. 1993. *External Exposure to Radionuclides in Air, Water, and Soil.* Federal Guidance Report No. 12. EPA 402-R-93–081. U.S. Environmental Protection Agency and Oak Ridge National Laboratory. Available at www.epa.gov/radiation/docs/federal/402-r-93–081.pdf.

Evans, R.D. 1955. *The Atomic Nucleus.* McGraw-Hill, New York.

ICRP (International Commission on Radiological Protection). 1975. *Report of the Task Group on Reference Man.* ICRP Publication 23. Pergamon Press, New York.

ICRP. 1977. "Recommendations of the International Commission on Radiological Protection." ICRP Publication 26. *Annals of the ICRP* 1(3).

ICRP. 1983. "Radionuclide Transformations: Energy and Intensity of Emissions." ICRP Publication 38. *Annals of the ICRP*, 11–13.

ICRP. 1987. "Data for Use in Protection against External Radiation." ICRP Publication 51. *Annals of the ICRP* 17(2/3).

ICRP. 1991. "1990 Recommendations of the International Commission on Radiological Protection." ICRP Publication 60. *Annals of the ICRP* 21(1–3).

ICRP. 1996. "Conversion Coefficients for Use in Radiological Protection Against External Radiation." ICRP Publication 74. *Annals of the ICRP* 26(3/4).

Jacob, P., and H.G. Paretzke. 1986. "Gamma-Ray Exposure from Contaminated Soil." *Nuclear Science and Engineering* 93(3): 248–261.

Jacob, P., H.G. Paretzke, H. Rosenbaum, and M. Zankl. 1986. "Effective Dose Equivalents for Photon Exposures from Plane Sources on the Ground." *Radiation Protection Dosimetry* 14(4): 299–310.

Jacob, P., H.G. Paretzke, H. Rosenbaum, and M. Zankl. 1988a. "Organ Doses from Radionuclides on the Ground. Part I. Simple Time Dependencies." *Health Physics* 54(6): 617–633.

Jacob, P., H.G. Paretzke, and H. Rosenbaum. 1988b. "Organ Doses from Radionuclides on the Ground. Part II. Non-trivial Time Dependencies." *Health Physics* 55(1): 37–49.

Kocher, D.C. 1980. "Effects of Indoor Residence on Radiation Doses from Routine Releases of Radionuclides to the Atmosphere." *Nuclear Technology* 48: 171–178.

Kocher, D.C. 1981a. *Radioactive Decay Data Tables*. DOE/TIC-11026. U.S. Department of Energy, Washington, DC.

Kocher, D.C. 1981b. *Dose-Rate Conversion Factors for External Exposure to Photons and Electrons*. NUREG/CR-1918, ORNL/NUREG-79. Oak Ridge National Laboratory, Oak Ridge, TN.

Kocher, D.C. 1983. "Dose-Rate Conversion Factors for External Exposure to Photons and Electrons." *Health Physics* 45(3): 665–686.

Kocher, D.C., and A.L. Sjoreen. 1985. "Dose-Rate Conversion Factors for Photon Emitters in Soil." *Health Physics* 48(2): 193–205.

Kocher, D.C., and K.F. Eckerman. 1988. *External Dose-Rate Conversion Factors for Calculation of Dose to the Public*. DOE/EH-0070. U.S. Department of Energy, Washington, DC.

Loevinger, R., and M. Berman. 1968. "A Schema for Absorbed-Dose Calculations for Biologically-Distributed Radionuclides." MIRD Pamphlet No. 1. *Journal of Nuclear Medicine* Supplement 1: 7–14.

NAS/NRC (National Academy of Sciences/National Research Council). 1964. *Studies in Penetration of Charged Particles in Matter*. NAS-NRC Publication 1133, ed. U. Fano. National Academy of Sciences—National Research Council, Washington, DC.

Negin, C.A., and G. Worku. 1992. *MicroShield, Version 4, User's Manual*. Grove 92–2. Grove Engineering, Rockville, MD.

NRC (Nuclear Regulatory Commission). 1977. *Regulatory Guide 1.109. Calculation of Annual Doses to Man from Routine Releases of Reactor Effluents for the Purpose of Evaluating Compliance with 10 CFR Part 50, Appendix I*. Nuclear Regulatory Commission, Washington, DC.

Schleien, B., L.A. Slayback, Jr., and B.K. Birky. 1998. "Interaction of Radiation with Matter." Chapter 5 in *Handbook of Health Physics and Radiological Health*, 3rd ed. Williams & Wilkins, Baltimore, MD.

Skrable, K., C. French, G. Chabot, and A. Major. 1974. "A General Equation for the Kinetics of Linear First Order Phenomena and Suggested Applications." *Health Physics* 27(1): 155–157.

Trubey, D.K. 1966. *A Survey of Empirical Functions Used to Fit Gamma-Ray Buildup Factors*. ORNL/RSIC-10. Oak Ridge National Laboratory, Oak Ridge, TN.

Zankl, M., N. Petoussi, and G. Drexler. 1992. "Effective Dose and Effective Dose Equivalent—the Impact of the New ICRP Definition for External Photon Radiation." *Health Physics* 62(5): 395–399.

# 11

# Estimating and Applying Uncertainty in Assessment Models

Thomas B. Kirchner

Since about 1980, quantitative risk assessments for chemical and radiological contaminants have largely supplanted qualitative risk assessments (Paustenbach 1995). Industry now insists on adequate scientific evidence to support potentially restrictive regulation (Bonin and Stevenson 1989). Indeed, the 1979 Supreme Court decision concerning the U.S. Environmental Protection Agency's proposed regulation of benzene concluded that stricter regulations would be appropriate only if it could be demonstrated that risks to people would be significantly reduced (Graham et al. 1988). This decision set quantitative assessments as the standard for enforceable regulations. Furthermore, as Morgan (1998) stated, risk implies uncertainty; to analyze risk invariably means that uncertainties must be considered.

Quantitative risk assessments require the use of mathematical models. Models that give quantitative, repeatable results are viewed as providing impartial assessments and can integrate the expertise of many scientists. Two types of models are generally distinguished: empirical models and process-level models. An empirical model is a mathematical representation of a set of data. The representation can be an arbitrary function, such as a polynomial, with coefficients chosen to provide a good fit to the data, or it can be a function based upon laws or principles, such as exponential decay of radionuclides. Empirical models are based upon the assumption that there exists a predictable relationship between one set of variables and another. In contrast, process-level simulation models are designed to represent the dynamics of a system rather than to describe a data set (figure 11.1).

Empirical models can be fit to data using direct numerical techniques, such as computing the coefficients to exactly fit a polynomial of order $n-1$ to $n$ data points or using iterative search techniques to find parameters for a function to approximate the

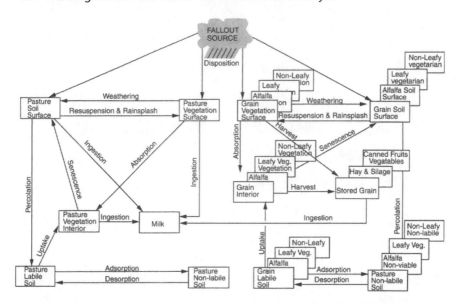

Figure 11.1 Process-level simulation models typically represent the system of interest as a series of compartments with flows of material or energy between the compartments. This diagram represents the PATHWAY simulation model (Whicker and Kirchner 1987).

data (Press et al. 1988). Statistical analyses rely upon the use of empirical models. The models are usually linear equations that can be fit using direct methodologies, such as minimizing the sums of squared differences between the model predictions and the data. For example, in most statistical regression models, data are divided into sets of one or more independent variables that are used to predict one or more dependent variables. Linear relationships are usually assumed to exist between the dependent and independent variables, although linearity may be achieved through a transformation of the variables. Nonlinear regression models usually are parameterized using iterative search techniques. The units associated with the parameters, such as meters or grams, and the distribution assumed to be exhibited by the parameter are implicitly defined by the data and the statistical methodology.

Empirical models are useful tools for describing data sets, particularly when interpolations are needed. One strength of statistical models is that inference within the domain of the data is strong and a probabilistic measure of uncertainty can be derived. Empirical models are not generally useful for extrapolation beyond the domain of the data used in their construction except when they are based upon laws or principles (Zar 1984). One exception is the time series model, which are empirical models that incorporate time as an independent variable (Chatfield 1980). Time series models can sometimes be used for forecasting if you can assume that the process giving rise to the time series of data continues to operate in the same way in the future.

Process-level simulation models can be simple mathematical equations much like those used for empirical models, but more often they are formulated as expressions designed to represent the change in a system through time (i.e., dynamic equations). A system is typically visualized as a series of compartments linked by flows. The compartments, or state variables, are abstract or idealized containers of matter or energy representing one or more state variables of the real system, such as the biomass of plants per unit area of land surface or the radionuclide concentration per unit mass of plant tissue. The rate of flow between compartments is represented using differential equations, difference equations, or discrete events. Process-level models can also use partial derivatives to represent the flow of material or energy in a continuous medium, such as the flow of water through soil. In any case, the approach requires the iterative solution of the rate equations or the simulation of events over the time span to be considered. The length of each time step is determined by the type of equations being solved and, with differential equations, the accuracy with which the equations are to be solved. The Pathway model (Whicker and Kirchner 1987) shown in figure 11.1 is an example of a model based on both differential equations and discrete events.

Some simulation models incorporate stochastic processes into their structure. Stochastic processes are defined in terms of probabilities and are implemented using pseudorandom number generators (Naylor et al. 1968). The processes represent stochastic events, such as precipitation or the occurrence of accidents. Models that make use of stochastic processes need to be run many times to generate enough realizations of its predictions to form a distribution of possible outcomes. Summaries, such as the mean and variance of a prediction or its 5th and 95th percentile, can also be used as the output variables of the stochastic model.

Because simulation models represent the important dynamics and general features of a system rather than a specific set of data, they tend to be used in risk assessment to a much greater extent than empirical models. Simulation models typically do not provide the strength of inference that can be obtained from well-designed statistical sampling efforts and analyses. However, as noted above, empirical models are usually constrained to interpolation within the range of available data and are not useful for extrapolation in time or space. Furthermore, risks associated with regulated radionuclides (or chemicals) often have to be estimated to meet regulatory approval before initiating activities that are needed to provide appropriate data. The applicability of a process-level model, its sensitivity to required input data, the quality of its predictions, and the uncertainty surrounding its predictions must be based upon extensive validation, sensitivity, and uncertainty analyses. In the absence of such analyses, the inference that you can draw from the model's predictions should be considered weak at best.

The level of uncertainty is an important factor to consider when validating a model's performance, either as an absolute criteria (e.g., is the level of uncertainty acceptable?) or as a relative criteria (e.g., do the observations to which the model is compared fall within the confidence or tolerance interval for the model predictions?). Sensitivity analysis is designed to identify the parameters that have the greatest effect on model predictions. This type of analysis can be used as a tool to help guide model development, to identify critical areas where research could improve model predictions or reduce uncertainty, and to build confidence in the performance of models. The relationship between sensitivity analysis, validation, and uncertainty is briefly reviewed below following the discussion of uncertainty analysis.

## Why Perform an Uncertainty Analysis?

The goal in performing an uncertainty analysis is usually to evaluate the predictive ability of a model. Because models are, by definition, simplified representations of real systems, their outputs will invariably reflect uncertainty due to assumptions made to simplify their structure, uncertainty in the values assigned to parameters of the model equations, uncertainty or variability in inputs to the model, and simulation of natural variability in the processes or among the components being represented in the model. An assessment of the uncertainty in a model prediction or output is required to fully evaluate the range of outcomes possible given these various sources of uncertainty. Uncertainty in model predictions may change through simulation time or across the domain of applicability for a model (Breshears et al. 1989). The level of uncertainty in predictions may also change as the quality of the input data for a model changes.

The results of uncertainty analyses can also be used to evaluate the performance of a model both during its development and as part of its validation. Indeed, the magnitude of uncertainty alone can be used as a criterion for accepting or rejecting a model. In building a model, you must make choices, such as which processes and states to include and which states to aggregate to simplify the model. The level of uncertainty can be used as a criterion for comparing various versions of a model. For example, it is often tempting to add greater detail to a model to improve its ability to represent the behavior of a system. Adding complexity to a model may improve, for example, its ability to fit a particular set of observations. However, the added complexity also may increase the uncertainty in model predictions. Eventually, adding complexity to a model is likely to increase uncertainty in model predictions to unacceptable levels. Uncertainty analysis, therefore, provides one criterion by which to judge whether a model is "improved" by adding or deleting processes or states.

This chapter focuses on parametric uncertainty. A parametric uncertainty analysis is the process by which uncertainty in model parameters is propagated through a model and the distributions of model predictions or outputs are described. *Parameter* in this context refers to initial values of state variables and model parameters. Uncertainty in model inputs, or driving variables, will also be considered parametric uncertainty because there is little or no difference when analyzing contributions to uncertainty from these two sources. *Output variable* refers to a state variable or function of state variables whose response is of particular interest. Structural uncertainty arising from alternative, plausible model formulations will not be considered. Structural uncertainty is problematic because it can be difficult, at best, to express it in a probabilistic manner, both in assigning probabilities to the various alternatives and in combining results that may not be equivalent (e.g., of a different scale or level of aggregation). In addition, the consideration of alternative formulations is often addressed during the construction and validation of a model.

## Describing Uncertainty

Uncertainty in parameters used in models can be classified into two types: aleatory and epistemic. *Aleatory uncertainty* involves parameters that are assumed to vary across space or time in a stochastic way. For example, precipitation may show seasonal patterns but appear to be a stochastic phenomenon with respect to any given day. *Epistemic uncertainty* involves parameters that are assumed to be constant throughout a simulation but have uncertainty associated with their values from a lack of knowledge about the particular assessment being performed. This uncertainty can be represented as a distribution of confidence over a range of possible values. For example, infiltration rates for a given soil type may be relatively constant, but uncertainty could be assigned to the rate if the soil type is unknown. A distribution of epistemic uncertainty is sometimes called a distribution of subjective probability because it reflects the modeler's degree of belief about the likelihood of a parameter taking on various values. Understanding the distinction between true variability and distributions of confidence in uncertainty analysis is essential when discussing uncertainties (Hattis and Burmaster 1994).

The terms type A and type B have also been applied to uncertainty in models (NCRP 1996; IAEA 1989) and metrology (ANSI 1997). The definitions of type A and type B uncertainty focus on characterizing the uncertainty in the results of calculations. The term "type B uncertainty" is applied to a distribution of results where the true result is a single but unknown value. The distribution in this case represents the lack of knowledge about the true value or, conversely, the confidence you have in a computed value being the true value. Type B uncertainty is thus closely aligned with epistemic uncertainty. The term "type A uncertainty" is applied to a distribution of results that represents aleatory uncertainty, or natural variability. Although the distinction between type A and type B uncertainties might appear to be clear-cut, it is often difficult to cleanly partition uncertainty into these two categories. A distribution of doses could be characterized as type A uncertainty if it is meant to represent the variability in doses of an exposed population, or could be characterized as type B uncertainty if it is meant to represent the possible doses that an exposed individual might have received. It is the interpretation of the model result (the end point) that determines whether its uncertainty represents type A or type B.

In the past, uncertainties due to natural variability and to lack of data were usually not distinguished in the methods used to propagate the uncertainties through models. Indeed, estimates of variability based on sparse data were often subjectively adjusted to account for the additional uncertainties associated with using the data in a different context. The goal in assigning subjectively derived distributions to parameters, thought to be truly constant, and in subjectively adjusting distributions representing natural variability was to increase confidence that decisions made using the model results would have a low likelihood of type II (false negative) errors. There is growing interest, however, among scientists and decision makers to produce realistic rather than conservative estimates of risk, to specifically identify the components of confidence versus expected frequency in the results from risk assessments, and to partition the contributions from subjective and objective uncertainties (Helton 1994; Hoffman and Hammonds 1994; MacIntosh et al. 1994).

**470** Radiological Risk Assessment and Environmental Analysis

It is also important to distinguish whether parameters are constant through time or variable through time, as demonstrated in the following example. Assume a cow is eating pasture that has been contaminated with a radionuclide. The concentration of the radionuclide in the pasture decays as a first-order process because of radioactive decay and physical processes that remove the contamination from the leaves. Loss from the cow is also a first-order process because of radioactive decay and biological elimination. The model for this system is

$$\frac{dQ_1}{dt} = -K_1 Q_1, \tag{11.1}$$

$$\frac{dQ_2}{dt} = IQ_1 - K_2 Q_2, \tag{11.2}$$

where

$Q_1$ and $Q_2$ = the pasture and cow compartments, respectively
$K_1$ and $K_2$ = the loss rates
$I$ = the ingestion rate of pasture by the cow

If the parameters for the model are assumed to be constant through time but are assigned distributions to represent variability such as that among pastures and animals, then the results are quite different than if the parameters are assumed to vary stochastically through time to represent, for example, day-to-day changes in consumption rates or spatial variability in concentrations in the pasture (figure 11.2).

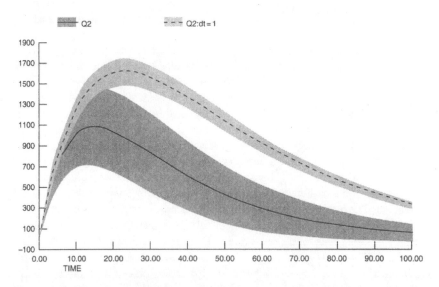

Figure 11.2 Assuming parameters are stochastic but constant through time ($Q_2$) produces much different results than assuming parameters are stochastically variable through time ($Q_2 : dt = 1$, representing simulations in which parameters change each day). The means ±1 standard deviation are plotted. $Q_2$ could also represent type B uncertainty.

Most uncertainty analyses must consider both epistemic and aleatory sources of uncertainty. The importance of stochastic variability in an analysis depends heavily on the scale of the assessment question being asked. Aleatory uncertainty is likely to be important when considering individuals as opposed to aggregates, such as estimating an individual dose as opposed to a collective dose. Aleatory uncertainty is also important if spatially or temporally defined random events significantly affect the system, such as disturbances to waste repositories due to natural events or human intrusion. Time-dependent uncertainty is generally not considered in simple risk assessment models. These models often assume conditions, such as a constant rate of input, under which time-dependent variability can be ignored. Dynamic simulation models can consider time-dependent uncertainty, although implementing such uncertainty requires care, as discussed below in the section titled "Assigning Distributions."

## Probability Distributions

A basic requirement for performing a quantitative uncertainty analysis of a model is to formulate a quantitative representation of the uncertainty associated with the model's parameters. A minimum requirement is to define an estimate of range for each of the relevant parameters. If the model is not complicated mathematically, then the interval of the output variables can be estimated by propagating the ranges of the parameters through the model. More often, available information indicates there is a greater chance that a parameter will take on one value as compared to another. This probabilistic viewpoint is often expressed by defining a probability distribution for the parameter. Even without data, subjective opinion can be used to weight the selection of parameter values toward likely values based on theoretical considerations or expert opinion. A limited amount of data can be combined with subjective knowledge to derive distributions for parameters. Under ideal circumstances, there is sufficient data to tabulate the frequency with which particular values of a parameter are likely to be observed. A standard statistical distribution can then be selected and its parameters fit to the data from these samples of observations.

## Descriptive Statistics

Probability distributions can be described in terms of their functional forms and parameters, their graphical forms, and various descriptive statistics such as the mean, variance, skewness, kurtosis, and percentiles. In risk assessment, there is often interest in representing the uncertainty in the mean of a distribution, using a confidence interval, or in the percentiles of the distribution of possible values, using tolerance intervals. Graphical methods for examining the properties of a distribution should not be overlooked because typical descriptive statistics often do not convey enough information to characterize the shape of a distribution, such as having two or more modes.

Two functional forms of probability distributions that provide useful information are the probability density function (PDF) and the cumulative distribution function (CDF) (figures 11.3 and 11.4).

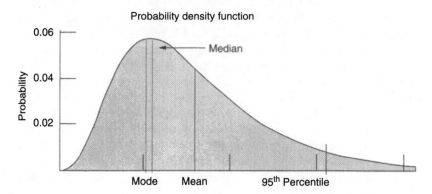

Figure 11.3 The probability density function shows the probability of the values of a random variable.

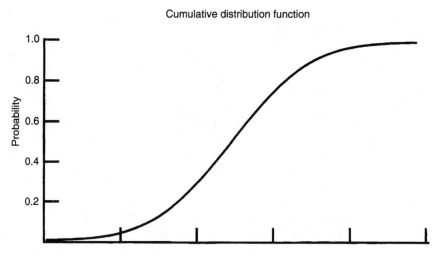

Figure 11.4 The cumulative distribution function shows the probability (on the abscissa) that a value is less than or equal to a point on the ordinate axis.

The PDF describes the likelihood of observing a particular value for a variable across a range, and it is given the functional representation $f(x)$. The CDF is obtained by integrating the PDF:

$$F(x) = \int_{-\infty}^{\infty} f(x)dx. \tag{11.3}$$

When the variable of interest can take on only discrete values, the frequency with which the various values are observed is represented by the probability mass function, $p(x_i)$, which has a CDF that has discrete steps (figures 11.5 and 11.6).

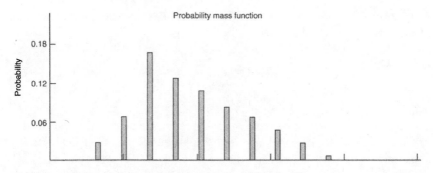

Figure 11.5 The probability mass function shows the probability for the various values of a random variable having a discrete distribution.

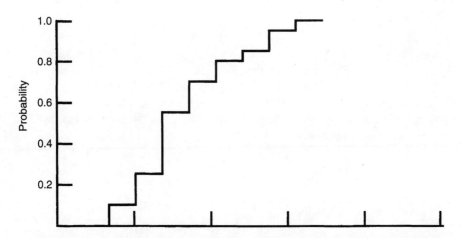

Figure 11.6 The cumulative distribution function for a discrete distribution has steps.

Several of the parameters typically used to describe a distribution are based on the concept of expectation. The expectation of a function of a random variable, $x$, is the product of that function and the probability associated with $x$; that is,

$$E[G(X)] = \int_{-\infty}^{\infty} G(x)f_x(x)dx \qquad (11.4)$$

for continuous distributions, or

$$E[G(X)] = \sum_{-\infty}^{\infty} G(x)f_x(x) \qquad (11.5)$$

for discrete distributions, where $G(X)$ is a function. For example, the expected value of a continuous random variable is

$$E[X] = \int_{-\infty}^{\infty} x f_x(x) dx, \qquad (11.6)$$

which is the definition of the mean of $x$.

Properties of a distribution that often are of interest are its mode, mean, median, and a measure of its dispersion or spread. The degree of skewness and kurtosis of a distribution can also be of importance when characterizing a distribution. The mode of a distribution is the value having the greatest likelihood of being observed, which is the peak of the curve of a PDF. Some distributions can have two or more such peaks, in which case the distributions are described as multimodal. The median value, $M$, is the value that satisfies

$$\int_{-\infty}^{M} f_x(x) dx = 0.5 \qquad (11.7)$$

For a set of observations, the median is estimated to be the point where, for $n$ observations ($n$ being odd), $(n-1)/2$ of the observations are less than the value (i.e., the median is the middle value in a list of observations ordered by value). If $n$ is even, the median is assumed to be the midpoint between the two middle values.

The mean of a distribution, $\mu$, is defined as

$$\mu = \int_{-\infty}^{\infty} x f(x) dx \qquad (11.8)$$

for continuous distributions, or

$$\mu = \sum x_i p(x_i) \qquad (11.9)$$

for discrete distributions, where $f(x)$ is the PDF of $x$ and $p(x_i)$ is the probability mass function of $x_i$. The mean of a distribution, $\mu$, is also called the first moment of the distribution. The central moments of a distribution are the expectations of powers of $(x-\mu)$. The second central moment of a distribution, the variance, $\sigma^2$, is a measure of the dispersion of the distribution and is defined as

$$\sigma^2 = \text{var}[X] = \int (x - \mu)^2 f(x) dx \qquad (11.10)$$

and

$$\sigma^2 = \text{var}[X] = \sum (x_i - \mu)^2 p(x_i) \qquad (11.11)$$

for continuous and discrete distributions, respectively. Central moments are defined relative to the mean of the distribution.

The third central moment gives information about the skewness of a distribution:

$$\mu_3 = \int (x - \mu)^3 f(x) dx \tag{11.12}$$

$$\mu_3 = \sum (x_i - \mu)^3 p(x_i). \tag{11.13}$$

Positive values of skewness indicate that the right tail of the distribution is stretched to the right or, equivalently, that the mode of the distribution is shifted left (figure 11.7). The left tail of a distribution having negative skewness is stretched to the left.

The fourth central moment provides information about the flatness of the distribution (figure 11.8):

$$\mu_4 = \int (x - \mu)^4 f(x) dx \tag{11.14}$$

$$\mu_4 = \sum (x_i - \mu)^4 p(x_i). \tag{11.15}$$

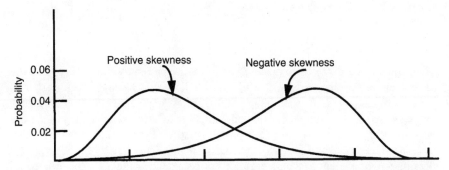

Figure 11.7 Positive skewness is characterized by the right tail being stretched to the right, while the left tail of a negatively skewed distribution is longer. A distribution with no skewness is symmetric.

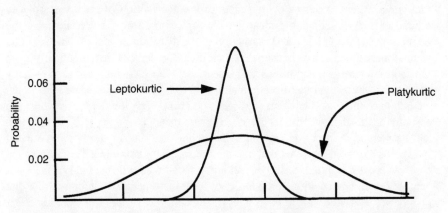

Figure 11.8 A platykurtic distribution is relatively flat, while a leptokurtic distribution has a sharp peak.

When comparing an observed distribution to an expected distribution, relatively flat distributions are often labeled *platykurtic*, and relatively peaked distributions are called *leptokurtic*.

In addition to the moments of a distribution, which measure the central tendency, dispersion, and shape of a distribution, the percentiles or fractiles of a distribution are often of interest. The $p$th fractile of a distribution is a value, $X$, for which the probability of observing values less than $X$ is $p$. The percentile of a distribution is equivalent to the fractile represented as a percentage. Thus, 95% of all values are expected to be less than the 0.95 fractile (figure 11.3).

## Statistical Intervals

Statistical intervals are used to express confidence about the estimates of parameters of distributions. Most scientists are familiar with confidence intervals on means. Confidence intervals can also be placed on standard deviations and percentiles. In risk assessment, there is often interest in reporting a level representing some upper percentile for the population, as opposed to a percentile for the mean. In this case, tolerance levels are of interest.

### Confidence Intervals

The parameters derived from the analysis of samples are only estimates of the true parameters of the distribution that was sampled. They are subject to uncertainty, which is often expressed as confidence intervals on the parameters. The confidence intervals placed on sample means and sample standard deviations are not truly objective measurements of variability; they simply represent the confidence you have in the method that is used to construct the interval. In other words, a frequency distribution equivalent to the distributions of confidence cannot be constructed by additional sampling of the data; thus, there is no objective method for testing the veracity of a distribution of confidence for a sample mean. For example, repeated sampling of a population will generate a frequency distribution of sample means that is equivalent, in terms of its shape, to the distribution of confidence placed around an individual sample mean but that is centered on the true mean of the population (figure 11.9). The construction of the distribution of confidence about a sample mean is objective in the sense that it can be derived mathematically from the data after certain assumptions are made. Unless those assumptions are explicitly verified, however, their validity is a matter of belief. Therefore, the degree to which a confidence interval is deemed credible is determined by the degree to which the assumptions associated with its construction are met.

The method used to construct the 95% confidence interval for a sample mean is expected to yield intervals that contain the true population mean in 95% of the cases where it is applied. Strictly speaking, a specific confidence interval does not enclose the true mean with a probability of 0.95; the true mean is either within that confidence interval or not. It is also not a statement that says if you were to collect more samples and compute the means of those samples, 95% of those sample means would be expected to fall within this particular interval. Rather, it is expected that

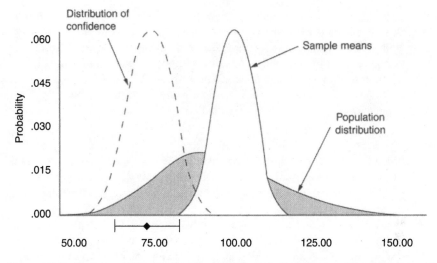

Figure 11.9 A lognormal distribution of a population (dark shading) and the distribution of sample means for a sample size of 10 (light shading). The distribution of confidence for one sample mean is shown with a dashed line, and the associated confidence interval is shown below the graph.

95% of the intervals constructed around various sample means would contain the true mean.

The confidence interval around the sample mean is derived from the assumption that a normal distribution of sample means would be attained if many samples from the population were collected and the mean of each sample computed. This assumption of normality is based upon the Central Limit Theorem of statistics and holds even if the distribution of the population being sampled is not normal. However, large sample sizes may be required to show convergence of the distribution of the mean to a normal distribution if the samples being averaged are drawn from a highly skewed distribution. For example, consider a simulation in which you collect samples of 10 items from a population distribution that was lognormal and that had a mean of 100 and a standard deviation of 20. The collection of 10,000 samples of size 10 was simulated, and the distribution of the means of those samples was computed. The results (figure 11.9) show that the distribution of the sample means is normal (probability = 0.0003 using a chi-square goodness-of-fit test), whereas the distribution for the population is lognormal.

The distribution of sample means constructed in the example above should not be confused with the distribution constructed using the estimate of the sample mean and its standard deviation for a single sample. The standard deviation of the sample mean is also called the standard error. The standard error, $s_e$, is equal to the standard deviation, $s$, for the sample divided by the square root of the sample size, $n$:

$$s_e = \frac{s}{\sqrt{n}}. \tag{11.16}$$

The uncertainty in the estimate of the mean decreases as the size of a sample increases. Notice that the standard error is estimated from a single sample, not from many samples, as was the case for the distribution of means shown in figure 11.9. This extrapolation from one sample to a distribution of means from many samples is based upon applying the Central Limit Theorem.

The confidence interval for the mean is constructed using the sample mean, $\overline{X}$, the standard error, $s_e$, and a parameter called the $t$-statistic:

$$\overline{X} - t_{(1-\alpha), n-1} s_e < \mu < \overline{X} + t_{(1-\alpha), n-1} s_e. \quad (11.17)$$

The value of the $t$-statistic depends upon the sample size, $n$, and the level of confidence, $\alpha$. A confidence interval can be constructed for every 1 of the 10,000 samples collected for the example of figure 11.9 to build the distribution of sample means. Because confidence intervals are centered on the sample mean rather than the true, but unknown, population mean, samples that had means in the extreme tails of the distribution are likely to yield confidence intervals that do not include the true mean of the population. For example, one sample yielded a mean of 79.55 and a standard error of 4.31. The 0.05 $t$-parameter for a sample size of 10 is 2.262. Thus, the confidence interval for that sample ranges from 69.8 to 89.29 and does not include the true mean, 100 (figure 11.9). Of the 10,000 samples that were run to construct figure 11.9, 94.4% of them had 95% confidence intervals that included the true mean of the population. The difference between the 94.4% observed and the 95% expected is due to sampling error.

The estimate of the confidence interval associated with the standard deviation of a distribution depends on the type of distribution. If the distribution being sampled is normal, then

$$U = \frac{vs^2}{\sigma^2} \quad (11.18)$$

has a chi-square distribution with $v = n-1$ degrees of freedom, where $n$ is the size of the sample, $s^2$ is the sample variance, and $\sigma^2$ is the true variance (Zar 1984). The confidence interval for $\sigma^2$ is

$$\frac{vs^2}{\chi^2_{(1-\alpha/2),v}} \leq \sigma^2 \leq \frac{vs^2}{\chi^2_{\alpha/2,v}}, \quad (11.19)$$

where $\alpha$ the is the confidence level (Zar 1984). The formula can be applied to lognormal distributions by applying it to logarithmic transforms of the data.

The confidence interval for the Poisson distribution's single parameter $\lambda$ deserves consideration because Poisson distributions are frequently encountered in the fields of epidemiology and radiation measurement. In the Poisson distribution, $\lambda$ is equal to both the mean and the variance of the distribution. If $n$ events are observed where the distribution of events is Poisson, then the $100(1 - \alpha)\%$ confidence interval for $\lambda$ can be estimated by finding the smallest and largest $\lambda$ for which $n$ events are expected with probabilities $1 - (\alpha/2)$ and $\alpha/2$, respectively (Bailar and Ederer 1964) (figure 11.10).

In this example, if the number of observations is 20, then the Poisson distributions with $\lambda$ as small as 12.21 and as large as 30.89 have the probability 0.025 or greater

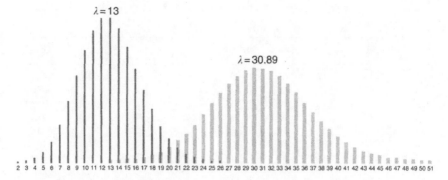

Figure 11.10  For *n* events, the 95% confidence interval for the Poisson parameter, *l*, is bounded by the Poisson distributions that have *n* as the 97.5 and 2.5 percentiles, respectively.

Table 11.1  Confidence intervals for the Poisson distribution's single parameter, $\lambda$.[a]

|  | Poisson, $\lambda = 12.21$ | | Poisson, $\lambda = 30.89$ | |
| --- | --- | --- | --- | --- |
| *n* | Probability | Cumulative probability | Probability | Cumulative probability |
| 13 | 0.11  | 0.658 | 0     | 0     |
| 14 | 0.102 | 0.752 | 0     | 0.001 |
| 15 | 0.088 | 0.828 | 0.001 | 0.001 |
| 16 | 0.072 | 0.886 | 0.001 | 0.002 |
| 17 | 0.055 | 0.928 | 0.002 | 0.005 |
| 18 | 0.04  | 0.957 | 0.004 | 0.009 |
| 19 | 0.027 | 0.975 | 0.006 | 0.015 |
| **20** | **0.018** | **0.975** | **0.01** | **0.025** |

[a] For an observation of 20 counts, the 95% confidence interval for $\lambda$ is 12.21 to 30.89.

for observing 20 events (table 11.1). The lower and upper confidence limits for $\lambda$ when there are *n* events can be found by searching for $\lambda$ values that satisfy

$$p = \sum_{x=0}^{n} \frac{e^{-\lambda} \lambda^x}{x!} \qquad (11.20)$$

for $p = 1 - \alpha/2$ and $p = \alpha/2$, respectively. Table 11.2 lists confidence limits for $\lambda$ given the number of observations *n*.

## Tolerance Intervals

In risk assessment, it is often not the mean of a distribution that is of interest but the location of various percentiles of the distribution. If the parameters for a given distribution are known, then the levels associated with any given percentile can usually be determined exactly. For example, the one-tailed 5% percentile for a standard normal distribution is 1.64σ, which means that 5% of the values in the population represented by the distribution would be greater than μ + 1.64σ.

Table 11.2 Ninety-five percent confidence levels for λ given the number of observed events, n

| n | Lower | Upper | n | Lower | Upper | n | Lower | Upper |
|---|---|---|---|---|---|---|---|---|
| 1 | 0.025 | 5.572 | 27 | 17.79 | 39.28 | 65 | 50.17 | 82.85 |
| 2 | 0.242 | 7.225 | 28 | 18.6 | 40.47 | 70 | 54.57 | 88.44 |
| 3 | 0.619 | 8.768 | 29 | 19.42 | 41.65 | 75 | 58.99 | 94.01 |
| 4 | 1.09 | 10.24 | 30 | 20.24 | 42.83 | 80 | 63.44 | 99.57 |
| 5 | 1.624 | 11.67 | 31 | 21.06 | 44 | 85 | 67.9 | 105.1 |
| 6 | 2.203 | 13.06 | 32 | 21.89 | 45.17 | 90 | 72.37 | 110.6 |
| 7 | 2.815 | 14.42 | 33 | 22.71 | 46.34 | 95 | 76.86 | 116.1 |
| 8 | 3.455 | 15.76 | 34 | 23.55 | 47.51 | 100 | 81.37 | 121.6 |
| 9 | 4.117 | 17.09 | 35 | 24.38 | 48.68 | 110 | 90.41 | 132.5 |
| 10 | 4.797 | 18.39 | 36 | 25.21 | 49.84 | 120 | 99.5 | 143.4 |
| 11 | 5.493 | 19.68 | 37 | 26.05 | 51 | 130 | 108.6 | 154.3 |
| 12 | 6.202 | 20.96 | 38 | 26.89 | 52.16 | 140 | 117.7 | 165.2 |
| 13 | 6.924 | 22.23 | 39 | 27.73 | 53.31 | 150 | 126.9 | 176 |
| 14 | 7.656 | 23.49 | 40 | 28.58 | 54.47 | 160 | 136.1 | 186.8 |
| 15 | 8.398 | 24.74 | 41 | 29.42 | 55.62 | 170 | 145.4 | 197.5 |
| 16 | 9.148 | 25.98 | 42 | 30.27 | 56.77 | 180 | 154.6 | 208.3 |
| 17 | 9.905 | 27.22 | 43 | 31.12 | 57.92 | 190 | 163.9 | 219 |
| 18 | 10.67 | 28.45 | 44 | 31.97 | 59.07 | 200 | 173.2 | 229.7 |
| 19 | 11.44 | 29.67 | 45 | 32.82 | 60.21 | 250 | 219.9 | 282.9 |
| 20 | 12.21 | 30.89 | 46 | 33.68 | 61.36 | 300 | 267 | 335.9 |
| 21 | 13 | 32.1 | 47 | 34.53 | 62.5 | 350 | 314.3 | 388.6 |
| 22 | 13.79 | 33.31 | 48 | 35.39 | 63.64 | 400 | 361.7 | 441.1 |
| 23 | 14.58 | 34.51 | 49 | 36.25 | 64.78 | 450 | 409.3 | 493.5 |
| 24 | 15.38 | 35.71 | 50 | 37.11 | 65.92 | 500 | 457.1 | 545.8 |
| 25 | 16.18 | 36.91 | 55 | 41.43 | 71.59 | 750 | 697.3 | 805.6 |
| 26 | 16.98 | 38.1 | 60 | 45.79 | 77.23 | 1,000 | 938.9 | 1,063 |

The usual situation, however, is that the parameters for the distribution are not known exactly. Instead, there will only be estimates of the parameters, and the quality of the estimates will depend on the size of the sample used to compute the estimates (figure 11.11). Estimates of percentiles have greater uncertainty than do estimates of means.

Three approaches can be used to estimate the percentiles for a distribution. In the first method, tabular values of the tolerance parameter, $g'_{(1-\alpha;p,n)}$, are used to estimate the percentiles. The tolerance parameter is used to compute the one-sided tolerance levels using the formula

$$t_p = \overline{X} + g'_{(1-\alpha;p,n)} s$$

or

$$t_p = \overline{X} - g'_{(1-\alpha;p,n)} s, \qquad (11.21)$$

where

$t_p$ = the upper $p$ tolerance limit
$\overline{X}$ = the estimate of the mean

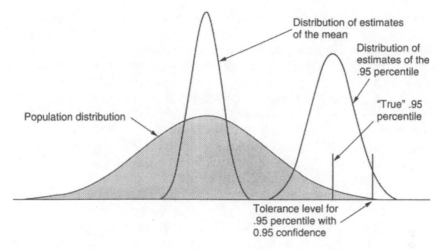

Figure 11.11 Tolerance levels reflect the uncertainty in the estimates of the mean and variance that are obtained from sampling a distribution.

$g'_{(1-\alpha;p,n)}$ = the tolerance parameter for the one-sided percentile $p$ and a confidence level of $P$

$s$ = the estimate of the standard deviation

Hahn and Meeker (1991) provide a table of values of $g\langle PR\rangle_{(1-\alpha;p,n)}$ for normal distributions. For example, suppose you have estimated the mean and variance for a normal distribution based upon 21 samples and you are interested in estimating the level of the upper 0.95 percentile with a confidence of 0.99. You would select $g\langle PR\rangle_{(1-\alpha;p,n)} = 2.766$ in row 21 from the column for the 0.99 confidence level and the 0.95 proportion of the distribution in Hahn and Meeker's table A.12d. The level is then computed as

$$t_{0.95} = \overline{X} + 2.766s, \quad (11.22)$$

where $t_{0.95}$ is the upper 0.95 tolerance limit, $\overline{X}$ is the estimate of the mean, and $s$ is the estimate of the standard deviation. Two-sided tolerance bounds on a population distribution or confidence bounds on a percentile can also be constructed using other tables in Hahn and Meeker (1991).

This approach works well if the distribution is normal and the probabilities of interest correspond to those of the table. A distribution-free approach can be used when it is not known whether the assumption of normality is appropriate. The distribution-free approach estimates the position of percentiles using order statistics. The order statistics are the data ordered from the smallest value, $x_{(1)}$, to the largest value, $x_{(n)}$. The distribution-free point estimate of the $p$th percentile lies between $x_{(np)}$ and $x_{(np+1)}$ (Hahn and Meeker 1991). Therefore, the $(np) + 1$ order statistic is used to estimate the location of the $p$th percentile, where $(np)$ is the integer part of the product of $n$ and $p$. For example, the point estimate of the 95th percentile for an $n$ of 50 would be $x_{(48)}$. One-tailed and two-tailed confidence intervals can be computed for percentiles if $n$ is sufficiently large.

The minimum sample size required to provide a one-tailed confidence interval for a percentile can be determined easily. The probability of randomly selecting $n$ samples from a distribution and having them all fall below the $100p$th percentile of the distribution is $p^n$. Thus, $x_{(n)}$ can be used to estimate the $p$th percentile of a distribution with confidence $1-\alpha$ if $p^n \leq \alpha$. For example, the minimum sample size to estimate with 95% confidence the 95th percentile of a distribution is 59. The probability of having 59 out of 59 samples fall below the 95th percentile is 0.0485. Therefore, any value that exceeds the largest of the 59 values is likely to fall above the true 95th percentile of the distribution with a confidence of about 95%. The value of $x_{(59)}$ represents the upper one-tailed confidence level on the 95th percentile, with $x_{(57)}$ being the point estimate of the 95th percentile.

It may seem reasonable to assume that the 95th percentile of a distribution can be estimated with as few as 19 samples because any subsequent sample that exceeded $x_{(19)}$ would be in the top 5% of the distribution of the sample. However, the probability of selecting 20 values at random from a distribution and having all of them fall below the 95th percentile is $0.95^{20} = 0.3585$. Thus, $x_{(20)}$ will be below the true 95th percentile in about 36% of samples of size 20. The largest percentile that can be estimated with a confidence of $(1-\alpha)$ in a sample of size $n$ using $x_{(n)}$ is $p = e^{\ln(\alpha)/n}$. Thus, for $n = 20$, $x_{(20)}$ is a valid estimator only of the 86th percentile with 95% confidence.

Computing one-tail confidence intervals for percentiles with a sample size greater than the minimum $n$ and for two-tailed confidence intervals requires the use of the binomial distribution. The cumulative probability for observing $j$ or fewer events in $n$ trials with a probability $p$ of an event occurring is

$$B(j, n, p) = \sum_{i}^{j} \binom{n}{i} p^i (1-p)^{n-i}, \tag{11.23}$$

where

$$\binom{n}{i} = \frac{n!}{i!(n-1)!}. \tag{11.24}$$

The location of the upper one-tailed estimate of the $100p$th percentile of a distribution is the value $x_{(j)}$, where $j$ is the smallest integer for which $B(j-1; n, p) \geq (1-\alpha)$. The location of the lower one-tailed estimate of the $100p$th percentile of a distribution is the value $x_{(j)}$, where $j$ is the largest integer for which $B(j-1; n, p) \leq \alpha$. The distribution-free estimate of the upper one-tailed 95% confidence level for the 95th percentile when $n = 100$ occurs at $x_{(99)}$, and the lower one-tailed 95% confidence level occurs at $x_{(91)}$. The 95% one-tailed lower and upper confidence levels for the 5th percentile are $x_{(2)}$ and $x_{(10)}$, respectively. A two-tailed distribution-free confidence interval can be constructed by solving $B(u-1; n, p) - B(l-1; n, p) \geq 1-\alpha$ for $u$ and $l$ such that $0 < l < u \leq n, 0 < p < 1$, and $u$ and $l$ are as close together as possible. When $n$ is sufficiently large, there can be more than one set of $u$–$l$ pairs that satisfy the conditions and you can choose whether to select confidence levels that are symmetric about $x_{(np+1)}$. Tables of order statistics for estimating one- and two-tailed tolerance intervals can be found in Hahn and Meeker (1991).

Estimating and Applying Uncertainty in Assessment Models   483

The third method for estimating tolerance intervals requires a significant investment in computation time. The approach is simply to run a Monte Carlo simulation on the model to generate a distribution for the output variable, and then use the $x_{(np+1)}$ order statistic to estimate the percentile of interest. The process is repeated many more times until a distribution for the estimates of the percentile is obtained. The estimate for the percentile that is used as the tolerance level, with confidence $P$, is the $(nP + 1)$ ordered value of the estimated percentiles. Thus, if 100 Monte Carlo simulations of a model were run, then the best estimate of the 0.95 percentile level would be the 96th value in the ordered list. Repeating this process 100 times would give 100 estimates of the 0.95 tolerance level, so you could sort these values from smallest to largest and select the largest value as the best estimate of the 0.95 tolerance level with a confidence level of 0.99.

## Typical Distributions

Constructing distributions for parameters is often the most difficult part of the analysis because there are often limited data in the risk assessment model. Several probability distributions are commonly used in uncertainty analyses (table 11.3). The choice of which distribution to use generally depends on the data available, the information about the process giving rise to the distribution, and the subjective assessment of the most likely shape.

The assumption is often made that distributions representing variability are normal or lognormal (figure 11.12), but such assumptions should be verified.

Table 11.3  Mean and variance of frequently encountered distributions[a]

| Distribution | Mean | Variance |
|---|---|---|
| Uniform | $\frac{a+b}{2}$ | $\frac{(b-a)^2}{12}$ |
| Log-uniform | $\frac{b-a}{\ln(b)-\ln(a)}$ | $(b-a)\frac{[\ln(b)-\ln(a)](b+a)-2(b-a)}{2[\ln(b)-\ln(a)]^2}$ |
| Normal | $\mu$ | $\sigma^2$ |
| Exponential | $\frac{1}{\lambda}$ | $\frac{1}{\lambda^2}$ |
| Lognormal | $\mu_x = e^{\mu_y + \frac{\sigma_y^2}{2}}$ | $\sigma_x^2 = e^{2\mu_y + 2\sigma_y^2} - e^{2\mu_y + \sigma_y^2}$ |
| Chi squared | $k$ | $2k$ |
| Poisson | $\lambda$ | $\lambda$ |
| Triangular | $\frac{a+b+c}{3}$ | $\frac{a^2-a(b+c)+b^2-bc+c^2}{18}$ |

[a] For the uniform and log-uniform distributions, $a$ and $b$ are the minimum and maximum values. For the triangular distribution, $a$ is the minimum, $b$ is the mode, and $c$ is the maximum.

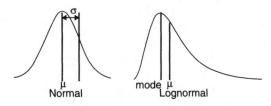

Figure 11.12 Normal and lognormal distributions are often used in uncertainty analyses. The mode and the mean, $\mu$, are equivalent in the normal distribution. In the lognormal distribution, the mean lies to the right of the mode. The standard deviation, $\sigma$, is a measure of the dispersion of the distribution.

The Central Limit Theorem states that a normal distribution will result from adding or averaging a number of values selected at random (i.e., independently) from identically distributed random variables. The normal distribution is frequently assumed in statistical analyses of experimental data because of the prevalence of testing assumptions about the means of samples, and the distribution of means is expected to be normal. Extending the Central Limit Theorem to cover the sum of random variables having different distributions, such as the sum of various model parameters or state variables, is somewhat questionable, and you should not simply assume that such a sum would be normal.

In a lognormal distribution, the logarithms of the data are normally distributed. Lognormal distributions arise from multiplying identically distributed random variables. A product is the antilogarithm of the sum of the logarithms of randomly distributed variables, and thus the Central Limit Theorem can be invoked. Lognormal distributions are positively skewed and are often described in terms of their geometric mean and geometric standard deviation. The geometric mean is the antilogarithm of the mean of the logarithms of the data, and the geometric standard deviation is the antilogarithm of the standard deviation of the logarithms of the data. The arithmetic mean for a lognormal distribution lies to the right of the mode. If the geometric standard deviation exceeds 3.28, then the arithmetic mean will be larger than the 95th percentile for the distribution (Edelmann and Burmaster 1997).

Poisson distributions are discrete distributions that arise from events that occur randomly through time with a constant average rate, such as the decay of radioactive elements when the number of atoms in a sample is large. The distribution of the number of events in an interval of time has a Poisson distribution. For large sample sizes, the Poisson distribution converges toward a normal distribution.

The exponential and Poisson distributions are related because the intervals between random events will have an exponential distribution. The single parameter for the exponential distribution, $\lambda$, is equal to the inverse of the mean of the intervals.

The chi-square distribution is one of a family of gamma distributions. It is well known because of its use in classical statistics for contingency tests. It also provides a basis for estimating the uncertainty on the variance of normal distributions (Mood et al. 1974).

Uniform, log-uniform, triangular, log-triangular, trapezoid, log-trapezoid, and empirical distributions are not usually associated with natural processes, but they are frequently used when data are scarce to represent uncertainty. The uniform distribution is often used to represent a case where a minimum and maximum for a parameter can be defined but where there is little or no information to argue that the probability of selecting a value from one portion of the range is any different than the probability of selecting a value from any other portion of the range. The triangular distribution is used when a range can be supplemented with a judgment about the most likely value. A trapezoid can be used when a subrange rather that a single value is thought to be more likely. Both the triangular and the trapezoid distributions can result from the sum of two uniform distributions. The log-uniform, log-triangular, and log-trapezoid distributions are often used when the range of the possible values covers several orders of magnitude.

## Correlations and Multivariate Distributions

Correlations among input parameters to a model can have large impacts on an analysis (Smith et al. 1992). Covariance is a measure of the linear relationship or correlation between two random variables. The covariance between variables $X$ and $Y$, denoted cov[$XY$], will be positive when $X-\mu_x$ and $Y-\mu_y$ tend to have the same sign with high probability. The covariance between $X$ and $Y$ will be negative when there is a high probability that $X-\mu_x$ and $Y-\mu_y$ will be of opposite signs. The magnitude of cov[$XY$] is not a good indicator of the degree of linear relationship between $X$ and $Y$ because it depends upon the variance of $X$ and $Y$. The correlation coefficient is better for judging the interdependency of $X$ and $Y$. The correlation coefficient, $\rho$, is computed by dividing the covariance of $X$ and $Y$ by the product of the standard deviations of $X$ and $Y$:

$$\rho = \frac{\text{cov}[XY]}{\sigma_X \sigma_Y} \quad (11.25)$$

$X$ and $Y$ are said to be independent if they have no correlation.

Our knowledge about the correlation between parameters is often limited, at best, but the effect of covariance between parameters on the variance of a model output can be very important. If a conservative assumption is warranted, then the correlation between the variables can be assumed to be perfect, so that $\rho = \pm 1$ and cov[$XY$] $= \pm\sigma_x\sigma_y$. However, this assumption can have a significant impact on the uncertainty of the model results.

The multivariate normal distribution provides one means of representing correlations among variables. The multivariate normal has the property that the marginal and conditional distributions are normal (figure 11.13). The marginal distribution is the distribution for one of the variates, and the conditional distribution is the distribution for one variate given values for the remaining variates. A conditional distribution can be thought of as the cross section of a slice made through a

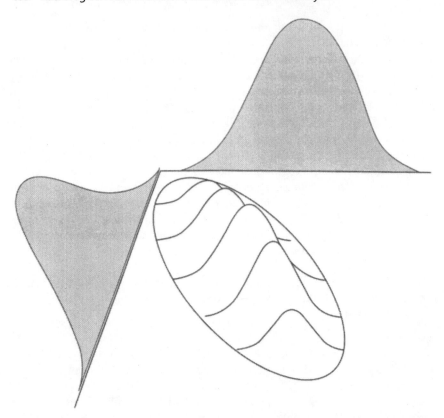

Figure 11.13 In a bivariate normal distribution, each of the marginal distributions is normal, as are the conditional distributions. A slice through the distribution at right angles to either axis will be a normal distribution. The angle of the main axis of the distribution is determined by the degree of correlation between the two variates.

multivariate distribution orthogonal to the axis of interest. Bivariate normal distributions are used frequently in statistics, such as in the statistical model underlying simple linear regression.

Multivariate lognormal distributions have a multivariate normal distribution of the logarithms of the variates, but the correlation between the logarithms of the variates is not equal to the correlation between the variates (IAEA 1989). The relationship between the correlation coefficient for the logarithms, $\rho\langle PR\rangle$, and the correlation coefficient for the untransformed variates, $\rho$, is given by

$$\rho' = \frac{\ln\left\{1 + \rho\sqrt{[\exp(\sigma_1^2) - 1]\exp(\sigma_2^2) - 1}\right\}}{\sigma_1 \sigma_2}. \qquad (11.26)$$

Multivariate discrete distributions also exist, such as the multinomial. Most of the univariate distributions, however, do not have multivariate analogs. For example, there is no multivariate triangular distribution. If Monte Carlo methods are used to propagate uncertainty, then methods exist for inducing correlations among distributions with nearly any type of distribution (discussed under "Latin Hypercube Sampling," below).

## Assigning Distributions

One of the most substantial challenges to performing an uncertainty analysis is developing the distributions of uncertainty associated with the inputs to a model (Cullen and Frey 1998). The uncertainty in a parameter may arise from natural variability in the parameter, measurement errors, or lack of knowledge about a parameter. Uncertainty may be introduced through assigning parameters based on approximations. Uncertainty also can reflect parameter values that are assigned using subjective judgment. Sometimes parameter uncertainty encompasses the bias introduced in a model because of aggregation of states or processes in formulating the model (Gardner et al. 1982; O'Neill and Rust 1979; Mosleh and Bier 1992).

An important aspect of assigning uncertainty to a parameter is to specify exactly what the uncertainty represents. The definition of the uncertainty on a parameter must be consistent with the definition and use of the parameter. For example, suppose a model uses a transfer coefficient to estimate the concentration of a radionuclide in cow's milk given the concentration of the radionuclide in forage. In assigning uncertainty to the transfer coefficient, you would need to consider whether the concentration in milk was expected to represent the concentration in milk coming from a dairy that obtains milk from several sources, the average concentration in a pool of milk obtained from a single herd of cows, or the concentration in milk from a single cow. Natural variability among cows and variability in the spatial distribution of the radionuclide would likely cause you to assign a higher uncertainty to the transfer coefficient when applied to estimating milk concentrations from a single cow than when applied to estimating milk concentrations from a dairy. The problem of properly specifying uncertainty so that a meaningful probability distribution can be assigned has led Howard and Matheson (1984) to suggest the clarity test as a conceptual means of ensuring proper specificity. The test involves the notion that a researcher could measure the parameter of interest if given sufficient resources. A properly specified parameter would give sufficient details to identify the conditions under which an experiment could be conducted to obtain the required data.

The problem of specificity of uncertainty is particularly problematic when assigning distributions to time-varying stochastic parameters in models. Simulating the variability of a parameter through time typically involves sampling the distribution of the parameter at various times as the simulation proceeds. In such simulations, it is critical that the length of time over which the parameter remains constant corresponds to the time interval assumed when assigning a distribution to the parameter. For example, assume that the consumption of contaminated milk by an individual is being simulated and that the milk consumption rate is assumed to vary from day to

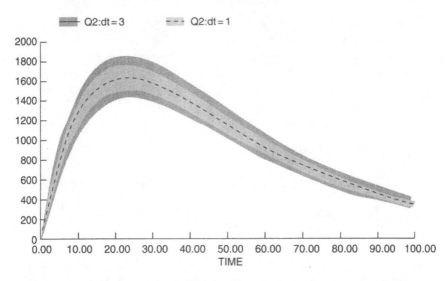

Figure 11.14 A simple stochastic model was run for two sets of 100 simulations using a time step of 1 and 3. The means ±1 standard deviation for one compartment are plotted. The difference between the uncertainties in the two sets of runs is due only to the time step used to solve the differential equations.

day. The model estimates the total intake of the contaminant by integrating the rate of milk consumption through time using numerical methods to solve differential equations. New values for the stochastic parameters are selected at the start of each time step. If the time step for solving the rate equations exceeds one day, then the variability in consumption through time is likely to be exaggerated. If the time step is less than one day, then the variability is likely to be depressed (figure 11.14).

Ensuring that a correct time step is used to sample the distributions of parameters is not necessarily a trivial problem. Data representing variability through time in different parameters may be based on different time steps, and the time step used in a model is often controlled by the design and, in the case of models employing differential equations, by the numerical accuracy required of the solution. Differential equations are frequently used to represent processes that occur continuously through time. Differential equations can be solved in computer simulation models using numerical methods that solve the equations over discrete intervals of time. The accuracy of the methods depends on the time step, the form of the equations, and the magnitude of the rate constants in the equations. Adaptive time-step methods, such as the Runge-Kutta-Felburg or Gear's methods, can be used to control the accuracy of the solution by adjusting the size of the time step throughout a simulation. For this reason, when using adaptive time-step methods to solve differential equations, the time step with which stochastic parameters are sampled and assigned throughout the simulation must be controlled independently of the solution time step.

Aggregating states that show natural variability in their dynamics can also be problematic. For example, consider the problem of predicting the average concentration of $^{131}$I in milk for a herd of dairy cows starting with rate constants for individual cows. If the cows produce equal volumes of milk, then the average concentration would correspond to the concentration expected if the milk from the herd were pooled together. Like many processes in models of radionuclide transport, the decline in the concentration ($Bq^{-1}$) of $^{131}I$ in milk can be well described using first-order kinetics:

$$\frac{dQ}{dt} = kQ, \tag{11.27}$$

where $Q$ is the concentration of $^{131}I$ in milk and $k$, the rate of loss, is negative (Sasser 1965). The solution to this differential equation is

$$Q_t = Q_0 e^{kt}, \tag{11.28}$$

where $Q_0$ is the initial value of $Q$, and $Q_t$ is the value at time $t$. The rate of decline in the concentration of $^{131}I$ in milk is controlled by the combination of physical decay and biological elimination. Sasser (1965) measured half-times for the rate of loss in six cows ranging from 0.60 to 0.85 days. A simulation model was constructed to represent the decline in milk concentration from each of the six cows and from a hypothetical herd having an average rate of loss equal to the mean of the six observed rate constants. The dynamics of the herd were assumed to be identical to the dynamics for an individual cow; that is,

$$\frac{dQ_h}{dt} = k_h Q_h. \tag{11.29}$$

The model also computed the average concentration of $^{131}I$ in milk across the six cows, as shown in figure 11.15:

$$Q_a = \sum_1^6 Q_i/6. \tag{11.30}$$

For this example, $Q_a$ (figure 11.15, solid line) can be viewed as the true solution. Plotting the logarithm of $Q_h$ yields a straight line, whereas the logarithm of $Q_a$ is a concave curve. The difference between $Q_a$ and $Q_h$ is model bias, the error introduced by using an inadequate representation of the system being modeled. Given the analytical solution for the model and the condition that $k$ is a random variable, an approximation for the function for the mean can be written as

$$\overline{Q}_t \approx Q_0 e^{\overline{k}t + (t^2\sigma^2)/2}, \tag{11.31}$$

with the corresponding differential equation

$$\frac{d}{dt}\overline{Q}\left(\overline{k} + t\sigma^2\right); \quad k < 0. \tag{11.32}$$

Thus, the rate of decline in mean $^{131}I$ concentration in milk is expected to decrease as time increases. Suppose it was assumed that the dynamics of a population could be accurately modeled using the dynamics derived for an individual. The difference between the aggregated mean of the model prediction (dashed line) and the

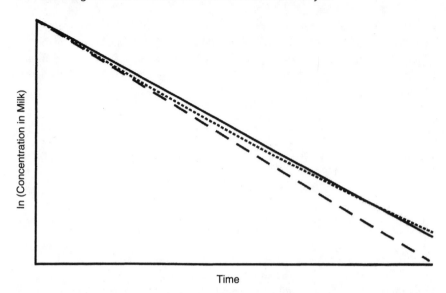

Figure 11.15 The dynamics of $^{131}$I in milk are simulated using the average of six rate constants measured by Sasser (1965) (dashed line) and by simulating milk concentrations for each of the six cows and averaging the milk concentrations (solid line). The dotted line is an approximation derived for the average concentration. The logarithm of the average concentration is a nonlinear function of time. The logarithm of the concentration computed using the average rate constant is a linear function of time.

observations (solid line) at some point in time might be assumed to be due to bias in the parameter values that must be accounted for by inflating the uncertainty in $k$. If the model had been designed to simulate individual cows in agreement with the scale of the experimental data, then there would have been no perceived need to inflate the uncertainty of $k$ to account for model bias. Likewise, if the modeler had recognized that the calculus of random variables could not be assumed to be the same as the calculus of simple variables, then a proper formulation for simulating the aggregate variable might have been derived and the model bias eliminated. If the appropriate design for a model cannot be implemented, then it may become necessary to inflate uncertainties to account for model bias.

### Deriving Distributions from Data

The parameters for an appropriate distribution can be estimated if sufficient data are available. The mean and variance are typical parameters of interest, but it is important to consider the type of distribution to be fit to the data. For example, some distributions, such as the Poisson and exponential distributions, have only a single parameter. Others, such as the hypergeometric or truncated normal, may require three or more parameters. One approach to choosing a distribution is to simply select the distribution that best fits the data. It often takes a large number of observations to distinguish the quality of fit of different distributions to a collection of data, especially when the coefficient of variation is large (Haas 1997). Sampling errors can also influence selecting a distribution. Hattis and Burmaster (1994) argue that it is important to evaluate what type of distribution is expected based on the processes that give rise to the variability in the model parameter. In this way, the appropriate distribution can be selected even when data are scarce.

## Estimating Parameters of a Distribution

There are several possible approaches to fitting a distribution to data. The three most frequently used methods are moment matching, maximum likelihood estimation, and least squares minimization. The method of moment matching is based on computing the moments for the data and then using those moments as estimates for the moments of the assumed distribution. For example, estimating the mean and variance of the data provides estimates of the parameters for a normal distribution, where the mean alone is sufficient for a Poisson. The types of distribution to fit can sometimes be narrowed down by using the third and fourth moments, that is, the "shape" parameters $\beta_1$ and $\beta_2$ ($\beta_1$ and $\beta_2$ are the third and fourth central moments normalized to the variance of the distribution). Hahn and Shapiro (1967) present a graph showing the regions defined by $\beta_1$ and $\beta_2$ to which some typical distributions are limited.

The method of maximum likelihood estimation is the most robust of the methods to fit a distribution to data, but it can be difficult to implement. The maximum likelihood estimator for a distribution having one parameter, $\Theta$, is defined to be

$$L(\Theta) = \prod_{i=1}^{n} f(x_i), \qquad (11.33)$$

where the $x_i$ are the data and $f(x_i)$ is the frequency of the data in the sample. The method involves choosing $\Theta$ to maximize $L(\Theta)$, which is equivalent to maximizing the probability of observing the data in the sample (Pooch and Wall 1993).

The method of least squares involves choosing the parameters for the distribution that minimize the sum of the squared differences between the sample data and the values expected for the distribution. The method of least squares could be used to fit a PDF or a CDF to a set of data (Pooch and Wall 1993).

Estimating the parameters of a distribution is only the first step in considering the uncertainty associated with the data. Other uncertainties that could influence the interpretation of the data should also be evaluated. For example, systematic errors are not identified by the replicated sampling used in most experiments, but they could possibly be estimated by experts. In small samples, low-frequency events may have never been observed but could be accounted for by choosing an appropriate distribution.

## Using Limited Data

When data are very sparse, you may need to assume a distribution for a variable. A triangular distribution is frequently used when the true distribution is unknown

 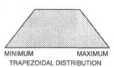 

Figure 11.16 Triangular, uniform, trapezoidal, and empirical distributions can be used when the exact form of a distribution is unknown.

but the extremes and a mean or modal value of the distribution can be estimated (Hodak 1994) (figure 11.16). The trapezoidal distribution is similar to the triangular distribution but does not have the emphasis of a single modal value (Seiler and Alvarez 1996). When only the maximum and minimum can be estimated, a uniform distribution is the least biased assumption. All values within a uniform distribution are equally likely. When uncertainties are large, log-triangular, log-trapezoidal, and log-uniform distributions should be considered. In these distributions, it is the logarithms of the data that are triangularly, trapezoidally, or uniformly distributed.

Seiler and Alvarez (1996) make a valid argument for using the beta distribution, or similar flexible distribution, in preference to the triangular, trapezoidal, or uniform choice when data are sparse. Their argument centers on the fact that these distributions are hardly equivalent to implying that no knowledge or extremely limited knowledge about the distribution exists. This is because (1) the endpoints must be selected, (2) points of discontinuity must be identified in the case of the triangular and trapezoidal distributions, and (3) the PDF based on the endpoints and discontinuities specifies the probability of specific values. The beta distribution can be parameterized to extend from a uniform distribution to a monomodal peaked distribution. Chan (1993) proposed another flexible function, the Bezier curve, to model distributions.

Occasionally, sufficient data exist to show that the distribution for a parameter cannot be well categorized into a typical statistical distribution. It is possible to generate samples from empirical distributions. Caution should be used, however, when sampling empirical distributions. One obvious danger in using empirical data is that the completeness of the distribution is correlated with sample size. Extreme but infrequent values may not be observed, yet such values may have profound influences on model predictions. In addition, empirical distributions may suffer from bias inherent in the data collection procedure.

When data are scarce and distributions are assigned on a subjective basis, assumptions about the form or the parameters for a distribution may be questioned. One strategy for evaluating those questions is to examine the impact of using alternative distributions. In many cases, particularly with complex models, the shape of the input distribution will have little impact on the form of the output distributions. In some cases, even substantial changes in the parameters of a distribution, such as the mean and variance, will have little impact simply because the model is not sensitive to that variable.

## Using Expert Elicitation

The most controversial method of assigning distributions to model inputs is undoubtedly the use of expert elicitation to identify parameters and their uncertainties when data are scarce or unavailable. Critics question whether opinions, even those of experts, deserve consideration in a quantitative assessment. Certainly, the use of expert opinion can lead to biases, over- or underestimation of uncertainties, or other errors. The alternative to using expert opinion is often to reject performing a quantitative assessment of risks or to ignore the uncertainty in some model inputs and, thus, overstating confidence in model results.

A basic conflict exists between most scientists and risk managers with regard to estimating and interpreting doses and their uncertainties. The goal of seeking the truth (accuracy) leads scientists to focus on minimizing type I errors (rejecting a null hypothesis when it is true), often without regard to type II errors (not rejecting a null hypothesis when it is false) (Eberhardt and Thomas 1991). Type I errors are also called false positive errors, and type II errors are called false negative errors. Scientists are likely to focus on type I errors because their probability can often be evaluated objectively; type II errors must often be evaluated subjectively (Zar 1984). On the other hand, risk managers often are concerned with minimizing the chance that an assessment will falsely conclude that there is no significant risk, that is, with minimizing type II errors (Eberhardt and Thomas 1991). Risk assessors must often try to satisfy the proponents of both viewpoints when extrapolating from experimentally determined uncertainties to the environmental uncertainties required for projecting potential risks. Scientific credibility requires that dose estimates be as accurate as possible; public acceptability of a study depends on establishing confidence that doses are not underestimated. Conservative dose estimates may be acceptable to the public, unacceptable to scientists because they are likely to result in false positive estimates, and unacceptable to risk managers because they may result in unnecessary destructive and costly cleanup activities.

Scientific training tends to focus on an experimentalist view of statistics, in which scientists test hypotheses about the means of treatments and controls, and variability is an inconvenience that they try to minimize through careful experimental design. Uncertainty in the mean can be reduced by increasing the number of replicates in the experiment. Systematic error, or bias, cannot similarly be reduced. Hence, eliminating bias is an important concern to most scientists. Focusing on eliminating systematic error during model construction tends to increase model complexity at the expense of model uncertainty (figure 11.17) (Gardner et al. 1980; Mosleh and Bier 1992). To ensure a low probability of making type II errors, researchers tend to use conservative (large) estimates of the uncertainty for parameters, even though the likelihood is small that all the parameter uncertainties would be maximized (but see Schlyakhter 1994; Henrion and Fischhoff 1986). When eliciting information from scientific experts, you should be aware of the inherent differences in goals that can exist between the scientific and the risk assessment communities.

**494** Radiological Risk Assessment and Environmental Analysis

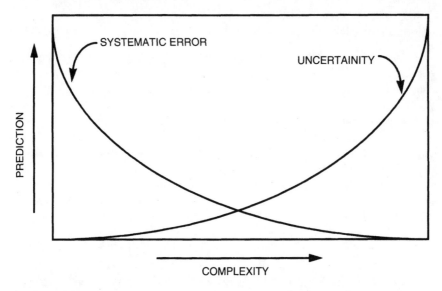

Figure 11.17 Increasing the complexity of a model initially improves its accuracy. However, uncertainty increases with complexity as less well-known parameters and processes are added to the model.

A thorough review of the literature on expert elicitation and decision analysis is beyond the scope of this chapter. You can find reviews of this subject area in Tversky and Kahnemann (1974), Morgan and Henrion (1992), Kahnemann et al. (1982), Meyer and Booker (1991), and Wright and Ayton (1994). Because of the prevalence of using subjective methods in risk assessment, a few examples of some of the potential pitfalls that can be encountered when using subjective methods for assigning uncertainty are presented here.

You should recognize that both the way questions are posed and the information that experts have accumulated can introduce bias into elicited responses. For example, Tversky and Kahnemann (1971) show that people tend to expect the behavior observed in large samples to be reflected in small samples, as well (representativeness bias). Thus, what appear to be unusual patterns in data (e.g., heads, heads, heads, tails, tails, tails in a coin toss experiment) can be elevated in importance by people even though the patterns are not statistically significant. This tendency, called "the law of small numbers" by Tversky and Kahnemann, is found even among technical people with statistical training.

Another tendency among people is to weight the estimates of frequency of events by their perceived importance (motivational bias) or the ease with which they can recall previous occurrences of the event or can imagine the event to occur (availability bias) (Morgan and Henrion 1992). For example, unusual events are likely to be reported more widely than common events and thus be subject to inflated frequencies. Tversky and Kahnemann (1973) and Lichtenstein et al. (1978) have conducted experiments that quantitatively demonstrate that familiarity biases perceived frequencies. Furthermore, if a known frequency of an event (the anchor point) is presented when eliciting estimates of frequencies of similar events, people tend to use those estimates to help scale their responses (anchoring and adjustment bias). Errors in the anchor point tend to bias the estimated frequencies in the same direction, and frequencies are also biased toward the anchor point (Kahnemann and Tversky 1973). Even when the questions are posed to technically trained people, researchers have found that base rates tend to be ignored or misapplied when eliciting frequencies of events. For example, Eddy (1982) posed the following question to physicians:

> The prevalence of breast cancer is 1% (in a specified population). The probability that a mammogram is positive if a woman has breast cancer is 79%, and 9.6% if she does not. What is the probability that a woman who tests positive actually has breast cancer? _____ %.

About 95% of the physicians estimated the answer to be about 75%. However, the correct answer is about 8%, which can be more easily seen if the problem is converted from single event probabilities to frequencies (Gigerenzer 1994). Out of 1,000 women, 10 are expected to have cancer, of which eight will test positive. However, 95 women will test positive who do not have cancer. Hence, only 8/103 or about 8% of the women testing positive will really have cancer.

Many people seem to be subject to an error in logic called the conjunction fallacy. An example problem that has been used to demonstrate this error is called the Linda problem (Wright and Ayton 1994):

> Linda is 31 years old, single, outspoken, and very bright. She majored in philosophy. As a student, she was deeply concerned with issues of discrimination and social justice and also participated in antinuclear demonstrations. Which of these two alternatives is more probable?
>
> (a) Linda is a bank teller.
> (b) Linda is a bank teller and active in the feminist movement.

In previous experiments, 80–90% of the subjects of the experiment answered (b), regardless of the fact that (b) is the product of the probability of (a) and the probability that Linda is active in the feminist movement and thus cannot be larger than (a) (Gigerenzer 1994). This tendency has been postulated to be associated with the representative heuristic, which is a nonstatistical decision-making device people employ. Using the representative heuristic, people judge alternatives by seeking the greatest similarity. Gigerenzer (1994) points out that the proportion of subjects that commit the conjunction fallacy is greatly diminished if the problem is restated to add:

> There are 100 people who fit the description above. How many of them are:
>
> (a) bank tellers
> (b) bank tellers and activists in the feminist movement

Apparently stating the problem in this form, as a frequency problem, negates the impact of the representative heuristic and helps people identify the inconsistencies that would arise if the conjunction fallacy were to be committed.

The risk assessment literature over the last few years has expressed concern that scientists generally tend to underestimate the true uncertainties when reporting their results (overconfidence bias) (e.g., Schlyakhter 1994; Henrion and Fischhoff 1986). Although it appears that scientists do tend to overestimate their confidence in their data, it also appears that the problem has been overstated for some studies and based on improper interpretation of published uncertainties in others. For example, it has been pointed out that over the years, physicists' measurements of the speed of light have routinely fallen outside the confidence intervals from previous experiments (Henrion and Fischhoff 1986). In this case, the published values of uncertainty represented the precision of the measurements made using a particular methodology and did not (and could not) represent the potential systematic error in the measurements. Another example involves uranium levels near a federal facility for which uncertainties were purportedly underestimated (Schlyakhter 1994). In this case, the author mistakenly assumed that the standard deviations reported for the data were accurate values of the population standard deviations, rather than estimates based on small sample sizes. Had the author used tolerance intervals to take into account sample size effects, he would not have been able to conclude that the uncertainties were underestimated.

The critics imply that appropriate estimates of uncertainty should be made sufficiently large so as to represent the combination of systematic and random errors. Although scientists can objectively assess measurement and random errors, estimating potential systematic error is generally considered subjective, speculative, and inappropriate for most scientific publications. Hence, risk assessors must often use meta-analysis and elicitation methods in addition to published data to derive estimates of uncertainty that are appropriate for their objectives. Furthermore, they must recognize that there are confidence intervals associated with published estimates of both means and variances that should not be ignored when using data for risk assessment.

Although research into the psychology of statistical reasoning and decision making by people has identified many potential problems, it is also true that much research has been aimed at the general population rather than at scientists or experts in particular fields. Motivated people who were knowledgeable in the subject area of the questions tended to perform better than the general public in experiments designed to test for biases and the use of nonstatistical decision-making rules (Beach and Braun 1994). Thus, you should be cautious when dealing with expert opinion but should not automatically reject the use of expert elicitation.

Formal protocols for eliciting expert judgment can help to ensure that the potential for cognitive biases is considered during the elicitation process. One such protocol involves 10 steps (Hora and Iman 1989; NRC 1990; NCRP 1996):

1. Selection of issues and associated parameters
2. Selection of experts
3. Preparation of background information
4. Training of experts in the elicitation method
5. Presentation of issues, questions, and background information to experts
6. Analysis of issues, questions, and background information by experts

7. Discussion of analyses of issues and background information by experts
8. Elicitations of judgment from the experts
9. Consolidation (and possibly aggregation) of the elicitation results
10. Review and communication of the elicitations

These steps include the features of other formal elicitation protocols, such as the Stanford/SRI Assessment Protocol (for a discussion of this and other protocols, see Morgan and Henrion 1992). The five stages of the Stanford/SRI protocol are motivation, structuring, conditioning, encoding, and verifying. The motivational stage introduces the nature of the problem and the expectations for the experts, including a discussion of potential motivational bias. The structuring phase proceeds to establish unambiguous definitions for the information to be elicited. The conditioning phase attempts to illuminate the cognitive processes of the experts for the benefit of both the expert and the analyst. This phase would include discussions about the various cognitive biases that could affect the experts' responses. The encoding phase seeks to generate a subjective probability distribution using the experts' knowledge. The verifying phase involves discussing the results with the experts to check for consistency.

## Methods of Propagation

Uncertainty is propagated through models using three common methods: analytical methods using mathematical statistics, mathematical approximation techniques, and Monte Carlo methods. Analytical and mathematical approximation techniques are usually feasible only for relatively small models of limited complexity and are generally used only when there is no time-dependent stochasticity. The mathematics of the analytical method can be tedious and difficult. Discontinuities in model functions or the use of time-dependent driving variables makes the analytical and mathematical approximation techniques difficult or impossible to use. Analytical methods are frequently used in metrology, and understanding these methods can give considerable insight into the kinds of problems that can arise in Monte Carlo analyses. Monte Carlo methods are generally easy to implement, and several commercial applications exist that can facilitate such analyses.

### Analytical Methods

Analytical methods are generally applied to relatively simple functions of random variables. Analytical methods can be used to find the expectations of functions of random variables or to find the distribution of functions of random variables. The expectations of functions, particularly the mean and variance, are usually of most interest, and methods for computing these moments are presented below. Finding the distributions of functions of random variables, that is, the CDF or PDF of the distribution, can be mathematically more challenging. There are three techniques that can be used: the CDF technique, the moment-generating function technique, and the transformation technique. These methods go beyond the scope of this chapter but can be found in most texts on statistical theory (e.g., Mood et al. 1974).

Equation 11.4 can be used to compute the expectations of random variables. As an example, consider computing the mean for the exponential of a uniformly distributed random variable. The mean is

$$E[e^X] = \int_a^b e^x \frac{1}{b-a} dx = \frac{e^b - e^a}{b-a}, \tag{11.34}$$

where $1/(b-a)$ is the PDF of a uniformly distributed variable having a minimum of $a$ and a maximum of $b$ and the limits of integration represent the range of $f(x)$. The function of the mean would be entirely different if the random variable had another distribution.

### Sum and Difference of Random Variables

Analytical methods to propagate uncertainty give exact solutions for the expectation and variance for the sum, difference, and product of random variables, independent of the distribution of those variables, but they can only give estimates for quotients. The formulas for these arithmetic operations are found in Mood et al. (1974):

$$E\left[\sum_{i=1}^n X_i\right] = \sum_{i=1}^n E[X_i] \tag{11.35}$$

and

$$\text{var}\left[\sum_{i=1}^n X_i\right] = \sum_{i=1}^n \text{var}[X_i] + 2\sum_{i=1}^{n-1}\sum_{j=i+1}^n \text{cov}(X_i, X_j) \tag{11.36}$$

for a sum, and

$$\text{var}\left[\sum_{i=1}^n X_i\right] = \sum_{i=1}^n \text{var}[X_i] - 2\sum_{i=1}^{n-1}\sum_{j=i+1}^n \text{cov}(X_i, X_j) \tag{11.37}$$

for a difference. If the variables in the sum or difference are independent, the formula for the variance reduces to

$$\text{var}\left[\sum_{i=1}^n X_i\right] = \sum_{i=1}^n \text{var}[X_i] \tag{11.38}$$

(i.e., the variance of a sum of independent random variables is equal to the sum of the variances).

## EXAMPLE 11.1

Let $I_{\text{milk}}$ be the integrated intake of $^{137}$Cs in milk and $I_{\text{eggs}}$ be the integrated intake of $^{137}$Cs in eggs. Assume that you can express the uncertainty in the integrated intakes as a mean and variance and that the integrated intakes can be assumed to be independent because the foods came from different geographic locations. The expected value of the total integrated intake from these two foods, $T$, would be

$$E[T] = E[I_{\text{milk}}] + E[I_{\text{eggs}}] + \mu_{I_{\text{milk}}} + \mu_{I_{\text{eggs}}}, \quad (11.39)$$

and its variance would be

$$\text{var}[T] = \text{var}[I_{\text{milk}}] + \text{var}[I_{\text{eggs}}] = \sigma^2_{I_{\text{milk}}} + \sigma^2_{I_{\text{eggs}}}. \quad (11.40)$$

### Product of Random Variables

The expected value of a product depends not only on the mean values of the terms but also on their covariance:

$$E[XY] = \mu_X \mu_Y + \text{cov}[X, Y]. \quad (11.41)$$

The variance depends on the means, variances, and covariances of the terms:

$$\begin{aligned}\text{var}[XY] &= \mu_Y^2 \text{var}[X] + \mu_X^2 \text{var}[Y] + 2\mu_X \mu_Y \text{cov}[X, Y] \\ &\quad - (\text{cov}[X, Y])^2 + E\left[(X-\mu_X)^2 (Y-\mu_Y)^2\right] \\ &\quad + 2\mu_Y E\left[(X-\mu_X)^2 (Y-\mu_Y)\right] + 2\mu_X E\left[(X-\mu_X)(Y-\mu_Y)^2\right].\end{aligned} \quad (11.42)$$

If $X$ and $Y$ are independent,

$$E[X, Y] = \mu_X \mu_Y. \quad (11.43)$$

and

$$\text{var}[X, Y] = \mu_Y^2 \text{var}[X] + \mu_X^2 \text{var}[Y] + \text{var}[X]\text{var}[Y]. \quad (11.44)$$

## EXAMPLE 11.2

Let $I_{\text{milk}}$ be the integrated intake of $^{137}$Cs and DCF be the dose conversion factor for a whole-body dose from ingested $^{137}$Cs. If both $I_{\text{milk}}$ and DCF are assumed to have uncertainties that can be expressed as means and variances, then the uncertainty on dose would be

$$E[D_{\text{whole body}}] = \mu_{I_{\text{milk}}} \times \mu_{\text{DCF}}, \quad (11.45)$$

and the uncertainty in the expected value would be

$$\text{var}[D_{\text{whole body}}] = \mu_{\text{DCF}}^2 \sigma^2_{I_{\text{milk}}} + \mu_{I_{\text{milk}}}^2 \sigma^2_{\text{DCF}} + \sigma^2_{\text{DCF}} \sigma^2_{I_{\text{milk}}}. \quad (11.46)$$

## Quotient of Random Variables

The mean and variance for the quotient of random variables do not have exact solutions. An approximation of the solution, obtained using a Taylor series expansion, is

$$E\left[\frac{X}{Y}\right] \approx \frac{\mu_X}{\mu_Y} - \frac{1}{\mu_Y^2}\text{cov}(X,Y) + \frac{\mu_X}{\mu_Y^3}\text{var}[Y] \tag{11.47}$$

and

$$\text{var}\left[\frac{X}{Y}\right] \approx \left(\frac{\mu_X}{\mu_Y}\right)^2 \left(\frac{\text{var}[X]}{\mu_X^2} + \frac{\text{var}[Y]}{\mu_Y^2} - \frac{2\,\text{cov}[X,Y]}{\mu_X\mu_Y}\right). \tag{11.48}$$

If $X$ and $Y$ are independent,

$$E\left[\frac{X}{Y}\right] \approx \frac{\mu_X}{\mu_Y} + \frac{\mu_X}{\mu_Y^3}\text{var}[Y] \tag{11.49}$$

and

$$\text{var}\left[\frac{X}{Y}\right] \approx \left(\frac{\mu_X}{\mu_Y}\right)^2 \left(\frac{\text{var}[X]}{\mu_X^2} + \frac{\text{var}[Y]}{\mu_Y^2}\right). \tag{11.50}$$

### EXAMPLE 11.3

Suppose that you want to estimate the initial concentration of $^{137}$Cs on vegetation following a deposition event. You are given the total deposition and the fraction of the total deposition remaining on the surface of the plants. Using the formula for a product, given above, you have computed the mean and variance for the amount of $^{137}$Cs on the surface of the plants, $\mu_{\text{surface}}$ and $\sigma^2_{\text{surface}}$. To estimate the concentration, you must divide the amount of $^{137}$Cs by the biomass of the vegetation, which is also expressed as a mean and variance. Assume the amount of $^{137}$Cs intercepted by the vegetation is independent of the biomass. The estimate for the concentration of $^{137}$Cs in vegetation is

$$E[C_{\text{veg}}] = \frac{\mu_{\text{surface}}}{\mu_{\text{biomass}}} + \frac{\mu_{\text{surface}}}{\mu_{\text{biomass}}^3}\sigma^2_{\text{biomass}}, \tag{11.51}$$

with the uncertainty on the concentration estimated by

$$\text{var}[C_{\text{veg}}] = \left(\frac{\mu_{\text{surface}}}{\mu_{\text{biomass}}}\right)^2 \left(\frac{\sigma^2_{\text{surface}}}{\mu_{\text{surface}}^2} + \frac{\sigma^2_{\text{biomass}}}{\mu_{\text{biomass}}^2}\right). \tag{11.52}$$

The propagation equations for these fundamental operations show how important covariance (correlation) among the variables can be. Covariance can affect both the mean and the variance of computed results. What these formulas do not express is the type of distribution that results from such operations, that is, the PDF or CDF for the distribution. As discussed above, the PDF and CDF can sometimes be derived using other mathematical techniques.

## Formulas for Normal and Lognormal Distributions

The formulas for estimating the expected values and variances for the arithmetic functions of random variables shown above do not require assumptions about the types of distributions associated with the variables. Often, however, the parameters will be assigned to have normal or lognormal distributions. These assumptions are sometimes made because experiments show the distribution is reasonable or because of theoretical or computational considerations. The following theorems are useful in propagating uncertainties when dealing with normal or lognormal distributions.

### Linear Operations

Let $X$ be a normally distributed random variable with mean $\mu_X$ and variance $\sigma_X^2$. This can be denoted as

$$X \sim N\left(\mu_X, \sigma_X^2\right). \tag{11.53}$$

Linear combinations of normally distributed random variables are also normally distributed:

$$\sum_1^n a_i X_i \sim N\left(\sum_1^n a_i \mu_i, \sum_1^n a_i^2 \sigma_i^2\right). \tag{11.54}$$

Adding a constant to normally distributed values shifts the mean but has no impact on the variance:

$$C + X \sim N\left(\mu_X + C, \sigma_X^2\right). \tag{11.55}$$

### Geometric Means and Standard Deviations

Assume $X$ has a lognormal distribution and let $m$ equal the mean of the logs of the $x_i$:

$$m = \mu_{\ln(x)}. \tag{11.56}$$

Let $v$ equal the variance of the logs of the $x_i$:

$$v = \sigma^2_{\ln(x)}. \tag{11.57}$$

The geometric mean for the distribution of $X$, $GM_X$, is

$$GM_X = e^m = e^{\mu_{\ln(x)}}, \tag{11.58}$$

and the geometric standard deviation, $GSD_X$, is

$$GSD_X = e^{v^{1/2}} = e^{\sigma_{\ln(x)}}. \tag{11.59}$$

The formulas for the calculation of expected values always deal with arithmetic means and variances. To convert from geometric means and standard deviations to arithmetic means and variances, use the following formulas:

$$\mu_X = e^{\left\{ \ln(GM_X) + \frac{[\ln(GSD_X)]^2}{2} \right\}}, \qquad (11.60)$$

$$\sigma_X^2 = e^{(2\{\ln(GM_X)+[\ln(GSD_X)]^2\})} - e^{\{2\ln(GM_X)+[\ln(GSD_X)]^2\}}, \qquad (11.61)$$

$$GM_X = e^{\left[\ln(\mu_X) - \frac{1}{2}\ln(\sigma_X^2/\mu_X^2 + 1)\right]} = \frac{\mu_X}{\sqrt{\frac{\sigma_X^2}{\mu_X^2} + 1}}, \qquad (11.62)$$

$$GSD_X = e^{\sqrt{\ln(\sigma_X^2/\mu_X^2 + 1)}}. \qquad (11.63)$$

### Mathematical Approximation Techniques

Approximations of the expected value and variance for a function can be computed from the means and variances of the parameters and the partial derivatives of the function at the mean values of the parameters. This method, sometimes called the delta method, is based upon a Taylor series expansion of the functions.

### Mean

The approximation for the mean of a function is

$$E[f(X_1, X_2, \ldots, X_n)] \approx f(\mu_1, \mu_2, \mu_3, \ldots, \mu_n) +$$

$$\frac{1}{2}\sum_{j=1}^{n}\left[\frac{\partial^2 f}{\partial X_j^2}\right] \cdot \text{var}[X_j] + \sum_{j=1}^{n}\sum_{j'>j}^{n}\left[\frac{\partial^2 f}{\partial X_j \partial X_{j'}}\right] \cdot \text{cov}[X_j, X_{j'}]. \qquad (11.64)$$

If all of the variables are independent, then

$$E[f(X_1, X_2, \ldots, X_n)] \approx f(\mu_1, \mu_2, \mu_3, \ldots, \mu_n) + \frac{1}{2}\sum_{j=1}^{n}\left[\frac{\partial^2 f}{\partial X_j^2}\right] \cdot \text{var}[X_j]. \qquad (11.65)$$

$\partial^2 f/\partial X_j^2$ is the second derivative of $f$ with respect to $X_j$ evaluated at $\mu_j$.

### Variance

The approximation for the variance of a function is

$$\text{var}[f(X_1, X_2, \ldots, X_n)] \approx \sum_{j=1}^{n}\left[\frac{\partial f}{\partial X_j}\right]^2 \cdot \text{var}[X_j] + 2\sum_{j=1}^{n}\sum_{j'>j}^{n}\left[\frac{\partial}{\partial X_j}\right] \cdot \text{cov}[X_j, X_{j'}]. \qquad (11.66)$$

Estimating and Applying Uncertainty in Assessment Models   503

If $X_j$ and $X_{j'}$ are independent, then $\text{cov}[X_j, X_{j'}] = 0$:

$$\text{var}\,[f\,(X_1, X_2, \ldots, X_n)] \approx \sum_{j=1}^{n}\left\{\left[\frac{\partial f}{\partial X_j}\right]^2 \cdot \text{var}\,[X_j]\right\}. \qquad (11.67)$$

This method of approximation is often used to estimate the mean and variance of functions of random variables (table 11.4).

## Example 11.4

Consider the function $e^Y$ where $Y$ is a distribution having mean $\mu_Y$ and variance $\sigma_Y^2$. The partial derivatives are

$$\frac{\partial e^Y}{\partial Y} = e^Y, \qquad (11.68)$$

$$\frac{\partial^2 e^Y}{\partial Y^2} = e^Y, \qquad (11.69)$$

Table 11.4  Approximations of the mean and variance for several frequently used mathematical functions[a]

| Function | Mean | Variance |
|---|---|---|
| $x^* = e^y$ | $\mu_x = e^{\mu_y + \frac{\sigma_y^2}{2}}$ | $\sigma_x^2 = e^{2\mu_y + 2\sigma_y^2} - e^{2\mu_y + \sigma_y^2}$ |
| $x = \ln(y)$ | $\mu_x = \ln(\mu_y) - \frac{\sigma_y^2}{2\mu_y^2}$ | $\sigma_x^2 = \frac{\sigma_y^2}{\mu_y^2}$ |
| $x^{**} = \ln(y)$ | $\mu_x = \ln(\mu_y) - \frac{1}{2}\ln\left(\frac{\sigma_y^2}{\mu_y^2} + 1\right)$ | $\sigma_x^2 = \ln\left(\frac{\sigma_y^2}{\mu_y^2} + 1\right)$ |
| $x = \sin(y)$ | $\mu_x = \sin(\mu_y) \cdot \left(1 - \frac{\sigma_y^2}{2}\right)$ | $\sigma_x^2 = \sigma_y^2 \cdot [\cos(\mu_x)]^2$ |
| $x = \cos(y)$ | $\mu_x = \cos(\mu_y) \cdot \left(1 - \frac{\sigma_y^2}{2}\right)$ | $\sigma_x^2 = \sigma_y^2 \cdot [\sin(\mu_x)]^2$ |
| $x = \sqrt{y}$ | $\mu_x = \sqrt{\mu_y} - \frac{1}{8}\mu_y^{-\frac{3}{2}} \cdot \sigma_y^2$ | $\sigma_x^2 = \frac{\sigma_y^2}{4\mu_y}$ |
| $x = c^y$ | $\mu_x = c^{\mu_y} + \frac{1}{2}\sigma_y^2 \ln(c)^2\, c^{\mu_y}$ | $\sigma_x^2 = \sigma_y^2 c^{2\mu_y}\ln(c)^2$ |
| $x^* = c^y$ | $\mu_x = e^{\ln(c)\mu_y + \frac{\ln(c)^2 \sigma_y^2}{2}}$ | $\sigma_x^2 = e^{2\ln(c)\mu_y + 2\ln(c)^2\sigma_y^2}$ $-e^{2\ln(c)\mu_y + \ln(c)^2\sigma_y^2}$ |
| $x = y^c$ | $\mu_x = \mu_y^c + \frac{1}{2}c\,(c-1)\,\mu_y^{c-2}\sigma_y^2$ | $\sigma_x^2 = \sigma_y^2 c^2 \mu_y^{2(c-1)}$ |

[a] The equations for the expressions marked with * or ** assume that the distribution for y is normal or lognormal, respectively, and give exact results for that case. The remaining equations are first-order approximations derived using a Taylor series.

so the estimators for the mean and variance are

$$\mu_X = e^{\mu_Y} + \tfrac{1}{2}e^{\mu_Y}\sigma_Y^2 = e^{\mu_Y}\left(1 + \tfrac{1}{2}\sigma_Y^2\right), \qquad (11.70)$$

$$\sigma_X^2 = e^{2\mu_Y}\sigma_Y^2. \qquad (11.71)$$

Compare these estimators with those in table 11.4, which were derived by assuming that $Y$ has a normal distribution.

## Propagation Using Interval Estimates

Occasionally it may be necessary to propagate uncertainty through mathematical expressions where the uncertainty is defined in terms of an interval, such as $X \pm \delta X$. Uncertainties in measurements are often reported in this format. The formulas for propagating such uncertainties through sums, differences, products, and quotients are similar to those for propagating means and variances (Taylor 1982).

### Sum and Difference

If several independent variables $X, \ldots, W$ are estimated with uncertainties expressed as the interval $\pm \delta X, \ldots, \pm \delta W$, then the sum and difference of the variables,

$$q = X + \cdots + Z - (U + \cdots + W), \qquad (11.72)$$

would have the uncertainty

$$\delta q = \sqrt{(\delta X)^2 + \cdots + (\delta Z)^2 + (\delta U)^2 + \cdots + (\delta W)^2}. \qquad (11.73)$$

If the variables show complete correlation, then the formula for propagation is simplified to

$$\delta q = \delta X + \cdots + \delta Z + \delta U + \cdots \delta W. \qquad (11.74)$$

### Products and Quotients

If several independent variables $X, \ldots, W$ are associated with small intervals of uncertainty $\pm \delta X, \ldots, \pm \delta W$, then the uncertainty on a product or quotient of the variables,

$$q = \frac{X \times \cdots \times Z}{U \times \cdots \times W}, \qquad (11.75)$$

is given by

$$\frac{\delta q}{|q|} = \sqrt{\left(\frac{\delta X}{|X|}\right)^2 + \cdots + \left(\frac{\delta Z}{|Z|}\right)^2 + \left(\frac{\delta U}{|U|}\right)^2 + \cdots + \left(\frac{\delta W}{|W|}\right)^2}. \qquad (11.76)$$

If the variables are completely correlated, then

$$\frac{\delta q}{|q|} = \frac{\delta X}{|X|} + \cdots + \frac{\delta Z}{|Z|} + \frac{\delta U}{|U|} + \cdots + \frac{\delta W}{|W|} \qquad (11.77)$$

## Other Functions

Uncertainty in the function of a variable, $q(X)$, where uncertainty in $X$ is denoted by the range $X \pm \delta X$, can be estimated by

$$\delta q(X) = \left|\frac{dq}{dX}\right| \delta X, \tag{11.78}$$

where $dq/dX$ is the derivative of $q$ with respect to $X$. In the case where the function is a power of $X$ (i.e., $q = X^n$), the fractional uncertainty in $q(x)$ is $|n|$ times the fractional uncertainty in $X$:

$$\frac{\delta q}{|q|} = |n| \frac{\delta x}{|x|} \tag{11.79}$$

(Taylor 1982). If $q$ is a function of two or more independent variables, then

$$\delta q(X,\ldots,Z) = \sqrt{\left(\frac{\partial q}{\partial X}\delta X\right)^2 + \cdots + \left(\frac{\partial q}{\partial Z}\delta Z\right)^2}. \tag{11.80}$$

### *Covariance and the Order of Operations*

One error that can be made when propagating uncertainties analytically is to ignore the covariance between two or more terms that results from having the same parameter in several terms. This situation can occur by simply doing legal algebraic operations on an expression. For example, take the case for computing

$$Y = A(X + Z). \tag{11.81}$$

Assume that $A, X$, and $Z$ are independent of one another. One way to compute the variance in $Y$ would be as follows:

1. Apply the formula for a sum to compute the variance of $X + Z$. Because the variables are independent, the covariance term can be ignored.
2. Apply the formula for a product using var[$A$] and var[$X + Y$], also assuming independence.

Notice what happens if the equation is rearranged:

$$Y = AX + AZ \tag{11.82}$$

It may seem reasonable to compute var[$AX$] and var[$AY$] using the formula for products, and then to apply the formula for a sum of independent variables to compute var[$Y$]. However, the terms $AX$ and $AY$ are no longer independent; they share the common factor $A$. The correct solution can be achieved when solved in this way by performing the operations as listed in steps 1 and 2 or by computing the covariance between $AX$ and $AZ$ and then using in the formula for a sum.

## Monte Carlo Methods

Monte Carlo methods can be applied to models of any size, but since they involve large numbers of simulations, they may be costly for large models. Given the relative ease of use and the power of inexpensive computers, Monte Carlo methods are probably the most frequently used for uncertainty analysis. The Monte Carlo method involves choosing parameter values using a random, or stochastic, selection scheme. Typically, parameter values are selected before the start of each simulation. Numerous simulations are run, and the values of the output variables for each of those simulations are saved for later analysis (figure 11.18). The values of the output variables may be saved one or more times during the simulation.

### EXAMPLE 11.5

Take a simple model, biological elimination of a long-lived radionuclide, and go through a few iterations of the Monte Carlo method. Ignoring radioactive decay for the moment, the differential equation for the model can be written as

$$\frac{dA}{dt} = -r(A), \tag{11.83}$$

for which you can obtain the analytical solution

$$A_t = A_0 e^{-rt}, \tag{11.84}$$

where $A$ is the amount of activity in the organism, and $r$ is the rate of elimination of the radionuclide from the body. Assume that $r$ is normally distributed with a mean of 0.2 and a standard deviation of 0.04. You should be aware that $r$ could not be normally distributed because the organism must always have a positive $r$. However, the chance of randomly choosing $r \leq 0$ is less than 1 out of 1,000.

The procedure starts by selecting 200 values of $r$ using the normal pseudorandom number generator. The first five values generated are

0.19471
0.23075

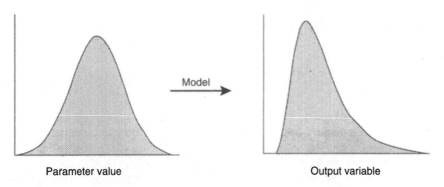

Figure 11.18 A simulation model maps the distributions of its parameters into a distribution of output variables.

0.23355
0.16258
0.16667

You must next choose the initial amount of activity, $A_0$, and the time, $t$, at which to sample the activity. For this example, an initial amount of activity of 1 μCi and a time of 10 days are chosen. Using the values of $r$ selected above, the analytical solution of the model at time 10 days yields

0.1426912
0.0995163
0.0096768
0.1967563
0.1888717

The model maps the domain of the parameter into the range of the state variable. There is a one-to-one correspondence between each value of $r$ and a value of $A_{10}$. After 200 simulations, the following statistics for the distribution were calculated:

Mean = 0.1432          Geometric mean = 0.1337

Variance = $3.033 \times 10^{-3}$   Geometric standard deviation = 1.447

To verify this solution, the analytical solution for the model is used to compute the uncertainty on $A_{10}$ given the distribution of $r$. The analytical solution is

$$A_t = A_0 e^{-rt}, \qquad (11.85)$$

which can be transformed to

$$\ln(A_t) = \ln(A_0) - rt. \qquad (11.86)$$

Remember that $r$ is normally distributed with $\mu = 0.2$ and $\sigma^2 = 0.0016$. Using the formulas for linear combinations of normal distributions,

$$\ln(A_t) \sim N[\ln(A_0) - \mu_r t, t^2 \sigma_r^2]. \qquad (11.87)$$

Because $\ln(A_t)$ is normally distributed, $A_t$ is lognormally distributed with a geometric mean of $e^{\ln(A_0) - \mu_r t}$ and a geometric standard deviation of $e^{t\sigma}$. Using the formulas to convert from geometric mean and geometric standard deviation to $\mu$ and $\sigma^2$,

$$\mu_{A_t} = e^{\left[(\ln A_0 - \mu_r t) + \frac{t^2 \sigma_r^2}{2}\right]}, \qquad (11.88)$$

$$\sigma_{A_t}^2 = e^{[2 \times (\ln A_0 - \mu_r t + t^2 \sigma^2)]} - e^{[2 \times (\ln A_0 - \mu_r t) + t^2 \sigma_r^2]} \qquad (11.89)$$

substituting the actual values gives

$$\mu_{A_t} = e^{\left\{[0-(0.2 \times 10)] + \frac{[100 \,(0.0016)]}{2}\right\}} = e^{-1.92} = 0.146607, \qquad (11.90)$$

$$\sigma_{A_t}^2 = e^{2(0-(0.2 \times 10)+100(0.0016))} - e^{2(0-(0.2 \times 10))+100(0.0016)}$$

$$= e^{-3.68} - e^{-3.84} = 3.72937 \times 10^{-3}, \qquad (11.91)$$

which are comparable to the Monte Carlo estimates.

## Generating Random Numbers

Pseudorandom number generators usually create a series of numbers to be used as input to a Monte Carlo uncertainty analysis. The numbers appear to be random in terms of statistics used to test for randomness. They are called pseudorandom numbers because they are generated using deterministic rules; hence, they give repeatable results. Random number generators have the property that they will always give the same sequence of numbers, and the sequence of random numbers will repeat after a certain number have been generated. This interval between repetitions of the sequence of numbers is often called the *period* for the generator. Good random number generators have long periods, a core uniform random number generator that yields a true rectangular distribution, and low levels of autocorrelation among the numbers they produce (Barry 1996). An $i$th-order autocorrelation is the correlation between values separated by $i-1$ values in the sequence.

## Potential Problems with Monte Carlo Methods

Monte Carlo methods are generally robust and relatively easy to implement. Circumstances can arise, however, that can lead to unpredictable results. One of these situations involves the use of distributions with infinite tails, such as the normal distribution. The presence of an infinite tail means it is possible, even if unlikely, that a value will be generated that is near or even equal to zero, that a negative value will be generated, or that a large positive or negative value will be produced. If the parameter cannot logically take on such values, then problems might arise. For example, a negative loss rate or a near-zero value in a denominator can easily produce undesired behavior. Truncating the input distribution at known limits can help to avoid this problem.

It is possible to specify input distributions that yield output distributions that cannot be usefully characterized. For example, the ratio of two variables having standard normal distributions results in a Cauchy distribution, which has tails that never converge to zero and hence has no moments (i.e., no mean or variance). Sample mean and variance can be computed from samples taken from a Cauchy distribution, but the mean and variance are unlikely to converge as sample size increases. In fact, variance may increase as sample size increases.

Random number generators on computers typically use a uniform number generator to produce one or more values that are then transformed into the appropriate distribution. Many of the uniform random number generators available on computer systems are written for fast execution rather than for generating good sequences of numbers. The length of the cycle on one common generator, the linear congruential method, is limited by the word size of the computer. Autocorrelation is also a problem with many standard random number generators supplied with complier libraries. Press et al. (1988) recommends a random number generator that uses shuffling to break up sequential correlations.

## Sampling Designs

An uncertainty analysis based upon Monte Carlo simulations requires that a sampling design be implemented. Simple random sampling is the least complicated design, but it is often the least efficient. Stratified designs, such as Latin hypercube (see below), can significantly reduce the number of simulations required, which can be important when dealing with complicated models or numerous scenarios.

### Simple Random Sampling

Monte Carlo simulations frequently employ simple random sampling of the distributions of the parameters. In simple random sampling, the distributions of the parameters are assumed to be independent, and values are chosen completely at random from within the distributions. Example 11.5, propagating uncertainty through a model of the biological elimination of a radionuclide, made use of simple random sampling. In some instances, you may know that parameters are correlated. Simulating correlated parameters is possible, but it requires the use of multivariate pseudorandom number generators or the introduction of correlations by manipulating the ordering of the values within samples of generated values (Iman and Conover 1982).

### Latin Hypercube Sampling

A problem with simple random sampling is that it usually takes many simulations to adequately sample most distributions. One approach to reducing the number of simulations is to use a stratified sampling scheme, such as the Latin hypercube design (McKay et al. 1979). Latin hypercube sampling has been shown to significantly reduce the number of simulations required to estimate the mean and variance of output variables. The design stratifies the distribution of each parameter into $n$ quantiles, each of which contains $100/n\%$ of the distribution. Thus, a value chosen at random from the distribution would have an equal probability of occurring in each quantile. Latin hypercube sampling is based upon a stratified sampling scheme in which, for $n$ simulations, each quantile of each simulation is sampled only once. Values are chosen at random from within the range defined for a selected quantile.

As an example of a Latin hypercube design, consider a Monte Carlo simulation that involves five parameters. Five quantiles are used in this example, as well. The process starts by constructing a five-by-five matrix. The columns of the matrix will identify simulations, and the rows will be parameters. Within each row, a number from 1 to 5 will be assigned to each cell (figure 11.19).

These numbers are assigned randomly without replacement so that a given number appears only once in a row. The numbers in the cells identify which quantile is to be sampled at the start of each simulation. For example, a 1 indicates that sampling is to be restricted to the quantile for 0 to 20%. After setting up such a matrix, a pseudorandom number generator is used to create a corresponding matrix of parameter values to be used in the simulations. New matrices are constructed before each set of five simulations. To avoid bias, the total number of simulations

|  | Simulation | | | | |
|---|---|---|---|---|---|
|  | 1 | 2 | 3 | 4 | 5 |
| $P_1$ | 3 | 5 | 4 | 2 | 1 |
| $P_2$ | 2 | 5 | 3 | 1 | 4 |
| Parameter $P_3$ | 5 | 4 | 1 | 3 | 2 |
| $P_4$ | 4 | 3 | 1 | 2 | 5 |
| $P_5$ | 1 | 2 | 5 | 3 | 4 |

Figure 11.19 Latin hypercube sampling design for five parameters and five simulations.

run should be a multiple of the number of quantiles (i.e., all of the simulations in every matrix should be run). Tests have been conducted comparing the results from simple random sampling and Latin hypercube sampling (McKay et al. 1979). These tests show that Latin hypercube sampling can produce good estimates of the mean and variance of the output distribution with significantly fewer simulations than simple random sampling.

A variation of the Latin hypercube design uses the midpoint of all quantiles rather than selecting a value at random from within the quantile (Morgan and Henrion 1992). This additional condition on sampling reduces the variance in the results, potentially reducing the number of simulations needed to adequately estimate the mean or variance of the distributions of outputs. The variation can introduce errors into the results if the model has periodic behavior with the period or frequency being controlled by a sampled parameter. This case is unlikely in most models used in risk assessments, with the possible exception of models that simulate the dynamics of natural populations.

The Latin hypercube sampling scheme has also been expanded to enable correlated sampling, even using marginal distributions of different types (Iman and Conover 1982). Spearman correlations between the variates cannot be ensured, but the method preserves rank order correlations. The modification is based on using a multivariate normal distribution to select the various quantiles for the variates and then mapping the normal deviates into the marginal distributions using an inverse transform function.

## Importance Sampling

Often in risk analysis, researchers are most concerned about the results that fall within one of the tails of the output distribution. Importance sampling is a highly stratified design that can be used to minimize the number of simulations by concentrating the sampling to those regions of the input parameter distributions that give rise to the extreme values in the results (Clark 1961; Fishman 1996). To increase the representation of results in the tails of the output distribution, the probability of sampling from within the critical regions of the input parameters is artificially increased. The importance weight is the degree to which the probability of sampling the critical input values is inflated. The importance weight of the output value is the product of the importance weights of all the inputs associated with that value. The output value can be restored to its actual probability by dividing by its importance weight. The design is complicated by the requirement that the weighting factors, as well as the values, must be propagated through the model. In addition, it can sometimes be difficult to identify the critical regions of the input distributions if the models are complex.

## Sampling Designs to Partition Variability and True Uncertainty

Hoffman and Hammonds (1994) argue that, for some questions, it is important to distinguish uncertainty due to natural variability from uncertainty due to lack of knowledge. Uncertainty can be partitioned between epistemic and aleatory uncertainties, presuming that the uncertainty on each input parameter can be attributed to either natural variability or lack of knowledge about a fixed but unknown value. In this case, the sampling scheme for the Monte Carlo simulations would nest the sampling for the parameters associated with natural variability within a sampling scheme for the parameters associated with lack of knowledge. For example, assume parameters A, B, and C have natural variability and parameters D, E, and F have constant but unknown values. A set of 100 Monte Carlo simulations could be run selecting only values of A, B, and C for each combination of values selected for D, E and F. If 100 sets of values for D, E, and F were selected, then a total of 10,000 simulations would need to be run, and 100 families of distributions for the output variable would be generated (figure 11.20). Hoffman and Hammonds also point out that, in some cases, the means and variances for the distributions of A, B, and C can also be classified as constant but unknown values and assigned to that category in the sampling scheme. This technique enables confidence bounds to be placed around the CDF for the output distribution, where the uncertainty in the CDF arises from the lack of knowledge about the fixed parameters. The nested sampling design required for these analyses can be cumbersome and time-consuming. Bogen (1995) suggests some innovative approaches for simplifying the joint assessment of uncertainty and variability.

### *Number of Simulations*

The choice of the number of simulations depends on the goals for the analysis and on practical considerations, such as the time required to perform the analyses. Fewer simulations are usually required to provide a good estimate of the mean of the output variables than the variance of their distributions. A good representation of the tails of an output distribution can take a very large number of simulations unless a method such as stratified sampling or importance sampling can be employed. The tails of a distribution are usually represented by low-frequency events, and thus a large sample size is needed to ensure good representation in the tails. For example, suppose that you are interested in the upper 5% of a distribution. Using simple

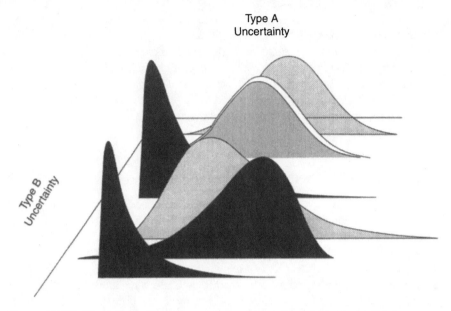

Figure 11.20 Monte Carlo simulation can be used to partition the total uncertainty among type A (natural variability) and type B (lack of knowledge) categories. In this example, there is a lack of knowledge about the appropriate distribution for a model input.

random sampling, the number of simulations, $n$, required to observe one value in the upper 5% of the distribution with 95% confidence is

$$0.95^n = 0.05; \tag{11.92}$$

$$n = \frac{\ln(0.05)}{\ln(0.95)} \cong 59. \tag{11.93}$$

This can be interpreted as saying that if you want to determine the position of the upper 5% quantile, with a 5% chance of the value being wrong, then you could perform 59 simulations, order the results, and use the largest value as the estimate of the 95th percentile. Although the largest value is expected to lie within the upper 5% quantile, it could occur anywhere within that quantile (i.e., it might lie within the upper 0.01% quantile). To reduce the chance to 1% of underestimating the 95th percentile requires 90 simulations.

Another method for determining sample size is to plot the output statistic of interest against the cumulative number of simulations (figure 11.21). For example, suppose that you are concerned with estimating the mean and variance of the output distribution to establish a confidence interval. In most situations, such estimates will stabilize as the number of simulations increases, with the mean usually stabilizing sooner than the variance. You can set up a series of sets of simulations, with $n$ simulations per set, and run enough sets to achieve a reasonably stable value of the variance. If the variance or mean fails to stabilize, you should critically examine the model structure and the parameter distributions to ensure that

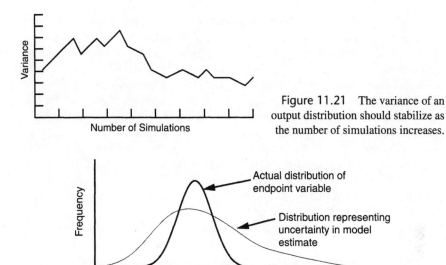

Figure 11.21  The variance of an output distribution should stabilize as the number of simulations increases.

Figure 11.22  The actual value of an endpoint if it is a single value, or the true distribution for an endpoint, should fall within the distribution representing the uncertainty in model predictions. The endpoint is $X$.

unreasonable parameter values are not being generated, such as values very near zero that are used in denominators, or negative values for parameters that logically must be positive.

### *Interpretation of the Output Distributions*

Although uncertainty analyses create distributions of outputs of models, interpreting those distributions is problematic and must be approached with caution. If true uncertainty is represented in the inputs, then the output distributions do not represent variability that you would expect to see in measurements of the states in the real system, nor are they guaranteed to encompass the range of responses of the real-world variables (figure 11.22). If the uncertainty on the parameters is estimated well, then it may be possible to claim that the uncertainty on an output should encompass either the true value or the variability expected in the real states, depending on whether the output represents a single value or a distribution, respectively. It must also be demonstrated, however, that the model simulates the real system adequately in terms of the objectives for the model (i.e., that the model is validated). Uncertainty analysis is not a substitute for model validation, but uncertainty can be used as a criterion for evaluating a model.

Three cases should be considered when evaluating distributions of model outputs: (1) that the distribution represents only variability (aleatory uncertainty), (2) that the distribution represents only true uncertainty (epistemic uncertainty), and (3) that the distribution results from an aggregation of both types of uncertainty.

The first case is extremely unlikely to be encountered in practice, but it establishes a frame of reference for the other two cases. For an example of aleatory uncertainty, consider computing the average tissue concentration of a herd of cattle in which each individual ingested a fixed, known amount of a contaminant as part of an experiment. If the masses of the cattle and the kinetics of uptake, degradation, and elimination of the contaminant in cattle are known, then a model could predict the tissue concentrations through time. In this hypothetical case, the interpretation of the distribution is straightforward and would be equivalent to the interpretation placed on distributions of measurements. The distribution would be equivalent to a frequency distribution, and thus you could make conclusions about the proportion of cattle showing various levels of tissue burdens. The mean, median, and other percentiles of the distribution would correspond to what would be expected from the distribution of actual measurement.

The second case is much more likely to be encountered in risk assessment. In this case, if there were no uncertainty, then the endpoint of the assessment would be represented by a single value rather than a distribution. For example, consider estimating the inhalation dose to an individual. Factors that would affect the dose estimate (e.g., environmental air concentrations, individual specific activity patterns, breathing rates, and clearance rates) would all have epistemic uncertainties regardless of whether those uncertainties were purely subjective estimates from experts or were derived from frequencies of actual measurements in people. The mean would be a probability-weighted estimate of the individual's dose. The uncertainty associated with the predicted dose would represent the assessor's level of confidence for the dose being a particular value. The distribution would not represent variability that would be expected in the doses of a population of similar individuals, nor would it guarantee that the range of responses in the real world would be encompassed by the distribution of estimated doses (figure 11.22). Hence, it would be inappropriate to conclude, for example, that 50% of people who have similar characteristics to the specific case would have doses less than the median. The median would be more appropriately identified as the dose estimate that the assessor thinks will exceed the true dose with 50% confidence. It is also inappropriate to conclude that 50% confidence means that the assessor expects the median estimate to exceed the true value in 50% of the cases where an estimate is made using this methodology. The subjective distributions associated with the inputs are not distributions representing random values.

The third case, with inputs having both aleatory and epistemic uncertainties, is also common in risk assessment. Aggregating variability and confidence (i.e., treating these sources of uncertainty as being equivalent) produces a distribution of risk or dose that confounds confidence and frequency. This distribution is therefore difficult to interpret. For example, the median value of a distribution is often used as a best estimate of dose or risk under the assumption that 50% of the estimates will lie below the median and 50% will lie above. If the distribution resulted from adding a distribution of confidence to a frequency distribution, then values near the median could be generated, for example, from adding an infrequent but large value to a small value associated with a low confidence of occurrence.

To illustrate this point in more detail, consider an example in which the total dose, $D$, is equal to the sum of two components, $A$ and $B$:

$$D = A + B. \tag{11.94}$$

The first component, $A$, is a constant having a value of 120. The second component, $B$, varies among individuals and is represented by a normal distribution having a mean of 50 and a standard deviation of 15. The true distribution of the sum has a mean of 170 and a standard deviation of 15 (figure 11.23).

Now suppose that the constant component is not known precisely but is estimated by an expert to be between 70 and 130, with 100 being the most likely value. A triangular distribution is used to represent the distributions of subjective uncertainty (figure 11.24).

The true value of 120 falls just below the 95th percentile of the triangular distribution having 70, 100, and 130 as its minimum, mode, and maximum, respectively. Thus, this example would be expected to bias the results toward lower values than the true values, but it would be acceptable in scientific terms because the subjective distribution includes the true value of the constant within the 95% tolerance interval.

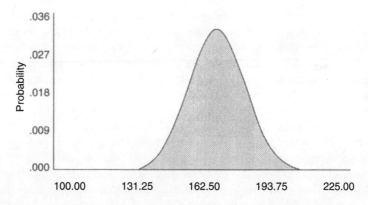

Figure 11.23  The "real" distribution of doses is a normal distribution with a mean of 170.

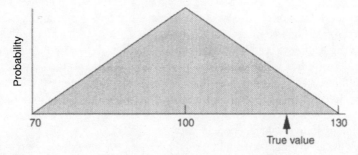

Figure 11.24  A triangular distribution was assigned to represent the subjective uncertainty in the constant A.

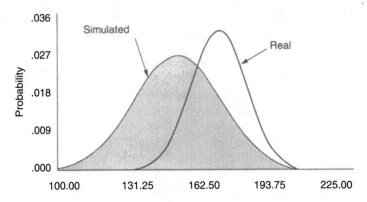

Figure 11.25 The distribution of simulated doses (shaded) lies to the left of the "real" distribution of doses.

The distribution of estimated doses simulated using a Monte Carlo method shows the impact of the uncertainty in $A$ (figure 11.25).

Because the distribution is symmetric, the mean and median of the simulated distribution are expected to be near the mode, which is to the left of the mode of the true distribution. The simulated distribution has greater variability (i.e., it is broader) than the true distribution because the uncertainty in $A$ is incorporated into the predictions. It is difficult to state what the modal value of the simulated distribution represents because it confounds the natural variability in factor $B$ with the confidence in the value assigned to $A$. Because this simple example is a linear equation and both distributions were symmetric, you could conclude that the mode of the true distribution is equally likely to be greater than or less than the mode of the simulated distribution. This is not equivalent to inferring that 50% of the doses lie on each side of the median of the simulated distribution.

The value associated with the 95th percentile of the simulated distribution is expected to result from a value of $A$ and $B$ near the upper tails of their respective distributions. Thus, you should have greater confidence that this value does not underestimate the true 95th percentile than in stating that the median of the simulated distribution does not underestimate the median of the true distribution (figure 11.26).

This example represents a worst-case but scientifically credible prediction because the most likely value of $A$ was assumed to be lower than its true value. Had the distribution assigned to $A$ ranged from 110 to 170 with a mode of 140, such that the true value was just within the lower bound of the 95% tolerance interval, then the distribution of simulated results would have been to the right of the true results (figure 11.27). Furthermore, had nested sampling been used (Hoffman and Hammonds 1994), you could have generated a family of CDFs, for example, to illustrate the impact of the epistemic uncertainty on the distributions of possible doses among individuals in a population.

Conclusions about risk should be drawn based on scientifically credible predictions of true values. You should strive to have a high degree of confidence that the value or range of values selected from a simulated distribution does not

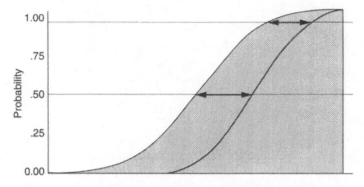

Figure 11.26 The difference between the cumulative frequency distribution for the simulated doses (shaded) and the true doses (line) decreases as cumulative probability increases.

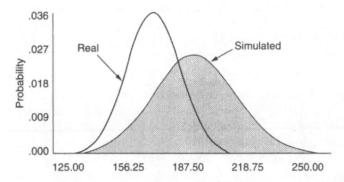

Figure 11.27 If the distribution of uncertainty on factor $A$ were biased toward larger values, then the distribution of simulated doses (shaded) would lie to the right of the real doses.

underestimate the true risks. When distributions of model inputs represent true uncertainty or confidence, you must endeavor to ensure that the distributions of results are properly interpreted. In particular, you should recognize that the mean or median of the output distribution is likely to be associated with only a moderate (about 50%) level of confidence.

## Sensitivity Analysis

Sensitivity analysis is used to assess the sensitivity of a state variable or a function of state variables to changes in a parameter's value. The results of sensitivity analysis may be used to

- Help identify the parameters that most influence the model outputs, thus requiring scrutiny when preparing an assessment
- Help identify processes or factors that can be safely ignored when constructing models

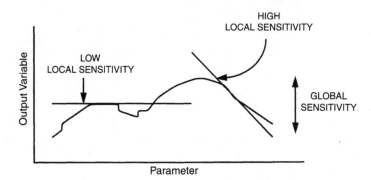

Figure 11.28 Local sensitivity represents the effect of a small change in a parameter around a particular value. Global sensitivity represents the effect of the parameter across its entire range.

- Build confidence that the model responds to perturbations in much the same way as the system being modeled (e.g., that it is not excessively sensitive or insensitive to its inputs)
- Help identify ways to reduce the uncertainty in the model outputs, thus improving its performance

Sensitivity analyses are typically conducted at two levels: local and global (Campolongo et al. 2000). Local sensitivity measures the impact of a parameter on model output in the region near a specific value (Miller 1974; Tomovic 1963; Tomovic and Vukobratovic 1972). Global sensitivity measures the impact of a parameter across its range of values (figure 11.28). Thus, an output variable can have a high local sensitivity and a low global sensitivity to a parameter if the parameter has a strong influence on the output variable but has relatively low variance or uncertainty.

## Local Sensitivity Analysis

Local sensitivity analysis is used to assess the effect of small perturbations in the values of a parameter around a particular value. For models represented by a single equation, you can quantify the local sensitivity of a parameter by finding the partial derivative of the function relative to the parameter and then substituting the nominal values of all parameters into the resulting equation. The nominal values chosen for the analysis are usually those considered to give the best model predictions for a particular situation.

The magnitude of the solutions of the differential equations depends upon the magnitude of the parameters. Normalization of the partial differentials can be used to adjust the sensitivity coefficient to help remove the effect of differences in magnitude among the various parameters, thus producing a sensitivity index or ratio; that is,

$$S_i = \frac{\partial y/y}{\partial x_i/x_i} = \frac{x_i}{y}\frac{\partial y}{\partial x_i}, \qquad (11.95)$$

where $y$ is the model output and the $x_i$ are the model parameters (EPA 2001).

Local sensitivity can also be estimated using numerical methods based on the assumption that the response surface of the output variable is effectively linear within the small region of the parameter space that is explored by the perturbations. The sensitivity index is computed as

$$S_i \cong \frac{x_i}{y} \frac{\Delta y}{\Delta x} \tag{11.96}$$

(EPA 2001). Although the sensitivity ratio is based on the partial derivative at a point, equivalent to using a small value of $\Delta$ in the numerical approximation, the difference equation form of $S_i$ has also been applied using the plausible range of a parameter as the value of $\Delta$ (EPA 2001). The validity of using the potential range of a parameter in assessing sensitivity depends strongly on there being a linear response between the parameter and output.

Because of the complexity of most simulation models, local sensitivity analyses usually rely on numerical techniques using a factorial sampling design to map the response surface around the nominal values of the parameters (Steinhorst et al. 1978; Rose 1983). A common choice for the sampling design for a local sensitivity analysis is a full factorial design at three levels (nominal, nominal $\pm n\%$). Such a design is usually sufficient to identify and rank all first-order effects, to identify important interactions among parameters, and to indicate where nonlinear relationships between parameters and output variables may be important. The number of simulations that need to be run using a full factorial design is $n = l^p$, where $n$ is the number of simulations (combinations), $l$ is the number of levels of each parameter to be used, and $p$ is the number of parameters to be evaluated. Thus, sampling four parameters at each of three levels will require 81 simulations.

Fractional factorial designs can also be used to reduce the number of simulations. In a fractional factorial design, all possible combinations are not considered. Instead, combinations are eliminated that produce information about interactions that are considered negligible relative to the first-order effects and lower order interactions (Law and Kelton 1982). For example, all interactions of order 3 and higher could be eliminated. The disadvantage of fractional factorial designs is that the higher order interactions that are not explicitly considered are confounded with the lower order interactions. If it becomes apparent that ambiguities exist in the analysis, however, another fractional factorial design can be designed to eliminate the ambiguity.

One extreme of the fractional factorial design is the independent parameter perturbation (IPP) scheme (Rose 1983), also know as the one-factor-at-a-time (OAT) approach (Saltelli 2000; Daniel 1958). The IPP method consists of sampling each parameter at each level while using only the nominal value for the remaining parameters. This analysis provides a ranking of first-order effects, that is, only the effect of each parameter on the output variables. Interactions among the parameters cannot be identified. After ranking the parameters with regard to first-order effects, you can then refine the analysis by running a full factorial design on a subset of the highest ranked parameters from the IPP analysis. The danger in this approach is that parameters with small first-order (main) effects may have significant higher order effects that will not be identified.

Although local sensitivity analyses are frequently conducted using discrete intervals and factorial designs, other methods for performing local sensitivity analysis are available. For example, you can use differential analysis, based on a Taylor series expansion of the model equations (Iman and Helton 1988; Campolongo et al. 2000), or the Fourier amplitude sensitivity test (FAST), creating a Fourier series representation of the model and then analyzing its behavior (Helton et al. 1991; McRae et al. 1982). These methods are often difficult to implement for models of even moderate complexity, but they can be useful when analyzing relatively simple models.

### Global Sensitivity Analysis

A global sensitivity analysis assesses the effect of a parameter on an output variable over the range of values that may be likely to be exhibited. The range of values is often based upon a frequency distribution that has been assigned to each parameter. Variance-based global sensitivity analysis attempts to partition the variance observed in an output variable among the parameters. Because of statistical and computational limitations, partitioning often is limited to producing an ordered list of parameters to which the outputs are sensitive.

The global sensitivity of a model prediction to parameters may be of interest when the prediction is represented by a distribution of possible outcomes. For example, suppose the purpose of the model is to make estimates of the exposure of an individual to radiation following a reactor accident. The model takes into account variation in people's activities, food consumption rates, and so forth. Uncertainties in the various parameters used in the model are propagated, and the predictions of the model are the geometric mean and geometric standard deviation of doses. A global sensitivity analysis may be used to identify the parameters that contributed the most to overall dispersion in estimated dose. Improvements in the predictive ability of the model may then be made by reducing the uncertainty in the values of the parameters of highest sensitivity.

Sampling techniques like those for local sensitivity analysis may also be used for global sensitivity analysis. For example, EPA (2001) suggests estimating the global sensitivity of an output to a parameter by denormalizing the sensitivity index by multiplying $S_i$ by the coefficient of variation, $\sigma_{x_i}/\mu_{x_i}$, or normalized range, $x_{i,max} - x_{i,min}/\bar{x}$, of the $i$th parameter. This approach may work if the model output has a linear response to all of the parameters. In the absence of such linearity, the probable range of values that each parameter may take must be sampled more thoroughly than for the local analysis. For the factorial and IPP designs, sampling can involve many more than three levels of each parameter. These designs become impractical for anything but very small models. Global sensitivity analysis of complex models is most often based on sampling-based techniques know as variance-based methods (Chan et al. 2000). These methods require the sampling of parameters and the generation of outputs using Monte Carlo methods. Two sampling designs used for uncertainty analysis, simple random sampling and Latin hypercube sampling (McKay et al. 1976, 1979), can be used for some global sensitivity analysis methods, as well. Other methods, such as correlation ratios, require more extensive sampling designs than those used for uncertainty analysis.

## Statistics for Ranking Parameters

If a model is simply the sum or product of two or more independent parameters, then the contribution to the total uncertainty of each parameter can be computed exactly. The variance of a sum of independent terms is equal to the sum of the variances of the terms, so each parameter contributes $\frac{\text{var}[p_i]}{\text{var}[Total]} \times 100\%$ of the total uncertainty. The contribution of parameters to the total uncertainty of a model that consists of a chain of independent multiplicative factors can be computed similarly. In this case, each parameter contributes $\frac{\text{var}[\ln(p_i)]}{\text{var}[\ln(Total)]} \times 100\%$ because the equation can be transformed from a product of parameters to the sum of the logarithms of the parameters.

Models more complex than simple additive or multiplicative chains usually require numerical and statistical methods for estimating their sensitivities to parameters. The variance-based methods for sensitivity analysis include analysis of variance (ANOVA), regression-based techniques, correlation ratios, Sobol's sensitivity index, and FAST (Campolongo et al. 2000). There are problems with and limitations to these statistical analyses. Nevertheless, they are frequently used with good success. In addition, some of these methods require sampling designs that go beyond those required for uncertainty analysis, potentially increasing the computational effort.

ANOVA can be used to rank the parameters and to identify important interactions among the variables (Steinhorst et al. 1978). In using ANOVA, you should recognize that this application of the method is not statistically valid. Because the simulation model has no truly random processes, it should be possible to allocate all of the variation in the output variables to the variation imposed by the input variables. This will lead to having an error mean square (EMS) term of zero, so that all the $F$-ratios will be infinite. Any value appearing in the EMS term arises from violating the assumptions of the ANOVA test. Nevertheless, the mean squares from an ANOVA are useful in ranking the contributions of the parameters to the variation in the output variables. ANOVA may be effectively used with local analyses because the different levels of each parameter that are sampled can be represented as factors in the analysis.

Regression-based methods are popular for sensitivity analyses because they are relatively easy to implement, although they require assumptions to be made, such as linear relationships between the parameters and model outputs (McKay 1995). Correlation coefficients can be used to rank the effects of parameters on output variables (e.g., Robinson and Hurst 1997). Kohberger et al. (1978) suggested using a model containing linear, quadratic, and all first-order interaction terms. Partial correlation coefficients (PCCs) have also been used as measures of the importance of a parameter to an output variable (Gardner et al. 1980). PCCs represent the correlation between a parameter and an output variable after removing all the *linear* effects of the other variables. PCCs are useful because sampling designs that require fewer simulations than a full factorial design, such as simple random sampling or

Latin hypercube sampling, can be used. Note, however, that the test does not identify interactions among parameters (Rose 1983).

A modification of the PCC analysis has been suggested for use in sensitivity analysis. The modification uses the rank of each parameter in the statistical analysis rather than its actual value (Iman et al. 1981). The use of ranks can significantly decrease the confounding influence of nonlinearity in the response of an output variable to a parameter.

Correlation ratios provide an estimate of the importance of each parameter based on the ratio of the variance correlation expectation (VCE) to the total variance (McKay 1995). The VCE is the variance in the expectation of the output variable, $Y$, conditional on a fixed value of a parameter, $X$, across all values of $X$:

$$\frac{\text{var}_{x_i}[E[Y|X_j = x_j]]}{\text{var}[Y]} = \frac{\text{VCE}}{\text{var}[Y]} \tag{11.97}$$

The calculation of correlation ratios is computationally expensive because it involves sampling all of the other parameters and running many simulations at each of many values of the parameter of interest. Estimation of the correlation ratio can be best accomplished using Latin hypercube sampling (Chan et al. 2000). The method involves creating $r$ baseline Latin hypercube sample matrices each of size $mk$, where $k$ is the number of parameters and $m$ is the number of samples per parameter. The results obtained from running the $rm$ baseline simulations using these input values is used to estimate the total variance in $Y$. The contribution of a parameter to the total variance is then computed using a series of modified parameter matrices. Unfortunately, the calculation of the correlation ratio has the undesirable property of sometimes returning negative values. McKay (1997) suggests an alternative expression that produces biased estimates but remains positive and converges to analytical values as the number of replicates, $r$, increases.

A measure of sensitivity that is equivalent to the correlation ratio can be computed using FAST (Campolongo et al. 2000; Chan et al. 2000). The measure is based on analyzing the behavior of a Fourier series created to represent the model (Cukier et al. 1973, 1978; Helton et al. 1991; McRae et al. 1982; Chan et al. 2000). This method transforms a multidimensional surface over all of the uncertain model parameters into a one-dimensional integral based on a Fourier series. Hamby (1995) and Campolongo et al. (2000) discuss some additional indices and methods that can be used to judge sensitivity.

## Uncertainty and Model Validation

Uncertainty analysis alone is not sufficient to establish that a model is capable of providing reasonable estimates. Validation of a model is also required. Although model validation is addressed in chapter 14, it is helpful to discuss the interplay between these two topics. Validation is an analysis that attempts to delineate the domain of applicability of a model (Kirchner and Whicker 1984). The domain of applicability of a model refers to the set of conditions under which a model can be assumed to adequately represent the system of interest. Thus, while uncertainty analysis is analogous to determining the imprecision in measurements from a piece of laboratory equipment, validation is analogous to determining the accuracy of the measurements and the range of input factors over which accurate measurements can be made.

The results of uncertainty analysis can play an important role in model validation. The level of uncertainty in model predictions can be used as a criterion for judging whether the model results compare favorably with observations. For example, one may regard simulations as unsatisfactory if their levels of uncertainty are above some arbitrary level, whether or not the mean predictions compare favorably with the associated observations. Thus, a high level of uncertainty could reduce confidence in the model predictions. The confidence or tolerance interval on model predictions can also be used to define a level of practical significance for use in comparing model results to observations, assuming that the levels of uncertainty are acceptable (Kirchner and Whicker 1984).

Model comparison experiments can be used to examine the effect of using different assumptions about model structure or parameter values. Model comparison experiments can also be used to test whether the results from two or more different models correspond. The efficiency and quality of comparisons between models can be enhanced by using specialized sampling schemes to reduce variance in the results. Two common techniques are antithetic sampling and common random numbers (Law and Kelton 1982). These sampling designs force correlation between input variables—negative correlation in the case of antithetic sampling and positive correlation in the case of common random numbers.

Antithetic sampling is designed to reduce the variance in the estimate of the mean of an output from a model. The antithetic sampling scheme requires that simulations be run in pairs. For each pair, sampling is done to ensure that a high value of a parameter for the first simulation is paired with a low value from the second simulation. The outputs of the pair of simulations are averaged to produce a single estimate. Antithetic sampling can be valuable when cost or time constraints limit the number of simulations that can be run for an analysis, but it has no advantage in the absence of these constraints.

The common random numbers sampling scheme is designed to reduce the variance in the difference between paired samples from simulations involving two models or from one model running under two different sets of conditions. Random numbers are synchronized so that a random number used for a particular purpose in one system is used for the identical purpose in the other system.

Both of these sampling schemes require that attention be given to creating the input values and saving the results. This is sometimes problematic because of the nature of the pseudorandom number generators used to create the parameter values. Frequently, all of the random number generators used in a model draw on a single uniform random number function. While this function will always return the same sequence of values, the values assigned to a parameter in, for example, the first simulation of one model may be different from the value assigned to that same parameter in the second model if the two models sample a different number of parameters. A common method to ensure that the input values are paired correctly (in the case of antithetic sampling) and that parameters are assigned the same sequence of values (in the case of common random numbers) is to generate the parameter values before all simulations. The values are then imported into the models as required.

## Summary

With the exception of simple, conservative screening-level models, most models used in risk assessment will have uncertainties associated with their outputs. These uncertainties arise from aleatory (stochastic) and epistemic (lack-of-knowledge) uncertainties attached to the model structure, parameters, and inputs. The evaluation and analysis of uncertainty in model predictions is an important part of establishing the credibility of a risk assessment. The evaluation of parametric model uncertainty involves propagating uncertainty in the parameters and inputs through the model. Analytical methods and mathematical approximation techniques can be used to propagate uncertainty through relatively simple models. Monte Carlo methods can be used with models of almost any complexity given sufficient computer resources. Although Monte Carlo methods are widely used, understanding the analytical methods can give insight into what may sometimes be seen as counterintuitive behavior in model responses under uncertainty, and into the kinds of problems that can arise when propagating uncertainty through some mathematical expressions.

Traditional descriptive statistics can be used to characterize the uncertainty in model outputs. When model uncertainties arise from the presence of both aleatory uncertainty and epistemic uncertainty, it is often useful to employ nested sampling designs to allow separation of these sources of uncertainty in the model outputs. In addition, risk assessment is often interested in the potential extremes of the predictions, rather than the mean responses. In those cases, tolerance intervals, rather than confidence intervals, will be of interest.

Sensitivity analysis is used to identify the parameters or inputs to which a model output is most sensitive. Local sensitivity analysis helps to define the impact of small perturbations in the values of parameters and is analogous to finding the partial derivative of the model "function" with respect to the parameters at a point. Global uncertainty analysis takes into account the entire range of values of parameters and inputs, which are usually defined in terms of distributions. Global uncertainty analysis attempts to partition the contributions to the uncertainty in outputs among the parameters and inputs to the model. Some techniques for global sensitivity analysis, such as the regression-based methods, can use the same sampling designs as those used for uncertainty analysis. Other techniques require specialized sampling designs that may require considerably more simulations than are required for uncertainty analysis.

The validation or benchmarking of models cannot be conducted without recognizing that there is uncertainty associated with model outputs. The degree of uncertainty alone can be used as a criterion for judging model performance. Uncertainty should also be considered when comparing one model to another or comparing a model to data. Specialized sampling designs can reduce computational effort when conducting model comparisons.

## Problems

1. The dose rate at a distance $r$ from a point source of gamma radiation is equal to the product of four factors divided by the square of the distance from the source:

$$\dot{D} = \frac{AE(\text{Ci} - \text{MeV}) \times 3.7 \times 10^{10}(\sec^{-1} \text{Ci}^{-1}) \times 1.6 \times 10^{-6}(\text{erg} - \text{MeV}^{-1}) \times C(\text{cm}^{-1})}{1.293 \times 10^{-3}(\text{gm} - \text{cm}^{-3}) \times 4\pi r^2 \times 100(\text{erg} - \text{rad}^{-1})},$$

where $1.293 \times 10^{-3}$ is the density of air at standard temperature and pressure, and it is assumed to be appropriate for this example. Combining the constants gives

$$\dot{D} = \frac{3.64 \times 10^4 \times A \times E \times C}{r^2} \frac{\text{rad}}{\sec}.$$

The other parameters are uncertain and have the following means, standard deviations, and variances:

| Variable | Definition | Mean | Standard deviation | Variance | Units |
|---|---|---|---|---|---|
| A | Activity | 100 | 10 | 100 | Ci |
| E | Average total photon energy | 1 | 0.2 | 0.04 | MeV per disintegration |
| C | Average energy absorption coefficient | $5 \times 10^{-5}$ | $1 \times 10^{-5}$ | $1 \times 10^{-10}$ | cm$^{-1}$ |
| r | Distance from source | 1,000 | 100 | 10,000 | cm |

   Assuming that these parameters are distributed normally, what are the mean dose rate and its standard deviation? Would you expect the distribution of dose rate to be normal? What is your estimate of the 95th percentile for the distribution of dose rate?

2. Using the parameter values from problem 1, estimate the mean, standard deviation, geometric mean, geometric standard deviation, and 95th percentile, assuming that the distributions of the parameters are lognormal. Would you expect the distribution to be lognormal? Did the assumption of the shape of the input distribution have much impact on the results?

3. How much does parameter $C$ contribute to the total uncertainty in problem 2?

4. Rather than do a probabilistic assessment, a regulation has asked that you estimate the dose rate using the 90th percentiles of $A, E$, and $C$ and the 5th percentile of $r$, assuming that the parameters are lognormal. How does this estimate of dose compare to the mean computed in problem 2? How does it compare to the 95th and 97.5th percentiles?

5. The model for estimating dose rate does not take into account other factors that could influence the dose to an exposed person, such as size and orientation. If you were to compute dose to an individual by adding factors to represent these features, would the total uncertainty increase or decrease? If you increased the mean of $E$ from 1 to 2 but did not change its standard deviation, would the variance increase, decrease, or stay the same? Would the geometric standard deviation increase, decease, or stay the same?
6. Is a detailed, realistic model always preferable to a simple model? Why or why not?
7. An assessment makes use of a dose assessment model for which some parameters are defined by distributions based on site-specific measurements and others are based on expert opinion. Can uncertainties on these two classes of parameters be combined in an assessment? How would you propose to propagate the uncertainties? If you wanted to reduce the uncertainty in the dose estimate, would it always be appropriate to concentrate your efforts on improving the estimate of uncertainty for the parameter contributing the most to total uncertainty? Why or why not?
8. A model is developed to estimate doses for one radionuclide at a time. You are modeling external doses due to exposure to an accidental release of reactor waste. There is uncertainty in both the total activity deposited in an area and the proportions of the various radionuclides in the waste. Can you reliably estimate the mean total dose as the sum of the mean doses for each radionuclide? Justify your answer.
9. A dose assessment is being conducted that makes use of badge readings as part of an epidemiological study. Badges that show no response are assigned values equal to the detection level for the badge. How will a change in the frequency for reading badges from monthly to weekly affect the total dose estimate? Would it be better to assign a dose of zero to measurements below the detection level? Is there another method that would be better? Would your answer be any different if the data are being used for estimating doses for the purpose of providing protection?
10. Several commercial applications can be used to perform Monte Carlo analyses on models that can be expressed as algebraic equations. Can you rely on these "off-the-shelf" tools to always give reliable results? Give an example of a situation where the estimated percentiles for a distribution based on Monte Carlo methods can be unreliable.

References

ANSI (American National Standards Institute). 1997. *U.S. Guide to the Expression of Uncertainty in Measurement.* ANSI/NCSL Z540–2-1997. National Conference of Standards Laboratories, Boulder, CO.

Bailar, J.C., III, and F. Ederer. 1964. "Significance Factors for the Ratio of a Poisson Variable to Its Expectation." *Biometrics* 20: 639–643.

Barry, T.M. 1996. "Recommendations on the Testing and Use of Pseudo-random Number Generators Used in Monte Carlo Analysis for Risk Assessment." *Risk Analysis* 16: 93–105.

Beach, L.R., and G.P. Braun. 1994. "Laboratory Studies of Subjective Probability: A Status Report." In *Subjective Probability*, ed. G. Wright and P. Ayton. Wiley & Sons, New York.

Bogen, K.T. 1995. "Methods to Approximate Joint Uncertainty and Variability in Risk." *Risk Assessment* 15: 411–419.

Bonin, J.J., and D.E. Stevenson, eds. 1989. *Risk Assessment in Setting National Priorities*. Plenum Press, New York.

Breshears, D.D., T.B. Kirchner, M.D. Otis, and F.W. Whicker. 1989. "Uncertainty in Predictions of Fallout Radionuclides in Foods." *Health Physics* 57: 943–953.

Campolongo, F., A. Saltelli, T. Sorensen, and S. Tarantola. 2000. "Hitchhiker's Guide to Sensitivity Analysis." In *Sensitivity Analysis*, ed. A. Saltelli, K. Chan, and E.M. Scott. John Wiley & Sons, New York.

Chan, K., S. Tarantola, A. Saltelli, and I.M. Sobol. 2000. "Variance Based Methods." In *Sensitivity Analysis*, ed. A. Saltelli, K. Chan, and E.M. Scott. John Wiley & Sons, New York.

Chan, S.Y.J. 1993. "An Alternative Approach to the Modeling of Probability Distributions." *Risk Assessment* 13: 97–102.

Chatfield, C. 1980. *The Analysis of Time Series: An Introduction*. Chapman & Hall, New York.

Clark, C.E. 1961. "Importance Sampling in Monte Carlo Analyses." *Operations Research* 9: 603–620.

Cukier, R.I., C.M. Fortuin, K.E. Schuler, A.G. Petschek, and J.H. Schaibly. 1973. "Study of the Sensitivity of Coupled Reaction Systems to Uncertainties in Rate Coefficients. I. Theory." *Journal of Chemical Physics* 59: 3873–3878.

Cukier, R.I., H.B. Levin, and K.E. Schuler. 1978. "Nonlinear Sensitivity Analysis of Multiparameter Model Systems." *Journal of Computer Physics* 26: 1–42.

Cullen, A.C., and H.C. Frey. 1998. *Probabilistic Techniques in Exposure Assessment: A Handbook for Dealing with Variability and Uncertainty in Models and Inputs*. Plenum, New York.

Daniel, C. 1958. "On Varying One Factor at a Time." *Biometrics* 14: 430–431.

Eberhardt, L.L., and J.M. Thomas. 1991. "Designing Environmental Field Studies." *Ecological Monographs* 61: 53–73.

Eddy, D.M. 1982. "Probabilistic Reasoning in Clinical Medicine: Problems and Opportunities." In *Judgment under Uncertainty: Heuristics and Biases*, ed. D. Kahnemann, P. Slovic, and A. Tversky. Cambridge University Press, New York.

Edelmann, K.G., and D.E. Burmaster. 1997. "Are All Distributions of Risk with the Same 95th Percentile Equally Acceptable?" *Human and Ecological Risk Assessment* 2: 223–234.

EPA (U.S. Environmental Protection Agency). 2001. *Risk Assessment Guidance for Superfund*: Vol. 3. *Part A, Process for Conducting Probabilistic Risk Assessment*. Appendix A. U.S. Environmental Protection Agency, Washington, DC.

Fishman, G.S. 1996. *Monte Carlo Concepts, Algorithms, and Applications*. Springer, New York.

Gardner, R.H., D.D. Huff, R.V. O'Neill, J.B. Mankin, J. Carney, and J. Jones. 1980. "Application of Error Analysis to a Marsh Hydrology Model." *Water Resources Research* 16: 659–664.

Gardner, R.H., W.G. Cale, and R.V. O'Neill. 1982. "Robust Analysis of Aggregation Error." *Ecology* 63: 1771–1779.

Gigerenzer, G. 1994. "Why the Distinction Between Single-Event Probabilities and Frequencies Is Important for Psychology (and Vice Versa). In *Subjective Probability*, ed. G. Wright and P. Ayton. Wiley & Sons, New York; pp. 130–161.

Graham, J.D., L. Green, and M.J. Roberts. 1988. *In Search of Safety: Chemicals and Cancer Risks*. Harvard University Press, Cambridge, MA; pp. 80–114.

Haas, C.N. 1997. "Importance of Distributional Form in Characterizing Inputs to Monte Carlo Assessments." *Risk Assessment* 17(1): 107–113.

Hahn, G.J., and W.Q. Meeker. 1991. *Statistical Intervals: A Guide for Practitioners*. Wiley Interscience, New York.

Hahn, G.J., and S.S. Shapiro. 1967. *Statistical Models in Engineering*. Wiley & Sons, New York; p. 355.

Hamby, D.M. 1995. "A Comparison of Sensitivity Analysis Techniques." *Health Physics* 68: 195–204.

Hattis, D., and D.E. Burmaster. 1994. "Assessment of Variability and Uncertainty for Practical Risk Assessment." *Risk Assessment* 14: 713–730.

Helton, J.C. 1994. "Treatment of Uncertainty in Performance Assessment for Complex Systems." *Risk Analysis* 14: 483–511.

Helton, J.C., J.W. Garner, R.D. McCurley, and D.K. Rudeen. 1991. *Sensitivity Analysis Techniques and Results for Performance Assessment at the Waste Isolation Pilot Plant*. SAND90-7103. Sandia National Laboratories. National Technical Information Service, Springfield, VA.

Henrion, M., and B. Fischhoff. 1986. "Assessing Uncertainty in Physical Constants." *American Journal of Physics* 54: 791–798.

Hodak, D.G. 1994. "Adjusting Triangular Distributions for Judgmental Bias." *Risk Assessment* 14: 1025–1031.

Hoffman, F.O., and J.S. Hammonds. 1994. "Propagation of Uncertainty in Risk Assessments: The Need to Distinguish Between Uncertainty Due to Lack of Knowledge and Uncertainty Due to Variability." *Risk Analysis* 14: 707–712.

Hora, S.C., and R.L. Iman. 1989. "Expert Opinion in Risk Analysis. The NUREG-1150 Methodology." *Nuclear Science and Engineering* 102: 323–331.

Howard, R.A., and J.E. Matheson. 1984. *Readings in the Principles and Practices of Decision Analysis*. Strategic Decision Systems, Menlo Park, CA.

IAEA (International Atomic Energy Agency). 1989. *Evaluating the Reliability of Predictions Made Using Environmental Transport Models*. Safety Series No. 100. International Atomic Energy Agency, Vienna, Austria.

Iman, R.L., and W.J. Conover. 1982. "A Distribution-free Approach to Inducing Rank Correlation among Input Variables." *Communications on Statistics: Simulation and Computing* 11: 311–334.

Iman, R.L., and J.C. Helton. 1988. "A Comparison of Uncertainty and Sensitivity Analysis Techniques for Computer Models. *Risk Analysis* 8: 71–90.

Iman, R.L., J.C. Helton, and J.E. Cambell. 1981. "An Approach to Sensitivity Analysis of Computer Models, Part I: Introduction, Input Variable Selection, and Preliminary Variable Assessment." *Journal of Quality Technology* 13(3): 174–183.

Kahnemann, D., and A. Tversky. 1973. "On the Psychology of Prediction." *Psychology Review* 80: 237–251.

Kahnemann, D., P. Slovic, and A. Tversky, eds. 1982. *Judgment under Uncertainty: Heuristics and Biases*. Cambridge University Press, New York.

Kirchner, T.B., and F.W. Whicker. 1984. "Validation of PATHWAY, a Simulation Model of the Transport of Radionuclides Through Agroecosystems." *Ecological Modeling* 22: 21–44.

Kohberger, R.C., D. Scavia, and J.W. Wilkinson. 1978. "A Method for Parameter Sensitivity Analysis in Differential Equation Models." *Water Resources Research* 14: 25–28.

Law, A.M., and W.D. Kelton. 1982. *Simulation Modeling and Analysis.* McGraw-Hill, New York.

Lichtenstein, S., P. Slovic, B. Fischhoff, M. Layman, and B. Combs. 1978. "Judged Frequency of Lethal Events." *Journal of Experimental Psychology: Human Learning and Memory* 4: 551–578.

MacIntosh, D.L., G.W. Suter II, and F.O. Hoffman. 1994. "Uses of Probabilistic Exposure Models in Ecological Risk Assessments of Contaminated Sites." *Risk Assessment* 14: 405–419.

McKay, M.D. 1995. *Evaluating Prediction Uncertainty.* NUREG/CR-6311. U.S. Nuclear Regulatory Commission and Los Alamos National Laboratory, Washington, DC.

McKay, M.D. 1997. "Nonparametric Variance-based Methods for Assessing Uncertainty Importance." *Reliability Engineering System Safety* 57: 267–279.

McKay, M.D., W.J. Conover, and D.E. Whiteman. 1976. *Report on the Application of Statistical Techniques to the Analysis of Computer Codes.* LANUREG-6526-MS. Los Alamos Scientific Laboratory, Los Alamos, NM.

McKay, M.D., W.J. Conover, and R.J. Beckman. 1979. "A Comparison of Three Methods for Selecting Values of Input Variables in the Analysis of Output from a Computer Code." *Technometrics* 21: 239–245.

McRae, G.J., J.W. Tilden, and J.H. Seinfeld. 1982. "Global Sensitivity Analysis—A Computational Implementation of the Fourier Amplitude Sensitivity Test (FAST)." *Computers and Chemical Engineering* 6: 15–25.

Meyer, M., and J. Booker. 1991. *Eliciting and Analyzing Expert Judgment: A Practical Guide.* Academic Press, New York.

Miller, D.R. 1974. "Sensitivity Analysis and Validation of Simulation Models." *Journal of Theoretical Biology* 48: 345–360.

Mood, A.M., F.A. Graybill, and D.C. Boes. 1974. *Introduction to the Theory of Statistics.* McGraw-Hill, New York.

Morgan, M.G. 1998. "Uncertainty Analysis and Risk Assessment." *Human and Ecological Risk Assessment* 4: 25–39.

Morgan, M.G., and M. Henrion. 1992. *Uncertainty: A Guide to Dealing with Uncertainty in Quantitative Risk and Policy Analysis.* Cambridge University Press, New York.

Mosleh, A., and V. Bier. 1992. "On Decomposition and Aggregation Error in Estimation: Some Basic Principles and Examples." *Risk Analysis* 12: 203–214.

Naylor, T.H., J.L. Balintfy, D.S. Burdick, and K. Chu. 1968. *Computer Simulation Techniques.* John Wiley & Sons, New York.

NCRP (National Council on Radiation Protection and Measurements). 1996. *A Guide for Uncertainty Analysis in Dose and Risk Assessments Related to Environmental Contamination.* NCRP Commentary No. 14. National Council on Radiation Protection and Measurements, Washington, DC.

NRC (U.S. Nuclear Regulatory Commission). 1990. *Severe Accident Risks: An Assessment for Five U.S. Nuclear Power Plants.* NUREG-1150. National Technical Information Division, Springfield, VA.

O'Neill, R.V., and B. Rust. 1979. "Aggregation Error in Ecological Models." *Ecological Modeling* 7: 91–105.

Paustenbach, D.J. 1995. "The Practice of Health Risk Assessment in the United States (1975–1995); How the U.S. and Other Countries Can Benefit from That Experience." *Human and Ecological Risk Assessment* 1: 29–80.

Pooch, U.W., and J.A. Wall. 1993. *Discrete Event Simulation: A Practical Approach*. CRC Press, Boca Raton, FL.

Press, W.H., B.P. Flannery, S.A. Teukolsky, and W.T. Vetterling. 1988. *Numerical Recipes in C*. Cambridge University Press, Cambridge.

Robinson, R.B., and B.T. Hurst. 1997. "Statistical Quantification of the Sources of Variance in Uncertainty Analyses." *Risk Analysis* 17: 447–453.

Rose, K.A. 1983. "A Simulation Comparison and Evaluation of Parametric Sensitivity Methods Applicable to Large Models." In *Analysis of Ecological Systems: State-of-the-Art in Ecological Modelling*, ed. W.K. Lauenroth, G.V. Skogerboe, and M. Flug. Elsevier Scientific, New York; pp. 129–140.

Saltelli, A. 2000. "What Is Sensitivity Analysis?" In *Sensitivity Analysis*, ed. A. Saltelli, K. Chan, and E.M. Scott. John Wiley & Sons, New York.

Sasser, L.B. 1965. "The Uptake and Secretion of Iodine-131 by Lactating Dairy Cows Resulting from a Contaminated Pasture." MS thesis, Colorado State University, Fort Collins, CO.

Schlyakhter, A.I. 1994. "An Improved Framework for Uncertainty Analysis: Accounting for Unsuspected Errors." *Risk Analysis* 14: 441–447.

Seiler, F.A., and J.L. Alvarez. 1996. "On the Selection of Distributions for Stochastic Variables." *Risk Analysis* 16: 5–18.

Smith, A.E., P.B. Ryan, and J.S. Evans. 1992. "The Effect of Neglecting Correlations When Propagating Uncertainty and Estimating the Population Distribution of Risk." *Risk Assessment* 12: 467–474.

Steinhorst, R.K., H.W. Hunt, G.S. Innis, and K.P. Haydock. 1978. "Sensitivity Analysis of the ELM Model." In *Grassland Simulation Model*, ed. G.S. Innis. Ecological Studies 26. Springer-Verlag, New York; pp. 231–255.

Taylor, J.R. 1982. *An Introduction to Error Analysis*. University Science Books, Oxford University Press, Mill Valley, CA.

Tomovic, R. 1963. *Sensitivity Analysis of Dynamic Systems*. McGraw-Hill, New York.

Tomovic, R., and M. Vukobratovic. 1972. *General Sensitivity Theory*. American Elsevier Publishing, New York.

Tversky, A., and D. Kahnemann. 1971. "The Belief in the Law of Small Numbers." *Psychology Bulletin* 76: 105–110.

Tversky, A., and D. Kahnemann. 1973. "Availability: A Heuristic for Judging Frequency and Probability." *Cognitive Psychology* 4: 207–232.

Tversky, A., and D. Kahnemann. 1974. "Judgment under Uncertainty: Heuristics and Biases." *Science* 185: 1124–1131.

Whicker, F.W., and T.B. Kirchner. 1987. "PATHWAY: A Dynamic Food-Chain Model to Predict Radionuclide Ingestion after Fallout Deposition." *Health Physics* 52: 717–737.

Wright, G., and P. Ayton, eds. 1994. *Subjective Probability*. John Wiley & Sons, New York.

Zar, J.H. 1984. *Biostatistical Analysis*. Prentice-Hall, Englewood Cliffs, NJ.

# 12

# The Risks from Exposure to Ionizing Radiation

Roger H. Clarke

Radiation exposures can damage living cells, causing death in some of them and modifying others. Most organs and tissues of the body are not affected by the loss of even considerable numbers of cells. However, if the number lost is large enough, there will be observable harm to the organ or tissue and, therefore, to the individual. This harm from radiation will occur only if the radiation dose is large enough to kill a large number of cells. This type of harm occurs in all individuals who receive a dose in excess of the threshold for the effect and is called *deterministic* (see figure 12.1). Other radiation damage may occur in cells that are not killed but are modified, in which case the damage in the viable cell is usually repaired. If this repair is not perfect, the modification will be transmitted to daughter cells and may eventually lead to cancer in the tissue or organ of the exposed individual.

If the cells are involved in transmitting genetic information to the descendants of the exposed individual, hereditary disorders may arise. These effects in individuals or their descendants are called *stochastic*, meaning of a random nature.

In short, deterministic (threshold) effects will occur only if the radiation dose is substantial, for example, in an accident. Stochastic effects (cancer and hereditary effects) may be caused by damage in a single cell. As the dose to the tissue increases from a low level, more and more cells are damaged and the probability of stochastic effects occurring increases, as shown in figure 12.2.

Since this book is concerned with environmental risk assessment, which does not normally involve deterministic risks, the emphasis of this chapter is on stochastic effects of exposure—cancer and hereditary risks. A review of the deterministic effects of radiation exposure can be found in ICRP (2007). This chapter is based on the most recent international reviews of radiation effects in the reports of the U.N.

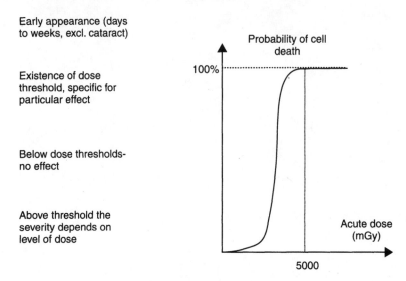

Figure 12.1  Principle characteristics of deterministic effects of radiation.

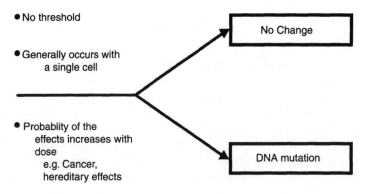

Figure 12.2  Summarizing the stochastic effects of radiation.

Scientific Committee on the Effects of Atomic Radiation (UNSCEAR 2000, 2001) and the International Commission on Radiological Protection (ICRP 2005, 2007).

Radiation exposure has been associated with most forms of leukemia and with cancers of many organs, such as lung, breast, and thyroid, but not with some other organs such as prostate. However, a small addition of radiation exposure (e.g., of the order of the global average level of natural radiation exposure) would produce an exceedingly small increase in the chances of developing an attributable cancer. Moreover, radiation-induced cancer is difficult to detect. It may manifest only decades after the exposure occurred and does not differ from cancers arising spontaneously or attributable to other factors. (In the developed world, more than 20% of deaths are attributable to cancer.) The major long-term evaluation of populations exposed to radiation is provided by the study of the approximately 80,000 survivors

The Risks from Exposure to Ionizing Radiation   533

of the atomic bomb attacks on Hiroshima and Nagasaki. It has revealed an excess of only a few hundred cancer deaths in the studied population. Since approximately half of this population is still alive, additional study is necessary in order to obtain the full picture.

Radiation exposure also has the potential for causing hereditary effects in the offspring of persons exposed to radiation. Such effects were once thought to threaten the future of the human race by increasing the rate of natural mutation to an inappropriate degree. However, hereditary effects have yet to be detected in human populations exposed to radiation, although they are known to occur in other species.

## Radiobiological Effects after Low Doses of Radiation

Both UNSCEAR (2000) and ICRP (2005) have reviewed the broad field of experimental studies of radiation effects in cellular systems and in plants and animals. Many of these responses and the factors modifying them form a basis for the knowledge of human radiation effects and can often be evaluated in more detail than studies of humans. Furthermore, fundamental radiobiology today includes the field of molecular radiobiology, which is contributing to an understanding of the mechanisms of radiation response.

### Biophysical Aspects of Radiation Action on Cells

Important advances and judgments are given in ICRP Publication 92 (ICRP 2003) and in an ICRP report on low-dose risks (ICRP 2005). The understanding of the early postirradiation biophysical processes in cells and tissues has advanced substantially. The following paragraphs briefly highlight some major points of development. Knowledge of the fine structure of energy deposition from radiation tracks in DNA dimensions has grown, largely through the further development of Monte Carlo track structure codes. Coupled with radiobiological information, track structure data have greatly affected scientific thinking on the nature of biologically critical damage to DNA (see figure 12.3).

In particular, it has been recognized that a high proportion of radiation-induced damage to DNA is represented in complex clusters of chemical alterations. Such clustered damage can arise via a combination of damages induced by the main tracks, secondary electrons, and secondary reactive radical species. Double-strand breaks (DSBs) and single-strand breaks (SSBs) in the DNA sugar-phosphate backbone plus a variety of damaged DNA bases can combine together in clusters, with a substantial fraction of total damage being closely spaced. There is also evidence that both the frequency and complexity of complex clustered damage depend on the linear energy transfer (LET) of the radiation. When DSBs, SSBs, and base damages are considered together, complex clustered damage may constitute as much as 60% and 90% of total DNA damage after low- and high-LET radiation, respectively. This is illustrated schematically in figure 12.4. These data highlight a major difference between DNA lesions induced by radiation and those arising spontaneously

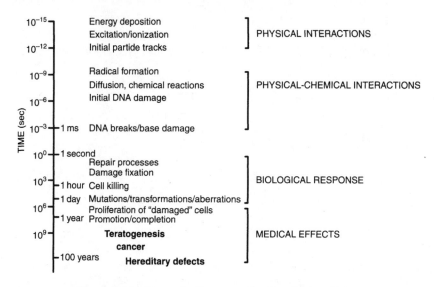

Figure 12.3  Events leading to stochastic radiation effects.

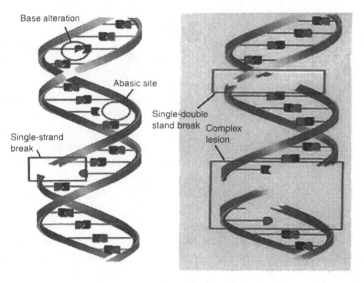

Figure 12.4  Types of radiation damage to DNA.

via oxidative attack by reactive chemical radicals: whereas the former are predominantly complex and clustered, the latter are randomly distributed and simple in their chemical structure. The different repair characteristics of simple and complex DNA lesions is an important factor in the development of judgments on health effects after low doses of radiation.

In addition to improvements in understanding of the induction of complex DNA damage by radiation, there have been other advances in radiation biophysics. For example, radiation-induced damage has been investigated at the level of chromosome structure, and this work has been paralleled by the biophysical modeling of the induction of gene/chromosomal mutations. There have also been valuable technical innovations, including the development of single particle irradiation systems (microbeams) and imaging methods for the cellular visualization of DNA–protein interactions during DNA damage response (ICRP 2005; Cherubini et al. 2002).

## *Chromosomal DNA as the Principal Target for Radiation*

Damage to DNA in the nucleus is the main initiating event by which radiation causes long-term damage to organs and tissues of the body. DSBs in DNA are regarded as the most likely candidate for causing critical damage (UNSCEAR 2000; ICRP 2007). Single radiation tracks have the potential to cause DSBs and, in the absence of 100% efficient repair, could result in long-term damage, even at the lowest doses. Damage to other cellular components (epigenetic changes) may influence the functioning of the cell and progression to the malignant state.

Numerous genes are involved in cellular response to radiation, including those for DNA damage repair and cell-cycle regulation. Mutation of these genes is reflected in several disorders of humans that confer radiation sensitivity and cancer proneness on the individuals. For example, mutation of one of the many *checkpoint* genes may allow insufficient time to repair damage because the cell loses its ability to delay progression in the cell cycle following radiation exposure.

Cells have a number of complex pathways capable of recognizing and dealing with specific forms of damage. This subject is reviewed in annex F of UNSCEAR (2000). One gene that plays a key role is the tumor suppressor p53, which is lost or mutated in more than half of all human tumors. The p53 protein produced by the gene controls both arrest of the cell cycle and one pathway of apoptosis (the programmed cell death that is instrumental in preventing some damaged cells from progressing to the transformed, malignant growth stage). Some of these processes are also implicated in stress response or adaptation processes that limit the extent or outcome of damage. Even with such beneficial processes induced and acting, it is clear that the misrepaired damage that follows radiation interactions gives the potential for progression to cancer induction or hereditary disease (see figure 12.5).

The research findings on the adaptive responses to radiation in cells and organisms are reviewed by the UNSCEAR (1993, 2000), the National Council on Radiation Protection and Measurements (NCRP 2001), and the ICRP (2005) and are not reviewed here. So far, however, there appears to be no generally reproducible reduction in tumor induction following low-dose irradiation.

## *Epigenetic Responses to Radiation*

A major feature of radiobiological research since 1990 has been a range of studies that provide evidence of postirradiation cellular responses that appear to result in genomic change and/or cellular effects without an obvious requirement for directly

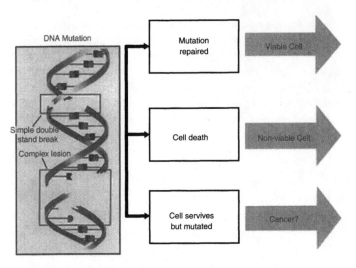

Figure 12.5    Consequences of DNA radiation damage.

induced DNA damage (Cherubini et al. 2002; ICRP 2005). In a broad sense, these processes may be termed "epigenetic" and contrast with the well-established radiobiological concept of direct DNA targeting by ionizing radiation tracks that has underpinned much of the post-1990 developments in biophysics and DNA damage response. Although there are elements of overlap, these epigenetic effects may be placed in two categories: (a) radiation-induced *genomic instability* and (b) postirradiation *bystander signaling* between cells.

The bystander effect relates to the expression of cell death/apoptosis, gene/chromosomal mutation, genomic instability, and/or changing patterns of protein abundance in cells not directly intersected by radiation tracks. While conventional DNA damage response is known to result in the expression of genomic damage within the first or second postirradiation cell cycles, the term "induced genomic instability" broadly describes a set of phenomena where genomic damage and its cellular consequences are expressed persistently over many postirradiation cell cycles.

The phenomena of induced genomic instability and bystander effects when expressed in vitro may show some common stress-related mechanisms. However, there are few data and some controversies on the relative contribution of bystander signaling to cellular effects overall and the extent to which this is dose dependent. Studies on bystander effects in vivo are in their infancy, although there are some positive data relating to factors affecting the breakage of chromosomes.

The multistage nature of tumorigenesis is considered in annex G of UNSCEAR (2000). Much knowledge about this process remains to be learned. Although the concept of sequential, interacting gene mutations as the driving force for neoplasia is more firmly established, there is a lack of understanding of the complex interplay between these events and the consequences for cellular behavior and tissue homeostasis. Uncertainty also exists regarding the contribution made to malignant development of nonmutational (epigenetic) cellular events such as gene silencing and cellular communication changes.

### *Effects at Low Doses of Radiation*

The basic premise of radiation response is that any radiation interaction with DNA results in damage that, if not repaired or if incorrectly repaired, may represent an initiating event in the tumorigenesis pathway. The mutation of genes commonly results in modulation of their expression, with loss of gene products (proteins) or alteration in their properties or amounts. The biochemical balance of the cell may then be disrupted, compromising the control of cell signaling or the proliferation and differentiation schedules. In this way, mutated cells, instead of being checked or killed, may be allowed to proceed to clonal growth. Some nonmutational (epigenetic) events or damage may be involved or contribute to these changes. In some cases the genome may be destabilized, allowing further mutations to accumulate, which may promote the tumorigenesis progression.

Mechanistic considerations can be used to determine if there is a threshold exposure level below which a biological response does not occur. Specifically, there is a need to know whether the repair processes at very low doses are more efficient and perhaps enhanced by the adaptive response, preventing any damage to the cellular components. Such a threshold could occur only if repair processes were totally effective in that dose range or if a single track were unable to produce an effect. The absence of consistent indications of significant departures from tumorigenic response at low doses in cellular end points (e.g., chromosome aberrations, gene mutation, and cell transformation), the activity of well-characterized error-prone DNA repair pathways, and the evidence on spontaneous DSBs in mammalian cells argue against adaptive or other processes that might provide for a dose threshold for radiation effects. The cellular processes such as apoptosis and cellular differentiation that can protect against later phases of tumorigenesis are judged to be efficient but can be bypassed; there is no reason to believe that these defenses act differently on spontaneous and radiation-induced tumors or have specific dose dependencies.

Thus, even at low doses, radiation may act as a mutational initiator of neoplasia and antitumorigenic defenses may be unlikely to show low-dose dependency. The dose response does not appear to be a complex function of increasing dose. The simplest representation is a linear relationship, which is consistent with most of the available mechanistic and quantitative data. There may be differences in response for different types of tumors, and statistical variations in each data set are inevitable. A departure from linearity is noted for leukemia data, for which a linear-quadratic function is used. Skin cancer and some cancers induced by alpha emitters may have virtual thresholds. Because of the multistep nature of the tumorigenesis process, linear or linear-quadratic functions are used for representational purposes only in evaluating possible radiation risks. The actual response may involve multiple and competing processes that cannot yet be separately distinguished, as summarized in table 12.1.

Table 12.1 Summary of conclusions on mechanisms of radiation carcinogenesis

---

DNA damage is the main initiating event by which radiation causes cell damage, cancer, and hereditary effects

Single radiation tracks have the potential to cause double-strand breaks and, in the absence of 100% efficient repair, could result in long-term damage even at the lowest doses

Fundamental judgments broadly support linear dose response

Cellular, animal, and epidemiological data are at their limits of resolution

Current challenges to linearity—
- For a threshold: error-free DNA repair and other postirradiation defenses, adaptive response
- For supralinearity: genomic instability, bystander effects, etc.— are not sufficient to alter these conclusions

---

## Dose and Dose-Rate Effectiveness Factor

For reasons related to statistical power, the dose-specific statistical estimates of radiation-related risk upon which this report is based reflect observed cancer excesses at equivalent doses greater than about 200 mSv, mainly delivered acutely. However, many of the more contentious issues in radiation protection involve risks from continuous exposures, or fractionated exposures with acute fractions of a few millisieverts or less. Experimental investigations tend to show that fractionation or protraction of dose is associated with reduced dose-specific risk, suggesting that dose-specific estimates based on high-dose, acute exposure data should be divided by a dose and dose-rate effectiveness factor (DDREF) for applications to low-dose, continuous, or fractionated exposures. The magnitude of DDREF is uncertain and has been treated as such in a number of recent reports based on quantitative uncertainty analysis (e.g., NCRP 1997; EPA 1999; NCI/CDC 2003). However, the mean of the probabilistic uncertainty distribution for DDREF employed in those analyses differs little from the value of 2 recommended by the ICRP (1991) and UNSCEAR (1993). A DDREF of 2 is also generally compatible with the animal data noted in ICRP (2007). For these reasons, the ICRP recommends that a DDREF of 2 continues to be used.

## Genetic Susceptibility to Cancer

The issue of interindividual genetic differences in susceptibility to radiation-induced cancer was reviewed in ICRP Publication 79 (ICRP 1998) and UNSCEAR (2000, 2001). There has been a remarkable expansion in knowledge since 1990 of the various single-gene human genetic disorders where excess spontaneous cancer is expressed in a high proportion of gene carriers, known as the *high penetrance* genes. There is also a growing recognition and some data on variant genes of lower penetrance where gene–gene and gene–environment interactions determine a far more variable expression of cancer.

Studies with cultured human cells and genetically altered laboratory rodents have also contributed much to knowledge and, with more limited epidemiological/clinical data, suggest that a high proportion of single-gene, cancer-prone disorders will show increased sensitivity to the tumorigenic effects of radiation.

Recently, good progress has been made in demonstrating experimentally the complex interactions that may underlie the expression of cancer predisposing genes of lower penetrance; however, this work is in its infancy.

## Heritable Diseases

Views on the risks of induction of heritable diseases by radiation exposure of the gonads were developed by extrapolating to humans quantitative data on dose response for germ cell mutations in experimental animals (predominantly mice) ICRP Publication 60 (ICRP 1991). Although extended follow-ups of mortality and cancer incidence in the offspring of the Japanese A-bomb survivors have been published (Izumi et al. 2003a, 2003b), these data do not alter the conclusions of previous analyses. In addition, few new quantitative data on mutation induction in mice have become available. However, since 1990 there have been significant developments in our understanding of the mutational process and new concepts for genetic risk estimation in human populations (UNSCEAR 2001).

The application of molecular genetic techniques has provided detailed knowledge of the molecular basis of naturally occurring mutations that cause heritable diseases in humans and of radiation-induced gene (specific locus) mutations in mouse germ cells. There is now strong evidence that large multilocus deletions of the genome constitute the predominant class of radiation-induced mutation. It is judged that only a proportion of such multigene loss events will be compatible with embryonic/fetal developmental and live birth. These findings have led to the concept that the principal adverse genetic effect in humans is likely to take the form of multisystem developmental abnormalities rather than single-gene diseases.

Another conceptual change based upon new human genetic information is the development of methods to assess the responsiveness of the frequency of chronic multifactorial diseases (e.g., coronary heart disease and diabetes) to an increase in mutation rate. This has allowed an improved estimate to be made of the risks associated with this large and complex class of disease where expression requires the interaction of genetic and environmental factors.

These human genetic, experimental, and conceptual advances have been integrated to form a new and more robust framework for the estimation of genetic risks (UNSCEAR 2001).

There have also been developments in the estimation of radiation-induced mutation rates in mice and humans using expanded simple-tandem-repeat DNA loci in mice and mini-satellite loci in humans. These DNA repeats are highly mutable with the mutations manifesting as changes in the number of tandem repeats. This increased mutability is expressed spontaneously and after radiation. Attention has been given to the mutational mechanisms involved, including the untargeted and transgenerational effects of radiation (UNSCEAR 2000, 2001; CERRIE 2004). However, since current knowledge indicates that mutations at these DNA repeat sequences are only rarely associated with genetic disorders, there is no good reason to include quantitative mutational data for these loci in the estimates of genetic risk.

## Cancer Epidemiology

Radiation-associated cancer in humans is studied in population groups that have been exposed to radiation doses so that cancer cases in excess of the normal background incidence may be identified. Estimates of risk may be derived from populations for which individual doses can be reasonably estimated. These populations include survivors of the atomic bombings, medically irradiated patients, those occupationally exposed, individuals exposed to radionuclides released into the environment, and some people exposed to elevated levels of natural background radiation (see table 12.2).

It is now known that radiation can cause cancer in almost any tissue or organ in the body, although some sites are much more likely to develop cancer than are others. A clearer understanding of physiological modifying factors, such as sex and age, has developed over the last few years. Although differences in the absolute risk of tumor induction by sex are not large and vary by site, the absolute risk is higher in women than in men for most solid cancers. People who were young at the time of radiation exposure have higher relative and absolute risks than older people do, but again this varies by site.

Further follow-up of radiation-exposed cohorts has demonstrated that excess cancers continue to occur for a long time after radiation exposure, and therefore, large uncertainties can arise in the projection of lifetime risks. Data for the Japanese atomic bomb survivors are consistent with a linear or linear-quadratic dose response over a wide range of doses, but quantifying risks at low doses is less certain because of the limitations of statistical precision, the potential residual biases or other methodological problems, and the possibility of chance findings due to multiple statistical testing. Longer follow-up of cohorts with a wide range of doses, such as the atomic bomb survivors, will provide more essential information at low doses, but epidemiology alone will not be able to resolve the issue of whether there are low-dose thresholds. It should be noted, however, that the inability to detect increased risks at very low doses does not mean that these increases in risk do not exist (UNSCEAR 2000; ICRP 2005).

Table 12.2  Sources of data for epidemiological radiation risk assessments

| |
|---|
| Japanese A-bomb survivors |
| Medical studies |
| Occupational studies |
| Studies in the former Soviet Union |
| Groups exposed to high natural background radiation |

## Japanese A-Bomb Survivors

The studies of the Japanese survivors are particularly important because the cohort includes a large exposed population of both sexes, a wide distribution of doses, and the full range of ages. The results of this study provide the primary basis for estimating the risk of radiation-induced cancer. Among the 86,572 individuals in the Life Span Study cohort of survivors of the atomic bombings, there were 7,578 deaths from solid tumors from 1950 through 1990. Of these cancer deaths, 334 can be attributed to radiation exposure. In this same period, 87 of 249 leukemia deaths can be attributed to radiation exposure. In 1991, some 48,000 persons (56%) were still living, while at the time of the latest evaluation in 2000, 44% of the population was still living (see table 12.3).

The Life Span Study cancer incidence and mortality data are broadly similar, demonstrating statistically significant effects of radiation for all solid tumors as a group, as well as for cancers of the stomach, colon, liver, lung, breast, ovary, and bladder. The incidence data also provide evidence of excess radiation risks for thyroid cancer and nonmelanoma skin cancers. Statistically significant risks were not seen in either the incidence or the mortality data for cancers of the rectum, gall bladder, pancreas, larynx, uterine cervix, uterine corpus, prostate, and kidney or renal pelvis. An association with radiation exposure is noted for most types of leukemia, but not for lymphoma or multiple myeloma. A summary of the number of cancers and the excess due to radiation is given in table 12.4.

Table 12.3 Positive and negative indications of the epidemiology of the A-bomb survivors

*Positive*

The primary source of information for whole-body exposure
86,572 individuals of both sexes and all ages with good dose data
  over a wide range of doses

*Negative*

Japanese population with different baseline rates of cancer
44% still alive in 2000, so future deaths uncertain
Acute exposure, relatively high, whereas interest is in low chronic
  exposures

Table 12.4 Number of radiation-induced cancers in the Life Span Study

| | |
|---|---|
| Cohort | 86,572 |
| Alive in 2000 | ~38,000 |
| Deaths from cancer or leukemia | ~10,000 |
| Number of radiation-induced cancer deaths: | |
|     Leukemia ~100 | |
|     Solid cancers ~400 | |

The numbers of solid tumors associated with radiation exposure are not sufficient to permit detailed analysis of the dose response for many specific sites or types of cancer. For all solid tumors combined, the slope of the dose–response curve is linear up to about 3 Sv, but the dose–response curve for leukemia is best described by a linear-quadratic function. Statistically significant risks for cancer in the Life Span Study are seen at organ doses above about 100 mSv.

## Other Cohorts

Studies of populations exposed to medical, occupational, or environmental radiation provide information on issues that cannot be addressed by the atomic bomb survivor data, such as the effects of chronic low doses, alpha doses to the lung from radon, highly fractionated doses, and variability among populations. For some cancer sites, including leukemia, breast, thyroid, bone, and liver, very useful results come from investigations other than the Life Span Study. Risk estimates derived from those studies generally agree well with those from the Life Span Study.

Large studies of occupationally exposed persons are also contributing valuable data on low-dose effects. A combined analysis of data for a large number of nuclear workers indicates that the risk of leukemia increases with increasing dose. However, the statistical precision of such studies is still low in comparison with the results at high-dose rate from the atomic bomb survivors. As a result, it is difficult to arrive at a definitive conclusion on the effects of dose rate on cancer risks, particularly since these effects may differ among cancer types. However, the conclusion reached in the UNSCEAR (1993) report, based on both epidemiological and experimental evidence, which suggested a reduction factor of something less than 3 when extrapolating to low doses or low-dose rates, still appears to be reasonable in general. This supports the final choice of a DDREF of 2 based on mechanistic studies (see "Radiobiological Effects after Low Doses of Radiation," above).

Information on the effects of internal doses, from both low- and high-LET radiation, has increased in the last decade. In particular, an elevated risk of thyroid cancer in parts of Belarus, the Russian Federation, and the Ukraine contaminated as a result of the Chernobyl accident shows a link with radioactive iodine exposure during childhood. However, risk estimation associated with these findings is complicated by difficulties in dose estimation and in quantifying the effect of screening for the disease. Other studies in the former Soviet Union have provided further information on internal doses (e.g., an increased risk of lung cancer among workers at the Mayak plant). Leukemia was elevated in the population living near the Techa River. However, the different sources of radiation exposure (both external and internal) and, in the case of the Techa River studies, the potential effects of migration, affect the quantification of risks. Results from several case–control studies of lung cancer and indoor radon have been published in recent years that, in combination, are consistent with extrapolations from data on radon-exposed miners, although the statistical uncertainties in these findings are still large.

Particular attention has been paid in annex I of UNSCEAR (2000) to risks for specific cancer sites. Again, the new information that has become available in recent years has helped in the examination of some risks. However, there remain problems in characterizing risks for some cancer sites because of the low statistical precision associated with moderate or small excess numbers of cases. This can limit, for example, the ability to estimate trends in risk in relation to factors such as age at exposure, time since exposure, and gender. An exception is breast cancer, where a comparison of data on the Japanese atomic bomb survivors and women with medical exposures in North America points to an absolute transfer of risks between populations. There are some cancer sites for which there is little evidence for an association with radiation (e.g., non-Hodgkin's lymphoma, Hodgkin's disease, and multiple myeloma). While the evaluations for the lymphomas are affected in part by the small numbers of cases in several studies, they should be contrasted with the evaluations for leukemia (excluding chronic lymphocytic leukemia), which while also a rare disease, has clearly been related to radiation in many populations.

Lifetime risk estimates are sensitive to variations in background tumor rates, and the variability can lead to differences that are comparable to differences associated with the transport method across populations or the method of risk projection. The variability in these projections highlights the difficulty of choosing a single value to represent the lifetime risk of radiation-induced cancer. Furthermore, uncertainties in estimates of risk for specific types of cancer are generally greater than for all cancers combined.

One radiation-associated cancer of particular importance in children is cancer of the thyroid gland. There is strong evidence that the risk of thyroid cancer decreases with increasing age at exposure, so that the risk in children younger than 15 years of age is substantially larger than that in adults. Children 0–5 years of age are five times more sensitive than those 10–14 years of age. In view of this sensitivity, it is not surprising that large increases in thyroid cancer incidence have been observed in children in Belarus, the Russian Federation, and the Ukraine following the Chernobyl accident in 1986. The thyroid cancer incidence rate in children from these regions was 10-fold higher in 1991–1994 than in the preceding five years. This topic is reviewed extensively in annex J of UNSCEAR (2000).

In a recent assessment of thyroid cancers from 1986 to 2002 in children or adolescents at the time of the accident (Cardis et al. 2006), the number of thyroid cancer cases diagnosed in Belarus, the Ukraine, and the most contaminated regions of Russia is close to 5,000 in a population of 14,400,000. Of these, 15 are known to have been fatal up to now. Individual thyroid doses ranged up to several tens of grays, while mean doses range from a few tens of milligrays to several grays in particular population groups. Figure 12.6 shows the results in Belarus for children, adolescents, and young adults.

## *In Utero Exposures*

Cancer may be induced by prenatal exposure. In humans, the induction of childhood cancers, leukemia, and solid cancers because of exposure to X-rays was first reported in 1958, when the Oxford Survey established an increased incidence of childhood tumors in the first 15 years of life for those exposed to X-rays in utero compared with those who were not exposed. The attribution of this increase to radiation exposure

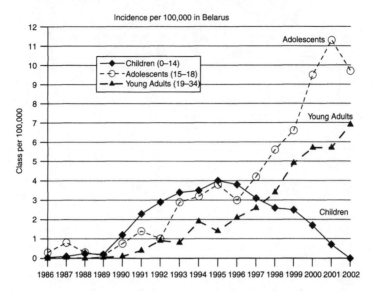

Figure 12.6 Annual incidence of childhood, adolescent, and young adult thyroid cancers in Belarus (from Cardis et al. 2006).

has been criticized by some on the grounds that the exposed women may have had medical or other conditions that were responsible for the increased cancer rates. Support for the causal role of radiation is found in some other studies, and the risk, if genuine, was estimated to be about 5% $Sv^{-1}$. No such effects were observed in survivors of the atomic bombings irradiated in utero.

Risks of induced cancer expressed in adulthood among those exposed in utero are more difficult to evaluate. Nevertheless, the fact that relative risks increase with decreasing age at exposure among the survivors of the atomic bombings causes concern about a potentially greater sensitivity to cancer induction for those exposed in utero than for those exposed at young ages. The atomic bomb survivors exposed in utero were 55 years old in the year 2000. Thus, it is especially important to evaluate their cancer risk experience late in life.

## Uncertainties in Risk Estimates Based on Mortality Data

Based on the available epidemiological data, UNSCEAR (2000) derived risk estimates for radiation-induced cancer. For a population of all ages and both genders with an acute dose of 1 Sv (low LET), it is suggested that lifetime risk estimates for solid cancer mortality might be taken as 9% for men and 13% for women. The uncertainties in these estimates may be of the order of a factor of 2 higher or lower. The estimates could be reduced by 50% for chronic exposures, again with an uncertainty factor of 2 higher or lower. Solid cancer incidence risks can be taken as being roughly twice those for mortality. Lifetime solid cancer risks estimates for those exposed as children might be twice the estimates for a population exposed at all ages. However, continued follow-up in studies of such groups will be important in determining lifetime risks. The experience of the Japanese atomic bomb survivors provides compelling evidence for linearity in estimating excess risks of solid cancers; therefore, as a first approximation, linear extrapolation of the estimates at 1 Sv could be used for estimating solid cancer risks at lower doses.

The estimates of lifetime risks for leukemia are less variable. The lifetime risk of death from leukemia may be taken as 1%, for either gender, following an acute dose of 1 Sv. The uncertainty in this estimate may be of the order of a factor of 2 higher or lower. In view of nonlinearity in the dose response, decreasing the dose 10-fold, from 1 Sv to 0.1 Sv, will result in a 20-fold decrease in the lifetime risk if the dose is acute. The risks of solid cancer and leukemia are broadly similar to those estimated in UNSCEAR (1993).

## Risk Coefficients for Cancer and Hereditary Effects

New data on the risks of radiation-induced cancer and hereditary effects have been used by the ICRP in risk modeling and disease detriment calculations in order to estimate nominal probability coefficients for use in risk assessment (ICRP 2007).

### Cancer Risk Coefficients

In ICRP Publication 60 (ICRP 1991), nominal cancer risks were computed based on mortality data; however, in the current report (ICRP 2007), risk estimates are based principally on incidence data. The reason for the change is that incidence data provide a more complete description of the cancer burden than do mortality data, particularly for cancers that have a high survival rate. In addition, cancer registry (incidence) diagnoses are more accurate and the time of diagnosis is more precise. It is recognized, however, that incomplete coverage of the A-bomb population because of migration from Hiroshima or Nagasaki introduces a factor of uncertainty on risk estimates based on these cancer incidence data. At the time of ICRP Publication 60, comprehensive incidence data were not available. Since then, a thorough evaluation of cancer incidence in the Life Span Study of Japanese atomic bomb survivors has been published (Thompson et al. 1994; Preston et al. 1994), and new analyses regarding the latest A-bomb cancer incidence data are expected soon. Site-specific risk estimates were taken from the most recent solid cancer incidence analyses of the atomic bomb survivor in the Life Span Study, with follow-up from 1958 through 1998, and adjusted to reduce bias in risk estimates due to uncertainty in individual dose estimates (Pierce et al. 1990). The newly implemented atomic bomb dosimetry system, DS02, is a considerable improvement over DS86. On average, the DS02 dose estimates are slightly greater than the DS86 estimates. Risk estimates using the two systems differ by less than 10% (Preston et al. 2004).

The resulting gender and age-averaged nominal risk coefficients derived from the ICRP (2007) are shown in table 12.5. The named organs and tissues are those for which the ICRP has felt able to quantify risk factors. The "Other Solid" represents

Table 12.5 Summary of gender-averaged nominal cancer risks for a population of all ages

| Tissue | Nominal incidence (% Sv$^{-1}$) | Lethality fraction | Nominal fatal risk (% Sv$^{-1}$) |
|---|---|---|---|
| Esophagus | 0.15 | 0.93 | 0.14 |
| Stomach | 0.79 | 0.83 | 0.66 |
| Colon | 0.65 | 0.48 | 0.31 |
| Liver | 0.30 | 0.95 | 0.29 |
| Lung | 1.14 | 0.89 | 1.01 |
| Bone | 0.07 | 0.45 | 0.03 |
| Skin | 10.00 | 0.002 | 0.02 |
| Breast | 1.12 | 0.29 | 0.32 |
| Ovary | 0.11 | 0.57 | 0.06 |
| Bladder | 0.43 | 0.29 | 0.12 |
| Thyroid | 0.33 | 0.07 | 0.02 |
| Bone Marrow | 0.42 | 0.67 | 0.28 |
| Other Solid$^a$ | 1.44 | 0.49 | 0.71 |
| **Total (rounded)** | **17** (7 excluding skin) | | **4** |

$^a$Other solid (14 total): adrenals, extrathoracic region, gall bladder, heart, kidneys, lymphatic nodes, muscle, oral mucosa, pancreas, prostate, small intestine, spleen, thymus, uterus/cervix.

a group of organs and tissues that the Commission considers to be radiosensitive, but the available data are insufficient to quantify individual risk. There are 14 of these organs or tissues, and the Commission has given a total risk factor that can be divided equally among the 14 to give a risk of 0.01 % Sv$^{-1}$ to each. The named organs and tissues are the adrenals, extrathoracic region, gall bladder, heart, kidneys, lymphatic nodes, muscle, oral mucosa, pancreas, prostate, small intestine, spleen, thymus, and uterus/cervix.

Table 12.5 first gives the nominal incidence rate and then the average lethality for cancer of the individual organ. The third column shows the consequent fatal nominal risk coefficient. These coefficients should be used in conjunction with the relevant organ equivalent dose to estimate risk. For whole-body exposure from low-LET radiations, a risk of 4% Sv$^{-1}$ should be applied.

In general for most environmental assessments low-LET radiation or alpha radiation is likely to be predominant. If the irradiation is from an alpha-emitting source, then a weighting factor of 20 should be used to convert the organ absorbed dose (Gy) to equivalent dose (Sv). If irradiation involves neutron exposures or other radiations such as protons, then reference should be made to the more detailed data in ICRP (2007).

## Hereditary Risk

The term "genetic risks" as used in this document denotes the probability of harmful genetic effects manifest in the descendants of a population that has sustained radiation exposures. These effects are expressed as increases over the baseline frequencies of genetic diseases in the population per unit dose of low-LET, low-dose/chronic irradiation. Since the publication of the 1990 recommendations of the ICRP (ICRP 1991), the 1990 report of the Committee on the Biological Effects of Ionizing Radiation (NAS/NRC 1990), and the UNSCEAR (1993) report, several important advances have been made in the prediction of genetic risks of exposure of human populations to ionizing radiation. On the basis of these, UNSCEAR (2001) revised its earlier risk estimates.

In its 2001 report, UNSCEAR completed a comprehensive review of the risk to offspring following parental exposure to radiation. For the first time, the review included an evaluation of those diseases that have both hereditary and environmental components, the so-called *multifactorial* diseases. The major finding is that the total hereditary risk is 0.3–0.5% $Gy^{-1}$ to the first generation following irradiation. This is less than one-tenth of the risk of fatal carcinogenesis following irradiation. Since it takes some hundreds of generations for defects to reach equilibrium, the risk to the first few generations is still about 10% of the carcinogenic risk.

There are some problems in comparing genetic risk coefficients with those for cancers. Cancer risk coefficients quantify the probability of harmful effects of radiation to the exposed individuals themselves, and genetic risk coefficients quantify the probability of harmful effects to the descendants of those exposed. In the case of genetic risk coefficients, the inclusion of risk up to two generations in the calculations can be justified on the basis that people are generally interested in the well-being of their children and grandchildren. The estimate restricted to the first postradiation generation has the advantage that it is more comparable to those for cancers and, therefore, deserves serious consideration.

Detriment adjusted probability coefficients for hereditary disease up to the second generation are recommended as $0.2 \times 10^{-2}$ $Sv^{-1}$ for the whole population and $0.1 \times 10^{-2}$ $Sv^{-1}$ for adult workers (ICRP 2007); the respective values from ICRP Publication 60 (ICRP 1991) are $1.3 \times 10^{-2}$ $Sv^{-1}$ and $0.8 \times 10^{-2}$ $Sv^{-1}$, but these relate to risks at a theoretical equilibrium and no longer seem justified.

## Overall Conclusions on Biological Effects at Low Doses

Although the mechanistic uncertainty remains, studies on DNA repair and the cellular/molecular processes of radiation tumorigenesis provide no good reason to assume that there will be a low-dose threshold for the induction of tumors in general. However, curvilinearity of the dose response in the low-dose region—perhaps associated with biochemical stress responses or changing DNA repair characteristics—cannot be excluded as a general feature. The mechanistic modeling of radiation tumorigenesis is at a relatively early stage of development, but the data available tend to argue against a dose threshold for most tumor types.

Until the above uncertainties on low-dose response are resolved, the international view is that an increase in the risk of tumor induction proportionate to the radiation dose is consistent with developing knowledge and that it remains, accordingly, the most scientifically defensible approximation of low-dose response.

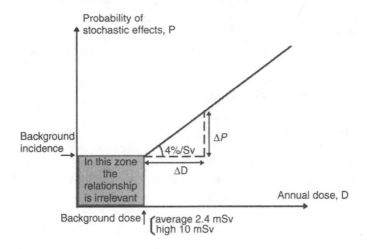

Figure 12.7  Region of the dose–response curve where risk factors apply.

This is illustrated in figure 12.7, where it is clear that the risk figures apply to the slope of the dose-response curve at the level of background, some few millisieverts per year. However, a strictly linear dose response should not be expected in all circumstances.

The dose response for the induction of heritable disease carries fewer low-dose biological uncertainties than that of multistage tumorigenesis, but the same uncertainties surrounding DNA damage response remain. An increase in the risk of germ-cell mutation that is proportionate to radiation dose is judged to be a scientifically reasonable approximation for the induction of heritable effects at low doses.

It must be recognized that ongoing and future studies in epidemiology and animal sciences, while remaining of great importance for qualitative risk assessment, will not resolve the uncertainties surrounding the effects in humans of low-dose radiation. Accordingly, there will be an increasing need for weight-of-evidence judgments based on largely qualitative data from cellular or molecular studies of the biological mechanisms that underlie health effects; the provision of such judgments demands strong support from biologically validated computational models of risk.

With ever-improving experimental technology, fundamental knowledge will continue to grow. On this basis, the emphasis is on the need for further work on the mechanisms of DNA damage response/cellular stress and studies of the consequences of these responses for neoplastic development. Current uncertainties on the role of epigenetic factors such as bystander effects or induced genomic instability may be reduced, but it may remain difficult to estimate their overall contribution to risk. However, the development of mechanistic models of radiation risk demands more than a simple improvement in the understanding of cellular or molecular processes.

## Problems

1. What is the main initiating event by which radiation causes long-term damage to organs and tissues of the body?
2. Why is the use of a simple proportionate relationship between increments of dose and increased risk a scientifically plausible assumption?
3. Explain the roles of induced genomic instability, bystander cell signaling, and adaptive response in the genesis of radiation-induced health effects for radiological protection purposes.
4. Why is genetic susceptibility to radiation-induced cancer involving strongly expressed genes assumed not to appreciably distort estimates of population risk? What about the potential impact of common but weakly expressing genes?
5. Define the dose and dose-rate effectiveness factor (DDREF). What value is recommended for radiological protection purposes? Why?
6. What is the gender averaged fatal cancer nominal risk coefficient for the whole population? Upon what is this value based?
7. What is the detriment-adjusted probability coefficient for hereditary disease up to the second generation for the whole population? Why does the respective previous value of $1.3 \times 10^{-2}$ $Sv^{-1}$ no longer seem justified?

### References

Cardis, E., G. Howe, E. Ron, V. Bebeshko, T. Bogdanova, A. Bouville, et al. 2006. "Cancer Consequences of the Chernobyl Accident: 20 Years On." *Journal of Radiological Protection* 26: 127–140.

CERRIE. 2004. *Report of the Committee Examining Radiation Risks of Internal Emitters (CERRIE)*. Committee Examining Radiation Risks of Internal Emitters, London. Accessed online at www.cerrie.org.

Cherubini, R., D.T. Goodhead, H.G. Menzel, and A. Ottolenghi, eds. 2002. "Proceedings of the 13th Symposium on Microdosimetry." *Radiation Protection Dosimetry* 99 (1–4).

EPA (U.S. Environmental Protection Agency). 1999. *Estimating Radiogenic Cancer Risks*. 402-R-00-003. Environmental Protection Agency, Washington, DC.

ICRP (International Commission on Radiological Protection). 1991. "1990 Recommendations of the International Commission on Radiological Protection." ICRP Publication 60. *Annals of the ICRP* 21(1–3).

ICRP. 1998. "Genetic Susceptibility to Cancer." ICRP Publication 79. *Annals of the ICRP* 28(1/2).

ICRP. 2003. "Relative Biological Effectiveness (RBE), Quality Factor (Q), and Radiation Weighting Factor ($w_R$)." ICRP Publication 92. *Annals of the ICRP* 33(4).

ICRP. 2005. "Low-Dose Extrapolation of Radiation Related Cancer Risk." ICRP Publication 99. *Annals of the ICRP* 35(4).

ICRP. 2007. "Biological and Epidemiological Information on Health Risks Attributable to Ionizing Radiation: A Summary of Judgments for the Purposes of Radiological

Protection of Humans." ICRP Publication 103. *Annals of the ICRP* 37(2–4), Annex A.

Izumi, S., A. Suyama, and K. Koyama. 2003a. "Radiation-Related Mortality among Offspring of Atomic Bomb Survivors after a Half-Century of Follow-up." *International Journal of Cancer* 107: 291–297.

Izumi, S., K. Koyama, M. Soda, and A. Suyama. 2003b. "Cancer Incidence in Children and Young Adults Did Not Increase Relative to Parental Exposure to Atomic Bombs." *British Journal of Cancer* 89: 1709–1713.

NAS/NRC (National Academy of Sciences/National Research Council). 1990. *The BEIR V Report*. National Academy Press, Washington, DC.

NCI/CDC (National Cancer Institute/Centers for Disease Control). 2003. *Report of the NCI-CDC Working Group to Revise the 1985 NIH Radioepidemiological Tables*. NIH Publication 03–5387. U.S. Department of Health and Human Services, National Institutes of Health, National Cancer Institute, Bethesda, MD.

NCRP (National Council on Radiation Protection and Measurements). 1997. *Uncertainties in Fatal Cancer Risk Estimates Used in Radiation Protection*. NCRP Report No. 126. National Council on Radiation Protection and Measurements, Bethesda, MD.

NCRP 2001. *Evaluation of the Linear-Non-Threshold Dose-Response Model for Ionizing Radiation*. NCRP Report No. 136. National Council on Radiation Protection and Measurements, Bethesda, MD.

Pierce, D.A., D.O. Stram, and M. Vaeth. 1990. "Allowing for Random Errors in Radiation Dose Estimates for the Atomic Bomb Survivor Data." *Radiat. Res.* 123: 275–284.

Preston, D.L., S. Kusumi, M. Tomonaga, S. Izumi, E. Ron, A. Kuramoto, et al. 1994. "Cancer Incidence in Atomic Bomb Survivors. Part III. Leukaemia, Lymphoma and Multiple Myeloma, 1950–1987." *Radiation Research* 137: S68–S97.

Preston, D.L., D.A. Pierce, Y. Shimizu, H.M. Cullings, S. Fujita, S. Funamoto, and K. Kodama. 2004. "Effect of Recent Changes in Atomic Bomb Survivor Dosimetry on Cancer Mortality Risk Estimates." *Radiation Research* 162: 377–389.

Thompson, D.E., K. Mabuchi, E. Ron, M. Soda, M. Tokunaga, and S.Ochikubo. 1994. "Cancer Incidence in Atomic Bomb Survivors. Part II: Solid Tumors, 1958–1987." *Radiation Research* 137: S17–S67.

UNSCEAR (U.S. Scientific Committee on the Effects of Atomic Radiation). 1993. *Sources and Effects of Ionizing Radiation*. 1993 Report to the General Assembly with Scientific Annexes. United Nations, New York.

UNSCEAR. 2000. *Sources and Effects of Ionizing Radiation*: Vol. 2. *Effects*. Report of the U.N. Scientific Committee on the Effects of Atomic Radiation. E.00.IX.4. United Nations, New York.

UNSCEAR. 2001. *Hereditary Effects of Radiation*. Report of the U.N. Scientific Committee on the Effects of Atomic Radiation. E.01.IX.2. United Nations, New York.

# 13

# The Role of Epidemiology in Estimating Radiation Risk: Basic Methods and Applications

Owen J. Devine
Paul L. Garbe

Epidemiology is the study of how disease risk and possible determinants of that risk vary in human populations (Last 1995). Much of our current knowledge about the health risks associated with certain activities and exposures is based on evidence from epidemiologic studies. For example, current public health activities related to reducing the level of cigarette smoking, decreasing childhood exposures to lead, and removing asbestos from buildings were started on the basis of the findings of epidemiologic investigations. In other chapters of this book, you will encounter estimates of the health risks associated with exposures to ionizing radiation. These estimates are almost all based on epidemiologic studies of populations, such as atomic bomb survivors, uranium miners, and medical patients, who have been exposed to radioactive materials (NRC/NAS 1990; UNSCEAR 1993; Lubin et al. 1994). In addition, differences of opinion concerning the health risks resulting from exposures to low levels of ionizing radiation are often based on varying interpretations of new and existing epidemiologic analyses (Cohen and Colditz 1994), and critical evaluation of these analyses often hinges on the underlying epidemiologic practice (Greenland 1992). As a result, a general understanding of the basic methods used in epidemiology is beneficial to professionals involved in the fields of radiation protection, risk assessment, and health physics.

This chapter provides an overview of the general methods used in epidemiologic investigations with a special emphasis on approaches pertinent to radiation-related epidemiology. Because epidemiologic investigations are based on interpreting the disease risk in populations, we begin with a description of how the risk and rate of disease are estimated on the basis of observed disease counts. We then examine measures of association that are commonly used to evaluate potential relationships between these risk and rate estimates and the exposure to a suspected causative factor. In addition, we provide a brief discussion on methods for evaluating the magnitude of an observed association between disease risk and a suspected causative exposure in light of the random variation expected in observable measures of disease burden. We conclude with an examination of issues specifically related to epidemiologic investigations of exposures to radioactive materials, including the critical nature of appropriate dosimetric practice.

## Measures of Disease Burden in Populations

The goal of epidemiologic analyses is to evaluate patterns of disease risk in populations and, if possible, identify predictors of that risk. The first step in accomplishing this goal is counting the number of individuals having disease in the population under study. These disease counts are generally based on one of two, not necessarily distinct, events: diagnosis of, or death due to, the disease of interest. A count of the number of new cases of the study disease is called *incidence*, while enumeration of disease-related deaths is referred to as *mortality*. For some diseases, these two measures will be similar; for other outcomes, specifically diseases that are not generally fatal, incidence and mortality counts can be quite different. For example, lung and thyroid cancer are often health outcomes of interest in epidemiologic studies on the impacts of exposure to radioactive materials. Because lung cancer is generally a fatal disease, mortality and incidence counts for this outcome are usually similar. Thyroid cancer, on the other hand, is rarely fatal, and as a result, incidence is likely a better measure of the potential impact of radiation exposure for this disease than is mortality. For the remainder of this chapter, we generally frame our discussion of estimators for the disease burden in a population in terms of incidence measures. However, similar approaches can be used to quantify mortality risk in the population under study.

### Estimating Disease Risk

We define *risk* as the conditional probability of developing the disease under study during a specified period of time. This probability is conditional because it depends upon a person being at risk, that is, alive and disease free, at the beginning of the period. The true risk for specific individuals within the population depends on characteristics such as age, sex, race, genetic susceptibility, and, potentially, other factors such as exposure to ionizing radiation or other contaminants. Because individual variability in disease susceptibility is associated with factors such as an individual's genetic background, which are currently not measurable, estimation and comparison of disease risk on an individual basis are not possible. As a result, epidemiologic studies generally assess how the *average* risk for disease among subgroups of a population is affected by differences in the prevalence of suspected causative factors. Unfortunately, the true underlying average risk for these subgroups is not observable but instead must be estimated from available information. As a result, inference about disease exposure relationships depends on developing meaningful estimators for the true risk in the study population.

One such estimator for the average risk in a population is based on the cumulative incidence of disease over a specified period of time. For example, suppose we are interested in estimating the risk of lung cancer in a specified population during the time interval $t_0$ through $t_1$. An estimator for the period-specific risk is

$$R(t_0, t_1) = \frac{D(t_0, t_1)}{N(t_0)}, \tag{13.1}$$

where $D(t_0, t_1)$ is the number of new cases of lung cancer that occur in the population during the observation period and $N(t_0)$ is the number of living, lung-cancer-free individuals in the population at time $t_0$.

The estimator for risk given in equation 13.1 is the most commonly used in epidemiologic studies and is appropriate for fairly short time intervals such as one to five years. For longer observation periods, however, the denominator may not reflect the true number of individuals who are at risk for disease development during the interval. To see why this is true, consider the fact that some portion of the study population is likely to die of causes other than the disease of interest during long observation periods. We have no way of knowing, in general, if those individuals who died from other causes would have developed the disease of interest had they survived. As a result, the estimator in equation 13.1 should be adjusted to account for the fact that some portion of the study group is not at risk for developing the disease under study for the entire period. One way to make this adjustment is to assume that withdrawals from the population at risk occur, on average, halfway through the observation period. This estimator, called the actuarial estimator by Kleinbaum et al. (1982), is given by

$$R(t_0, t_1) = \frac{D(t_0, t_1)}{\left[N(t_0) - \frac{W(t_0, t_1)}{2}\right]}, \tag{13.2}$$

where $W(t_0, t_1)$ is the number of individuals withdrawn from those at risk during the study period. Notice that use of the alternative risk estimator requires the analyst to identify all those removed from observation or *lost to follow-up* for any reason instead of just counting the number of cases of the disease under study. If only $N(t_0)$ and $D(t_0, t_1)$ are available, then risk must be estimated using equation 13.1.

An important point to remember when using equation 13.1 or 13.2 is that the accuracy of both estimators depends on an assumption that the true average risk during the interval $t_0$ through $t_1$ is constant. If the interval separating $t_0$ and $t_1$ is long, this assumption may become questionable. For example, remediation of a toxic waste site in the middle of a study period may significantly reduce cancer risks for subsequent years in the surrounding community. To increase the likelihood of providing an accurate risk estimate, time periods are generally limited to one to five years. However, we may want to develop average risk estimates for the study population that span longer periods. For example, suppose we want to estimate cancer risk over a period of 20 years. If we believe that the risk is fairly stable within shorter time intervals, one way to develop the 20-year risk estimate is to partition the interval into shorter periods and estimate the risk within each of the shorter intervals. These interval-specific risk estimates can then be combined to provide an estimator of risk across the entire study period. To combine these interval-specific risk estimates, we must assume that the risk for disease in a given time period is independent of the risk in any other time interval. Under this assumption, the risk in a combination of time intervals, say, $(t_0, t_1), (t_1, t_2) \ldots (t_{j-1}, t_j)$ can be estimated by

$$R(t_0, t_j) = 1 - \prod_{i=0}^{j-1} [1 - R(t_i, t_{i+1})], \qquad (13.3)$$

where $R(t_i, t_{i+1})$ is the interval-specific risk in the time period $t_i$ through $t_{i+1}$ estimated by using equation 13.1 or 13.2. Notice that the estimator is 1 minus the probability of surviving through time $t_j$ without developing disease or, equivalently, the probability of developing disease by time $t_j$.

The idea of combining interval-specific risk estimates can be expanded to incorporate changes in risk due to age as well as time. For example, when estimating risk over periods longer than five years, you may need to account for the fact that the population is increasing in age as well as experiencing a potential change in risk with time. As an illustration, suppose we wish to estimate the 20-year risk for lung cancer among a group of males who were 45 years old in 1975. Not only is it possible that the average annual lung cancer risk for all citizens has changed from 1975 to 1995, perhaps due to enhanced environmental policies, but it is also virtually certain that the study group's average annual risk increased as its members aged during the 20-year observation period. To address the possibility that risk can change with time and age, we can use equation 13.1 to estimate risk within age and time intervals and then combine these age- and time-specific estimates in a manner similar to equation 13.3. The result is an estimate of the average risk in a population that is age $a_j$ at time $t_j$ and was age $a_0$ at time $t_0$. This approach of combining age- and time-specific risk estimates across age/time strata is an integral tool in the statistical method known as survival analysis and constitutes the basic approach used to develop lifetime risk estimates (Thomas et al. 1992).

## Estimating Disease Rate

The risk estimates discussed so far provide a description of the average disease risk existing in a population under study. Epidemiologists, however, are often interested in the frequency at which new cases of the study disease occur. The usual measure of this frequency is the incidence rate. While the estimators discussed so far correspond to the risk, or the probability of developing disease among those who are at risk, the incidence rate reflects the occurrence of new cases per unit of time spent at risk by the study population. For example, suppose 10 individuals are observed for five years, during which three of the subjects develop the disease of interest. The average incidence rate, usually referred to as simply the incidence rate (IR), for this population over this period is

$$\text{IR} = \frac{D(t_0, t_1)}{\text{PY}}, \qquad (13.4)$$

where PY is the person-years at risk accumulated by this group over the five-year period. Using our example data, these person-years are given by

$$PY = 10 \text{ persons} \times 5 \text{ years} = 50 \text{ PY},$$

and the incidence rate is therefore

$$IR = \frac{3 \text{ cases}}{50 \text{ PY}} = 0.06 \text{ cases per PY}$$

Often incidence rates are multiplied by large round numbers for ease of expression. For example, the incidence rate calculated above could be multiplied by 1,000 and expressed as 60 cases per 1,000 PY at risk.

It is unlikely that everyone in a study group who develops disease does so at the last second of the last year of observation; rather, individuals will develop disease at some point during the study period. After these individuals develop disease or, in practice, when they are diagnosed, they should no longer contribute person-time at risk to the denominator of equation 13.4. In addition, individuals may stop contributing person-time at risk because they die from other causes, leave the study, or are simply lost to follow-up. If individual follow-up data are available for all members of the study population, then a more precise estimate of the incidence rate is given by

$$IR = \frac{D(t_0, t_1)}{\sum_{i=1}^{N(t_0)} PY_i}, \qquad (13.5)$$

where $PY_i$ is the observed time at risk for individual $i$ in the study population.

## Estimating Disease Prevalence

We now need to distinguish between measures of incidence and prevalence of disease. Whereas the incidence rate represents the frequency of the appearance of new cases per unit of time spent at risk, prevalence is defined as the ratio of the total number of cases at a given point in time to the total number of people in the population. Thus, prevalence at time $t_i$, $P(t_i)$, is given by

$$P(t_i) = \frac{D(t_i)}{N(t_i)}. \qquad (13.6)$$

Notice that the prevalence does not provide information about the time-dependent rate of disease occurrence in the population, as does the incidence rate, but rather reflects a snapshot of the overall burden of disease.

We now consider the relationship between the incidence rate in a given population and the disease risk. As stated above, we define risk as the conditional probability of developing disease during some specified period assuming that a person is alive and disease-free at the beginning of that interval. Because risk is a probability, it can range between 0 and 1 and is unitless. Incidence rates, however, can exceed 1 if, for example, we count recurrent diagnoses of symptoms of a chronic underlying illness. In addition, incidence rates are expressed in units of the population time spent at risk. Thus, the terms "risk" and "incidence rate" are not interchangeable, and there is often confusion about the difference in these measures, even among epidemiologists. To examine the difference between these metrics, suppose we have estimated the incidence rate (IR) of a disease in a population during some given period of time. Because the incidence rate reflects the instantaneous rate of disease occurrence, then the instantaneous rate of change in the number of people at risk for developing disease at any point in time, say, $N(t)$, is

$$\frac{dN(t)}{dt} = -\text{IR} \times N(t). \tag{13.7}$$

If we suppose that the time interval began at $t = t_0$ and ended at $t = t_1$, then solution of the above differential equation leads to a risk of developing disease, $r(t_0, t_1)$, of

$$r(t_0, t_1) = 1 - e^{[-\text{IR}*(t_1-t_0)]}.$$

If we expand equation 13.7 in a Taylor series about $(t_1 - t_0) = 0$, then it becomes clear that for rare diseases (i.e., diseases with low incidence rates) and for short time intervals, the risk in period $t_0$ through $t_1$ is approximately equal to the incidence rate. For longer time intervals and more common diseases, however, this approximate equality may not hold, and the association between risks and incidence rates is best described as in equation 13.7.

Epidemiologic investigations often depend on the comparison of incidence rates among population groups. For example, an investigator might compare the cancer incidence rate in a community that experienced exposure to radioactive material with the rate in another community whose citizens had no measurable exposure. The individuals within these communities will likely be of various ages, ranging from the very young to the elderly.

Because the incidence rates for most cancers are higher among older citizens, differences in the age composition of the two communities may mask differences in the rates due to differing levels of exposure. For example, suppose that the community that had no measurable exposure has a larger proportion of elderly citizens than the exposed community does. This larger proportion of elderly citizens will likely cause the incidence rate in the unexposed group to be higher than would be expected in another unexposed population whose age distribution more closely matched that of the exposed community. As a result, the increased rate in the comparison group may mask a real effect of radiation exposure.

To remove the effect of age before making comparisons of interest, epidemiologists will often age-standardize incidence rates in both the exposed and unexposed groups (Bishop et al. 1975). Age standardization is a normative procedure in which the observed incidence rates are adjusted to reflect what the observed rate would be if the study population had a standard age distribution (e.g., the age distribution of the U.S. population in the year 2000). In the most straightforward approach to age standardization, called the direct method, age-specific incidence rates are calculated for specified age classes within the study population. The age-standardized rate is then obtained as the weighted average of the observed age-specific rates, with the weights equal to the proportion of the reference population in each age class. Incidence rates can be standardized for other population variables in addition to age, although age standardization is the most common. For example, age and sex standardization is sometimes applied to incidence rates to match not only the age distribution, but also the proportion of males and females.

The goal of standardization is to remove the effects of differences among groups that are not of interest but may affect the observed incidence rate, thus allowing a more focused evaluation on differences of interest. Unstandardized incidence rates are often referred to as crude rates, while standardized rates are often referred to as adjusted incidence rates.

## Measures of Association between Disease Risk and Suspected Causative Factors

In some cases, the goal of obtaining epidemiologic measures is purely to provide a summary of the level of disease burden in a population. For example, the Centers for Disease Control and Prevention routinely collects information on disease incidence and mortality using a variety of surveillance systems. The goal of this surveillance, which is sometimes called descriptive (as opposed to analytic) epidemiology, is to characterize the disease experience in the target population. Epidemiologists, however, are often interested in going beyond simply describing the level of disease present in the population to assessing possible associations between disease risk and the presence or absence of suspected causative factors. For example, you might wish to compare the incidence of lung cancer in a community that has been exposed to ionizing radiation with the corresponding incidence in a community with no such exposure. A higher observed incidence in the exposed community could indicate a possible association between the risk of developing cancer and exposure to the radioactive material. Two questions should immediately come to mind if such an association is observed. First, how do you determine if a difference in the disease experience of the two communities differs meaningfully from the random variation you would expect in observed disease rates? Second, if such a meaningful difference is observed, does this fact establish a causative link between the exposure and the cancer risk? The answer to the first question involves methods for the statistical analysis of epidemiologic data that we address further below. The answer to the second question is a resolute *no*. Epidemiologic studies do not provide concrete answers on causative factors for disease, but rather they provide evidence of relationships based on observable associations.

### Risk Ratio

Given this caveat, we now examine the measures of association between exposure and disease burden that are used in most epidemiologic studies. Suppose we have designed an epidemiologic study with the goal of comparing the level of disease risk in two populations: one population is located adjacent to a nuclear facility, and one is located much farther away. For illustration, we refer to these populations as exposed and unexposed, respectively. We start with a hypothesis, called the null, that there is no difference in the risk of developing a particular cancer between the exposed and the unexposed groups. An alternative hypothesis is that the exposed population has an increased risk of developing cancer. Suppose the study period is five years, the number of individuals in the unexposed population at the beginning of this time frame is $N_0$, and the initial population size in the exposed group is $N_1$. In addition, assume that $D_0$ and $D_1$ cancers are observed in the unexposed and exposed populations, respectively, during the five-year study period. Table 13.1 is a two-by-two table representing the data observed in the study. Constructing this table is a first step in virtually every analysis of epidemiologic data. The rows in the table are arranged by exposure status. The first row contains the exposed population, and the second contains the unexposed. The columns subdivide each row into individuals who are observed to develop the disease during the study period and those who do not. Using equation 13.1, we can use this table to estimate the cancer risk in the exposed population as

$$R_1 = \frac{D_1}{N_1}. \tag{13.8}$$

Similarly, the estimate of cancer risk in the unexposed population is given by

$$R_0 = \frac{D_0}{N_0}. \tag{13.9}$$

The ratio of these two risk estimates is an intuitive measure for comparing the risk in the exposed and unexposed groups. This ratio, called the risk ratio or the relative risk (RR), is given by

$$\text{RR} = \frac{R_1}{R_0}. \tag{13.10}$$

Traditionally, the risk in the exposed group is put in the numerator of the risk ratio so that values greater than 1 are indicative of higher risk in the exposed group. Conversely, a risk ratio less than 1 could be an indication of a protective effect of the exposure. An alternative means of comparing the risk in the two groups is to calculate the risk difference (RD):

$$\text{RD} = R_1 - R_0. \tag{13.11}$$

Risk difference values greater than 0 can indicate excess risk in the exposed population, whereas values less than 0 can indicate a protective effect.

Table 13.1 Number of diseased, disease-free, exposed, and unexposed individuals in a hypothetical epidemiologic study

|  | Diseased | Disease free | Total |
| --- | --- | --- | --- |
| Exposed | $D_1$ | $N_1 - D_1$ | $N_1$ |
| Unexposed | $D_0$ | $N_0 - D_0$ | $N_0$ |
| Total | $D_1 + D_0$ | $N - D_1 - D_0$ | $N$ |

## Risk Odds Ratio

An alternative measure of association is based on the odds of developing disease given membership in either the exposed on unexposed group. If $R_1$ is the risk of disease in the exposed group, then the odds of developing disease among members of the exposed group are $R_1/(1 - R_1)$. Similarly, the odds of developing disease among members of the unexposed group are $R_0/(1 - R_0)$. The odds ratio (OR), sometimes more specifically referred to as the risk odds ratio, is a common measure of association in epidemiologic studies and is defined as

$$\text{OR} = \frac{R_1(1 - R_0)}{R_0(1 - R_1)}. \tag{13.12}$$

An important characteristic of the odds ratio is evident when the disease of interest is rare. If the risk for disease is small, say, approximately less than 0.10 (Kleinbaum et al. 1982), such that $R_1$ and $R_0$ are close to 1, then the odds ratio, as defined in equation 13.12, will be quite close to $R_1/R_0$. In other words, the odds ratio and the risk ratio are approximately equal for rare diseases.

## Exposure Odds Ratio

Before we move on to measures of association based on incidence, we consider an alternative means of defining the odds ratio that is important in certain types of epidemiologic studies. Suppose, at the beginning of the study, that we do not know which subjects are exposed or unexposed, but we do know who has been diagnosed with the disease under study and who has not. In this situation, instead of having exposed and unexposed comparison groups, we have a group with the disease, called cases, and a group without the disease, called controls. As an alternative to examining the risk of developing disease given exposure, we could use these data to determine if the case group is more likely to have experienced the exposure of interest than is the control group. To do this, instead of calculating the odds of developing disease, we would compute the ratio of the odds of exposure in the case and control groups. This ratio, sometimes called the exposure odds ratio, will have a value greater than 1 if the case group has larger odds of having been exposed than does the control population. Using the notation from table 13.1, the odds of exposure among the cases is given by

$$O_c = \frac{D_1}{D_0},$$

while the odds of having been exposed among the controls is

$$O_0 = \frac{N_1 - D_1}{N_0 - D_0}.$$

Thus, the exposure odds ratio (EOR) for comparing the odds of having been exposed among the case and control groups is

$$\text{EOR} = \frac{D_1(N_0 - D_0)}{D_0(N_1 - D_1)}, \tag{13.13}$$

which is exactly equivalent to the risk odds ratio of equation 13.12. Therefore, when we are interested in examining the exposure history of case versus control subjects, the odds ratio for exposure is equivalent to the odds ratio for developing disease in a study in which exposure status is known and no participants are lost to follow-up.

We now consider measures of association based on rates rather than risk estimates. Recall that the incidence rate, as defined in equation 13.5, provides a measure of the change in the population's disease status per unit of person-time at risk. Like the risk ratio, the ratio of incidence rates, often referred to as just the rate ratio, allows analysts to compare the rate of disease development in exposed and unexposed groups. Thus, if the incidence rates in the exposed and unexposed populations are designated by $IR_1$ and $IR_0$, respectively, the incidence rate ratio (IRR) is defined as

$$\text{IRR} = \frac{IR_1}{IR_0}. \tag{13.14}$$

In some situations, it may not be possible to follow the unexposed and exposed groups to determine the incidence of new cases. Instead, the investigator may have to rely on a comparison of the current disease burden in the two populations. In this case, the ratio of the disease prevalence, as defined in equation 13.6, may be of interest where, again using the subscript 1 to denote the exposed group and 0 to represent the unexposed, the prevalence ratio (PR) at time $t_i$ is given by

$$\text{PR}(t_i) = \frac{P_1(t_i)}{P_0(t_i)}. \tag{13.15}$$

Depending on the resolution of the available exposure data, it may be more informative in an epidemiologic study to subdivide the study group into more than just two populations. For example, suppose you want to examine the potential association between exposure to radioactive materials and cancer risk among workers at a nuclear power generating facility. Workers at such a plant are likely to have varying degrees of exposure depending upon their particular duties. Therefore, classification into just two categories may be difficult. To obtain a more meaningful exposure gradient, an investigator may need to subdivide the study population into a number of groups corresponding to potential dose received. The investigator could then select one category, usually the one with the lowest potential for exposure, as the reference with which the measures of disease impact in the other categories can be compared. For example, suppose we have three levels of possible exposure among the workers in a facility: no exposure to radioactive material, moderate levels of exposure, and heavy exposure. Using the no-exposure group as the referent and differentiating summary measures of disease among the groups by the subscripts 0, 1, and 2, we can express the odds ratio for comparing the moderately exposed and unexposed groups as

$$OR_{0,1} = \frac{D_1(N_0 - D_0)}{D_0(N_1 - D_1)}.$$

Similarly, the odds ratio for disease comparing the high exposure versus no exposure group is

$$OR_{0,2} = \frac{D_2(N_0 - D_0)}{D_0(N_2 - D_2)}.$$

Note that an observed value of $OR_{0,2}$ that is greater than $OR_{0,1}$ may be indicative of a dose–response trend in the data. If possible, further division of the study population into more refined exposure categories may shed more light on this possibility. This type of increased specificity in characterizing exposure may be carried out to the point where individual exposure or dose estimates are available for every member of the study population. In this situation, person-specific dose estimates could be used to fit a mathematical model relating dose and risk. One of the most commonly used models is the logistic model, which we discuss in more detail below. For now, it is sufficient to realize that the logistic model implies that the log odds of developing disease are linearly related to the amount of exposure or dose. For example, if individual radiation dose measurements were available for all employees in our example facility, then we could fit the logistic model

$$\ln\left[\frac{R_i}{(1-R_i)}\right] = \beta_0 + \beta_1 \cdot \text{dose}_i, \qquad (13.16)$$

where $R_i$ is the $i$th individual's risk of developing disease, $\text{dose}_i$ is that individual's measured dose, and the coefficients $\beta_0$ and $\beta_1$ are unknown but will be estimated from the data. Notice in the model that $\beta_1$ is the measure of association between the exposure of interest and the risk for disease. If $\beta_1$ equals 0, then the odds of developing disease is not related to exposure. On the other hand, if $\beta_1$ is greater than 0, then the estimate for this coefficient reflects the per unit increase in the log odds of disease per unit of exposure. Conversely, if $\beta_1$ is less than 0, then the coefficient represents the decrease in the log odds of disease per unit of exposure.

The standardized mortality ratio (SMR) is an alternative measure of disease–exposure association that has been commonly used in epidemiologic investigations of worker populations to investigate the possible impacts of occupational exposures. The SMR is defined as the ratio of the observed number of cause-specific deaths in an exposed population to the number of deaths that would be expected to be caused by the study disease if that population had no exposure. To illustrate, suppose we observe $D_i$ deaths due to kidney cancer among male workers of age $i$ who are employed at a particular nuclear weapons production facility. We will assume that there are $I$ age classes in the worker population such that $i = 1, 2, \ldots I$. The sum of the $D_i$ values across the $I$ ages gives us the total number of fatal kidney cancers observed in our target worker population that we need for the numerator of the SMR. To get the denominator of the SMR, that is, the number of kidney cancer deaths we would expect in this group if they had no exposure, we must use information on age-specific kidney cancer death rates observed in some unexposed population. To be useful in calculating the SMR, however, this population must be as similar to the study group as possible with the exception of having no potential for the exposure of interest. For example, if $IR_i$ is the incidence rate of fatal kidney cancer among all U.S. males in age group $i$, then we could calculate the expected number of age-specific kidney cancer deaths among our worker study group as

$$E[D_i] = IR_i \cdot PY_i,$$

where $PY_i$ is the number of person-years at risk spent by the workers when they were in age class $i$. Alternatively, we could decide that the rate of kidney cancer mortality among the entire population of U.S. males is not a good value to use in our estimation of the expected number of deaths among workers. One reason for this concern might be the so-called healthy worker effect, which implies that those healthy enough to be employed might, on average, be healthier than the population as a whole (Symons and Taulbee 1984). As an alternative, we could calculate the expected number of kidney cancer deaths if there was no exposure using the observed age-specific kidney cancer mortality rates among administrative workers at the same facility, as long as these administrative workers had no chance for exposure. Once we have decided on an appropriate comparison group from which to derive the age-specific rates for those without exposure, the SMR is calculated as

$$\text{SMR} = \frac{\sum_{i=1}^{I} D_i}{\sum_{i=1}^{I} E[D_i]},$$

or simply the ratio of the observed number of kidney cancer deaths among the workers under study to the expected number of such deaths in the absence of exposure.

Notice that the SMR is defined in terms of mortality as opposed to disease incidence. We make this distinction to emphasize the SMR's primary use in occupational studies. An incidence measure, called the standardized incidence ratio (SIR), can be calculated in a similar fashion based on incident cases as opposed to mortality. While the SMR and the SIR are common measures of disease impact in studies of worker populations, you should note that these measures are generally not comparable between populations. For example, if you were interested in the relative impact of exposure in the worker populations of two nuclear facilities, then comparison of the SMRs observed in each plant might be misleading, even if the expected number of deaths in each group were derived using the same comparison population (Symons and Taulbee 1984).

Before leaving this initial assessment of measures of association between exposure and disease, two questions must be addressed. The first is how to interpret an observed value for one of these measures given the random variation expected in the observed counts of disease on which they are based. The key issue in answering this question is to remember that true risk in the population being studied is not observable and that any estimate of the risk or odds ratio, or of a coefficient of a logistic regression model, is subject to sampling error. As an illustration, suppose we establish two sets of exposed and unexposed workers within a nuclear facility. The two exposed groups are selected to be as demographically similar as possible, as are the two unexposed populations. If we estimate two risk ratios by comparing the observed disease risk in the two sets of exposed and unexposed subjects, it is highly unlikely that these two estimates will be equal. The obvious reason for this difference is that, even though the two exposed groups experienced the same underlying increased risk and the two unexposed groups incurred the same background risk, the development of disease depends on factors other than the presence or absence of exposure. The actual risk for developing disease given a certain level of exposure is likely different for all individuals in the study because of other nonmeasurable risk factors, such as genetic susceptibility to disease. As a result, even estimates of the background risk for disease are subject to apparently random variation. Thus, to interpret the meaning of any estimate of an odds ratio, rate ratio, or coefficient of a logistic regression model, we need to determine how likely we would be to observe these values given the expected variation in the observed data that is not accounted for by exposure.

The second question is which measure of association should be used. The appropriate choice for a measure of association in an epidemiologic study depends on the plan used to collect the needed information. This plan is called the *study design*. Several factors affect the choice of the study design to be used in an epidemiologic analysis, and some of these are discussed in the following section. However, a decision must be made in the design phase as to what association measure most appropriately addresses the hypothesis of interest. For example, if we are planning a study with a multiyear period of observation, during which many participants are likely to be lost to follow-up, then estimating the risk ratio, even with the corrections of equation 13.2, may not be feasible. Instead, we may want to focus on estimating the incidence rate ratio using equation 13.14. Alternatively, if we have exposed and unexposed populations that we can follow completely through the study period, then the risk ratio estimate of equation 13.10 may be a more meaningful indicator of possible exposure effects. Much of the rationale for choosing a particular measure of association between exposure and disease will become evident in the ensuing discussion of common epidemiologic study designs.

## Study Designs Commonly Used in Epidemiologic Investigations

Most epidemiologic studies can generally be grouped into two categories based on the information available when the study begins. In one situation, we know the exposure status of individuals in the study population and, in the course of the study, will collect information on the incidence, prevalence, or other measures of disease impact. This type of study design is generally referred to as a *cohort study*. Alternatively, at the beginning of the study we may know the disease status of the participants, that is, who is disease-free and who has had a diagnosis of the disease of interest. In this situation, we would use a *case–control* study design to compare the level of exposure between the case subjects and the controls to determine if a history of exposure is more prevalent among the case group.

### Cohort Designs

In the most basic cohort study design, we compare the level of disease between two groups: a group that has experienced the exposure of interest and one that has not. These two groups are often referred to as the exposed and unexposed cohorts. A cohort could, however, refer to the entire study population. For example, if we are interested in fitting a dose–response model using individual dose measurements and disease information, we could design a study in which the entire study population could be considered as one cohort.

Cohort studies can be classified according to the time at which collection of disease information begins. For example, we may be interested in historical exposures to communities surrounding U.S. Department of Energy nuclear weapons production facilities. If we can establish the levels of possible exposure in these communities, then we would likely want to retrospectively collect disease incidence data because impacts of these past exposures may not be evident in the current community. This type of cohort study, in which disease incidence data are collected from a starting point in the past, is referred to as a *retrospective* cohort study. Alternatively, we may want to establish cohorts in the present time based on exposure status and monitor the disease incidence in these groups through some specified future study period. This type of design is referred to as a *prospective* cohort study.

An advantage of a prospective as opposed to retrospective cohort study is that more quality assurance can be placed on the information collected on subjects. Thus, a prospective design might be preferable in a study of lung cancer and exposure to radioactive materials because we would want to know, with as much certainty as possible, which subjects are cigarette smokers and which are not. In a prospective study, this type of information can be collected as the subjects are enrolled into the study population, whereas smoking information is notoriously difficult to gather retrospectively. In addition, a retrospective study may have to rely on historical medical records to establish which study participants developed disease. In many cases, these records are difficult to find or may not exist. This is especially true when some of the study subjects are deceased. Thus, accurate classification of disease status is likely to be more difficult in retrospective than in prospective studies. On the other hand, we may be interested in assessing the impact of exposures that happened in the past, and the disease history of the study cohort may be much more informative than the current or future health status of a prospective study group. This is especially true when the disease of interest is likely to be fatal and those most affected by the exposure may already be deceased and would therefore have no chance to be included in the prospective study. We can, however, design cohort studies that are both retrospective and prospective in that the collection of disease information can begin retrospectively yet continue until some future date.

The cohort designs discussed thus far are intended to provide estimates of disease incidence that allow analysts to estimate either risk or rate ratios or, perhaps, the coefficients of a mathematical model relating exposure to disease risk. Alternatively, we may want a snapshot of the current health of the cohort. For example, we can identify exposed and unexposed groups and design a study to canvas these populations to determine the proportion of each cohort that has been diagnosed with the disease of interest. The type of design used to accomplish this canvassing is called a *cross-sectional study*. Note that in a cross-sectional study, we obtain no information on disease incidence but only on disease prevalence. Thus, hypotheses on disease–exposure associations in cross-sectional studies must be based on comparisons of prevalence estimates such as the prevalence ratio of equation 13.15.

In general, cohort studies are expensive because of the need to identify and trace the entire cohort and to ascertain the past, present, or future disease status of the participants. On the other hand, a properly designed prospective cohort study, if feasible, is often the best means to evaluate hypotheses regarding disease exposure associations (Kelsey et al. 1996).

## Case–Control Designs

The second general design used in epidemiologic investigations is the case–control study. In a case–control design, the disease status of the participants is known at the study's inception and the purpose of the investigation is to compare the level of exposure in the case group with that in the control group. For example, you may wish to evaluate the hypothesis that past occupational exposures to radioactive materials affect people's risk for developing non-Hodgkin's lymphoma (NHL). Rather than identifying, tracing, and interviewing two cohorts of individuals, one with past occupational exposure and one without, analysts could use a case–control approach. In this example, a researcher could use a cancer registry to identify a suitable number of NHL cases that meet the study criteria. Using other data sources (e.g., driver's license, tax, and voting records), the study team could identify another group of individuals with no record of NHL diagnosis who could be recruited into the study as controls.

In designing such a study, researchers should attempt to match the control group with the case group as closely as possible on factors such as race, ethnicity, social economic status, gender, age, and other potential risk factors for the disease of interest (Schlesselman 1982). A study based on this type of recruiting is called a *matched* case–control study. To increase the ability to detect meaningful levels of association between exposure and disease, researchers often match more than one control to each case in a matched design. Once the cases and controls are recruited into the study, the analysis focuses on establishing the exposure history of the two groups. In our example, if previous exposure to radioactive materials is more prevalent among the cases than among the controls, we might suspect that an association between occupational radiation exposure and NHL exists in these data. Notice that the case–control design involves recruiting participants into the study on the basis of current disease status. As a result, we cannot estimate the disease risk under a case–control design. We can, however, estimate the exposure odds ratio using equation 13.13, which is equivalent to the risk odds ratio for a cohort study with no losses to follow-up.

The advantage of a case–control design is that, by focusing on a smaller number of individuals than a cohort study, researchers can collect more precise information on study participants for a fixed cost. In addition, in a cohort design, the number of cases observed in the exposed and unexposed groups is not known until the data are collected. For rare diseases, such as some cancers, the number of cases observed in a cohort study may be too small for researchers to meaningfully evaluate whether exposure is associated with disease. A case–control study, however, is designed to provide a specified number of cases, so researchers know they will have an adequate number of cases before collecting the data.

## Nested Designs

We have focused on a comparison of cohort and case–control study designs. Sometimes, however, it is advisable to design a study that combines features of both. For example, a cohort of potentially exposed individuals who resided near a nuclear weapons production facility may be established in which model-based dose estimates are obtainable for all participants. If we fit a model, such as the logistic regression of equation 13.16 to estimate the increase in the odds of disease per unit of dose, we are clearly performing a dose–response analysis within a cohort study design. In this case, the cohort is everyone who resided near the facility who meets the criteria for inclusion in the study. We may wish, however, to gain additional information on potential risk factors that could affect the relationship between estimated dose and risk in a more refined analysis, but cost constraints may prevent us from obtaining this additional information on all members of the cohort. We could collect, however, the additional information for a subset of the original study population, which would be a sampled collection of cases and noncases identified in the cohort portion of the study (Langholz and Clayton 1993).

In the second stage of the study, we will compare the doses received by the cases and the sampled controls. This second part of the study is called a nested case–control study because the cases and controls in the smaller study were selected from the original cohort study population.

## Assessing the Observed Level of Association between Disease and Exposure

We have already mentioned that the number of cases of disease observed in a population under study during any period should be thought of as a random variable. As an illustration, suppose we have designed a study to observe disease incidence among 10 groups of participants with each group made up of 10 individuals. In addition, we will assume that five of the groups experienced exposure to only background levels of radiation, while the other five received more than background exposure because of their proximity to some radiation source. Suppose that the study has been designed so that the 10 groups are as similar as possible on other factors that may affect the risk of developing the disease of interest. Figure 13.1 shows a graph of hypothetical yet plausible incidence rates that we might expect to see in such a study. The key point to note in this figure is the spread in the observed incidence rates among the groups with similar levels of exposure. Notice the variability in rates among the unexposed groups and the even greater within-cohort variation in the exposed set. This spread in the observed rates reflects the random variation you would expect in the observed number of cases among populations even if they were experiencing the same true underlying average risk.

As another example, if a population of 1,000 individuals has an average risk of 1 in 100 of developing disease during a five-year period, it does not mean that if we were to observe that group for those five years we would see exactly 10 diagnoses. On average, over a large number of such groups, we might see close to 10 cases per

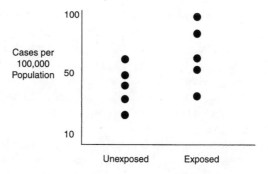

Figure 13.1 Possible incidence rates among five groups of 10 individuals, each of whom have been exposed to elevated levels of radiation, and among five groups of 10 individuals, each of whom have been exposed only to background radiation levels.

five years, but the observed number during any particular five-year interval in any given group would fluctuate.

Separating meaningful differences in underlying risk from the random variation that we expect in observed measures is the key issue in evaluating epidemiologic hypotheses of disease-exposure relationships. This section discusses analytic approaches that can be used to increase the chance of detecting true differences in risk on the basis of observable risk measures. Before beginning this discussion, however, we should realize that differences between the observed rates among groups with similar exposures (e.g., those in figure 13.1) could also reflect systematic, in addition to random, effects. For example, diagnostic practices could differ among groups, especially if they are recruited in different geographic areas, leading to potential misclassification of disease status. Alternatively, the method used to recruit participants into the study could differ both within and across exposure cohorts, introducing bias into any subsequent analysis. Misclassification and bias can cause differences in the measures of the disease risk among groups that appear to be associated with exposure but, in fact, have nothing to do with exposure status.

## *Interpreting Estimates of Disease Exposure Association*

We now consider how to evaluate hypotheses of association between exposure and measures of disease burden in epidemiologic studies. Suppose we are using a risk ratio to summarize the difference in estimated risks between an exposed and unexposed cohort. In general, we evaluate the observed magnitude of this measure by developing an estimate of the amount of variation expected in an observed risk ratio when no true association between disease and exposure exists and examining the magnitude of the observed measure relative to that variation. The assumption of no association between the disease and the exposure is called the *null hypothesis*, and we will reject this hypothesis only if the data provide striking evidence to the contrary. For simplicity, we symbolize the null hypothesis of no association using the term $H_0$. The measure of variability we will use is the standard error of the observed measure derived under the assumption that $H_0$ is true. You are probably familiar with methods of computing the standard error for random variables that follow a normal or lognormal distribution. Developing metrics for the standard error of estimates such as the risk, rate, and odds ratios is more complex. This complexity results from the fact that these measures are nonlinear functions of random variables, which are decidedly nonnormal.

Suppose, however, that we have an estimator for the standard error of a measure, $M$, of association between disease and exposure. We will call the estimator for the standard error of $M$ $\sigma_0$, where the 0 in the subscript of $\sigma_0$ indicates that this is the standard error we anticipate in observed values of $M$ under $H_0$. We will now construct what is called a *test statistic* to evaluate the likelihood of $H_0$ given our observed data. A general form for such a test statistic is

$$Z = \frac{M - E[M]}{\sigma_0}, \tag{13.17}$$

where $Z$ is the value of the statistic and $E[M]$ is value we would expect to see for $M$ under $H_0$. As an illustration, return to the risk ratio described in equation 13.10. For the risk ratio, we can state the null hypothesis in the form

$$H_0 : \text{RR} = \frac{R_1}{R_0} = 1;$$

that is, if there is no association between exposure and disease, then we expect the ratio of the estimated risks in the exposed and unexposed cohorts to be close to 1. It turns out that we can develop a test statistic for the null hypotheses by examining only the observed, $D_1$, and expected, $E[D_1]$, number of cases in the exposed group and the variance we would expect for this count under $H_0$. By assuming both the total number of diseased individuals and the size of the exposed and unexposed cohorts to be fixed, we can use an argument based on the hypergeometric distribution (Evans et al. 1993) to show that the expected value for $D_1$ under $H_0$ is

$$E[D_1] = (D_1 + D_0)\left(\frac{N_1}{N}\right),$$

while the standard error of $D_1$ under the null hypotheses can be estimated by

$$\text{Std}(D_1) = \sqrt{\frac{N_1 \cdot N_0 \cdot (D_1 + D_0) \cdot (N - D_1 - D_0)}{N^2(N - 1)}}.$$

Therefore, using the general approach of equation 13.17, we can develop a test statistic for the null hypotheses that the risk ratio equals 1, versus the alternative that the ratio exceeds 1, of

$$Z = \frac{D_1 - E[D_1]}{\text{Std}(D_1)}. \tag{13.18}$$

Under the null hypotheses, the expected value for $D_1$ is estimated by assuming that the proportion of cases in the exposed group is the same as the proportion of the total number of subjects who were exposed. If the observed number of cases in the exposed group differs greatly from this expectation, when normalized by the estimated standard error, then we might question the validity of the hypotheses of no association. The magnitude of the above test statistic can be evaluated by assuming that, for reasonably large $N$ and when the null hypothesis is correct, $Z$ in equation 13.18 approximately follows a standard normal distribution (i.e., a normal distribution with mean 0 and standard deviation 1). Therefore, we can evaluate the likelihood of the null hypothesis being true by using a standard normal distribution function to calculate the probability of observing a $Z$ value larger than the one seen in the study.

If this calculated probability is small, say, less than 0.05, then it begins to appear unlikely that the observed value of the test statistic is plausible under the null hypothesis. As a result, we might question the validity of the no-association hypothesis. Alternatively, if the probability of observing a $Z$ value like the one seen in the study is relatively large, we might conclude that there is not enough evidence in the data to reject the null hypothesis of no-exposure disease association.

The probability of obtaining a test statistic value as large as or larger than the one seen in the study under the null hypotheses is often referred to as the $p$ value. Thus, if the $p$ value is small, we begin to question the validity of the null hypotheses, whereas if the $p$ value is large, we should be less willing to question $H_0$. The $p$ values observed in studies are often compared with established comparison values to provide points of reference. Traditional values for these comparison probabilities, often called significance levels, are 0.01, 0.05, and 0.1, with 0.05 being the most common. When the observed $p$ value is less than one of these fixed comparison levels, the test is usually called statistically significant at that level. For example, if comparison of the observed value of $Z$ in equation 13.18 with a standard normal distribution produced a $p$ value of 0.03, the test could be said to be significant at the 0.05 level.

An alternative form of the test given by equation 13.18 is based on the result that the random variable produced by squaring a standard normal deviate follows a chi-squared distribution with one degree of freedom (Hogg and Craig 1978). Therefore, if the null hypothesis is true and the test statistic produced by equation 13.18 is approximately standard normal, then squaring this value and multiplying by a correction factor of $N/N_1$ produces a test statistic that can be assumed to approximately follow a chi-squared distribution with 1 degree of freedom. This test, which is commonly used in epidemiologic analyses, is called the Mantel-Haenszel chi-squared test. The Mantel-Haenszel test for the null hypothesis that the risk ratio is equal to 1 is given by

$$Z^2 = \frac{(N-1)[D_1(N_0 - D_0) - D_0(N_1 - D_1)]^2}{N_1 \cdot N_0 \cdot (D_1 + D_0) \cdot (N - D_1 - D_0)}. \tag{13.19}$$

In this case, the $p$ value is the probability of getting a $Z^2$ value larger than the one observed based on the central chi-squared distribution with 1 degree of freedom. Again, if this $p$ value is low, say, below 0.05, we might question the validity of the null hypothesis, whereas for large $p$ values, we could conclude that there is not enough evidence in the data to reject the no-association hypothesis.

## Confidence Intervals

An alternative to using a test statistic to evaluate the likelihood of the null hypothesis being true is to construct what are called confidence intervals surrounding the estimated association measures. The goal in constructing such intervals is to form a range, say, $[a, b]$, such that, given the variability we expect in the observed measure of interest, we have a certain amount of confidence that the true value of the measure is within the stated interval. For example, if we think of the observed risk ratio as an estimate for the unobservable ratio of the true average risks in the exposed and unexposed populations, then the confidence interval is constructed to provide a specified level of assurance that the derived interval contains the true risk ratio. Obviously, derivation of this interval depends on the value of the risk ratio estimate, the variability associated with the estimate, and the degree of confidence we wish to require on bounding the true value. If we determine $a$ and $b$ to provide a 95% level of confidence, then the resulting range is referred to as the 95% confidence interval. For illustration, an approximate 95% confidence interval for the estimated risk ratio of equation 13.10 has lower bound

$$a = \text{RR}\left\{\exp\left(-\Theta_{0.025}\sqrt{\frac{1-R_1}{N_1 R_1} + \frac{1-R_0}{N_0 R_0}}\right)\right\}$$

and upper bound

$$b = \text{RR}\left\{\exp\left(\Theta_{0.025}\sqrt{\frac{1-R_1}{N_1 R_1} + \frac{1-R_0}{N_0 R_0}}\right)\right\},$$

where $\Theta_{0.025}$ is the value associated with a cumulative probability of 0.0975 from the standard normal distribution. Notice that this confidence interval is referred to as approximate. This reflects the assumption that, for large values of $N_0$ and $N_1$, the natural log of the risk ratio divided by its approximate standard error under $H_0$ approximately follows a standard normal distribution.

We have made assumptions of approximate normality both in evaluating the magnitude of the test statistics and in constructing the confidence intervals. Although this assumption is generally acceptable in most circumstances, for studies with few participants or those investigating extremely rare diseases, tests based on the true distribution of the test statistic may be preferable. The forms for these tests, called *exact tests*, are available in a variety of epidemiologic and biostatistics publications (Agresti 1990; Feinberg 1987). In addition, just as a measure of association can be extended to situations in which there is more than one level of exposure, the inferential methods we have discussed can also be extended to multiple exposure levels. For example, a generalized form of the Mantel-Haenszel chi-squared statistic is often applied across multiple exposure levels (Kelsey et al. 1996).

As a final point in our overview of methods for evaluating hypotheses on exposure–disease relationships, we should consider the difference between confidence intervals, as discussed here, and uncertainty ranges discussed elsewhere in this book. The term *confidence interval* has a very specific and precise meaning in the epidemiologic and statistical context. It refers to the required level of confidence, given expected sampling error, that the specified range contains the true but unknown parameter. Given a specified level of confidence, the width of this interval depends on the sampling variability we would expect in the estimated value. An uncertainty range for an unknown parameter, as used in risk assessments, often reflects a range of possible values for a true but unknown quantity (IAEA 1989; Hoffman and Hammonds 1994). For example, the uncertainty in a dose estimate may reflect a probabilistic representation of the possible values for the mean true dose of a collection of individuals. However, given the traditional statistical interpretation of the term confidence interval, an uncertainty range and a confidence interval are not equivalent and should not be used interchangeably.

We now consider how to interpret the results of analyses like those described above. So far, we have based our evaluations only on comparisons of observed exposure levels and disease counts. We have not taken into account other factors that may mask the true association, or lack thereof, between exposure and disease. To illustrate this point, consider a hypothetical study focused on the potential association between a given type of cancer and occupational internal exposure to alpha emitters. The study design compares the five-year cancer risk for two worker populations, one with a history of internal exposure to alpha emitters and one with very low documented occupational exposure to radioactive materials. Suppose we have recruited 706 exposed and 2,058 unexposed workers to participate in the study and that all subjects are white males. In addition we will assume that the exposed and unexposed cohorts have similar age distributions. Table 13.2 presents the two-by-two table of hypothetical data for this study.

Using equation 13.10, we can estimate the risk ratio for bone cancer comparing the exposed and unexposed workers by

$$RR = \frac{\left(\frac{60}{706}\right)}{\left(\frac{90}{2058}\right)} = 1.94.$$

Thus, the observed risk among the exposed group of workers appears to be almost twice that in the unexposed group. We now will evaluate how likely an observed relative risk of this size is under the null hypothesis of no association given the

Table 13.2 Number of participants in a hypothetical worker study who have been diagnosed with cancer by presence or absence of occupational exposure to radioactive materials

| Occupational exposure | Cancer | No cancer | Total |
|---|---|---|---|
| Yes | 60 | 646 | 706 |
| No | 90 | 1,968 | 2,058 |
| Total | 150 | 2,614 | 2,764 |

Risk ratio = 1.94; Mantel-Haenszel $Z^2$ = 17.43; 95% confidence interval = [1.42, 2.66].

sampling error we might expect in the data. Using the Mantel-Haenszel chi-squared statistic of equation 13.19, we obtain

$$Z^2 = \frac{(2764 - 1)(60 \times 1968 - 90 \times 646)^2}{706 \times 2058 \times 150 \times 2614} = 17.43.$$

To evaluate the likelihood of $H_0$ being correct, we compute the probability of obtaining a $Z^2$ value greater than 17.43 from a central chi-squared distribution with one degree of freedom. The probability of obtaining a $Z^2$ at least as large as was observed in these data under the no-association hypothesis is less than 0.001, indicating that it is extremely unlikely to get a test statistic of this size by chance alone. We can also compute the 95% confidence interval for the above risk ratio estimate, which is [1.42, 2.66]. Notice that this interval does not contain the null value of 1. Given this simple analysis, we have a fair amount of evidence to conclude that there is a substantial increase in cancer risk among the exposed workers.

Now suppose we become aware that the smoking history of the study subjects is available. From this new information, we learn that 201 (28%) of the 706 exposed workers are current or former smokers, while only 51 (2%) of the 2,058 unexposed workers have ever smoked. Using this information, we can split, or stratify, the analysis to examine the effect of occupational exposure separately for smokers and nonsmokers. This stratified analysis is illustrated in table 13.3.

Notice that for both the smoking and nonsmoking categories, the estimated risk ratio for occupational exposure is virtually equal to the null value of 1. In addition, the smoking-specific Mantel-Haenszel test statistics are equal to zero to two decimal places in the stratified analysis, and both stratified confidence intervals include the null value of 1. Therefore, we have a strange situation in which separating the analysis based on a third factor altered the conclusions drawn from the combined analysis.

Table 13.3 Number of participants in a hypothetical worker study diagnosed with cancer stratified by smoking status

| Occupational exposure | Cancer | No cancer | Total |
|---|---|---|---|
| *Current or Former Smoker[a]* | | | |
| Yes | 40 | 161 | 201 |
| No | 10 | 41 | 51 |
| Total | 50 | 202 | 252 |
| *NonSmoker[b]* | | | |
| Yes | 20 | 485 | 505 |
| No | 80 | 1927 | 2007 |
| Total | 100 | 2412 | 2512 |

[a] Risk ratio = 1.01; Mantel-Haenszel $Z^2$ = 0.00; 95% confidence interval = [0.55, 1.89].
[b] Risk ratio = 0.99; Mantel-Haenszel $Z^2$ = 0.00; 95% confidence interval = [0.61, 1.61].

The situation we have presented is called *confounding*, and the factor added to the analysis, in this case smoking status, is called a confounder. A *confounder* is a factor that is associated with both the exposure and the outcome of interest but is not in the causal pathway between the two. As a result of the confounder's association with exposure, the prevalence of the confounding variable in the exposed group can differ from the prevalence of the confounder in the unexposed group. In addition, because of the confounder's association with the outcome, this difference can lead to an apparent association between the outcome and the exposure if the confounder is ignored. In our example, smoking is much more prevalent among the exposed group than among the cohort without occupational exposure. In addition, smoking is a risk factor for the cancer being studied. Table 13.3 shows that 201 out of 252 smokers developed cancer, while only 505 out 2,512 nonsmokers developed the disease. Therefore, when the strata are combined (i.e., when smoking status is ignored in the analysis), the increased number of cases in the occupationally exposed cohort due to a higher level of smoking results in the appearance of an association between cancer and radiation exposure. While such a high risk of disease in our hypothetical smoking group is not likely, it serves a purpose in the illustration of the potential effects of ignoring a confounding attribute.

Another potential problem associated with the analysis of epidemiologic data is effect modification or interaction. Whereas a confounder tends to mask the true level of association between disease and exposure, an *effect modifier* alters the exposure effect, resulting in differing levels of association between disease and exposure at different levels of the interacting effect. Perhaps the best example of effect modification in radiation epidemiology is the synergistic effect between exposure to radon decay products and smoking that has been observed in epidemiologic evaluations of cohorts of underground miners (Lubin and Steindorf 1995). In these data, the increased risk of lung cancer observed at a given level of exposure appears to be greater among those miners who smoked than among those who were nonsmokers. In other words, smoking appears to magnify the impact on lung cancer risk at a given amount of exposure to radon decay products. As a result, smoking can be considered as an effect modifier in these studies.

The key difference between a confounder and an effect modifier is that the confounder is associated with both the exposure and the disease, but an interacting factor is not associated with exposure per se but may alter the potential effect of the exposure. As an example, consider the data presented in table 13.4, which shows the results of a hypothetical study of lung cancer between two groups of workers: one with a history of occupational exposure to inhaled alpha emitters and one without such a radiation exposure.

In this hypothetical example, we have an exposed and unexposed cohort each with 3,000 subjects. We assume that 40% of the subjects in the unexposed cohort are current or former smokers, while only 10% of the exposed subjects have a smoking history. The first two parts of table 13.4 present the data stratified by smoking status. Examination of the smoking-specific risk ratios indicates a fairly strong interaction between smoking and the exposure of interest, with a risk ratio for exposure of 2.00 among smokers and no indication of excess risk among the nonsmokers. The Mantel-Haenszel chi-squared test statistic (equation 13.19)

Table 13.4 Number of participants in a hypothetical worker study diagnosed with cancer by occupatonal exposure to radioactive material stratified by and combined across smoking status

| Occupational exposure | Cancer | No cancer | Total |
|---|---|---|---|
| Part 1. History of cigarette smoking[a] | | | |
| Yes | 12 | 288 | 300 |
| No | 24 | 1,176 | 1,200 |
| Total | 36 | 1,464 | 1,500 |
| Part 2. No history of cigarette smoking[b] | | | |
| Yes | 54 | 2,646 | 2,700 |
| No | 36 | 1,764 | 1,800 |
| Total | 90 | 4,410 | 4,500 |
| Part 3. Combined across smoking history[c] | | | |
| Yes | 66 | 2,934 | 3,000 |
| No | 60 | 2,940 | 3,000 |
| Total | 126 | 5,874 | 6,000 |

[a] Risk ratio = 2.00; Mantel-Haenszel $Z^2$ = 4.10; 95% confidence interval = [1.01, 3.95].
[b] Risk ratio = 1.00; Mantel-Haenszel $Z^2$ = 0.00; 95% confidence interval = [0.66, 1.52].
[c] Risk ratio = 1.10; Mantel-Haenszel $Z^2$ = 0.29; 95% confidence interval = [0.78, 1.55].

has a $p$ value of 0.04 for the exposure effect among smokers and a $p$ value of 1.00 among the nonsmokers.

Part 3 of table 13.4 summarizes the results of an analysis in which the two smoking strata are combined. To illustrate the effect of ignoring an effect modifier, notice how the estimate of the risk ratio from the combined analysis is substantially less than the risk ratio derived from examination of smokers only. In fact, the $p$ value for the Mantel-Haenszel statistic based on the combined data is 0.59, which would lead us to accept the no-association hypothesis. Thus, by ignoring the interaction term, we are missing an indication of increased risk among a subgroup of the study population.

The need to adjust for confounders and potential interaction terms is the primary reason that analysis of two-by-two tables is usually just the first step in an epidemiologic analysis. In fact, risk, rate, and odds ratios based on these types of tables are often referred to as crude estimates to indicate their preliminary nature. As was shown in the preceding two examples, one way to adjust for the presence of confounders and effect modifiers is to stratify the data into categories based on the potential confounding or interacting variable and to estimate the association between exposure and disease within each stratum. In the case of confounding, a modification of the Mantel-Haenszel chi-squared test (Kelsey et al. 1996) can be used to combine the stratified information into an adjusted overall test for association between disease and exposure. An alternative way to adjust for confounding and interaction is to develop models relating the combined effects of the exposure of interest and other potentially important information to the risk for disease. We discussed one such model above in our consideration of the logistic model of equation 13.16.

To illustrate how this model can be extended to account for potential confounders using the previous example, define a variable, $D_i$, which has a value of 1 if the $i$th person in the study group is exposed or a value of 0 if the $i$th person is not exposed to an internal alpha emitter. In addition, define an additional variable, $S_i$, which has a value 1 if the $i$th person has a history of cigarette smoking and 0 otherwise. We will use these variables to model the log odds of developing lung cancer based on both exposure and smoking status using the logistic regression model

$$\ln(\text{odds}) = \beta_0 + \beta_1 \cdot D_i + \beta_2 \cdot S_i. \tag{13.20}$$

Now consider two study participants: both are nonsmokers, but only one has been exposed to radioactive material. For both of these individuals, $S_i$ has a value of 0. For the individual who has been exposed, however, $D_i$ has a value of 1, and using the model in equation 13.20, we can model his or her log odds of developing disease, which we will designate by $\ln(O_e)$, as

$$\ln(O_e) = \beta_0 + \beta_1.$$

For an unexposed individual who does not smoke, the log odds of disease, which we will designate by $\ln(O)$, is modeled as

$$\ln(O) = \beta_0.$$

Now consider the difference

$$\ln(O_e) - \ln(O) = \beta_1,$$

which, in terms of the odds ratio, is given by

$$\frac{O_e}{O} = e^{\beta_1}.$$

Thus, we have our desired measure of association, in this case the odds ratio, as a function of the parameters used in the logistic model. Now consider two other individuals: one who smokes but is not exposed whose odds for disease we will designate as $O_s$, and another who both smokes and is exposed and, as a result, is subject to odds of disease given by $O_{es}$. By using the correct values of 0 and 1 for $D_i$ and $S_i$ in equation 13.20 and by raising the differences to the base $e$, we can see that

$$\frac{O_e}{O} = \frac{O_{es}}{O_s} = e^{\beta_1}, \tag{13.21}$$

while

$$\frac{O_s}{O} = \frac{O_{es}}{O_e} = e^{\beta_2}.$$

Therefore, by using the model in equation 13.20, we have separated the effect of the exposure from the confounding effect of smoking. If the smoking term $\beta_2 \cdot S_i$ had been omitted from the model, the resulting estimated odds ratio would reflect some unknown convolution of smoking and exposure effects.

Now suppose we suspect that smoking is an effect modifier in that we hypothesize that a person's risk from radiation exposure is altered depending on whether he or she smokes. We can construct a logistic model to reflect this interaction as:

$$\ln(\text{odds}) = \beta_0 + \beta_1 \cdot D_i + \beta_2 \cdot S_i + \beta_3 \cdot D_i \cdot S_i. \tag{13.22}$$

Notice that we have added what is called an interaction term, $\beta_3 \cdot S_i \cdot D_i$, to the original model of equation 13.20. To see the impact of adding this term, again let $O$ represent the odds for disease for an unexposed nonsmoker, $O_e$ the odds for disease for an exposed nonsmoker, $O_s$ the disease odds for an unexposed smoker, and $O_{es}$ the odds for disease for an exposed smoker. Using the appropriate values of 0 and 1 for $D_i$ and $S_i$ to signify membership in the exposed/unexposed smoking/nonsmoking groups, equation 13.22 produces an estimated odds ratio for disease among nonsmokers of

$$\frac{O_e}{O} = e^{\beta_1}$$

and an odds ratio for the effect of exposure among smokers of

$$\frac{O_{es}}{O_s} = e^{\beta_1 + \beta_3}.$$

Therefore, as opposed to a confounding effect, where the odds ratio for exposure was the same in both the smoking and nonsmoking groups, if smoking is an effect modifier, then the impact of exposure to radioactive materials itself differs depending on a person's smoking status. In other words, smoking acts to change the risk due to exposure.

The obvious question is how to use the information on exposure, confounders, effect modifiers, and disease status to develop estimates of the $\beta$ values in equation 13.22. In most cases, these estimates are derived by using a technique known as *maximum likelihood* (ML). The main idea behind the ML approach is to express the probability of developing disease for the $i$th person in the study group as

$$\text{Prob(disease under the model)}^{Y_i} \cdot [1 - \text{Prob (disease under the model)}]^{(1-Y_i)},$$

where the variable $Y_i$ takes a value of 1 if the $i$th person has the disease and 0 if he or she does not. In addition, algebraic manipulation of the logistic model, for example, the model in equation 13.20, shows that the desired probability can be modeled as

$$\text{Prob(disease under the model)} = \frac{e^{\beta_0 \cdot \beta_1 \cdot D_i + \beta_2 \cdot S_i}}{\left(1 + e^{\beta_0 + \beta_1 \cdot D_i + \beta_2 \cdot S_i}\right)}.$$

If we designate this probability of disease under the model for individual $i$ as $P_i$, indicating individual $i$s particular values for $D_i$ and $S_i$, then the joint probability of seeing the collection of cases and noncases that was observed among our $N$ study subjects, which we will call $L$, is

$$L = \prod_{i=1}^{N} P_i^{y_i} \cdot (1 - P_i)^{(1-y_i)}. \tag{13.23}$$

This joint probability is called the *likelihood*, and it represents the probability of observing the collection of cases and noncases that we actually counted in our study. Notice that the likelihood is a function of the values of $P$, which in turn are functions of each individual's known exposure and smoking status and the unknown $\beta$ values we wish to estimate. In addition, if individual $i$ has the disease, then $y_i$ equals 1, and he or she contributes $P_i$ to the likelihood. Alternatively, if individual $i$ does not have the disease and $y_i$ equals 0, then he or she contributes $(1 - P_i)$. Therefore, observed disease status is incorporated into the likelihood by inclusion of $P_i$ or $(1 - P_i)$ depending on whether individual $i$ has the disease. The ML approach, as implied by the name, is to use multivariate calculus techniques to find those values for $\beta$ that maximize the likelihood. In other words, we will estimate the unknown values for $\beta_0$, $\beta_1$, and $\beta_2$ by using those values for these parameters that maximize the probability of seeing the collection of cases and noncases that were actually observed in the study. The resulting estimates are called *maximum likelihood estimates* (MLEs). Using the MLEs for the $\beta$ values, we can then develop ML estimates for the adjusted odds ratio due to exposure by substituting the estimated $\beta$ values for the unknown parameters in equation 13.20. An additional advantage of the ML approach is that it allows analysts to estimate the standard errors associated with the estimated parameters. These standard errors can then be used to evaluate hypotheses and construct confidence intervals for the estimates. Estimated odd ratios produced by these types of models are often called *adjusted odds ratios* to reflect the attempt to account for potential confounders and effect modifiers.

In the illustration of how logistic regression can be used to adjust for potential confounders and effect modifiers, we have not addressed the appropriateness of the assumed model for data collected in a case–control study. Because the model specifies disease odds as the outcome and exposure status as a known predictor, it appears that the model is not appropriate for case–control designs in which disease status is known and inferences are based on the level of exposure in the case and control groups. However, the logistic model is appropriate for case–control designs because both cohort and case–control designs result in similar likelihood functions under the logistic model (Prentice and Pyke 1979). As a result, the MLEs will be the same regardless of the time at which cases are identified.

The applicability of the logistic regression model under a cohort or case–control design has led to widespread use of this technique in epidemiologic studies in which exposure and disease information is available for all subjects. In some situations, however, the necessary information is not available for each study participant but only on collections of individuals with similar characteristics. For example, suppose we cannot identify the specific individuals within a given area who developed disease, but we do know the total population size and the total number of persons who developed disease in the region. In addition, suppose we have some additional information on the area such as an estimate of the average radiation exposure and the percentage of the population who smoke. Because we do not have individual information (e.g., we do not know if individual $i$ is a case or noncase or know his or her specific exposure level), we cannot develop the estimate of $P_i$ based on the logistic regression model. We can analyze this type of aggregated data, however, using a technique called Poisson regression.

If you are familiar with discrete probability distributions, you will recognize in equation 13.23 that the contribution of the $i$th individual to the likelihood function based on the logistic model is the distribution function of the Bernoulli random variable $Y_i$. This variable takes a value of 1 if the $i$th individual has the disease in question and 0 if he or she does not. In *Poisson regression*, we assume that the number of cases observed in a particular group is a Poisson random variable with an expected value that depends on the total number of people in the group and other measurable quantities such as the average radiation exposure. For example, suppose we have a measure of the average lung dose resulting from indoor exposure to radon and radon decay products for a collection of $K$ counties in the United States. In addition, we will assume that observed lung cancer incidence rates are available for a study period that has relevance to these dose estimates, for example, a 20-year incidence rate that begins 5–10 years after the exposure measurements were taken. We have summary measures of disease and exposure but lack individual information. If we redefine $Y_i$ to be a random variable representing the number of cases in county $i$, we can model the relationship for the expected value of $Y_i$, designated as $E[Y_i]$, and average exposure as

$$E[Y_i] = N_i e^{\beta_0 + \beta_1 \times D_i}, \qquad (13.24)$$

where $D_i$ is now the average radiation dose among residents of county $i$ and $N_i$ is the number of inhabitants of the county at the beginning of the study period. Notice the model implies that the disease risk in county $i$, which we will call $R_i$, is dependent on the average dose such that

$$\frac{E[Y_i]}{N_i} = R_i = e^{\beta_0 + \beta_1 D_i}. \qquad (13.25)$$

If we consider a categorization in which $D_i$ takes a value of 1 if county $i$ has an estimated average exposure level greater than some cut-off value and 0 if it does not, then using the notation that $R_1$ is the risk in any exposed area and $R_0$ is the risk in an unexposed region, the model in equation 13.25 implies that the risk ratio is given by

$$\frac{R_1}{R_0} = e^{\beta_1}. \qquad (13.26)$$

Therefore, if we can develop an estimate for $\beta_1$ in the model of equation 13.24, we can estimate the risk ratio attributable to exposure. Just as in the logistic regression approach, ML can be used to estimate the unknown model parameters. In this situation, however, the likelihood cannot be constructed by combination of Bernoulli distribution functions. Instead, we will assume that $Y_i$, the number of cases observed in county $i$, is a random variable that follows a Poisson distribution. We also will assume that the number of disease cases in any area does not depend on the number of cases observed in any other area. Under these assumptions, the likelihood function (i.e., the probability of getting the distribution of cases across counties that was seen) is given by

$$\text{Prob (getting the collection of } Y_i\text{s observed)} = \prod_{i=1}^{K} P(Y_i = y_i),$$

where $P(Y_i = y_i)$ is the Poisson distribution function for $Y_i$ given by

$$P(Y_i = y_i) = \frac{(N_i\mu_i)e^{-N_i\mu_i}}{y_i!}$$

and

$$\mu_i = e^{\beta_0 + \beta_1 D_i}. \tag{13.27}$$

Under ML, we will again estimate the parameters of interest by finding those values for $\beta_0$ and $\beta_1$ that maximize the likelihood of observing the collection of area-specific diseases that were actually counted in the study. Once we have the estimate for $\beta_1$, we can substitute this value into equation 13.26 to obtain an MLE for the risk ratio due to exposure. As in logistic regression, the ML approach under the Poisson model also yields estimates of the standard errors of the estimated parameters. These standard error estimates allow researchers to test hypotheses of the form $H_0 : \beta_1 = 0$, which implies a risk ratio of 1, and to create confidence intervals on the estimates.

The estimated odds ratios and risk ratios resulting from the use of logistic and Poisson regression are often reported relative to a unit of the measure of exposure. For example, the National Academy of Sciences evaluation of the association between a history of childhood exposure to $^{131}$I and risk for thyroid cancer is often summarized as a risk ratio of 8 at a thyroid dose of 1 gray (NRC 1990). In addition, this analysis can be conducted with continuous as opposed to categorical measures of the exposure of interest. If the exposure measure is continuous, then the $\beta_1$ term in the logistic model should be interpreted as the increase in log odds of disease per unit increase in exposure. Similarly, if a continuous measure of exposure is used in a Poisson regression, then the resulting estimate of $\beta_1$ represents the estimated increase in the log risk ratio per unit increase of exposure.

A final note should be made on Poisson regression approaches and other analytic methods when they are applied to aggregated data. These types of studies, in which information on disease status, exposure, and confounding or modifying factors is not available at the individual level but only in aggregations across individuals, are sometimes called *ecologic studies*. The problem with ecologic studies is that they are subject to bias in that factors that appear to affect the risk for disease in a certain manner at the aggregate level may, in fact, exert quite a different effect on disease risk at the unmeasured individual level. This potential bias in ecologic study findings is sometimes referred to as the *ecologic fallacy* and is a subject of debate in radiation epidemiology (Cohen and Colditz 1994; Greenland 1992). Therefore, ecologic-level studies should be thought of as preliminary investigations whose results may point the way for more focused individual-level cohort or case–control follow-up investigations.

So far, we have discussed analytic techniques for evaluating possible associations between potential causative factors and the risk of developing disease. Another group of analysis techniques focus on the length of time between an event, say, exposure to radioactive material, and some potentially related outcome, for example, diagnosis with cancer. The length of time between these two events is often referred to as survival time; hence, this analysis approach is called *survival analysis*.

To see the difference between survival analysis and the methods discussed so far, consider a hypothetical five-year study of the association between occupational exposure to uranium and development of kidney disease. At the end of the five-year period, we could count the cases of kidney disease that occurred in the group and use a logistic regression model to evaluate whether disease incidence is related to the level of uranium exposure adjusted for potential confounders and effect modifiers. Using this approach, however, we are potentially missing some important information. Suppose that uranium exposure also increases the risk of developing lung cancer in the exposed group. Because lung cancer is usually fatal, we would expect the exposed group to have more individuals removed from being at risk for developing kidney disease than the unexposed cohort because some in the exposed group would have died of lung cancer. Therefore, because we do not know how many exposed people would have developed kidney disease if they had not first developed lung cancer, analysis of these data using logistic regression may be misleading.

In survival analysis, we can address the problem of individuals being differentially removed from the study by adjusting risk estimates to reflect how the number of individuals at risk for disease, called the *risk set*, changes with time. For example, in the above illustration, if we knew the date of exposure and the dates of diagnosis for the kidney disease and lung cancer, we could alter the population at risk to account for individuals who die of lung cancer. Thus, we would be more likely to identify the association between uranium exposure and kidney disease. The methods and models used in survival analysis can be complex and are beyond the scope of this chapter. However, there is a large body of literature on the subject to which you can refer (e.g., Cox and Oakes 1984; Hosmer and Lemeshow 1999).

Given the fact that survival analysis techniques use more information than do logistic and Poisson regressions, you may ask why these approaches are not always used in model-based analyses of epidemiologic data. One reason is the theoretical and computational complexity involved in correct application of these methods. In addition, to apply even exploratory survival analysis methods, researchers must know individual follow-up times for all study participants from time of exposure to time of removal from the risk set. This level of information on study participants is rarely available in radiation epidemiology. For example, even determining exactly when an individual's exposure to radioactive materials began can be difficult because exposures often occurred long in the past and were distributed across many years. An important exception to this difficulty is in studies of atomic bomb survivors for whom the precise moment of exposure initiation is known. The health data on the atomic bomb survivor cohort have been extensively analyzed using survival analysis methods.

## Issues in Radiation Epidemiology

So far in this chapter, we have addressed general issues concerning the analysis of epidemiologic data. This section focuses on some specific issues related to the use of epidemiologic methods to investigate the human health effects resulting from exposures to radioactive materials. Perhaps the most important prerequisite for such use of epidemiologic approaches is developing relevant measures of exposure and dose.

The first requirement for dose estimates to be useful in an epidemiologic study is that the estimates must be temporally relevant to any subsequent risk for disease. For example, Lubin et al. (1994) assumed a five-year latency period between exposure to radon decay products and initiation of the time at risk for developing lung cancer related to that exposure. This latency period reflects empirical observations that increases in risk due to these exposures do not appear to be immediate but are delayed until a number of years after exposure.

Sometimes dosimetric practice itself can render dose estimates temporally unsuitable for use in epidemiologic investigations. For example, suppose you are interested in evaluating a possible association between the estimated bone surface dose resulting from internal exposure to uranium and the risk of developing bone cancer. Standard dosimetric practice for this type of alpha emitter would be to estimate an organ-specific committed dose equivalent for a specified commitment period, say, 50 years. The problem with this committed dose approach is that if we were to use that estimate in an epidemiologic study, we would be erroneously assuming that every individual is at risk of developing cancer at the 50-year committed dose for all times before the end of the commitment period. Figure 13.2 illustrates the problem with this assumption. The dashed line represents the 50-year committed dose estimate resulting from a single exposure, while the solid line illustrates the change in the exposed individual's true cumulative dose over time. Notice that for a large portion of our study interval, which begins after the latency period between

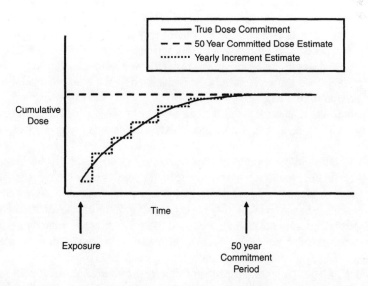

Figure 13.2 True organ-specific dose commitment resulting from a one-time exposure to radioactive material, estimates for that dose based on a 50-year committed dose approach, and an estimate based on yearly increments in accrued dose.

time of exposure and initiation of the resulting increased risk, use of the committed dose estimate results in a higher dose estimate than is justified on the basis of the person's actual accumulated dose. This overestimation of dose could lead to biased results in an epidemiologic study, especially if the material has a relatively long effective half-life in the organ of interest.

A preferable dose estimate for epidemiologic uses is the organ-specific dose estimate like the one illustrated by the dotted line in figure 13.2. For this estimate, each year's portion of the dose commitment is sequentially added to the current cumulative dose to better describe the true accumulation of dose over time. This approach, however, can be somewhat complicated for extended exposures to radionuclides with long effective half-lives. In this case, to produce a year-specific dose estimate, we would have to add the portion of the committed dose accrued in each year due to new exposures to the appropriate portion of the committed dose resulting from previous years' exposures (Killough et al. 1998).

In addition to being relevant in terms of time, radiation dose estimates must also be relevant in terms of the outcome of interest. For this reason, effective dose estimates are not likely to be useful for epidemiologic analyses. If we consider cancers to be the primary outcome of interest, then an epidemiologic study will likely focus on the risk of cancer occurring at a specific site such as the lungs, kidneys, or bone surface. In this situation, an organ-specific dose estimate is much more likely to be relevant in terms of possible association with increased risk for the cancer of interest.

Our discussion of issues related to developing dose estimates that are applicable to epidemiologic research is predicated on the assumption that some mechanism, for example, a dose reconstruction, is available to produce such estimates for everyone to be included in the study. In this context, the term "individual dose estimate" may be misleading because model-based estimations of dose provide individualized dose estimates only to the level of resolution of the model. For example, a dose reconstruction can provide a dose estimate for a given group of individuals who resided or worked at a specified location during a specified time and who share some set of common characteristics. This estimate, although often referred to as an individual dose estimate, is only individualized to the level of membership in the group. Interindividual variation in true dose within such groups is therefore not identifiable. In fact, in some cases, the best we may be able to do in an epidemiologic investigation is to develop crude metrics of exposure, such as proximity to a possible source.

This variation in the level of potential accuracy of dose and exposure estimates highlights the final requirement if dose estimates are to be useful in epidemiologic investigations: adequate representation of dose estimate uncertainty. All estimates of dose, even those based on laboratory assays, are subject to some level of uncertainty. Just as this uncertainty must be incorporated in estimates of associated risk, lack of precise knowledge about true exposure levels must be accounted for in the planning and analysis of epidemiologic studies (Thomas et al. 1993).

Epidemiologic studies of the effects of exposures to radioactive materials must be designed carefully. If you are using a cohort approach, for example, the definition of what constitutes an exposed as opposed to unexposed group is critical. Often, however, such decisions are based on arbitrarily defined zones or worker job titles that may introduce bias into results due to misclassification of true exposure status. The National Research Council and National Academy of Sciences (NRC/NAS 1995) have suggested a phased approach in which primary radionuclides and pathways at a given site are identified in a screening procedure. The results of these analyses are used to identify groups of individuals most likely to have been exposed.

In addition, the uncertainty inherent in estimates of exposure and dose must be considered when designing an epidemiologic study. Often the number of participants to be recruited for a study is the number necessary to achieve a desired level of statistical power. Statistical power is the probability that the study will detect an association between an exposure and disease if such an association truly exists. The power of an epidemiologic analysis depends on a variety of factors, such as the number of participants, the distribution of exposures across the study population, and the magnitude of the true association between disease risk and exposure. If researchers plan a study based on uncertain dose estimates, then they must account for this lack of precision when calculating the sample size needed to attain a specified level of power (Devine and Smith 1998; Devine 2003), because the study's power depends on the distribution of true exposures in the study population (McKeown-Eyssen and Thomas 1985). In general, to realistically address dose estimate uncertainty in planning an epidemiologic study, you need a larger study population than would be necessary if the true doses were observable.

The fact that we do not know each subject's true dose must be addressed in the analysis of data that are collected in epidemiologic studies on the effects of exposure to radioactive materials. Consider the logistic regression model in equation 13.16. If, instead of knowing $D_i$, we can only observe $Z_i$, where $Z_i$ is a surrogate measure for the unknown true dose, then we must account for the imprecision associated with $Z_i$. To make this accounting, we must know, or be able to reasonably estimate, the relationship between the true dose and the surrogate measure and incorporate that relationship into the analysis.

It is usually assumed that estimates of association between disease risk and imprecise measures of dose will be lower than the true association between risk and the actual dose. This reduction in the level of association is often referred to as *attenuation to the null*. This attenuation effect, however, is only certain for linear models relating disease risk and dose and for a specific type of relationship between $D_i$ and $Z_i$ called *classical error*. For nonlinear models relating risk and dose, such as the logistic and Poisson regression models, and in situations where the relationship between the true and estimated dose does not correspond to the classical error model, the difference between estimates of association based on the surrogate measure and those based on the true dose are often unknown (Thomas et al. 1993). In addition, the standard error of the association estimate should be increased when it is based on the surrogate dose measure to reflect lack of knowledge about the true dose.

As a result, epidemiologic studies of the impact of radiation exposure on human health must reflect, to the best extent possible, our lack of knowledge concerning true exposure and dose levels, and researchers should make every attempt to account for this uncertainty in the study planning and analysis. Studies in which uncertain dose values are used with no accounting for the uncertainty inherent in these estimates should be viewed with some skepticism. Finally, remember that uncertainty in dose estimates is likely to be present in virtually every study of the impact of radiation exposure on human health. Even studies of occupational exposures based on workers wearing dosimeter badges are likely to involve some degree of uncertainty about the accuracy of the measured doses because of factors such as variations in the precision of measurement devices over time.

Analysis techniques to account for using uncertain dose estimates in the analysis of epidemiologic data are beyond the scope of this chapter. Richardson and Gilks (1993) present methods for accounting for this type of uncertainty in analyses of occupational exposures. In addition, Kerber et al. (1993) and Thomas et al. (1991) provide examples of accounting for dose estimate uncertainty in studies of radiation effects in populations whose exposure estimates are based on environmental dose reconstruction techniques.

## Conclusion

This chapter provides a brief overview of the types of approaches used in epidemiologic analyses in general and in evaluations of the potential health effects of human exposure to radioactive material in particular. This material is not intended to serve as a guide in conducting epidemiologic investigations, but rather provides a basic understanding of epidemiologic concepts that can help you in critically evaluating the literature and in understanding the crucial role of dosimetric practice in radiation epidemiology. We have stressed that scientifically sound estimates of exposure and dose, along with realistic quantification of the uncertainty inherent in these estimates, are the cornerstones on which sound radiation epidemiologic evaluations must be based.

## Problems

1. You are reading about an epidemiologic study on a possible increase in risk for thyroid nodules among individuals who lived within two miles of a nuclear weapons production facility during its years of operation from 1960 through 1985. The researchers have identified all 212 persons who currently reside within the two-mile limit. To serve as a comparison group, they have identified 625 individuals who currently reside at least 10 miles from the plant. The data on the prevalence of thyroid nodules in the two groups are as follows:

   | Currently resides within two miles of the plant | Thyroid nodule present | Thyroid nodule not present | Total number of individuals |
   |---|---|---|---|
   | Yes | 21 | 191 | 212 |
   | No | 36 | 589 | 625 |
   | Total | 57 | 780 | 837 |

   a. Calculate the risk ratio for having a thyroid nodule for those who live within two miles of the plant relative to those who do not.
   b. Calculate the Mantel-Haenszel test statistic for evaluating the hypothesis that there is no association between risk and living within two miles of the former production plant.

c. The probability of observing a Mantel-Haenszel test statistic of the size seen in this study is 0.04, if the hypothesis of no association between where a person lives relative to the plant and their risk for thyroid nodules is true. How would you interpret this result in terms of the plausibility of the no-association hypothesis? Do these data provide proof that living near the plant causes thyroid nodules?

2. If the definition of the exposure and the disease data presented in problem 1 were all the information available to you in the summary of the study's findings, what questions might you pose in terms of the relevance of the definition of exposure used in the study to the true exposure of interest? Are there any possible confounding factors that might lead you to have questions about the study results?
3. Calculate the odds of disease among those who live within two miles of the plant and the odds of disease among those who do not. Using these values, calculate the risk odds ratio for comparing these two groups. Is this odds ratio similar to the risk ratio calculated in problem 1? Why or why not?
4. You read in a press release that workers at a nuclear power production facility who have been diagnosed with bone cancer are twice as likely to have worked in a certain area of the plant as are workers with no such diagnosis. Given this description, do you think the design used in this research was a cohort or case–control study? Why? How would the reporting of the study results likely be changed if the design were other than the one you selected?
5. The results of a cross-sectional study on the health of employees at a nuclear power production facility are reported in the newspaper. From the article, you learn that the all current and living former employees were examined in the study to determine the presence or absence of a specified cancer. The article concludes with a statement that the incidence rate for cancer development among these workers is 10 cancers per 100,000 person-years at risk. You immediately question either the validity of the study or the accuracy of the newspapers summary of the results. Why?
6. Suppose, as part of your health monitoring duties at a nuclear facility, you are charged with following the health of six randomly selected disease-free employees for a period of five years to determine the incidence rate for the development of a particular disease that might be related to an occupational exposure. The following table summarizes the follow-up time and the disease occurrence data for the six individuals over the five-year study:

| Subject number | Time (years) at which subject was diagnosed with study disease | Time (years) at which subject was removed from the study |
| --- | --- | --- |
| 1 | 1.5 | 1.5 |
| 2 | 2.0 | |
| 3 | 3.5 | 3.5 |
| 4 | 3.0 | |
| 5 | 0.5 | 0.5 |
| 6 | 5.0 | |

Notice that three individuals developed the study disease during the follow-up period and, as a result, were removed from the population at risk at the time of their diagnosis. In addition, two other individuals were removed from the study population prior to study completion due to other reasons after contributing two and three years of person-time at risk.

  a. Using equation 13.8, calculate the risk of developing the study disease among this group.
  b. Calculate the total person-years at risk among the six individuals during the five-year study period.
  c. Derive the incidence rate for the study disease based on the observed cases and the calculated person-years at risk.
  d. Explain the different aspects of disease burden described by the risk and the incidence rate in this study. Why are the magnitudes of these measures similar or different in this example?

7. You are designing the dosimetric methods to be used in an epidemiologic study of possible association between dose to the bone surface, resulting from an acute exposure to an alpha-emitting radionuclide, and the risk of developing bone cancer. Suppose the radionuclide of interest has an effective half-life in the bone surface of 60 months and the dose conversion factor you plan to use is designed to estimate the 50-year committed dose to the bone surface. Will this approach give estimates of dose that are appropriate for use in the epidemiologic investigation? Explain your answer. What if the radionuclide of interest had a one-month effective half-life? (Hint: Assume that the dose resulting from the exposure will be negligible after a period of 10 times the effective half-life.)

8. An environmental group reports that it has evidence that the prevalence of birth defects in a community surrounding a nuclear waste storage facility is two times larger than the national prevalence of birth defects. The group collected the data on which it based this statement by running advertisements in local newspapers requesting all persons who ever lived near the plant to write in and report whether they or their children suffer from any birth defects. When the group decided that a large enough number of responses to the advertisement had been received, they calculated the proportion of respondents who reported a birth defect, and it was this number they compared to the national average. What problems, if any, in the methods used in this analysis might lead you to question the reported results?

9. You have been asked to provide comments on the results of an epidemiologic investigation concerning the possible association between exposure to radioactive fallout from atmospheric weapons testing and the risk of breast cancer in women. The study population is composed of two cohorts, a group exposed to atmospheric fallout and a group with no such exposure. Because family history of breast cancer is a risk factor for development of this disease, you are glad to see that the investigators collected information from study participants on family medical history. You learn that 10% of the exposed group and 11% of the unexposed group have close relatives

who have been diagnosed with breast cancer. Should you be concerned that family history of breast cancer is a confounder in this analysis? Why or why not?

## Acknowledgment

We sincerely thank Dr. Judith Qualters of the U.S. Centers for Disease Control and Prevention for her many helpful comments and suggestions on this chapter.

References

Agresti, A. 1990. *Categorical Data Analysis*. New York: John Wiley & Sons.
Bishop, Y.M., S.E. Feinberg, and P.W. Holland. 1975. *Discrete Multivariate Analysis*. Cambridge, MA: MIT Press.
Cohen, B.L., and G.A. Colditz. 1994. "Tests of the Linear–No Threshold Theory for Lung Cancer Induced by Exposure to Radon." *Environmental Research* 64: 65–89.
Cox, D.R., and D. Oakes. 1984. *Analysis of Survival Data*. New York: Chapman & Hall.
Devine, O.J. 2003. "The Impact of Ignoring Measurement Error When Estimating Sample Size for Epidemiologic Studies." *Evaluation and Health Professions* 26: 315–339.
Devine, O.J., and J.M. Smith. 1998. "An Estimating Sample Size for Epidemiologic Studies: The Impact of Ignoring Exposure Measurement Uncertainty." *Statistics in Medicine* 17: 1375–1389.
Evans, M., N. Hastings, and B. Peacock. 1993. *Statistical Distributions*. New York: John Wiley & Sons.
Feinberg, S.E. 1987. *The Analysis of Cross-Classified Data*. Cambridge, MA: MIT Press.
Greenland, S. 1992. "Divergent Biases in Ecologic and Individual-Level Studies." *Statistics in Medicine* 11: 1209–1223.
Hoffman, F.O., and J.S. Hammonds. 1994. "Propagation of Uncertainty in Risk Assessments: The Need to Distinguish Between Uncertainty Due to Lack of Knowledge and Uncertainty Due to Variability." *Risk Analysis* 14(5): 707–712.
Hogg, R.V., and A.T. Craig. 1978. *Introduction to Mathematical Statistics*. New York: Macmillan.
Hosmer, D.W., and S. Lemeshow. 1999. *Applied Survival Analysis: Regression Modeling of Time to Event Data*. New York: John Wiley & Sons.
IAEA (International Atomic Energy Agency). 1989. *Evaluating the Reliability of Predictions Using Environmental Transfer Models*. Safety Series 100. Vienna: International Atomic Energy Agency.
Kelsey, J.L., A.S. Whittemore, A.S. Evans, and W.D. Thompson. 1996. *Methods in Observational Epidemiology*. New York: Oxford University Press.
Kerber, R.A., J.E. Till, S.L. Simon, J.L. Lyon, D.C. Thomas, S. Preston-Martin, M.L. Rallison, R.D. Lloyd, and W. Stevens. 1993. "A Cohort Study of Thyroid Disease in Relation to Fallout from Nuclear Weapons Testing." *Journal of the American Medical Association* 270: 2076–2082.
Killough, G.G., M.J. Case, K.R. Meyer, R.E. Moore, S.K. Rope, D.W. Schmidt, B. Shleien, W.K. Sinclair, P.G. Voilleque, and J.E. Till. 1998. *Radiation Doses and Risks to Residents from FMPC Operations from 1951–1988*, vols. 1 and 2. Task 6 Final Report:

Fernald Dosimetry Reconstruction Project. RAC Report No. 1-CDC-Fernald-1998-Final. Neeses, SC: Risk Assessment Corporation.

Kleinbaum, D.G., L.L. Kupper, and H. Morgenstern. 1982. *Epidemiologic Research, Principles and Quantitative Methods*. New York: Van Nostrand Company.

Langholz, B., and D. Clayton. 1993. *Sampling Strategies in Nested Case-Control Studies*. Technical Report 56. Los Angeles, CA: Division of Biostatistics, Department of Preventive Medicine, School of Medicine, University of Southern California.

Last, J.M. 1995. *A Dictionary of Epidemiology*. New York: Oxford University Press.

Lubin, J.H., and K. Steindorf. 1995. "Cigarette Use and the Estimation of Lung Cancer Attributable to Radon in the United States." *Radiation Research* 141: 79–85.

Lubin, J.H., J.D. Boice, C. Edling, R.W. Hornung, G. Howe, E. Kunz, R.A. Kusiak, H.L. Morrison, E.P. Radford, J.M. Samet, M. Timarche, A. Woodward, Y.S. Xiang, and D.A. Pierce. 1994. *Radon and Lung Cancer Risk; A Joint Analysis of 11 Underground Miners Studies*. NIH Publication No. 94–3644. Behtesda, MD: National Cancer Institute, National Institutes of Health, U.S. Department of Health and Human Services.

McKeown-Eyssen, G.E., and D.C. Thomas. 1985. "Sample Size Determination in Case-Control Studies: The Influence of the Distribution of Exposure." *Journal of Chronic Diseases* 38: 559–568.

NRC (National Research Council). 1990. *Health Effect of Exposure to Low Levels of Ionizing Radiation: BEIR V*. Washington, DC: National Academy Press.

NRC/NAS (National Research Council/National Academy of Sciences). 1990. *Health Effects of Exposure to Low Levels of Ionizing Radiation BEIR V*. Washington, DC: National Academy Press.

NRC/NAS. 1995. *Radiation Dose Reconstruction for Epidemiologic Uses*. Washington, DC: National Academy Press.

Prentice, R.L., and R. Pyke. 1979. "Logistic Disease Incidence Models and Case-Control Studies." *Biometrika* 66 (3): 403–411.

Richardson, S., and W.R. Gilks. 1993. "Conditional Independence Models for Epidemiologic Studies with Exposure Measurement Error." *Statistics in Medicine* 12: 1703–1722.

Schlesselman, J.J. 1982. *Case Control Studies, Design, Conduct, Analysis*. New York: Oxford University Press.

Symons, M.J., and J.D. Taulbee. 1984. "Statistical Evaluations of the Risk of Cancer Mortality Among Industrial Populations." In *Statistical Methods for Cancer Studies*, ed. R.G. Cornell. New York: Marcel Decker.

Thomas, D.C., J. Gauderman, and R. Kerber. 1991. *A Non-Parametric Monte Carlo Approach to Adjustment for Covariate Measurement Errors in Regression Analysis*. Technical Report 15. Los Angeles, CA: Division of Biostatistics, Department of Preventive Medicine, School of Medicine, University of Southern California.

Thomas, D., S.D. Darby, F. Fagnani, P. Hubert, M. Vaeth, and K. Weiss. 1992. "Definition and Estimation of Lifetime Detriment from Radiation Exposures: Principles and Methods." *Health Physics* 63 (3): 259–272.

Thomas, D., D. Stram, and J. Dwyer. 1993. "Exposure Measurement Error: Influence on Exposure–Disease Relationships and Methods of Correction." *Annual Review of Public Health* 14: 69–93.

UNSCEAR (U.N. Scientific Committee on the Effects of Atomic Radiation). 1993. *Sources and Effects of Ionizing Radiation*. UNSCEAR 1993 Report to the General Assembly, with Scientific Annexes. New York: United Nations.

# 14

# Model Validation

Helen A. Grogan

Numerous chapters in this book highlight the importance of mathematical models in all aspects of radiological assessment, from estimating the release and transport of radionuclides through the environment, to exposure pathway analysis, to estimating doses and human health risks. Different types of models are employed depending on the goal of the analysis, such as screening models, research models, and assessment models. Screening models are designed to identify the most significant radionuclides, pathways, or media in relation to a specific issue for the purposes of focusing the analysis and resources. These models generally use conservative assumptions designed to overestimate radionuclide concentrations and impacts and to limit false negative results. Research models are designed to provide insight and understanding about the processes and mechanisms controlling the behavior of radionuclides in some component of the environment. Such models tend to be quite detailed and use realistic assumptions for the parameter values. Assessment models are comprehensive in scope so that the consequences of radionuclides in the environment can be evaluated for decision purposes. Such models are often composed of smaller models or modules. In all cases, however, the model is a mathematical representation of a natural system. Although physical models of a system are also possible, such as a soil column in a laboratory or a field lysimeter, they are not the subject of this chapter. Rather, it is the mathematical models used to represent these systems.

A mathematical model may be solved using a calculator, or it may be coded into computer language. Implementing the model using a verified computer code allows rapid and repeated calculations with little or no computational error. Many modeled systems are quite complex, so a computer code is necessary to carry out the calculations. Computer code verification is the process of checking that the computer code has been constructed correctly so it conforms to specifications (i.e., it truly represents the conceptual model) and that there are no inherent numerical problems with obtaining a solution. Verification, however, says nothing about the suitability of the model to answer the specific assessment question. The mathematical model may range from a simple empirical relationship to a complex multidimensional model. Regardless of the model type or its complexity, the user should be interested in how the model performs, in other words, its validity in light of the assessment question it is intended to answer.

Extensive literature exists debating the term "validation," and there are many philosophical arguments about what it means and whether it is achievable. Much of this debate arises because different groups of scientists have attached different meanings to the term "model validation," as evidenced by a series of editorials on modeling contaminant transport through groundwater systems (Bredehoeft and Konikow 1993; McCombie and McKinley 1993; Bair 1994). At one extreme, all models are valid because they formalize and make explicit conceptual models in ways that facilitate critical evaluation (Forrester 1968), and at the other extreme, models can never be validated; they can only be invalidated (Oreskes et al. 1994). For the reader who is fascinated by such statements and the rationale behind them, Beck et al. (1997) provides a good review and entry into this topic. A key point to emerge from all the discourse is that model validation should not be viewed as a label that is attached to a model (Hassanizadeh and Carrera 1992), because the specified purpose of the model needs to be determined before its validity can be assessed (Caswell 1976). This is a vital point to remember for radiological assessments where public communication of the results is important. The purpose of this chapter is to provide a useful definition of model validation for radiological assessment, to identify the different components of a model validation process, and to lay out a framework for achieving this.

## Validation Process

*Validation* is an analysis that attempts to delineate the domain of applicability of a model (Peterson and Kirchner 1998; Kirchner 1994). The *domain of applicability* of a model refers to the set of conditions under which a model can be assumed to adequately represent the system of interest. The validation process determines the accuracy of the model and the range of input factors over which the model provides accurate estimates. Thus, as Kirchner clarifies in chapter 11 on uncertainty analysis, while uncertainty analysis is analogous to determining the imprecision in measurements from a piece of laboratory equipment, validation is analogous to determining the accuracy of the measurements and the range of input factors over which accurate measurements can be made. As noted above, the validity of a model cannot be judged without specifying the assessment question it is intended to answer. Beck et al. (1997) identifies several questions to determine a model's validity. For example, "Will the model perform its task reliably?" A positive answer to this question provides assurance that the risk of an undesirable outcome will be minimal. This is particularly important if there are large negative consequences associated with poor predictions. A second question may be, "Which among several candidate models is the most reliable instrument for performing the given task?" To answer this question, it is necessary to review the performance of a number of models over a range of conditions relevant to the task and identify the most suitable one.

From a practical standpoint, the process of model validation can be broken down into two main parts: the composition of the model and the performance of the model. These two concepts are discussed in the following sections.

## *Model Composition*

To evaluate the composition of the model, its composition and structure are scrutinized, as well as the manner in which its constituent hypotheses are assembled. Attention is given to the form of the equations and the parameter values used in the model. Understanding the formulation of a model is necessary to evaluate any potential limitations due to inherent constraints or assumptions. Models invariably contain assumptions, and it is important to review these. Law and Kelton (1982) refer to this process as establishing the face validity of a model and identify testing the assumptions as a second step in this process. Establishing the compositional validity of a model is important even when a model's performance can be evaluated quantitatively because model outputs are usually synthetic or aggregate variables and hence are subject to aggregation error (Gardner et al. 1982; O'Neill and Rust 1979; Mosleh and Bier 1992) and the possibility of compensatory errors.

The process of evaluating the compositional validity of a model is often achieved through peer review. Oftentimes, this review proceeds in stages, starting with internal review by peers within the same organization, followed by what tends to be a more formal external review by independent technical experts. Publishing details of the model in refereed journals is another example of the peer review process.

One aspect of achieving compositional validity is verification of the model code if it has been implemented on a computer. Verification is an essential stage in the development of a computer model and should not be confused with validation. *Verification* is the process of testing the internal correctness of a computer code that implements mathematical or numerical solutions dictated by a conceptual model of a real system. For example, a computer code is verified when sufficient tests have been performed to ensure that it is free from errors in programming and data entry and that the accuracy of numerical approximations to solutions of equations is acceptable. Verification should also ensure that the model is fully operational, in the sense that all possible pathways through the code have been exercised using the complete range of input parameters for which it was designed. Software testing procedures and tools for program or code verification are not addressed in this chapter.

One difficulty with the verification of model codes is that the scientific review process is typically associated with publications, and publications tend to describe the conceptual and mathematical representation of the model rather than the code for the model. Thus, model code verification is usually not part of the peer review process.

To facilitate review of a model, the assumptions should be explicit in the model documentation. Model sensitivity analysis is one useful tool for testing parameter assumptions. The degree of sensitivity of the model output to the input parameters can provide important insights. For example, if a model is insensitive to processes known to be important for the system being modeled, its validity will be questioned. Model assumptions can only be tested by altering the structure of the model. When possible, the assumptions should be tested empirically. When empirical tests cannot be conducted, it is frequently useful to examine the effects of substituting alternative mechanisms or structures. For example, it is often useful to demonstrate the effect of adding more or less structural complexity to a model. Such tests also contribute to understanding the uncertainty in the model due to structural complexity. In chapter 11, in the section "Uncertainty and Model Validation," Kirchner discusses specialized sampling schemes for use in model comparison experiments to provide answers to some of these questions.

A second stage of model testing involves the comparison of results obtained by different codes for a well-defined hypothetical system. This form of model testing is referred to by different names, including benchmarking and intercomparison exercises. Benchmarking allows comparison of the algorithms employed to solve a given set of equations and comparison of the relative importance of the various processes included in the models. Absolute answers are generally unavailable for these problems. The International Atomic Energy Agency (IAEA) Environmental Modelling for Radiation Safety (EMRAS) program launched in September 2003 (www-ns.iaea.org/projects/emras/) includes such intercomparison exercises. The EMRAS program continues some of the work of previous international programs in the field of radioecological modeling, such as BIOMOVS (Biospheric Model Validation Study), which was initiated by the Swedish National Institute for Radiation Protection in 1985 and followed by BIOMOVS II, as well as programs sponsored by the IAEA: VAMP (Validation of Model Predictions, 1988–1996) and BIOMASS (Biospheric Modelling and Assessment, 1996–2001). BIOPROTA (www.bioprota.com) is a separate international program designed to address the key uncertainties in long-term assessments of contaminant releases into the environment from radioactive waste disposal. It also evolved from these earlier programs and includes model comparison exercises.

To explain and understand the similarities and differences in model results, careful analysis of model structure, modeling assumptions, and parameter values is required. Frequently, it is only through such intercomparison exercises that inconsistencies in a model formulation come to light (Grogan 1989). When a model is reviewed in isolation, such inconsistencies can be difficult to identify and may go undetected. All these activities contribute to establishing model credibility. For models that are designed to predict the behavior of radionuclides released into the environment in the future, comparison of model predictions against independent measured concentrations clearly is not feasible. This is the case for radiological assessment models applied to geologic disposal of radioactive waste. For these situations, model intercomparison exercises such as BIOPROTA, BIOMASS, and its predecessors are most valuable. The challenges and an approach to this type of modeling in the context of the Swedish program are described in a special issue of *Ambio*, a journal of the human environment (Bradshaw et al. 2006).

## Model Performance

The quantitative comparison of model predictions with observations from the system is important for models that are to be used as predictive tools. Such comparisons are not confined to model-predicted estimates for present-day situations; they also apply to reconstruction of historical events.

Model *accuracy* is a measure of how close the model estimate is to a similar measured quantity. This is evaluated by comparing model estimates against an independent set of measurements of like quantities. The measurements are independent if they have not been used to develop or parameterize the model in any way. This process is often referred to as "blind testing." Model accuracy can be contrasted to model *precision*, which is the ability of the model to reproduce the same result upon repeated trials given uncertainty in the input parameters. Model precision is established by parameter uncertainty analysis, which is the topic of chapter 11.

The acceptable level of estimative accuracy should be identified early in the model testing process, as well as the decisions that will be made if this level is not achieved. These assessments must be made over the range of conditions in which the model may be applied. The acceptable level of accuracy is subject to judgment and varies depending on the specific problems or questions being addressed by the model (IAEA 1989). As a rule of thumb, however, an environmental transport model is performing very well when the predictions are within a factor of 2 of the observations. Typical goals for model performance are to predict measured concentrations within a factor of 3, 5, or 10, depending on the assessment question. Furthermore, it is beneficial to state the percentage of time that the model predicts within a specified factor of the observations (e.g., within a factor of 3 in 65% of the cases, and within a factor of 5 in 90% of the cases). When model predictions and observations differ by large factors, such as 100 or more, this should trigger a reevaluation of the model and the data used to address the assessment question.

A fundamental principle for validation is that the data against which model predictions are tested must not have been used in constructing the model. The independence of the validation data is necessary to avoid the criticism that the model should be expected to fit data to which it has been explicitly or implicitly calibrated. In a well-designed validation experiment, all data to be used for comparisons should be sequestered. Sometimes even seeing a particular pattern in a set of validation data is sufficient for a modeler to reconsider the values that should be used for some parameters or perhaps how the process should be modeled. In a blind validation experiment, the validation data are identified, verified, and protected by one or more people not directly associated with setting up the model for the test. These data archivists are also responsible for providing the modelers the information necessary to run the model experiment, such as specific input data or scenario information.

One of the important limitations in model evaluation is the lack of data to use in quantitative model validation experiments. One reason that models are popular for the analysis of behavior of systems is that experimental manipulation of many systems is impossible because of system complexity, the cost of obtaining data, or regulations prohibiting experiments. For these same reasons, data that can be used in validation experiments are often uncommon. Data that do exist are often incomplete in the sense that only a limited subset of model outputs may be represented, and only some or none of the required input data may have been collected. In the case of missing input data, modelers often substitute data collected spatially or temporally near the system of interest or use values based upon expert opinion. These substitutions often result in higher levels of uncertainty in model predictions than would be obtained had appropriate site-specific data been used. In addition, errors in data, particularly systematic errors, can lead to the wrong conclusions when model predictions are compared to the data (Peterson and Kirchner 1998). Frequently, when data are not available for the assessment end point, such as the dose to individuals, they are available for intermediate steps in the assessment, such as concentrations in vegetation and milk. Comparisons of model predictions for these quantities are often assumed to provide an adequate test of the model. Because compensating errors may occur, this assumption may not be valid.

The Chernobyl accident in April 1986 presented a unique opportunity to test radiological assessment models. The information gathered after the accident from measurement programs instituted in countries of the former Soviet Union and many European countries produced environmental data sets of sufficient quality and quantity for model testing purposes. Scenario A4, organized in the BIOMOVS study, was the largest and most comprehensive evaluation of agricultural food chain models ever undertaken. Scenario A4 examined the air–forage–milk/beef and air–grain pathways for $^{131}$I and $^{137}$Cs at multiple sites for six months following the Chernobyl accident (BIOMOVS 1991; Peterson et al. 1996; Peterson and Kirchner 1998). The IAEA's VAMP program used Chernobyl observations collected over several years following the accident to develop further test scenarios (IAEA 1995, 1996), while later phases of the program developed test scenarios that focused on process level modeling rather than entire environmental transport pathways (Garger et al. 1999; Konoplev et al. 1999; Kryshev et al. 1999; Thiessen et al. 2005b). In a separate project, the environmental contamination from the Chernobyl accident provided an opportunity to test models of external dose to people living in these regions (Golikov et al. 1999).

The historical dose reconstructions of the impact of environmental releases from the U.S. nuclear weapons complex have all generated data sets for model testing. Examples are provided in Rood et al. (2002) and Till et al. (2000). Data collected following an accidental release of $^{131}$I in 1963, and evaluated as part of the Hanford Environmental Dose Reconstruction Project, formed the basis of a test scenario used in an international model testing study (Thiessen et al. 2005a).

## *Calibration*

During the process of model development, model output values may be compared to a set of observational data, and the model parameters adjusted so that the model predictions match the observational data within some predefined range. This process is called model calibration. Even though a model's output may appear to perform poorly when compared to the observations, the model output may exhibit a strong correlation to the observations, indicating that the differences among observations can be accounted for by the model. Thus, the model output may be biased either high or low as compared to the observations. Under such circumstances, a model calibration factor may be applied to improve the model performance.

In contrast, weak correlations between model predictions and observations indicate that differences among the observations are controlled by factors unaccounted for by the model. Weak correlations can be the consequence of a number of factors, including a poor model structure, poor parameterization of the model, or high system variability (Hoffman and Gardner 1983).

### EXAMPLE 14.1 MODEL CALIBRATION

Figure 14.1 shows the calibration of a model used to predict the dispersion of radon and its decay products released to the atmosphere from the K-65 silos at the Feed Materials Production Center (FMPC) in Fernald, Ohio (Killough and Schmidt 2000). The storage silos had been in operation since the early 1950s and were filled with $^{226}$Ra-bearing ore from which uranium had been extracted. Two data sets for radon monitoring in air near the FMPC during the 1980s were available. One consisted of measurements taken by Mound Laboratory from July 2, 1984, through July 2, 1986, at 17 stations, most of which were within the site boundary. The other consisted of measurements of radon concentrations performed by FMPC staff at the seven site boundary stations (BS1–BS7) and two or more background locations during the years 1981 through 1987. The site boundary was about 1 km

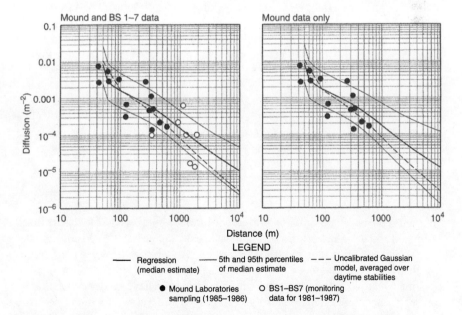

Figure 14.1 Calibration of Gaussian model for radon diffusion (data from Killough and Schmidt 2000).

from the center of the production area. The Gaussian model median predictions prior to calibration are represented by the dashed line in figure 14.1, and the calibrated median predictions are represented by the thicker solid line. The left-hand plot presents the calibration result using both data sets, and the right-hand plot presents the calibration results based on the Mound data set alone. Including the BS1–BS7 data reduces the uncertainty in the calibration curve at larger distances (1–10 km) from the silos.

## Tests of Model Performance

The types of quantitative tests that can be applied to the comparison of model results and empirical observations depend on the types and amount of data available. Frequently, metrics based upon statistical techniques are used to help perform these evaluations. Some of these tests are described in the sections below, and examples of their application can be found throughout this chapter. It is important to keep in mind, however, that tests to allow a statistical inference with regard to validation are not well developed for the kinds of data frequently available in radiological assessment. Furthermore, there is no universal test to determine the validity of a model.

In applying any quantitative technique for comparing results and observations, one should be aware of the distinction between statistical significance and practical significance (Law and Kelton 1982). Low variability in data and model predictions can make a small difference or bias statistically significant. Conversely, high variability in data and model predictions can result in large differences between model results and data that are shown to be statistically nonsignificant. Thus, it is also important to evaluate for differences of practical significance with regard to the objectives for the model when making comparisons between model predictions and observations.

### *Testing for Bias*

The predicted-to-observed ratio ($P/O$) provides a simple measure of the bias in the model-predicted estimates to the observed values:

$$= \frac{P}{O}, \tag{14.1}$$

where $O$ is the observed value, such as a concentration, and $P$ is the model-predicted estimate for that value. A ratio of 1 indicates exact agreement. $P/O$ ratios greater than 1 indicate that the model overpredicts the observed values, and ratios less than 1 indicate that the model underpredicts. In many cases, there is a distribution of $P/O$ ratios to evaluate. For example, suppose you have 100 air concentration measurements representing quarterly averages over 25 years. You might compare these measured concentrations to predicted concentrations of the same quantity. You would then have 100 $P/O$ ratios or a distribution of $P/O$ ratios. The distribution of

$P/O$ ratios can be used to make a statement about the model's ability to predict quarterly average concentrations. Common statistics such as the mean and standard deviation may be used to summarize the distribution of $P/O$ ratios. If the difference between the predicted estimates and the observed values tends to be large (a factor of 10 or more), then the geometric mean (GM) and geometric standard deviation (GSD) of the $P/O$ are more useful:

$$\text{GM} = \exp\left(\overline{\ln \frac{P}{O}}\right), \tag{14.2}$$

$$\text{GSD} = \exp\left[\sqrt{\frac{1}{n-1}\sum_{i=1}^{n}\left(\ln \frac{P_i}{O_i} - \overline{\ln \frac{P}{O}}\right)^2}\right]. \tag{14.3}$$

The overbars indicate averages over the sample. Again, a GM $P/O$ ratio less than 1 indicates model underprediction of the observed values, and a GM ratio greater than 1 indicates model overprediction.

When several of the predictions differ from the observations by a factor of 10 or more, then a log-transformed measure of model bias is appropriate because it provides a more balanced indicator (Hanna et al. 1991). The log-transformed measure of bias is the geometric mean bias (MG) (Cox and Tikvart 1990):

$$\text{MG} = \exp\left(\overline{\ln O} - \overline{\ln P}\right). \tag{14.4}$$

Geometric mean bias values of 0.5 and 2.0 indicate a factor of 2 model overprediction and underprediction, respectively.

Another measure of bias is the fractional bias (FB), which is a measure of the mean bias in the model. The fractional bias is the difference between the mean of the observed values and the mean of the model-predicted estimates normalized to the average of the two means:

$$\text{FB} = \frac{2\left(\overline{O} - \overline{P}\right)}{\overline{O} + \overline{P}}. \tag{14.5}$$

An FB value of 0 indicates perfect agreement between the average observed value and the average model-predicted estimate. Note that the FB provides a measure similar to the mean (or geometric mean) $P/O$ ratio. Negative FB values indicate model overprediction. An FB of $-0.67$ is equivalent to model overprediction by about a factor of 2, where, on average, the model-predicted values are twice the observed values. In contrast, positive values for FB indicate model underprediction, and an FB of 0.67 is equivalent to model underprediction by about a factor of 2. In this case, the mean $P/O$ ratio is 0.5. The FB index is sensitive to changes in the $P/O$ ratio below about 5. Figure 14.2 demonstrates that, when the model predictions are more than a factor of 5 larger than the observed values, this is not a very useful measure. Yu et al. (2006) have refined this metric to define a normalized mean bias factor, $B_{\text{NMBF}}$, to address some of these difficulties.

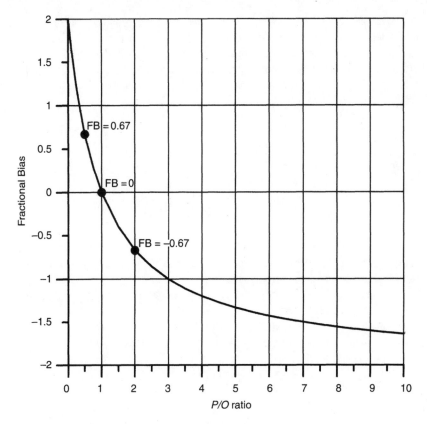

Figure 14.2   Relationship between FB and P/O ratio.

## *Measures of Scatter*

The normalized mean square error (NMSE) provides a measure of the variance or scatter in the model predictions from the observed values (Cox and Tikvart 1990):

$$\text{NMSE} = \frac{\overline{(O-P)^2}}{\overline{O}\cdot\overline{P}}. \tag{14.6}$$

A value of 0 indicates perfect agreement between predictions and observations. A value of 1 indicates that the typical difference between predictions and observations is approximately equal to the mean. The NMSE is a useful measure when the typical difference between predictions and observations is about a factor of 2.

When several of the predictions differ from the observations by a factor of 10 or more, then a log-transformed measure of model variance is appropriate (Hanna et al. 1991). The log-transformed measure of bias is the geometric mean variance

(VG) (Cox and Tikvart 1990):

$$\text{VG} = \exp\overline{\left[(\ln O - \ln P)^2\right]}. \tag{14.7}$$

A VG of 1.6 indicates a typical factor of 2 difference between model-predicted and observed data pairs. A perfect model would have FB and NMSE values of 0, and MG and VG values of 1.0. Note that MG and VG are similar to the GM and GSD of the distribution of $P/O$ ratios.

### Correlation and Regression

The correlation coefficient ($r$) is a popular test of statistical significance of the association between the model-predicted and observed values and can be used as a measure of model performance:

$$r = \frac{\sum (O_i - \overline{O})(P - \overline{P})}{\sqrt{\sum (O_i - \overline{O})^2 \sum (P_i - \overline{P})^2}}. \tag{14.8}$$

If the relation between the predicted and observed values is linear, then it can be expressed by the regression equation:

$$P = a + bO, \tag{14.9}$$

where $a$ is the y-intercept and $b$ is the slope.

When the differences between the predicted and observed values are large (exceeding a factor of 10), a log-transformed regression fit to the data can provide a more meaningful comparison. See, for example, figure 14.3, which is discussed in more detail below.

### Visual Display of Information

Apart from the various quantitative measures of model performance, it is often instructive to prepare visual displays of the predicted and observed model values. Such displays allow the user to assimilate a wealth of information about the model performance. Take, for example, figure 14.3, which shows the performance of five different atmospheric dispersion models for a range of test situations. The plots are shown on a log-log scale and include the ideal correlation line (i.e., $r = 1$). It is immediately apparent that of the five models, RATCHET performs best under the test conditions. By contrast, table 14.1 summarizes different quantitative measures of model performance for each of the five models under the same test conditions represented in figure 14.3. It takes much longer to assimilate the information.

The $r$-values (log-transformed correlation coefficients) confirm the results of the visual inspection that indicated RATCHET performs best in this case. The $r$-value also provides a quantitative measure of the difference in performance of the TRIAD and INPUFF models, which is less easily discerned by eye.

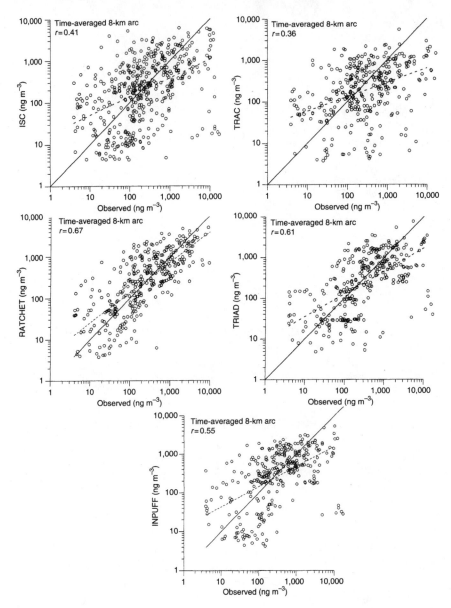

Figure 14.3 Comparison of the performance of five atmospheric dispersion models (from Rood 1999, figure 4, reprinted with permission).

## EXAMPLE 14.2 EVALUATION OF A SIMPLE DILUTION MODEL

The FMPC was part of the nuclear weapons complex and processed uranium into uranium ingots for machining or for extrusion into tubular form. The feed material included uranium concentrates, uranium compounds recycled from other stages of nuclear weapons production, and some uranium ores. The 1,000-acre site was

Table 14.1  Performance measure results for five atmospheric dispersion models (from Rood 1999, table 6)

| Performance measure (8-km results) | ISC | TRAC | RATCHET | TRIAD | INPUFF2 |
|---|---|---|---|---|---|
| Geometric mean $C_p/C_o$ | 1.1 | 0.64 | 0.86 | 0.88 | 0.91 |
| Geometric std $C_p/C_o$ | 6.2 | 6.4 | 4.4 | 4.7 | 5.3 |
| FB | 0.26 | 0.89 | 0.46 | 0.57 | 0.58 |
| FB confidence interval | 0.023–0.46 | 0.72–1.1 | 0.27–0.65 | 0.36–0.78 | 0.39–0.77 |
| NMSE | 5.0 | 12 | 7.8 | 7.7 | 7.8 |
| NMSE confidence interval | 3.3–6.6 | 8.3–15 | 4.7–11 | 5.1–11 | 4.8–11 |
| MG | 0.89 | 1.6 | 1.2 | 1.1 | 1.1 |
| log(MG) confidence interval | −0.29–0.097 | 0.29–0.67 | 0.041–0.29 | −0.016–0.32 | −0.052–0.29 |
| VG | 28 | 38 | 9.1 | 11 | 16 |
| log(VG) confidence interval | 2.7–3.9 | 3.0–4.2 | 1.8–2.6 | 1.9–2.9 | 2.2–3.3 |
| $r^a$ | 0.41 | 0.36 | 0.67 | 0.61 | 0.55 |
| r confidence interval | 0.31–0.51 | 0.26–0.45 | 0.60–0.72 | 0.53–0.67 | 0.47–0.63 |
| % within a factor of 5 | 67 | 69 | 79 | 72 | 72 |

[a] Calculated using a log-transformed regression.

located about 15 miles northwest of Cincinnati, Ohio, and operated from 1954 through 1988. The Fernald Dosimetry Reconstruction Project evaluated the doses and health impacts to the public from environmental releases of radionuclides from the FMPC (Meyer et al. 1996). Most of the uranium releases went to the atmosphere, with much smaller amounts released to surface water. Screening calculations showed that inhalation was the primary exposure pathway. Because the aquatic pathways were not an important contributor to dose, a simple dilution model, described by equation 14.10, was used. A key issue was to determine the adequacy of the model to calculate radionuclide concentrations in the receiving water ($C_m$ in Bq m$^{-3}$):

$$C_m = \frac{C_o}{S}, \tag{14.10}$$

where $C_o$ is the radionuclide concentration in effluent at the point of release (Bq m$^{-3}$) and $S$ is the dilution factor, a ratio of the receiving water body flow rate to the liquid effluent flow rate. The effluent concentration is the radionuclide release rate divided by the effluent flow rate:

$$C_o = \frac{W_o}{Q_o}, \tag{14.11}$$

where $W_o$ is the radionuclide release rate in effluent at the point of release (Bq s$^{-1}$) and $Q_o$ is the effluent flow rate at the point of release (m$^3$ s$^{-1}$).

The two receiving water bodies for liquid effluent discharges from FMPC were the Great Miami River and Paddy's Run Creek. The Great Miami is a large river that flows in a northerly direction to the east of the site, and Paddy's Run is a small creek with erratic, seasonal flow that begins north of the FMPC and flows southward

**602** Radiological Risk Assessment and Environmental Analysis

along the western edge of the site. Most of the liquid effluent releases were directly to the Great Miami River from a sewage treatment plant through a buried pipeline. Discharges to Paddy's Run came from runoff and spills. The available model input data were as follows:

- Records of liquid effluent discharges from FMPC
- Great Miami River flow rates measured by the U.S. Geological Survey on a daily basis since 1910 at several stations along the river
- Limited flow rate and volume estimates for Paddy's Run

Independent measurements of uranium concentrations in water samples taken from Paddy's Run and the Great Miami at specified locations were available to test the model performance. These measurements were independent because they had not been used in any way to develop or calibrate the model. The predicted and observed uranium concentrations in the Great Miami downstream of FMPC are plotted for a three-year period in figure 14.4. Because the FMPC discharge rate to the river during this time was relatively constant, at approximately one million gallons per day, the final uranium concentration in the river was inversely proportional to the flow rate. The *P/O* ratios for the two receiving water bodies are shown in figure 14.5. The model-predicted concentrations in the Great Miami in many cases are within a factor of 3 of the observed concentrations and are all within a factor of about 10. There also appears to be a temporal structure to the *P/O* ratios, which could warrant further investigation if this were a more significant exposure pathway. Figure 14.5b indicates that the model has a tendency to overpredict uranium concentrations in Paddy's Run, although most model predictions are within a factor of 10 of the observations. By taking the analysis a step further and plotting the *P/O* ratio against the observed concentration for Paddy's Run (figure 14.6), it is apparent that the model overprediction occurs when the uranium concentrations in the creek are very low. This provides further reassurance that the model is performing adequately for quantifying exposures for this minor exposure pathway at FMPC. In summary, the model testing confirmed that it was appropriate to use the simple dilution model

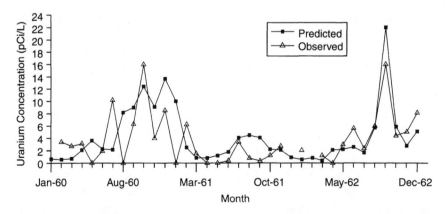

Figure 14.4 Predicted and observed uranium concentrations in the Great Miami River downstream of the FMPC.

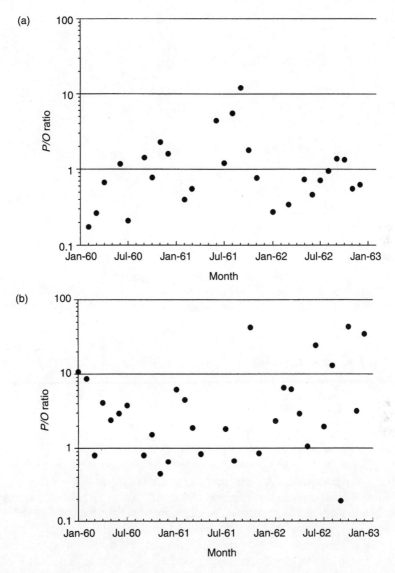

Figure 14.5 *P/O* ratios for the two receiving water bodies: Great Miami River (a) and Paddy's Run Creek (b).

to reconstruct radionuclide concentrations and exposures associated with aquatic releases from FMPC.

## Reasons for Poor Model Performance

The reasons that a model does not perform well may be numerous and varied, but for convenience, they can be divided into three general categories: the user, the model

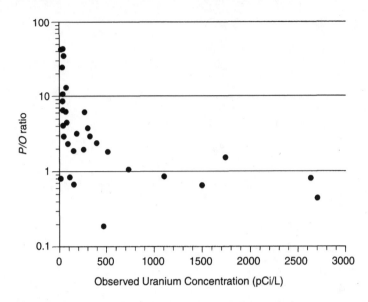

Figure 14.6 *P/O* ratio against the observed uranium concentration for Paddy's Run Creek.

itself, and the assessment question, sometimes called the scenario. The following sections discuss these in turn.

## *User Error*

A model does not exist in isolation; it has to be applied by someone—the user—and in the process, the user often makes many decisions that affect the model's performance. The significance of the user as a determinant of model performance is often overlooked. An inexperienced user may obtain very different results for an assessment question compared to those of an experienced user of the same model. One reason for this can be that the inexperienced user is unable to assess the face validity of the model results. In addition, an inexperienced user may be more susceptible to applying an "off-the-shelf" computer model as a "black box" and in the process may be unaware of the model's limitations with regard to the system the user seeks to describe. In a similar vein, inexperience on the part of the user may lead to the use of generic parameter values that are assigned as default values but do not apply to the site-specific situation for the model application. Alternatively, the user may misunderstand and/or misinterpret information that is used directly or synthesized in some way for input to the model.

Regardless of the user's experience, the user is often required to make judgments when applying the model. Some judgments may relate to the assessment question and how best to answer it using a model. Others may involve deciding what processes (physical, biological, chemical, etc.) to include in the model and the appropriate parameter values to use as input for them. Still other judgments may relate to assigning uncertainty estimates to some or all of the input parameters. Consequently, it has been observed that different users of the same model will produce different results, and conversely, the same user of different models can produce results that are nearly identical (see Peterson et al. 1996).

Finally, human error is a factor that should not be ignored. Implementing quality assurance/quality control methods to check input parameter values, units, and so forth, may be time-consuming, but they are important steps to avoiding such problems. Clearly documented, transparent computer models that are well structured with input parameter values and distributions specified in a single location, and automatically updated when revised, may not be essential for achieving good model performance, but they greatly increase the ease with which peer review can be accomplished and any errors detected.

## The Model

There are many aspects to a model that could explain why it performs poorly when applied to a specific assessment question. Models are designed with different objectives, and these will affect their applicability to address a specific assessment question. The emphasis in screening and compliance models is on not underestimating the assessment end point, whether it is concentration in a specific media and pathway, or dose (underestimation is possible with generic screening models). For this reason, the extent of overestimation of predicted values compared to observed values may be considerable. Many radiological assessment models are thought to provide best estimates of the doses to members of a critical group from various exposure pathways associated with an environmental source of radioactivity. However, oftentimes a number of conservative assumptions are contained in these models. Model comparison exercises and blind testing are valuable tools to highlight such situations. At the same time, however, it is important to recognize that three different strategies are available to a user to address a specific assessment question: A new model can be developed to tackle the problem, an existing model can be tailored to answer the question, or an existing model can be used without modification. User experience and modeling proficiency are key determinants of the approach taken and the type of model selected. The importance of the assessment question also should determine the time and resources allocated to it, and influence the decision of which strategy to adopt.

The model structure and the way it is implemented (i.e., mathematical representation) have implications for its performance. Flexible model structures allow site-specific applications to be implemented more easily. Many environmental transport and exposure pathway models used for radiological assessment are similar with regard to the conceptualization of the environment and the calculation methods. Consequently, differences in predicted values are often associated with differences in parameter values and their distributions. Typically, the predicted radionuclide concentrations or doses are controlled by a few key parameters. The availability of site-specific data to quantify these parameters is important for a model to perform well.

Two types of error observed in models are compensatory errors and compounded errors. For example, comparison of model predictions for subsequent steps in a model calculation may reveal that concentrations in the first step are under- or overpredicted, whereas the concentrations in the final prediction are in good agreement with observations of the system. In this case, the errors have the net effect of canceling each other out. This situation has been observed for the air–forage–milk pathway, where the milk distribution factor that describes the transfer from the pasture into milk is overestimated, but the initial concentration in pasture is underestimated (Peterson et al. 1996). It is only through in-depth analysis of a model that such issues are identified. This demonstrates that improving the quantification of a specific parameter does not guarantee improved model performance overall, at least not initially. It also highlights the difficulty associated with assuming that comparison of predicted and observed concentrations for an intermediate value in the model calculation provides an adequate test of the model performance with regard to the assessment endpoint. Despite considerable data being available on radionuclide concentrations in the environment, measurement data on the doses to individuals tend to be sparse.

Compounded errors occur when a prediction error in one step of the model calculation becomes magnified in subsequent steps as additional errors accumulate.

Agreement between modelers about the magnitude of a specific parameter does not necessarily mean the parameter is well known. Take, for example, figure 14.7, which shows the assumed technetium distribution factors for milk for 15 different models that participated in a BIOMOVS model intercomparison exercise involving long-term irrigation of agricultural land with contaminated groundwater (Grogan 1989). The empirical distribution factor relates the equilibrium concentration of the radionuclide in milk to the daily concentration in the feed. At that time, with the exception of four models, a $^{99}$Tc distribution factor of $10^{-2}$ d L$^{-1}$ was used. This was

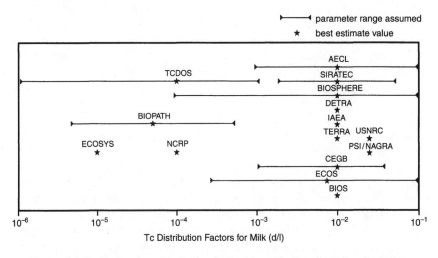

Figure 14.7  Technetium distribution factors for milk (data from Grogan 1989).

because the same literature source (IAEA 1982) was used, in which it is stated that this value was selected in the absence of relevant data on technetium. The selected value assumes that the chemical behavior of technetium is similar to that of iodine. The remaining four models used significantly lower values that were based upon limited experimental findings for $^{99}$Tc uptake by livestock that were reported in the literature.

## *The Assessment Question*

The specific assessment question that a model is required to answer can have a significant impact on model performance. Models tend to perform better for assessment questions that focus on the risks from relatively continuous releases of radioactivity into the environment, such as routine emissions from stacks or routine liquid effluent discharges. In these cases, quasi-steady-state conditions exist in the environment. Model predictions are based on processes averaged over longer temporal and spatial scales. Assessment questions that focus on episodic releases tend to present a greater challenge, especially if the assessment end point is a prediction of the location of maximum impact as opposed to the magnitude of this impact. Episodic releases are more dynamic, so the temporal and spatial scales to capture them are relatively short. The historical dose reconstruction for the Rocky Flats Site provides an example of this. A major fire occurred in Building 771 on September 11, 1957, which resulted in a significant release of plutonium to the atmosphere (Voillequé 1999; Rood and Grogan 1999; Rood et al. 2002). The fire began at 10:00 P.M. and ended at 2:00 A.M. the following morning. Although the first 45 min of the fire was modeled in 1-min intervals to account for the rapidly changing conditions in the building, the plutonium releases during this time were summed to give 15-min totals, and the remainder of the fire was modeled in 15-min intervals, corresponding to the resolution of the available meteorological data. There was a significant shift in wind direction approximately 45 min after the start of the fire. Because the size of plutonium particles released during the fire is not known, this uncertainty, combined with uncertainties associated with the time resolution of the meteorological conditions, meant that although the maximum likely inhalation exposure could be estimated within about a factor of 10 (5th to 95th percentile of distribution), the location of that exposure was subject to much greater uncertainty. In contrast, the inhalation risks from routine releases of plutonium from 1953 through 1989 could be estimated at a specific location within well-defined uncertainty bounds (Rood et al. 2002).

The presence or absence of site-specific information that can be used as input data affects model performance. The types of information required to answer the assessment question include details about agricultural practices such as the dates of sowing and harvesting, storage times for crops, and human dietary habits, to name a few.

Apart from the availability of input data, the quality of the input data is another factor that affects the performance of a model. In the case of independent data sets used to test a model, it is important to recognize that these data sets are also subject to uncertainties and biases. Careful consideration needs to be given to the reliability, completeness, and suitability of data to be used for model testing. Measurement data may be biased due to the instruments or procedures used to make the measurements. One example is air sampling of particulates to determine radionuclide concentrations, a topic discussed in chapter 2 on source terms. The temporal and/or spatial scales associated with different data sets need to be considered to ensure they are compatible with the assessment question and the models to be used.

## Conclusions

Model validation tends to be a contentious topic. Rather than present a philosophical discussion on what it means and whether it is achievable, this chapter takes a more pragmatic approach. Model validation for radiological assessment is an analysis that attempts to delineate the domain of applicability of a model; that is, those situations in which a model can be relied upon to make predictions that are sufficiently accurate for the purpose at hand (Peterson and Kirchner 1998).

The process of model validation is broken down into two main parts: an analysis of the composition of the model and an analysis of the performance of the model. Model validation is incomplete if either is ignored. At the outset, the specific assessment question that the model is required to address must be stated. Embedded in this assessment question is the concept of establishing the correct degree of complexity required for a model for a particular purpose (Beck et al. 1997).

Rigorous testing is fundamental to establishing model credibility for assessments. Furthermore, a model's success or failure can only be determined after in-depth testing. Where possible, this should include tests of intermediate steps within the model because compensatory errors are not uncommon. Blind testing is the best way to determine the predictive accuracy of a model; however, independent data sets for achieving this may not exist. In the absence of independent test data, the predictive accuracy of a model can never be fully established, and the applicability of a model may be questioned. A model that is to be applied over a range of situations also needs to be tested over a similar range. For example, if a model is designed for use at a range of sites at various locations or for a spectrum of radionuclides, then the model needs to be tested under these conditions. The importance of evaluating a model's performance from a number of different angles and using a spectrum of tools, including visual displays of information, cannot be overstated. The reader is cautioned that although statistical tests of model performance are available, the procedure for applying them to the task of model validation is not well established.

## Problems

1. Discuss the criteria you would use to identify an appropriate environmental transport model for an assessment of doses to representative individuals from ongoing routine releases to the atmosphere at a nuclear facility. Explain any site-specific characteristics that might affect your decision. Would you expect your decision to change if the assessment were of historical releases at the facility? What if it were of potential future releases?

Table 14.2 Comparison of doses predicted by four models for continuous deposition of a radionuclide onto agricultural land for 30 years (based on Hoffman and Gardner 1983 table 11-7)

| Pathway | Dose per unit activity deposited per unit area for 30-y deposition | | | |
| --- | --- | --- | --- | --- |
|  | Model A | Model B | Model C | Model D[a] |
| Leafy vegetables | 1.05 | 0.20 | 0.26 | 0.27 (0.03–2.4)[a] |
| Non-leafy vegetables | 5.29 | 2.62 | 2.09 | 0.34 (0.025–4.7) |
| Milk | 0.20 | 0.28 | 0.102 | 0.15 (0.011–2.1) |
| Meat | 0.045 | 0.029 | 0.031 | 0.055 (0.003–0.97) |
| All pathways combined | 6.59 | 3.12 | 2.46 | 1.2 (0.21–6.9) |

[a]GM (95% range).

2. Design a field experiment to evaluate the performance of an environmental transport model to assess the media concentrations of two contrasting radionuclides, for example, $^{99}$Tc and $^{137}$Cs, released into the environment.
3. If your model's predictions have not been compared against independent measurements of the system, what measures could you take to improve confidence in your model?
4. Under what situations will a simple model be more realistic than a more sophisticated, complex model?
5. Of the different models described in this book, which are afflicted with the greatest uncertainty? Of these, which are amenable to validation under field conditions?
6. Table 14.2 provides a comparison of the results obtained from four different models used to predict doses from a continuous uniform deposition of a radionuclide onto agricultural land for 30 years. Models A, B, and C provide deterministic estimates of the dose, whereas model D provides an estimate of the geometric mean dose and 95% range using parameter uncertainty analysis. Models A, B, and C are stated to have a tendency to overpredict dose. What conclusions would you reach about the performance of the different models based on a comparison of the model results? Would your conclusions change if you were informed that there are close similarities in both the mathematical formulation and the databases from which parameter values are derived for all four models? Do the uncertainty analysis results for model D affect your conclusions?

References

Bair, E.S. 1994. "Model (In)Validation—A View from the Courtroom" [guest editorial]. *Ground Water* 32(4): 530–531.
Beck, M.B., J.R. Ravetz, L.A. Mulkey, and T.O. Barnwell. 1997. "On the Problem of Model Validation for Predictive Exposure Assessments." *Stochastic Hydrology and Hydraulics* 11(3): 229–254.
BIOMOVS (Biospheric Model Validation Study). 1991. *Multiple Model Testing Using Chernobyl Fallout Data of I-131 in Forage and Milk and Cs-137 in Forage, Milk, Beef*

*and Grain.* BIOMOVS Technical Report 13. Swedish Radiation Protection Institute, Stockholm, Sweden.

Bradshaw, C., T. Lindborg, and U. Kautsky, eds. 2006. "Special Issue: Transdisciplinary Ecosystem Modeling for Safety Assessments of Radioactive Waste Disposal." *Ambio* 35(8): 417–523.

Bredehoeft, J.D., and L.F. Konikow. 1993. "Ground-Water Models: Validate or Invalidate" [editorial]. *Ground Water* 31(2): 178–179.

Caswell, H. 1976. "The Validation Problem." In *Systems Analysis and Simulation in Ecology*, vol. 4, ed. B.C. Patten. Academic Press, New York; pp. 313–325.

Cox, W.M., and J.A. Tikvart. 1990. "A Statistical Procedure for Determining the Best Performing Air Quality Simulation Model." *Atmospheric Environment* 24A: 2387–2395.

Forrester, J.W. 1968. *Principles of Systems*. Wright-Allen Press, Cambridge.

Gardner, R.H., W.G. Cale, and R.V. O'Neill. 1982. "Robust Analysis of Aggregation Error." *Ecology* 63: 1771–1779.

Garger, E.K., F.O. Hoffman, K.M. Thiessen, D. Galeriu, A.I. Kryshev, T. Lev, C.W. Miller, S.K. Nair, N. Talerko, and B. Watkins. 1999. "Test of Existing Mathematical Models for Atmospheric Resuspension of Radionuclides." *Journal of Environmental Radioactivity* 42(2): 157–175.

Golikov, V., M. Balanov, V. Erkin, and P. Jacob. 1999. "Model Validation for External Doses Due to Environmental Contaminations by the Chernobyl Accident." *Health Physics* 77(6): 654–661.

Grogan, H.A., ed. 1989. *Irrigation with Contaminated Groundwater*. BIOMOVS Technical Report 6—Scenario B2. National Institute of Radiation Protection, Stockholm, Sweden.

Hanna, S.R., D.G. Strimaitis, and J.C. Chang. 1991. *Hazard Response Modeling Uncertainty (A Quantitative Method)*. Vol. 1: *User's Guide for Software for Evaluating Hazardous Gas Dispersion Models*. Air Force Engineering and Service Center, Tyndall Air Force Base, FL.

Hassanizadeh, S.M., and J. Carrera. 1992. "Special Issue on Validation of Geo-hydrological Models" [editorial]. *Advanced Water Resources* 15(1): 1–3.

Hoffman, F.O., and R.H. Gardner. 1983. "Evaluation of Uncertainties in Environmental Radiological Assessment Models." In *Radiological Assessment: A Textbook on Environmental Dose Assessment*, ed. J.E. Till and H.R. Meyer. NUREG/CR-3332. U.S. Nuclear Regulatory Commission, Washington, DC; pp. 11–15.

IAEA (International Atomic Energy Agency). 1982. *Generic Models and Parameters for Assessing the Environmental Transfer of Radionuclides from Routine Releases: Exposures of Critical Groups*. Safety Series No. 57. International Atomic Energy Agency, Vienna, Austria.

IAEA. 1989. *Evaluating the Reliability of Predictions Made Using Environmental Transport Models*. Safety Series No. 100. International Atomic Energy Agency, Vienna, Austria.

IAEA. 1995. *Validation of Multiple Pathways Assessment Models Using Chernobyl Fallout Data of $^{137}Cs$ in Central Bohemia (CB) of Czech Republic—Scenario CB*. Report of the first test exercise of the VAMP Multiple Pathways Assessment Working Group. IAEA-TECDOC-735. International Atomic Energy Agency, Vienna, Austria.

IAEA. 1996. *Validation of Models Using Chernobyl Fallout Data from Southern Finland—Scenario S*. Second Report of the VAMP Multiple Pathways Assessment Working Group. IAEA-TECDOC-904. International Atomic Energy Agency, Vienna, Austria.

Killough, G.G., and D.W. Schmidt. 2000. "Uncertainty Analysis of Exposure to Radon Released from the Former Feed Materials Production Center." *Journal of Environmental Radioactivity* 49(2): 127–156.

Kirchner, T.B. 1994. "Data Management and Modeling." In *Environmental Information Management, Proceedings of a Symposium*. Albuquerque, NM., ed. W.K. Michener, J.W. Brunt, and S.G. Stafford. Taylor and Francis, Bristol, PA; pp. 357–375

Konoplev, A.V., A.A. Bulgakov, F.O. Hoffman, B. Kanyár, G. Lyashenko, S.K. Nair, A. Popov, W. Raskob, K.M. Thiessen, B. Watkins, and M. Zheleznyak. 1999. "Validation of Models of Radionuclide Wash-off from Contaminated Watersheds Using Chernobyl Data." *Journal of Environmental Radioactivity* 42: 131–141.

Kryshev, I.I., T.G. Sazykina, F.O. Hoffman, K.M. Thiessen, B.G. Blaylock, Y. Feng, D. Galeriu, R. Heling, A.I. Kryshev, A.L. Kononovich, and B. Watkins. 1999. "Assessment of the Consequences of the Radioactive Contamination of Aquatic Media and Biota for the Chernobyl NPP Cooling Pond: Model Testing Using Chernobyl Data." *Journal of Environmental Radioactivity* 42: 143–156.

Law, A.M., and W.D. Kelton. 1982. *Simulation Modeling and Analysis*. McGraw-Hill, New York.

McCombie, C., and I. McKinley. 1993. "Validation—Another Perspective" [guest editorial]. *Ground Water* 31(4): 530–531.

Meyer K.R., P.G. Voillequé, G.G. Killough, D.S. Schmidt, S.K. Rope, B. Shleien, R.E. Moore, M.J. Case, and J.E. Till. 1996. "Overview of the Fernald Dosimetry Reconstruction Project and Source Term Estimates for 1951–1988." *Health Physics* 71(4): 425–437.

Mosleh, A., and V. Bier. 1992. "On Decomposition and Aggregation Error in Estimation: Some Basic Principles and Examples." *Risk Analysis* 12: 203–214.

O'Neill, R.V., and B. Rust. 1979. "Aggregation Error in Ecological Models." *Ecological Modeling* 7: 91–105.

Oreskes, N., K. Shrader-Frechette, and K. Belitz. 1994. "Verification, Validation, and Confirmation of Numerical Models in the Earth Sciences." *Science* 263: 641–646.

Peterson, S.R., and T.B. Kirchner. 1998. "Data Quality and Validation of Radiological Assessment Models." *Health Physics* 74(2): 147–157.

Peterson, S.R., F.O. Hoffman, and H. Köhler. 1996. "Summary of the BIOMOVS A4 Scenario: Testing Models of the Air-Pasture-Cow Milk Pathway Using Chernobyl Fallout Data." *Health Physics* 71(2): 149–159.

Rood, A.S. 1999. *Performance Evaluation of Atmospheric Transport Models*. Revision 1. 3-CDPHE-RFP-1999-FINAL (Rev. 1). Prepared for the Colorado Department of Public Health and Environment by Radiological Assessments Corporation, Neeses, SC.

Rood, A.S., and H.A. Grogan. 1999. *Estimated Exposure and Lifetime Cancer Incidence Risk from Plutonium Released from the 1957 Fire at the Rocky Flats Plant*. 2-CDPHE-RFP-1999-FINAL. Prepared for the Colorado Department of Public Health and Environment by Radiological Assessments Corporation, Neeses, SC.

Rood, A.S., H.A. Grogan, and J.E. Till. 2002. "A Model for a Comprehensive Assessment of Exposure and Lifetime Cancer Incidence Risk from Plutonium Released From the Rocky Flats Plant, 1953–1989." *Health Physics* 82(2): 182–212.

Thiessen, K.M., B.A. Napier, V. Filistovic, T. Homma, B. Kanyár, P. Krajewski, A.I. Kryshev, T. Nedveckaite, A. Nényei, T.G. Sazykina, U. Tveten, K.L. Sjöblom, and C. Robinson. 2005a. "Model Testing Using Data on $^{131}$I Released from Hanford." *Journal of Environmental Radioactivity* 84(2): 211–224.

Thiessen, K.M., T.G. Sazykina, A.I. Apostoaei, M.I. Balonov, J. Crawford, R. Domel, S.V. Fesenko, V. Filistovic, D. Galeriu, T. Homma, B. Kanyár, P. Krajewski, A.I. Kryshev, I.I. Kryshev, T. Nedveckaite, Z. Ould-Dada, N.I. Sanzharova, C. Robinson, and K.L. Sjöblom. 2005b. "Model Testing Using Data on $^{137}$Cs from Chernobyl Fallout in the Iput River Catchment Area of Russia." *Journal of Environmental Radioactivity* 84(2): 225–244.

Till, J.E., G.G. Killough, K.R. Meyer, W.S. Sinclair, P.G. Voillequé, S.K. Rope, and M.J. Case. 2000. "The Fernald Dosimetry Reconstruction Project." *Technology* 7: 270–295.

Voillequé, P.G. 1999. *Estimated Airborne Releases of Plutonium During the 1957 Fire in Building 71*. 10-CDPHE-RFP-1999-FINAL. Prepared for the Colorado Department of Public Health and Environment by Radiological Assessments Corporation, Neeses, SC.

Yu, S., B. Eder, R. Dennis, S.-H. Chu, and S.E. Schwartz. 2006. "New Unbiased Symmetric Metrics for Model Evaluation of Air Quality Models." *Atmospheric Science Letters* 7: 26–34.

# 15

# Regulations for Radionuclides in the Environment

David C. Kocher

Many environmental radiological assessments are performed to evaluate compliance with regulations or other guidance that essentially define acceptable levels of radionuclides in the environment for particular exposure situations. Compliance with standards for radionuclides in the environment can be important in planning for future activities (e.g., developing new operating facilities or waste disposal sites, undertaking remediation of environmental contamination from past practices, or planning for responses to radiation accidents), evaluating current practices (e.g., releases from operating nuclear facilities), or evaluating past practices (e.g., historical releases from nuclear facilities).

The primary purpose of this chapter is to present information on current standards in the United States for controlling (limiting) routine and accidental radiation exposures of the public. Radiation exposures of the public are regulated under several federal laws that are concerned with protection of public health and the environment. Many of these laws address exposure to hazardous chemicals as well as radionuclides.

This chapter is divided into several sections that discuss the following topics:

The principal U.S. laws that provide authority for regulating exposures of the public to radionuclides and hazardous chemicals

The roles of the most important U.S. governmental institutions with responsibilities for regulating radiation exposures of the public and the roles of national and international advisory groups in radiation protection

**614**  Radiological Risk Assessment and Environmental Analysis

Regulations, recommendations, and guidance for controlling routine radiation exposures of the public

Regulations and recommendations that are concerned with establishing exempt levels of radionuclides in the environment

Guidance on responses to accidental releases of radionuclides to the environment

Because radiation exposures of the public often are regulated under environmental laws that also are concerned with control of exposures to hazardous chemicals, discussions in this chapter emphasize limits on health (cancer) risk that are embodied in laws and regulations, especially when different standards are compared. Risk is the only basis for comparing potential impacts of exposure to radionuclides and other hazardous substances, especially carcinogens.

## Principal Laws for Regulating Exposures to Radionuclides and Hazardous Chemicals in the Environment

The principal federal laws that provide authority for regulation of radionuclides and hazardous chemicals in the environment include the following:

Atomic Energy Act and Uranium Mill Tailings Radiation Control Act (UMTRCA)
Safe Drinking Water Act
Clean Water Act
Clean Air Act
Resource Conservation and Recovery Act (RCRA)
Comprehensive Environmental Response, Compensation, and Liability Act (CERCLA)
Toxic Substances Control Act (TSCA) and Indoor Radon Abatement Act

The basic concerns of these laws are summarized in table 15.1. Discussions in later sections describe the different approaches to protection of public health and the environment that are embodied in these laws and how laws other than the Atomic Energy Act that are concerned with hazardous chemicals apply to radionuclides or otherwise influence regulation of radionuclides. Other aspects of these laws and their implementing regulations, such as how they apply to noncarcinogenic hazardous chemicals, are discussed in Kocher (1999). Mandates for regulation contained in different environmental laws also are discussed in Overy and Richardson (1995).

In considering how the laws listed in table 15.1 apply to radionuclides, it is useful to distinguish between radionuclides that arise from operations of the nuclear fuel cycle and non-fuel-cycle radioactive materials. The former are regulated by the U.S. Environmental Protection Agency (EPA), U.S. Nuclear Regulatory Commission (NRC), or U.S. Department of Energy (DOE) under the Atomic Energy Act. Non-fuel-cycle radioactive materials include naturally occurring radioactive material (NORM) other than source material (e.g., mining wastes or radium that

Table 15.1  Basic concerns of principal federal laws for regulating radionuclides and hazardous chemicals in the environment

| Law | Basic concerns |
|---|---|
| Atomic Energy Act; Uranium Mill Tailings Radiation Control Act (UMTRCA) | Development, use, and control of atomic energy for defense and peaceful purposes; regulation of source, special nuclear, and byproduct materials to protect public health and safety[a,b] |
| Safe Drinking Water Act | Protection of U.S. drinking water supplies and resources |
| Clean Water Act | Restoration and maintenance of chemical, physical, and biological integrity of U.S. surface waters |
| Clean Air Act | Protection and enhancement of quality of U.S. air resources to promote public health and welfare and productive capacity of its population |
| Resource Conservation and Recovery Act (RCRA) | Protection of human health and environment from disposal of hazardous and nonhazardous solid waste; reduction or elimination of generation of hazardous waste; material recycling and recovery to conserve natural resources |
| Comprehensive Environmental Response, Compensation, and Liability Act (CERCLA) | Federal response and compensation for unpermitted and uncontrolled releases, including threats of release, of hazardous substances to the environment |
| Toxic Substances Control Act (TSCA); Indoor Radon Abatement Act | Regulation of toxic substances in commerce to protect human health and environment; mitigation of indoor radon to protect public health |

[a] Source, special nuclear, and byproduct materials are defined in the Atomic Energy Act and its implementing regulations in Title 10, parts 30, 40, and 70, of the Code of Federal Regulations. Source material is natural material (uranium or thorium) from which nuclear fuel is made; special nuclear material is fissionable material (plutonium, $^{233}$U, or $^{235}$U) used in nuclear weapons or reactors; and, until 2005, when the definition was amended (see text), byproduct material included (1) any radioactive material except special nuclear material, resulting from the production or use of special nuclear material (i.e., fission and activation products produced when nuclear fuel is used in reactors) and (2) uranium or thorium mill tailings.

[b] The Atomic Energy Act does not include requirements for protection of public health and safety in uses of source, special nuclear, or byproduct materials, nor does it specify approaches to be taken in establishing standards for radiation protection of the public. Standards and other requirements are defined only in guidance and regulations established by the U.S. Environmental Protection Agency, U.S. Nuclear Regulatory Commission, and U.S. Department of Energy under authority of the Atomic Energy Act.

does not occur in uranium or thorium mill tailings) and radionuclides produced in accelerators, which historically were not regulated under the Atomic Energy Act unless they were the DOE's responsibility. These non-fuel-cycle radioactive materials are referred to as naturally occurring and accelerator-produced radioactive material (NARM).

The distinction between radioactive materials that arise from operations of the nuclear fuel cycle and NARM has become less important as a result of a provision of the Energy Policy Act of 2005 that expanded the definition of byproduct material in the Atomic Energy Act to include the following materials that are produced, extracted, or converted after extraction for use in commercial, medical, or research activities: (1) discrete sources of $^{226}$Ra, (2) radioactive material produced in an accelerator, and (3) any discrete source of NORM other than source material that the NRC determines would pose a threat to public health and safety or to the common defense and security similar to the threat posed by a discrete source of $^{226}$Ra. Those non-fuel-cycle radioactive materials, which do not include diffuse sources of NORM other than source material (e.g., mining wastes or wastes from extraction of energy resources), that are not DOE's responsibility are now subject to regulation by the NRC under the Atomic Energy Act.

The applicability of the federal laws listed in table 15.1 to regulation of radionuclides in the environment is summarized in table 15.2. Radionuclides that arise from operations of the nuclear fuel cycle are subject to regulation under the Atomic Energy Act and other laws. NARM also is subject to regulation under several laws. In the absence of regulations that could be established under TSCA or RCRA, however, management and disposal of NARM in the commercial and non-DOE governmental sectors that is not subject to regulation by NRC under the Atomic Energy Act and is not regulated under other laws currently are regulated only by the states.

Table 15.2  Summary of applicability of federal environmental laws to radionuclides$^a$

| Law | Radionuclides from nuclear fuel cycle$^b$ | NARM$^c$ |
|---|---|---|
| Atomic Energy Act; UMTRCA | Y | N$^d$ |
| Safe Drinking Water Act | Y | Y |
| Clean Water Act | N$^e$ | Y |
| Clean Air Act | Y$^f$ | Y |
| RCRA | N$^g$ | Y$^h$ |
| CERCLA | Y | Y |
| TSCA; Indoor Radon Abatement Act | N$^g$ | Y$^i$ |

$^a$Y = law applies to specified radioactive materials; N = law does not apply, except as noted.
$^b$Source, special nuclear, and byproduct materials as defined under the Atomic Energy Act.
$^c$Naturally occurring and accelerator-produced radioactive material.
$^d$NARM is regulated under the Atomic Energy Act when such material is the responsibility of the DOE. In 2005, the definition of byproduct material in the Atomic Energy Act was amended to include accelerator-produced radioactive material and certain discrete sources of $^{226}$Ra and other naturally occurring radioactive material, excluding source material (see text). Those materials in commercial and non-DOE governmental sectors are now subject to regulation by the NRC under the Atomic Energy Act.
$^e$Discharges of radionuclides to surface waters permitted under the Atomic Energy Act are exempt from regulation under the Clean Water Act, except discharges of high-level radioactive waste into surface waters are banned.
$^f$The EPA has exempted facilities regulated by the NRC under the Atomic Energy Act from standards for airborne emissions of radionuclides established under the Clean Air Act.
$^g$Radioactive materials defined under the Atomic Energy Act are specifically excluded from regulation under RCRA and TSCA, but in practice, such radioactive materials that are mixed with other hazardous materials regulated under RCRA or TSCA are managed primarily in accordance with regulations established under those laws.
$^h$The EPA has not established regulations specifically for NARM under RCRA and cannot do so until NARM waste is declared to be hazardous in RCRA or its implementing regulations.
$^i$The EPA may establish regulations specifically for NARM under TSCA but has not done so. Under the Indoor Radon Abatement Act, the EPA has established guidance on mitigation of indoor radon but is not authorized to establish enforceable standards.

## Institutional Responsibilities for Radiation Protection of the Public

This section describes the roles of the most important U.S. governmental institutions with responsibilities for radiation protection of the public. The roles of the most important national and international advisory groups in radiation protection also are described.

### Responsibilities of U.S. Governmental Institutions

The most important federal agencies with responsibilities for radiation protection of the public include the EPA, NRC, and DOE. The states also have important responsibilities.

#### U.S. Environmental Protection Agency

The EPA has several responsibilities for radiation protection of the public as a result of the many different legislative mandates under which it operates. These responsibilities include the following:

- Development of federal guidance on radiation protection of the public, which is not enforceable by the EPA but is intended for use by all federal agencies with regulatory responsibilities for radiation protection, especially the NRC and DOE.
- Establishment of generally applicable standards for radionuclides in the environment that apply to specific sources or practices that are regulated under the Atomic Energy Act. All such sources and practices also are regulated by the NRC or DOE. The NRC or DOE, not the EPA, usually enforces these EPA standards.
- Establishment of standards for radionuclides in the environment under laws other than the Atomic Energy Act that also are concerned with nonradiological hazards. The EPA enforces these standards and may delegate some enforcement authorities to the states.
- Development of guidance on mitigation of indoor radon under the Indoor Radon Abatement Act.
- Development of guidance on responses to radiation accidents, which is intended to be used by state and local authorities.

#### U.S. Nuclear Regulatory Commission

The NRC is an independent regulatory authority established in 1974 that assumed many of the regulatory responsibilities previously assigned to the U.S. Atomic Energy Commission (AEC). Under authority of the Atomic Energy Act, the NRC is responsible for protecting public health and safety in peaceful uses of source, special nuclear, and byproduct materials. The NRC's regulatory responsibilities are carried out by means of its authority to license all commercial and non-DOE governmental activities involving the use of those radioactive materials. The NRC also has licensing authority over certain facilities operated by the DOE, including facilities for storage and disposal of high-level radioactive waste.

The NRC is an enforcement authority for environmental radiation standards that the EPA establishes under the Atomic Energy Act. The NRC enforces any such standards that apply to licensed commercial or non-DOE governmental activities, as well as standards that apply to DOE activities over which the NRC has licensing authority.

## U.S. Department of Energy

The DOE, which was established in 1977, is responsible for atomic energy defense activities that had been assigned to AEC and many other energy research, development, and demonstration activities. Under the Atomic Energy Act, the DOE is responsible for protecting public health and safety in carrying out its authorized activities.

The DOE is an enforcement authority for environmental radiation standards that the EPA establishes under the Atomic Energy Act. The DOE enforces any such standards that apply to the activities of DOE or its contractors, except when the EPA or NRC is the enforcement authority. The DOE also may establish its own standards, usually by issuing DOE Orders, but occasionally by rulemaking, and the DOE enforces all such standards. DOE standards also apply to NARM for which the DOE is responsible, as well as to radioactive materials defined in the Atomic Energy Act, because the EPA has not established standards for management and disposal of NARM under TSCA or RCRA.

## State Governments

State governments have two important responsibilities in radiation protection of the public. First, under the Agreement State provisions of the Atomic Energy Act, the NRC may transfer portions of its licensing authority over commercial uses of source, special nuclear, and byproduct materials to states that agree to assume such authority. States may not assume licensing authority over activities of the DOE or other federal agencies that are licensed by the NRC.

Second, in the absence of EPA regulations for NARM that could be established under TSCA or RCRA, many of these radioactive materials have been subject to regulation only by the states. Exceptions include that (1) the EPA regulates some sources and practices involving NARM under various environmental laws (see table 15.2), (2) the DOE is responsible for regulating any such materials under its control, and (3) the NRC can now regulate some (but not all) of these materials in the commercial and non-DOE governmental sectors under the Atomic Energy Act.

In addition, Federal Facilities Agreements established under the Federal Facilities Compliance Act give states important technical oversight of activities at DOE sites involving radioactive waste management under the Atomic Energy Act and environmental restoration under CERCLA. In some cases (e.g., the Hanford Site in Richland, WA), states have entered into formal agreements with the DOE and EPA that provide for a direct state role in decision making in waste management and environmental restoration activities.

## Role of Advisory Organizations

The most influential advisory organizations in radiation protection are the International Commission on Radiological Protection (ICRP) and the National Council on Radiation Protection and Measurements (NCRP). The International Atomic Energy Agency (IAEA) has been increasingly influential in recent years in countries other than the United States.

The ICRP is an association of scientists from many countries, including the United States, and the NCRP is a nonprofit corporation chartered by Congress. Both organizations develop recommendations on many aspects of radiation protection, but their recommendations are not binding on national authorities. Regulatory authorities in the United States have not always adopted recommendations of the ICRP and the NCRP.

The IAEA is an intergovernmental organization that operates under the auspices of the United Nations. One of the IAEA's statutory objectives is to develop radiation protection standards. The IAEA's current radiation protection standards, which are based mainly on recommendations of the ICRP (1991, 1993a), are contained in its basic safety standards (IAEA 1996). Those standards specify requirements on radiation protection that are recommended for use by member states (including the United States), but member states, not the IAEA, are responsible for implementation and enforcement. IAEA standards have not influenced development of radiation protection standards in the United States beyond the influence of ICRP recommendations that are incorporated in the standards.

## Standards for Controlling Routine Radiation Exposures of the Public

This section discusses regulations, recommendations, and guidance for controlling routine exposures of the public to radionuclides in the environment that have been developed by regulatory authorities in the United States and by international and national advisory groups. This section also compares the various standards by expressing them in terms of limits on lifetime cancer risk, and it discusses the important issue of consistency of standards (i.e., "risk harmonization").

### Basic Approaches to Regulating Exposure to Radionuclides in the Environment

Regulations for radionuclides in the environment that have been developed under different federal laws are varied and diverse. Nonetheless, all laws and regulations incorporate one of two basic approaches, referred to as the radiation paradigm and the chemical paradigm, in controlling radiation exposures of the public (Kocher and Hoffman 1991, 1992; Kocher 1999). This section also discusses the assumed dose–response relationship that underlies all regulations for any carcinogens.

## Radiation Paradigm for Risk Management

The radiation paradigm for risk management is embodied in all regulations that apply only to radionuclides that have been developed under the Atomic Energy Act (Kocher and Hoffman 1991, 1992; Kocher 1999). The radiation paradigm is based on the system of radiation protection for practices recommended by the ICRP (1977, 1991) and NCRP (1987a, 1993). This system contains three basic elements:

1. *Justification* of practices, meaning that any practice that increases radiation exposure should result in a positive net benefit to society
2. *Optimization* of exposures, meaning that exposures should be reduced to levels that are as low as reasonably achievable (ALARA), taking into account technical and economic factors and other societal concerns
3. *Limitation* of dose to individuals in critical population groups

The principle of optimization (ALARA) is emphasized above the principle of dose limitation for individuals; that is, exposures should be optimized even if all individual doses are below the limit. ICRP Publication 26 (ICRP 1977) emphasizes optimization of collective (population) dose using cost–benefit analysis. ICRP Publication 60 (ICRP 1991), however, emphasizes the importance of source constraints, also referred to as dose constraints, which are constraints on individual dose for specific practices that should be set at a fraction of the dose limit, in applying the principle of optimization. Both interpretations have been incorporated in regulations developed by federal agencies. The ALARA principle should still be applied in reducing doses from any practice below an applicable source constraint.

Given that the principle of justification has been satisfied, the radiation paradigm for risk management, which is depicted in figure 15.1, has the following basic elements:

1. A *limit* on radiation exposure, corresponding to an *upper bound* on acceptable risk
2. A requirement to *reduce* exposures below the limit to levels that are ALARA

Exposures above the upper bound are regarded as unacceptable, meaning *intolerable*, and exposures normally must be limited to levels below the upper bound regardless of cost. As indicated in figure 15.1, exposures just below the limit are regarded as barely tolerable, and application of the ALARA principle normally is expected to result in exposures well below the limit (ICRP 1991; NCRP 1993).

The ICRP's 1990 recommendations on radiation protection distinguished between *practices* and *interventions* (ICRP 1991). A practice is any activity that increases radiation exposure (e.g., operations of nuclear facilities or disposal of radioactive waste), whereas an intervention is any activity that decreases radiation exposure for a preexisting situation (e.g., cleanup of contaminated land after a nuclear accident, removal of naturally occurring radionuclides from drinking water, or mitigation of indoor radon).

The ICRP intends that the system of radiation protection involving the principles of justification, optimization, and dose limitation described above should be applied

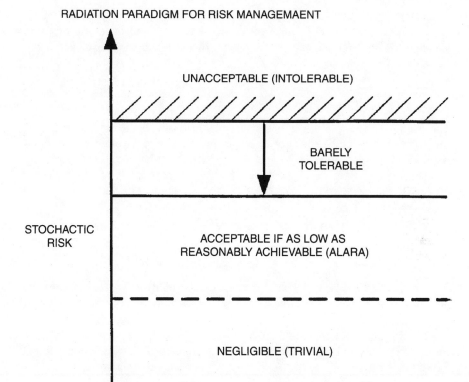

Figure 15.1 Radiation paradigm for risk management incorporating upper bound on acceptable (barely tolerable) risk and reduction of risks below upper bound as low as reasonably achievable (ALARA).

to practices (ICRP 1991). For interventions, the ICRP recommends, first, that any action to reduce exposures should have a positive net benefit, taking into account monetary costs as well as nonradiological risks and societal impacts. That is, any intervention should do more good than harm. Second, given that an intervention is justified, the principle of optimization should be applied to determine the appropriate action that maximizes the net benefit. The main difference between practices and interventions is that the dose limit for practices does not apply to interventions because applying the dose limit might entail unreasonable costs or otherwise do more harm than good. Instead of applying a dose limit to interventions, the ICRP and NCRP have developed recommendations on levels of exposure for particular situations (e.g., exposure to indoor radon) above which action to reduce exposures should almost always be justified.

The depiction of the radiation paradigm in figure 15.1 also indicates that there are levels of radiation exposure so low that the associated risks generally can be regarded as negligible (trivial). At such low levels of exposure, efforts at further reduction of exposures based on the ALARA principle generally would be unwarranted (NCRP 1993). The concept of a trivial exposure that could be applied to any exposure situation has not been incorporated explicitly in any regulations developed under the Atomic Energy Act.

## Chemical Paradigm for Risk Management

The chemical paradigm for risk management, which is embodied in all regulations for radionuclides and chemical carcinogens that have been developed under environmental laws other than the Atomic Energy Act, is fundamentally different from the radiation paradigm described above (Kocher and Hoffman 1991, 1992; Kocher 1999). Given that the principle of justification has been satisfied, the chemical paradigm, which is depicted in figure 15.2, has the following basic elements:

1. A *goal* for acceptable risk
2. Allowance for an *increase* (relaxation) in risks above the goal, based primarily on considerations of technical feasibility and cost

The chemical paradigm for risk management essentially is the opposite of the approach to regulation of radionuclides under the Atomic Energy Act depicted in figure 15.1.

The chemical paradigm is exemplified by approaches to risk management for radionuclides and other carcinogens under the Safe Drinking Water Act and CERCLA, which are described further below. In each case, a goal for acceptable risk, which may be unattainable by any means, is established in law or regulations. Regulatory requirements then may be set above the goal based primarily on considerations of technical feasibility and cost. It is important to emphasize that risks above statutory or regulatory goals are not necessarily unacceptable, meaning intolerable. More generally, the chemical paradigm does not yet incorporate the concept of an intolerable risk, which is a basic element of the radiation paradigm.

There is another important difference between the radiation and chemical paradigms for risk management. In the radiation paradigm, the upper bound on acceptable risk, which is represented by the dose limit in the system of radiation protection, is intended to apply to all controlled sources of exposure combined and all exposure pathways. There is no analog to the dose limit in the chemical paradigm. Rather, all standards for radionuclides and hazardous chemicals that have been developed under laws other than the Atomic Energy Act apply only to specific substances, environmental media (e.g., air or water), or sites.

## Linear, Nonthreshold Dose–Response Hypothesis

Both approaches to controlling exposures of the public to radionuclides and chemical carcinogens in the environment discussed above are based on an assumption that the probability of induction of a stochastic health effect (cancers or severe hereditary effects) is a linear function of dose, without threshold. This assumption is referred to as the linear, nonthreshold dose–response hypothesis. The assumptions that any dose, no matter how small, imposes some risk and that the risk per unit

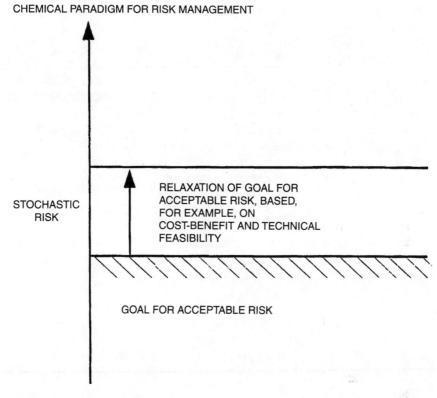

Figure 15.2 Chemical paradigm for risk management incorporating goal for acceptable risk and allowance for increases (relaxation) in risks above the goal based primarily on considerations of technical feasibility and cost.

dose is independent of dose underlies all regulations for radionuclides and chemical carcinogens.

The validity of the linear, nonthreshold dose–response hypothesis for radiation has been much debated, in part because there is a lack of direct scientific evidence of risks from radiation exposure at levels of natural background and below (National Research Council 1990, 2006), which are dose levels of concern in radiation protection of the public. Annual doses to the whole body from natural background, excluding indoor radon, are about 1 mSv (100 mrem) or below (NCRP 1987b). Nonetheless, based in part on direct evidence of increased risks of cancer at acute doses above about 100 mSv (10 rem) and fits to data on dose–response in humans using a linear-quadratic model that is nearly linear at doses below about 2 Sv (200 rem) (National Research Council 1990, 2006), the linear, nonthreshold dose–response hypothesis is regarded by authoritative advisory groups (ICRP, NCRP, and IAEA) and regulatory authorities (EPA, NRC, DOE, and the states) as a prudent assumption for purposes of risk *management* (control of exposures), even though it may not be valid for purposes of risk *assessment* (estimation) in cases of actual exposures at doses well below limits of epidemiological detection.

## Radiation Protection Standards for the Public

This section describes radiation protection standards for the public that have been developed by federal agencies and the states. These standards were based mainly on recommendations by the ICRP and NCRP. Current ICRP and NCRP recommendations on radiation protection standards for the public and the IAEA's basic safety standards for protection of the public also are discussed.

Radiation protection standards specify limits on routine exposure that are regarded as necessary for protection of human health. These standards are intended to apply to all controlled sources of exposure combined, except indoor radon and deliberate medical practices; they do not apply to uncontrollable sources of natural background radiation. In contrast, the many environmental radiation standards described in the next section apply only to specific practices or sources. Radiation protection standards must be met, except in cases of accidents or emergencies, regardless of cost; that is, routine exposures above limits specified in the standards are considered to be intolerable.

Radiation protection standards for the public are concerned primarily with limiting risks of stochastic health effects, mainly cancers but also including severe hereditary effects. Dose limits in current standards are sufficiently low to ensure that deterministic effects, which usually occur only at annual doses above about 0.15 Sv (15 rem), would be prevented in all organs and tissues except, perhaps, in the lens of the eye and the skin (ICRP 1991).

### Guidance of the U.S. Environmental Protection Agency

Federal guidance on radiation protection of the public was first developed by the Federal Radiation Council (FRC 1960, 1961). FRC guidance, although it has not been officially superseded by the EPA, is now outdated and no longer presents the EPA's views on basic requirements for radiation protection of the public.

In 1994, the EPA issued proposed federal guidance on radiation protection of the public (EPA 1994a) to replace the earlier FRC guidance. Provisions of the proposed guidance of interest to this discussion are listed in table 15.3. The EPA's proposed guidance was largely consistent with recommendations in ICRP Publication 26 (ICRP 1977) and NCRP Report No. 91 (NCRP 1987a). More recent ICRP and NCRP recommendations differ mainly in their use of the effective dose rather than the effective dose equivalent (ICRP 1991; NCRP 1993). To ensure compliance with the dose limit and the ALARA objective, the proposed guidance emphasizes that authorized limits for specific practices or sources should be set at a fraction of the annual dose limit of 1 mSv (100 mrem). This provision is consistent with a recommendation by the NCRP (1987a, 1993), which states that whenever the dose from a single site could exceed 25% of the annual dose limit, the site operator should assure that doses from all controlled sources combined would not exceed the dose limit on a continuous basis. This part of the proposed guidance also agrees with current ICRP (1991) recommendations on the use of source constraints.

Table 15.3  U.S. Environmental Protection Agency's proposed federal guidance on radiation protection standards for the public (EPA, 1994a)[a]

---

Doses to individuals and populations should be maintained as low as reasonably achievable (ALARA)
Dose limits
  Annual effective dose equivalent to individuals of 1 mSv (100 mrem) from all controlled sources combined, including sources not associated with operations of nuclear fuel cycle but excluding indoor radon
  Annual effective dose equivalent to individuals up to 5 mSv (500 mrem) may be permitted, with prior authorization, in unusual, temporary situations
  Continued exposure over substantial portions of a lifetime at or near 1 mSv (100 mrem) per year should be avoided
Authorized limits for specific sources or practices
  Authorized limits should be established to ensure that annual dose limit of 1 mSv (100 mrem) for all controlled sources combined and the ALARA objective are satisfied; authorized limits normally should be set at a fraction of the annual dose limit

---

[a] Proposed guidance was intended to replace guidance of the FRC (1960, 1961), but the EPA has not issued its guidance in final form.

The EPA's proposed federal guidance on radiation protection of the public (EPA 1994a) has not been issued in final form because various federal agencies could not reach a consensus. The EPA has since drafted new proposed guidance but, at the time of this writing, has not issued a proposal for public comment.

### Radiation Protection Standards of the U.S. Nuclear Regulatory Commission

The NRC's current radiation protection standards for the public in 10 CFR part 20 were issued in 1991 (56 *FR* 23360)[1] and went into effect for all licensees at the beginning of 1994. These standards, which are based mainly on recommendations in ICRP Publication 26 (ICRP 1977), are summarized in table 15.4. The annual dose limits apply at locations of exposure of members of the public, but the concentration limits apply anywhere in unrestricted areas.

The hourly limit on external dose and limits on concentrations of radionuclides in air and water in unrestricted areas provide standards expressed in terms of quantities that can be directly measured to ensure compliance with the annual dose limit of 1 mSv (100 mrem) for all release and exposure pathways combined. Concentration limits in air and water were calculated using dose coefficients for inhalation and ingestion of radionuclides given in ICRP Publications 30 and 48 (ICRP 1979, 1986).

The NRC's radiation protection standards differ from the EPA's proposed federal guidance (see table 15.3) in two related respects. First, the NRC's primary dose limit of 1 mSv (100 mrem) per year applies to individual licensees, but the proposed federal guidance applies this dose limit to all controlled sources combined. Second, the proposed federal guidance emphasizes use of authorized limits (source constraints) at a fraction of the dose limit to ensure compliance with the dose limit and ALARA objective. With the exception of dose constraints for airborne emissions of radionuclides and for decontamination and decommissioning of licensed facilities discussed further below, the NRC has relied on site-specific applications

**Table 15.4** Radiation protection standards for the public established by federal agencies

---

*U.S. Nuclear Regulatory Commission (10 CFR part 20)*

Doses to the public shall be maintained as low as reasonably achievable (ALARA)
Dose limits
  Annual effective dose equivalent to individuals of 1 mSv (100 mrem), excluding dose from permissible releases of radionuclides into sanitary sewerage
  Annual effective dose equivalent to individuals up to 5 mSv (500 mrem) may be permitted with prior NRC authorization
  Effective dose equivalent in any unrestricted area from external sources of 0.02 mSv (2 mrem) in any hour
Limits on annual average concentrations of radionuclides in air and water in unrestricted areas based on annual effective dose equivalent of 0.5 mSv (50 mrem) from inhalation and ingestion

*U.S. Department of Energy (DOE 1990)*

Doses to the public shall be maintained as low as reasonably achievable (ALARA)
Dose limits
  Annual effective dose equivalent to individuals of 1 mSv (100 mrem) from all routine DOE activities combined
  If avoidance of annual dose above 1 mSv (100 mrem) is impractical, annual effective dose equivalent to individuals up to 5 mSv (500 mrem) may be permitted temporarily with prior DOE authorization
Concentration guides for radionuclides in air and water in uncontrolled areas based on annual effective dose equivalent of 1 mSv (100 mrem) from inhalation and ingestion; concentration guides for noble gases and short-lived radionuclides in air based on annual effective dose equivalent of 1 mSv (100 mrem) from external exposure

---

of the ALARA objective to reduce doses from individual sources to levels far below the dose limit. Site-specific applications are difficult to quantify in regulations.

### Radiation Protection Standards of the U.S. Department of Energy

The DOE's current radiation protection standards for the public are contained in DOE Order 5400.5 (DOE 1990). These standards, which are based mainly on recommendations in ICRP Publication 26 (ICRP 1977), are summarized in table 15.4.

The DOE has been developing revised radiation protection standards for the public (DOE 1993, 1995). In addition to existing provisions in DOE Order 5400.5 (DOE 1990), revised standards may include an annual effective dose equivalent of 0.3 mSv (30 mrem) from all DOE sources combined as a criterion to be used in evaluating compliance with the annual dose limit of 1 mSv (100 mrem). This dose criterion takes into account that sources not under DOE control may contribute to exposures of the public. If an estimated annual effective dose equivalent to individual members of the public from all DOE sources combined exceeds 0.3 mSv (30 mrem), the DOE site operator would need to evaluate the dose from all non-DOE sources at or near the site to ensure that the annual limit of 1 mSv (100 mrem) from all sources combined will be met.

DOE Order 5400.5 also contains standards for controlling exposures of the public to specific practices or sources, which are discussed further below.

## State Radiation Protection Standards

State governments are the only authorities that may develop enforceable radiation protection standards that apply to all controlled sources of exposure of the public except indoor radon, including sources that are not currently regulated by a federal agency, such as diffuse sources of NORM that have been technologically enhanced (i.e., TENORM) but do not arise from operations of the nuclear fuel cycle and are not regulated by the DOE.

In 1968, the Conference of Radiation Control Program Directors (CRCPD) was established to provide a forum for state radiation control programs to communicate with each other and with federal agencies. Since 1990, the CRCPD has issued "Suggested State Regulations for the Control of Radiation," which generally are consistent with federal requirements and contain additional provisions for those aspects of radiation protection that are addressed only by the states.

Current radiation protection standards for the public developed by the CRCPD are contained in part D of the suggested state regulations (CRCPD 2005). Those standards are consistent with the EPA's proposed federal guidance (see table 15.3) and current NRC and DOE standards (see table 15.4). Part N of the suggested state regulations (CRCPD 2005) emphasizes that the annual dose limit of 1 mSv (100 mrem) that applies to all licensed sources combined includes exposures to TENORM that is not currently regulated by a federal agency.

## Current Recommendations of the ICRP, NCRP, and IAEA

The ICRP's current recommendations on radiation protection are contained in Publication 60 (ICRP 1991). Dose limits for the public recommended by the ICRP are as follows:

Annual effective dose of 1 mSv (100 mrem), except higher effective doses could be allowed in a single year, in special circumstances, provided that the average effective dose over five years does not exceed 1 mSv (100 mrem) per year.
Annual equivalent dose to lens of the eye of 15 mSv (1.5 rem).
Annual equivalent dose to skin of 50 mSv (5 rem).

Limits on equivalent dose to the lens of the eye and skin are intended to prevent deterministic effects in those tissues in sensitive population groups, such as infants and children.

The ICRP also emphasizes the use of source (dose) constraints for individual sources or practices at a fraction of the annual dose limit of 1 mSv (100 mrem) in optimizing exposures and ensuring that the annual dose limit, which applies to all controlled sources combined excluding indoor radon and deliberate medical practices, will be met (ICRP 1991). A dose constraint of about 0.3 mSv (30 mrem) per year is considered appropriate in most cases.

At the time of this writing, the ICRP is developing revised recommendations on radiation protection. It is not expected that the basic radiation protection standard for the public of 1 mSv (100 mrem) per year and an emphasis on source (dose) constraints for individual practices or sources at a fraction of the dose limit will be changed.

The NCRP's current recommendations on radiation protection are contained in Report No. 116 (NCRP 1993). For the most part, they are similar to ICRP's recommendations (ICRP 1991) summarized above. In particular, the NCRP recommends annual limits on effective dose of 1 mSv (100 mrem) for continuous exposure and 5 mSv (500 mrem) for infrequent exposures. Other NCRP recommendations are discussed further below.

The IAEA's current recommendations on radiation protection are contained in the basic safety standards (IAEA 1996). These standards include dose limits for the public that are the same as the dose limits in ICRP Publication 60 (ICRP 1991) given above. Other provisions of the basic safety standards are discussed further below.

When an annual dose limit of 1 mSv (100 mrem) was first recommended by the ICRP (1977) and the NCRP (1987a), the objective was to limit lifetime cancer risks to individual members of the public to about 1 in 1,000 ($10^{-3}$), based on an assumed risk of fatal cancers of about $10^{-2}$ per Sv. The fatal cancer risk for members of the public now is assumed to be about $5 \times 10^{-2}$ per Sv (ICRP 1991; NCRP 1993), and the lifetime risk corresponding to the dose limit is assumed to be about $4 \times 10^{-3}$. Given this level of risk associated with the dose limit, use of a dose constraint for individual sources at 25% of the dose limit (NCRP 1987a, 1993) is potentially important in ensuring that the lifetime fatal cancer risk would not exceed about $10^{-3}$. Similarly, the ICRP (1991) assumed that use of dose constraints for individual sources at a fraction of the dose limit, coupled with an expectation that individuals rarely would be members of the critical group for more than one source, would ensure that the lifetime risk from all controlled sources combined would not exceed about $10^{-3}$. The magnitude and variability of natural background radiation were additional factors in deciding not to reduce the recommended dose limit for the public in spite of the increase in the cancer risk.

## Summary of Radiation Protection Standards for the Public

Radiation protection standards of all federal agencies, as well as the CRCPD representing the states, specify an annual dose limit for individual members of the public of 1 mSv (100 mrem). This dose limit is in accordance with ICRP and NCRP recommendations and is intended to apply to all controlled sources of exposure combined, excluding indoor radon and deliberate medical practices. However, no single federal agency is responsible for regulating all controlled sources of exposure. As a consequence, dose constraints on individual practices or sources at a fraction of the dose limit and state standards are important in ensuring compliance with the dose limit for the public.

It cannot be overemphasized that compliance with the annual dose limit of 1 mSv (100 mrem) does not, by itself, constitute adequate protection of the public. All radiation protection standards emphasize compliance with the ALARA objective in reducing doses as far below the annual dose limit as practicable. That is, the standard for radiation protection is compliance with the annual dose limit *and* the ALARA objective. The importance of the ALARA objective in controlling exposures of the public is considered further below.

## Standards for Specific Practices or Sources

This section describes environmental radiation standards to limit routine exposures of the public to specific practices or sources (also called source constraints, dose constraints, or authorized limits) that have been developed by the EPA, NRC, and DOE. Selected state standards and selected recommendations of the ICRP, NCRP, and IAEA also are discussed. Standards that do not include exposures to natural background are important in that they provide a practical means of ensuring compliance with radiation protection standards for all controlled sources combined.

A useful perspective on environmental radiation standards and radiation protection standards for the public is provided by data on the average annual effective dose equivalent to adults in the United States from exposure to natural background and man-made sources. Data on such exposures are listed in table 15.5 (NCRP 1987b). For an assumed risk of fatal cancers of $5 \times 10^{-2}$ per Sv (ICRP 1991; NCRP 1993), the average risk from exposure to natural background, including radon, over a lifetime of 70 years is about $10^{-2}$.

Table 15.5  Average radiation doses to adults from exposure to natural background and other sources: U.S. population (NCRP 1987b)

| Source of exposure | Annual effective dose equivalent | |
|---|---|---|
|  | mSv | mrem |
| Natural sources |  |  |
|   Indoor radon | 2.0 | 200.0 |
|   Other[a] | 1.0 | 100.0 |
| Nuclear fuel cycle[b] | 0.0005 | 0.05 |
| Consumer products[c] | 0.1 | 10.0 |
| Medical procedures |  |  |
|   Diagnostic X-rays | 0.39 | 39.0 |
|   Nuclear medicine | 0.14 | 14.0 |
| Total | 3.6 | 360.0 |

[a] Other natural sources include cosmic rays, cosmogenic and terrestrial radionuclides, and radionuclides in the body.
[b] Estimated dose is average value within 80 km (50 miles) of operating nuclear facilities.
[c] Consumer products exclude tobacco; estimated average annual dose equivalent to bronchial epithelium from smoking is 0.16 Sv (16 rem).

**630** Radiological Risk Assessment and Environmental Analysis

Environmental radiation standards or guidance have been developed or considered by the EPA, NRC, or DOE for the following categories of practices or sources (laws providing authority for the standards or guidance are given in parentheses):

Operations of uranium fuel-cycle facilities (Atomic Energy Act)
Radioactivity in drinking water (Safe Drinking Water Act)
Radioactivity in liquid discharges to surface water (Clean Water Act)
Uranium and thorium mill tailings (UMTRCA)
Other residual radioactive material (Atomic Energy Act; CERCLA)
Radioactive waste management and disposal (Atomic Energy Act; TSCA)
Airborne emissions of radionuclides (Clean Air Act)
Indoor radon (Indoor Radon Abatement Act)

Federal standards and guidance in these categories are described in the following sections. These discussions emphasize dose constraints or other numerical criteria that apply to an exposure situation of concern. It is important to recognize, however, that when a practice or source is regulated under the Atomic Energy Act or UMTRCA or when a particular standard developed under another law applies to NRC-licensed or DOE facilities (e.g., standards for airborne emissions of radionuclides under the Clean Air Act), the standard for radiation protection is compliance with applicable dose constraints or other numerical criteria and the ALARA objective, not compliance with the numerical criteria alone.

## Operations of Uranium Fuel-Cycle Facilities

Standards for operations of particular uranium fuel-cycle facilities have been established by the EPA, and the NRC has established design objectives for limiting releases from nuclear power reactors.

*U.S. Environmental Protection Agency Standards* EPA standards for operations of uranium fuel-cycle facilities in 40 CFR part 190, which were established in 1977 (42 *FR* 2858), are listed in table 15.6. These standards were based on judgments about doses and releases from operating facilities that were reasonably achievable using best-available effluent control technologies; they were not based on considerations of a limit on acceptable dose or risk. The dose and release limits were already being met by operating facilities, and there was little, if any, additional cost to comply. Use of dose constraints for the whole body or the critical organ was consistent with the formulation of radiation protection standards in ICRP Publication 2 (ICRP 1959) and existing federal guidance on radiation protection (FRC 1960, 1961).

The EPA's uranium fuel-cycle standards were the first authorized limits for specific practices or sources involving man-made radionuclides developed under the Atomic Energy Act. The same dose constraints were used later in regulating other practices or sources. As information on risks from radiation exposure was developed after promulgation of the standards, an annual dose constraint of about 0.25 mSv (25 mrem) for specific practices or sources came to be regarded as necessary

Table 15.6  U.S. Environmental Protection Agency standards for operations of uranium fuel-cycle facilities (40 CFR part 190)

*Limits on annual dose equivalent to individuals from all radionuclides, except radon and its decay products*

0.25 mSv (25 mrem) to whole body
0.75 mSv (75 mrem) to thyroid
0.25 mSv (25 mrem) to any other organ

*Limits on releases of radionuclides per gigawatt-year of electrical energy produced*

2 PBq (50,000 Ci) of $^{85}$Kr
0.2 GBq (5 mCi) of $^{129}$I
0.02 GBq (0.5 mCi) of $^{239}$Pu plus other alpha-emitting transuranic radionuclides with half-lives greater than 1 year

Standards apply to normal operations in milling of uranium ore, chemical conversion of uranium, isotopic enrichment of uranium, fabrication of uranium fuel, electricity generation in light-water-cooled nuclear power plants using uranium fuel, and reprocessing of spent uranium fuel. They do not apply to mining operations, transportation of radioactive material, operations at waste disposal sites, and reuse of recovered nonuranium special nuclear material and byproduct material.

to ensure that risks to members of the public from all controlled sources combined would not exceed acceptable (barely tolerable) levels (NCRP 1987a, 1993), even though that dose constraint was not based originally on considerations of an acceptable risk.

*U.S. Nuclear Regulatory Commission's Design Objectives for Nuclear Power Reactors*  NRC regulations in 10 CFR part 50 require that releases of radionuclides from commercial light-water-cooled nuclear power reactors shall be maintained ALARA. In appendix I of 10 CFR part 50, which was promulgated in 1975 (40 *FR* 19439), the NRC established design objectives for nuclear power reactors in the form of numerical guides to meet the requirement to maintain releases ALARA. These design objectives, which strictly apply only for the purpose of issuing licenses to construct nuclear power reactors, are listed in table 15.7.

The numerical guide for implementing the ALARA requirement essentially quantifies an optimization of collective (population) dose using cost–benefit analysis. This guide applies only to the design of nuclear power reactors. A dollar value of collective dose, such as $1,000,000 per person-Sv ($10,000 per person-rem) avoided, also has been widely used in the nuclear industry in applying the ALARA objective to protection of workers and the public.

## Radioactivity in Drinking Water

Standards for radioactivity in public drinking water supplies have been established by the EPA. In addition, the DOE established standards that apply to drinking water supplies provided by the DOE or its contractors and to discharges from DOE facilities that affect private or public water systems.

Table 15.7 U.S. Nuclear Regulatory Commission's design objectives for commercial light-water-cooled nuclear power reactors (10 CFR part 50, appendix I)

*Liquid effluents, all exposure pathways*

Limits on annual dose equivalent or committed dose equivalent to individuals
   0.03 mSv (3 mrem) to whole body
   0.1 mSv (10 mrem) to any organ

*Gaseous effluents*

Limits on annual absorbed dose in air
   0.1 mGy (10 mrad) from gamma radiation
   0.2 mGy (20 mrad) from beta radiation
Limits on annual dose equivalent to individuals
   0.05 mSv (5 mrem) to whole body
   0.15 mSv (15 mrem) to skin

*Atmospheric releases of radioactive iodine and radionuclides in particulate form, all exposure pathways*

Limit on annual dose equivalent or committed dose equivalent to individuals of 0.15 mSv (15 mrem) to any organ

*Implementation of requirement to maintain releases of radioactivity ALARA*

Additional effluent controls shall be used if the cost of controls is less than $100,000 per person-Sv ($1,000 per person-rem) avoided to whole body or thyroid in the population within 80 km (50 mi) of reactor

Design objectives expressed in terms of dose equivalent or absorbed dose are not standards for operating reactors. However, the NRC requires the reporting of releases and remedial action plans if design objectives are exceeded during routine operations, so they provide de facto operating standards.

*Approach to Setting Drinking Water Standards* The EPA establishes standards for radionuclides and other contaminants in drinking water under authority of the Safe Drinking Water Act. These standards apply at the tap, rather than the source, and take into account the effects of water treatment by a municipal water supply in removing contaminants. For many contaminants, including radionuclides, drinking water standards are concerned primarily with reducing levels that occur naturally in sources of drinking water. In accordance with the EPA's strategy for protecting groundwater (EPA 1991a), drinking water standards also have been applied under other environmental laws (e.g., RCRA and CERCLA) to protect potential sources of drinking water that might be contaminated by human activities.

The Safe Drinking Water Act specifies that the EPA must establish maximum contaminant level goals (MCLGs) and maximum contaminant levels (MCLs) for contaminants in drinking water. MCLGs are *nonenforceable* health *goals* that must correspond to levels at which *no* known or anticipated health effects would occur and which allow an adequate margin of safety in protecting public health. The MCLG for radionuclides, and other known or probable human carcinogens, must be *zero*, given the assumption of a linear, nonthreshold dose–response relationship. This goal is not attainable at any cost.

MCLs are the *legally enforceable standards* for contaminants in drinking water. The Safe Drinking Water Act specifies that the MCL for any contaminant must be set as close to the MCLG as possible, taking into account technical feasibility and cost. In accordance with this provision, MCLs for radionuclides and other carcinogens have been based primarily on best available technology for water treatment systems, rather than judgments about acceptable health risks to the public. MCLs for radionuclides and other carcinogens established by the EPA correspond to lifetime cancer risks in the range of about $10^{-4}$ to $10^{-6}$.

MCLs are established for individual contaminants only. For radionuclides and other carcinogens, neither the Safe Drinking Water Act nor its implementing regulations specify a limit on acceptable risk from all contaminants in drinking water combined.

*U.S. Environmental Protection Agency's Interim Standards* In 1976, the EPA established interim standards for radioactivity in community drinking water systems in 40 CFR part 141 (41 *FR* 28402). Those standards addressed both naturally occurring and man-made radionuclides, but they were concerned primarily with reducing background levels of naturally occurring radionuclides, especially radium. Interim standards for radium, gross alpha-particle activity, and man-made beta- or gamma-emitting radionuclides are listed in table 15.8.

The EPA required that standards (MCLs) for individual man-made beta- or gamma-emitting radionuclides, except $^3$H and $^{90}$Sr, be calculated from the specified dose criterion on the basis of the assumptions noted in table 15.8. The specified dose coefficients were calculated using internal dosimetry models in ICRP Publication 2 (ICRP 1959), which became outdated shortly after the interim standards

Table 15.8 U.S. Environmental Protection Agency standards for radioactivity in community drinking water systems (40 CFR part 141)[a]

---

*Limits on concentrations of alpha-emitting radionuclides (MCLs)*

0.2 Bq L$^{-1}$ (5 pCi L$^{-1}$) for $^{226}$Ra plus $^{228}$Ra

0.6 Bq L$^{-1}$ (15 pCi L$^{-1}$) for gross alpha-particle activity, including $^{226}$Ra but excluding radon and uranium

30 μg L$^{-1}$ for uranium[b]

*Limit on annual dose equivalent to individuals from man-made, beta- or gamma-emitting radionuclides*[c]

0.04 mSv (4 mrem) to whole body or any organ

---

[a] Except as noted, limits were contained in interim standards promulgated in 1976 and were retained in revised standards promulgated in 2000.
[b] MCL was first established in revised standards promulgated in 2000. This standard is expected to correspond to an activity concentration of natural uranium of about 1 Bq L$^{-1}$ (27 pCi L$^{-1}$) (EPA 2000b).
[c] MCLs for individual radionuclides, except $^3$H and $^{90}$Sr, are calculated assuming a daily consumption of drinking water of 2 L and dose coefficients for ingestion of radionuclides derived from data in National Bureau of Standards Handbook 69 (1963); concentration limits for $^3$H and $^{90}$Sr are specified in 40 CFR part 141.

were promulgated. However, the interim standards did not allow MCLs for beta or gamma emitters to be recalculated using newer dose coefficients.

Radium was considered to be the most important radionuclide in drinking water. In accordance with a requirement of the Safe Drinking Water Act discussed above, the MCL for radium was based on an analysis of the costs of reducing existing, natural levels in community drinking water systems in relation to health risks (cancers) averted in the population. For a daily ingestion of drinking water of 2 L, the MCL for radium corresponds to an annual effective dose equivalent of about 0.05 mSv (5 mrem) or an annual effective dose (ICRP 1991) between about 0.04 mSv (4 mrem) and 0.1 mSv (10 mrem), depending on the relative abundances of the two radium isotopes. The interim standard for gross alpha-particle activity was intended mainly for use as a screening tool to detect high levels of naturally occurring $^{226}$Ra and was not expected to be important if the radium standard was met.

The interim standard for man-made beta- or gamma-emitting radionuclides was based mainly on maximum observed levels of $^{90}$Sr and $^{137}$Cs in surface waters due to fallout from atmospheric testing of nuclear weapons. Other sources of man-made radionuclides were not expected to be important in public drinking water supplies. The EPA anticipated that there would be no additional costs in meeting the standard for beta or gamma emitters.

*Revisions of U.S. Environmental Protection Agency Standards* In 1991, the EPA issued proposed revisions of the interim standards for radioactivity in community drinking water systems (EPA 1991b). These revisions included increases in existing MCLs for radium and beta- or gamma-emitting radionuclides and new MCLs for uranium and $^{222}$Rn. Before the revised drinking water standards for radionuclides were promulgated, the Safe Drinking Water Act Amendments of 1996 were passed. The amendments contained two provisions that influenced the EPA's efforts to revise the interim drinking water standards.

First, the amendments specified that any revision of an existing drinking water standard "shall maintain, or provide for greater, protection of the health of persons." The EPA concluded that this requirement means that MCLs for radionuclides in the interim standards promulgated in 1976 cannot be relaxed (increased) (EPA 2000a). In effect, the amendments established a national policy that existing federal drinking water standards, including the interim standards for radionuclides, now defined minimally acceptable standards for health protection of the public from contaminants in drinking water, even though existing standards for radionuclides and other carcinogens had not been based on considerations of a limit on acceptable risk.

On the basis of the policy that existing drinking water standards cannot be relaxed, the EPA promulgated revised drinking water standards for radionuclides in 2000 (65 *FR* 76708). Interim standards for radium, gross alpha-particle activity, and man-made beta- or gamma-emitting radionuclides established in 1976 were retained, including the specification that outdated dose coefficients for ingestion of radionuclides (National Bureau of Standards 1963) must be used to calculate MCLs for beta- or gamma-emitting radionuclides. In addition, a new MCL for uranium listed in table 15.8 was added. The MCL for uranium was based on an assessment of the chemical toxicity of uranium and a cost–benefit analysis.

Second, the Safe Drinking Water Amendments Act of 1996 directed the EPA to withdraw the proposed MCL for radon in drinking water (EPA 1991b), which the EPA did (EPA 1997). The amendments also directed the EPA to promulgate a new standard for radon to be based on results of a study by the National Research Council (1999). The EPA was directed to adopt a multimedia strategy to limit risks from exposure to radon, which recognizes that radon in drinking water is not the most important source of radon in indoor air.

In 1999, the EPA issued proposed regulations to address the requirement of the Safe Drinking Water Amendments Act to promulgate standards for radon in drinking water (EPA 1999a). The proposed regulation included a stringent standard that would apply if a state chose not to develop a multimedia radon mitigation program and a more relaxed standard that would apply if a state developed such a program. At the time of this writing, the EPA had not promulgated final standards for regulating radon in drinking water.

*Importance of Drinking Water Standards*  EPA standards for radionuclides in public drinking water supplies are among the most important of all environmental radiation standards. These standards apply to drinking water that is consumed by a majority of the U.S. population. In addition, these standards have been applied to protection of potential *sources* of drinking water, especially groundwater, that may be contaminated by past or future practices (e.g., radioactive waste disposal); these applications are consistent with the EPA's strategy for protection of groundwater resources (EPA 1991a). An important issue is whether standards that were based originally on judgments about concentrations of radionuclides that could be achieved in public drinking water supplies at a reasonable cost would also be reasonably achievable in other situations that were not envisioned when the standards were first developed. This issue is important even though current law specifies that drinking water standards are minimally acceptable standards for protection of the public from contaminants in drinking water.

*U.S. Department of Energy Standards*  The DOE's current requirements for radiation protection of the public and the environment in DOE Order 5400.5 (DOE 1990) include the following standards for radionuclides in drinking water:

Limit on annual effective dose equivalent to individuals of 0.04 mSv (4 mrem) for drinking water supplies operated by or for the DOE, excluding dose from naturally occurring radionuclides

Limits on liquid effluents from DOE activities such that intakes to private or public water systems downstream of facility discharges will not exceed EPA drinking water standards in 40 CFR part 141

These standards are consistent with current EPA standards.

### Radioactivity in Liquid Discharges

In 1982, the EPA established standards under authority of the Clean Water Act that apply to liquid discharges of naturally occurring radionuclides from mines or mills

Table 15.9  U.S. Environmental Protection Agency standards for liquid discharges of naturally occurring radionuclides (40 CFR part 440)

*Limits on concentrations in daily effluents*

0.4 Bq L$^{-1}$ (10 pCi L$^{-1}$) for dissolved $^{226}$Ra
1.1 Bq L$^{-1}$ (30 pCi L$^{-1}$) for total $^{226}$Ra
4 mg L$^{-1}$ for uranium

*Limits on average concentrations in daily effluents for 30 consecutive days*

0.11 Bq L$^{-1}$ (3 pCi L$^{-1}$) for dissolved $^{226}$Ra
0.4 Bq L$^{-1}$ (10 pCi L$^{-1}$) for total $^{226}$Ra
2 mg L$^{-1}$ for uranium

Standards apply to mines or mills used to produce or process uranium, radium, or vanadium ores.

that produce or process uranium, radium, or vanadium ores in 40 CFR part 440 (47 *FR* 54598). These standards are listed in table 15.9. They were based on best-available effluent control technologies, rather than consideration of potential health risks to the public from ingestion of surface water into which radionuclides are discharged.

The EPA has not developed other standards for discharges of naturally occurring radionuclides under the Clean Water Act. As noted in the following section (see table 15.10), standards in table 15.9 also apply to liquid discharges from active uranium or thorium processing sites that are regulated by the EPA under UMTRCA.

## Uranium and Thorium Mill Tailings

Standards for management and disposal of uranium and thorium mill tailings have been established by the EPA. In addition, the NRC has established standards for operations of licensed uranium mills and disposal of uranium mill tailings at licensed sites. DOE requirements for management and disposal of mill tailings generally conform to EPA standards; they are not discussed in this chapter.

*U.S. Environmental Protection Agency Standards*  The EPA's current standards for uranium and thorium mill tailings in 40 CFR part 192 were first established in 1983 (48 *FR* 590, 48 *FR* 45926), and provisions for groundwater protection were revised in 1995 (60 *FR* 2854). Only naturally occurring radionuclides are found in mill tailings, and the most important radionuclides regarding protection of public health are radium, radon, and their decay products. Toxic heavy metals and organic compounds also are a concern at mill tailings sites, and such contaminants are included in standards for groundwater protection.

The EPA's mill tailings standards are contained in several subparts of 40 CFR part 192 that apply to different aspects of management, disposal, or remediation.

Table 15.10  U.S. Environmental Protection Agency standards for uranium and thorium mill tailings (40 CFR part 192)

*Subpart A: Control of residual radioactive materials at inactive uranium processing sites managed by the DOE*

Limit on emissions of radon
Annual average release rate of $^{222}$Rn to the atmosphere of 0.7 Bq m$^{-2}$ (20 pCi m$^{-2}$) per second, or annual average concentration of $^{222}$Rn in air outside disposal site of 20 Bq m$^{-3}$ (0.5 pCi L$^{-1}$) above background

Limits on concentrations in groundwater[a]
0.2 Bq L$^{-1}$ (5 pCi L$^{-1}$) for $^{226}$Ra plus $^{228}$Ra
0.6 Bq L$^{-1}$ (15 pCi L$^{-1}$) for gross alpha-particle activity, excluding radon and uranium
1.1 Bq L$^{-1}$ (30 pCi L$^{-1}$) for $^{234}$U plus $^{238}$U

Controls for limiting radon emissions and releases to groundwater shall be designed to be effective for up to 1,000 years to the extent reasonably achievable and, in any case, for at least 200 years

*Subpart B: Cleanup of land and buildings contaminated with residual radioactive materials from inactive uranium processing sites managed by the DOE*

Limits on concentrations of $^{226}$Ra in soil above background in land averaged over any area of 100 m$^2$
0.2 Bq g$^{-1}$ (5 pCi g$^{-1}$) averaged over the first 15 cm below the ground surface
0.6 Bq g$^{-1}$ (15 pCi g$^{-1}$) averaged over any 15-cm thick layers more than 15 cm below the ground surface

Concentrations of radon decay products in air (including background) in any occupied or habitable building
Limit of $6 \times 10^{-7}$ J m$^{-3}$ (0.03 WL)[b]
Objective for remedial action of $4 \times 10^{-7}$ J m$^{-3}$ (0.02 WL)

Limit on gamma radiation level above background in any occupied or habitable building of 20 μR h$^{-1}$
Compliance with groundwater protection standard in subpart A

*Subpart D: Management of uranium byproduct materials at active uranium processing sites licensed by the NRC or Agreement State*

During processing operations and before the end of the closure period, compliance with
Groundwater protection standard in subpart A, except for the concentration limit for $^{234}$U plus $^{238}$U
Flux standard for $^{222}$Rn from tailings piles in subpart A, but not the concentration standard outside the site
Individual dose constraints in EPA's uranium fuel-cycle standards (40 CFR part 190)
Limits on radioactivity in liquid discharges to surface waters in EPA's 40 CFR part 440

After the closure period and for a period specified in subpart A, compliance with
Flux standard for $^{222}$Rn from tailings piles in subpart A, but not the concentration standard outside the site; flux standard does not apply to any portion of a site containing concentrations of $^{226}$Ra in soil above background less than cleanup standard in subpart B

*Subpart E: Management of thorium byproduct materials at active thorium processing sites licensed by the NRC or Agreement State*

Standards for uranium, $^{222}$Rn, and $^{226}$Ra in subpart D apply to thorium, $^{220}$Ra, and $^{228}$Ra, respectively; flux standard for $^{222}$Rn during uranium processing operations and before the end of the closure period does not apply to releases of $^{220}$Rn from thorium processing sites during that same period

During thorium processing operations and before the end of the closure period, individual dose constraints in the EPA's uranium fuel-cycle standards (40 CFR part 190) apply; dose from $^{220}$Rn and its short-lived decay products is excluded

[a] Groundwater protection standards were based on the EPA's interim drinking water standards in 40 CFR part 141, except the standard for uranium was consistent with that for drinking water added in 2000 (see table 15.8). Standards also include concentration limits for certain other heavy metals and organic compounds and provisions for establishing alternative concentration limits.

[b] 1 WL (working level) = $2.08 \times 10^{-5}$ J of potential alpha-particle energy per m$^3$ ($1.3 \times 10^5$ MeV L$^{-1}$) of air from short-lived radon decay products. When short-lived decay products are present and in activity equilibrium, 1 WL = $3.7 \times 10^3$ Bq m$^{-3}$ (100 pCi L$^{-1}$) of $^{222}$Rn.

These standards are listed in table 15.10. Some aspects of these standards, including subpart C, are described below.

Standards in subpart B, except the groundwater protection standards, were based primarily on background levels of radionuclides in areas of the western United States where uranium and thorium ore deposits exist and residual radioactive materials were produced (EPA 1982). These standards essentially require that, to the extent reasonably achievable, control and cleanup of mill tailings should result in health risks to the public no greater than risks from unmined ore from which tailings are produced.

Standards in subpart B, except the groundwater protection standards, also are interrelated in that compliance with the concentration limit for $^{226}$Ra in surface soil of 0.2 Bq g$^{-1}$ (5 pCi g$^{-1}$) was intended to ensure that concentrations of indoor radon decay products would be less than the objective for remedial action of $4 \times 10^{-7}$ J m$^{-3}$ (0.02 working level [WL])$^2$ and the indoor gamma radiation level above background would be less than 20 μR h$^{-1}$ (EPA 1982). The concentration limit for $^{226}$Ra in subsurface soil of 0.6 Bq g$^{-1}$ (15 pCi g$^{-1}$) was not based on limitation of risk, but was intended only to provide a standard that would allow detection of mill tailings in subsurface soil by measurement of photon radiation in the field (EPA 1982).

Subpart C is concerned with implementing standards in subparts A and B; it does not contain additional numerical criteria. This subpart specifies that operations and closure at mill tailings sites must conform to requirements for management of hazardous waste established by the EPA under RCRA. These requirements emphasize the importance of protecting groundwater in accordance with drinking water standards. RCRA requirements are appropriate for mill tailings because, as indicated by the groundwater protection standards, these materials contain substantial concentrations of toxic heavy metals and organic compounds.

Except for the individual dose constraints during processing operations and before the end of the closure period, mill tailings standards are expressed in terms of measurable quantities. Standards in subparts A and B can be converted to estimates of dose to individuals who are assumed to reside on contaminated land near a site. The most important contributors to doses corresponding to those standards are the limits on levels of indoor radon and external gamma radiation indoors, the latter including external exposure to radium in soil (National Research Council 1998).

The dose corresponding to the standard for indoor gamma radiation level of 20 μR h$^{-1}$ can be estimated by assuming that an exposure of 1 R corresponds to an effective dose equivalent of about 7 mSv (0.7 rem) (ICRP 1987a) and by assuming an indoor residence time of 85% (ICRP 1987b). The estimated annual effective dose equivalent from all external sources during indoor residence is about 1 mSv (0.1 rem).

The dose corresponding to the limit on concentrations of indoor radon decay products of $6 \times 10^{-7}$ J m$^{-3}$ (0.03 WL) can be estimated on the basis of the mean annual effective dose equivalent per unit exposure for an indoor residence time of 85% recommended by the ICRP (1987b). The estimated annual effective dose equivalent from exposure to indoor radon is about 8 mSv (0.8 rem).

On the basis of the calculations described above, the EPA's mill tailings standards correspond to a limit on annual effective dose equivalent of nearly 10 mSv (1 rem), and the contribution from all sources of exposure other than indoor radon is about 1 mSv (0.1 rem). The annual dose excluding indoor radon is about the same as the primary dose limit of 1 mSv (100 mrem) per year from all controlled sources combined in the EPA's proposed federal guidance on radiation protection of the public (see table 15.3). This similarity indicates that compliance with a recommendation in the proposed federal guidance that the dose from individual practices or sources normally should be limited to a fraction of the primary dose limit may be impractical at contaminated properties in the vicinity of uranium mill tailings sites.

Dose calculations described above apply to individuals who are assumed to reside on contaminated land near uranium mill tailings sites; they do not apply to individuals who are assumed to reside on a tailings pile itself at some time in the future. Since undiluted mill tailings typically contain concentrations of $^{226}$Ra of about 10–40 Bq g$^{-1}$ (300–1,000 pCi g$^{-1}$) (DOE 1996), which are considerably higher than the cleanup standard for $^{226}$Ra at specified depths in soil of 0.2 or 0.6 Bq g$^{-1}$ (5 or 15 pCi g$^{-1}$), permanent residence on an exposed tailings pile would result in doses from external exposure and indoor radon that are roughly 100 times higher than estimated doses during residence on contaminated land near a site given above. Doses of this magnitude clearly would be unacceptable under any circumstances. The intent under UMTRCA is that perpetual federal control will be maintained over mill tailings piles to prevent individuals from living on them.

Some criteria in EPA mill tailings standards have been applied to other exposure situations. For exposure to indoor radon decay products, the objective for remedial action of $4 \times 10^{-7}$ J m$^{-3}$ (0.02 WL) corresponds to a radon concentration of about 150 Bq m$^{-3}$ (4 pCi L$^{-1}$), which is the current EPA guideline for mitigation of radon in homes and schools discussed further below. In addition, the standard for cleanup of radium in surface soil has been widely applied by the states, particularly in state standards for TENORM that is not currently regulated by a federal agency (National Research Council 1998; CRCPD 2005).

*U.S. Nuclear Regulatory Commission Standards*   NRC standards for operation of uranium mills and disposal of mill tailings at licensed sites in 10 CFR part 40, appendix A, were established in 1985 (50 *FR* 41862) and have been amended several times since. They conform in most respects to EPA mill tailings standards in subpart D of 40 CFR part 192, except that external photon exposures from tailings or wastes should be reduced to background levels and exposure to indoor radon decay products is not addressed. NRC standards also include (1) technical criteria for siting and design of disposal facilities and for complying with requirements for groundwater protection and (2) a requirement that airborne effluents from milling operations shall be maintained ALARA.

## Other Residual Radioactive Material

This section describes various standards and recommendations that have been developed to control residual radioactive material other than uranium or thorium mill tailings.

*U.S. Department of Energy Standards for Residual Radioactive Material* Current DOE standards for control of residual radioactive material are contained in Order 5400.5 (DOE 1990). These standards address the following:

Release of contaminated property for unrestricted use by the public
Interim storage of residual radioactive material
Long-term management of uranium, thorium, and their decay products

DOE standards for control of residual radioactive material are in the form of a hierarchy of requirements, guidelines, and authorized limits. This hierarchy includes:

1. Requirements for radiation protection of the public
2. Guidelines (either generic or derived) for acceptable levels of residual radioactive material that can be used if requirements for radiation protection of the public are met
3. Authorized limits on acceptable levels of residual radioactive material that are used if generic or derived guidelines would not ensure compliance with requirements for radiation protection of the public

The various requirements, guidelines, and procedures for establishing authorized limits are summarized in table 15.11. Revisions of DOE standards for control of residual radioactive material may be included in revised radiation protection standards for the public and the environment (DOE 1993, 1995).

*U.S. Nuclear Regulatory Commission Standards for Decontamination and Decommissioning of Licensed Facilities* In 1997, the NRC established standards for decontamination and decommissioning of many licensed nuclear facilities in 10 CFR part 20, subpart E (62 *FR* 39058). This regulation is referred to as the License Termination Rule (LTR), and it applies to all licensed facilities except those that are subject to NRC mill tailings standards in 10 CFR part 40, appendix A, and uranium solution extraction (recovery) facilities. The LTR specifies standards to permit unrestricted or restricted use by the public of land and structures that contain residual source, special nuclear, or byproduct material. These standards are listed in table 15.12. Previous guidelines for land contaminated with residual thorium or uranium (NRC 1981), discussed further below, also may be used.

The LTR does not apply to sale or release of contaminated equipment. Previous guidelines for unrestricted use by the public of contaminated buildings and equipment (NRC 1974, 1982) still apply in those cases. Those guidelines were based primarily on levels of residual radioactive material that can be measured routinely, rather than consideration of doses to the public that would result from unrestricted use. However, annual doses that correspond to the contamination guidelines for buildings and equipment were expected to be no more than a few tens of microsieverts (few mrem) and, thus, in compliance with the dose criterion for unrestricted use in the LTR.

In 1999, the NRC established standards for decontamination and decommissioning of contaminated lands and structures at licensed thorium mills and uranium recovery facilities in 10 CFR part 40, appendix A (64 *FR* 17506). These standards are

**Table 15.11  U.S. Department of Energy standards for control of residual radioactive material (DOE 1990)**

*Requirements for radiation protection of the public*

Limit on annual effective dose equivalent to individuals of 1 mSv (100 mrem) from all routine DOE activities and exposure to residual radioactive material; higher doses from acute exposure are permitted provided that the annual effective dose equivalent averaged over a lifetime is not expected to exceed 1 mSv (100 mrem)

Levels of residual radioactive material shall be reduced ALARA, with administrative controls used to reduce exposures ALARA if residual radioactive material is not reduced to levels that would permit unrestricted use by the public

Release of land and structures for unrestricted use by the public shall be governed by guidelines for residual radioactive material established in the Formerly Utilized Sites Remedial Action Program (FUSRAP) and the remote Surplus Facilities Management Program (SFMP)

*Guidelines for acceptable levels of residual radioactive material to permit unrestricted use by the public, which incorporate guidelines for FUSRAP and remote SFMP sites*

Limits on residual concentrations of radium and thorium in soil, airborne radon decay products in occupied or habitable structures on private property, and external gamma radiation inside buildings or habitable structures as given in subpart B of EPA's mill tailings standards (40 CFR part 192)

Limits on residual concentrations of radionuclides other than radium and thorium in soil shall be derived on the basis of the dose limit in radiation protection requirements and prescribed site-specific procedures and data

Compliance with dose limit in radiation protection requirements for residual radioactive material in soil and external gamma radiation, including exposure on open lands

Limits on residual radioactive material on surfaces of structures and equipment, based on NRC guidelines (NRC 1982) and NRC Regulatory Guide 1.86 (NRC 1974)

Compliance with applicable federal and state standards for radionuclides in air and water

*Authorized limits on residual radioactive material at each site and at vicinity properties*

Limits shall be established to ensure that, as a minimum, dose limit in radiation protection requirements will not be exceeded under worst-caset, plausible-use scenarios

Limits shall be set equal to generic or derived guidelines, except where site-specific data show that guidelines are not appropriate (e.g., when dose limit in radiation protection requirements would be exceeded)

*Guidelines for interim storage of residual radioactive material at FUSRAP and remote SFMP sites*

Limits on concentrations of $^{222}$Rn in air above background
  3,700 Bq m$^{-3}$ (100 pCi L$^{-1}$) at any point within a site
  1,100 Bq m$^{-3}$ (30 pCi L$^{-1}$) averaged over a year and over a site
  110 Bq m$^{-3}$ (3 pCi L$^{-1}$) averaged over a year at any location outside a site

Limits on release rate of $^{222}$Rn above background of 0.7 Bq m$^{-2}$ (20 pCi m$^{-2}$) s$^{-1}$

Limits on radionuclide concentrations in groundwater or quantities of residual radioactive material as established in federal or state standards

Control and stabilization features should be designed to ensure, to the extent reasonably achievable, an effective life of 50 years with a minimum life of at least 25 years

*Guidelines for long-term management of uranium, thorium, and their decay products*

Limits on release rate of $^{222}$Rn to the atmosphere and $^{222}$Rn concentration in air outside the boundary of contaminated areas as given in subpart A of EPA's mill tailings standards (40 CFR part 192)

Protection of groundwater in accordance with applicable DOE Orders and federal and state standards

Design lifetime for control and stabilization features as given in subpart A of EPA's mill tailings standards (40 CFR part 192)

Control of access to sites and prevention of misuse of onsite residual radioactive material by appropriate administrative controls and physical barriers designed to be effective, to the extent reasonable, for at least 200 years

**Table 15.12** U.S. Nuclear Regulatory Commission standards for decontamination and decommissioning of licensed nuclear facilities (10 CFR part 20, subpart E)

---

*Conditions for unrestricted use by the public of contaminated land and structures*

Annual effective dose equivalent to individuals would not exceed 0.25 mSv (25 mrem), including dose from groundwater sources of drinking water

Levels of residual radioactive material have been reduced ALARA

*Conditions for license termination of facilities under restricted conditions if dose constraint for unrestricted use is not reasonably achievable or would result in net harm*

Provisions for legally enforceable institutional controls would provide reasonable assurance that annual effective dose equivalent to individuals will not exceed 0.25 mSv (25 mrem)

Annual effective dose equivalent is ALARA and would not exceed 1 mSv (100 mrem) under conditions of unrestricted use if institutional controls were no longer in effect, or would not exceed 5 mSv (500 mrem) if compliance with 1 mSv (100 mrem) is not achievable, is prohibitively expensive, or would result in net harm

*Standards apply for 1,000 years after decommissioning*

---

Standards do not apply to uranium or thorium mill tailings sites or uranium recovery facilities regulated under 10 CFR part 40, appendix A.

listed in table 15.13. They are concerned mainly with cleanup of naturally occurring radionuclides other than radium that are not addressed in EPA mill tailings standards in 40 CFR part 192 (see table 15.10) and the previous 10 CFR part 40, appendix A. Radionuclides of greatest concern are uranium and $^{230}$Th. NRC standards, which are based on EPA mill tailings standards, essentially specify that doses to the public from exposure to residual radionuclides other than radium under conditions of unrestricted release of contaminated sites should be limited to levels that are consistent with the maximum allowable doses from exposure to radium.

*U.S. Environmental Protection Agency's Standards for Cleanup of Radioactively Contaminated Sites* The EPA has not established standards specific to cleanup of residual radioactive material other than mill tailings under authority of the Atomic Energy Act. Rather, the EPA has chosen to regulate cleanup of radioactively contaminated sites under existing regulations that were established in 1990 under authority of CERCLA. Cleanup standards, which apply to sites that are contaminated with hazardous chemicals as well as radionuclides, were established in the National Contingency Plan (NCP) in 40 CFR part 300 (55 *FR* 8666).

Requirements for remediation of contaminated sites under CERCLA and the NCP are discussed in detail in NCRP (2004). CERCLA and the NCP specify that *goals* for remediation of contaminated sites shall be developed taking into account:

- Applicable or relevant and appropriate requirements (ARARs) established under other federal or state environmental laws, with EPA drinking water standards established under the Safe Drinking Water Act, including the

Table 15.13  U.S. Nuclear Regulatory Commission standards for decontamination and decommissioning of licensed thorium mills and uranium recovery facilities (10 CFR part 40, appendix A)

*Conditions for unrestricted public use of contaminated land and structures*

> Annual effective dose equivalent to individuals would not exceed dose resulting from cleanup of radium-contaminated land to standards specified in subpart B of the EPA's 40 CFR part 192, excluding dose from radon[a]
>> Maximum annual dose corresponding to cleanup standards for radium in soil (dose benchmark) shall be calculated over 1,000 years using site-specific parameters[b]
>
> Concentrations of radionuclides other than radium in soil and surface activity on remaining structures are reduced ALARA
> If radium dose benchmark before application of ALARA requirement exceeds 1 mSv (100 mrem) per year, decommissioning plan requires NRC approval

[a] EPA standards are given in table 15.10.
[b] At most sites, the annual effective dose equivalent from external exposure to radium in soil at concentrations specified in EPA standards should be no more than 0.6 mSv (60 mrem), and doses from inhalation and ingestion are expected to be considerably less (EPA 1982).

standards for radionuclides listed in table 15.8, specified as ARARs for remediation of groundwater or surface waters that are current or potential sources of drinking water
- An upper bound on lifetime cancer risk of $10^{-6}$ to $10^{-4}$ from all substances and all exposure pathways combined at specific sites
- A hazard index (i.e., ratio of a daily intake to a maximum daily intake judged to be safe) of 1.0 or less for noncarcinogenic hazardous substances (including uranium)

CERCLA and the NCP also specify that compliance with ARARs and upper bounds on lifetime cancer risk and intakes of noncarcinogens can be waived under certain circumstances, including, for example, that (1) compliance would result in greater risk to human health and the environment than noncompliance, (2) compliance is technically infeasible or impractical, or (3) compliance would not balance the cost of a response against the benefit in protecting human health and the environment. These provisions mean that goals specified in CERCLA and the NCP are used to define levels of contamination above which the *feasibility* of remediation must be considered, but compliance with the goals in site cleanups is required only when it is feasible and cost-effective.

Following promulgation of the NCP, the EPA issued various guidance documents to interpret the regulations. Guidance issued in 1991 emphasized that remedial action usually is not warranted when the lifetime cancer risk under conditions of unrestricted use would be less than $10^{-4}$, provided that drinking water standards are met in water resources, although lower risk goals can be used if warranted by site-specific conditions (Clay 1991). The 1991 guidance also emphasized that the upper boundary of the risk range of $10^{-6}$ to $10^{-4}$ is not precisely $1 \times 10^{-4}$, but that an estimated risk of *around* $10^{-4}$ (e.g., up to about $3 \times 10^{-4}$) may be considered acceptable if justified by site-specific conditions.

In 1997, the EPA developed guidance specific to establishing remediation goals at radioactively contaminated sites under CERCLA and the NCP (Luftig and Weinstock 1997). The 1997 guidance indicated that ARARs at radioactively contaminated sites that were expressed as limits on annual effective dose equivalent above 0.15 mSv (15 mrem) would not be adequately protective for the purpose of establishing goals for remediation. The EPA's view that this dose criterion was consistent with a lifetime cancer risk of around $10^{-4}$ provided compatibility with standards developed by the EPA in other radiation control programs, including standards for disposal of spent fuel, high-level waste, and transuranic waste in 40 CFR part 191 discussed further below, and it provided a means of comparison with NRC cleanup standards, including those discussed above in this chapter and listed in table 15.12.

The intended role of the annual dose criterion of 0.15 mSv (15 mrem) in 1997 guidance (Luftig and Weinstock 1997) was somewhat ambiguous. It was not clear whether remediations that comply with the annual dose criterion would be generally acceptable, or whether the annual dose criterion should be used only when an ARAR is expressed in terms of dose, rather than risk, and the goal for remediation of radioactively contaminated sites generally should be to limit lifetime cancer risks to $10^{-4}$. This ambiguity was potentially important because a lifetime risk of no more than $1 \times 10^{-4}$ could require that annual doses be limited to levels substantially lower than 0.15 mSv (15 mrem). For example, if a risk of cancer incidence of $7.6 \times 10^{-2}$ $Sv^{-1}$ ($7.6 \times 10^{-4}$ $rem^{-1}$) is assumed (EPA 1994b), the risk over a 70-year lifetime that corresponds to the annual dose criterion is about $8 \times 10^{-4}$. Cancer incidence is used by the EPA in risk assessments for all carcinogens under CERCLA.

Guidance issued in 1999 indicated that the EPA does not use an annual effective dose equivalent of 0.15 mSv (15 mrem), in addition to a lifetime cancer risk of around $10^{-4}$, in establishing goals for remediation of radioactively contaminated sites under CERCLA and the NCP (EPA 1999b). That is, EPA emphasizes the use of risk, rather than dose, in establishing remediation goals. Assessments of dose should be conducted only when necessary to evaluate compliance with ARARs that are expressed in terms of dose (e.g., NRC standards for disposal of low-level radioactive waste discussed further below).

*Dispute between U.S. Environmental Protection Agency and U.S. Nuclear Regulatory Commission over Cleanup Standards*   NRC standards for decontamination and decommissioning of licensed nuclear facilities in 10 CFR part 20, subpart E (the LTR), listed in table 15.12, differ from the EPA's preferred approach to regulating cleanup of radioactively contaminated sites described above. As indicated in its 1997 guidance (Luftig and Weinstock 1997), for example, the EPA believes that NRC cleanup standards are not adequately protective of human health and the environment in two respects.

First, in the EPA's view, the NRC's annual dose constraint of 0.25 mSv (25 mrem) for unrestricted use of contaminated sites does not comply with the goal established under CERCLA and the NCP (40 CFR part 300) of limiting lifetime cancer risks to around $10^{-4}$.

Second, NRC cleanup standards do not include a separate provision to address groundwater protection in accordance with drinking water standards. As a consequence, compliance with the NRC's annual dose constraint for unrestricted use, which applies to all exposure pathways combined, could result in radionuclide concentrations in groundwater in excess of drinking water standards. The EPA believes that a separate provision is needed to conform to its groundwater protection strategy (EPA 1991a) and to the requirement of CERCLA that EPA drinking water standards (MCLs) are ARARs for cleanup of contaminated groundwater at Superfund sites.

The two agencies attempted to resolve their disagreements over cleanup standards at radioactively contaminated sites by means of a Memorandum of Understanding (MOU) (EPA/NRC 2002). However, the MOU did not eliminate the possibility that the EPA could intervene in decisions by the NRC to terminate licenses under the LTR if the EPA believes that cleanups that would be acceptable under the LTR are not adequately protective of human health and the environment in accordance with standards developed under CERCLA. NRC decisions to terminate licenses are particularly vulnerable to challenge by the EPA if drinking water standards would not be met in groundwater that is a potential source of drinking water.

The dispute between the EPA and the NRC is discussed in more detail by the National Research Council (1998) and the NCRP (2004). The issue is not whether NRC cleanup standards in the LTR are adequately protective of human health and the environment, because it clearly is not the case that risks would be intolerable under the NRC's preferred approach to regulation. The difference between the NRC's annual dose constraint for unrestricted use of 0.25 mSv (25 mrem) and the EPA's goal for lifetime cancer risk of around $10^{-4}$ is not significant in regard to limiting health risks to the public, especially when the requirement in the LTR to reduce doses ALARA is taken into account. Furthermore, as emphasized above, drinking water standards were not based on an a priori judgment about levels of contamination that are required to protect public health without regard for the feasibility of achieving the standards. It also is the case that allowable excursions above drinking water standards under the LTR would not be large. Rather, the dispute between the two agencies is basically a matter of differences of opinion about suitable approaches to risk management (i.e., differences between the radiation paradigm used by the NRC and the chemical paradigm used by the EPA under CERCLA). In addition, as emphasized in NCRP (2004), differences in assumptions used to estimate dose or risk at contaminated sites are much more important than any differences between the NRC's annual dose criterion and the EPA's lifetime risk criterion.

*U.S. Environmental Protection Agency Guidance on Transuranium Elements in the Environment* In 1987, the EPA issued draft interim recommendations on acceptable levels of transuranium elements in the environment above background levels due to fallout from atmospheric testing of nuclear weapons (EPA 1987). These recommendations were developed to address responses to accidental releases of transuranium elements to the environment, and they do not apply to any other radionuclides. In addition, the recommendations do not provide general criteria for decontamination and decommissioning of sites or facilities, they do not apply to the transient period during and immediately following an accident, and they should not be used as limits for planned releases to the environment. The recommendations include the following:

- To the extent practicable, annual alpha absorbed dose to individuals from transuranium elements should be limited to 0.01 mGy (1 mrad) to pulmonary lung or 0.03 mGy (3 mrad) to bone, endosteal bone surfaces, or red bone marrow (range I).
- Doses above range I and less than annual effective dose equivalent of 1 mSv (100 mrem) from all sources, excluding natural background and medical practices, are acceptable provided risks to the population are justified, general surveillance and routine monitoring are implemented, and doses are maintained ALARA (range II).
- Doses above range II and less than annual effective dose equivalent of 5 mSv (500 mrem) from all sources, excluding natural background and medical practices, on an intermittent basis require continual monitoring and evaluation of individuals and limitations on access or use pending remedial actions (range III).
- Remedial actions should accomplish permanent reductions in risks to the public, and occupancy or land-use restrictions should not be relied on to protect future generations.
- Remedial actions should assure compliance with applicable environmental standards (e.g., drinking water standards and groundwater protection requirements).
- Soil contamination level of 7 kBq m$^{-2}$ (0.2 µCi m$^{-2}$), to depth of 1 cm and for particle sizes less than 2 mm, and air concentration of 0.04 mBq m$^{-3}$ (1 fCi m$^{-3}$) provide screening levels of alpha-emitting transuranium elements for use in demonstrating compliance with range I recommendations but do not define limits for use in implementing range I recommendations.

Range I recommendations are expressed in terms of absorbed dose. Since the EPA uses an average quality factor of 20 to calculate dose equivalent from alpha particles (Eckerman et al. 1988), range I recommendations correspond to annual dose equivalents from alpha particles of 0.2 mSv (20 mrem) to the lung or 0.6 mSv (60 mrem) to bone, bone surfaces, or red marrow.

The EPA has not indicated that it intends to issue the draft interim guidance in final form.

*State Standards for Cleanup of Naturally Occurring and Accelerator-Produced Radioactive Material* States are responsible for regulating NARM that is not currently regulated by a federal agency. The most important NARM of concern to the states is TENORM, and the most important radionuclides in such materials are radium and radon.

Current suggested state regulations (CRCPD 2005) include a provision that land may not be released for unrestricted use when the concentration of $^{226}$Ra or $^{228}$Ra, averaged over any area of 100 m$^2$ and to a depth of 15 cm, exceeds 0.2 Bq g$^{-1}$ (5 pCi g$^{-1}$). However, this concentration limit may be relaxed if a site-specific assessment shows that concentrations of indoor radon would not exceed 150 Bq m$^{-3}$

(4 pCi L$^{-1}$) and the annual effective dose equivalent to individuals, excluding indoor radon, would not exceed 0.15 mSv (15 mrem). The concentration limit for radium in surface soil is consistent with EPA mill tailings standards in subpart B of 40 CFR part 192 (see table 15.10), the criterion for indoor radon is the same as the EPA's recommended action level discussed further below, and the annual dose criterion is the same as the dose criterion discussed in 1997 EPA guidance on cleanup of radioactively contaminated sites under CERCLA and the NCP (Luftig and Weinstock 1997).

Several states have established cleanup standards for $^{226}$Ra in soil (National Research Council 1998). Most state standards are expressed as limits on concentrations of $^{226}$Ra that are consistent with EPA cleanup standards in 40 CFR part 192, subpart B (see table 15.10). In some cases, the cleanup standard for radium depends on the radon emanation rate from residual radioactive material, with higher levels of $^{226}$Ra up to 1.1 Bq g$^{-1}$ (30 pCi g$^{-1}$) allowed when the radon emanation rate is low. An exception to the approach of establishing a limit on concentrations of $^{226}$Ra in soil is the cleanup standard established by the State of New Jersey, which specifies a constraint on annual effective dose equivalent to individuals of 0.15 mSv (15 mrem) from all exposure pathways, excluding radon.

*NCRP Recommendations on Remediation of Natural Sources* The NCRP has developed a recommendation on levels of public exposure to natural sources other than radon that should require remedial actions (NCRP 1987b, 1993). The recommended remedial action level for indoor radon is discussed further below.

The NCRP has recommended that remediation of natural sources should be undertaken when the annual effective dose from continuous exposure, excluding the dose from radon and its decay products, is expected to exceed 5 mSv (500 mrem). External exposure normally should be the primary concern, but significant internal exposure to natural sources other than radon should be included. If remedial actions are undertaken, the ALARA principle should be applied to reduce doses below the recommended action level.

The NCRP recommendation on remediation of natural sources other than radon applies to natural background and to TENORM, including mill tailings. In contrast to the EPA's approach in proposed federal guidance on radiation protection of the public (EPA 1994a), standards for uranium and thorium mills tailings (40 CFR part 192), and guidance on remediation of radioactively contaminated sites (Luftig and Weinstock 1997), the NCRP essentially regards TENORM, including mill tailings, as an enhanced form of natural background that should be treated separately from man-made radionuclides for purposes of radiation protection. That is, the NCRP recommendation does not require a distinction between undisturbed background and TENORM.

The NCRP recommendation on remediation of natural sources other than radon has not been adopted by any federal agency or state.

*ICRP Recommendations on Remediation of Radioactively Contaminated Sites* ICRP recommendations on remediation of radioactively contaminated sites, which the ICRP refers to as an intervention, are contained in Publication 82 (ICRP 1999). These recommendations, which are based on the principles of justification and optimization (ALARA) and apply to other protective actions as well, are summarized as follows:

> An existing annual effective dose that approaches or exceeds about 100 mSv (10 rem) will almost always justify intervention and may be used as a generic reference level for establishing protective actions under nearly any conceivable circumstance; an existing annual equivalent dose to a specific organ or tissue that exceeds a threshold for deterministic effects also will almost always require intervention.
>
> An existing annual effective dose less than about 10 mSv (1 rem) may be used as a generic reference level below which intervention is not likely to be justifiable for some prolonged exposure situations.

The ICRP also emphasizes that intervention might be justifiable when an existing annual effective dose is below the generic reference level of 10 mSv (1 rem) and that intervention may possibly be necessary when an existing annual effective dose is well below the generic reference level of 100 mSv (10 rem), depending on the particular circumstances.

The recommended dose below which intervention is not likely to be justified is consistent with EPA standards for management and disposal of uranium and thorium mill tailings, DOE standards for control of residual radioactive material, and NRC standards for decontamination and decommissioning of thorium mills and uranium recovery facilities discussed above. The recommended dose at which intervention will almost always be justified is well above levels that correspond to those standards. In considering such comparisons, it should be appreciated that ICRP recommendations are intended to be applied to a wide variety of exposure situations and in countries with much more limited economic resources than the United States. It is necessary that ICRP recommendations not be unduly restrictive and that they encourage flexibility in their application under a wide variety of circumstances.

### Radioactive Waste Management and Disposal

Federal agencies have established standards for management and disposal of radioactive wastes other than mill tailings. The EPA has established standards for management and disposal of spent nuclear fuel, high-level waste, and transuranic waste. The NRC has established licensing criteria for disposal of radioactive wastes in deep geologic repositories (principally spent fuel and high-level waste) and near-surface facilities (principally low-level waste). The DOE has established standards for disposal of low-level waste and NARM. The EPA also may develop standards for management and disposal of NARM, and the CRCPD has developed suggested state standards for management and disposal of NARM. In addition, EPA standards for management and disposal of hazardous chemical wastes are important when radioactive waste is mixed with hazardous chemical waste.

*U.S. Environmental Protection Agency Standards for Spent Fuel, High-Level Waste, and Transuranic Waste* EPA standards for management (except for transportation), storage, and disposal of spent nuclear fuel, high-level waste, and transuranic waste were first established in 40 CFR part 191. These standards were promulgated in 1985 (50 *FR* 38066) and revised in 1993 (58 *FR* 66398)

Table 15.14  U.S. Environmental Protection Agency standards for management and disposal of spent fuel, high-level waste, and transuranic waste (40 CFR part 191)[a]

*Subpart A: Management and storage*

Facilities regulated by the NRC or Agreement State
  Limits on annual dose equivalent to individuals, including dose from all operations of uranium fuel-cycle facilities covered by 40 CFR part 190
    0.25 mSv (25 mrem) to whole body
    0.75 mSv (75 mrem) to thyroid
    0.25 mSv (25 mrem) to any other organ
  Facilities operated by DOE and not regulated by NRC or Agreement State
  Limits on annual dose equivalent to individuals
    0.25 mSv (25 mrem) to whole body
    0.75 mSv (75 mrem) to any organ
  Upon application for alternative standard, limits on annual dose equivalent from all sources, excluding natural background and medical practices
    1 mSv (100 mrem) for continuous exposure
    5 mSv (500 mrem) for infrequent exposure

*Subparts B and C: Disposal*[b]

Reasonable expectation of compliance with standards for 10,000 years after disposal[c]
Containment requirements[d]
  Cumulative releases of radionuclides to accessible environment shall have a likelihood of
    Less than 1 chance in 10 of exceeding limits specified in appendix A, table 1, of the standard
    Less than 1 chance in 1,000 of exceeding 10 times the specified limits[e]
Individual protection requirements
  Undisturbed performance of disposal system (releases of radionuclides absent human intrusion) shall not cause annual effective dose equivalent to individuals in accessible environment from all potential exposure pathways to exceed 0.15 mSv (15 mrem)
Groundwater protection requirements
  Undisturbed performance of disposal system shall not cause levels of radioactivity in any underground source of drinking water in the accessible environment to exceed limits (MCLs) specified in 40 CFR part 141 as they exist when implementing agency (NRC or DOE) determines compliance with requirements

[a] Standards do not apply to the management and disposal of spent fuel and high-level waste at the Yucca Mountain Site in Nevada; in response to congressional directive, the EPA has developed separate standards for Yucca Mountain in 40 CFR part 197 (see following section and table 15.15).
[b] The EPA may substitute for any provisions of subparts B and C if requiring compliance appears to be inappropriate.
[c] Provision takes into account that there are unquantifiable uncertainties in assessing the performance of waste disposal systems over long periods of time, and that absolute assurance of compliance with quantitative criteria in subparts B and C cannot be obtained.
[d] The accessible environment is any location more than 5 km from the outer boundary of the original location of disposed waste.
[e] Limits on cumulative releases of radionuclides are expressed as activities per 1,000 metric tons of heavy metal for disposal of spent fuel or high-level waste, or activities per 37 PBq ($10^6$ Ci) of alpha-emitting transuranium radionuclides with half-lives greater than 20 years for disposal of transuranic waste.

to address deficiencies in the 1985 standards that had been noted in a court of appeals ruling on the standards. Current standards in 40 CFR part 191 are listed in table 15.14. As described later in this section, these standards do not apply to the proposed facility for disposal of spent fuel and high-level waste at Yucca Mountain, Nevada.

In subpart A, differences between the dose constraints at facilities licensed by the NRC or an Agreement State and the dose constraints at facilities operated by the DOE provide a clear example of setting standards on the basis of doses that the EPA judged to be reasonably achievable. The EPA judged that doses that were reasonably achievable at non-DOE waste management and storage facilities were somewhat lower than doses that were reasonably achievable at similar DOE facilities.

Cumulative release limits in the containment requirements for disposal correspond to an estimated risk to the U.S. population over 10,000 years of about 1,000 health effects per repository. On the basis of analyses of disposal systems in different geologic media using foreseeable technology, the EPA judged that this population risk is reasonably achievable. The containment requirements are probabilistic. That is, predictions of cumulative releases must be expressed as probability distributions that take into account uncertainties in model parameters and the likelihood of occurrence of different scenarios for release of radionuclides, including releases caused by inadvertent human intrusion into a disposal facility (e.g., by drilling). The standard also includes guidance on the frequency and severity of inadvertent intrusion into a repository to be assumed in demonstrating compliance with the containment requirements.

The individual protection and groundwater protection requirements limit allowable releases of radionuclides to the accessible environment in any year due to natural processes and events. The individual protection standard was based on the EPA policy of limiting lifetime cancer risks from waste management and disposal activities to around $10^{-4}$, and the groundwater protection standard was based on the policy of protecting potential sources of drinking water in accordance with standards for community drinking water systems (EPA 1991a). The EPA judged that both of these requirements were reasonably achievable.

For several reasons, containment requirements for disposal in 40 CFR part 191 were controversial. First, the use of limits on cumulative releases of radionuclides over a specified time deviated from the usual approach of specifying a limit on annual dose to individuals. Limits on cumulative releases of radionuclides do not provide limits on rates of release to the accessible environment and, thus, limits on individual dose. It was only when an individual protection requirement was added to the standards when they were promulgated in 1985 (50 *FR* 38066) that a limit on annual dose to individuals was also specified. Second, the containment requirements corresponded to a very low risk of about 0.1 health effects per year in the U.S. population. Third, based on guidance in the standard on the frequency and severity of inadvertent human intrusion into a geologic repository to be assumed in performance assessments, projected releases to the accessible environment at well-chosen sites were due almost entirely to human intrusion rather than natural processes and events. As a consequence, decisions about acceptable waste disposals could be based on evaluations of highly speculative human behavior. Finally, the containment requirements applied for 10,000 years after disposal but not beyond. This time limit for applying the standard was based mainly on a judgment by the EPA that changes in geologic and climatologic conditions over longer times would be significant but largely unpredictable. Analyses indicated, however, that most releases of radionuclides to the accessible environment at well-chosen sites would occur well beyond 10,000 years, so the standards would not apply at times of largest potential impacts on public health and the environment.

The groundwater protection requirements in 40 CFR part 191 also were controversial. The NRC and DOE objected to the inclusion of separate groundwater protection requirements that were consistent with drinking water standards. Those objections focused on two concerns. First, if an annual dose constraint to individuals from all exposure pathways, including consumption of drinking water, is sufficiently low (e.g., 0.25 mSv), groundwater resources would be adequately protected with no need of further restrictions. With such a dose constraint, possible exceedances of drinking water standards for radionuclides would not be large, only small volumes of groundwater that could affect only small populations would be contaminated above drinking water standards, and the impact on public health would not be significant. Second, the Safe Drinking Water Act required that drinking water standards be evaluated periodically by reconsidering their costs and benefits. Since those standards could be changed between the time a disposal site is selected and a final decision about compliance is made, the possibility of significant changes in drinking water standards would make it difficult to plan for acceptable disposals.

Finally, the individual protection requirements for disposal in 40 CFR part 191 also were contentious. The NRC and DOE preferred a somewhat higher annual dose constraint of 0.25 mSv (25 mrem) to be consistent with other standards for specific practices or sources that had been established under the Atomic Energy Act.

*U.S. Environmental Protection Agency Standards for Yucca Mountain Site*
In 1992, Congress directed the EPA to issue new standards, to be based on recommendations of the National Academy of Sciences (NAS), that would apply to disposal of spent fuel and high-level waste at the Yucca Mountain Site in Nevada. Congress indicated a strong preference for a standard that would be expressed in terms of the maximum annual effective dose equivalent to individuals due to releases to the accessible environment, with no time limit for applying the standard. The congressional directive was developed in response to controversies over containment requirements in EPA standards in 40 CFR part 191 discussed above, including the importance of scenarios for inadvertent human intrusion in determining projected releases.

NAS recommendations on disposal standards for spent fuel and high-level waste at the Yucca Mountain Site were issued in 1995 (National Research Council 1995). In 2001, in response to the 1992 congressional directive and NAS recommendations, the EPA established standards for management and disposal of spent fuel and high-level waste at the Yucca Mountain Site in 40 CFR part 197 (66 *FR* 32074). These standards are listed in table 15.15. They retained many of the existing standards in 40 CFR part 191 that apply at any other site, including the individual dose constraint for undisturbed performance, a separate groundwater protection requirement that was based on drinking water standards for radionuclides in 40 CFR part 141 (see table 15.8), and a time limit of 10,000 years for applying the standards. None of these provisions was recommended in the NAS study. The individual dose constraint that applies to releases caused by inadvertent human intrusion was new and was based upon a similar recommendation in the NAS study. The new standards do not include the containment requirements in 40 CFR part 141, which conforms to a recommendation by the NAS that a standard to limit population risk is not needed.

Table 15.15  U.S. Environmental Protection Agency standards for management and disposal of spent fuel and high-level waste at Yucca Mountain Site in Nevada (40 CFR part 197)[a]

---

*Subpart A: Management and storage*

For management and storage of spent fuel and high-level waste outside the Yucca Mountain repository but within the Yucca Mountain Site and for storage of spent fuel and high-level waste inside the Yucca Mountain repository combined
  Limit on annual effective dose equivalent to individuals in the general environment of 0.15 mSv (15 mrem)

*Subpart B: Disposal*[b]

Reasonable expectation of compliance with standards for 10,000 years after disposal
Individual protection standard[c]
  Limit on annual effective dose equivalent to individual members of the public at a specified location with a specified diet and living style due to releases from undisturbed Yucca Mountain disposal system (releases absent human intrusion) of 0.15 mSv (15 mrem) from all potential transport and exposure pathways
Human intrusion standard[c]
  Limit on annual effective dose equivalent to individual members of the public at a specified location with a specified diet and living style due to releases caused by a human intrusion scenario involving inadvertent drilling through a degraded waste package of 0.15 mSv (15 mrem) from all potential environmental transport and exposure pathways
Groundwater protection standards
  Releases from undisturbed Yucca Mountain disposal system will not cause levels of radioactivity in a specified representative volume of groundwater at a specified location to exceed
    0.2 Bq L$^{-1}$ (5 pCi L$^{-1}$) for $^{226}$Ra and $^{228}$Ra combined
    0.6 Bq L$^{-1}$ (15 pCi L$^{-1}$) for gross alpha-particle activity, including $^{226}$Ra but excluding radon and uranium
    Annual dose equivalent to whole body or any organ of 0.04 mSv (4 mrem) from all beta- or gamma-emitting radionuclides combined, assuming a daily intake of drinking water of 2 L

---

[a] Standards also are included in NRC's licensing criteria for the Yucca Mountain Site in 10 CFR part 63.
[b] Disposal standards include other provisions not given in this table (see text).
[c] The EPA issued proposed revisions of individual protection and human intrusion standards (see text), but revised standards had not been promulgated at the time of this writing.

The groundwater protection standard for beta- or gamma-emitting radionuclides at the Yucca Mountain Site differs from the groundwater protection standard in 40 CFR part 191 in an important way. The latter standard specifies that existing MCLs in EPA drinking water standards in 40 CFR part 141 shall be used as groundwater protection standards, which means that MCLs for beta or gamma emitters must be based on outdated dose coefficients for ingestion of radionuclides (National Bureau of Standards 1963). The groundwater protection standard at Yucca Mountain does not include such a provision, which means that updated dose coefficients for ingestion can be used to relate projected concentrations of radionuclides in groundwater to estimates of annual dose equivalent to the whole body or any organ. In addition, the groundwater protection standard at the Yucca Mountain Site does not include the MCL for uranium that was added in revising the interim drinking water standards in 2000 (see table 15.8).

EPA standards for disposal at the Yucca Mountain Site include other provisions of importance in demonstrating compliance. First, compliance with dose criteria shall be based upon the mean of a probability distribution of projected doses over 10,000 years that is obtained by accounting for uncertainties in projected performance. Second, changes in society, the biosphere (other than climate), human biology, or human knowledge or technology should not be projected over the next 10,000 years but should be assumed to remain constant as they are when a license application is submitted. Factors related to changes in geology, hydrology, and climate should be projected over that time frame. Finally, the standards address how unlikely features, events, and processes that could affect performance of the Yucca Mountain Site should be taken into account in performance assessments.

Standards for disposal at Yucca Mountain also specify that the peak dose to an individual member of the public that would occur beyond 10,000 years after disposal and within the period of geologic stability shall be calculated. No regulatory standard was applied to the results of such an analysis, but the results and their bases provide an indicator of the long-term performance of the disposal system. The period of geologic stability at Yucca Mountain is expected to be about $10^6$ years (National Research Council 1995). A similar provision was applied to the human intrusion standard if exposures of members of the public that result from the specified drilling intrusion scenario are projected to occur beyond 10,000 years after disposal.

In 2004, a court of appeals ruled that the 10,000-year period of compliance in the EPA's individual protection standard for disposal at Yucca Mountain was not based upon, and consistent with, findings in the NAS study (National Research Council 1995), as directed by Congress. The NAS found that (1) such a standard should apply to the peak dose within the period of geologic stability of about $10^6$ years and (2) there was no scientific basis for limiting application of an individual protection standard to 10,000 years. The court of appeals vacated the disposal standards for Yucca Mountain to the extent that they specify a 10,000-year compliance period. This ruling applied only to the individual protection and human intrusion standards. The groundwater protection standards in 40 CFR part 197 were upheld, partly on the grounds that the NAS made no finding or recommendation for or against such a standard. Therefore, consistency with NAS findings, as directed by Congress, is not at issue.

In 2005, the EPA issued proposed revisions of standards for disposal at Yucca Mountain in 40 CFR part 197 to address the findings by a court of appeals noted above (EPA 2005). A major challenge to the EPA is the need to develop a standard beyond 10,000 years that is reasonably achievable, while still providing adequate protection of the public. Any such standard must account for projected releases and annual doses during that period, which are likely to be much higher than during the first 10,000 years. At the time of this writing, the EPA had not promulgated final revisions of disposal standards.

*U.S. Environmental Protection Agency's Certification Criteria for Waste Isolation Pilot Plant Site* EPA standards for disposal of spent fuel, high-level waste, and transuranic waste in 40 CFR part 191 discussed above and summarized in table 15.14 were intended to be generally applicable to any disposal site. However, those standards currently apply only to disposal of the DOE's defense transuranic waste at the Waste Isolation Pilot Plant (WIPP) facility in New Mexico. WIPP is the only authorized facility for disposal of defense transuranic waste, and no other facility has been authorized for any other DOE or non-DOE transuranic waste.

In establishing the DOE program for disposal of defense transuranic waste at the WIPP facility, Congress stipulated that the facility would not be licensed by the NRC, which means that the DOE is the implementing agency for EPA standards in 40 CFR part 191 at that site. However, the WIPP Land Withdrawal Act of 1992 gave the EPA authority to certify the DOE's demonstration of compliance with 40 CFR part 191 for disposal of transuranic waste at the WIPP facility. This is the only instance where the EPA is authorized to certify compliance with any of its environmental standards developed under the Atomic Energy Act.

In 1996, the EPA established criteria for certifying disposal of defense transuranic waste at the WIPP facility in 40 CFR part 194 (61 *FR* 5224). The certification criteria include (1) allowance for use of expert judgment and peer review in demonstrating compliance with disposal standards; (2) requirements for considering both natural and human-initiated processes and events in demonstrating compliance with containment requirements in the disposal standards, including specification of future site conditions; and (3) specifications for generating probability distributions of projected cumulative releases to the accessible environment, doses to individuals, and radionuclide concentrations in groundwater.

In 1998, the EPA certified that disposal of defense transuranic waste at the WIPP facility would comply with standards in 40 CFR part 191 (63 *FR* 27354), and disposal began shortly thereafter. The WIPP Land Withdrawal Act requires that the EPA recertify disposal at the WIPP facility every five years. The EPA issued the first such recertification in 2006 (71 *FR* 18010).

*U.S. Nuclear Regulatory Commission Standards for Geologic Repositories*
The NRC is the licensing authority for geologic repositories that are developed under the Nuclear Waste Policy Act and are intended for disposal of spent fuel and high-level waste, including the repository at Yucca Mountain.

The first NRC standards for disposal of radioactive waste in geologic repositories were established in 10 CFR part 60. These standards were promulgated in 1983 (48 *FR* 28194) and amended in 1996 (61 *FR* 64257). Although these standards were intended to apply primarily to spent fuel and high-level waste, they also apply to any other waste containing source, special nuclear, or byproduct material that would be sent to a repository licensed under 10 CFR part 60, including, for example, high-activity low-level waste that is not acceptable for near-surface disposal. As noted above, the NRC does not have licensing authority over the WIPP facility for disposal of defense transuranic waste.

NRC standards for geologic repositories in 10 CFR part 60 include three criteria related to performance, siting, and design (i.e., to the performance of particular engineered and natural barriers) that are intended to help provide "reasonable assurance" that the containment requirements in EPA disposal standards in 40 CFR part 191 (see table 15.14) would be met. These criteria include the following:

Substantially complete containment of waste within waste packages for 300–1,000 years

Limit on the release rate of any radionuclide from the engineered barrier system following the containment period of $10^{-5}$ per year of the inventory of that radionuclide at 1,000 years following permanent closure or $10^{-5}$ per year of the inventory of all radionuclides placed in a disposal facility that remains after 1,000 years of decay

A groundwater travel time of at least 1,000 years prior to waste emplacement for the fastest path of likely radionuclide travel from the edge of the disturbed zone to the accessible environment

Currently, there are no authorized or planned facilities for disposal of radioactive waste in a geologic repository to which NRC licensing criteria in 10 CFR part 60 would apply. These standards do not apply to Yucca Mountain, and there are no plans to develop another geologic repository that might be used for disposal of spent fuel or high-level waste, transuranic waste that is not acceptable for disposal in the WIPP facility, or high-activity low-level waste that is not acceptable for disposal in a near-surface facility.

In parallel with the EPA's development of standards for disposal of spent fuel and high-level waste at the Yucca Mountain Site in 40 CFR part 197, the NRC established licensing criteria in 2001 that apply only to that site in 10 CFR part 63 (66 *FR* 55732). In contrast to the NRC's licensing criteria in 10 CFR part 60 as they related to the EPA's generally applicable environmental standards for disposal of spent fuel and high-level waste in 40 CFR part 191, the NRC's licensing criteria for the Yucca Mountain Site include the same standards for management, storage, and disposal as the EPA standards for that site (see table 15.15). The NRC also issued proposed revisions of 10 CFR part 63 (NRC 2005) that conform to the EPA's proposed revisions of 40 CFR part 197 that were developed to address findings by a court of appeals discussed above concerning deficiencies in the time limit of 10,000 years for complying with the individual protection and human intrusion standards.

*U.S. Environmental Protection Agency Standards for Low-Level Waste*
The EPA is authorized under the Atomic Energy Act to establish generally applicable environmental standards for management, storage, and disposal of low-level waste. The EPA, however, has not developed such standards. NRC and DOE standards, which are described below, apply to disposal of commercial or (non-DOE governmental) low-level waste and to DOE low-level waste, respectively.

*U.S. Nuclear Regulatory Commission Standards for Near-Surface Disposal Facilities* Current NRC standards for near-surface land disposal of commercial or non-DOE governmental radioactive waste containing source, special nuclear, or byproduct material in 10 CFR part 61 were established in 1982 (47 *FR* 57446) and amended in 1993 to apply to above-ground disposal (58 *FR* 33886). These standards are intended to apply primarily to disposal of low-level waste, but they do not stipulate that only wastes classified as low-level waste are acceptable for near-surface disposal in licensed facilities. Performance objectives and technical

Table 15.16  U.S. Nuclear Regulatory Commission's licensing criteria for near-surface disposal of radioactive waste (10 CFR part 61)[a]

---

*Performance objectives for near-surface disposal facilities*

  Limits on annual dose equivalent to individual members of the public beyond the facility boundary[b]

  0.25 mSv (25 mrem) to whole body
  0.75 mSv (75 mrem) to thyroid
  0.25 mSv (25 mrem) to any other organ

  Releases beyond facility boundary should be maintained ALARA
  Design, operation, and closure of the disposal facility must ensure protection of inadvertent intruders onto the disposal site after active institutional controls are removed (assumed to occur at 100 years after facility closure)

*Technical requirements for near-surface disposal*

  Requirements on site suitability and design, facility operation and site closure, and waste characteristics
  Limits on concentrations of radionuclides that are generally acceptable for near-surface disposal for class A, B, and C wastes, as specified in tables 1 and 2 of the standard, and requirements for disposal of waste in the different classes

---

[a] Standards are intended to apply primarily to disposal facilities for commercial or non-DOE governmental low-level waste. Standards do not apply to (1) disposal of certain wastes by materials licensees in accordance with requirements in NRC's 10 CFR part 20, (2) disposal of commercial or DOE high-level wastes in geologic repositories, (3) uranium or thorium mill tailings, and (4) wastes not regulated under authority of the Atomic Energy Act.
[b] NRC staff has recommended that a limit on annual effective dose equivalent of 0.25 mSv (25 mrem) can be used instead of a limit on annual dose equivalent to the whole body or a critical organ (NRC 2000).

---

requirements in NRC standards for near-surface disposal of radioactive waste are summarized in table 15.16. Disposals to which these standards do not apply are noted in the table.

Dose constraints for individuals beyond the facility boundary are the same as in EPA uranium fuel-cycle standards in 40 CFR part 190 (see table 15.6). Limits on concentrations of radionuclides in class A, B, and C wastes and requirements for disposal of waste in each class constitute the waste classification system for near-surface disposal, which is intended to ensure protection of inadvertent intruders at any site, as stipulated in the performance objectives. Those concentration limits were based in part on assessments of dose in assumed intrusion scenarios and limits on annual dose equivalent to an inadvertent intruder of 5 mSv (0.5 rem) to the whole body and bone or 15 mSv (1.5 rem) to any other organ, which were consistent with dose limits in radiation protection standards for the public at the time 10 CFR part 61 was developed.

The NRC did not specify a time period for compliance with performance objectives in 10 CFR part 61, especially the dose constraints for members of the public beyond the facility boundary. NRC staff has recommended that performance objectives should be applied for 10,000 years, but that calculations should be extended beyond that time if projected doses for particular radionuclides are increasing at 10,000 years (NRC 2000). Results of such calculations could be used to establish site-specific limits on inventories of long-lived radionuclides, especially if projected

Table 15.17   U.S. Department of Energy's performance objectives for disposal of low-level waste in DOE M 435.1-1, chapter (DOE 1999)

---

Protection of members of the public beyond facility boundary

   Limit on annual effective dose equivalent to individuals of 0.25 mSv (25 mrem) from all release and exposure pathways, excluding dose from radon and its decay products in air
   Limit on annual effective dose equivalent to individuals of 0.1 mSv (10 mrem) from releases to air, excluding dose from radon and its decay products[a]
   Limit on average release rate of radon to air of 0.74 Bq m$^{-2}$ s$^{-1}$ (20 pCi m$^{-2}$ s$^{-1}$) or, alternatively, limit on concentration of radon in air at facility boundary of 19 Bq m$^{-3}$ (0.5 pCi L$^{-1}$)[a]

Performance objectives apply for 1,000 years after disposal

---

Standards replace those in DOE Order 5820.2A, chapter III (DOE 1988a). Standards apply to disposals after September 26, 1988, except as specified in existing contracts under DOE Order 5820.2A; standards include other provisions not given in this table (see text).
[a] Performance objective is consistent with EPA standards for airborne emissions of radionuclides in 40 CFR part 61.

doses exceed the annual dose limit for the public in 10 CFR part 20 (see table 15.4) of 1 mSv (100 mrem) (NRC 2000).

Non-DOE waste with concentrations of radionuclides greater than class C limits specified by the NRC is not generally acceptable for near-surface disposal. In a 1989 revision of 10 CFR part 61 (54 *FR* 22578), the NRC stipulated that waste greater than class C must be sent to a geologic repository (currently intended to be the Yucca Mountain facility) unless disposal elsewhere is approved by the NRC. Disposal of waste greater than class C would not be subject to licensing requirements in 10 CFR part 61. If waste greater than class C were sent to Yucca Mountain, the applicable EPA standards would be those in 40 CFR part 197 (see table 15.15), and the applicable licensing requirements would be those in 10 CFR part 63.

*U.S. Department of Energy Standards for Low-Level Waste Disposal*   Current DOE requirements for disposal of low-level waste are given in chapter 4 of the implementing manual (M 435.1–1) of DOE Order 435.1 (DOE 1999). Performance objectives for disposal of low-level waste are listed in table 15.17.

DOE M 435.1–1 (DOE 1999) includes several requirements on performance assessments in addition to the requirement to demonstrate compliance with performance objectives. First, performance assessments must demonstrate that projected releases of radionuclides shall be maintained ALARA. Second, an assessment of projected impacts on water resources must be performed to establish limits on inventories of long-lived and mobile radionuclides for disposal. DOE M 435.1–1 does not specify how inventory limits for such radionuclides shall be determined (i.e., criteria for protection of water resources). Rather, each facility must meet requirements for protection of the environment in DOE Orders 5400.1 (DOE 1988b) and 5400.5 (DOE 1990). Third, an assessment of potential impacts on a hypothetical inadvertent intruder must be performed for the purpose of establishing limits on concentrations of radionuclides for disposal. Those limits are to be based on performance measures for chronic and acute exposure scenarios of 1 mSv (100 mrem) per year and 5 mSv (500 mrem), respectively.

DOE M 435.1–1 also requires that a site-specific composite analysis be prepared that considers the effects of other interacting sources of potential public exposure at a DOE low-level waste disposal site (e.g., releases from other disposal sites, operating facilities, or waste storage areas). Performance measures that apply to potentially overlapping sources shall be consistent with DOE requirements for radiation protection of the public and the environment, which presently are the requirements in DOE Order 5400.5 (DOE 1990) listed in table 15.4, and shall be evaluated for 1,000 years after disposal.

*Federal Standards for Naturally Occurring and Accelerator-Produced Radioactive Material* The EPA may develop standards for management and disposal of waste containing NARM under authority of TSCA (Cameron 1996). The EPA, however, has not developed such standards.

The DOE regulates all NARM waste for which it is responsible as part of its authorized activities under the Atomic Energy Act. The DOE manages larger volumes of lower-activity NARM waste (i.e., NORM waste) consistent with EPA requirements for uranium and thorium mill tailings in 40 CFR part 192 (DOE 1990; see table 15.11), and smaller volumes of higher activity NARM waste may be managed as low-level waste (DOE 1999). The DOE manages and disposes of NARM waste based on its properties (i.e., its resemblance to mill tailings or low-level waste), and standards for similar wastes are applied.

Prior to 2005, the NRC could not regulate NARM. The NRC is now authorized under the Atomic Energy Act to regulate NARM and certain discrete sources of $^{226}$Ra and NORM other than source material. The NRC intends to regulate disposal of waste containing those materials in accordance with requirements for near-surface disposal of low-level waste in 10 CFR part 61 (NRC 2007; see table 15.16).

*State Standards for Naturally Occurring and Accelerator-Produced Radioactive Material* States are responsible for regulating NARM waste that is not currently regulated by the NRC or DOE. Suggested state regulations (CRCPD 2005) include provisions that apply to disposal of NARM waste, principally wastes containing TENORM. Those regulations specify that disposal methods for uranium mill tailings that are regulated under the EPA's 40 CFR part 192 are generally acceptable for TENORM. Other disposal methods might also be suitable, provided that basic criteria for radiation protection of the public are met. Those criteria specify a limit on annual effective dose equivalent, excluding the dose from radon and its decay products, of 1 mSv (100 mrem) from all licensed sources combined. As a consequence, allowable doses from disposal of NARM would be a fraction of the dose limit that applies to all sources combined. Suitable disposal methods might include down-hole disposal of some oilfield wastes, disposal in industrial or sanitary landfills, and onsite disposal with institutional controls. Cost–benefit analysis may be used to evaluate alternative disposal options.

*Management and Disposal of Mixed Radioactive and Hazardous Chemical Waste* The definition of hazardous waste in RCRA specifically excludes source, special nuclear, and byproduct materials regulated under the Atomic Energy Act.

As a consequence, waste that contains radioactive material defined in the Atomic Energy Act and hazardous waste defined under RCRA, which is referred to as "mixed waste," is subject to dual regulation, with the radionuclides regulated in accordance with requirements for disposal of radioactive waste developed under the Atomic Energy Act and the rest of the waste regulated under RCRA.

The approach to protecting public health and the environment in disposal of hazardous chemical waste under RCRA is different from the approach under the Atomic Energy Act, as embodied in EPA, NRC, and DOE regulations described above. First, the goal under RCRA is *zero* release of hazardous substances to the environment. Second, instead of specifying limits on health risk or other quantitative criteria (e.g., acceptable levels of hazardous substances in the environment that apply over long time periods) and requiring performance assessments to demonstrate compliance with those limits at specific disposal sites, EPA regulations to implement RCRA emphasize compliance with detailed and prescriptive technical standards for obtaining operating permits at hazardous waste treatment, storage, and disposal facilities. Those standards are technology based and apply at all such facilities. Particularly important are the universal treatment standards (UTSs) for hazardous waste in 40 CFR part 268. In addition to UTSs, RCRA and implementing regulations in 40 CFR part 264 emphasize groundwater protection standards—either background levels of contamination, limits on contamination similar to MCLs in drinking water established by the EPA under the Safe Drinking Water Act, or alternative limits established by the EPA—and a requirement for corrective actions if groundwater protection standards are exceeded. Compliance with groundwater protection standards is based on monitoring, rather than assessments of the long-term performance of waste disposal facilities.

RCRA precludes any hazardous waste regulation for mixed waste that is inconsistent with the Atomic Energy Act. No irreconcilable technical inconsistencies between requirements of the two laws and their implementing regulations have been found. Thus, for example, mixed waste that contains low-level radioactive waste normally would be managed in accordance with technical requirements for hazardous chemical waste established under RCRA even though radionuclides in the waste are not subject to regulation under RCRA. As noted above, RCRA requirements also are included in EPA regulations for disposal of uranium or thorium mill tailings in 40 CFR part 192.

The exclusion of radioactive material from regulation under RCRA does not apply to NARM,[3] so NARM waste not subject to regulation under the Atomic Energy Act could be regulated under RCRA without regard for the presence of other hazardous materials. However, the definition of hazardous waste in implementing regulations in 40 CFR part 261 does not include NARM waste, and the EPA cannot regulate any waste material under RCRA unless it is declared to be hazardous in law or regulations. Furthermore, many important wastes containing NORM are specifically excluded from the definition of hazardous waste in 40 CFR part 261, including wastes generated in mining and processing of ores and minerals, production of oil and natural gas or geothermal energy, and combustion of fossil fuels. Regulation of NARM waste under RCRA thus would require changes in the definition of solid hazardous waste.

TSCA also does not apply to source, special nuclear, and byproduct materials that are regulated under the Atomic Energy Act. If such radionuclides are mixed with toxic substances that are regulated under TSCA (e.g., PCBs, dioxins, and asbestos), the radionuclides are managed mainly in accordance with TSCA requirements in a manner similar to the case of mixed radioactive and hazardous chemical waste that is managed mainly in accordance with RCRA requirements.

### Airborne Emissions of Radionuclides

The EPA regulates airborne emissions of radionuclides and other hazardous air pollutants under authority of the Clean Air Act. Standards established under the Clean Air Act apply to radionuclides and other hazardous substances individually. Neither the Clean Air Act nor its implementing regulations specify a limit on acceptable risk from all hazardous air pollutants combined.

Current EPA standards for airborne emissions of radionuclides, which are included in the National Emission Standards for Hazardous Air Pollutants (NESHAPs) in 40 CFR part 61, were first established in 1989 (54 *FR* 51654) and amended in 1991 (56 *FR* 65934), 1992 (57 *FR* 23305), 1994 (59 *FR* 36280), 1995 (60 *FR* 46206), and 1996 (61 *FR* 68972). These standards are listed in table 15.18. The standard for phosphogypsum used in agriculture is intended to limit the risk from external exposure to $^{226}$Ra in soil as well as airborne emissions of $^{222}$Rn. All other standards are concerned only with limiting risks due to airborne emissions of radionuclides.

NESHAPs for radionuclides were established in response to a mandate by a court of appeals regarding an existing standard for vinyl chloride. The court required that EPA, in accordance with the Clean Air Act, base NESHAPs on determinations of (1) an acceptable risk to individuals or populations and (2) an ample margin of safety below the acceptable risk for protection of public health. This ruling meant that NESHAPs had to be based on considerations of acceptable health risks to the public, but they could not be based on technical feasibility and cost. In response to the court order, NESHAPs for radionuclides and other carcinogens were set such that the lifetime risk to individuals would not exceed about $10^{-4}$ and the lifetime risk to the greatest number of individuals in exposed populations (i.e., the average individual risk) would not exceed about $10^{-6}$.

NESHAPs for radionuclides issued in 1989 also applied to facilities licensed by the NRC or an Agreement State, except for users of radionuclides in the form of sealed sources only. In subsequent amendments, the EPA rescinded standards for commercial facilities licensed by the NRC in response to the 1990 Clean Air Act Amendments and an MOU with the NRC. The EPA agreed that existing NRC regulations limit airborne emissions of radionuclides to an extent consistent with or more restrictive than EPA standards in 40 CFR part 61. Applicable NRC regulations include (1) 10 CFR part 40, appendix A, for inactive uranium mill tailings disposal sites, (2) 10 CFR part 50, appendix I, for nuclear power reactors, and (3) a 1996 amendment to radiation protection standards for the public in 10 CFR part 20 (61 *FR* 65120), which established an annual dose constraint of 0.1 mSv (10 mrem), excluding the dose from radon, for airborne emissions from licensed facilities other than nuclear power reactors.

Table 15.18  U.S. Environmental Protection Agency standards for airborne emissions of radionuclides (40 CFR part 61)[a]

*Limit on annual effective dose equivalent to individuals of 0.1 mSv (10 mrem) for*

DOE facilities emitting any radionuclide other than radon, except at disposal facilities for spent fuel, high-level waste, or transuranic waste subject to EPA's 40 CFR part 191, subpart B, or disposal facilities for mill tailings subject to EPA's 40 CFR part 192, and excluding dose from $^{222}$Rn

Non-DOE federal facilities not licensed by the NRC, except any dose from $^{222}$Rn is excluded and limit does not apply at:

    Disposal facilities for spent fuel, high-level waste, or transuranic waste subject to EPA's 40 CFR part 191, subpart B

    Inactive uranium mill tailings disposal sites subject to EPA's 40 CFR part 192

    Low-energy accelerators, except limit on annual effective dose equivalent is 0.03 mSv (3 mrem) from isotopes of iodine

    Emissions of $^{222}$Rn from underground uranium mines with production greater than specified values

*For elemental phosphorus plants*

Limit on annual emissions of $^{210}$Po from all calciners and nodulizing kilns of 0.07 TBq (2 Ci); or Limit on total annual emissions from any plant of 0.17 TBq (4.5 Ci) when specified scrubbers are installed

*For phosphogypsum distributed in commerce for uses in agriculture*

Limit on average concentration of $^{226}$Ra of 0.4 Bq g$^{-1}$ (10 pCi g$^{-1}$)

*Limit on emission rate of $^{222}$Rn of 0.7 Bq m$^{-2}$s$^{-1}$ (20 pCi m$^{-2}$s$^{-1}$) from*

DOE facilities for storage and disposal of material containing radium
Inactive phosphogypsum stacks (waste piles from phosphate mining)
Operating and inactive uranium mill tailings piles, except for inactive disposal sites licensed by the NRC under 10 CFR part 40, appendix A

[a] Standards do not apply to NRC licensees because the EPA has agreed that existing NRC regulations limit airborne emissions of radionuclides to an extent consistent with or more restrictive than EPA standards (see text). Standards also do not apply to Agreement State licensees when an Agreement State has developed regulations to limit airborne emissions that are compatible with NRC regulations.

Although the NRC agreed that applying EPA's NESHAPs for radionuclides to licensed facilities represented an unnecessary duplication of its existing requirements, the NRC resisted the EPA's insistence that it establish a separate dose constraint for airborne emissions from licensed nuclear facilities other than nuclear power reactors. The NRC's preferred approach to radiation protection relies on the annual dose limit of 1 mSv (100 mrem) for all release and exposure pathways combined and vigorous application of the ALARA principle in reducing doses at specific sites, rather than dose constraints for particular release or exposure pathways at a fraction of the dose limit. This dispute was similar to the disagreement between the two agencies concerning the need for a separate groundwater protection requirement in standards for cleanup of radioactively contaminated sites and radioactive waste disposal.

The following additional points are noteworthy. First, standards were not established for certain sources of radionuclide emissions to air that were judged by the EPA to be insignificant, including operating surface uranium mines, coal-fired boilers, and facilities for management of spent fuel, high-level waste, and transuranic waste. For surface uranium mines and coal-fired boilers in particular, estimated risks were less than the assumed levels of acceptable risk to individuals and populations of about $10^{-4}$ and $10^{-6}$, respectively, although the average risk due to emissions from coal-fired boilers was higher than the average risk due to airborne emissions from DOE facilities or nuclear power plants.

Second, in establishing the limit on annual effective dose equivalent of 0.1 mSv (10 mrem) for many sources, EPA ignored a statement by NCRP (1984a) that (1) a limit on annual effective dose equivalent of 1 mSv (100 mrem) from all controlled sources combined provides an upper bound on acceptable risk to individuals and (2) an authorized limit for specific practices or sources at 25% of the dose limit and further application of the ALARA objective at specific sites would provide an adequate margin of safety for exposed individuals and populations.

Finally, although NESHAPs for radionuclides had to be based on considerations of acceptable risk, they were shown to be reasonably achievable with existing effluent control technologies (EPA 1989a, 1989b). The feasibility of emission controls also was considered in determining an ample margin of safety in NESHAPs for radionuclides and other carcinogens, that is, in deciding that the average individual risk should not exceed about $10^{-6}$.

## Indoor Radon

Indoor radon is the single source of radiation exposure that poses the greatest health risk to the public (see table 15.5). Guidance on mitigation of indoor radon has been established by the EPA and the U.S. Department of Health and Human Services (DHHS). Recommendations on mitigation of indoor radon also have been developed by the NCRP, ICRP, and IAEA. The historical development of guidance and standards for protection against indoor radon is discussed by Harley (1996).

*U.S. Environmental Protection Agency and U.S. Department of Health and Human Services Guidance*   In 1986, the EPA and DHHS established federal guidance on mitigation of indoor radon (EPA and DHHS 1986). That guidance was revised in 1994 (EPA and DHHS 1994) in response to a provision of the Indoor Radon Abatement Act of 1988, which established a national goal of reducing indoor radon concentrations to background (outdoor) levels. Concentrations of radon in outdoor air average about 7 Bq m$^{-3}$ (0.2 pCi L$^{-1}$) but are highly variable (NCRP 1987b), and the average concentration in homes is about 50 Bq m$^{-3}$ (1.3 pCi L$^{-1}$) (EPA and DHHS 1994).

Current federal guidance on mitigation of indoor radon is listed in table 15.19. This guidance applies to schools as well as homes, even though residence times in schools are less than in homes, based primarily on a judgment that the mitigation level is reasonably achievable in schools.

The recommended mitigation level for indoor radon was based on two considerations: (1) risks to individuals who receive the highest exposures and (2) the cost–benefit of reducing risks in the whole population, given the distribution of levels of radon in homes. For exposure at the recommended mitigation level of 150 Bq m$^{-3}$ (4 pCi L$^{-1}$), the estimated lifetime risk of fatal lung cancer is $2 \times 10^{-3}$ for persons who have never smoked and $3 \times 10^{-2}$ for smokers (EPA and DHHS 1994). For former smokers, the risk might be intermediate.

Federal guidance on indoor radon is not an enforceable standard for limiting public exposure to radon in homes and schools. It has been widely used, however, as a de facto standard in the real estate and home insurance industries.

*Recommendations of the NCRP, ICRP, and IAEA* The mitigation level for radon in homes recommended by the NCRP (1984b, 1993) is an annual exposure to short-lived radon decay products of 2 working level months (WLMs).[4] The recommended mitigation level was based originally on an assumption that the excess lifetime risk of fatal lung cancer should not exceed 0.02 (NCRP 1984b). It was retained in current recommendations (NCRP 1993) based on consideration of the feasibility of remediation at that level and an assumption that the excess lifetime risk should not exceed 10 times the risk corresponding to the average radon level in homes. If radon decay products are assumed to be in 50% activity equilibrium with the parent, the NCRP's recommended mitigation level corresponds to a radon concentration of 370 Bq m$^{-3}$ (10 pCi L$^{-1}$), or slightly more than twice the current federal guidance (see table 15.19). The relationship between the two recommendations is discussed further in Harley (1996). The NCRP has not developed separate recommendations on mitigation of radon in schools or workplaces.

The ICRP (1993a, 1999) has recommended a mitigation level of 200–600 Bq m$^{-3}$ (5–16 pCi L$^{-1}$) for radon in homes. The upper end of this range was based on an assumption that efforts to avoid annual effective doses that exceed 10 mSv (1 rem) should almost always be justified, and the lower end was based on a judgment that optimization of exposures (application of the ALARA objective) should suggest an action level not less than about one-third of that dose (ICRP 1993a).

The mitigation level for radon in schools and workplaces recommended by the ICRP (1993a, 1999) is 500–1,500 Bq m$^{-3}$ (14–40 pCi L$^{-1}$). The increase by a

Table 15.19 Federal guidance on mitigation of indoor radon: homes and schools (EPA and DHHS 1994)

| Radon concentration in air, Bq m$^{-3}$ (pCi L$^{-1}$) | Recommended action |
| --- | --- |
| >150(4)[a] | Mitigation should be undertaken |
| 70–150 (2–4) | Mitigation should be considered, especially when concentration can be reduced below 70 Bq m$^{-3}$ (2 pCi L$^{-1}$) |

[a] Mitigation level corresponds to concentration of short-lived radon decay products in air of about $4 \times 10^{-7}$ J m$^{-3}$ [0.02 Working Level (WL)].

factor of 3 compared with the recommended mitigation level in homes was based on the different residence times.

The IAEA (1996) adopted the mitigation level of 200–600 Bq m$^{-3}$ (5–16 pCi L$^{-1}$) for radon in homes that had been recommended by ICRP (1993a). The IAEA's recommended mitigation level in schools and workplaces is 1,000 Bq m$^{-3}$ (27 pCi L$^{-1}$), which is the average of the range of mitigation levels recommended by the ICRP (1993a).

NCRP, ICRP, and IAEA recommendations on mitigation levels for indoor radon are all somewhat higher than current federal guidance. This difference is due primarily to the greater emphasis in federal guidance on reducing risks in the whole population rather than risks to individuals who receive the highest exposures (National Research Council 1998).

## Risks Associated with Radiation Standards for the Public

All standards for radionuclides in the environment are concerned with limiting cancer risks to the public. An interesting perspective on the many and diverse radiation standards for the public is obtained by comparing estimates of lifetime risk associated with criteria in the standards.

Table 15.20 gives estimates of cancer risks from chronic exposure over an average lifetime of 70 years associated with (1) radiation protection standards for the public, which apply to all controlled sources combined, and selected authorized limits and guidance for specific practices or sources and (2) exposures to natural background radiation. Most risk estimates assume a fatal cancer risk per unit effective dose equivalent of $5.0 \times 10^{-2}$ Sv$^{-1}$ ($5.0 \times 10^{-4}$ rem$^{-1}$) and tissue weighting factors ($w_T$ values) currently recommended by the ICRP (1991). EPA risk estimates are used in cases of exposure to indoor radon and the containment requirements for disposal of spent fuel, high-level waste, and transuranic waste.

Use of lifetime risks of fatal cancers in table 15.20 is based on the historical emphasis in radiation protection on fatal cancer as the primary health effect of concern (ICRP 1977, 1991). In recent years, however, cancer incidence has assumed greater importance as the primary health effect of concern in protection of the public. The EPA uses cancer incidence in applying its standards developed under CERCLA to cleanup of sites that are contaminated with radionuclides or chemical carcinogens, and the EPA generally uses cancer incidence in regulating chemical carcinogens under authority of any environmental laws. For uniform irradiation of the whole body, the lifetime risk of cancer incidence is about 50% higher than the risk of fatal cancer (EPA 1994b). The ratio of the two risks in specific organs or tissues is highly variable and ranges from factors of about 500 and 10 for skin and thyroid cancer, respectively, to close to 1.0 for acute leukemia and cancers of the esophagus, stomach, liver, and lung (EPA 1994b).

Cancer risks in table 15.20 also are intended to be best (central) estimates of risk. This approach is based on the historical emphasis on use of best estimates of risk in radiation protection, without consideration of their uncertainties (ICRP 1977, 1991). In contrast, the EPA generally regulates chemical carcinogens in the environment using upper 95% credibility limits of estimated risks that account for uncertainties

Table 15.20  Lifetime risks of fatal cancers associated with selected radiation exposures and standards or guidance for controlling radiation exposures of the public[a]

| Risk | Exposure, standard, or guidance |
|---|---|
| $4 \times 10^{-2}$ | Cleanup standards at uranium mill tailings sites, including indoor radon (EPA's 40 CFR part 192, subpart B) |
| $0.2–3 \times 10^{-2}$ | Concentration of radon in air of 150 Bq m$^{-3}$ (4 pCi L$^{-1}$) in federal guidance on mitigation of radon in homes (EPA and DHHS 1994)[b] |
| $2 \times 10^{-2}$ | Annual dose equivalent to whole body of 5 mSv (500 mrem) in existing federal guidance on radiation protection of the public (FRC 1960)[c] |
| $1 \times 10^{-2}$ | Average annual effective dose equivalent of 3 mSv (300 mrem) from exposure to natural background, including indoor radon (NCRP 1987a) |
| $0.7–9 \times 10^{-3}$ | Average concentration of radon in air of 50 Bq m$^{-3}$ (1.3 pCi L$^{-1}$) in homes (EPA and DHHS 1994)[b,d] |
| $4 \times 10^{-3}$ | Annual effective dose equivalent of 1 mSv (100 mrem) from all controlled sources combined, excluding indoor radon, in proposed federal guidance on radiation protection of the public (EPA 1994b)[e] |
| $4 \times 10^{-3}$ | Indoor gamma radiation level of 20 $\mu$R h$^{-1}$ in cleanup standards at uranium mill tailings sites (EPA's 40 CFR part 192, subpart B), assuming indoor residence time of 85% |
| $2 \times 10^{-3}$ | Concentrations of $^{226}$Ra in soil of 0.2 and 0.6 Bq g$^{-1}$ (5 and 15 pCi g$^{-1}$) in cleanup standards at uranium mill tailings sites (EPA's 40 CFR part 192, subpart B), assuming continuous external exposure indoors and outdoors |
| $9 \times 10^{-4}$ | Annual dose equivalent to whole body or effective dose equivalent of 0.25 mSv (25 mrem)[f] |
| $5 \times 10^{-4}$ | Annual effective dose equivalent of 0.15 mSv (15 mrem)[g] |
| $4 \times 10^{-4}$ | Annual effective dose equivalent of 0.1 mSv (10 mrem) from airborne emissions (EPA's 40 CFR part 61) |
| $2 \times 10^{-4}$ | Concentration of uranium of 30 $\mu$g L$^{-1}$ in drinking water (EPA's 40 CFR part 141)[h] |
| $2 \times 10^{-4}$ | Concentration of $^{226}$Ra of 0.2 Bq L$^{-1}$ (5 pCi L$^{-1}$) in drinking water (EPA's 40 CFR part 141) |
| $1 \times 10^{-4}$ | Goal for cleanup of radioactively contaminated sites (EPA's 40 CFR part 300; Clay 1991; Luftig and Weinstock 1997; EPA 1999b) |
| $1 \times 10^{-4}$ | Annual dose equivalent to whole body or effective dose equivalent of 0.04 mSv (4 mrem) from beta- or gamma-emitting radionuclides in drinking water (EPA's 40 CFR part 141) |
| $1 \times 10^{-4}$ | Annual dose equivalent to thyroid of 0.75 mSv (75 mrem) from ingestion of $^{131}$I[f] |
| $1 \times 10^{-4}$ | Annual dose equivalent to lungs of 0.25 mSv (25 mrem) from inhalation of insoluble natural uranium[f] |
| $9 \times 10^{-5}$ | Annual dose equivalent to bone of 0.25 mSv (25 mrem) from inhalation of insoluble $^{239}$Pu[f] |
| $4 \times 10^{-5}$ | Annual dose equivalent to bone of 0.25 mSv (25 mrem) from ingestion of soluble natural uranium[f] |
| $1 \times 10^{-5}$ | Annual dose equivalent to bone of 0.04 mSv (4 mrem) from $^{90}$Sr in drinking water (EPA's 40 CFR part 141) |
| $7 \times 10^{-6}$ | Annual dose equivalent to thyroid of 0.04 mSv (4 mrem) from $^{129}$I in drinking water (EPA's 40 CFR part 141) |
| $3 \times 10^{-8}$ | Containment requirements for disposal of spent fuel, high-level waste, and transuranic waste (EPA's 40 CFR part 191, subpart B)[i] |

[a]Estimated risks assume continuous exposure over 70 years and, unless otherwise noted, a risk of fatal cancers per unit effective dose equivalent of $5 \times 10^{-2}$ Sv$^{-1}$ ($5 \times 10^{-4}$ rem$^{-1}$) and tissue weighting factors ($w_t$ values) currently recommended by the ICRP (1991).

*continued*

Table 15.20 (Continued)

[b] Lower bound applies to individuals who have never smoked, and upper bound applies to smokers. For former smokers, the risk may lie in between.
[c] Dose limit in existing federal guidance is outdated and now applies only to occasional or unusual exposure situations that do not occur over long time periods.
[d] Average annual effective dose equivalent from indoor radon estimated by the NCRP (1987b) is 2 mSv (200 mrem) and the associated lifetime risk is $7 \times 10^{-3}$.
[e] Dose limit is included in current radiation protection standards for the public developed by the NRC (10 CFR part 20) and DOE (1990).
[f] Dose constraint in EPA's uranium fuel-cycle standards (40 CFR part 190) and other environmental standards.
[g] Dose constraint in EPA standards for management and disposal of spent fuel, high-level waste, and transuranic waste (40 CFR part 191, subpart B; 40 CFR part 197).
[h] Mass concentration is assumed to correspond to an activity concentration of 1 Bq L$^{-1}$ (27 pCi L$^{-1}$); effective dose equivalent per unit activity intake is approximately the same for all naturally occurring isotopes of uranium (Eckerman et al. 1988).
[i] Average risk in U.S. population.

in data in humans or animals from which risk estimates are derived. The EPA has used the same approach in regulating cleanup of radioactively contaminated sites under CERCLA. For uniform irradiation of the whole body, an upper 95% credibility limit of risk, either fatal cancer or cancer incidence, is about a factor of 2 higher than central estimates (NCRP 1997; EPA 1999c). The uncertainty in the cancer risk in specific organs or tissues is greater and is highly variable (National Research Council 2006).

Risk estimates in table 15.20 lead to the following observations. First, excluding the containment requirements for disposal of spent fuel, high-level waste, and transuranic waste in 40 CFR part 191, risks associated with the different standards for specific practices or sources vary by nearly 4 orders of magnitude.

Second, the risk associated with cleanup standards at uranium mill tailings sites, which is due primarily to indoor radon, and the average risk to smokers and nonsmokers associated with federal guidance on radon in homes exceed the risk associated with the annual dose limit of 1 mSv (100 mrem) in current radiation protection standards for all controlled sources of exposure combined, excluding indoor radon.

Third, average risks in the U.S. population from exposure to natural background, including indoor radon, and exposure to indoor radon only are greater than the risk associated with current radiation protection standards for all controlled sources combined, excluding indoor radon. The average risks from natural background and indoor radon only also are greater by about an order of magnitude or more than risks associated with all standards for specific practices or sources except standards for mill tailings.

Finally, risks associated with some standards for specific practices or sources are comparable to or less than the risk associated with a negligible annual individual dose of 0.01 mSv (1 mrem) for any practice or source recommended by the NCRP (1993). The NCRP recommendation is discussed further below.

### Consistency of Radiation Standards for the Public

The issue of consistency of standards—that is, the extent to which different standards for radionuclides and hazardous chemicals in the environment correspond to similar health risks to the public—has been considered by many investigators (e.g., Travis et al. 1987; Kocher 1988; Kocher and Hoffman 1991, 1992; EPA-SAB 1992; Brown 1992; GAO 1994; Taylor 1995; Overy and Richardson 1995; National Research Council 1998; Kocher 1999; NCRP 2004). Comparisons of risks associated with different standards, similar to those in table 15.20, have been presented in some cases (Travis et al. 1987; Kocher 1988; GAO 1994; National Research Council 1998; Kocher 1999). Given the wide range of risks associated with different radiation standards, the Government Accounting Office concluded that a consensus on acceptable radiation risk is lacking (GAO 1994).

The desire for consistency in radiation standards for the public in limiting health risks is understandable. However, there are several important reasons why such a consistency should not be expected (National Research Council 1998).

First, radiation standards for the public have been developed under various laws that mandate different approaches to setting standards. The most important consideration is the difference between the radiation paradigm for risk management under the Atomic Energy Act (see figure 15.1) and the chemical paradigm for risk management under other environmental laws (see figure 15.2). Radiation standards that were based on the chemical paradigm include drinking water standards developed under the Safe Drinking Water Act and standards for cleanup of radioactively contaminated sites developed under CERCLA. To reiterate, the radiation paradigm involves a limit (upper bound) on acceptable (barely tolerable) risk and reduction of risks below the limit using the ALARA principle. In contrast, the chemical paradigm involves a goal for acceptable risk and allowance for an increase (relaxation) in risks above the goal based primarily on technical feasibility and cost.

The concept of a limit in the radiation paradigm is fundamentally different from the concept of a goal in the chemical paradigm. A goal for acceptable risk does not define a limit on acceptable (barely tolerable) risk that must be met without regard for cost or other relevant factors. Consequently, it can be quite misleading to compare standards in the form of limits with standards in the form of goals. For example, it is not particularly meaningful to compare the lifetime risk associated with the annual dose limit of 1 mSv (100 mrem) in radiation protection standards for the public with the goal for lifetime cancer risk of around $10^{-4}$ in cleanup of contaminated sites under CERCLA.

Second, some radiation standards were based primarily on judgments about acceptable health risks to the public, while other standards were based primarily on judgments about doses, releases, or environmental levels (i.e., risks) that are reasonably achievable. There is no a priori reason to expect that risks associated with these two types of standards would be consistent. For example, the annual dose limit of 1 mSv (100 mrem) in radiation protection standards for the public was based on a judgment about a maximum tolerable risk, while drinking water standards for radionuclides, which have much lower associated risks, were based essentially on cost–benefit analyses for removal of radionuclides from public drinking water supplies. Standards with such different bases are not directly comparable.

Third, standards may differ significantly in their applicability. For example, radiation protection standards for the public developed under the Atomic Energy Act apply to all controlled sources combined, excluding indoor radon and medical exposures. All other standards for radionuclides in the environment developed under the Atomic Energy Act and other laws apply only to specific practices or sources. These two types of standards are not directly comparable. Indeed, risks associated with standards for specific practices or sources, except for indoor radon, should be substantially less in most cases than the risk associated with the annual dose limit of 1 mSv (100 mrem) in radiation protection standards (EPA 1994a). Furthermore, various standards for specific practices or sources apply to different exposure situations. Most of those standards were based primarily on judgments about doses, releases, or environmental levels (and, therefore, risks) that are reasonably achievable. In those cases, there is no a priori reason to expect that risks judged to be reasonably achievable for one exposure situation (e.g., releases from nuclear fuel-cycle facilities) would be consistent with risks judged to be reasonably achievable for a different situation (e.g., radioactive waste disposal). It is primarily in the interest of achieving some degree of consistency that standards for different exposure situations often are about the same. Finally, standards developed under the Atomic Energy Act, except for some standards for mill tailings (see table 15.10), apply to all release and exposure pathways combined for exposure situations of concern, while standards developed under other laws (e.g., standards for airborne emissions and drinking water) often apply only to particular release or exposure pathways. It should not be expected that these two types of standards would be consistent.

Fourth, some standards are concerned primarily with protection of individuals who are assumed to receive the highest exposures, while other standards are concerned primarily with protection of whole populations (i.e., individuals who receive an average exposure). Because maximum individual risks generally are higher than average risks in a population, risks associated with the two types of standards could differ substantially and may not be directly comparable. For example, the uranium fuel-cycle standards are concerned primarily with protection of maximally exposed individuals, while drinking water standards are concerned primarily with protection of entire populations.

Fifth, some standards are concerned with exposures to naturally occurring radionuclides, but others are not. Given the high doses and risks associated with natural background radiation, risks associated with different standards can vary substantially depending on whether the standards apply to naturally occurring radionuclides. For example, standards for mill tailings and federal guidance on indoor radon are concerned only with exposures to naturally occurring radionuclides that have been increased by human activities. The associated risks are relatively high as a direct result of the relatively high doses and risks from undisturbed natural background and the impossibility of reducing doses to levels below background. Other standards that do not include exposures to natural background (e.g., the uranium fuel-cycle standards) reasonably can have much lower associated risks, and the two types of standards are not directly comparable. Drinking water standards also are concerned primarily with naturally occurring radionuclides, but natural levels of radioactivity in sources of drinking water usually are relatively low, so the standards have lower associated risks.

Finally, some differences in risks associated with various radiation standards are explained by two other factors. First, different standards were developed at different times, and judgments about acceptable doses and risks have changed considerably over time. For example, the annual dose limit in radiation protection standards for the public has been reduced from 5 mSv (500 mrem) (FRC 1960) to its current value of 1 mSv (100 mrem), based primarily on a substantial increase in the assumed risk per unit dose. In addition, a judgment by the EPA that a lifetime risk of around $10^{-4}$ is an upper bound on acceptable risk for specific practices or sources is a relatively recent development that has resulted in increasingly stringent radiation standards including, for example, annual dose constraints of 0.1 mSv (10 mrem) for airborne emissions and 0.15 mSv (15 mrem) for management and disposal of radioactive wastes. Second, dosimetric quantities used in standards have changed over time. Early standards were expressed in terms of the dose equivalent to the whole body or critical organ, while standards in the United States that are more recent are expressed in terms of the effective dose equivalent. For inhalation or ingestion of radionuclides, the risk associated with a given dose to the critical organ can be substantially less than the risk associated with the same effective dose equivalent (Eckerman et al. 1988).

The important conclusion from these discussions is that risks associated with different radiation standards should not be compared unless the bases for the standards and their applicability are understood and the standards are interpreted properly. Otherwise, inappropriate and misleading conclusions about the significance of differences in risks associated with different standards can result (National Research Council 1998).

## Importance of ALARA Objective to Consistent Regulation

In addition to identifying important reasons for apparent inconsistencies in various radiation standards for the public, as described above, the National Research Council (1998) considered the important issue of the relationship between standards and doses (risks) that are experienced by exposed individuals and populations. These considerations are summarized as follows.

Radiation standards for the public include quantitative criteria that define limits or goals on acceptable doses (risks) for exposure situations of concern. However, doses (risks) that are experienced by individuals and populations are not, in most cases, determined by those criteria. Rather, application of the ALARA objective usually is the most important factor in determining actual doses and risks, irrespective of the particular law or regulation under which a source of exposure is controlled. It is important to recognize that the ALARA objective defines a *process* for control of exposures and that the outcome of that process cannot be specified in advance in regulations. In most cases, limits or goals specified in standards, although they represent important statements of principle and define upper or lower bounds on dose (risk) in applying the ALARA objective, are relatively unimportant in determining actual doses (risks).

The importance of the ALARA objective in controlling exposures of the public is illustrated by the following examples. First, in operations of nuclear facilities that are regulated under the Atomic Energy Act, the average annual individual dose is only about 0.05% of the annual dose limit of 1 mSv (100 mrem) for all controlled sources combined in radiation protection standards for the public (see table 15.5), and doses to individuals who receive the highest exposures normally are no more than about 10% of the dose limit and often are substantially less (EPA 1989a, 1989b). This example shows that in the important case of operating nuclear facilities, doses (risks) to most members of the public are determined largely by vigorous application of the ALARA objective. The dose limit in radiation protection standards and even, in some cases, authorized limits for specific practices or sources at a fraction of the dose limit (e.g., EPA's uranium fuel-cycle standards in 40 CFR part 190 and standards for airborne emissions in 40 CFR part 61) are relatively unimportant in determining actual doses (risks).

Second, in considering acceptable risks at radioactively contaminated sites that are subject to remediation under CERCLA and its implementing regulations in EPA's 40 CFR part 300, considerable attention is given to the goal for limiting lifetime cancer risk of around $10^{-4}$ (EPA 1999b; Luftig and Weinstock 1997; Clay 1991). However, far less attention has been given to the outcome that negotiated cleanup levels at some sites, as incorporated in Records of Decision, had associated risks substantially above the goal of $10^{-4}$ (Baes and Marland 1989; EPA 1994c). Cleanup levels that are judged acceptable at any site are based on a decision process defined in CERCLA and 40 CFR part 300 that is similar to applications of the ALARA objective under the Atomic Energy Act. That is, acceptable risks in cleanup of contaminated sites may be determined primarily by site-specific applications of the ALARA objective rather than the risk goal specified in EPA regulations. This important conclusion also was reached by the NCRP (2004) in a study of similarities and differences between the EPA's approach to cleanup of radioactively contaminated sites under CERCLA, which is based on the chemical paradigm for risk management, and the NRC's approach under the Atomic Energy Act, which is based on the radiation paradigm. The NCRP study concluded that application of the ALARA objective was the critical factor in providing consistency to cleanup decisions in the two approaches to risk management.

Finally, EPA drinking water standards for radionuclides (40 CFR part 141), developed under the Safe Drinking Water Act, are important because they apply to drinking water systems that are used by most of the U.S. population. In addition, they generally are applied to protection of groundwater resources at new waste disposal sites and at contaminated sites that undergo remediation under other laws. Drinking water standards specify limits on concentrations of radionuclides, but those limits were based originally on judgments about levels of radionuclides that were reasonably achievable, given existing levels in sources of drinking water and the effectiveness of available methods of water treatment, rather than on an a priori judgment about acceptable risks from radionuclides in drinking water. That is, even though drinking water standards are now considered to define minimally acceptable health risks from radionuclides in drinking water, they were based on application of the ALARA objective.

These considerations indicate that all standards for controlling radiation exposures of the public developed under any laws have as their unifying principle the objective that exposures from any practice or source should be ALARA. To the extent that the ALARA objective is applied consistently to all practices and sources, those standards will be consistent with regard to the doses (risks) actually experienced and risk harmonization will be achieved, irrespective of risks associated with quantitative criteria in the standards. In judging whether the ALARA objective has been applied reasonably consistently, it is important to appreciate that doses (risks) that are ALARA can vary greatly depending on the particular practice or source. It is not the goal of ALARA to achieve similar risks for all practices or sources.

## Exemption Levels for Radionuclides in the Environment

Preceding discussions in this chapter are concerned with standards for limiting radiation exposures of the public on the basis of a general concern about limiting health risks. This section discusses recommendations, regulations, and guidance on establishing exemption levels for radionuclides in the environment. In contrast to standards discussed above, discussions in this section address the concern that there are levels of radiation exposure so low that they need not be regulated (controlled), either because doses and risks are trivial or because the costs of reducing doses and risks far outweigh the associated benefits in improved health protection. This section describes general concepts for exemption of practices or sources and reviews recommendations developed by national and international advisory organizations and regulations and guidance established by the NRC.

### Concepts of Exemption

Two important general concepts are used in establishing exemption levels for radionuclides in the environment. The first is referred to as a de minimis level, and the second is referred to as an exempt level or level that is below regulatory concern (BRC).

### De Minimis Level

A de minimis level, which usually is expressed in terms of an annual dose, is generally applicable to any practice or source, and it defines a level below which efforts by regulatory authorities to control exposures generally are unwarranted. A de minimis dose is based primarily on a judgment that the associated risk is negligible (trivial), largely without regard for whether such a dose is reasonably achievable. If all individual doses were below the de minimis level, no further reductions in exposure based on application of the ALARA objective would be attempted. It should be emphasized, however, that achieving a de minimis dose is not the goal of ALARA, and that doses that are ALARA for specific exposure situations can be well above a de minimis level. The concept of a generally applicable de minimis dose is illustrated, for example, by the dashed line at the bottom of figure 15.1.

### Exempt or Below Regulatory Concern Level

An exempt or BRC level differs from the concept of a de minimis level in that (1) it applies only to a specific practice or source, rather than any practice or source and

(2) it is based on ALARA considerations rather than considerations of a negligible (trivial) risk for any exposure situation. That is, it is a level, which may be expressed in terms of a quantity of radioactive material (e.g., total activity or concentrations of radionuclides) or annual dose, that is judged by regulatory authorities to be ALARA for a specific practice or source (e.g., waste disposal) at any site.

An exempt or BRC level may vary greatly depending on the particular practice or source. Furthermore, an exempt or BRC level may be substantially higher than levels that correspond to a generally applicable de minimis dose.

## *Recommendations of Advisory Organizations*

The NCRP and IAEA have developed recommendations on exemption of practices or sources from regulatory control. These recommendations essentially are concerned with defining and using a generally applicable de minimis dose.

### Recommendations of the NCRP

Current NCRP recommendations on radiation protection of the public (NCRP 1993) include a recommendation that an annual effective dose to an individual from any practice or source of 0.01 mSv (1 mrem) or less is negligible. This recommendation was based on considerations of the magnitude of the dose, the difficulty in detecting and measuring dose and associated health effects at this level, and risks associated with the mean and variance of natural background radiation. In essence, the NCRP regards individual doses from any practice or source at a small fraction of the largely unavoidable dose from natural background as negligible. For continuous exposure over a 70-year lifetime, an annual effective dose of 0.01 mSv (1 mrem) corresponds to a lifetime risk of fatal cancers of about $4 \times 10^{-5}$.

### Recommendations of the IAEA

The IAEA has developed recommendations on exemption principles that could be applied to any practice or source (IAEA 1988, 1996). Individual practices or sources could be exempted from regulatory control under the following conditions:

The annual effective dose to individuals receiving the highest exposures would be on the order of 0.01 mSv (1 mrem) or less *and*

Either the collective effective dose from one year of an unregulated practice would be no more than about 1 person-Sv (100 person-rem), or an assessment of optimization of protection indicates that exemption is the optimum option

The IAEA used these criteria to derive exempt amounts of radionuclides for purposes of use and disposal; these exemption levels are given in the basic safety standards (IAEA 1996). Exemption levels for naturally occurring radionuclides apply only to their use in consumer products, as radioactive sources, or for their elemental properties.

## Exemptions Established by the U.S. Nuclear Regulatory Commission

The NRC has established many exemptions for radioactive material in several of its regulations and in other guidance. These exemptions represent applications of the concept of an exempt or BRC level for specific practices or sources, rather than a generally applicable de minimis dose. The NRC also attempted to establish more general exemption principles for several practices and sources that would be expressed in terms of annual individual and collective (population) doses that are BRC (NRC 1990), but this effort did not succeed (NRC 1993).

### Exemptions in U.S. Nuclear Regulatory Commission Regulations

The NRC has established many exemptions for radioactive material in its regulations, as summarized below:

Radiation protection standards for the public in 10 CFR part 20 specify limits on amounts of radionuclides in waste materials that are exempt from further regulatory control, including (1) concentrations or annual releases of radionuclides for discharge into sanitary sewer systems, with any excreta from individuals undergoing medical treatment with radioactive material exempt without limits, and (2) concentrations of 2 kBq $g^{-1}$ (0.05 μCi $g^{-1}$) for land disposal of scintillation materials and animal carcasses containing $^3$H or $^{14}$C.

Regulations in 10 CFR part 30 include (1) exempt quantities and concentrations of byproduct material and (2) exemptions for many specific items, mostly consumer products, containing small quantities of byproduct material.

Regulations in 10 CFR part 40 include exemptions for specific consumer products and other products or materials containing small quantities of source material.

Regulations in 10 CFR part 71 include (1) an exemption from all requirements for packaging and transportation of any materials containing concentrations of radionuclides less than 74 Bq $g^{-1}$ (2 nCi $g^{-1}$) and (2) exemption from certain requirements for packages containing fissile material or americium and plutonium only.

All exemptions in NRC regulations were based on considerations of associated doses to the public, including doses to workers at unlicensed facilities, and costs and benefits of regulating practices or sources of concern. In all cases, the NRC judged that the exemptions would have an insignificant adverse impact on public health and safety. These exemptions are not expressed in terms of dose, however, and they are not easily related to doses to the public. Furthermore, individual doses associated with these exemptions are highly variable and often are well above the de minimis level of 0.01 mSv (1 mrem) per year recommended by the NCRP (1993) and the IAEA (1988, 1996) (Schneider et al. 2001).

## U.S. Nuclear Regulatory Commission Guidance on Disposal of Thorium or Uranium

In 1981, the NRC issued guidance on disposal of residual thorium or uranium from past processing operations (NRC 1981). This guidance includes the following concentration limits for disposal of these materials with no restrictions on burial method:

0.4 Bq g$^{-1}$ (10 pCi g$^{-1}$) of natural thorium or uranium with their decay products present and in activity equilibrium
1.3 Bq g$^{-1}$ (35 pCi g$^{-1}$) of depleted uranium
1.1 Bq g$^{-1}$ (30 pCi g$^{-1}$) of enriched uranium

The concentration limit for unrestricted disposal of natural thorium or uranium was based on the soil cleanup standard for $^{226}$Ra at specified depths of 0.2 or 0.6 Bq g$^{-1}$ (5 or 15 pCi g$^{-1}$) in EPA standards for mill tailings (40 CFR part 192, subpart B). Concentration limits for depleted and enriched uranium were based on a limit on annual alpha absorbed dose of 0.01 mGy (1 mrad) to the lungs or 0.03 mGy (3 mrad) to bone, as given in proposed guidance on dose limits for exposure to transuranium elements in the environment (EPA 1977), discussed above, and an analysis of exposure pathways from unrestricted disposal of these materials.

The 1981 NRC guidance on unrestricted disposal of thorium or uranium has been used mainly to provide criteria for remediation of contaminated sites to permit unrestricted use by the public. NRC standards for decontamination and decommissioning of licensed nuclear facilities in 10 CFR part 20, subpart E, discussed above, specify that the 1981 guidance may be used to define acceptable cleanup levels of thorium or uranium. This guidance also has been used to define unrestricted disposals of waste containing uranium at DOE sites (see, e.g., Kocher and O'Donnell 1987; Lee et al. 1995, 1996).

Annual doses to individuals that correspond to concentration limits in the NRC guidance on disposal of thorium or uranium are well above the de minimis level of 0.01 mSv (1 mrem) recommended by the NCRP (1993) and the IAEA (1988, 1996). However, the guidance applies only to naturally occurring radionuclides, for which the dose from natural background also is well above the de minimis level (NCRP 1987b) (see table 15.5). A de minimis dose of 0.01 mSv (1 mrem) per year often would apply only to man-made radionuclides (NCRP 1987a).

## Protective Action Guides for Accidents

Preceding discussions in this chapter are concerned with standards, including exemption levels, that apply to routine exposures of the public. This section discusses protective action guides (PAGs) for undertaking responses to radiation accidents. PAGs are distinct from radiation protection standards for the public, which do not apply to accidents. Enforceable limits on dose (or risk) are inappropriate for accidents, which are uncontrolled situations. Rather, it is more appropriate to design engineered safety systems to reduce the probability of accidents and their severity and to develop plans for responding to accidents in the unlikely event they occur.

This section describes the concept of a PAG and the different time phases after an accident for defining PAGs. It then reviews PAGs developed by the EPA and U.S. Food and Drug Administration (FDA), proposed PAGs developed by the U.S. Department of Homeland Security for responding to incidents involving radiological dispersal devices (RDDs) or improvised nuclear devices (INDs), reactor siting criteria in NRC regulations, and recommendations of the ICRP and IAEA.

## *Purpose and Scope of Protective Action Guides*

The purpose of a PAG is to prevent the occurrence of unacceptable exposures of the public in the event of an accident. PAGs define *projected* doses or measured levels of radioactivity in environmental media (e.g., animal feed and human foods) following an accident above which countermeasures to avoid exposures of the public are warranted. Practical countermeasures include sheltering, evacuation, administration of stable iodine to block uptake of $^{131}$I by the thyroid, control of foodstuffs and water, control of access to contaminated areas, decontamination of persons and areas, and relocation of populations.

PAGs normally are recommendations, rather than requirements, because the appropriate response often depends on the nature of an accident and demographic and environmental conditions during an accident. PAGs do not define limits on deliberate releases, even though some PAGs are consistent with radiation protection standards for the public.

## *Time Phases for Defining Protective Actions*

PAGs normally are defined for three time phases of an accident, often referred to as the early phase, intermediate phase, and recovery phase, as described below.

1. Early phase: time from hours to days after initiation or occurrence of an accident, when immediate decisions about responses (e.g., sheltering, evacuation, and administration of stable iodine) are required. These decisions often are based mainly on predictions of meteorological and other conditions at the site of an accident, early knowledge of the nature of an accident, and initial projections of dose.
2. Intermediate phase: time from weeks to months after an accident is brought under control, when environmental measurements can be used to assess the need for additional protective actions (e.g., control of foodstuffs and drinking water, control of access to contaminated areas, and relocation of populations).
3. Recovery phase: time from months to years, when actions to reduce levels of radionuclides in the environment to allow permanent residence at the site under normal conditions may be undertaken (e.g., decontamination of areas).

### Protective Action Guides Established by Federal Agencies

PAGs for responding to nuclear accidents have been established by the EPA, and the FDA has established PAGs for responding to accidental contamination of foods and animal feed. These PAGs are intended to be used by state and local authorities in responding to accidents.

### Recommendations of the U.S. Environmental Protection Agency

In 1992, the EPA issued its current recommendations on PAGs for nuclear incidents, except nuclear war (EPA 1992). These recommendations are intended to apply to (1) accidents at nuclear power plants, DOE or Department of Defense facilities, foreign reactors, and research facilities; (2) accidental contamination of materials at steel mills or scrap metal recycling facilities; and (3) accidents in transport of radioactive materials. PAGs for the different time phases of an accident developed by the EPA are listed in table 15.21. Except as noted, PAGs are intended to limit cancer risks. Recommended dose limits for emergency workers provide a graded approach to protection.

### Recommendations of the U.S. Food and Drug Administration

EPA recommendations on accidents described above also include PAGs for radionuclides in food and animal feed that were established by the FDA in 1982 (47 FR 47073). The FDA has revised its recommendations, and they now apply to contamination of tap water used for drinking as well (FDA 1998). The FDA's revised PAGs were based on values developed by international consensus, especially recommendations of the ICRP (1984).

Current FDA recommendations on PAGs for foods, including tap water used for drinking, are listed in table 15.22. PAGs expressed in terms of dose are the same as recommendations of the ICRP (1984) on lower bounds of doses for which countermeasures are likely to be warranted. The PAG expressed in terms of the effective dose equivalent is intended to limit the risk of stochastic effects, primarily cancers, and the PAG expressed in terms of the dose equivalent to any organ or tissue is intended to prevent deterministic effects. In any accident situation, the more limiting of the two doses should be applied.

Current FDA recommendations in table 15.22 also include derived intervention levels (DILs), which are measurable quantities of radionuclides in food and tap water that are intended to correspond to the PAGs. DILs were developed for groups of radionuclides that are expected to be the most important in nuclear accidents; they were derived using age- and gender-specific intake rates of foods and tap water and age-specific dose coefficients for ingestion of radionuclides given in ICRP Publication 56 (ICRP 1989). If there is concern that food will continue to be contaminated significantly beyond the first year after an accident, the long-term circumstances need to be evaluated to determine whether DILs should be continued or if other guidance may be more appropriate.

Table 15.21  U.S. Environmental Protection Agency recommendations on protective action guides for responses to nuclear accidents (EPA 1992)

*Early time phase of hours to days after an accident*

Evacuation or sheltering at[a]

Effective dose equivalents above 10 mSv (1 rem)
Committed dose equivalents to thyroid above 50 mSv (5 rem)
Dose equivalents to skin above 500 mSv (50 rem)[b]

With state medical approval, administration of stable iodine at committed dose equivalents to thyroid above 250 mSv (25 rem)

Dose limits for workers that perform emergency services

50 mSv (5 rem) for all activities
100 mSv (10 rem) while protecting valuable property, provided lower dose is not practicable
250 mSv (25 rem) for life saving or protection of large populations, provided lower dose is not practicable
>250 mSv (25 rem) for life saving or protection of large populations, but only on a voluntary basis for persons who are fully aware of risks involved

*Intermediate time phase during the first year after an accident[c]*

Relocation of the general population at

Effective dose equivalents above 20 mSv (2 rem)
Dose equivalents to skin from beta irradiation above 1,000 mSv (100 rem)[b]

Application of simple dose reduction techniques at effective dose equivalents below 20 mSv (2 rem)

*Recovery phase after the first year following an accident[d]*

Effective dose equivalents in any year not to exceed 5 mSv (0.5 rem)
Effective dose equivalents over 50 years, including the first and second years, not to exceed 50 mSv (5 rem), or an average of 1 mSv (0.1 rem) per year

---

[a] Under unusual circumstances, doses for initiating evacuation or sheltering may be increased by a factor of 5.
[b] Dose criterion is intended to prevent deterministic effects in skin.
[c] Recommendations for the intermediate time phase include PAGs for radionuclides in food established by the FDA in 1982 (47 FR 47073). However, the FDA has since issued new recommendations (see table 15.22). EPA recommendations for this time phase do not include PAGs for radionuclides in drinking water.
[d] PAGs for the recovery phase are longer-term objectives. For reactor accidents, the EPA expects that objectives will be met if a PAG of 20 mSv (2 rem) for the intermediate time phase is met during the first year.

Current FDA recommendations do not include limits on concentrations of radionuclides that should be permitted in animal feeds. The recommendations do include protective actions for animal feeds that are intended to reduce or prevent subsequent contamination of food.

## Proposed Recommendations of the U.S. Department of Homeland Security

In 2006 (71 *FR* 174), the U.S. Department of Homeland Security issued proposed PAGs for responding to incidents involving an RDD or an IND. The EPA's PAGs in table 15.21 do not apply to those situations.

Table 15.22  Current FDA recommendations on protective action guides for responding to radioactive contamination of human foods (FDA 1998)[a]

*Protective actions should be undertaken for*
  Committed effective dose equivalents above 5 mSv (0.5 rem)
  Committed dose equivalents to any organ or tissue above 50 mSv (5 rem)

*Derived intervention levels (DILs) for groups of radionuclides*[b]

| Radionuclides | Bq kg$^{-1}$ (pCi g$^{-1}$) | Age group[c] |
|---|---|---|
| $^{90}$Sr | 160 (4.3) | 15 years |
| $^{131}$I | 170 (4.6) | 1 year |
| $^{134}$Cs + $^{137}$Cs | 1,200 (32) | Adult |
| $^{238}$Pu + $^{239}$Pu + $^{241}$Am | 2 (0.054) | 3 months |
| $^{103}$Ru[d] | 6,800 (180) | 3 months |
| $^{106}$Ru[d] | 450 (12) | 3 months |

[a] Recommendations replace 1982 FDA recommendations that are included in the current EPA guidance (EPA 1992) (see table 15.21). Recommendations apply to tap water used for drinking as well as foods consumed by humans.
[b] DILs correspond to doses above which protective actions should be undertaken; they apply during the first year after an accident and to all components of a diet in foods prepared for consumption, including tap water used for drinking.
[c] Critical age group for which a DIL was derived.
[d] For $^{103}$Ru and $^{106}$Ru combined, individual concentrations divided by their respective DIL and then summed must be less than 1.

Proposed protective actions and associated PAGs in the different time phases after incidents involving an RDD or IND are summarized below.

- Early time phase:
  - Limiting exposures of emergency workers to an effective dose equivalent of 50 mSv (5 rem), or greater under exceptional circumstances when doses above that level may be unavoidable.
  - Sheltering of the public at projected effective dose equivalents above 10–50 mSv (1–5 rem), with sheltering normally beginning at 10 mSv (1 rem); sheltering may begin at lower doses if advantageous.
  - Evacuating the public at projected effective dose equivalents above 10–50 mSv (1–5 rem), with evacuation normally beginning at 10 mSv (1 rem).
  - Administrating potassium iodide in response to incidents involving radioiodine in accordance with guidance of FDA (2001).

- Intermediate time phase:
  - Limiting worker exposures to an annual effective dose equivalent of 50 mSv (5 rem).
  - Relocating the general public at projected effective dose equivalents above 20 mSv (2 rem) during the first year and 5 mSv (0.5 rem) during subsequent years.

- Interdiction of food or drinking water at projected annual effective dose equivalents above 5 mSv (0.5 rem).
- Late time phase:
  - Final cleanup actions based on optimization of exposures.

Most of the proposed PAGs are based on those developed previously by the EPA and FDA for use in other accident situations. Exceptions are the proposed PAGs for limiting worker exposures during the intermediate time phase and for final cleanup actions during the late time phase. The former is based on current federal guidance on radiation protection standards for workers developed by the EPA in 1987 (52 *FR* 2822). The latter is based on the consideration that conditions during the late time phase no longer present an emergency situation, but that an appropriate response is better viewed in terms of objectives of site restoration and cleanup. Given the broad range of long-term impacts that may occur from RDDs or INDs, a numerical guideline to define responses during the late time phase is not recommended. Rather, a process of optimizing final cleanup decisions on the basis of case-specific conditions is appropriate; such an approach is consistent with ICRP recommendations discussed below.

The process of optimizing final cleanup actions in the late time phase in the event of an incident involving an RDD or IND could be similar to that used in cleanup of contaminated sites under CERCLA. However, the goal of achieving a lifetime cancer risk of around $10^{-4}$ at CERCLA sites is not likely to be an appropriate starting point for reaching final cleanup decisions in the event of a serious incident involving an RDD or IND. Legal frameworks that could be applied to such cleanups, concerns raised by their application, and options for addressing those concerns are discussed in Elcock et al. (2004).

### *U.S. Nuclear Regulatory Commission's Reactor Siting Criteria*

The NRC has established criteria for siting of nuclear power reactors in 10 CFR part 100. In evaluating proposed reactor sites, the extent of an exclusion area and a low population zone are determined on the basis of projected doses to individuals that result from a postulated severe accident. The exclusion area is defined in terms of the projected dose at any point on the boundary of the area during a period of two hours immediately following the postulated accidental release. The low population zone is defined in terms of the projected dose at any point on the outer boundary during the entire period of passage of the radioactive cloud.

Reactor siting criteria that apply to site applications after January 10, 1997, were established in 1996 (61 *FR* 65157). Boundaries of the exclusion area and low-population zone are defined by a maximum projected effective dose equivalent of 250 mSv (25 rem). This dose does not define limits on acceptable dose to the public in the event of a nuclear accident, nor does it define an intervention level. The NRC uses this dose only for the purpose of evaluating design features of nuclear power plants with respect to postulated reactor accidents to assure that risks to the public in the event of such accidents would be low.

## ICRP Recommendations on Responses to Accidents

The ICRP's first comprehensive recommendations on responses to radiation accidents were issued in 1984 (ICRP 1984). As noted in the preceding section, those recommendations were used in developing the FDA's current PAGs for responding to accidental contamination of foods.

Current ICRP recommendations on responses to radiation accidents were issued in 1993 (ICRP 1993b). Recommendations on intervention levels during the early and intermediate time phases of an accident include (1) dose levels above which intervention is almost always justified and (2) ranges below the justified intervention levels within which optimized intervention levels are expected to be found.

ICRP recommendations on countermeasures during the early time phase include the following:

- Justified intervention levels of:

  - 50 mSv (5 rem) effective dose for sheltering
  - 500 mSv (50 rem) equivalent dose to the thyroid for administration of stable iodine
  - 500 mSv (50 rem) effective dose or 5,000 mSv (500 rem) equivalent dose to skin for evacuation during the first week

- Range of optimized intervention levels that should be not more than a factor of 10 lower than justified intervention levels.

ICRP recommendations on countermeasures during the intermediate time phase include the following:

- For relocation:

  - Justified intervention level of 1,000 mSv (100 rem) effective dose
  - Range of optimized intervention levels of 5–15 mSv (0.5–1.5 rem) effective dose per month for prolonged exposure

- For restriction of a single foodstuff:

  - Justified intervention level of 10 mSv (1 rem) effective dose in a year
  - Range of optimized intervention levels, expressed as radionuclide concentrations in foodstuffs, of 1,000–10,000 Bq kg$^{-1}$ (27–270 pCi g$^{-1}$) for beta- or gamma-emitting radionuclides and 10–100 Bq kg$^{-1}$ (0.27–2.7 pCi g$^{-1}$) for alpha-emitting radionuclides

The ICRP (1993b) also discusses available countermeasures during the long-term recovery phase and optimization of the dose level to permit withdrawal of those countermeasures on the basis of cost–benefit analysis. Recommendations on dose levels that would warrant intervention during this time phase are not given.

## IAEA Guidelines for Intervention Levels in Emergency Exposure Situations

The IAEA's basic safety standards (IAEA 1996) include guidelines for intervention levels in emergency exposure situations. These guidelines are intended to provide starting points for judgments that are required in selecting the most appropriate intervention levels for particular emergency exposure situations. Higher or lower intervention levels may be appropriate in some situations. The IAEA's recommended intervention levels are summarized as follows:

- Dose levels to the most exposed individuals (i.e., critical groups) at which intervention is expected to be undertaken under any circumstances:
  - For acute exposure, projected absorbed dose in less than two days of 1 Gy (100 rad) to whole body and bone marrow, 6 Gy (600 rad) to lungs, 3 Gy (300 rad) to gonads and skin, 5 Gy (500 rad) to thyroid, or 2 Gy (200 rad) to lens of the eye; the possibility of deterministic effects at doses greater than about 0.1 Gy (10 rad) delivered over less than two days to the fetus should be considered in the justification and optimization of actual intervention levels for immediate protective action.
  - For chronic exposure, annual equivalent dose of 0.2 Sv (20 rem) to gonads, 0.1 Sv (10 rem) to lens of the eye, or 0.4 Sv (40 rem) to bone marrow.

- Generic optimized intervention levels of doses averaged over suitably chosen samples of the population at which urgent protective actions are warranted:
  - For sheltering, 10 mSv (1 rem) of avertable effective dose in a period of no more than two days
  - For temporary evacuation, 50 mSv (5 rem) of avertable effective dose in a period of no more than one week
  - For administration of stable iodine (iodine prophylaxis), 100 mGy (10 rad) of avertable committed absorbed dose to thyroid due to radioiodine

- Generic action levels for contaminated foodstuffs:
  - In foods destined for general consumption, 100 or 1,000 Bq kg$^{-1}$ (2.7 or 27 pCi g$^{-1}$) for beta- or gamma-emitting radionuclides and 10 Bq kg$^{-1}$ (0.27 pCi g$^{-1}$) for alpha-emitting radionuclides
  - In milk, infant foods, and drinking water, 100 or 1,000 Bq kg$^{-1}$ (2.7 or 27 pCi g$^{-1}$) for beta- or gamma-emitting radionuclides and 1 Bq kg$^{-1}$ (0.027 pCi g$^{-1}$) for alpha-emitting radionuclides
  - For administration of stable iodine (iodine prophylaxis), 100 mGy (10 rad) of avertable committed absorbed dose to thyroid due to radioiodine

- Generic optimized intervention levels of doses averaged over suitably chosen samples of the population at which temporary relocation or permanent resettlement are warranted:
  - For initiating and terminating temporary relocation, effective doses of 30 mSv (3 rem) in a month and 10 mSv (1 rem) in a month, respectively; if

effective dose accumulated in a month is not expected to fall below 10 mSv (1 rem) within a year or two, permanent resettlement should be considered
- Permanent resettlement should also be considered if a lifetime effective dose is projected to exceed 1 Sv (100 rem)

## Conclusions

This chapter presents information on current and proposed standards, guidance, and recommendations for controlling routine and accidental radiation exposures of the public. These discussions indicate that radiation protection of the public in the United States has become rather complex during the last few decades. This complexity has resulted from such factors as an increased public concern about health risks from exposure to radionuclides and other hazardous substances in the environment, the different responsibilities of various federal agencies and state governments for radiation protection of the public, and the several environmental laws under which radiation exposures of the public may be regulated.

Discussions in this chapter emphasize the bases for the many different standards for controlling exposures to radionuclides in the environment. Two important themes have emerged. First, standards for controlling exposures to radionuclides in the environment have increasingly been influenced by standards that apply to other hazardous substances. Given this influence, it is important to understand the fundamental difference in approaches to health protection of the public that are embodied in standards for radionuclides developed under the Atomic Energy Act (the radiation paradigm) and standards for radionuclides and other hazardous substances developed under other environmental laws (the chemical paradigm). It is especially important to understand the difference between a limit, as embodied in standards developed under the Atomic Energy Act, and a goal, as embodied in many standards developed under other environmental laws.

The second important theme is that application of the principle that exposures should be maintained ALARA (optimized) is the single most important factor in determining health risks to the public for any controlled sources of exposure to radionuclides or hazardous chemicals, essentially without regard for limits or goals on dose or risk that are specified in laws or regulations. The key to harmonizing public health protection standards for radionuclides and hazardous chemicals lies in reasonably consistent application of the ALARA principle to all exposure situations, rather than imposing consistent limits or goals on dose or risk in all situations.

### Notes

1. For each regulation published in the Code of Federal Regulations (CFR), reference to the *Federal Register* (*FR*) notice containing the promulgated regulation is given, in the form "[volume] *FR* [page]." These notices provide important supplementary information on the basis for regulations.

2. See table 15.10, footnote b, for definition of working level (WL).

3. When RCRA was promulgated, this exclusion applied only to radioactive materials associated with operations of the nuclear fuel cycle. Now that the definition of byproduct material in the Atomic Energy Act has been amended to include some NARM, wastes containing those materials presumably are now excluded from regulation under RCRA. The exclusion, however, does not apply to diffuse NORM.

4. An exposure of 1 working level month (WLM) is a cumulative exposure to short-lived radon decay products equivalent to an exposure to 1 WL for a working month (170 h).

References

Baes, C.F., III, and G. Marland. 1989. *Evaluation of Cleanup Levels for Remedial Action at CERCLA Sites Based on a Review of EPA Records of Decision.* ORNL-6479. Oak Ridge National Laboratory, Oak Ridge, TN.

Brown, S.L. 1992. "Harmonizing Chemical and Radiation Risk Management." *Environmental Science and Technology* 26(12): 2336–2338.

Cameron, F.X. 1996. "The Odyssey of the Good Ship NORM: The Search for a Regulatory Safe Harbor." In *NORM/NARM: Regulation and Risk Assessment. Proceedings of the 29th Midyear Topical Meeting of the Health Physics Society*, ed. K.L. Mossman and K.B. Thiemann. Health Physics Society, McLean, VA; 13–18.

Clay, D.R. 1991. *Role of the Baseline Risk Assessment in Superfund Remedy Selection Decisions.* Directive 9355.0–30. Office of Solid Waste and Emergency Response (OSWER), U.S. Environmental Protection Agency, Washington, DC.

CRCPD (Conference of Radiation Control Program Directors, Inc.). 2005. *Suggested State Regulations for the Control of Radiation.* Available online at www.crcpd/org/free_docs.asp.

DOE (U.S. Department of Energy). 1988a. *Radioactive Waste Management.* DOE Order 5820.2A. U.S. Department of Energy, Washington, DC.

DOE. 1988b. *General Environmental Protection Program.* DOE Order 5400.1. U.S. Department of Energy, Washington, DC.

DOE. 1990. *Radiation Protection of the Public and the Environment.* DOE Order 5400.5. U.S. Department of Energy, Washington, DC.

DOE. 1993. "10 CFR Part 834—Radiation Protection of the Public and the Environment." Proposed Rule. *Federal Register* 58(56): 16268–16322.

DOE. 1995. "10 CFR Parts 830 and 834—Nuclear Safety Management and Radiation Protection of the Public and the Environment." Notice of Limited Reopening of Comment Periods. *Federal Register* 60(169): 45381–45385.

DOE. 1996. *Integrated Data Base Report—1995: U.S. Spent Nuclear Fuel and Radioactive Waste Inventories, Projections, and Characteristics.* DOE/RW-0006, Rev. 12. U.S. Department of Energy, Washington, DC.

DOE. 1999. *Radioactive Waste Management Manual.* DOE M 435.1–1. U.S. Department of Energy, Washington, DC.

Eckerman, K.F., A.B. Wolbarst, and A.C.B. Richardson. 1988. *Limiting Values of Radionuclide Intake and Air Concentration and Dose Conversion Factors for Inhalation, Submersion, and Ingestion.* Federal Guidance Report No. 11, EPA-520/1-88-020. U.S. Environmental Protection Agency and Oak Ridge National Laboratory, Washington, DC.

Elcock, D., G.A. Klemic, and A.L. Taboas. 2004. "Establishing Remediation Levels in Response to a Radiological Dispersal Event (or 'Dirty Bomb')." *Environmental Science and Technology* 38(9): 2502–2512.

EPA (U.S. Environmental Protection Agency). 1977. *Proposed Guidance on Dose Limits for Persons Exposed to Transuranium Elements in the General Environment.* EPA/520/4-77-016. U.S. Environmental Protection Agency, Washington, DC.

EPA. 1982. *Final Environmental Impact Statement for Remedial Action Standards for Inactive Uranium Processing Sites (40 CFR 192) Vol. 1.* EPA 520/4-82-013-1. U.S. Environmental Protection Agency, Washington, DC.

EPA. 1987. *Interim Recommendations on Doses to Persons Exposed to Transuranium Elements in the General Environment.* Draft Guidance. U.S. Environmental Protection Agency, Washington, DC.

EPA. 1989a. *Risk Assessment Methodology, Environmental Impact Statement for Proposed NESHAPs for Radionuclides, Vol. I.* EPA 520/1-89/005. U.S. Environmental Protection Agency, Washington, DC.

EPA. 1989b. "40 CFR Part 61—National Emission Standards for Hazardous Air Pollutants: Radionuclides." Final Rule. *Federal Register* 54(240): 51654–51715.

EPA. 1991a. *Protecting the Nation's Ground Water: EPA's Strategy for the 1990s.* The Final Report of the EPA Ground-Water Task Force. U.S. Environmental Protection Agency, Washington, DC.

EPA. 1991b. "40 CFR Parts 141, 142—National Primary Drinking Water Regulations: Radionuclides." Proposed Rule. *Federal Register* 56(138): 33050–33127.

EPA. 1992. *Manual of Protective Action Guides and Protective Actions for Nuclear Incidents.* EPA 400-R-92-001. U.S. Environmental Protection Agency, Washington, DC.

EPA. 1994a. "Federal Radiation Protection Guidance for Exposure of the General Public." Proposed Recommendations. *Federal Register* 59(246): 66414–66428.

EPA. 1994b. *Estimating Radiogenic Cancer Risks.* EPA 402-R-93-076. U.S. Environmental Protection Agency, Washington, DC.

EPA. 1994c. "40 CFR Part 196—Environmental Protection Agency Radiation Site Cleanup Regulation." Draft Proposed Rule. U.S. Environmental Protection Agency, Washington, DC. Available online at www.epa.gov/radiation/cleanup/html/replacmt.htm.

EPA. 1997. "40 CFR Parts 141 and 142—Withdrawal of Proposed National Primary Drinking Water Regulations for Radon-222." *Federal Register* 62(15): 42221–42222.

EPA. 1999a. "40 CFR Parts 141 and 142—National Primary Drinking Water Regulations: Radon-222." Notice of Proposed Rulemaking. *Federal Register* 64(211): 59246–59294.

EPA. 1999b. *Radiation Risk Assessment at CERCLA Sites: Q & A.* Directive 9200.4-31P. EPA/540/R/99/006. U.S. Environmental Protection Agency, Washington, DC.

EPA. 1999c. *Estimating Radiogenic Cancer Risks. Addendum: Uncertainty Analysis.* EPA 402-R-99-003. U.S. Environmental Protection Agency, Washington, DC.

EPA. 2000a. "40 CFR Parts 141 and 142—National Primary Drinking Water Regulations: Radionuclides. Notice of Data Availability." *Federal Register* 65(78): 21575–21628.

EPA. 2000b. "40 CFR Parts 9, 141, and 142—National Primary Drinking Water Regulations: Radionuclides." Final Rule. *Federal Register* 65(236): 76708–76753.

EPA. 2005. "40 CFR Part 197—Public Health and Environmental Radiation Protection Standards for Yucca Mountain, Nevada." Proposed Rule. *Federal Register* 70(161): 49014–49065.

EPA and DHHS (U.S. Environmental Protection Agency and U.S. Department of Health and Human Services). 1986. *A Citizen's Guide to Radon.* OPA-86-004. U.S. Government Printing Office, Washington, DC.

EPA and DHHS. 1994. *A Citizen's Guide to Radon.* EPA ANR-464, DHHS 402K92-001.

EPA/NRC (U.S. Environmental Protection Agency/U.S. Nuclear Regulatory Commission). 2002. "Memorandum of Understanding between the Environmental Protection Agency and the Nuclear Regulatory Commission. Consultation and Finality on Decommissioning and Decontamination of Contaminated Sites." OSWER No. 9295.8–06A. U.S. Environmental Protection Agency, Washington, DC.

EPA-SAB (U.S. Environmental Protection Agency–Science Advisory Board). 1992. *Commentary on Harmonizing Chemical and Radiation Risk Reduction Strategies.* EPA-SAB-RAC-COM-92–007. U.S. Environmental Protection Agency, Washington, DC.

FDA (U.S. Food and Drug Administration). 1998. *Accidental Radioactive Contamination of Human Food and Animal Feeds: Recommendations for State and Local Agencies.* U.S. Department of Health and Human Services, Rockville, MD.

FDA. 2001. *Potassium Iodide as a Thyroid Blocking Agent in Radiation Emergencies.* U.S. Department of Health and Human Services, Rockville, MD.

FRC (Federal Radiation Council). 1960. "Radiation Protection Guidance for Federal Agencies." *Federal Register* 25(97): 4402–4403.

FRC. 1961. "Radiation Protection Guidance for Federal Agencies." *Federal Register* 26(185): 9057–9058.

GAO (U.S. General Accounting Office). 1994. *Nuclear Health and Safety: Consensus on Acceptable Radiation Risk to the Public Is Lacking.* GAO/RCED-94–190. General Accounting Office, Washington, DC.

Harley, N.H. 1996. "Radon: Over- or Under-Regulated?" In *NORM/NARM: Regulation and Risk Assessment. Proceedings of the 29th Midyear Topical Meeting of the Health Physics Society*, ed. K.L. Mossman and K.B. Thiemann. Health Physics Society, McLean, VA; 39–45.

IAEA (International Atomic Energy Agency). 1988. *Principles for the Exemption of Radiation Sources and Practices from Regulatory Control.* Safety Series No. 89. International Atomic Energy Agency, Vienna, Austria.

IAEA. 1996. *International Basic Safety Standards for Protection Against Ionizing Radiation and for the Safety of Radiation Sources.* Safety Series No. 115-F. International Atomic Energy Agency, Vienna, Austria.

ICRP (International Commission on Radiological Protection). 1959. *Recommendations of the International Commission on Radiological Protection. Report of Committee II on Permissible Dose for Internal Radiation.* ICRP Publication 2. Pergamon Press, Oxford.

ICRP. 1977. "Recommendations of the International Commission on Radiological Protection." ICRP Publication 26. *Annals of the ICRP* 1(3). Pergamon Press, Oxford.

ICRP. 1979. "Limits for Intakes of Radionuclides by Workers." ICRP Publication 30, Part 1. *Annals of the ICRP* 2(3/4). Pergamon Press, Oxford.

ICRP. 1984. "Protection of the Public in the Event of Major Radiation Accidents: Principles for Planning." ICRP Publication 40. *Annals of the ICRP* 14(2). Pergamon Press, Oxford.

ICRP. 1986. "The Metabolism of Plutonium and Related Elements." ICRP Publication 48. *Annals of the ICRP* 16(2/3). Pergamon Press, Oxford.

ICRP. 1987a. "Data for Use in Protection against External Radiation." ICRP Publication 51. *Annals of the ICRP* 17(2/3). Pergamon Press, Oxford.

ICRP. 1987b. "Lung Cancer Risk from Indoor Exposures to Radon Daughters." ICRP Publication 50. *Annals of the ICRP* 17(1). Pergamon Press, Oxford.

ICRP. 1989. "Age-Dependent Doses to Members of the Public from Intake of Radionuclides." ICRP Publication 56. *Annals of the ICRP* 20(2). Pergamon Press, Oxford.

ICRP. 1991. "1990 Recommendations of the International Commission on Radiological Protection." ICRP Publication 60. *Annals of the ICRP* 21(1–3). Pergamon Press, Oxford.

ICRP. 1993a. "Protection against Radon-222 at Home and at Work." ICRP Publication 65. *Annals of the ICRP* 23(2). Pergamon Press, Oxford.

ICRP. 1993b. "Principles for Intervention for Protection of the Public in a Radiological Emergency." ICRP Publication 63. *Annals of the ICRP* 22(4). Pergamon Press, Oxford.

ICRP. 1999. "Principles for the Protection of the Public in Situations of Prolonged Exposure." ICRP Publication 82. *Annals of the ICRP* 29(1/2). Pergamon Press, Oxford.

Kocher, D.C. 1988. "Review of Radiation Protection and Environmental Radiation Standards for the Public." *Nuclear Safety* 29(4): 463–475.

Kocher, D.C. 1999. "Regulation of Public Exposures to Radionuclides and Hazardous Chemicals—Seeking Common Ground." *Radiation Protection Management* 16(1): 23–50.

Kocher, D.C., and F.O. Hoffman. 1991. "Regulating Environmental Carcinogens: Where Do We Draw the Line?" *Environmental Science and Technology* 25(12): 1986–1989.

Kocher, D.C., and F.O. Hoffman. 1992. "Reply to Weisberger Regarding 'Regulating Environmental Carcinogens.'" *Environmental Science and Technology* 26(5): 845–846.

Kocher, D.C., and F.R. O'Donnell. 1987. *Considerations on a De Minimis Dose and Disposal of Exempt Concentrations of Radioactive Wastes*. ORNL/TM-10388. Oak Ridge National Laboratory, Oak Ridge, TN.

Lee, D.W., J.C. Wang, and D.C. Kocher. 1995. *Operating Limit Study for the Proposed Solid Waste Landfill at Paducah Gaseous Diffusion Plant*. ORNL/TM-13008. Oak Ridge National Laboratory, Oak Ridge, TN.

Lee, D.W., D.C. Kocher, and J.C. Wang. 1996. "Operating Limit Evaluation for Disposal of Uranium Enrichment Plant Wastes." In *NORM/NARM: Regulation and Risk Assessment. Proceedings of the 29th Midyear Topical Meeting of the Health Physics Society*, ed. K.L. Mossman and K.B. Thiemann. Health Physics Society, McLean, VA; 91–98.

Luftig, S.D., and L. Weinstock. 1997. *Establishment of Cleanup Levels for CERCLA Sites with Radioactive Contamination*. Directive 9200.4–18. Office of Solid Waste and Emergency Response (OSWER), U.S. Environmental Protection Agency, Washington, DC.

National Bureau of Standards. 1963. *Maximum Permissible Body Burdens and Maximum Permissible Concentrations of Radionuclides in Air and Water for Occupational Exposure*. NBS Handbook 69. U.S. Department of Commerce, Washington, DC.

National Research Council. 1990. *Health Effects of Exposure to Low Levels of Ionizing Radiation. BEIR V*. National Academy Press, Washington, DC.

National Research Council. 1995. *Technical Bases for Yucca Mountain Standards*. National Academy Press, Washington, DC.

National Research Council. 1998. *Evaluation of Guidelines for Exposures to Technologically Enhanced Naturally Occurring Radioactive Materials*. National Academy Press, Washington, DC.

National Research Council. 1999. *Risk Assessment of Radon in Drinking Water*. National Academy Press, Washington, DC.

National Research Council. 2006. *Health Risks from Exposure to Low Levels of Ionizing Radiation*. National Academies Press, Washington, DC.

NCRP (National Council on Radiation Protection and Measurements). 1984a. "Control of Air Emissions of Radionuclides." Statement No. 6. National Council on Radiation Protection and Measurements, Bethesda, MD.

NCRP. 1984b. *Exposures from the Uranium Series with Emphasis on Radon and Its Daughters*. NCRP Report No. 77. National Council on Radiation Protection and Measurements, Bethesda, MD.

NCRP. 1987a. *Recommendations on Limits for Exposure to Ionizing Radiation*. NCRP Report No. 91. National Council on Radiation Protection and Measurements, Bethesda, MD.

NCRP. 1987b. *Exposure of the Population of the United States and Canada from Natural Background Radiation.* NCRP Report No. 94. National Council on Radiation Protection and Measurements, Bethesda, MD.

NCRP. 1993. *Limitation of Exposure to Ionizing Radiation.* NCRP Report No. 116. National Council on Radiation Protection and Measurements, Bethesda, MD.

NCRP. 1997. *Uncertainties in Fatal Cancer Risk Estimates Used in Radiation Protection.* NCRP Report No. 126. National Council on Radiation Protection and Measurements, Bethesda, MD.

NCRP. 2004. *Approaches to Risk Management in Remediation of Radioactively Contaminated Sites.* NCRP Report No. 146. National Council on Radiation Protection and Measurements, Bethesda, MD.

NRC (U.S. Nuclear Regulatory Commission). 1974. *Termination of Operating Licenses for Nuclear Reactors.* Regulatory Guide 1.86. U.S. Nuclear Regulatory Commission, Washington, DC.

NRC. 1981. "Disposal or Onsite Storage of Residual Thorium or Uranium (Either as Natural Ores or Without Daughters Present) from Past Operations." Branch Technical Position, SECY-81-576. U.S. Nuclear Regulatory Commission, Washington, DC.

NRC. 1982. *Guidelines for Decontamination of Facilities and Equipment Prior to Release for Unrestricted Use or Termination of Licenses for Byproduct, Source, or Special Nuclear Material.* U.S. Nuclear Regulatory Commission, Washington, DC.

NRC. 1990. "Below Regulatory Concern: Policy Statement." *Federal Register* 55(128): 27522–27537.

NRC. 1997. "10 CFR Part 2—Withdrawal of Below Regulatory Concern Policy Statements." *Federal Register* 58(162): 44610–44611.

NRC. 2000. *A Performance Assessment Methodology for Low-Level Radioactive Waste Disposal Facilities.* NUREG-1573. U.S. Nuclear Regulatory Commission, Washington, DC.

NRC. 2005. "10 CFR Part 63—Implementation of a Dose Standard after 10,000 Years." Proposed Rule. *Federal Register* 70(173): 53313–53320.

NRC. 2007. "10 CFR Parts 20, 30, 31, 32, 33, 35, 50, 61, 62, 72, 110, 150, 170, and 171—Requirements for Expanded Definition of Byproduct Material." Final Rule. *Federal Register* 72(189): 55864–55937.

Overy, D.P., and A.C.B. Richardson. 1995. "Regulation of Radiological and Chemical Carcinogens: Current Steps toward Risk Harmonization." *Environmental Law Reporter* 25(12): 10655–10708.

Schneider, S., D.C. Kocher, G.D. Kerr, P.A. Scofield, F.R. O'Donnell, C.R. Mattsen, S.J. Cotter, J.S. Bogard, and C. Wiblin. 2001. *Systematic Radiological Assessment of Exemptions for Source and Byproduct Materials.* NUREG-1717. U.S. Nuclear Regulatory Commission, Washington, DC.

Taylor, J.M. 1995. *U.S. Nuclear Regulatory Commission and U.S. Environmental Protection Agency Risk Harmonization Issues and Recommendations.* SECY-95-249. U.S. Nuclear Regulatory Commission, Washington DC.

Travis, C.C., S.A. Richter, E.A.C. Crouch, R. Wilson, and E.D. Klema. 1987. "Cancer Risk Management: A Review of 132 Federal Regulatory Decisions." *Environmental Science and Technology* 21(5): 415–420.

# Index

Boldface entries denote in-depth coverage of a topic.

absorbed fraction, 404, 420, 457–459
accident exposure regulations, **673–682**
activity mean thermodynamic diameter (AMTD), 438–439
activity median aerodynamic diameter (AMAD), 409–411, 416–417, 438–439
advisory organizations, and environmental regulations, **619**
AERMOD atmospheric model, **137–138**
age groups, dosimetry, **13**, **430–444**, **454**
airborne emissions of radionuclides, and environmental regulations, **660–663**
ALARA (as low as reasonably achievable), 390, 394, 396, 620–622, 624, 682
  and consistent regulation, **669–672**
  and specific applications, 629–632, 639, 641–643, 645–648, 656–657, 661–663
aleatory uncertainty, and assessment models, 469, 471, 511, 513–515, 524
AMAD (activity median aerodynamic diameter), 409–411, 416–417, 438–439
AMTD (activity mean thermodynamic diameter), 438–439

animals, in terrestrial food chain pathways, 267, **318–321**
annual limit on intake (ALI), and internal dosimetry, 401, 408, 430
antithetic sampling, and uncertainty, 523
applicable or relevant and appropriate requirements (ARARs), 642–645
aquatic boundary layer, 176–177
aquatic dispersion, 148–150, **182–183**, 190
  and transport models, **160–163**, 167, **168–170**, **178**, 179–180
aquatic food chain, 261–262, **340–371**
aquatic screening models, coastal, **179–181**
aquatic systems, and food products, 147, 187, 194, 200, **340–371**
aquatic transport models, 159, **160–163**, 167, **168–170**, **178**, 179–180
  and Chernobyl, 149, **166–167**, **192–200**
ARARs (applicable or relevant and appropriate requirements), 642–645
assessment models, and uncertainty, **465–525**
assessment question error, and model validation, **607–608**

atmospheric boundary layer (ABL), 81, 106, 112, 113, 115, 124
and turbulence, 82, 90, 93
atmospheric composition, **80–89**
atmospheric dispersion, 81, 89, 103
and transport models, 79–80, 82–83, 88, **108–139**
atmospheric stability classes, **116**
atmospheric transport, **79–139**
diffusion and deposition models, 79–80, 82–83, 88, **108–139**
and dry deposition, 128, **129–130**, 137
Gaussian models, 133–134
and model validation/selection, **132–137**
multicomponent models, **125–126**
atmospheric turbulence, **81–108**, 109, 112–113, **116**, 118–139, 292
seabreeze effects, 84–87
and suspension, 296–301
wind direction, **116**
Atomic Energy Act, 614–618, 630, 655, 658–660, 667–669, 682
averaging time, in atmospheric transport models, **116–118**

below regulatory concern (BRC), and regulatory exemptions, 671–672
benchmarking, and model validation, 524, 592
Bessel functions, and transport models, 157, 158
bias, and model validation, **596–598**
bioaccumulation
and aquatic food chain pathways, **344–371**
bioconcentration factor (BCF), 345–353
and cesium, 232, 252, 269, 330–331, 349, 353
biomass
animal, 267–268, 362
and Chernobyl, **356–360**, **368–369**
distributions, **360–362**, 365–368
PATHWAY model, 295
and terrestrial food chain pathways, 294–295
and vegetation, 266–268, 279, 294–295, 305, 308–309, 370–371
BIOPROTA, and model validation, 592
boating activities, and external dosimetry, **456**

Borel's theorem, and terrestrial food chain pathways, **285–286**
boundary layer, 103, 114, 137
aquatic, 176–177
atmospheric boundary layer (ABL), 81, 106, 112, 113, 115, 124
and turbulence, 82, 90, 93
thermal internal boundary layer (TIBL), 86–87
box models, *see* compartment models
buildings
radiation shielding effects, **455**, 462
wake effects, 87–89, **114–115**, 128, 138, 286, 300

calibration, and model validation, 226, **251–252**, **594–596**
CALMET atmospheric model, 138
CALPUFF atmospheric model, **138**, 292
cancer risk, 9–10, 139, **540–545**, 557, 614, 664, 666
and Chernobyl, 542, 543
and chronic radiation exposure, 664–666
and Clinch River, Tennessee, 352
coefficients, 14, 139, 422, **545–546**
in utero, **543–544**
Japanese A-bomb survivors, 14, 539–540, **541–542**, 543, 545
lifetime, 544–545, 619, 628, 633, 643–645, 667, 670
CAP88 atmospheric model, **138–139**, 270
case–control designs, and epidemiology, **565–566**
CDF, *see* cumulative distribution function
central limit theorem, and uncertainty, 477–478, 484
CERCLA (Comprehensive Environmental Response, Compensation, and Liability Act), 614–616, 622, 630, 664, 666, 667
consistency of standards, 670–671
and radiation contamination cleanup, 642–645
cesium
and bioaccumulation, 232, 252, 269, 330–331, 349, 353
and Chernobyl, 132, 153, 191–196, 354–371, 594
and Clinch River, Tennessee, 191, 348–352
nuclides, 33, 35, 36, 45, 49, 452

Index    691

protective regulations, 634, 678
releases, 152, 211, 230, 263, 264, 291
source terms, 37, 49, 51, 53–54, 60, 61
transport processes, 306, 308, 310, 315, 319–320, 322–323
CFR, *see* Code of Federal Regulations
chemical paradigm for risk management, **622**
chemical transport models, and groundwater, **213**
Chernobyl, 58, **200**, 594
  and aquatic transport, 149, **166–167**, **192–200**, **354–364**, 367, **368–371**
  and atmospheric transport, 84, 120
  and bioaccumulation, **354–364**, 367, **368–369**
  and cancer epidemiology, 542, 543
  and cesium, 132, 153, 191–196, 354–371, 594
  cooling pond analyses, **354–360**, 363, **366–371**
  and Dnieper River, 191–200
  and irrigation, **197–200**
  and phytoplankton, 361, 364–365, 368
  and population dynamics, **360–362**, 364–365, 369
  and Pripyat River, 166–167, 191, **192–198**, 200
  and radioiodine releases, 37, 132, 192, 194–195, 199, 594
  and reservoir contamination, **197–200**
  and surface deposition, 302, 318
  and wet deposition, 130, 132
  and zooplankton, **361–362**, 364–365, 369
chi-square distribution, 485, 569–570, 572–574, 682
clay, *see* soil
Clean Air Act, 26, 39, 139, 614–616, 630, 660–662
Clean Water Act, 614–616, 630, 636, 660
Clinch River, Tennessee, 342–343, 348–352, 352
Clinton Pile, Oak Ridge facility, 42
coastal waters, flow characteristics, **179–181**
Code of Federal Regulations
  Environmental Protection Agency
    40 CFR 61, airborne radionuclide emission standards, 660–661
    40 CFR 141, drinking water standards, 633, 635
    40 CFR 190, uranium fuel-cycle standards, 38, 56, 630–631, 656, 670
    40 CFR 191, high-level waste disposal standards, 644, 648–651, 653–657
    40 CFR 192, uranium mill tailings standards, 637, 639, 641, 647, 658, 665, 674
    40 CFR 197, Yucca Mountain high-level waste standards, 652, 653, 657
    40 CFR 264, groundwater protection, 659
    40 CFR 268, radioactive waste universal treatment standards (UTSs), 659
    40 CFR 440, liquid discharge standards, 636
  Nuclear Regulatory Agency
    10 CFR 20, radiation protection standards for the public, 394, 625–626, 640, 642, 657, 673
    10 CFR 40, uranium and thorium mill tailings standards, 639, 640, 643
    10 CFR 50, design objectives for nuclear power reactors, 631–632
    10 CFR 60, geologic repository radioactive waste disposal standards, 655
    10 CFR 61, licensing criteria for radioactive waste near-surface disposal, 656, 657
    10 CFR 100, nuclear power reactor siting criteria, 57, 679
cohort designs, and epidemiology, **563–565**, 566
colloids, and groundwater transport models, **232–233**
colon, *see* gastrointestinal tract
Columbia River, 10–11, 26, 42
committed dose equivalent, 581, 632, 677–678
  and internal dosimetry, 396–397, 399–401, **402–408**, 416–417, 421, 427
communication, risk, **21–24**, 497, 590
compartment models, 125–126, 148–149, 191, 411

**692** Index

compartment models (*continued*)
  multicompartment models, **125–126**, 272, **324–333**
  single-compartment models, 273, **273–287**, 324
  and terrestrial food chain pathways, 272, 273, **275–287**, 279, **324–333**
compensatory errors, and model validation, 591, 606, 608
Comprehensive Environmental Response, Compensation, and Liability Act, *see* CERCLA
Conference of Radiation Control Program Directors (CRCPD), 627, 628, 648
confidence intervals, in epidemiology, **570–580**
confounding, 514, 516, 519, 522, 573–577, 579–580
contaminated sites, cleanup standards, **642–645, 647–648**
continuous point and line sources, in atmospheric models, **110, 111–112**
continuous transport processes, and terrestrial food chain pathways, **288**
conversion to dose, **12–13**
convolution integral, and terrestrial food chain pathways, **284–285**
correlation and covariance, **231, 485–487**, 499, 500, **505, 599**
CRCPD (Conference of Radiation Control Program Directors), 627, 628, 646–648
cumulative distribution function (CDF), and uncertainty, 471–474, 491, 497, 500, 511, 516
currents, ocean, and radionuclide transport, 176–177

Darcy's Law, and groundwater flow, 217
daughter products, *see* radionuclides, decay products
decay products, *see* radionuclides, decay products
decay rate constant, and terrestrial food chain pathways, 287, 288
de minimis level exception, and environmental regulations, **671**
deposition, *see* dry deposition; wet deposition
deposition velocity ($V_d$, $V_w$, and $V_T$), 129–132, 139, 184, 279–280, **289–295**, 303

derived air concentration (DAC), and internal dosimetry, 401, 408
descriptive statistics, and assessment model uncertainty, **471–476**
deterministic approach, and surface water transport, 190
deterministic effects, and ionizing radiation exposure, 531–532, 624, 627, 648, 676, 681
deterministic models, and terrestrial food chain pathways, 271, 274
deterministic values, and uncertainty analysis, 4, 15–17
diffusion model, Gaussian, **109–120**, 121, 122–123, **222**
discharge, *see* effluent
discrete fracture transport models, and groundwater, 215
discrete processes, and terrestrial food chain pathways, **288**
disease–exposure association, and epidemiological assessment, **566–580**
dispersion
  aquatic, 148–150, 154–155, 159, 167, **178**, 190
  estuaries, **168–171**
  lakes, **182–183**
  rivers, **160–163**
  atmospheric, 37–38, 57, 81–83, 103
  modeling, 109, 114–116, 118, 127–128, 133–134, 137–138
dispersion coefficient
  and groundwater, 220, 221, 223, **224**, 236
  and transport models, **160–163, 168–174, 178**, **182–183**, 185
distribution coefficient ($K_d$), 151–152, 187–189, 229–231, 310–311, 315
DNA, 37, 532, 533–534, **535**, 536–539, 547–548
Dnieper River, and Chernobyl, 191–200
DOE, *see* U.S. Department of Energy
domain of applicability, in models, 468, 522, 590, 608
dose
  conversion to risk, **13–15**
  dose-rate effectiveness factor, **538**
dose coefficients
  age, **13**, 433, 434, **454**
  and electrons, 423, 453, 454, 461
  and external dosimetry, **448–461**

ingestion, 324, 331, 427–428, 430–432, 436, 437
inhalation, 441–442
and internal dosimetry, **402–408**, **437**
and photons, 404, 423, 450–461, 639
and uncertainty, 12, 15
dose equivalents, and radioiodine, 665, 677
dose–response curve, 542, 548, 561, 564, 566, **622–624**
dosimetry, 12–13
  external, 12–13, 389–390, **447–461**
  internal, 12–13, **389–444**
    age-dependent doses, **430–443**
    bone model, **419–421**
    committed dose equivalent, 396, 397, 399–401, **402–408**, 416–417, 421
    defined, 391–392
    dose conversion coefficients, 427, 431, 432, 434, 435, 442–443
    dose equivalents, **402–408**
    and effective dose ($E$), 424
    exposure regulations, **394–444**
    vs. external exposure, **389–394**
    and food products, 392
    gastrointestinal tract, 406–407, 417–419, 424–426, 433–434, 436–437, 442
    ICRP publications, **394–444**
    internal dose control, **392–394**
    and radioactive clouds, **421–422**
    recent recommendations, **422–444**
    source tissue, 402–403, 407
  and radionuclide decay products, 406, 432, 442, 448, 451–452
  and uncertainties in risk factors, 15
drinking water, environmental regulations, **631–635**
dry deposition, and terrestrial food chain pathways, 128, **129–130**, 137, **290–292**, 460
dual continuum transport models, and groundwater, 215
dynamic multicompartment models, and terrestrial food chain pathways, **324–332**
dynamic vs. steady-state models, and food chain transport, 272

ecologic studies, in epidemiology, 579
eddy diffusion coefficients, and Richardson number, **102–103**

effect modifier, in epidemiology, 573, 574, 576, 577, 580
effective dose ($E$), and internal dosimetry, 424
effective loss rate, 276, 279, 280, 302, 307
effective release height, in atmospheric transport models, **113–115**
effluent, 65–66, 88, 120, 121, 122
  discharge of, 164, 172, 183, 601–602, 607, 635–636
  and fuel processing plants, 59–61
  and nuclear reactors, 44–46, 49, 52, 55, 56, 62–63, 632
  and uranium mills, 39, 40, 41, 639
egg products, *see* food products
electrons
  and dose coefficients, 423, 453, 454, **461**
  and radiation exposure, 448, 451, 453–454, 457, 533
energy, and terrestrial food chain pathways, 264, 267–268
Environmental Modelling for Radiation Safety (EMRAS), and model validation, 592
Environmental Protection Agency (EPA), 137–138, 382, 387, 614, **617**, 618
  accident exposure regulations, 676–677, 679
  history, 38–39, 40, 59–60
  routine exposure regulations, **624–625**, **630–639**, **642–647**, **648–654**, **655**, **660–663**
environmental regulations, **613–682**
  accidents, **674–682**
  airborne emissions, **660–663**
  and below regulatory concern (BRC) level, **671–672**
  consistency of standards, **666–671**
  and Department of Homeland Security, 675, **677–679**
  and DOE, 618, **626–627**, **640–642**
  and dose–response hypothesis, **622–624**
  drinking water, **631–635**
  and EPA, **617**, **624–625**, **630–631**, **636–639**, **662–663**, 676
  exemptions, **671–675**
  and FDA, 675, **676–678**
  federal government responsibilities, **617–619**
  and IAEA, **664**, **672**, **681–682**

environmental regulations (*continued*)
  and ICRP, **627–628, 663–664, 680**
  improvised nuclear devices (INDs), 675
  indoor radon, **662–665**
  liquid discharges, **635–636**
  and NCRP, **628, 663–664**, 672
  and NRC, **617–618,625–626, 631, 639, 640, 674, 679**
  protective action guides (PAGs) for accidents, **674–682**
  public radiation protection standards, **619–672**
  radiological dispersal devices (RDDs), 675, 677–679
  and risk management
    chemical paradigm, **622**
    radiation paradigm, **620–622**
  specific practice/source standards, **629–672**
  and state governments, **618, 627**
  uranium, **630–631, 636–639**
  waste management and disposal, **649–661**
environmental transport, **5–8, 79–139, 147–201, 208–253, 260–333, 340–371**
EPA, *see* Environmental Protection Agency
epidemiology, and case-control designs, **565–566**
  cause-effect association, **557–563**
  cohort designs, **563–565**, 566
  confidence intervals, **570–580**
  confounding, 573–576, 577, 579, 580
  disease burden, **552–557**
  disease prevalence, **555–557**
  disease-exposure association, **566–580**
  disease risk, **552–554**
  dose-response models, 563–564, 566
  effect modifiers, 573, 574, 576, 577, 580
  incidence rate, **554–555**
  nested studies, **566**
  null hypothesis (H0), 567–571
  odds ratios, **559–563**, 562
  relative risk, **557–559**
  risk estimation, **551–584**
  study designs, **563–566**, 582–583
  and uncertainty, 15, 583–584
epigenetic responses, and risk from radiation, **535–537**

epistemic uncertainty, in assessment models, 469, 471, 511, 513–515, 524
equivalent continuum transport models, and groundwater, 215
equivalent dose ($H_T$), 423–424
estuaries, and surface water transport, **167–176**
exact tests, and epidemiology, 570
expert opinion, and uncertainty analysis, **493–497**
exposure-disease association, in epidemiology, **566–580**
exposure factors, defined, **8–11**
exposure regulations, **394–444, 614–616, 619–670, 673–682**
external dosimetry, 12–13, 389–390, **447–461**

fall leaf drop, 308–309
fallout, 128, **129**
federal guidance, and external dosimetry, **450–452**
Fernald Feed Materials Production Center (FMPC), 4, 18–21, 41, 66, 595, 600–604
Fick's law of diffusion, and groundwater, 222, 224, 234
finite plumes, in atmospheric transport models, **119**
first-order loss systems, and terrestrial food chain pathways, **275–284**
flow characteristics
  estuaries, **167–168**
  lakes, **181–182**
  oceans and coastal waters, **176–177**
  rivers, **159–160**
fluid flow compartments, and terrestrial food chain pathways, **286–287**
FMPC, *see* Fernald Feed Materials Production Center
food chain, **260–331, 340–371**
food products, 208, 325–326, 329–332, 392, 520, 594
  and aquatic systems, 147, 187, 194, 200, **340–371**
  and protective action guides, 675–681
  and terrestrial systems, **260–333**, 380, 383, 387
food web, **260–331, 340–371**

Fourier amplitude sensitivity test (FAST), 520, 522
fractional bias, *see* bias
Froude number, in atmospheric transport models, 118–119
fuel processing plants, **41**, **58–61**, 64, 208

gastrointestinal tract, 12, 383–384, 541, 546
and internal dosimetry, 400, 406–407, **417–419**, 424–426, 433–434, 436
Gaussian diffusion model, 95–97, **109–120**, 122–123, 128, 133–134, 460
genetic risk, 398, **538–539**, 540, 546–547
geochemical groundwater transport models, **231**
geologic repositories, NRC standards, **654–655**
geometric means and standard deviations, and uncertainty, **501–502**
gravitational effects, in atmospheric transport models, **113**
gravitational settling, and terrestrial food chain pathways, **289–290**
ground roughness, effects in external dosimetry, **455–456**
groundwater
  dating of, 208–209
  and dispersion coefficients, 223, **224**, 236
  and distribution coefficients ($K_d$), 229–231
  equations for flow, 217
  and Fick's law of diffusion, 222, 224, 234
  flow equations, **216–222**
  and food products, 208
  and mill tailings waste, **215–216**
  and near-surface disposal, **210**
  and nuclear power plant accidents, **211**
  and radioactive decay coefficients, 251
  and radioactive waste, **209–210**, **211–215**
  and radionuclide migration models, **209–253**
  and retardation coefficients ($R_d$), 220, 221, 229, 236
  and soil partition coefficients, 251, 252
  transport models
    chemical, **213**
    and colloids, **232–233**
    convective-dispersive equations, **236–243**
    dispersion in porous media, 220, **222–226**, 227
    equations, **219–222**, **236–243**
    flux model, **240–242**
    geochemical models, **231**
    and hydraulic conductivity ($K$), **226–228**
    instantaneous models, **243**
    and intrinsic permeability ($k$), 226–227
    and low-level waste, **250–251**
    methods for minimum dilution, **243–246**
    misuse of, **253**
    numerical methods, **233–236**
    parameters for, **222–233**
    point concentration model, **237–240**
    and population doses, **246–250**
    and porosity, 220, **222–226**, 227
    and radionuclides, **208–253**
    retardation factor, **228–232**
    solutions, **233–250**
    and sorption, **228**
    and tailings ponds, 210–211
    thermal-hydrologic, **212–213**
    thermomechanical, **212**
    types, **211–216**
    validation, **251–252**
  and uranium mining and milling, **210–211**

Hanford site
  exposure scenario, **10–11**
  and fuel processing, 41,42, 58–59
  Hanford Environmental Dose Reconstruction Project, 12, 24, 26, 59, 594
  and model validation, 593
  and Native Americans, 24, 26
  and radioiodine releases, 10, 12, 42, 59, 64, 594
  RATCHET model, 7, 291–292, 599–601
  and waste management, 618
heat transfer transport models, and groundwater, **212**
Henry's law, and wet deposition, 131
hereditary risk, and radiation, **539**, **546–547**, 548
high-level waste standards, **209–210**, **211–215**, **648–650**

historical assessment, 2, 4, 6–8, 9, 18
Hanford Environmental Dose
   Reconstruction Project, 12, 24, 26,
   59, 594
home construction, and radiation
   shielding, **455**
hydraulic conductivity ($K$), and
   groundwater transport models,
   **226–228**, 313
hydrodynamics, *see* surface water transport;
   groundwater, transport models
hydrologic cycle, **340–371**

IAEA (International Atomic Energy
   Agency), 42, 59, 135, 149, 592, 594
   accident exposure regulations, **680–682**
   routine exposure regulations, 619–621,
      624–630, 633, 647–648, **662–664**,
      **672–675**
   and terrestrial food chain pathways, 306,
      307, 319, 322
ICRP (International Commission on
   Radiological Protection), 13,17, 319,
   393–444
   Publications
      No. 2, permissible internal radiation
         doses, 397, 399
      No. 26, recommendations on
         radiological protection, **394–399**,
         451, 626
      No. 30, worker radionuclide intake
         limits, 394, 396, **399–422**
      No. 56, age-dependent radionuclide
         doses, **432–434**
      No. 60, 1990 radiological protection
         recommendations, **422–430**, 453,
         539, 547, 627, 628
      No. 67, age-dependent ingestion dose
         coefficients, **434–437**
      No. 69, age-dependent ingestion dose
         coefficients, **437**
      No. 71, age-dependent inhalation dose
         coefficients, **437–443**
      No. 72, compiled age-dependent
         ingestion/inhalation dose
         coefficients, **442–443**
      No. 79, genetic cancer susceptibility,
         538
      No. 88, fetal doses from maternal
         intake, **444**
      No. 89, basic radiological protection
         reference values, **443–444**
      No. 92, radiation weighting factors,
         533
      No. 95, infant doses from maternal
         intake, **444**
   and remediation of radioactively
      contaminated sites, **647–648**
   and risks for ionizing radiation exposure,
      533, 538, 539, 545–547
Idaho National Laboratory, 42, 59, 134,
   285–286
importance sampling, and uncertainty,
   **510–511**
improvised nuclear devices (INDs),
   environmental regulations, 675
Indoor Radon Abatement Act, 614–616,
   617, 630, 662
INEEL, *see* Idaho National Laboratory
ingestion, and dose coefficients, 324, 331,
   427–428, 430–432, 436, 437
inhalation, and internal dosimetry, 409–411,
   416–417, 438–439, 441–442
instantaneous point source, in atmospheric
   models, **109–110**
intermedia exchange and transfer, in surface
   water transport models, **151–154**
internal dosimetry, *see* dosimetry, internal
International Atomic Energy Agency,
   *see* IAEA
International Commission on Radiological
   Protection, *see* ICRP
intestines, *see* gastrointestinal tract
in utero exposures, and cancer
   epidemiology, **543–544**
in vivo studies, 37, 536
iodine, *see* radioiodine
irrigation, and Chernobyl, **197–200**

Japanese A-bomb survivors, 14, 533,
   539–540, **541–542**, 543, 545

kerma, and point-kernel method in external
   dosimetry, 461

lakes, and surface water transport, **182–187**
lanthanum, radioactive (RaLa), 59
Latin hypercube sampling, and uncertainty,
   **509–510**, 520, 522

Index 697

leaching, 215, 251, 309, **310–318**, 329
leptokurtic distributions, and uncertainty, 476
line source, in atmospheric models, **110**
linear operations, and uncertainty, **501**
LOCA (loss of coolant accident), 56–58
logic errors, and uncertainty, 493–497
long-term radionuclide release transport models, **156–190**
Los Alamos National Laboratory, 26–27, 41, 59
loss of coolant accident (LOCA), 56–58
low-dose radiation, risks from, 532, **533–540, 547–549**
low-level waste disposal standards, 215, **250–251, 655, 657–658**

Mantel-Haenszel test, in epidemiology, 569, 570, 572, 574
Marshall Islands, and nuclear weapons testing, 37
mass transport, and dispersion coefficient, 220, 221
mass transport equation, *see* groundwater, transport models
maximum contaminant levels (MCLs), 632–635
maximum likelihood (ML), and epidemiology, 576–579
maximum permissible body burden (MPBB), and internal dosimetry, 393, 397
maximum permissible concentration (MPC), and internal dosimetry, 393, 397, 401
maximum permissible dose equivalent (MPDE), and internal dosimetry, 393, 397
mean, mathematical, and uncertainty, **502–503**
measures of radiation dose, in epidemiology, 581–583
meat, *see* food products
mechanistic models, and food chain transport, **270–275**
MILDOS model, and food chain transport, 270, 297
milk products, *see* food products
mill tailings, and groundwater, **215–216**

mixing height, and atmospheric transport, **106–108, 115**
model composition, evaluating validity of, **591–593**
model error, and model validation, **605–607**
model performance, evaluating validity of, **593–594**
model uncertainty, and confounding, 514, 516, 519, 522
model validation, **18–21, 589–608**
  and assessment question error, **607–608**
  and atmospheric transport, **132–136**
  and benchmarking, 524, 592
  and bias, **596–598**
  and BIOPROTA, 592
  and calibration, 226, **251–252, 594–596**
  and Chernobyl, 594
  and compensatory errors, 591, 606, 608
  and correlation, **599**
  defined, 18, 590
  and Environmental Modelling for Radiation Safety (EMRAS), 592
  and environmental transport, 18, 21, 62
  and Gaussian atmospheric transport model, 133–134
  and groundwater transport, **251–252**
  and Hanford site, 594
  and input data, 607–608
  and model error, **605–607**
  and poor model performance, **603–608**
  and process, **590–596**
  and sensitivity analysis, 592
  and source terms, 18
  and tests of model performance, **596–603**
  and uncertainty, **522–524**
  validation vs. verification, 591
Monte Carlo statistical methods, and uncertainty, 16, 483, 497, **506–517**, 520, 524
multicompartment models, **125–126**, 272, **324–333**
multivariate distributions, **485–487**

NARM (naturally occurring and accelerator-produced radioactive material), 615–616, 618, 646, 648, **658–659**; *see also* NORM, TENORM
National Academy of Sciences (NAS), 61, 64, 579, 583, 651, 653–655

National Contingency Plan (NCP), 642–645, 647
National Council on Radiation Protection and Measurements, *see* NCRP
National Emission Standards for Hazardous Air Pollutants (NESHAPs), 39, 40, 139, 660–662
National Reactor Testing Station, *see* Idaho National Laboratory
Native Americans, exposures of, 10–11, 24, 26
NCRP (National Council on Radiation Protection and Measurements), 38, 61–62, 127–128, 321, 393–394, 420–421
    routine exposure regulations, 619–621, 624, 628, **647**, **663–664**, **672**
near-surface disposal facilities, **210**, **655–657**
NESHAPs (National Emission Standards for Hazardous Air Pollutants), 39, 40, 139, 660–662
Nevada Test Site, and dose assessments, 37, 225, 295
non-first-order loss systems, and terrestrial food chain pathways, **284–287**
non-Gaussian treatments, 134, 137
NORM (naturally occurring radioactive material), 614–616, 627, 658; *see also* NARM, TENORM
normal/lognormal distributions, and uncertainty, **501–502**
NRC, *see* U.S. Nuclear Regulatory Commission
nuclear reactors, **42–56**, 62–63, 632
    and accidents, 56–58, **211**
Nuclear Regulatory Commission, *see* U.S. Nuclear Regulatory Commission

Oak Ridge facilities, 40–43, 58–59
oceans, and flow characteristics, **176–181**
odds ratios, **559–563**, 562, 577

PAGs (protective action guides), for nuclear accidents, **674–681**
particle-in-cell models, of atmospheric diffusion, **126–127**
Pasquill-Gifford stability categories, **95–100**, 118, 119–120, 128, 133

PATHWAY model, 290, 295, 304–305, 308–309, 325–328, 466
PDF, *see* probability density function
percolation, **221–222**, 302, 309, **310**, 311, 330
permeability intrinsic ($k$), and groundwater transport models, 226–227
photons, and dose coefficients, 404, 423, 450–461, 639
    and radiation exposure, 119, 421, 448, **459**, 538
phytoplankton, 342, 356–358, **360–361**, 364–370
platykurtic distributions, and uncertainty, 475–476
plowing, 266, **318**
plumes
    and aquatic transport, 156–157, 164–165, 171, 173, 180, 186
    and atmospheric transport, **79–139**
    and external dosimetry, 453, 457, 460
    Gaussian models, 95–97, 102, **109–120**, 122–123, 128, 460
plutonium, 32–33, 35, 36, 60, 64, 150–151
    and bioaccumulation, 306, 315, 321, 322, 349, 353
    and Chernobyl, 193–194, 200
    and Hanford, 42, 58, 59
    and Los Alamos, 41, 59
    protective regulations, 434, 631, 665, 673, 678
    and Rocky Flats, **6–8**, 10, 25, 41, 303, 607
point-kernel method, and external dosimetry, **457–461**
point source, in atmospheric models, **109–112**
Poisson distributions, and uncertainty, 478–479, 484
Poisson regression, in epidemiology, 577–579, 583
poor model performance, in model validation, **603–608**
population dynamics, and Chernobyl, **361–362**, 364–369
porosity, and groundwater transport models, 220, **222–226**, 227
precipitation scavenging, *see* wet deposition
Pripyat River, and Chernobyl, 166–167, 191, **192–198**, 200

probabilistic risk assessment, 12, 17, 57–58
probability density function (PDF), and uncertainty, 471–474, 492, 497–498, 500
probability distributions, and uncertainty, **471–472**
progeny, radionuclides, *see* radionuclides, decay products
propagation, mathematical, and uncertainty, **497–506**
protective action guides (PAGs), for nuclear accidents, **674–681**
protective regulations, **613–682**
public radiation protection standards, **617–672**
puff transport and diffusion models, 107, **120–124**, 135, **138**, 292
pulmonary absorption particle classes, 440–443
pulmonary clearance classes, 414–416
p values, 569

radiation contamination cleanup, and ARARs, 642–645
radiation dose, average U.S. adult, 629
radiation exposure, and biological effects, **531–549**
radiation paradigm for risk management, 602, **620–622**
radiation shielding effects, and building structures, 455, 462
radiation weighting factor ($w_R$), and internal dosimetry, 423–424
radioactive cloud, submersion in, and internal dosimetry, **421–422**
radioactive lanthanum (RaLa), 59
radioactive waste
  disposal regulations, **648–661**
  and groundwater, **209–210**, **211–215**, **250–251**
  high-level, **209–210**, **211–215**
  low-level, **215**, **250–251**
  solid, **61–62**
  spent fuel, 644, 648–654, 666, 657, 659
radiocesium, *see* cesium
radioiodine
  and bioaccumulation, 306, 315, 318, 319, 322, 349
  dose equivalents, 665, 677
  and Hanford, 291
  nuclides, 35, 36, 45, 51, 53–54
  releases, 139, 489–490, 632
    Chernobyl, 37, 132, 192, 194–195, 199, 594
    Hanford, 10, 12, 42, 59, 64
  and terrestrial food chain pathways, 274, 290–292, 294, 295
  source terms, 45–46, 48–49, 56, 57, 59, 60–61
  and thyroid, 57, 268, 282, 542, 579, 675
radiological assessment process, **1–27**
radiological dispersal devices (RDDs), 675, 677–679
radionuclides
  as tracers, 262–264
  decay products, 32, 36, **153**, 393
    and dosimetry, 406, 432, 442, 448, 451–452
    and groundwater transport, **220–221**, 230, 251
    and leaching, 314
    and surface water transport, **153**
  exposure pathways, 79
  ingestion rates, and Chernobyl, **362**
  masses, 32, 34–35
  migration models, and groundwater, **209–253**
  progeny, *see* radionuclides, decay products
  properties, 33–36
  releases, accidental, in transport models, **154–156**
  sources, 32–36, **38–56**
  transport models
    atmospheric, **79–139**
    and Bessel functions, 157, 158
    and Chernobyl, **194–200**
    and coastal water screening, **179–181**
    and dispersion coefficients, **160–163**, **168–174**, **178–180**, **182–183**, 185
radon, 39, 139, 153, 542, 581, 629
  and Fernald Feed Materials Production Center, 4, 18, 595
  nuclides, 45, 51, 53, 54, 316
  protective regulations, 422, 616, 620, 623, 625, 674; *see also* Indoor Radon Abatement Act

radon (*continued*)
  standards for specific applications, 635–639, 646–647, 657–658, 660–661, **662–665**, 666
rainfall, 221–222, 292–295; *see also* wet deposition
rainout, 131, 132, 266, 292–293
RaLa (radioactive lanthanum), 59
random numbers, 467, 506, **508**, 509, 523–524
random variables, **498–501**
RATCHET (Regional Atmospheric Transport Code for Hanford Emission Tracking) model, 7, 291–292, 599–601
RCRA, *see* Resource Conservation and Recovery Act
Reactor Safety Study, 57, 58, 302
reference individuals, 401–402, 404, 408, 432, 447–448
Regional Atmospheric Transport Code for Hanford Emission Tracking (RATCHET) model, 7, 291–292, 599–601
regression statistical analysis, 466, 521, 575–579, 580, 583, **599**
regulatory models, for atmospheric diffusion, **137–139**
relative risk, and epidemiology, **557–559**
reservoirs, and Chernobyl, **197–200**
Resource Conservation and Recovery Act (RCRA), 614–616, 618, 630, 638, 658–660
respiratory system, internal dosimetry model for, **409–417**
RESRAD code, 230, 270, 312
resuspension, *see* suspension
retardation coefficient ($R_d$), and groundwater, 221, **228–232**, 248, 311
Richardson number, and atmospheric stability, **99–107**
risk, defined, 14, 552
risk coefficients, **14–15**, 399–400, 422, **545–547**
risk communication, **21–24**, 497, 590
risk management, 602, **620–622**
risk from radiation
  cancer epidemiology, **540–545**
  and dose, **13–15**, 533–540, 547–549
  epigenetic responses, **535–537**
  hereditary factors, **538–539**, **546–547**, 548
  Japanese A-bomb survivors, 14, 539–540, **541–542**, 544, 545
  risk coefficients, **545–547**
risk ratio, **557–559**
rivers, and surface water transport, **159–167, 194–200**
Rocky Flats Environmental Technology Site, 6–10, 24–26, 41, 317–318
root uptake, and terrestrial food chain pathways, **303–307**
routine exposure regulations, **619–670, 672–675**

Safe Drinking Water Act, 614–616, 622, 630, 632–635, 651, 670–671
sand, *see* soil
Savannah River Site, 24, 42, 59, 269–270
  and atmospheric turbulence, 90, 92, 133–134
scatter, measures of, **598–599**
screening models
  atmospheric, **127–128**
  coastal waters, **179–181**
  estuaries, **171–176**
  lakes, **183–187**
  rivers, **163–167**
  soil, 270
sector averaging, in atmospheric transport models, **112–113**
sediment effects, in transport models, **187–190**
Sellafield nuclear site, 60, 150–151
senescence, and terrestrial food chain pathways, **308–309**
sensitivity analysis, and uncertainty, **517–522**, 592
shallow land burial, **210, 655–657**
shielding, from radiation, **455**, 462
shorelines, and external dosimetry, **456–457**
similarity theory, and Richardson number, **103–107**
simple random sampling, and uncertainty, **509**, 520
single-compartment models, 273, **273–287**, 324
sink compartments, and terrestrial food chain pathways, **277**

site conceptual exposure models (SCEMs), 9, **376–387**
soil
  column transport, **309–318**
  and dry deposition, **289–295**
  partition coefficients, and groundwater, 251, 252
  and terrestrial food chain pathways, 266
  transport
    to animals, **320–312, 318–324**
    to vegetation, **294–309, 310–318**
    vertical migration, **309–318**
solid radioactive waste disposal, **61–62**
sorption, and groundwater transport models, **228**
source characteristics, in atmospheric transport models, **113–120**
source compartments, and terrestrial food chain pathways, **275–277**
source terms, **3–5**, 18, 21, **31–66, 213**, 231
  and cesium releases, 37, 49, 51, 53–54, 60, 61
  and low-level waste, **250–251**
  and nuclear reactor accidents, **56–58**
  and prospective analyses, **62–64**
  and retrospective analyses, **64–66**
specific effective energy (SEE), 403–405, 407, 427–428, 432
stability categories, atmospheric, **94–106**
stakeholder participation, 21, **24–27, 377–378**
standardized mortality ratio, 561–562
state governments, and environmental regulations, **618, 627, 646–647, 658–659**
statistical intervals, and uncertainty, **476–483**
statistical models, and food chain transport, **269–270**
stochastic effects, 395–397, 531–549
stochastic models, 17, **135–136**, 271, **288–289**, 506–507
stochastic parameters, 487–488
stochastic processes, **288–289**, 467, 470–471
stomach, *see* gastrointestinal tract
strontium, 45, 51, 53–54, 60–61, 152, 211
  and bioaccumulation, 232, 252, 306, 315, 319, 322, 348–353
  and Chernobyl, 153, 166, 192–200
  and Clinch River, Tennessee, 348–352
  dose equivalents, 407, 434–436, 442, 443
  and internal dosimetry, 405–406, 432–433, 435–436, 441
  protective regulations, 633, 634, 665, 678
study design
  case-control, **565–566**
  cohort, **563–565**, 566
  and epidemiology, **563–566**, 582–583
  nested, **566**
  and random sampling, **509**, 520
  and uncertainty, **509–511**, 520
surface water transport, **147–200**
survival analysis, in epidemiology, 579–580
suspension
  and inhalation, 321, 330
  and sediment, 150–161, 187–189, 195–196, **296–303**

target tissue, and internal dosimetry, 402–403, 407
Techa River, Russia, 542
TENORM (technically enhanced naturally occurring radioactive material), 627, 639, 646, 647, 658, 659; *see also* NORM, NARM
terrain effects, and atmospheric motion, 84–86, **118–119**
terrestrial food chain pathways, **260–333**, 380, 383, 387
thermal internal boundary layer (TIBL), 86–87
thermal transport models, and groundwater, **212–213**
Three Mile Island, 57–58, 91
threshold effects, *see* deterministic effects
thyroid, 12, 64, **398–399**, 543–544, 632, 680–681
  and radioiodine, 57, 268, 282, 542, 579, 675
tides, in transport models, 168
tillage, 266, **318**
tissue weighting factors, and internal dosimetry, **399–400**
tolerance intervals, and uncertainty, **479–483**
Toxic Substances Control Act (TSCA), 614–616, 618, 630, 658, 659, 660
tracer kinetics, and terrestrial food chain pathways, **273–275**

transport and diffusion calculations, atmospheric, **89–108**
TSCA (Toxic Substances Control Act), 614–616, 618, 630, 658, 659, 660
type A and B uncertainty, in assessment models, 469, 512
type I and II errors, and null hypothesis, 493

U.S. Department of Energy (DOE), 40–41, 59, 61, 64, 230–231, 614–616
   low-level waste disposal standards, **657–658**
   routine exposure regulations, **618**, 626–627, 631, **635, 640–641**, 657–658
U.S. Nuclear Regulatory Commission (NRC), 88, 127, 455, 457, 614, **616–618**
   accident exposure regulations, **673–674, 679**
   history, 40, 52, 57–58, 61–64
   routine exposure regulations, **625–626**, 631, **639–645**, 651, **654–657**, 658
UMTRCA (Uranium Mill Tailings Radiation Control Act), 614–616, 630, 636
uncertainty analysis, 4, **12, 15–18, 133–136, 465–525**, 583–584
universal treatment standards (UTSs) for radioactive waste, 659
UNSCEAR (U.N. Scientific Committee on the Effects of Atomic Radiation), 66, 533, 535–536, 538, 543–544, 547–548
uranium, 40–42, 192, 652
   and bioaccumulation, 306, 315, 316–317, 322, 349, 353
   and Fernald Feed Materials Production Center, 4–5, 19–21, 66, 595, 600–602, 604
   fuel fabrication, **41–42**, 48, **58–60**
   mines and mills, 39, 40, 41, **210–211, 636–639**
   nuclides, 32–33, 35–36, 153, 209, 230, 262
   protective regulations, 38, **614–616, 630–646, 649, 658–661, 665–666**; see also Atomic Energy Act
   releases, 4, 18–19, 40, 231, 297
Uranium Mill Tailings Radiation Control Act (UMTRCA), 614–616, 630, 636
user error, in model validation, **604–605**

validation, *see* model validation
variability, vs. uncertainty, **511–512**
variance, and uncertainty, **502–504**, 513
vegetation
   and radiation shielding, 455–456
   and terrestrial food chain pathways, 266–267, **295–307**, 309, **310–318**
vegetation-to-animal transport, **318–320**
vegetation-to-soil transport, **307–309**
verification, vs. validation, for models, 591
visual display of information, in model validation, **599–603**

wake effects, buildings, *see* buildings, wake effects
washout, 131–132, 266, 292–293
waste disposal, environmental regulations, **649–661**
Waste Isolation Pilot Plant (WIPP) site, **653–655**
weapon and fuel fabrication, **41**
weathering, 279–280, **307–308**
wet deposition, 128, **130–132**, 137, **292–295**, 460
wind, *see* atmospheric turbulence
WIPP (Waste Isolation Pilot Plant) site, **653–655**
workplace exposure, and internal dosimetry, 389–391, 393
World Health Organization, environmental health criteria for radionuclides, 59

X-10 plant, Oak Ridge facilities, 58–59
X-rays, 543–544, 629

Y-12 plant, Oak Ridge facilities, 41
Yucca Mountain Site, 210, 212, 252, **651–653**

zooplankton, 343, 352, 356, 357, **360–363**, 364–370